Design of Municipal Wastewater Treatment Plants Volume I: Chapters 1–12

WEF Manual of Practice No. 8

ASCE Manual and Report on Engineering Practice No. 76

Prepared by Joint Task Force of the Water Environment Federation (formerly the Water Pollution Control Federation) and the American Society of Civil Engineers, with grant assistance from the U.S. Environmental Protection Agency.

WEF Facilities Development Subcommittee of the Technical Practice Committee and ASCE Environmental Engineering Division Committee on Water Pollution Management.

1992

Water Environment Federation
601 Wythe Street
Alexandria, VA 22314-1994

and the

American Society of Civil Engineers
345 East 47th Street
New York, NY 10017-2398

The material presented in this publication has been prepared in accordance with generally recognized engineering principles and practices and is for general information only. This information should not be used without first securing competent advice with respect to its suitability for any general or specific application.

The contents of this publication are not intended to be and should not be construed to be a standard of the Water Environment Federation (WEF) or the American Society of Civil Engineers (ASCE) and are not intended for use as a reference in purchase specifications, contracts, regulations, statutes, or any other legal document.

No reference made in this publication to any specific method, product, process, or service constitutes or implies an endorsement, recommendation, or warranty thereof by WEF or ASCE.

WEF and ASCE make no representation or warranty of any kind, whether express or implied, concerning the accuracy, completeness, suitability, or utility of any information, apparatus, product, or process discussed in this publication and assumes no liability.

Anyone using this information assumes all liability arising from such use, including but not limited to infringement of any patent or patents.

Library of Congress Cataloging-in-Publication Data
Design of municipal wastewater treatment plants/prepared by a Joint Task Force of the Water Environment Federation and the American Society of Civil Engineers.—2nd ed.
p. cm.—(WEF manual of practice; no. 8) (ASCE manual and report on engineering practice; no. 76)
Includes bibliographical references and indexes.
ISBN 0-943244-83-8 (set).—ISBN 0-943244-84-6 (v.1).—
ISBN 0-943244-85-4 (v.2)
ISBN 0-87262-834-5
1. Sewage disposal plants—Design and construction. 2. Sewage—Purification. 3. Sewage sludge. I. Joint Task Force of the American Society of Civil Engineers and the Water Environment Federation. II. Series: Manual of practice; no. 8. III. Series: ASCE manuals and reports on engineering practice; no. 76
TD201.W337 no.8a
[TD746]
628.1'68 s—dc20   91-30528
[628.3]   CIP

Library of Congress Catalog Card No. 91-30528
Printed in the United States of America by Book Press, Inc., Brattleboro, Vermont.

Contents, Volume I

Contents, Volume II

Joint Task Force

The Joint Task Force for this manual comprises Planning Group members, principal authors, contributing authors, and reviewers. Listed according to these categories of participation, the individual task force members are gratefully acknowledged as follows:

Planning Group

Lee Glueckstein	Michael C. Mulbarger
Joseph F. Lagnese, *Chairman*	Robert W. Okey
Lam Lim	Sherwood Reed (ASCE)
Anton F. Miorin	Donald E. Schwinn

Principal Chapter Authors

Orris E. Albertson	12	Michael C. Mulbarger	1, 2, 3
Max Augustus	10	Brooks W. Newbry	18
John Calmer	14	Robert E. Pailthorp	6
Ronald W. Crites	13	Roderick D. Reardon	15
James C. Dowell	5	Joseph C. Reichenberger	4
Albert C. Gray	16	James F. Stahl	19
John R. Harrison	12	John J. Stankunas	8
Dail C. Hollopeter	4	Richard Stratton	9
Elizabeth M. Huning	5	Rudy J. Tekippe	11
Shin-Joh Kang	15	Michael J. Wallis	17
Don Kincannon	12	Thomas E. Weiland	10, 19
Terry L. Krause	7	A. Gordon Wheler	12
Richard D. Kuchenrither	20	Kenneth L. Zacharias	7

Contributing Chapter Authors

Kashi Banerjee	16	Keith A. Danzinger	8
Edwin Barth	11	Allan J. DeLorme	15
Robert H. Beyer	8	Jamel Demir	17
Brandon Braley	9	Dennis M. Diemer	17
John A. Burgh	17	Gary S. Dodson	17
Donald Campbell	6	John F. Donovan	20
Arthur Condron	11	Michael D. Doran	12
Mike A. Crawford	20	Joseph A. Drago	14, 16
Kenneth Cundari	8	David N. Ellis	8
Glen Daigger	11	Ann E. Farrell	17

Alvin Firmin	9	Gary A. Rice	17
Jane B. Forste	20	Mike G. Rivard	20
Richard V. Frykman	7	Michael Robson	14
Walter Garrison	11	Robert A. Ryder	14
Donald Gates	14	Russel M. Sanchez Adams	14
Mark Gould	18	Perry L. Schafer	20
Shashi Gupta	8	Donald E. Schwinn	1, 2, 3
J. Ronald Hargreaves	19	Thomas S. Selfridge	17
Robert Hegg	11	Michael Selna	10
James Heidman	11	Almon M. Shen	17
Russ Huisinga	7	Karen Speer	18
Blair Jones	14	George Sprouse	11
James M. Kelly	19	Roger Stephenson	11
Timothy Koch	4	Rao Surampalli	12
Joseph F. Lagnese	1, 2, 3	Berrin Tansel	20
Kris Lindstrom	20	Christine Tokar	18
David Lounsbury	18	Karen L. Van Aacken	9
Jane Madden	9	Carlos E. Vargas	7
Douglas W. Mair	5	Orest Walchuk	8
William J. Martin	20	Thomas K. Walsh	18
Sean P. McNamara	8	Edward Wetzel	11
Seong W. Min	19	George C. White	14
Harry Q. Minter	8	Brian Wilson	11
Randall C. Naef	17	Thomas E. Wilson	11, 15
Daniel Norris	4	James Wren-Jarvis	20
Robert W. Okey	1, 2, 3	Ronald Wukasch	11
Joseph G. Rabosky	16	John F. Yost	7
Kerwin Rakness	11		

Chapter Reviewers

Thomas A. Allbaugh	11	Robert B. Brobst	16, 19
James E. Alleman	11	Donald S. Brown	18
Carl M. Anderson	17	Carl A. Brunner	17, 20
Robert B. Barbour	4, 15	Curtis W. Bryant	12, 13
Peter J. Barthuly	5	Roger J. Byrne	11
Allen Baturay	19	James T. Canaday	5
Aivars Bergs	5	Gerald T. Carey	17
Jerry C. Bish	7	James H. Clark	12, 17
Harry E. Bostian	19	Craig S. Coker	17, 19, 20
William C. Boyle	11, 12	Holly A. Cornell	2, 3, 4
Kenneth O. Brannan	15	John Crosse	10
Richard Brenner	11, 12	Gordon L. Culp	16

Glen T. Daigger	12	William J. O'Donnell	2–5, 7
Kurt A. Dally	6	James T. Olsta	13, 20
Atal E. Eralp	3, 9	Edward J. Opatken	12
Charles G. Farley	10, 18, 20	Douglas M. Owen	15
Joseph B. Farrell	17–20	Thomas M. Pankratz	10
A.A. Freidman	17–19	Denny S. Parker	12
Thomas P. Gilligan	11	John Piccirillo	19
Daniel L. Glasgow	4	Nicholas L. Presecan	9, 10
John Glorioso	19	Movva Reddy	15
Jack T. Goldberg	6, 17	George C. Riek	10
Blair Hansen	10	Ronald Saikowski	2, 3, 8
John Hawk	8	F. Michael Saunders	11, 12, 15
Jonathan C. Hay	17, 18	Perry L. Schafer	4, 7
James Heidman	11, 15	Kenneth P. Schnaars	7, 15
Gene Heyer	7	Ralph B. Schroedel, Jr.	5, 18
Steve Hinton	12	J. Holland Scott	15, 17, 18
Stephen Hoffman	6	Elmer Seegmueller	6
Terry L. Johnson	18	Robert E. Selleck	14
Thomas M. Keinath	11, 12	Jeannette A. Semon	6, 7, 17–19
James Kenney	8	Peter A. Shack	2, 3, 4
Kenneth K. Kienow	8	Timothy J. Shea	19
Kenneth L. Knickerbocker	15	Ronald B. Sieger	7, 8, 10
John S. Kontor	7	Edwin D. Simmons	11
J. Robert Krebs	5, 17	Gary J. Swanson	2, 3, 10
James F. Kreissl	9, 13	Ronald A. Tsugita	2, 3, 4
Robert W. Lecey	11, 15	Warren R. Uhte	10–13
Donald G. Linn	2, 3, 19	Ashok Varma	12
Bruce Logan	12	Albert D. Venosa	14
Manuel Madriz	8	T. Viraraghavan	10, 13
James J. McKeown	11, 14	Charles Von Dreusche	19
William F. McTernan	16	Malcolm R. Walker	13
J. Carrell Morris	14, 16	James Wheeler, EPA	All
Lawrence Mullen	7	Walter Zachritz	13
Peter G. Nicol	17		

WEF Staff Assistance

Jodi L. Levy (Production Coordinator)
Lorna E. Johnson (Technical Editor)
Norma Jean Reck (Editorial Assistant)
Myron F. Tiemens (Staff Liaison and Managing Editor)

Water Environment Federation

The Water Environment Federation is a nonprofit, educational organization composed of member and affiliated associations throughout the world. Since 1928, WEF has represented water quality specialists, including civil, design, and environmental engineers, biologists, bacteriologists, local and national government officials, treatment plant operators, laboratory technicians, chemists, industrial technologists, students, academics, and equipment manufacturers and distributors.

For information on membership, publications, and conferences contact

Water Environment Federation
601 Wythe Street
Alexandria, VA 22314-1994
(703) 684-2400

American Society of Civil Engineers

The American Society of Civil Engineers (ASCE) offers civil engineering professionals many opportunities for technical advancement, networking, and leadership and technical skill training. Also available to members are major savings on educational seminars, conferences, conventions, and publications. First class, low-cost insurance programs are among the most competitive available.

Members may participate on a national level, networking with colleagues on forums to advance the profession. This participation affords the opportunity to develop leadership skills and to expand personal contacts. On the local level, chapters (called Sections and Branches) act as advocates in the public interest on local issues and present seminars and programs relevant to the needs of the local community.

Manuals of Practice for Water Pollution Control

(As developed by the Water Environment Federation)

The WEF Technical Practice Committee (formerly the Committee on Sewage and Industrial Wastes Practice of the Federation of Sewage and Industrial Wastes Association) was created by the Federation Board of Control on October 11, 1941. The primary function of the committee is to originate and produce, through appropriate subcommittees, special publications dealing with technical aspects of the broad interests of the Federation. These manuals are intended to provide background information through a review of technical practices and detailed procedures that research and experience have shown to be functional and practical.

Water Environment Federation
Technical Practice Committee Control Group

P.T. Karney, *Chairman*
F.D. Munsey, *Vice-Chairman*
A.J. Callier
L.J. Glueckstein
C. Lowery
T. Popowchak

Authorized for Publication by the Board of Control
Water Environment Federation, 1992
Quincalee Brown, *Executive Director*

ASCE Manuals and Reports on Engineering Practice

(As developed by the ASCE Technical Procedures Committee July 1930 and revised March 1935, February 1962, and April 1982)

A manual or report in this series consists of an orderly presentation of facts on a particular subject, supplemented by an analysis of limitations and applications of these facts. It contains information useful to the average engineer in his everyday work, rather than the findings that may be useful only occasionally or rarely. It is not in any sense a "standard," however, nor is it so elementary or so conclusive as to provide a "rule of thumb" for nonengineers.

Furthermore, material in this series, in distinction from a paper (which expresses only one person's observations or opinions), is the work of a committee or group selected to assemble and express information on a specific topic. As often as is practicable, the committee is under the general direction of one or more of the Technical Divisions and Councils, and the product evolved has been subjected to review by the Executive Committee of that Division or Council. As a step in the process of this review, proposed manuscripts are often brought before the members of the Technical Divisions and Councils for comment, which may serve as the basis for improvement. When published, each work shows the names of the committee by which it was compiled and indicates clearly the several processes through which it has passed in review so its merit may be definitely understood.

In February 1962 (and revised in April 1982), the Board of Direction voted to establish

> A series entitled "Manuals and Reports on Engineering Practice" to include the Manuals published and authorized to date, future Manuals of Professional Practice, and Reports on Engineering Practice. All such Manual or Report material of the Society would have been refereed in a manner approved by the Board Committee on Publications and would be bound, with applicable discussion, in books similar to past Manuals. Numbering would be consecutive and would be a continuation of present Manual numbers. In some cases of reports on joint committees, bypassing of Journal publications may be authorized.

American Society of Civil Engineers

A.S. Vernick, *Chair,* Environmental Engineering Division, 1991
W.F. McTernan, *Chair,* Water Pollution Management Committee
S.C. Reed, *Chair,* Manual Review Panel

Authorized for Publication by the Publications Committee of the American
Society of Civil Engineers
Edward O. Pfrang, *Executive Director,* ASCE

Design of Municipal Wastewater Treatment Plants

ABSTRACT: This manual emphasizes contemporary design practices for municipal wastewater treatment plants and includes the results of surveys of plant designers about design practices and of plant superintendents concerning process performance, operational problems, and design deficiencies.

Chapter 1 describes the purpose and scope of the manual, including a historical overview of major developments in treatment plant design. Chapter 2 presents overall design considerations, particularly those that precede detailed design, and Chapter 3 discusses the basic principles of integrated facility design, which are the key to successful design of individual unit processes. Chapters 4 through 8 present the basic subjects of site selection and plant layout, plant hydraulics and wastewater pumping, occupational safety and health, support systems, and selection of construction materials. Design of liquid unit processes is discussed in Chapters 9 through 16. Chapters 17 through 20 cover solids and residuals and present handling, processing, and beneficial use and disposal options.

This manual is intended to be the most comprehensive and useful single source of design practice information for the design engineer of municipal wastewater treatment plants.

Preface

This manual of practice, which replaces the wastewater treatment plant design manual published in 1977, is intended to be the most comprehensive and state-of-the-art reference source for the practicing design engineer of municipal wastewater treatment plants. Accordingly, it emphasizes contemporary design practice and is primarily directed to users already having general knowledge of basic design principles and procedures. Approximately 180 leading practitioners representing a national cross section of contemporary design experience have contributed to this manual's preparation. In addition, the manual contains many of the results of three WPCF surveys, one of lead process engineers from 15 major design firms gathering information about current design practices, another of more than 1000 plant superintendents concerning technologies and design deficiencies, and the third of approximately 130 plant superintendents collecting data on process performance and other plant information.

WEF and the ASCE share responsibility and jointly participated in a task force for the preparation of this almost completely revised third edition of the manual. Under a cooperative agreement with WEF, EPA provided grant assistance for WEF staff and administrative support. The EPA project officer for the cooperative agreement was Charles Vanderlyn.

This manual comprises an introductory chapter and 19 others grouped into two volumes:

Volume I: Chapters 1–12
Volume II: Chapters 13–20

The following persons responded to the 1989 WPCF Design Practice Survey:

Henry H. Benjes
Robert Bonner
Glenn Daigger
Dale A. DeCarlo
John Jenks
John Keegan
Albert Pincince

Arthur H. Plautz
Joseph C. Reichenberger
John Stukenberg
Rudy Tekippe
David Walrath
James White
Thomas E. Wilson

The Joint Task Force members are or have been affiliated with the following companies or organizations:

Alexandria Sanitation Authority (VA)
American Colloid Company
R.V. Anderson and Associates, Ltd.
Baker/TSA Inc.
Bio Gro Systems, Inc.
Black & Veatch
Brown & Caldwell
Buchart-Horn, Inc.
Bureau of Sanitation, Stamford (CT)
Burgess & Niple Ltd.
Burns McDonnell
CH2M Hill
Camp Dresser & McKèe, Inc.
Capital Controls Co., Inc.
Carlson and Associates
Carson Joint Water Pollution Control Plant (CA)
Central Contra Costa Sanitary District
The Citadel (SC)
Clark County Sanitation District (NV)
Clemson University (SC)
Consoer, Townsend & Associates, Inc.
County Sanitation Districts of Los Angeles County
Daniels Laboratory (GA)
Delta Diablo Sanitation District (CA)
Donohue & Associates, Inc.
E-T Technologies
EIMCO
ERM Technologies, Inc.
East Bay Municipal Utility District (CA)
Eastern Environmental Systems
Engineering-Science, Inc.
Envirex, Inc.
Enviro Enterprises, Inc.
Enviro-Gro Technologies

Environmental Protection Agency
Gannett Fleming Affiliates, Inc.
Gore & Storrie, Ltd.
Greeley and Hansen
Green Bay Metropolitan Sewerage District (WI)
G.S. Dodson & Associates
Havens & Emerson, Inc.
Hazen and Sawyer
HDR Engineering, Inc.
Heritage Environmental Services, Inc.
Howard, Needles, Tammen & Bergendoff
Hycor Corp.
Hyperion Wastewater Treatment Plant (CA)
JACA Corporation
John Carollo Engineers
Jones & Henry Engineers, Inc.
Kennedy, Jenks, Chilton Inc.
LOTEPRO Corp.
Louisville Metropolitan Sewerage District (KY)
Malcolm Pirnie, Inc.
McNamee Advanced Technology, Inc.
McNamee, Porter & Seeley, Inc.
Metcalf & Eddy, Inc.
Metropolitan Denver Sewage Disposal District No. 1 (CO)
James M. Montgomery Consulting Engineers, Inc.
Mullen and Associates
George S. Nolte and Associates
O'Brien & Gere Engineers, Inc.
Oklahoma State University
P.C. Peer Consultants
Planning, Design, and Research Engineers
Process Applications, Inc.

Purdue University (IN)
Rosemount, Inc., Varec Division
Schrieber Corp., Inc.
Sitech Engineering
Solidur Plastic's Company
Southwest Technology
   Development Institute (NM)
Stearns & Wheler
Stottler, Stagg & Associates
Sverdrup Corporation
Syracuse University (NY)
Toltz, King, Duvall, Anderson &
   Associates

Tufts University (MA)
University of Arizona
University of Arkansas
University of California, Berkeley
University of Regina (Sask., Canada)
University of Wisconsin
Larry Walker Associates
Westin Engineering, Inc.
Weyerhauser Company
Whitman and Howard, Inc.
Wilson & Company
Zimpro Passavant

Figure Contributors

Cherry Burrell Fluid Handling
  Division
611 Sugar Creek Road
Delavan, WI 53115

CPC Engineering Corporation
441 Main Street
P.O. Box 36
Sturbridge, MA 01566

Disposable Waste Systems, Inc.
(Marketing Company—JWC
  Environmental)
16802 Afton Street, Suite 200
Irvine, CA 92714

Dorr-Oliver Incorporated
612 Wheeler's Farm Road
P.O. Box 3819
Milford, CT 06460-8719

Dresser Industries, Jeffrey Division
P.O. Box 387
Woodruff, SC 29388

Fairbanks Morse Pump Corporation
3601 Fairbanks Avenue
Kansas City, KS 66101-0906

Gorman Rupp-Ramparts Division
15601 19th Place West
P.O. Box 697
Sand Springs, OK 74063

Hydraulic Institute
30200 Detroit Road
Cleveland, OH 44145

Infilcro Degremont, Inc.
2924 Emerywood Parkway
Richmond, VA 23229

ITT Marlow
1150 Tennessee Avenue
Cincinnati, OH 45229

John Meunier, Inc.
6290 Perinault Street
Montreal, Quebec, Canada H4K 1K5

Mono Group, Inc.
847 Industrial Drive
Bensenville, IL 60106

Parkson Corporation
P.O. Box 408399
Fort Lauderdale, FL 33340

Robbins and Myers, Inc.
Fluids Handling Group
1895 W. Jefferson Street
P.O. Box 960
Springfield, OH 45501

Robbins and Myers, Inc.
Moyno Environmental Products

Rosemount Inc.
Varec Division
10800 Valley View Street
Cypress, CA 90630

Screw Conveyor Corporation
700 Hoffman Street
Hammond, IN 46327

Serpentix Conveyor Corporation
9085 Marshall Court
Westminster, CO 80030

Smith & Loveless, Inc.
14040 Santa Fe Trail Drive
Lenexa, KS 66215

Vaughan Company, Inc.
364 Monte-Elma Road
Montesano, WA 98563

Walker Process Equipment
840 North Russell Avenue
Aurora, IL 60506

WEMCO Division, Envirotech
  Corporation
P.O. Box 15619
Sacramento, CA 95852-1619

Wiesmann Engineering, Inc.
P.O. Box 10037
Largo, FL 34643

Chapter 1
Introduction

This manual, an update of the second edition published in 1977, aims to be the principal reference of contemporary practice for the design of municipal wastewater treatment plants. This manual was written for design professionals familiar with wastewater treatment concepts, the design process, and the regulatory basis of water pollution control and is not intended to be a primer for either the inexperienced or the generalist.

As a consensus publication, the manual focuses on current plant design practices of consulting firms, including performance information gained from typical examples of several thousand U.S. plants or plant modifications designed and built during the past 15 years. Reflecting work experiences of authors and reviewers (approximately 200 design professionals from across the U.S. and Canada), together with information carefully gleaned from more than one thousand literature sources, this manual constitutes a major overhaul of the second edition. Therefore, in this third edition, technologies that are no longer considered current practice are deleted; pertinent practices are updated; and new, unproven advances in the wastewater treatment field are touched on as emerging technologies. In some instances, alternate design methodologies or criteria are presented for issues that lack consensus.

This manual also serves to update and verify regulatory design requirements. However, because of the continuing evolution of environmental law at the local, state, and federal levels, some practice and design criteria presented in this manual may be inconsistent with requirements of authorizing agencies. Therefore, design engineers should check with appropriate regulating agencies for the most current requirements.

Some of the surveys included in this manual were conducted by WPCF, ASCE, and EPA to assess current practices of the design of municipal wastewater treatment plants. In addition to some limited special purpose surveys, WPCF conducted the following three surveys to quantify the prevailing practice of major consulting engineers in the field and the actual performance of integrated treatment works and unit processes contained in wastewater treatment plants: a survey of technology and design deficiencies at publicly owned treatment works—correlations of problems with plant parameters (1990), a wastewater treatment plant data survey (1989), and a design practice survey (1989).

In this manual, industrial and institutional wastes are discussed only to the extent that they affect the treatment of municipal domestic wastewaters. For detailed discussions of the administration, operation, and maintenance of municipal wastewater treatment plants, the reader may refer to "Operation of Municipal Wastewater Treatment Plants," WPCF's recently updated Manual of Practice No. 11.[1]

This two-volume manual consists of twenty chapters. Each chapter focuses on a particular subject or unit-processing objective. The sequence of the discussions presented is based on the premise that everything is connected to everything else. In other words, successful design of a municipal wastewater treatment plant is based on consideration of each unit process, upsystem and downsystem effects of that unit's place, and performance in the overall scheme of the treatment works.

Chapter 1 presents a brief historical overview of major developments that have shaped the design of municipal wastewater treatment plants. These developments range from simple pit privies and drainage ditches of the early 1800s to technologically complex, environmentally regulated, and intensively reviewed facilities of today. Also included is a discussion of present-day issues on which the future direction of plant designs may be based. Chapter 2 discusses design as a responsibility shared by several concerned entities and the variables that influence cost and decision making that are beyond the designer's control. Chapter 3 discusses opportunities for cost effectiveness of an overall design and introduces decision-making considerations that precede detailed consideration of any unit process.

Chapters 4 through 8 present basic subjects of site selection and plant layout, plant hydraulics and wastewater pumping, occupational safety and health, support systems, and construction materials. Chapters 9 through 16 discuss unit processes in the liquid-processing train, beginning with preliminary and primary treatment and ending with add-on processes for advanced wastewater treatment. These chapters trace an imaginary flow as if it were traveling through a typical treatment facility that incorporates primary and secondary treatment. Design of coarse screening, grit removal, and other preliminary treatment processes are discussed in Chapter 9, and design of primary treatment methods, including sedimentation and fine screens, is described in Chapter 10. Chapter 11 describes the design of suspended-growth systems; Chapter 12 covers attached growth and dual biological systems.

Chapter 13 discusses natural treatment systems that may be designed for nutrient and effluent suspended solids control. Chapter 14 summarizes design principles for several methods of disinfection and refers to comprehensive disinfection books for design details. Both traditional and updated methods of advanced waste treatment for nutrient control are discussed in Chapters 15 and 16. Chapter 15 limits discussion to integrated processes for biological nutrient removal in either one or more than one unit process. Chapter 16 discusses add-on processes for advanced wastewater treatment objectives, including physical–chemical processes for nutrient removal.

Chapters 17 through 20 present waste residue handling, processing, and disposal concepts. Chapter 17 discusses waste residue transport (both slurry and cake) and sludge-thickening and dewatering technologies, including re-lated conditioning considerations. Chapter 18 includes complete treatment of biological and chemical sludge stabilization, and Chapter 19 covers thermal sludge processing. Chapter 20 discusses the ultimate disposal of waste solids, noting that the ultimate disposal decision can significantly impact sludge-processing selections of preceding unit processes. The success or failure of any treatment facility can depend on the reliability of its ultimate solids dis-posal method.

The practice of designing municipal wastewater treatment plants, an im-perfect science, has no hard and fast rules that dictate successful design. Regulators and owners, recognizing that experience is critical to successfully determine design, should also apply common sense and remain open to change and innovation. Arriving at the most economical and attractive design for an integrated facility also requires designing for consistency with regu-latory guidelines of unit processes. Owners and designers working together can provide for flexibility of sludge-processing and disposal options capable of both recycling and safely disposing of waste solids. Furthermore, owners and regulatory personnel working together can continue to develop ultimate sludge disposal options. Failure to do so could make the operation of treat-ment works more difficult than it need be. Besides a sound design, the suc-cessful operation of wastewater facilities depends on other considerations such as preventive and routine maintenance, budgetary reserves that provide for replacing equipment, and responsible operation and management.

Published standards for specific design and loading criteria are referenced throughout the manual. Caution should be exercised when using such stan-dards and references for specific design or loading criteria. The original time and conditions on which the guidance was developed should be considered.

Some standards may remain universally applicable, some may not. For example, the "Recommended Standards for Sewage Works," commonly called the "ten state standards,"[2] and its several revisions, which have served as useful references for regulators and designers, were first developed in 1951 and were last revised in 1991. These guidelines continue to be a reference for the north central U.S.

WPCF's recently prepared Manual of Practice FD-17, "Combined Sewer Overflow Pollution Abatement,"[3] discusses concepts for peak flow mitigation and elimination of excessive infiltration/inflow through innovative collection systems and stormwater management. Accordingly, this design manual does not include opportunities and techniques for stormwater management in the collection system that may reduce the hydraulic peaks that dictate the sizing of many processes found in the plant's liquid-processing train. Also, the possibility of seasonal or alternate points of release are not detailed. Note that both of these considerations in plant design could result in significant savings in initial or life cycle cost.

Finally, because successful design can be achieved by a variety of approaches, some latitude for professional judgment and special circumstances must be preserved. This idea is clearly reflected in the results of a survey of 15 major design firms, summarized in Table 1.1. Survey results indicate that design professionals working independently of one another and exercising a choice of options achieved different facility designs that operate successfully. Any effort to require conformity of design criteria, whether by regulatory agency or peer directive, might limit the variety of successful design options.

*H*ISTORICAL OVERVIEW

This section provides a brief historical overview of the way in which municipal wastewater treatment has evolved in response to the public's changing perception of the sometimes conflicting needs of nature and society.

NINETEENTH CENTURY. In the early 1800s, the continental U.S. was a sparsely populated, undeveloped country. By the late 1800s, the population had grown from 5 to 76 million inhabitants; the occupied land mass had increased from 2.3 to 7.8 million km^2 (0.9 to 3 million sq miles). Then, as now, governing bodies responded to the public's negative reaction to sights and odors perceived to be unpleasant by improving waste management. Pit privies and open drainage ditches were replaced by buried sewers that conveyed not only wastes but also stormwater to other locations where it would impact less on the users. By 1860, the "sewered population" had increased to 1 million.[4]

Table 1.1   Summary of 1989 WPCF design practice survey.

Survey topics	Findings
Hydraulics (Chapter 5)	
Weir trough and launders	Majority (92%) preferred free flow under all design conditions.
Minimum design velocity for raw wastewater to prevent solids deposition	Majority (73%) preferred 2 ft/sec.[a]
Minimum design velocity for return activated sludge to prevent solids deposition	No majority . . . 33% preferred 2 ft/sec.
Raw wastewater pump selection	No majority . . . 42% preferred centrifugal, nonclog or screw, and 42% preferred centrifugal, open impellor.
Return activated sludge pump selection	No majority . . . 50% preferred centrifugal, open impellor.
Raw wastewater flow meter	No majority . . . 50% preferred a Parshall Flume.
Return activated sludge flow meter	Majority (67%) preferred a Magmeter.
Construction materials (Chapter 8)	
Coating for anaerobic digesters	No majority . . . 42% preferred coal tar epoxy.
Coating location in digester	Majority (62%) preferred under the roof and above and 1 ft below the sidewall water line.[b]
Preliminary treatment (Chapter 9)	
Preferred method and equipment for odor control	No majority . . . 40% preferred wet or carbon scrubbers.
Preferred grit removal system	No majority for transport system and technology . . . 28% preferred recessed impellor pumps with cyclone classifiers.
Use of trash racks ahead of coarse screens	Majority (67%) responded no.
Primary treatment (Chapter 10)	
Design surface overflow rate	No majority . . . 33% relied on ten state standards[2] (controlling condition of 1000 or 1500 gpd/sq ft at design average day or peak hourly flows, respectively).[c]
Configuration	Majority (69%) responded circular.
Suspended-growth biological treatment (Chapter 11)	
Configuration for 5-mgd plant[d]	No majority . . . 27% preferred complete mix.
Preferred sizing equation for aeration basin	No majority . . . 27% preferred the concepts of McKinney.
Preferred method for *Nocardia* foam control	Majority preferred return sludge chlorination (87%) and water spray (80%).

Table 1.1   Summary of 1989 WPCF design practice survey (continued).

	Survey topics	Findings
Suspended-growth biological treatment (Chapter 11) (continued)	BOD load (lb/hr[a]) used to calculate oxygen requirements for complete mix system with 8-hour detention time and diurnal hydraulic peak to average flow ratio of 2:1	No majority . . . 43% elected average day for the maximum month.
	Preferred air activated sludge aeration system for 10-mgd plant	Majority (57%) preferred fine bubble.
	Design SVI for sizing final clarifiers with an applied MLSS of 2000 mg/L	Majority (57%) preferred 150 mL/g.
	Average dry weather flow surface overflow rate that would be used to size final clarifiers	No majority . . . 47% preferred 600 gpd/sq ft.
Attached growth biological treatment (Chapter 12)	Preferred approach for biotowers	No majority . . . 30% preferred modified Velz equation and 30% preferred volumetric loading allowances.
	Minimum wetting rate for biotowers	No majority . . . 33% preferred 0.75 gpm/sq ft.[f]
	Use of soluble BOD, total BOD, or both for process design	No majority . . . 42% preferred soluble BOD.
	Maximum first-stage loading for design of rotating biological contactor	No majority . . . 38% preferred 2.5 lb soluble BOD_5/d/1000 sq ft.[g]
Effluent disinfection (Chapter 14)	Preferred method for secondary treatment system with 200-MPN/100 mL fecal coliform standard and stringent residual chlorine limit	Majority (77%) preferred chlorination/dechlorination.
	Assuming a design flow of greater than 1 mgd, the same coliform standard, no allowable residual chlorine, UV light disinfection is considered reliable and cost effective	No majority . . . 46% replied filter effluent and 46% thought it would not be appropriate for either a filtered or high-quality secondary effluent.

[a] ft/sec × 0.304 8 = m/s.
[b] ft × 0.304 8 = m.
[c] gpd/sq ft × 0.040 74 = $m^3/m^2 \cdot d$.
[d] mgd × 3785 = m^3/d.
[e] lb/hr × 0.453 6 = kg/h.
[f] gpm/sq ft × 0.679 02 = $L/m^2 \cdot s$.
[g] lb/d/1000 sq ft × 4.882 = $g/m^2 \cdot d$.

By the late 1800s, the total population had grown from 5 million (the early 1800s) to approximately 76 million, while the sewered population had increased from 1 (in 1860) to approximately 25 million. The drastic increase in the sewered population reflects the public awareness of the link between human disease and waste disposal practices. In response to public concern for health, municipalities improved waste management practices.

During the 1800s, wastewater disposal strategies were primarily a local responsibility. Some communities found that simple wastewater collection and dilution at the nearest watercourse were inadequate. By the early 1900s, approximately 60 towns and cities, totaling about one million inhabitants, had some form of municipal wastewater treatment. An objective of early treatment was the removal of visible settling and floating solids. By the end of the 1800s, several important wastewater management practices had been developed, which include the following:

- In 1886, the development of standards for wastewater treatment began with the establishment of the Lawrence, Massachusetts, experimental station.
- In 1887, the first formal biological waste treatment, an intermittent sand filter, was tried at Medford, Massachusetts.
- Rudolph Hering's recommendation of 0.11 m^3/s (4 cfs) per 1000 persons for the Chicago Drainage Canal was the first attempt to define assimilative capacity of a receiving body for wastewater.
- The first specific federal water pollution control legislation, which, through the Rivers and Harbors Appropriations of 1899 (often called the Refuse Act), prohibited dumping of solids to navigational waters without a permit from the U.S. Army Corps of Engineers.[5,6]

TWENTIETH CENTURY. Figure 1.1 summarizes the municipal wastewater treatment experience in the U.S. in the twentieth century. Figure 1.1 shows the changes that have occurred in this century.[7]

During the early 1900s, an era of rapid urbanization and industrial development, scores of people migrated from farmlands to the cities, thereby creating additional demands to sewer the growing population centers. Until the 1930s, the rate of increase of total population kept pace with the rate of increase of sewered population. However, by 1930, the nearly 50% increase in the total population outstripped the sewered population, which only increased by 50% of the small sewer-based population in 1900. Though sewer service lagged in the early 1900s, wastewater treatment technologies such as the following continued to evolve.

- In 1901, the first trickling filter was placed in operation in Madison, Wisconsin.
- In 1909, the first Imhoff tank was installed.

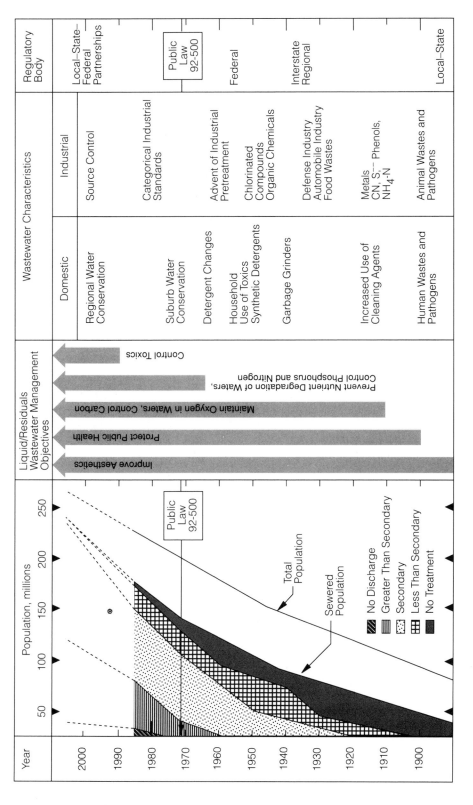

Figure 1.1 Historical trends for municipal wastewater management practices in the U.S.[1,4,5]

- In 1914, liquid chlorine was first applied for plant-scale disinfection.
- In 1916, the first activated sludge plant began operation in San Marcos, Texas.

Advanced forms of treatment were intended to protect the oxygen resources of the receiving stream. Initially, the oxygen demands of both carbon and nitrogen were a concern. However, in the 1930s, treatment strategies were only concerned with controlling the 5-day oxygen demand of biodegradable carbon. Practically, control of oxygen-demanding nitrogenous substances did not reemerge as a treatment objective until the 1960s.

Advanced forms of treatment increased the volume of waste solids or residue for disposal, thereby creating additional handling and disposal problems. Residual management practices trailed the development of liquid-processing alternatives. In the 1920s, separate, heated digestion and use of digestion gas evolved. In 1921, the first vacuum filters and centrifuges were used in Milwaukee. In the early thirties, the first large-scale dewatering and drying or incineration of sludges was used in Chicago.[8] The development and use of gravity thickening and chemical conditioning resulted from the need to mechanically dewater sludges.

The mathematical modeling of stream assimilative capacity was derived from the work of H.B. Streeter and E.B. Phelps in the 1920s. Also during the twenties, a few engineers, scientists, and public works officials organized and founded the Federation of Sewage Works Associations. The Federation underwent several name changes until, in 1991, it was renamed The Water Environment Federation (WEF).

Water pollution abatement programs gradually evolved from local public works and public health programs and remained the responsibility of local governing bodies during the early 1900s; later, the states began to oversee the programs. Federal involvement was inconsequential until the depression years of the thirties. At that time, federal expenditures and recovery programs financed local public works programs that built many of the major interceptors still found in today's older cities. As a result of federally financed programs, the number of people receiving some form of wastewater treatment increased dramatically. Also, cooperative watershed-planning agencies that emerged at this time addressed problems and needs on a regional basis rather than on a local, state, or national basis.

During World War II, the nation's water pollution control efforts were put on hold, though urban and industrial growth continued unabated. After World War II, the nation's waters again became the focus of attention, resulting in the passage of the Federal Water Pollution Control Act of 1948, the first comprehensive federal program aimed at controlling water pollution. In 1952, the extension of the act and other legislation that provided more appropriate funding at nearly 5-year intervals for the next 15 years increased federal involvement and stimulated the construction of municipal wastewater treatment facilities. The Clean Water Restoration Act of 1966 authorized expenditures of more than $1 billion, with a matching federal grant program of more than 30%.

Wastewater technology in common use from 1935 to 1955 was described in the first edition of this design manual.[9] Secondary treatment, at that time, was the exception rather than the rule. The norm then of conventional and chemically enhanced primary treatment soon gave way to biological treatment, often designed for less than prevailing secondary treatment equivalency. By the late 1950s, the national population had grown to 120 000 000, with nearly 60 000 000, or 50% of the entire population, being recipients of some form of wastewater treatment.

Biological treatment of municipal wastewater is susceptible to upset from industrial waste. A number of ordinances were promulgated that controlled the discharge of industrial wastes to municipal wastewater. Increased use of secondary treatment produced more biological sludge, which is more difficult to process. In the 1950s and 1960s, dissolved air flotation was advanced for both main stream solids–liquid separation and waste sludge thickening. Organic polyelectrolytes for sludge conditioning and enhancement solids-liquid separation also evolved from the postwar chemical industry as an alternative to and enhancement of inorganic chemical-conditioning strategies.

In the sixties, heightened public awareness of and dissatisfaction with the condition of the nation's water resources resulted in the development of water conservation programs and reformulation of synthetic detergents. By 1970, liquid-processing technologies for advanced levels of suspended solids, organics, phosphorus and nitrogen control, and solids-processing technologies using elevated temperatures and pressures to oxidize or condition waste sludges began to emerge.

In 1977, WPCF and ASCE again cosponsored the publication of the second edition of this design manual.[10] The second edition described the existing technologies from 1955 to 1975, commonly secondary treatment concepts. It also described expectations of that period for broad-scale application of then newly emerging, advanced physical–chemical treatment. Initially, many of the more sophisticated forms of advanced physical–chemical treatment did not gain general acceptance. Some contributing factors were operating difficulties, a sudden increase in energy costs (in the early seventies), a greater understanding of and concern for the environment, and an understanding of the limitations of the technology.

The Federal Water Pollution Control Act Amendments of 1972 (Public Law 92-500) spurred changes already taking place. The law provided ambitious goals for the nation's waterways, broadened enforcement powers under the National Pollutant Discharge Elimination System (NPDES) permit program, and provided $18 billion for construction grant authorization (in the form of 75% participating funding grants) for the first 3 years of the program.

Almost every component of the community's physical wastewater management system qualified for funding under the law's definition of treatment works.

The law also created a new agency to interpret and implement the desires of Congress—the U.S. Environmental Protection Agency (EPA). EPA actively participated in all aspects of water pollution control planning, including the design, construction, and operation of wastewater collection conduits and treatment facilities.

During the next 15 years, the federal government spent approximately $50 billion for construction grants. Changes to original legislation added, eliminated, expanded, or contracted deadlines, implementing regulations and guidelines, goals, and definitions. Funding authorizations and appropriations also changed. Many states who developed sufficiently sophisticated regulatory organizations took over management and execution of the federal program. Design was influenced by greater public participation in the decision-making process, more stringent industrial pretreatment programs and sewer use ordinances, and delays in construction of improvements because of increased regulatory involvement in funding and technical review.

Figure 1.1 shows that during the last two decades, the number of people receiving secondary or higher levels of treatment has doubled from 70 to 140 million. Additionally, untreated wastewater releases from sewered population centers has almost been eliminated.[7]

This third edition of the manual focuses on the technology successfully and routinely used for wastewater treatment in the U.S. from 1970 to 1990. During this 20-year period, contemporary design has emphasized secondary and more stringent treatment objectives. More stringent treatment objectives have been largely limited to high-level control of carbonaceous and nitrogenous oxygen-demanding pollutants, suspended solids, and phosphorus. Nitrogen elimination and zero discharge land-based treatment strategies have, for the most part, been limited to small communities and highly restricted waterways. Toxic control, especially for organic chemicals, by plant-based practices or user controls has only recently emerged as a design consideration.

Since the publication of the second edition of the manual in 1977, the high cost of traditional wastewater treatment approaches has led to the rediscovery and advancement of natural treatment alternatives in communities with an ample land resource. Today's newer, natural treatment technologies are now in the same position as sophisticated forms of advanced physical–chemical treatment were when they were described in the manual's second edition. Sound design and operating criteria for natural treatment technologies are less certain than those for well-established conventional technologies because practical experience has only begun to provide a basis for design.

In the last 20 years, technical advances in wastewater treatment include

- The recent introduction of vortex separators for grit removal,
- Greater understanding and application of processing enhancements to control activated sludge bulking,
- Processing improvements that promote biologically enhanced nitrogen and phosphorus removal,
- Pure oxygen activated sludge treatment,
- Sequential batch reactors, and
- Renewed interest in attached growth biological treatment technologies, with greater understanding of their expanded capabilities.

Technologies that have emerged include

- Belt filters for thickening and dewatering applications,
- Isolated application of high-pressure dewatering presses, and
- Use of both composting and chemical solids stabilization to achieve waste residue more acceptable for beneficial recycling.

In many instances, these changes resulted from increasing consideration of foreign technologies, particularly those of Europe. Technologies largely abandoned or found wanting during this 20-year period include vacuum filtration, independent physical–chemical treatment with granular activated carbon adsorption, high-pressure and temperature sludge conditioning, and nozzle centrifugation for waste activated sludge thickening. Legislative closure of ocean sludge disposal in 1990 marked the end of the 20-year period.

E*PA NEEDS SURVEY*

EPA's biannual Needs Survey provides a means for assessing the present status and future direction of the nation's water pollution control efforts for POTWs. Tables 1.2 and 1.3 summarize the 1986 Needs Survey Report to Congress.

PRESENT CONDITIONS. Table 1.2 classifies the municipal wastewater treatment plants in operation by their rated average day design flow capacity. As shown, nearly 80% of the plants are designed for < 3785 m^3/d (1 mgd); only 4% of the plants provide approximately 65% of the nation's wastewater treatment capacity. Data also show that the largest percentage increases, both in terms of number of plants and total design flow, are expected to occur in the small, <380 m^3/d (0.1 mgd), and in the large, >37 850 m^3/d (10 mgd), installation categories.

Table 1.2   Summary of treatment plants by flow range.[7]

Present and projected number of plants and their total design flows	Flow range, mgd[a]				
	0.01 to 0.10	0.11 to 1.00	1.1 to 10.0	≥ 10.1	Total
Plants, number (%)					
1986	4 960 (32)	7 003 (45)	2 898 (19)	577 (4)	15 438 (100)
2005	5 664 (33)	7 430 (44)	3 191 (19)	695 (4)	16 980 (100)
Increase	704 (14)	327 (6)	293 (10)	118 (20)	1 542 (10)
Design flow, mgd (%)					
1986	251 (1)	2 671 (7)	9 372 (26)	24 383 (66)	36 677 (100)
2005	287 (1)	2 763 (6)	10 389 (24)	29 584 (69)	43 023 (100)
Increase	36 (14)	92 (3)	1 017 (11)	5 000 (24)	6 346 (17)

[a] mgd × 3785 = m^3/d.

Table 1.3 Summary of treatment plants by level of treatment.[7]

Present and projected number of plants, total design flows, and populations served	Level of treatment					
	Less than secondary	Secondary	More stringent than secondary	No discharge[a]	Other[b]	Total
Plants, number (%)						
1986	2 112 (14)	8 403 (55)	3 115 (20)	1 762 (11)	46 (<1)	15 438 (100)
2005	45[c] (<1)	9 675 (57)	4 906 (29)	2 273 (13)	81 (<1)	16 980 (100)
Increase (decrease)	(2 067) (98)	1 272 (15)	1 791 (57)	511 (29)	35 (43)	1 542 (10)
Design flow, mgd (%)						
1986	5 529 (15)	15 714 (43)	14 373 (39)	973 (3)	88 (<1)	36 677 (100)
2005	387[c] (1)	18 844 (44)	21 996 (51)	1 686 (4)	110 (<1)	43 023 (100)
Increase (decrease)	(5 142) (93)	3 130 (20)	7 623 (53)	713 (73)	22 (25)	6 346 (17)
Average flow, mgd						
1986	2.6	1.9	4.6	0.6	1.9	2.4
2005	8.6	1.9	4.5	0.7	1.4	2.5
Population served, mil						
1986	28.8	72.3	54.9	5.7	10.5	172.2
2005	2.9	107.4	85.8	10.8	36.8	243.7
Increase (decrease), %	(90)	48	56	89	250	42

[a] Evaporation/percolation lagoons, land application systems, and other plants with effluent reuse.
[b] This grouping provides closure of categorical numbers with totals. Available information does not disclose plant type.
[c] Plants with tentative approval of secondary treatment waiver application as of 1986.

Design of Municipal Wastewater Treatment Plants

Table 1.3 compares the number of wastewater treatment plants and their rated average day hydraulic capacity with the design level of treatment.

FUTURE DIRECTIONS. Tables 1.2 and 1.3 present EPA's projections. Table 1.3 shows that, in the next 20 years, EPA expects that plant performance levels of lower than secondary treatment will be limited to fewer than 50 coastal communities, and the majority of the nation's plant capacity will be devoted to performance levels above secondary treatment equivalency.

In the 1986 Needs Survey, EPA estimated that approximately 23% of the major (1 mgd or larger) and 17% of the minor (< 1 mgd) wastewater treatment plants in the U.S. needed to comply with their final NPDES discharge requirements and needed some form of construction to achieve compliance. Approximately 50% of these facilities will require construction of some form of advanced treatment facilities.

EPA needs surveys through 1988 show that a high financial commitment will be required to bring the nation's wastewater treatment plants into compliance with the then prevailing standards. Considering pending issues that will likely be resolved in favor of more stringent standards and a broader area of regulatory control, these projected financial burdens can be expected to increase. Future economic challenges loom even greater when considering other aspects of wastewater management beyond the treatment plant such as combined sewer overflow and stormwater discharge.

Congress, while expanding its emphasis on standard setting and enforcement activities, is moving away from its historic grant programs to programs offering low interest rates on loans and a newly defined state and federal partnership. The success of EPA in achieving its goals remains to be demonstrated. Nevertheless, expected economic pressure and competition for limited public revenue can result in an increased burden on the designer for solutions that offer greater economy and require more extensive consideration of alternative technologies and management practices.

In the future, water conservation will become more common in water-limited areas. Water conservation needs will undoubtedly result in (if not force) beneficial recycling of treated wastewater for cooling, irrigation, and agricultural use. Industries and communities may continue to share the POTW for compatible waste control. These considerations suggest that wastewater received at treatment plants is likely to increase in strength. Increased efforts to control the discharge of toxics to the nation's waterways will continue into the future.

THE DESIGNER'S ROLE. Ideally, designers translate regulatory requirements, public goals, financial constraints, and technology into wastewater treatment plants that operate reliably, economically, and unobtrusively and that meet discharge standards. Designers of future facilities may have to change from the past emphasis of relying on short-term convenience to a focus on achieving long-term gain. Designers aware of past experiences but amenable to change can meet future challenges of new standards and goals as defined by the public's changing perception of need and national priorities.

Water pollution abatement resulting in unbalanced atmospheric deterioration and land degradation is not a viable solution. To control the cost of treatment and address environmental impacts, the designer should explore opportunities in the service area and its collection system. In designing a municipal wastewater treatment plant, the designer walks a careful line between providing facilities that can respond to uncertainties of the future and excessive overdesign, which can result in the misuse of public monies for superfluous facilities. When in doubt, the simplicity of proven technology may serve the designer and his client better than new, unproven technology.

*R*EFERENCES

1. "Operation of Municipal Wastewater Treatment Plants." Manual of Practice No. 11, Water Pollut. Control Fed., Alexandria, Va. (1990).
2. "Recommended Standards for Sewage Works." Great Lakes—Upper Miss. River Board of Sanit. Eng. (1991).
3. "Combined Sewer Overflow Pollution Abatement." Manual of Practice No. FD 17, Water Pollut. Control Fed., Alexandria, Va. (1989).
4. Babbitt, H.E., "Sewerage and Sewage Treatment." 6th Ed., John Wiley & Sons, New York, N.Y. (1947).
5. Galli, G., "100 years of construction news." *Eng. News-Rec.*, **192**, 433 (1974).
6. Longest, H.L., "The Past is Prologue: Looking Back, Looking Ahead at the U.S. EPA's Construction Grants Program." Paper presented at the 9th U.S.-Jap. Conf. Sew. Treat. Technol., Tokyo (1983).
7. "1986 Needs Survey Report to Congress." EPA 430/9-87-001, U.S. EPA, Office Munic. Pollut. Control (1987).
8. Lue-Hing, C., *et al.*, "Treatment Technologies and Effluent Quality for the Future, 2000+." *Proc. Public Waste Manage. Ocean Choice*, Mass. Inst. Technol., Cambridge (1986).
9. "Sewage Treatment Plant Design." Manual of Practice No. 8, Water Pollut. Control Fed., Washington, D.C. (1959).
10. "Wastewater Treatment Plant Design." Manual of Practice No. 8, Water Pollut. Control Fed., Washington, D.C. (1977).

Chapter 2
Overall Design Considerations

*I*NTRODUCTION

Many factors and considerations beyond process selection, design criteria, and their application influence the magnitude and costs of a treatment plant project. This chapter describes the factors and considerations representing today's practices.

---

*P*ROJECT PARTICIPANTS

Decision making for any municipal wastewater treatment plant design involves many participants: the public, the regulator, the legal counsel, the owner, the designer, the financier/investment banker, the operator, and the contractor. The owner serves as the focus of all of the project's activities. The design professional, as a member of the design team under the owner's direction, responds to the project's design needs. The design team consists of principal design engineers and supporting specialists.

All projects begin with an identification of a problem by the regulator, the public, legal counsel, or owner. The design professional then enters the project during the idea generation and evaluation phase of the problem-solving activity. Thereafter, the design professional or firm generally participates actively in all of the project's activities, typically until the end of the first year of operation.

Figure 2.1 shows cost-saving opportunities that most projects provide. As shown, major opportunities for cost savings occur during the initial phases of project development. Federal regulations for the Construction Grants and State Revolving Loan Programs provide for independent, third-party value engineering of the detailed design (typically at the 30 and 60% completion phase) if project costs exceed $10 million to stimulate cost savings. Although not included in Figure 2.1, operator input as part of the process selection and detail design effort can and should be an important contribution to the cost-saving effort, particularly for projects involving operating plants.

*D*ESIGN ACTIVITIES AND STANDARDS

Today's design practice for municipal wastewater treatment plants reflects rule-making and guidance documents of EPA, the requirements of the appropriate state regulatory agency, responsiveness to needs and codes of the implementing owner community and national building organizations, the design guidelines or criteria found in readily available reference documents, and conventions and practices of organizations and individuals who furnish equipment and design facilities. Regulatory, community, and national standards are always subject to change; therefore, the designer should ensure that the standards are up to date and relevant before they are applied to the design.

Table 2.1[1] describes the chronological evolution, typical of the late 1980s, of municipal wastewater treatment plant design. Typically, even with timely project progress and approval, elapsed time from the onset of facilities planning through the completion of the plant's first year of operation has been 7 to 9 years. The lack of future federal participating grants may accelerate the

---

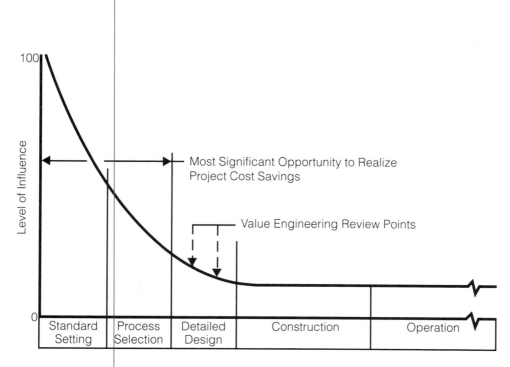

Figure 2.1 Opportunities to influence project cost during project phases.

Points of Opportunity to Influence Project Costs									
Project Participant	Standard Setting		Process		Construction		Aesthetic Impact Mitigation		Plant Operation
	Emissions	Design Criteria	Selection	Detailed Design	Materials	Methods	Const.	Oper.	
Public	●		●				●	●	
Regulator	●	●	●	●	●	●	●	●	●
Owner		●	●	●	●	●	●	●	●
Designer		●	●	●	●	●	●	●	●

implementation time through the elimination of many federal approval steps and the delegation of recently acquired federal responsibilities back to the states.

One example of the control that was imposed by EPA through the Construction Grants program concerns equipment, process, and power redundancy for reliable wastewater treatment works as summarized in Tables 2.2 and 2.3.[2] Reliability classes are based on beneficial use of receiving waters and the probable adverse impact on them of inadequately treated discharge. These classes are

- Class I—Treatment works that discharge to navigable waters that could be permanently or unacceptably damaged by effluent that

---

Overall Design Considerations   **19**

is degraded in quality for only even a few hours (for example, discharges near drinking water sources, shellfish waters, or contact sports).

- Class II—Treatment works that discharge to navigable waters that would not be permanently or unacceptably damaged by short-term effluent quality degradation but could be damaged by continued (several days) effluent quality degradation (for example, discharges to recreational waters).
- Class III—Treatment works not otherwise designated as class I or II.

Some states may have more stringent reliability requirements that may vary as a function of the rated capacity of the plant.[3] Some of the above reliability measures may also be appropriate at key points in the processing scheme if the plant effluent is injected to groundwater or directly provided for beneficial reuse.

Table 2.1 Chronology of activities in the 1980s related to plant design.[1]

Steps	Activities	Review
Facilities planning	• Define current conditions and problems. • Develop and assess alternate solutions for — Implementation, — Environmental benefits and liabilities, and — Cost effectiveness. • Select proposed solution. • Perform preliminary design. • Estimate costs and manner of funding. • Develop advance implementation schedule.	Review by public; review and approval by owner, state, and regional federal regulatory agencies; and special EPA review and approval if incremental AWT costs exceed $2 million.
Design	• Review and set design concept. • Prepare appropriate process instrumentation and control diagrams.	

Table 2.1 Chronology of activities in the 1980s related to plant design (continued).[1]

Steps	Activities	Review
Design (continued)	• Perform detailed design evaluations as needed. • Prepare detailed plans and specifications to describe and locate proposed improvements in the contract documents in sufficient detail to allow installation by contractor-chosen methods. • Prepare plans of operation for proposed improvements and existing treatment works during construction.	Review of design using internal quality control/quality assurance procedures; external value engineering review if project costs exceed $10 million and response to review comments; review and approval by owner, state, and federal regulatory agencies.
Construction	• Receive funding notification and respond to design review comments. • Receive and evaluate bids and select lowest responsive bidder. • Secure funding and award project. • Provide resident inspection and office construction administration services, including review and approval of shop drawings, equipment submissions, invoices and change orders, and verification of work performed in accordance with contract documents. • Prepare operation and maintenance manuals and provide start-up assistance and training. • Approve warranty testing and beneficial acceptance of improvements. • Provide performance certification after 1 year of operation.	Review and approval of owner, state and federal agencies occurs at each request for payment by the contractor. Final federal audit and grant payment occurs after the 1-year certification of performance.

Table 2.2   EPA Construction Grants Program guidance for minimum equipment and process reliability for the liquid-processing train.[2]

Component	Reliability classification					
	I		**II**		**III**	
	Treatment system	Power source	Treatment system	Power source	Treatment system	Power source
Holding basin	Adequate capacity for all flows	Optional	Not applicable	No	Not applicable	No
Degritting						Yes
Primary sedimentation	Multiple units[a]	Yes	Same as class I	Optional	Two minimum[a]	No
Trickling filters	Multiple units[b]	Yes	Same as class I	Optional	No backup	No
Aeration basins	Two minimum w/equal volume	Yes	Same as class I	Optional	Single unit permissible	No
Blowers or mechanical aerators	Multiple units[c]	Yes	Same as class I	Optional	Two minimum[c]	No
Diffusers	Multiple sections[d]		Same as class I	Optional	Same as class I	
Final sedimentation	Multiple units[b]	Yes	Multiple units[a]	Optional	Two minimum[a]	No
Chemical flash mixer	Two minimum or backup[e]	Optional	No backup	Optional	Same as class II	No
Chemical sedimentation	Multiple units[b]	Optional	No backup	Optional	Same as class II	No
Flocculation	Two minimum	Optional	No backup	Optional	Same as class II	No
Disinfection basins	Multiple units[b]	Yes	Multiple units[a]	Yes	Same as class II	No

[a] Remaining capacity with largest unit out of service must be for at least 50% of the design maximum flow.
[b] Remaining capacity with largest unit out of service must be for at least 75% of the design maximum flow.
[c] Remaining capacity with largest unit out of service must be able to achieve design maximum oxygen transfer; backup unit need not be installed.
[d] Maximum oxygen transfer capability must not be measurably impaired with largest section out of service.
[e] If only one basin, backup system must be provided with at least two mixing devices (one may be installed).

Table 2.3 **EPA Construction Grants Program guidance that is common to all reliability classifications for minimum equipment and process reliability.[2]**

Component	Guidance
Power source	Two separate and independent electric power sources from either two separate utility substations or one substation and one standby generator.
Pumps	Sufficient capacity with largest unit out of service to handle peak flow. Sludge backup pumps need not be installed and electrical redundancy may be restricted to main pumps.
Critical lighting and ventilation	Electrical redundancy for all operating stations.
Liquid-processing train	Minimum facilities, including trash removal or comminution, grit removal (except treatment works that do not pump or dewater sludge), provision for removal of settled solids from channels, pump wells, and piping before degritting or primary sedimentation. Unit operation bypass allowed around comminution, regardless of the number of units, unless two or more units are provided and operating unit can handle peak flow. Mechanically cleaned bar screens or comminutors must have backup bar screen and power source.
Solids-processing train	Provisions for preventing contamination of treated wastewater and alternative method of sludge disposal, treatment, or both if unit operations lack a backup capability.
Holding tanks	Permissible as alternative to backup capability if capacity of tanks is sufficient for estimated time of repair.
Anaerobic digestion	At least two tanks. Mixing equipment backup or flexibility must ensure that total mixing capability is not lost with one piece of equipment out of service. Backup equipment need not be installed.
Aerobic digestion	Backup basin not required. Include at least two blowers or mechanical aerators; less than design oxygen transfer is permissible with one unit out of service. Backup equipment need not be installed.
Dewatering	Multiple units with capacity to dewater design sludge flow with largest capacity unit out of service.

$P_{ROCUREMENT}$

CONTRACTUAL APPROACHES. Design- and construction-related
activities described in Table 2.1 represent what may be termed the classical
architect/engineer approach. With this approach, vendor-furnished equipment
is procured according to performance or prescriptive equipment specifications through contractors who are bidding from plans and specifications
prepared by a consulting engineer.

All funding and ownership of facilities rest with the owner in the classical
architect/engineer approach. The owner may also elect to procure program
administrative, design, and construction management services in complicated
projects.

Alternate contractual approaches may incorporate turnkey or a combination of turnkey and architect/engineer procurement of desired facilities. The
term "turnkey" applies if a single entity assumes total responsibility for plant
design, construction, start-up, and sometimes financing. The approaches and
their relative merits are presented in Table 2.4.[4] All involve the activities of a
design team in some capacity.

In a turnkey approach, the owner cannot escape the risks associated with
changing regulations and wastewater variability. Legal counsel should be
sought early and used throughout the preparation of turnkey contracts. Con-
flict in turnkey projects results most often from the owner's failure to define
clearly the expectations in terms of performance and quality of goods and services.

Privatization, the involvement of nonpublic and entrepreneurial interests
in project development and system operation, provides another option for
procurement of project needs. In the most comprehensive of three general alternative approaches to privatization, the private enterprise provides all required engineering, construction, funding, and operational services. These
are covered by contractual arrangements between the municipal owner and
the developer as to specific responsibilities and remuneration. This approach
would critically depend on existence of favorable federal tax incentives for
the developer. Such incentives do not presently exist. As another option, a
municipality could design, build, and sell the facility or enter into a sale or
leaseback agreement with a private corporation. As the third approach involves more limited private sector participation, private sector operating services may be obtained for a facility designed and built by the municipality or
municipal agency.

SPECIFICATIONS. Two general types of specifications are used to
procure goods and services for construction of a municipal project under any
procurement option: the prescriptive and the performance specification.

Table 2.4   Relative merits of contractual approaches.[4]

Consideration	Turnkey[a]	Architect and engineer[b]	Combination[c]
Financial	Theoretically, the most costly method because contractor must assume all risks associated with performance and possibly financing. Nonetheless, savings may result because of timeliness and single-source responsibilities.	Theoretically, less costly than turnkey approach because a variety of contractors can bid on complete plans and specifications.	Theoretically, less costly than turnkey approach.
Time	Probably one of the quicker methods of completing a project.	Potentially the slowest way to implement a project, particularly if equipment is prepurchased.	Between A&E and turnkey. Time saved by having vendor prepare balance of plant specifications is offset by bidding time required by general contractors.
Engineering	Somewhat less engineering control and review because of rapid schedule.	Probably good engineering/quality control levels.	Reasonably good engineering control.
Competition	Severely limited number of competitive bids, especially if central process is proprietary.	A construction competition would be achieved, A&E selection may be limited where process is proprietary.	—

Table 2.4 Relative merits of contractual approaches (continued).[4]

Consideration	Turnkey[a]	Architect and engineer[b]	Combination[c]
Risk	Maximum for contractor, minimum for owner with clearly defined scope. Quality of performance, system, equipment, and service materials may be less than anticipated unless adequately defined with reinforcing consequential damages.	Reasonable; equipment/process vendor will likely have most liabilities; pre-qualification may help.	—
Other	Least trouble-free approach for owner with contracts that clearly define scope.	Possible coordination difficulties if owner furnishes equipment.	—

[a] Total responsibility for engineering design; construction; startup; and, in some cases, financing assumed by one entity.
[b] Complete plans and specifications prepared by consulting engineer for issuance to general contractors for bidding and construction. This process may include installation of vendor-furnished equipment that has been prepurchased by owner through request-for-proposal process or earlier prescriptive equipment specifications. If so, vendor drawings are usually furnished in the final bid package.
[c] Combination of the other two options so that some of the plant is furnished by the vendor and the remainder is designed by the architect and engineer; combined plans and specifications are then issued to and bid by general construction contractors.

A prescriptive specification sets forth explicit criteria governing the processes (or services) that are to be provided. Its explicit nature complicates its preparation but offers the owner maximum protection in terms of protecting the quality of the installation and easing bid comparison. It typically delineates acceptable manufacturers and suppliers, normally providing for the owner's consideration of or "equal products" under a format or procedure for the designer evaluation.

A performance specification defines the input conditions and the desired objective. Vendors favor this type of procurement because it allows them greater latitude in the use of their products. This latitude tends to diminish the owner's control of the quality of the installation and complicate the analysis of bids. These risks, however, may be mitigated by prequalification. The prequalification procedure entails providing an opportunity, before the bidding process, for equipment manufacturers to submit qualifying information as a basis for the engineer's determination of whether the product conforms to the specifications and can thus be considered for bidding by construction contractors. In some cases, the procurement of the equipment and the procurement of its installation constitute separate stages, with the equipment procurement preceding completion of final detailed engineering.

Award of contracts is normally based on first cost if all items covered can be evaluated equally on a first cost basis. If not, an evaluated life cycle cost is most equitable. Evaluated life cycle bids should realistically reflect the cost of all consumable products (including replacement parts), adjusting for inflation during the anticipated service life of the installation. Further, the evaluation should reflect the cost of subsequent processing of all products and sidestreams if tangible differences between alternatives are anticipated. Bonus or penalty factors may be used to emphasize features that are important to the owner. The evaluation methodology should be discussed with all vendors before preparing final specifications to ensure the methodology's fairness.

Other critical aspects of specifications are the provisions for protecting the owner against contract violation of guarantee and completion time stipulations. These provisions may be for liquidated damages, consequential damages, or both types; the total, generally, should not exceed the amount of the contract.

Liquidated (or delay) damages are penalties executed for failure to complete the project in the time specified in the contract. Such damages may include any identifiable loss to the owner due to delay and such considerations as owner inspection costs, delay expenses from other ongoing dependent contracts, additional electric power or chemical costs experienced while awaiting operation of the new process, and penalties for failure to meet emissions standards.

Consequential (or performance) damages are losses resulting from the failure of the process to work as specified. They should accurately reflect the cost for operating with or correcting the process deficiency. Performance damages should only be triggered as a last resort after all other remedies to correct the deficiencies have been exhausted. Performance damages may include the cost of installing a compliant system, the cost of added technology to bring the original system into compliance, and any additional costs for power and chemical consumption, corrected for inflation. Although performance damages may be large, they also assure the owner that only responsible vendors will be inclined to bid the job. Vendors deserve explicit contract language defining and limiting consequential damages.

DEFINING OVERALL OBJECTIVES

The determination of performance objectives and influent waste characteristics precedes functional plant design. In general, the designer has little control over either, although their proper definition and evaluation remain a critical responsibility of the designer. This section considers design objectives both from the broad-based management perspective and the more narrow perspective of compliance standards that are routinely encountered in today's practice.

MANAGEMENT OBJECTIVES. Wastewater treatment inevitably produces liquid discharge (which must be disposed of) and air emissions. Air emissions will become more important in the future, perhaps significant enough to sometimes dictate the liquid and solids treatment choices or, even more fundamentally, the location of the plant.

Treatment merely renders liquid and residue products suitable for their respective disposal objectives. The broadly defined management objectives for disposal of these two products are the same, namely, to first consider and implement reuse opportunities that are beneficial and economically attractive and, secondly, to dispose of remaining liquid and solids fractions so that they do not interfere with beneficial uses of the receiving media (air, land, or water). If interference is encountered, the selected disposal strategy should cause the least negative environmental impact for the three receiving media considered as a whole. Simply expressed, management objectives for municipal wastewater treatment can be listed in order of priority:

1. Reduce impact on receiving waters,
2. Maximize beneficial reuse, and
3. Minimize impact in other receiving media (air and land).

Beneficial reuse applications of wastewater liquids vary regionally to reflect groundwater recharge, agricultural reuse, and a myriad of industrial and recreational applications. Beneficial properties of wastewater sludges include their energy value, soil-building properties, and nutrient content. Tables 2.5 and 2.6[4] provide example target quality characteristics for representative applications of wastewater treatment liquid products and Table 2.7[5] provides similar information for solids. As shown, regulatory agencies may also define acceptable technologies (with attendant design criteria) as a substitute for, or supplement to, target quality characteristics for the product liquid or solids management program. Chapter 20 of this manual provides additional information regarding controlling regulations for the ultimate disposal of wastewater sludges. Detailed information regarding state-of-the-art beneficial reuse of the liquid and solid products from wastewater treatment is found in other WPCF publications.[4,6]

In February 1989, after nearly a decade in preparation, EPA proposed 40 CFR 257 and 503 criteria for the disposal of wastewater sludge.[7] Scientists, engineers, and operators reviewed these proposed rules and submitted comments, which EPA is considering in its revision. Criteria in the proposed rule should not be used before promulgation.

Inspection of Tables 2.6[4] and 2.7[5] shows that beneficial use of both liquid and solid products from wastewater treatment may require controlling biodegradable and refractory pollutants, heavy metals, and other selected inorganics. Many of these constituents are more correctly and economically controlled before they enter the collection system. Thus, one of the wastewater management objectives is to control the release of those substances to the collection system, thereby avoiding adverse effects on the design and operation of the wastewater treatment plant. Specifically, management may consider several controls for its service area:

- Enforce local sewer-use ordinances to protect the collection system, health and safety of plant personnel, and the public;
- Control releases of inorganic and organic toxics (hazardous materials) sufficiently to prevent processing upset, pass-through, or accumulation, any of which could otherwise impair the plant's liquid or solid disposal schemes or violate receiving media criteria;
- Ensure industrial compliance with federal categorical pretreatment standards;
- Control industrial release practices that could cause plant upset and performance standards violations; and
- Cost-effectively control groundwater and stormwater entering the collection system to avoid the needless reduction of collection system and treatment works capacity.

General Pretreatment Regulations[8] require all publicly owned treatment works (POTWs) designed to accommodate flows of more than 18 900 m^3/d (5 mgd) and smaller POTWs with significant industrial discharges to establish a local pretreatment program.

Table 2.5   Federal drinking water standards.[4]

Type of standard or parameter	Concentration, mg/L
Primary	
Total coli, MPN/100 mL	1
Turbidity, units	1
Radioactivity, PCi/L	
Gross alpha	
Particle activity	15
Radium 226/228	15
Tritium	20 000
Strontium 90	8
Arsenic	0.05
Barium	1
Cadmium	0.01
Chromium	0.05
Fluoride versus T	1.4–2.4
Lead	0.05
Mercury	0.002
Nitrate as N	10
Selenium	0.01
Silver	0.05
Endrin	0.002
Lindane	0.000 4
Methoxychlor	0.1
Toxaphene	0.005
2,4-D	0.1
2,4,5-TP silvex	0.01
Trihalomethanes	0.1
Secondary	
Color, units	15
Corrosivity	None
Odor, TON	
pH, units	6.5–8.5
Chloride	250
Copper	1
Foaming agents	0.5
Hydrogen sulfide	0.05
Iron	0.3
Manganese	0.05
Sulfate	250
TDS	500
Zinc	5
Hardness as $CaCO_3$	100

Table 2.6   Example criteria for wastewater reuse applications.[4]

Reuse application and parameter	Example criteria, mg/L
Groundwater recharge[a]	—
Agricultural reuse[b]	
Aluminum	5.0
Arsenic	0.1
Beryllium	0.1
Boron	5.0
Cadmium	0.01
Chloride	<200
Chromium	0.1
Cobalt	0.05
Copper	0.20
Fluoride versus T	1.0
Iron	5.0
Lead	5.0
Lithium	2.5
Manganese	0.2
Molybdenum	0.01
Nickel	0.2
Selenium	0.02
Sodium adsorption ratio, no units	<9
Sulfate	<400
TDS	<2000
Zinc	2.0
Vanadium	0.1
Industrial reuse[c]	
Alkalinity as $CaCO_3$	100–350
pH, units	8.2–10
Aluminum	0.01–5
Copper	0.05–0.5
Foaming agents	1
Hardness as $CaCO_3$	1–350
Iron	0.3–1.0
Silica dioxide	1–3
Suspended solids	5–10
TDS	500–700
Zinc	0.01
Recreational reuse[d]	—

[a] Approval is typically on a case-by-case basis, with separate standards set for surface spreading and direct injection.

[b] Limits vary depending on the crop and the potential for public contact. Representative values are presented.

[c] Criteria are specifically tailored for each industrial application. Some cooling water applications only require the addition of chlorine, acid, and inhibitors. Typical criteria for boiler feedwater are presented for low-pressure (<150 psig) and intermediate-pressure (150–700 psig) steam generation.

[d] Standards vary depending on passive or active (primary and secondary) uses according to state water quality criteria.

Table 2.7   Massachusetts solids quality targets for land application.[5]

Stabilization

Use one of the following acceptable processes for pathogen reduction (acceptability of process is specifically defined in terms of one or more of the following: time/temperature, pH/time, depth/time/temperature relationships, and volatile solids reductions):

- Aerobic digestion,
- Air drying,
- Anaerobic digestion,
- Low-temperature composting, or
- Lime stabilization.

If (1) the solids are to be used in a type I application, (2) applied where a crop intended for direct human consumption is to be planted within 24 months and the sludge will come into direct contact with the edible portion of the crop, or (3) the solids will be not mixed with the soil within 48 hours after application to the land, one of the following acceptable processes for further pathogen reduction must be incorporated (acceptability of process is specifically defined in terms of one or more of the following: time/temperature, time/solids content/temperature relationships, volatile solids reductions, and electron radiation/gamma ray irradiation dosages):

- High-temperature composting,
- Heat drying,
- Heat treatment,
- Thermophilic aerobic digestion,
- Electron radiation,
- Gamma ray irradiation, or
- Pasteurization.

Heavy metal or chemical limits for use classifications

	Type I Unregulated distribution following initial approval not to exceed, mg/kg[a]	Type II Regulated distribution with site-specific approval not to exceed, mg/kg[a]	Type III Site-specific and non-food-chain use approval; if greater than, record with deed, mg/kg[a]
Cadmium	2	25	25
Lead	300	1000	1000
Nickel	200	200	200
Zinc	2500	2500	2500
Copper	1000	1000	1000
Chromium (total)	1000	1000	1000
Mercury	10	10	10
Molybdenum	10	10	10
Boron (water soluble)	300	300	300
PCBs			
Solids as fertilizer	2		
Solids as soil builder	2		

[a] Dry weight basis.

During the past 20 years, concerns with atmospheric emissions of municipal wastewater treatment plant pollutants have been largely limited to aesthetics (odor) and the products of combustion from sludge incineration, boilers, and engines. Pathogens in aerosols from gaseous products have also received some attention. However, recent concern has arisen regarding the release (stripping) of hazardous organics from municipal sewers, aeration tanks, biotowers, anaerobic digesters, or composting systems.[9] The anticipated increased concern for air emission control throughout the U.S. is expected to result in stricter industrial discharge regulations and more careful scrutiny and application of current treatment technologies and plant-siting criteria.

COMPLIANCE STANDARDS. Effluent National Pollutant Discharge Elimination System (NPDES) discharge standards are intended to protect and preserve beneficial uses of the receiving water body based on water quality criteria, technology-based limits, or both. The receiving water quality criteria are typically established for the 7-day (consecutive), 10-year, low-flow regime. These criteria generally define the allowable release of conservative and nonconservative pollutants based on dilution and, if appropriate, the reaction kinetics of nonconservative pollutants in the natural environment. Conservative pollutants are typically defined as those pollutants that are not subject to change by biological or chemical action. If limits based on water quality criteria are less stringent than technology-based limits, technology-based limits are controls and the pollutants so limited are considered to be "effluent limited."

National minimums, termed secondary treatment equivalency, for municipal wastewater dischargers are defined in Table 2.8.[10] Note that secondary treatment regulations[10] include some exceptions and allow states to establish more stringent effluent quality requirements. As noted in the previous subsection, minimum national criteria for use of municipal wastewater sludges are presently being developed.

If secondary treatment equivalency standards would result in a violation of the receiving water's quality criteria, the discharger's impact on the receiving stream is termed "water quality limited," thus requiring advanced waste treatment (AWT). Sometimes, the macronutrient (total nitrogen and phosphorus) restrictions reflect technology-based effluent limitations (rather than water quality criteria) as regulatory agencies respond to state, regional, and international legislation and agreements calling for nutrient removal from municipal wastewater releases. Regulatory agencies often set stringent effluent standards to protect public water supplies or highly valued natural resources.

In the mid-1980s, limits for whole-effluent toxicity began to be included in some NPDES permits. Whole effluent toxicity is determined by bioassay procedures that measure both the acute and chronic toxicity of the plant's

---

Table 2.8   Minimum national performance standards for POTWs (secondary treatment equivalency).[10]

Parameter	30-day average	7-day average
Conventional secondary treatment processes		
5-day BOD,[a] the most stringent of		
Effluent, mg/L	30	45
Percent removal[b]	85	—
5-day CBOD,[a] the most stringent of		
Effluent, mg/L	25	40
Percent removal[b]	85	—
Suspended solids (SS), the most stringent of		
Effluent, mg/L	30	45
Percent removal[b]	85	—
pH, units	Within the range of 6.0–9.0 at all times	
Whole effluent toxicity	Site specific	
Fecal coliform bacteria, MPN/100 mL[c]	200	400
Trickling filters and stabilization ponds (equivalent of secondary treatment)		
5-day BOD,[a] the most stringent of		
Effluent, mg/L	45	65
Percent removal[b]	65	—
5-day CBOD,[a] the most stringent of		
Effluent, mg/L	40	60
Percent removal[b]	65	—
Suspended solids (SS),[d] the most stringent of		
Effluent, mg/L	45	65
Percent removal[b]	65	—
pH, whole effluent toxicity and fecal coliform bacteria remain unchanged		

[a] Chemical oxygen demand (COD) or total organic carbon (TOC) may be substituted for the 5-day BOD when a long-term BOD:COD or BOD:TOC correlation has been demonstrated.

[b] Percent removal requirements may be waived on a case-by-case basis for combined sewer service areas and for separate sewer areas not subject to excessive inflow and infiltration where the base flow plus infiltration is ≤120 gpd/cap and the base flow plus infiltration and inflow is ≤275 gpd/cap.

[c] Not defined in federal secondary treatment equivalency regulations but permits typically include cited levels, often only on a seasonal basis.

[d] The state may adjust the SS limits for ponds subject to EPA approval.

effluent to aquatic specimens, usually fathead minnows and daphnia. The inclusion of this parameter will likely be expanded to most permits reissued hereafter. Should a violation of a toxicity permit limit occur, a POTW may be required to conduct a toxicity reduction evaluation (TRE) to identify the sources of effluent toxicity, the causative pollutants, and the effectiveness of pollution control options in reducing the toxicity. If specific causative chemicals are identified, the POTW's permit may be revised to include specific limitations on these pollutants.

MUNICIPAL WASTEWATER CHARACTERISTICS: SOURCE AND PHASE

The wastewater quality and quantity characteristics of a plant's influent typically reflect the nature of the contributing area, water uses, and conditions of the conveyance system. Except for infiltration/inflow and industrial discharge conditions, these characteristics are not usually subject to alteration by the engineering planning required for the wastewater treatment plant project. In general, the plant designer determines the wastewater characteristics and develops an end-of-the pipe solution responsive to compliance standards and other wastewater management objectives. This section identifies wastewater characteristics critical to design.

CHARACTERISTICS OF COMMUNITY WATER SUPPLY. The nonconsumptive portion of the water used in a wastewater plant's service area constitutes most of the wastewater routinely received at a plant. This component of wastewater reflects the character of the raw water supply, the water treatment processes, and the history of beneficial water use.

Typically, groundwater contains scaling compounds and is highly buffered. At the other extreme, lake waters are usually slightly mineralized, with little or no buffer. Softening, demineralization, or both may be practiced with or without accompanying changes in background alkalinity. Simple raw water coagulation and clarification with aluminum or iron salts will add anions and deplete alkalinity. Soft, unstabilized waters will aggressively solubilize metal from the water system and customer distribution piping, with potential adverse effects on sludge quality and disposal scheme (for example, copper) or enhanced background phosphorus removals (for example, iron).

The magnitude of the available buffer (alkalinity) is important when designing one or more of the following processes: nitrification, metal salt or lime addition for phosphorus removal, pH adjustment, and closed system oxidation. The scaling nature of both the water supply and the wastewater can impair equipment if processes include boilers, steam, cooling, or water

---

seals. Chloride, sulfate, sodium, and other inorganics pass through wastewater treatment plants and can affect some disposal strategies. High chlorides also influence material selections for elevated temperature-processing schemes. Under anaerobic conditions, high sulfates can result in concrete corrosion, malodors, and toxic air quality.

INFILTRATION AND INFLOW. The second most significant component of the wastewater received at the treatment plant is infiltration (water seepage through the collection system pipes and vaults) and inflow (surface and subsurface stormwater entering the collection system, mirroring the character of the precipitation event). Inflow can be immediate or delayed; delayed inflow is the runoff associated with the melt of an accumulated snow cover.

Infiltration, a function of regional groundwater tables, will vary seasonally and annually. Old sewer systems, constructed before 1970 using mortar or mastic jointing materials, can substantially contribute to infiltration problems.

"Recommended Standards for Sewage Works" (commonly called the ten state standards) defines an allowable infiltration (exfiltration) rate of 200 gpd/in.dia/mile for new pipe construction.[11] Acceptable infiltration values before replacement or rehabilitation becomes appropriate in older, existing sewers can be ten or more times higher. This determination depends on a case-specific economic analysis for each sewer system. EPA guidelines consider infiltration to be excessive if the plant flow for nonrunoff conditions and adjusted for nondomestic use is greater than 120 gpd/cap. That guidance suggests that inflow for storm-induced flow conditions would be excessive if plant flow is greater than 275 gpd/cap.[1]

Infiltration water quality matches that of the native groundwater. Occasionally, native groundwater contributes to the background iron in raw wastewater, thereby enhancing phosphorus removal.

Inflow is high in communities with combined sewer systems. Although combined sewer service may be only a small fraction of a plant's service area, inflow derived from the combined sewer service area will often dominate the design and operation of the treatment works. Precipitation-induced inflow will reflect low-buffered, often acidic rainwater and the additional pollutants derived from rooftops, roadways, and land use of the service area.

The designer faces special design issues when the plant serves a combined sewer service area because oversized combined sewers and interceptors serve as traps for sediment and settleable solids. Often, elevated masses of influent screenings, grit, and suspended solids received at the treatment plant during or following a storm event will reflect the extent of past accumulations in the sewer and pollutants introduced with the stormwater.

In some combined sewer systems, special consideration is needed for regulators or overflow structures provided to direct combined wastewater flow in excess of sewer or plant capacities in the receiving stream. These systems can result in nonintended reverse-flow conditions where the receiving water elevation varies with the tidal pulse. Malfunctioning tidegates can allow seawater to enter the collection system during both dry and wet weather conditions. The transient or endemic receipt of seawater at the treatment plant may dictate special material selections to minimize maintenance (corrosion) and may impose inhibitory stresses on some unit processes (for example, sodium and sulfide toxicity in anaerobic digestion).

DOMESTIC WASTES. The most reproducible, if not predictable, component of municipal wastewater treatment is the character of its domestic waste. This waste reflects the character and practices of the served population. As the collection system and service population base expand, the impact of the domestic population becomes less pronounced in terms of peak to average and minimum ratios of both flow and pollutant masses.

Typically, domestic wastewater flow variations follow daily and weekly lifestyle patterns of the serviced residential customers. Figure 2.2[12] illustrates the classical variation of wastewater flows and pollutants. Minimum domestic flows and pollutant concentrations are observed during the early morning hours; peak flows and pollutant concentrations are typically experienced in the late morning or early afternoon. Fifty percent of the plant's domestic pollutant load during a weekday could arrive at the plant during an 8-hour period; the plant's maximum hourly load can be more than double the plant's average hourly load (8% of average daily load).

In general, peak nitrogen concentrations will precede the flow peak, whereas peak phosphorus concentrations will coincide with or lag behind the flow peak. In residential and college communities, phosphorus concentrations will reflect the weekend-washing habits of the users.

The "ten state standards" recommend that new treatment plants be designed for a domestic load contribution of at least 0.08 kg (0.17 lb) of BOD per capita per day and 0.09 kg (0.20 lb) of suspended solids per capita per day unless available information justifies other design criteria.[11] Historically, the 0.08-kg (0.17-lb) per capita per day value has been used to define the population equivalent of industrial waste. Further, those standards also recommend that if garbage grinders are used in the service area, the design domestic loads can be increased to 0.09 kg (0.20 lb) of BOD and 0.10 kg (0.23 lb) of suspended solids per capita per day.[11]

The ASCE/WPCF Manual of Practice, "Gravity Sewer Design and Construction," provides detailed information on estimating flow rates from a variety of sanitary, domestic, and commercial fixtures and sources.[13] The "ten state standards" recommend the use of an average daily flow value of 100 gallons per capita per day for new domestic service in undeveloped areas unless water-use data or better flow estimates are available.[11] This suggested per capita flow value includes an allowance for moderate sewer infiltration and commercial/institutional use. Sewer rehabilitation, enforcement of inflow control ordinances, and water conservation ordinances may reduce domestic

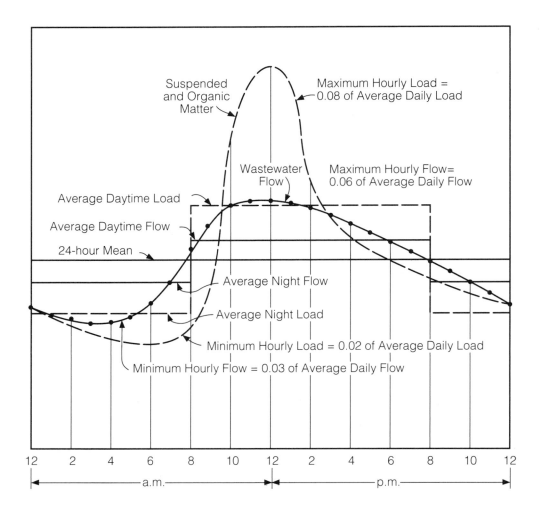

Figure 2.2   Hourly variation in flow and strength of municipal wastewater.[12]

waste flows. Current practice for projecting flow conditions relies less on such aggregate per capita allowances and more on evaluative procedures that separately consider water use and infiltration/inflow.

Domestic loads from an area's transient population should also be considered in designing a plant if its service area includes transients who work and visit the area during the day but maintain their permanent residence elsewhere. Readily available information from New York City suggests a nominal worker and transient flow contribution of 56.8 to 114 L (15 to 30 gal) per capita per day. Corresponding pollutant (BOD and suspended solids) load contributions from these sources range from 0.009 to 0.023 kg (0.02 to 0.05 lb) per capita per day. Plumbing codes, architectural standards, or state criteria can be used to develop site-specific estimates for restaurants and hotels.

Table 2.9 delineates the typical major pollutant composition of domestic wastewater using, in part, guideline recommendations of "Recommended Standards."[11] Table 2.10 provides additional detail for a variety of other wastewater characteristics. Typically, normal residential use of water will add approximately 200 to 300 mg/L of dissolved inorganic solids.[12,14-16]

Table 2.9   Typical major pollutant composition of domestic wastewater.

Parameter	Concentration by phase, mg/L[a]		
	Soluble[b]	Particulate	Total
Suspended solids[c]			
Volatile[d]			190
Inert			50
Total			240
5-day BOD[e]	65	135	200
Chemical oxygen demand[f]	130	260	400
Total nitrogen[g]	20	10	30
Total phosphorus[h]	5	2	7

[a] Based on assumed flow of 378 L/cap·d (100 gpd/cap).
[b] As typically defined with coarse filter, not necessarily soluble (see Table 2.11).
[c] Based on assumed 0.09 kg/cap·d (0.20 lb/cap/d) of suspended solids.
[d] Eighty percent volatile.
[e] Based on assumed 0.08 kg/cap·d (0.17 lb/cap/d) of BOD_5 and 30% soluble.
[f] Twice the BOD.
[g] 0.02 kg/cap·d (0.04 lb/cap/d) of total nitrogen and 65% soluble, with ammonia nitrogen composing most of the soluble content.
[h] 0.003 kg/cap·d (0.006 lb/cap/d) of phosphorus and 65% soluble.

INDUSTRIAL, COMMERCIAL, AND INSTITUTIONAL WASTE CONTRIBUTIONS. Quantities of industrial, commercial, and institutional components of municipal wastewaters may sometimes be the most difficult to estimate for design purposes, particularly projected future contributions. Occasionally, the industrial or institutional component can dominate the plant design.

Industrial contributions in any municipal wastewater may range from insignificant to many times the domestic contribution. Standard commercial, industrial, and institutional flow allowances used by some U.S. cities are presented in other publications.[13] Seasonal, weekly, holiday, and daily variations of industrial releases should be expected unless information to the contrary exists. The type of waste—simple carbohydrates or complex proteins

Table 2.10   Other characteristics of domestic wastewater.

Characteristic	Description
Physical	
Solids	
Screenings	Highly variable, depending on screen size.
Grit	Highly variable; nominal design values are 5 mg/L and 10% volatile.
Floating debris, grease, and oil slicks	Indicate untreated or ineffectively treated waste. The simplest forms of treatment are generally effective in removing floatables. Nominal design values for raw wastewater floatables are 5 mg/L and 90% volatile with approximately 30 200 kJ/kg (13 000 Btu/lb) of volatile matter.
Suspended solids	Typically, 50% of settled solids; volatile matter is approximately 23 200 kJ/kg (10 000 Btu/lb).
Dissolved solids	Highly variable amounts but will constitute approximately 65% of the total solids; approximately 65% of dissolved solids will be volatile (see chemical characteristics below).
Temperature	Wastewater temperatures depend largely on the climate and, to a small degree, whether groundwater or surface water is the source of the service area's water supply. Temperatures can vary from less than 10°C to more than 20°C during a year. Temperature affects biochemical reaction rates and gas solubility.
Color	The color of domestic wastewater indicates its age. Fresh wastewater is normally gray. As wastewater ages, it becomes black due to pronounced septicity.
Odors	Principal odors are generally associated with various reduced compounds of sulfur such as hydrogen sulfide and mercaptans.
Chemical	
Organics	Principal organic substances of domestic wastewater are proteins (40–60%), carbohydrates (20–40%), and lipids (10–20%). Additionally, wastewater contains measurable quantities of foaming substances (2–5 mg/L) and small concentrations of hazardous compounds that exhibit a wide range of sorbability, biodegradability, and stripability (see Table 2.14).
Inorganics	Common inorganic constituents in normal wastewater include chloride, hydrogen ions, alkalinity-causing compounds, nitrogen, phosphorus, sulfur, and heavy metals. The following are average incremental (mg/L) increases of inorganic solids associated with domestic water use after secondary treatment:[16,a]

Sodium	70	Silica	15
Potassium	10	Calcium	15
Ammonium (N)	15	Hardness ($CaCO_3$)	70
Nitrate (N)	5	Alkalinity ($CaCO_3$)	85
Chloride	75	Phosphorus	5
Sulfate	30	Magnesium	5

Table 2.10   Other characteristics of domestic wastewater (continued).

Gases	Gases commonly found in raw wastewater include nitrogen, oxygen, carbon dioxide, hydrogen sulfide, ammonia, and methane. Hydrogen sulfide creates an immediate oxygen demand and odor at treatment works; odorless methane is explosive in critical ratios with air.
Biological	In raw wastewater, an excellent medium for organism growth, selected concentrations for microorganisms of interest are

Total coliform	100–1000 million MPN/L,
Fecal coliform	10–100 million MPN/L, and
Total virus	1000–10 000 infectious units/L.

[a] Concentrations rounded to nearest 5 mg/L.

and fats, soluble or particulate matter, organics or inorganics, and nutrient enriched or nutrient poor matter—can influence the selection of treatment processes and the performance of the treatment plant.

When industrial wastes are dominant, bench-scale or pilot-plant evaluations may be necessary to develop or ensure the use of appropriate design criteria whether or not new technologies are applied. The designer should recognize that industrial operations and waste may be continuous or batch; vary daily, weekly, and seasonally for any single industrial facility; and vary from one industrial facility to another for the same type of industry. Failure to properly anticipate plant influent load under the most restrictive processing conditions and appropriate discharge limitations may result in noncompliance with the plant's NPDES permit conditions.

Flow and conventional pollutant load contributions from commercial sources are generally considered within the allowance for domestic sources. This consideration becomes less appropriate for smaller service areas. In small service areas, such commercial operations as coin-operated laundries may substantially affect the character of an area's wastewater. Often, pollutant load estimates from nondomestic sources can be developed by the use of numerical guidelines; where an industrial or institutional load is or may become significant, specific sampling programs and interviews are necessary to establish the impact of present operations and anticipated future changes.

OTHER WASTE CONTRIBUTIONS. The municipal wastewater treatment plant often provides an outlet for the septage generated in the surrounding, unsewered area and for sewer cleanings. In the future, more plants can be expected to serve as a convenient outlet for landfill leachate; waste solids from the water treatment plant; and, in some instances, groundwater contaminated with hazardous materials. Septage and leachate are characterized below. Hazardous materials or toxics are covered in the concluding section of this chapter.

Table 2.11[17,18] provides representative characterization of septage and landfill leachate. As septage includes solid residue that has somewhat stabilized over several years of storage, residual organics may exhibit low biodegradability. Landfill leachate characteristics vary widely, reflecting the character and age of the material placed in the landfill and the amount of water that infiltrates the landfill from ground and surface sources.

Published literature contains no representative characterizations for sewer cleanings. Sewer cleanings are expected to exhibit highly variable characteristics of organically enriched grit. The soluble component is somewhat less than that of septage, reflecting the in-place stabilization of settled organic material.

Waste solids from a water treatment plant can be expected to exhibit the characteristics of the suspended solid in the raw water supply and any solid (for example, powdered activated carbon) or solid-forming material (for example, alum addition and the resultant hydroxide precipitate) added during the course of treatment. The soluble pollutant phase of these waste solids will

Table 2.11   Typical major pollutant composition of septage and sludge landfill leachate.[17,18]

Parameter	Concentration by phase, mg/L			
	Volatile	Inert	Soluble[a]	Particulate
Septage[b]				
Suspended solids	10 000	5 000		
Total solids	25 000	15 000		
5-day BOD			3 000	2 000
Chemical oxygen demand (COD)			25 000	15 000
Total nitrogen as N			200	500
Total phosphorus as P			100	150
Sludge landfill leachate[c]				
Suspended solids	Insignificant			
5-day BOD			400	Negligible
Chemical oxygen demand (COD)			1 200	Negligible
Total organic carbon			800	Negligible
Total nitrogen as N			200	Negligible
Total phosphorus as P			40	Negligible

[a] As typically defined with coarse filter, not necessarily soluble.
[b] All values except soluble- and particulate-phase distribution reflect simplified expressions of published data.[17]
[c] All values reflect 5% of constituent range maximum except for assumed concentrations of SS, BOD, and phosphorus.[18]

reflect organics removed from the raw water supply and the time of storage at the water treatment plant. The aluminum (or iron) added during water treatment will enhance phosphorus removal at the wastewater treatment plant but not as much as would freshly added chemicals. Smaller plants may experience some problems related to flow and solids surging unless water plant discharges, particularly filter-backwashing wastes, are equalized.

PARTICLE SIZE DISTRIBUTION: TREATMENT IMPLICATIONS.
Wastewaters, treatment objectives, and design specifications are typically defined in terms of nonspecific parameters such as BOD and SS. Only recently have they also been defined in terms of their soluble and particulate phases. The division between these phases is typically defined as 1.2 μm, the pore size of the glass fiber filter for the SS test. For reference purposes, a 0.45-μm filter is used for millipore bacterial determinations. This subsection, using the information published by Levine et al.,[19] presents a more fundamental perspective to help the designer gain insight regarding the limits and applicability of treatment technologies and the parameters used to describe them.

Rates of sedimentation, mass transfer, adsorption, diffusion, and biochemical reaction are all influenced by particle size distribution. Table 2.12[19] shows size classifications, organic constituents and distribution, and biochemical oxidation rates reported in three wastewater characterization studies from the late 1950s and the early 1970s. The more rapid rate of biodegradation with particles having a size of 1.0 μm or less is readily apparent. Thus,

Table 2.12 Composition of organic materials in wastewater.[19]

Item	Classification			
	Soluble	Colloidal	Supra-colloidal	Settleable
Size range, μm	<0.08	0.08–1.0	1–100	>100
COD, % of total	25	15	26	34
TOC, % of total	31	14	24	31
Organic composition, % of total solids				
Grease	12	51	24	19
Protein	4	25	45	25
Carbohydrates	58	7	11	24
Biochemical oxidation rate k, day^{-1} (log base 10)	0.39	0.22	0.09	0.08

classical distinction between soluble and particulate solids seems to be supported by a significant decline of immediate biochemical stabilization capability of particles with sizes exceeding 1.0 μm. However, such distinctions do not apply to a true colloidal suspension.

Wastewater organic constituent characterization and technology assessment described by Levine *et al.* is provided in Figure 2.3.[19] Data associated with this work support several attendant observations or suggestions:

- Primary sedimentation at surface overflow rates of 32.6 to 48.8 m^3/m^2·d (800 to 1200 gpd/sq ft) will remove a minimum particle size of approximately 50 μm.

- The percentage of organic matter larger than 0.1 μm contained in settled municipal wastewater will be highly variable (ranging from 30 to 85% of the total as measured at four California plants and one Nevada plant) due to the nature of the community; the length, flow regime, and wastewater temperature in the collecting sewers (shear, solubilization, and enzymatic hydrolysis act to reduce the larger sized particles to smaller sizes); prior pumping and preliminary treatment system; the performance of the primary sedimentation tanks; and the presence of any recycled loads.

- Granular-media, primary effluent filtration (pulse bed, 0.45-mm sand size) will effectively remove particles larger than approximately 3 μm and some smaller material is removed by chance contact and entrapment or adsorption.

- Coagulation and flocculation can be used to aggregate wastewater constituents in the size range from less than 0.1 to 10 μm for eventual separation by sedimentation, filtration, or both. Sheet-mixing and pipeline flocculation concepts promote the most effective particle aggregation, especially for those particles smaller than 1 μm (which are not effectively measured by either the SS test or optical turbidity measurements typically used to monitor and study coagulation and flocculation processes).

- Aerobic biological treatment processes are unique in that they can effectively treat biologically degradable compounds in all of the size ranges encountered in wastewater treatment. During biological treatment, particle size distributions change because of new cell synthesis, flocculation, adsorption, enzymatic breakdown of macromolecules (one thousand to one million atomic mass units amu), and biochemical oxidation. Particulate matter, polymers, and macromolecules cannot be transported across bacterial membranes but can be removed from solution by adsorption to biofilm surfaces or entrapment in biological flocs. Proteolytic, lipolytic, and cellulolytic enzymes, each synthesized within bacterial cells, hydrolyze the macromolecules into smaller subunits that can be transported across the cellular membrane and metabolized. Molecules with a mass of less than 1000 amu can be taken up by bacteria. The biodegradability of low molecular mass compounds is controlled by molecular structure. Some compounds, such as humic and fulvic acids and other refractory or slowly degradable compounds, are stable under normal treatment conditions because of surface transfer biokinetic limitations. Removal of the difficult-to-treat particles before application of processes where adsorption is an important or dominant mechanism (such as in the activated sludge, trickling filtration, and activated carbon processes) can significantly enhance their performance.

- Anaerobic digestion studies (35°C, complete mix, pilot scale, 10-day detention time) of primary sludge show little or no reduction of particle sizes greater than 100 μm (approximately 45% of the total solids), size reduction by 65% for the material between 1.2 and 100 μm, and size reduction by approximately 85% for material with a particle size of less than 1.2 μm. As an explanation, the rate of particle hydrolysis depends directly on the particles' available surface area because enzymatic hydrolysis is primarily a surface phenomenon. Thus, smaller material is more readily degraded whereas degradation of larger material is surface limited by system kinetics.

*H*AZARDOUS SUBSTANCES CONTROL

During the 1980s, by implication and statute, the subject of hazardous or toxic substances in hazardous or toxic amounts became of great interest in municipal wastewater treatment. This section provides a brief primer on these substances and their effects on the design and practice of municipal wastewater treatment.

The following subsections summarize the intent and implications of the regulations impacting and potentially impacting POTWs, describe the principles regarding the measurement and treatability of hazardous organic substances, and provide representative limits for hazardous or deleterious materials at POTWs based on readily available references as of the end of 1989. Note that the subject of hazardous wastes is ever changing and becoming more specific. Further, state and local regulations control some hazardous wastes and not others. The material presented herein should not be applied unless the user independently updates and validates the material of interest. Chapter 6 provides additional information regarding hazardous substances and design measures that will reduce or prevent adverse effects on the occupational health and safety of plant workers and neighbors.

STATUTES AND GUIDANCE. Figure 2.4[20] describes POTW activities and sources of pollutants that are potentially governed by EPA regulations

---

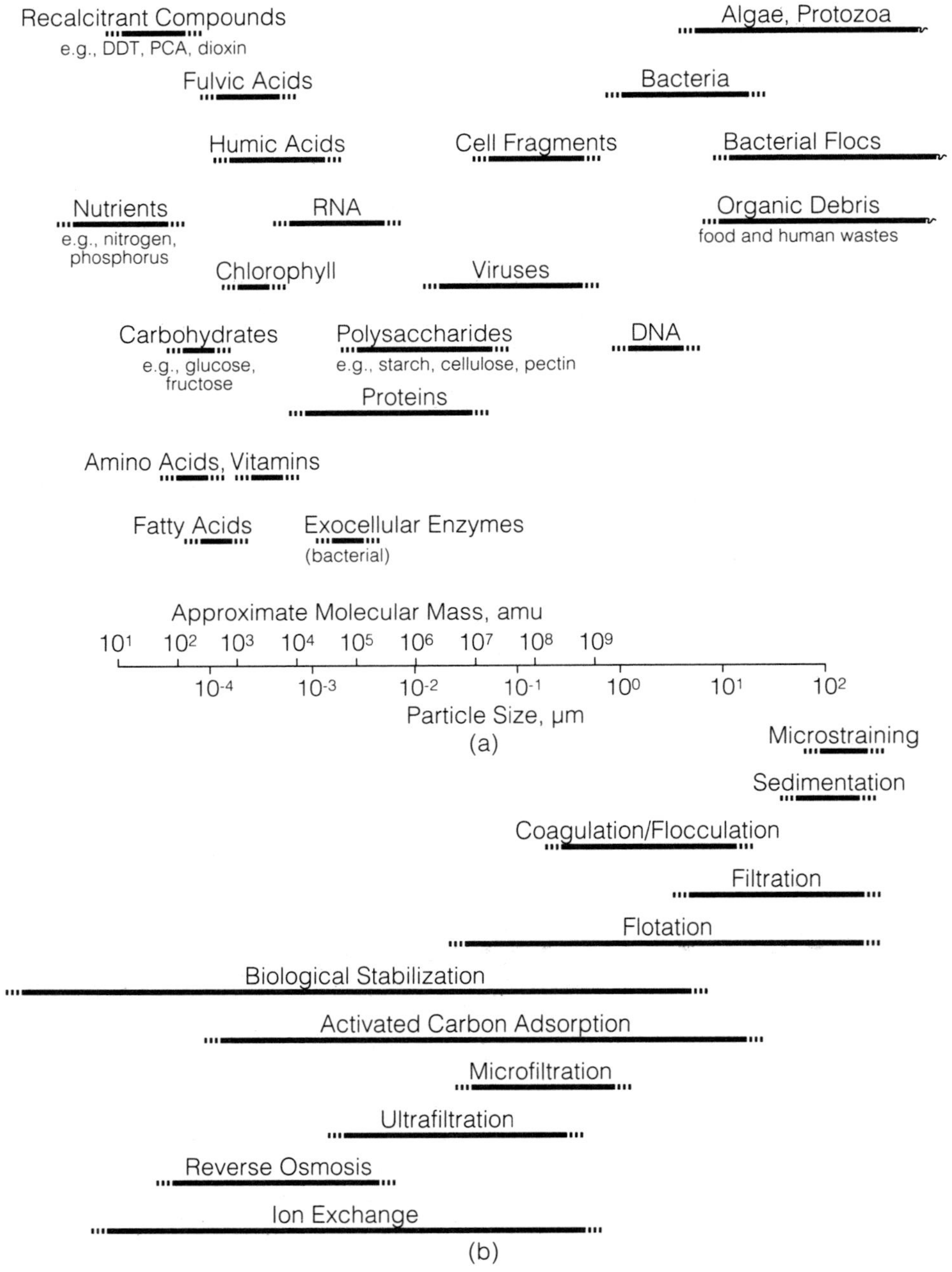

Recalcitrant Compounds
e.g., DDT, PCA, dioxin

Algae, Protozoa

Fulvic Acids

Bacteria

Humic Acids

Cell Fragments

Bacterial Flocs

Nutrients
e.g., nitrogen,
phosphorus

RNA

Organic Debris
food and human wastes

Chlorophyll

Viruses

Carbohydrates
e.g., glucose,
fructose

Polysaccharides
e.g., starch, cellulose, pectin

DNA

Proteins

Amino Acids, Vitamins

Fatty Acids

Exocellular Enzymes
(bacterial)

Approximate Molecular Mass, amu

10^1 10^2 10^3 10^4 10^5 10^6 10^7 10^8 10^9

10^{-4} 10^{-3} 10^{-2} 10^{-1} 10^0 10^1 10^2

Particle Size, μm

(a)

Microstraining

Sedimentation

Coagulation/Flocculation

Filtration

Flotation

Biological Stabilization

Activated Carbon Adsorption

Microfiltration

Ultrafiltration

Reverse Osmosis

Ion Exchange

(b)

Figure 2.3 Particle size organic characterization and related treatment technologies: (a) particle sizes and approximate molecular masses of wastewater organics and (b) treatment technologies.[19]

regarding hazardous substances. These statutes, their implementing regulations, and their legal ramifications for POTWs are described in the following paragraphs.[20-22]

What constitutes a hazardous waste disposal (and treatment and storage) site is strictly defined by law. Congressional directives regulating hazardous wastes originate in six statutes:

- Clean Water Act (CWA);
- Clean Air Act (CAA);
- Resource Conservation and Recovery Act (RCRA);
- Comprehensive Environmental Response, Compensation, and Liability Act (CERCLA), which is also known as the "Superfund" Act;
- Superfund Amendments and Reauthorization Act (SARA); and
- Toxic Substances Control Act (TSCA).

Statutory coverages depend on how a waste enters and leaves a system.

Simply expressed, the federal government regulates POTWs under CWA. Within this act, it regulates, specifically, the 126 most hazardous substances (termed "Priority Pollutants") through General Pretreatment Regulations under a locally enforced and implemented pretreatment program, broadly, through the NPDES permit program. The NPDES permit is a liability shield for all circumstances identified and authorized through its issuance. The general objectives of the pretreatment program are to prevent toxics in amounts that would

- Interfere with liquid and residual management operations,
- Pass through the treatment works, and
- Retard opportunities for beneficial recycling.

Additionally, pretreatment regulations specifically prohibit the introduction of pollutants that are explosive, corrosive, obstructive, excessively variable, and excessively hot. Although categorical production-based discharge standards have been established for 34 industries, the program relies heavily on each municipality identifying its own problems and proposing solutions to account for site-specific factors.

The CAA provides for state implementation plans (SIPs) to ensure and maintain national ambient air quality standards (NAAQS) for six pollutants (carbon monoxide, particulates, lead, nitrogen dioxide, ozone, and sulfur oxides) and national emission standards for both mobile and stationary sources of pollutants. This program impacts POTWs that operate sludge incinerators and dryers, engines, and boilers through EPA programs for new

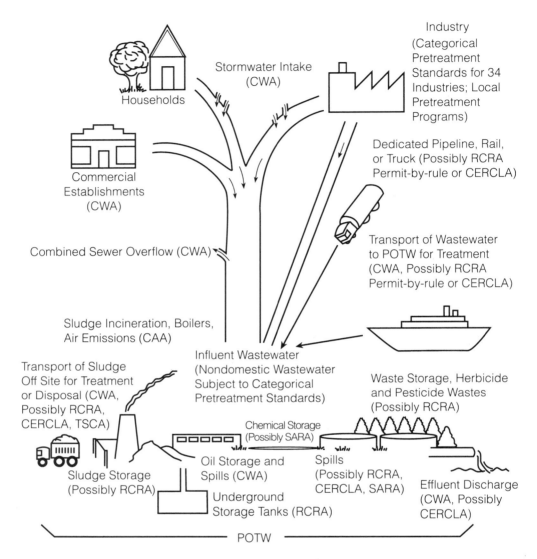

Figure 2.4 Activities and pollutant sources potentially governed by EPA's hazardous substance regulations.[20]

source review and prevention of significant deterioration (PSD), new source performance standards (NSPS), and national emission standards for hazardous air pollutants (NESHAPs).

The PSD program applies to any significant (more than 100 ton/yr) new source of any of the six pollutants and requires the use of the best available control technology in areas meeting NAAQS. More stringent emission restrictions may be required in nonattainment areas. NSPS for wastewater sludge incinerators apply to sludge incineration rates of 2200 lb/day and presently restrict particulates (1.3 lb/ton of dry sludge input) and opacity (20%). POTW NESHAPs limit mercury to 1.6 g/d per source and no more than

3.2 g/d per site and also set daily emission limits for beryllium if the POTW incinerates beryllium-containing waste. Implementing regulations of the CAA are the vehicle that regulatory bodies may use to restrict the allowable emission of strippable hazardous organics, as discussed earlier.

RCRA, concerned with 366 extremely hazardous substances (including the 126 priority pollutants of the CWA), regulates hazardous substances from the point of generation, through transport, storage and treatment, and disposal and nonhazardous wastes in underground storage tanks and municipal solid waste landfills. RCRA exempts domestic wastewater and POTWs from its jurisdiction through the domestic wastewater exclusion unless the POTW uses underground storage tanks or a municipal landfill for disposal of its sludges.

The domestic wastewater exclusion applies to industrial wastes as long as they mix with domestic wastewater before entering the POTW boundary headworks. If this does not occur and hazardous wastes are found to be either consciously or inadvertently received by rail, truck, vessel, or pipeline, RCRA's "permit by rule" procedures apply; the POTW will then automatically become a RCRA-permitted treatment, storage, or disposal facility (TSDF).

CERCLA and SARA are primarily concerned with prior disposal of hazardous substances. CERCLA POTW liabilities, with CWA and RCRA regulatory compliance, accompany wastewater exfiltration and overflow in the collection system, storage leaks, and effluent and solids disposal, or every aspect of a POTW operation. If prior POTW practices, whether conscious or not, pose a threat to public health, welfare, or the environment, the POTW is liable.

Liability may be shared with the permitting authority only if the POTWs NPDES permit identifies all release points, the conditions under which they operate, the expectation of contravention of any discharge standard, and the substances likely to be received by the POTW and, in turn, found in its sludges and effluent. Protection is not afforded to those who fail to make full disclosure, and full disclosure does not provide total protection if all hazardous substances were not identified (for example, CWA priority pollutant versus RCRA hazardous substances determinations), permitted (including any previously unpermitted constituents that are later determined to pose a threat), or both. Protection is not afforded to those who allow an industry to violate repeatedly their pretreatment obligations.

Title III of SARA established a new program that requires facilities, including treatment plants, to notify state and local officials if they have extremely hazardous substances at their facilities in excess of certain threshold amounts. Some facilities must also provide material safety data sheets (MSDS) on the hazardous chemicals stored on the site or lists of chemicals for which these data sheets are maintained, annually report the inventory and release to all media (air, land, and water) of those toxic chemicals that fall within Standard Industrial Codes 20 to 39, and meet certain threshold limits. Table 2.13[20] lists chemicals frequently encountered at POTWs and their threshold limits.

TSCA regulates the manufacture, use, and disposal of toxic substances. Under its enabling legislation, EPA is authorized to control the risks from more than 65 000 existing chemical substances and the use of new chemicals. Treatment plants are regulated under TSCA if they accept wastes contaminated with polychlorinated biphenyls (PCBs) or certain other toxic chemicals, which once over certain threshold concentrations require specific disposal procedures (for example, a waste with more than 50 mg/kg PCBs must be disposed of by approved U.S. EPA methods). If the plants experience a spill or leak of the same substances, TSCA governs their cleanup, conforming with specific stringent standards.

Plant liability for hazardous substances can be minimized by several actions:

- Avoid RCRA "permit by rule." Septage should not be accepted or should only be accepted from prequalified haulers with certification that the wastes are only from households and with advance approvals and testing of commercial and industrial wastes as appropriate.
- Enforce the pretreatment program. Do not allow the release of any pollutant at any rate not identified by the industrial user.
- Disclose fully and promptly all hazardous substances (name and mass), practices (including the receipt of septage), and conditions in which hazardous substances may be released to the environment under the NPDES permit.
- Keep permitting agency(ies) informed with written notification. Inform permitting agency of any significant change in hazardous substances in influent, effluent, and sludge and if problems result, solicit their advice.

Should a decision be made to receive RCRA or CERCLA cleanup wastes, the POTW should consider securing a modification of its NPDES permit (completely describing wastes, cleanup technology and performance, frequency of analysis, and sampling procedure), discharger indemnification, and regulatory approval of the action, including the cleanup technology and emissions. It is also advisable that wastes be initially accepted on a staged, provisional basis for no longer than the period of the plant's NPDES permit and that the compliance analyses include the RCRA hazardous substance list until information is secured to show that the more expansive determination is inappropriate.

RCRA or CERCLA cleanup standards can vary from site to site and are termed applicable or relevant and appropriate requirements (ARARs). Applicable requirements are cleanup standards; standards of control; and other

Table 2.13 **Extremely hazardous substances (EHS) that trigger SARA reporting if found at plants.[20]**

Substance	Threshold planning quantity (TPQ), lb[a]	Title III reportable quantity (RQ), lb[a]
Reporting to state emergency response commission (SERC) and local emergency planning (LEPC)		
Ammonia	500	100
Chlorine	100	10
Hydrogen chloride, anhydrous	500	5000
Hydrogen peroxide, concentration >52%	1000	1
Nitric acid	1000	1000
Ozone	100	1
Sulfur dioxide	500	1
Sulfuric acid	1000	1000

Additional hazardous chemicals requiring materials safety data sheets (MSDSs) per Occupational Safety and Health Administration regulations and SERC and LEPC reporting if present at POTW in quantities greater than 10 000 lb.[a]

Alum

Calcium hypochlorite

Carbon dioxide and other compressed gases

Cleaning solvents

Copper sulfate

Ferric chloride

Gasoline

Lime (CaO)

Methanol

Phosphoric acid

Pickle liquor

Polymers

Potassium permanganate

Sodium hypochlorite

Strong acids and bases

Examples of additional hazardous substances stored or generated by the POTW that are subject to MSDS reporting requirements

Chlorine dioxide

Methane

[a] lb × 0.453 6 = kg.

substantive environmental protection requirements, criteria, or limitations promulgated under federal and state law that specifically address a hazardous substance, remedial action, location, or other circumstance at the site. Relevant and appropriate requirements, while not legally applicable, address problems or situations sufficiently similar to those encountered at the site.

TREATABILITY UNDERSTANDINGS. Table 2.14[23-26] lists the hazardous pollutants most frequently encountered in plant sludges and the parameters that are used to characterize their fate during treatment or disposal in the natural environment. Inspection of this table shows that hazardous heavy metals are more routinely encountered at higher concentrations than are hazardous organics in wastewater sludges. Metals, as conservative substances, merely accumulate in plant waste and effluent suspended solids. On the other hand, hazardous organics can act as either conservative or nonconservative compounds, depending on the compound and processes in the plant. Aeration systems and flow-over weirs may strip organics if plant influent contains hazardous materials; their disposition in the liquid and solid phases and effect on effluent and sludge management should be studied.

Note that hazardous substance determinations often reach the limits of analytical capability and special care should be taken to guard against the reporting of false positives due to the laboratory or sampling procedure. The presence of bis(2-ethylhexyl) phthalate, a common plasticizer observed in all the samples in Table 2.14, may merely reflect the use of plastic tubing or containers for sample collection and storage. Similarly, any analysis that results in a tenfold change over previous observations or values less than ten times the detectable limit of any pollutant or blank should be regarded as suspect until confirmed by repetitive analyses.

A compound's chemical formula, molecular weight, solubility, Henry's law constant, organic carbon, and octanol–water partition coefficients serve to allow some prediction of the compound's relative strippability, adsorbability, and biodegradability. In general, the lower the molecular weight and the simpler the compound structure, the higher the compound's biodegradability. Biodegradability will also decline with the extent of halogenation (for example, the ratio of the weight of chlorine to that of the total compound). High solubilities tend to support high biodegradability and low stripping and adsorption. EPA reports experience that leads to the following conclusions:[9]

- Octanol–water partition coefficient, log K_{ow}: if this is greater than 3.5, the organic substance is highly adsorptive (for example, readily partitioning with the sludge). If the coefficient is less than 3.5, the substance is more likely to be removed by stripping or biodegradation. Others have suggested a coefficient of 2.0 as a better approximation of the strippability threshold.[27]

Design of Municipal Wastewater Treatment Plants

- Henry's law constant: if this is greater than 0.024, the compound is easily stripped from solution with strippability also increasing with a lower affinity for adsorption and a higher extent of halogenation.
- Volatile organic compounds: most VOC mass is not accounted for in sludges and effluents. Stripping and biodegradation are likely removal mechanisms; adsorption is small and may not be measurable.
- Base-neutral compounds: these have highly mixed removal mechanisms with compounds not partitioning to sludges more generally biodegradable. Volatilization and stripping are not likely to be significant removal mechanisms.
- Acid extractables: removal mechanism is dominated by biodegradation; many of these compounds are potentially formed during chlorination.
- Pesticides (PCBs): these are strongly adsorbed with little tendency to degrade in an anaerobic environment.
- Metals: these are largely concentrated in sludges.

The removal of hazardous substances by POTWs not only depends on their form but also the levels received in the influent wastewater (long-term and event-specific), the processes and opportunities for volatilization (large surface areas), mixing, stripping (aeration and mixing intensity and duration), recycling (closed pure oxygen systems will recycle stripped VOCs, allowing for greater biodegradation), biodegradation (older sludge cultures promise greater capabilities under continuous inoculums of organic hazardous waste), chemical uses (chlorine, ozone, and organic polymers), and effluent suspended solids. As would be expected, removal efficiencies of POTW hazardous substance are highly variable and, considered alone, meaningless for predicting performance of a given plant without detailed understanding of site-specific conditions for the plant's test. Nonetheless, data may help to reach a broad-scale understanding of treatment plant effects on hazardous substances. This understanding for hazardous organics and secondary treatment processes shows no complete pass-through. Representative removals range from 40 to 80% and some extreme removal efficiencies of 98% or higher are attained in each of the hazardous substance groups.[28]

The engineer and treatment plant operating management need to realize that most toxic materials, including pesticide-like substances, can partition strongly to the solid phase and, with metals, will be in the waste sludge. For a plant with toxic waste in the influent, the sludge then may often be more dangerous than the plant effluent.

SEWER USE/POTW GUIDANCE. The quandary facing POTWs is that although it is responsible for establishing local limits for hazardous wastes released to its sewer system (and the liability of environmentally sound treatment and disposal), specific regulatory standards for establishing such limits

Table 2.14   Hazardous pollutants most frequently encountered[a] in plant sludges.[23-26]

Pollutant	Percent of time detected	Median dry weight detected, mg/kg	Chemical formula	Molecular weight, g/mole	Water solubility, mg/L	Biochemical/physical/chemical (fate)[b,c]			Partition coefficients	
						Specific gravity, unitless	Relative biodegradability[b]	Henry's law constant, atm-m³/mol	Organic carbon, K_{oc} mL/g	Octanol-water, log K_{ow}
Metals and cyanide										
Arsenic	100	6	As	75		Metals are usually received at plants as ions and, to varying degrees, will be removed in the sludge. Toxic concentrations will range from <1.0 mg/L for mercury to >10 mg/L for zinc, nickel, and lead.				
Beryllium	61	1.2	Be	9						
Cadmium	100	30	Cd	112						
Chromium	100	400	Cr	52						
Copper	100	600	Cu	64						
Lead	100	400	Pb	207						
Mercury	98	3	Hg	201						
Nickel	100	130	Ni	59						
Selenium	100	3	Se	79						
Zinc	100	1400	Zn	65						
Cyanide	100	700	CN	26	2 000+					-0.25+
Volatile compounds										
Benzene	93	1.5	C_6H_6	78	1 800	0.88	D	0.005 6	83	2.12
Chlorobenzene	41	1.2	C_6H_5Cl	113	470	1.11	D	0.003 7	330	2.84
Chloroform	52	0.8	$CHCl_3$	119	8 200	1.49	A	0.002 9	31	1.97
1,2-Dichloroethane	23	20	$CLCH=CHCl$	97	8 500	1.27	B	0.000 98	14	1.2
Dichloromethane	95	9	CH_2CL_2	85	20 000	1.32	D	0.002 0	9	1.3
Tetrachloroethylene	70	3	$CL_2C=CCl_3$	166	150	1.63	A	0.026	360	2.6
Toluene	100	1700	$C_6H_5CH_3$	92	530	0.87	D	0.000 64	300	2.7
Trichloroethylene	84	9	$CHCl=CCl_2$	131	1 100	1.46	A	0.000 91	126	2.38
Vinyl chloride	14	40	$CH_2=CHCl$	62	2 700	0.97	No test	0.082	57	1.38

Table 2.14 Hazardous pollutants most frequently encountered[a] in plant sludges (continued).[23–26]

| | | | | | | | Biochemical/physical/chemical (fate)[b,c] | | Partition coefficients | |
Pollutant	Percent of time detected	Median dry weight detected, mg/kg	Chemical formula	Molecular weight, g/mole	Water solubility, mg/L	Specific gravity, unitless	Relative biodegrad-ability[b]	Henry's law constant, atm-m^3/mol	Organic carbon, K_{oc} mL/g	Octanol–water, log K_{ow}
Acid compounds										
Pentachlorophenol	41	10	C_6Cl_5OH	266	14	1.98	A	0.000 002 8	53 000	5
Phenol	89	20	C_6H_5OH	94	9 300	1.06	D	0.000 000 45	14	1.46
Base/neutral compounds										
Benzo(a)anthracene	57	9	$C_{18}H_{12}$	228	0.005 7		No test	0.000 001 2	1 380 000	5.6
Benzo(a)pyrene	11	300	$C_{20}H_{12}$	252	0.001 2		No test	0.000 001 6	5 500 000	6.06
Benzo(b)fluoranthene	18	2	$C_{19}H_{11}$	252	0.014		No test	0.000 012	550 000	6.06
Bis(2-ethylhexyl)phthalate	100	160	$C_{26}H_{30}O_4$	391	0.001 8	1.27	A			7.45
Chrysene	66	8	$C_{18}H_{12}$	228			A/N	0.000 001	200 000	5.61
Hexachlorobenzene	16	1.2	$C_6H_6Cl_6$	285	0.006		N	0.000 68	3 900	5.23
Phenanthrene	95	6	$C_{14}H_{10}$	178	1.0	1.06	D	0.000 16	14 000	4.46
Pyrene	93	7	$C_{16}H_{10}$	202	0.13	1.27	D/N	0.000 005 0	38 000	4.88

a Five or more plants of the 44 surveyed.
b Biological degradability data obtained using static slash procedures with 7-day initial incubation of 5 and 10 mg/L or organic to settled wastewater, followed by three more sub-cultures each at 7 days (28 days of acclimation total) where D = significant degradation/rapid acclimation; A = significant degradation/gradual acclimation; N = no significant degradation under test; and B = some degradation with volatilization.[24]
c Physical/chemical data where water solubilities are for neutral pH and temperatures of 20–30°C. Henry's law and organic carbon and octanol-water partition coefficients, respectively, are measures of a chemical's stripability, tendency for adsorption on soils and sediment, and affinity for uptake (bioaccumulation) in fatty tissue.[25,26]

currently include slightly more than 10% of the 126 priority pollutants (essentially the metals shown in Table 2.14,[23-26] the seven PCBs, phenols, and cyanide). The proposed 503 regulations include limits for 10 of the 13 metals, 15 of the 26 PCBs/pesticides, four of the 46 base-neutrals, one of the volatiles, none of the 11 acid extractables, none of the phenols, and no cyanide. Accordingly, the final regulations once promulgated may not cover more than 75% of the priority pollutants.

Table 2.15[22] summarizes the guidance prepared for one POTW that was faced with this dilemma and provides an example of one approach that could be used to develop limits to guide a POTWs acceptance of hazardous materials. Based on the principles and observations presented in the previous subsection, the following comments offer an additional understanding of the guiding assumptions used in the table's preparation:

- Hazardous substances that readily strip will be eventually subject to regulatory control in the sewer and treatment works (for example, those now under formal consideration in Southern California) and should be limited in anticipation of these controls.
- The POTW may receive hazardous substances that can degrade in higher concentrations than those that can strip or persist.
- Acclimation is most likely with a significant commitment to equalization (ideally, all significant sources of hazardous substances should provide a week or more of continuously mixed storage with delivery of 15% or less of the stored volume in any one day).
- Acceptable limits for biodegradable organics may be 10 or more times higher than the cited values. However, prudence would only permit higher numbers pending successful demonstration at lower values. Equilibrium considerations (see Chapter 3) suggest that successful demonstration should be based on values found in the plant's net waste solids at two or more times the solids residence time of the entire plant.

Table 2.15 General guidance for the acceptance of hazardous substances when no standards exist.[22]

Substances	Guidance	
Organics in public sewer	All substances with a Henry's law constant (H) >0.000 2 atm-m^3/mole should have no more than 10% of compounds water solubility at low flow in sewer receiving the waste.	
Organics in plant influent and no sludge standards	Volatile organic compounds, Henry's law constant (H): >0.02 atm-m^3/mole	
	Without equalization	≤0.000 1 mg/L
	With equalization	≤0.001 mg/L
	= 0.002 – 0.02 atm-m^3/mole	
	Without equalization	≤0.000 1 mg/L
	With equalization	≤0.01 mg/L
	<0.002 atm-m^3/mole	
	Without equalization	≤0.01 mg/L
	With equalization	≤0.1 mg/L
	>0.02 atm-m^3/mole and nonhalogenated	
	Without equalization	≤0.1 mg/L
	With equalization	≤1.0 mg/L
	Base-neutral and acid extractable compounds, octanol–water partition coefficient, K_{ow}	
	>10 000	
	Without equalization	≤0.000 1 mg/L
	With equalization	≤0.001 mg/L
	= 1000 – 10 000	
	Without equalization	≤0.001 mg/L
	With equalization	≤0.01 mg/L
	<1000	
	Without equalization	≤0.01 mg/L
	With equalization	≤0.1 mg/L
	Pesticide and PCB compounds	≤0.000 01 mg/L
Metals and organics in plant influent with sludge standards	Case-by-case analysis based on allowable hazardous substance dry weight km/kg concentration in sludge and net sludge production of POTW on a mg/L equivalent basis. For example, (allowable hazardous substance level in sludge, mg/kg) × [(net waste solids from plant, lb) / 8.34 (average daily influent flow, mgd)] = allowable hazardous substance in plant influent, mg/L.[a,b,c]	

[a] This equation conservatively allows no credit for biodegradation or stripping of organics and implicitly assumes no soluble-phase pollutant in plant effluent.
[b] lb × 0.453 6 = kg.
[c] mgd × 3785 = m^3/d.

REFERENCES

1. "Construction Grants 1985." EPA 430/9-81-020, U.S. EPA Office of Water Program Oper., Washington, D.C. (1985).
2. "Design Criteria for Mechanical, Electrical and Fluid System and Component Reliability." Suppl. to "Federal Guidelines: Design, Operation and Maintenance of Wastewater Treatment Facilities." EPA 430-99-74-001, U.S. EPA, Office of Water Program Oper., Washington, D.C. (1974).
3. "Guidelines for the Design of Wastewater Treatment Works." Tech. Advisory Board of the New England Interstate Water Pollut. Control Comm., TR-16, Boston, Mass. (1980).
4. "Water Reuse." Manual of Practice No. SM-3, Water Pollut. Control Fed., Alexandria, Va. (1983).
5. "Regulations for Land Application of Sludge and Septage." 310 CMR 32.00, Dep. Environ. Qual. Eng., Mass. (1983).
6. "Beneficial Reuse of Waste Solids." Manual of Practice No. FD-15, Water Pollut. Control Fed., Alexandria, Va. (1989).
7. "Standards for the Disposal of Sewage Sludge; Proposed Rule, 40 CFR Parts 257 and 503." *Fed. Regist.*, 5746 (1989).
8. U.S. Code of Federal Regulations, 40 CFR 403 (1991).
9. "Report to Congress on the Discharge of Hazardous Wastes to Publicly Owned Treatment Works (The Domestic Sewage Study)." EPA/530-SW-86-004, U.S. EPA, Office of Water, Washington, D.C. (1986).
10. U.S. Code of Federal Regulations, 40 CFR, Part 133 (1991).
11. "Recommended Standards for Sewage Works." Great Lakes—Upper Miss. River Board of State Sanit. Eng. Health Educ. Serv. Inc., Albany, N.Y. (1978).
12. Fair, G.M., and Geyer, J.C., "Water Supply and Wastewater Disposal." John Wiley & Sons, New York, N.Y. (1961).
13. "Gravity Sewer Design and Construction." Manual of Practice No. FD-5, Water Pollut. Control Fed./Am. Soc. Civ. Eng., Alexandria, Va. (1983).
14. "Wastewater Treatment Plant Design." Manual of Practice No. 8., Water Pollut. Control Fed., Alexandria, Va. (1977).
15. Metcalf & Eddy, Inc., "Wastewater Engineering." McGraw–Hill, Inc., New York, N.Y. (1972).
16. Weinberger, L.W., *et al.*, "Solving Our Water Problems—Water Renovation and Reuse." Ann. N.Y. Acad. Sci., 136 (1966).
17. "Handbook: Septage Treatment and Disposal." EPA-625/6-84-009, U.S. EPA, Munic. Environ. Res. Lab., Cincinnati, Ohio (1984).

18. "Environmental Regulations and Technology: Use and Disposal of Municipal Wastewater Sludge." EPA 625/10-84-003, U.S. EPA, Washington, D.C. (1984).

19. Levine, A.D., *et al.,* "Characterization of the size distribution of contaminants in wastewater: treatment and reuse implications." *J. Water Pollut. Control Fed.,* **57,** 805 (1985).

20. "Overview of Selected EPA Regulations and Guidance Affecting POTW Management." EPA 430/09-89/008, U.S. EPA, Office of Water (WH-595), Washington, D.C. (1989).

21. Zorc, J.M., *et al.,* "Minimizing liabilities facing POTWs." *J. Water Pollut. Control Fed.,* **60,** 29 (1988).

22. Mulbarger, M.C., "Special Wastes Acceptance Guidance: Baseline Understandings." Prepared for Danville, Ill., Sanit. Dist. (1989).

23. "Fate of Priority Pollutants in Publicly Owned Treatment Plants." EPA 440/1-82/303, U.S. EPA, Cincinnati, Ohio (1982).

24. Tabak, H.H., *et al.,* "Biodegradability studies with organic priority pollutant compounds." *J. Water Pollut. Control Fed.,* **53,** 1503 (1981).

25. "Superfund Public Health Evaluation Manual." EPA 540/1-86/060, U.S. EPA, Cincinnati, Ohio (1986).

26. "Hazardous Chemical Data Book." G. Weiss (Ed.), Noyes Data Corporation, Park Ridge, N.J. (1986).

27. Okey, R.W., Salt Lake City, Utah, Personal Communication (1991).

28. Engineering Science, Inc. "Introduction to Toxicity Reduction at Publicly Owned Treatment Works." Prepared for U.S. EPA, Office Munic. Pollutant Control, Office of Water, Washington, D.C. (1989).

Chapter 3
Principles of Integrated Facility Design

INTRODUCTION

This chapter discusses design considerations that rely principally on the insight and fundamental understanding of the designer who is accordingly accountable. These are important considerations because they provide great opportunity for maximizing cost effectiveness and minimizing fundamental design errors. The guiding principles discussed herein apply to the design of both new and upgraded facilities.

*F*OCUS ON PAST EXPERIENCE

Some measure of the limitations of completed designs and the occurrence of errors and omissions by designers can be found in some of the nationwide surveys of recent years. A brief review of the findings of two of these surveys provides a good focus for conveying the intent of this chapter.

Information presented in Chapter 2 shows that approximately 75% of the wastewater treatment plants in the U.S. are designed to treat an average daily flow of 3785 m^3/d (1 mgd) or less. Nearly 60% of these plants are sized between 378 and 3785 m^3/d (0.1 and 1.0 mgd). All are more likely than larger plants to be underfunded, understaffed, and not reached by the professional networks that can offer troubleshooting advice.

Since 1982, EPA's On-site Operator Assistance Program, created under Section 104(g)(1) of the Clean Water Act, has provided technical assistance to small plants to help operators and managers solve debilitating problems. A summary of EPA findings for 150 problem plants in its 10 regional jurisdictions is given in Table 3.1.[1]

Table 3.1 shows that successful wastewater treatment depends on operator understanding, responsible administration, and sound design. Shortcomings in any of these functions may result in processing upset and eventual failure. In small plants and larger installations, designers can reduce the risk of failure by selecting tolerant, flexible treatment processes with conservatively designed, responsive sludge-processing and disposal schemes. Designers can also promote operator training and work with administrators to ensure adequate support for the needs of the facility.

Results of a broader 1000-plant survey, which attempted to identify technology and design deficiencies at POTWs, was reported by WPCF in 1989.[2] Table 3.2 characterizes plants responding to the survey and their reported problem areas.

Unlike the EPA small plant survey, which focused on plants that reportedly had debilitating problems and relied on independent examination of causative problem areas, the WPCF survey relied on self-examination and did not single out plants with problems. In general, facilities surveyed are a representative cross section of the nation's plants. Approximately 72% have capacities of 18 900 m^3/d (5 mgd) or smaller. Most operate significantly below their average, day-rated hydraulic capacity (75% of plants at 75% capacity or less) and, although old (nearly 49% have been in operation for over 20 years), many of the older installations had been recently improved with expanded, modernized liquid- and solids-processing capabilities (21% between 10 and 5 years ago and 31% in the last 5 years). The most commonly used processes were activated sludge (59%), gravity sludge thickening (47%), anaerobic digestion (42%), sand bed dewatering (33%), and ultimate sludge disposal by land application (51%).

Table 3.1   Results of EPA survey of 150 small plants with debilitating problems.

Summary of most frequently-occurring performance-limiting factors (PLFs)	No. of EPA regions
Poor operator understanding/application of process control	10
Solids handling and sludge disposal	9
Infiltration/inflow	8
Staffing (too few staff, low pay, and high turnover)	6
Laboratory capability for process/NPDES testing	6
Process design errors	4
Support from municipality (administrative and technical)	4
Preventative maintenance program	3

Survey conclusions	Survey recommendations
Activated sludge may not be a good design choice for many small plants.	Give fuller consideration to simpler, more tolerant treatment processes (for example, fixed media and natural systems) that are less dependent on highly skilled operators. Select treatment technology based on realistic appraisal of all costs (including conservative estimates of sludge quantity and concentration for sludge treatment and disposal, staff salary, recruitment and training, equipment maintenance and replacement, and administrative costs).
Plant inflexibility undermines operability.	Designers should conscientiously build flexibility into small systems (for example, piping configurations, redundant unit processes, variable-speed pumps for wasting and return, aeration equipment, and consideration of equalization tanks for I/I extremes).
Small plants have front- and back-end problems with process design.	Pumps, piping, and aeration systems should be designed to accommodate increased solids and rags in the system where primary treatment is not provided.

Table 3.1 Results of EPA survey of 150 small plants with debilitating problems (continued).

Survey conclusions	Survey recommendations
Small plants have front- and back-end problems with process design (continued).	Operators should be made aware of the need to remove floating debris that passes primary screening and designers should provide practical facilities to facilitate their permanent removal.
	Solids-handling facilities should always consider (have) the capability of properly stabilized liquid sludge removal and disposal.
Heavy loads may confound unskilled operators.	Community administrators and design engineers should frankly discuss and agree on realistic loadings for the facility in the planning process (a conservative design approach should take into account I/I and industrial loadings).
Staffing difficulties aggravate poor performance.	Administrators should seek to attract and maintain a better staff through enhanced operator status and visibility using at least one (or nearly so) full-time position, with a salary comparable to other critical municipal functions (for example, the police chief), with reasonable authority for budgeting, purchasing, hiring and firing, and provide reasonable opportunities for training and certification.
Plant budgets and user charges may be too low.	Better fiscal management must start with a separate budget for the treatment plant that includes a sinking fund to cover replacement of major equipment and supports adequate staff salaries, training, and required certification courses.
Municipal support is a subtle but vital need.	Apply outreach and information transfer to increase community support; consider making the treatment plant into a multiple-use facility that accommodates recreational facilities and shares offices and building space with other community agencies and organizations.

Table 3.2 Results of WPCF 1000-plant survey of problem areas (1989).[2]

	Responding plants characteristics					Plants not responding, %
Average daily flow, mgd	<0.1	0.11-1	1.01-5	5.01-10	>10	
Respondents, %	8	32	32	8	20	6
Flow capacity, %	<75	75-100	100-125	125-150	>150	
Respondents, %	75	17	5	2	2	36
Age of plants, years	<6	6-10	11-15	16-20	>20	
Respondents, %	15	12	11	13	49	7

Problem category	Survey findings
Preliminary treatment	Most common major continuous problems were grease, grit and rag removal, corrosion, and grease buildup.
Secondary treatment equivalency Suspended growth systems	Major, periodic problems were limited to scum and foam buildup, filamentous growth, and solids settleability.
Attached or fixed-film growth systems	Many minor, periodic problems were reported, most often solids settleability and rising sludge. Filter flies were ranked as serious, periodic problems at 36% of the plants. Rising sludge was a minor, periodic problem at 35% of the plants.
Lagoons and stabilization ponds	Odor was a minor, periodic problem at 30% of the plants.
Effluent filtration	Extensive backwashing, rapid head loss development, and influent flow rate variation were prominent, periodic problems. Media loss was reported as a continuous, minor problem at 17% of the plants.
Disinfection	Residual chlorine and influent flow variation were a significant problem. Dechlorination was a minor problem at 15% of the plants.

Table 3.2   Results of WPCF 1000-plant survey of problem areas (1989) (continued).[2]

Problem category	Survey findings
Sludge concentration (thickening and dewatering)	Odor was commonly reported as a problem, ranging in severity from major, continuous to minor, periodic.
	Centrifuge wear was reported as a major, continuous problem.
	Gravity thickening and dissolved air flotation problems (beyond odor) were rarely reported to have operating problems.
Sludge stabilization and destruction	Supernatant quality, grit accumulation, and odors were commonly reported for anaerobic digestion.
	Solids settleability was a significant problem at 31% of the plants with aerobic digestion.
	Corrosion, equipment breakdown, and odors were the most commonly reported composting problems.
	Incineration problems were most commonly temperature and air pollution control.
Sludge transport and disposal	Landfilling and ocean disposal appeared to have the least problems; land availability was reported to be a major, continuous problem for 14% of the plants with land application schemes.
	Equipment breakdown was a significant to minor problem.
	Odors were reported to be a problem for many of the plants.
	Problem frequency and severity[a]
	Weather conditions 7
	Land availability 6
	Disposal site transport 4
	Landfilling 4
	Land application 4

Table 3.2 Results of WPCF 1000-plant survey of problem areas (1989) (continued).[2]

Problem category	Survey findings
Sludge transport and disposal (continued)	
Disposal site location	3
Covering practices	2
Cover material availability	2
Sludge barging transport	2
Operation, maintenance, and administration	
Problem frequency and severity[a]	
Infiltration/inflow	10
Hydraulic overload/underload	8
Equipment breakdown	8
Equipment age	7
Spare parts availability	7
Weather conditions	7
Sludge treatment train	6
Facility maintenance	6
Staffing level	6
Organic overload/underload	6
Staff morale	6
Process controllability	6
Operator knowledge/experience	5

[a] Relative frequency and severity indexed to most frequent and severe problem—infiltration/inflow—with an index number of 10.

Nearly 30% of responding plants reported minor, periodic problems in achieving their NPDES permit BOD and SS limitations. Plants reported little or no difficulty in achieving permitted pH and nutrient restrictions (nearly 33% practiced chemical addition).

The major findings of the WPCF survey are as follows:

- I/I and hydraulic overload problems are significant even though the majority of plants are below their rated averaged daily flow capacity;
- A large number of problems related to equipment breakdown, age, and spare parts were indicated, although most of the plants have been modernized or placed into operation in the last 10 years;
- Weather conditions (or inadequate design considerations of climatic conditions) were ranked ahead of sludge treatment train and land availability problems; and
- Odor issues were indicated as significant in every aspect of the residuals management train.

These findings yield to the certain conclusion that future designs must focus more on odor prevention and control. Frequently reported equipment and I/I-induced hydraulic problems suggest that present design, operating, and administrative practices result in less capacity and reliability than would be reasonably anticipated. The high ranking of weather suggests its likely influence on liquid processing (precipitation induced inflow), the availability of residual disposal sites (wet or frozen grounds), and the plant appurtenances necessary to deal with local weather conditions.

Equipment problems are related to one or more of the following: intolerant equipment, inappropriate application, inadequate design, poor manufacture, insensitive operation, or inadequate maintenance funding. I/I and hydraulic overload problems may also be linked to improper management of internal solids inventories.

Examples of improper management of internal solids inventories due to design or operational shortcomings include excessive operating levels of aerator solids linked with improper operating targets and inadequate wastage or improper return rates for actual plant flows. Deficiencies will be exacerbated by excessive recycles due to overflowing storage tanks or poor solids captures in the solids-processing train and improper solids removals from the plant due to undersized equipment, short operating schedules, or lack of a solids disposal site. These design or operational deficiencies can result in excessive effluent suspended solids even with the best final sedimentation tank design.

The two surveys show that, regardless of the size of the plant, a successful design is responsive to operating needs, representing a full range of operating and environmental conditions, for processing reserve (redundancy and flexibility) and tolerant, rugged equipment (reliability). However, the information also suggests that even the perfect design will not perform adequately without informed operation and responsible administration and, most importantly, reliable solids disposal based on the daily needs of the plant.

*F*ACILITY DESIGN REQUIREMENTS

A design may be functionally correct but fall short of expectations if it fails to account for start-up conditions, potential future expansion of the plant, the convenience and safety of the plant employees, and the plant's impacts on its surroundings.

PRESENT AND FUTURE DESIGN REQUIREMENTS. Reasonably accommodating the needs during the initial years of plant operation must be properly balanced with those of the future. In most cases, completely accommodating the objectives of any part of this time span will compromise those of another part. Experience has proved that the design should primarily accommodate the design year projected conditions, with allowances for (1) proper operation when loading conditions may be significantly less than design year loadings and (2) expansion or rehabilitation to handle loadings reasonably anticipated beyond the design year. Achieving the proper balance between the design period and the future sometimes creates a dilemma. In many cases, disregard for the future beyond the design year has resulted in abandonment of the original facility at great cost to the community. In other cases, an overly intensive design focus on an uncertain future beyond the design period has resulted in facilities with operations, maintenance, or performance shortcomings during the design period.

Because the reliability of loading projections declines as the time span of the projection increases, a facility process or layout commitment to an uncertain distant future deserves careful scrutiny if it would significantly compromise system operation during the first 15 to 20 years after start-up. As another consideration that reduces the reliability of commitments to the future, future changes in regulatory requirements or treatment technologies could invalidate assumptions underlying the future commitment. Nonetheless, the design of any treatment works should consider the likelihood that most plants will eventually be rehabilitated, upgraded, or expanded, regardless of the design period or the anticipated service lives of plant facilities. Recognition and reasonable accommodation of inevitable change and replacement is a key responsibility of the owner and the design engineer. With prudence and foresight, future plant modifications can be made easily and economically. Table 3.3[3] summarizes considerations involved in planning for the future beyond the design period. Chapters 4, 5, 7, and 8 provide additional relevant information.

Table 3.3   Considerations associated with designing for the future.[3]

Topics	Considerations
Design documentation	Standards and criteria for existing facilities, designed improvements, and future expansion planning (including equipment sizing; ultimate capacity layouts; and hydraulic, structural, and electrical details) should be documented as appropriate at the beginning of a project and properly archived by the owner and designer at its end. This documentation should include the hydraulic profile; process and instrumentation diagram; processing schematic(s); mass balances; design wastewater characteristics (including infiltration and inflow); performance objectives; peaking factors; service population; and significant sources of industrial, commercial, institutional or specialty wastes.
Plant layout	
Ultimate plant capacity	Definition of the ultimate plant capacity is the first step in designing for the future and can be estimated from service area boundaries, population projections, land-use plans, and the available site (or needed site) for various planning periods.
Future standards	Estimates of future compliance requirements for the plant's emissions (effluent, residuals, and atmospheric) should be used to allot space for future treatment needs and other process modifications as future standards evolve.
Adjacent land use	Planned uses of adjacent land represent an external variable to be used in formulating the site needs for the future.
Site planning considerations	The first step in developing the site plans for the current design should be based on the ultimate plant capacity and needs for the future, including • Fundamental layouts for the ultimate plant capacity showing operating and processing areas, site access, roadways, buffer zones, and drainage; • The ultimate capacity hydraulic profile to identify future hydraulic constraints and preserve and allocate hydraulic head to accommodate future expansion and distribution (pipelines and channels that are common to future expansion configurations must be carefully sized to avoid low velocity conditions and solids deposition problems in the early low-flow years of operation); and • An ultimate electrical load analysis to form the basis for the primary electrical distribution plan.

Table 3.3   Considerations associated with designing for the future (continued). [3]

Topics	Considerations
Design details	
Isolation and diversion during future construction	The design of any component of the treatment facility should anticipate repairs and phased expansion while the existing plant is in full operation. Assurance of operating ease during these conditions can be incorporated in the current design through the use of such isolation and diversion features as gates, valves, blind flanges, and stop log keys.
Structures	Often, certain structures are more appropriately designed for their ultimate capacity. Control rooms and operating centers should be sized to minimize expansion disruptions. Other buildings and facilities may be designed to accommodate alternate future uses or to facilitate future expansion. Common structural details used to accommodate future expansion include knockout panels, common wall designs, wall expansions, access allowances, rebar dowel splices, and key ways.
Equipment sizing	Unit process equipment sizing reflects collective consideration of capacity with units out of service, economies of scale in construction and operation, current design criteria, and phased and ultimate capacity planning. The attractiveness of increased operational flexibility through multiple, small units should be balanced against likely savings derived through fewer, larger units. The designer should compare the future replacement of smaller units by larger capacity equipment with the progressive expansion of the current equipment; in both cases, the design should account for allocation of future spatial needs.
Electrical and instrumentation	An often overlooked consideration is expansion planning for the electrical and instrumentation systems. Space capacity should be provided in the electrical duct bank, switchgear, and control system components to minimize future installation costs and operating inconvenience.

Principles of Integrated Facility Design   71

Designs of plants having a large disparity between their initial operating and design conditions deserve special consideration to ensure adaptability of processes. In this regard, normal redundancy of process units will usually provide the capability for efficient operation during initial periods of reduced loading. The use of variable delivery capability within the range of predicted operation for pumps, feed systems, blowers, and other equipment will also help accommodate a wide range of loading conditions. Hydraulic transport of solids-containing process flows is often a demanding consideration, requiring either mitigation by multiple piping or special provisions that permit interim periods of flushing or resuspension of low-flow deposition.

The designer must ensure that the facility can be operated and maintained without significant difficulties during the construction period. Achieving this objective sometimes requires rigorous analysis and special design accommodations, in some cases even the installation of temporary treatment facilities. The contractor alone cannot be expected to develop measures for maintaining compliance with the plant's permit during periods of disruptive construction. The contractor's responsibilities should be limited to safely carrying out construction procedures and work orders consistent with the contract documents prepared by the designer.

COMPATIBILITY WITH SURROUNDINGS. As noted in Chapter 1, the first stimulus for wastewater management was aesthetics. Aesthetics is now, and will likely remain, the foremost concern of the general public. Regardless of the plant's success in meeting its numerical performance standards, the public will judge the treatment facility to be unsuccessful if it is aesthetically offensive to its neighbors.

Mitigation of the visual, odor, and sound impacts of the plant should concern the owner and design engineer as much as meeting its performance objectives; mitigation of adverse aesthetic effects is, therefore, an essential goal for any facility improvement. If facility designs focus on possible aesthetic impacts and their control by process selection and source and site mitigation, plants will likely best serve the public and avoid the controversy and expense of remedial actions.

Alleviating odor is the most important consideration. Planned odor mitigation strategies seek to avoid open air turbulent mixing (and stripping) of raw and partially stabilized wastewater and minimize open air exposure of solids-processing recycles and residue before stabilization. Such strategies favor selecting processing concepts that will avoid generation of odors if odor-free alternatives are available.[4]

Sound is the second most important aesthetic consideration. Such concerns are mostly limited to noise generated by large, powered equipment (including engine generators, blowers, fans, and mechanical aerators). Variation, pulse, and tone of the noise can affect the listener as much as or more than the decibel energy of the sound wave. Mitigation strategies focus on equipment selection, acoustical architectural techniques, and the use of barriers or other sound wave attenuation measures within buildings, surrounding structures, and plant grounds. Noise impacts of construction and operation can be significant.

The third aesthetic issue concerns appearance. Building and landscape architects can enhance the visual appeal of plant structures and grounds. Sensitive landscaping and architectural design, usually incorporating low profile structures and structural shields of supporting auxiliary equipment, and multiple-use recreational planning for the plant's buffer zone around the plant can mitigate visual impacts and help reduce sound impacts on the surrounding population.

No discussion of aesthetic issues is complete without some discussion of truck traffic to which the public reacts because of concerns for public safety, disruption of neighborhood activities, and aesthetics. Because such concerns during construction and operation are expected, the designer should anticipate possible conflicts between public expectations and the needs of the facility and take steps to reduce these conflicts as much as practical. As examples, the designer may consider establishing service schedules noncoincidental with rush hours and presence of school children on the streets and specific truck routes to the plant that avoid residential streets and neighborhood commercial centers.

SENSITIVITY TO OPERATIONS. Operability is another critical aspect of design that deserves, but often lacks, the designer's attention. Operability may be considered as the extent to which a design affords opportunities for the operator to achieve the designer's efficiency and performance objectives. In this regard, the designer's challenge is two-fold: (1) providing for a work place that is safe, convenient, and pleasant and (2) developing a process design that tolerates reasonable degrees of loading and environmental variation for efficiency in process control and maintenance.

Experience is required in properly accommodating operability considerations with all the other considerations involved in an optimum design. Short of actual operating experience, the designer may best gain the necessary knowledge of operability by taking every opportunity to observe plant operations, thus gaining an operator's perspective of design adequacy. The design process must include operator review, ideally by operating staff of the plant being designed. If these operators cannot be consulted, other operators with experience in systems and processes similar to those of the design should be consulted. Direction, review, and feedback should be sought in the beginning of and throughout the design process.

Operability entails not only providing the plant components and features needed by the operator in normal circumstances but also providing the elements necessary to cope with unusual circumstances. The designer's attention must also encompass the myriad of non-process-related items that support plant operations. New features and options require review in the formative stage to allow an understanding of their effects on the design and gain the owner's commitment to provide, as required by changes, expanded staffing and other operations and maintenance needs.

The appropriateness of process complexity, automation, and flexibility varies both with the size of the plant and the level of required performance. The initial screening process for evaluating alternative systems and processes should include careful consideration of the different levels of staffing required.

F*ACTORS CRITICAL TO PROCESS DESIGN*

This section describes some of the major factors that are critical to the proper design of the plant processes. Designs that are developed with an understanding of these factors can tolerate changing conditions without exceeding process design objectives.

LOADING VARIABILITY. Loading variability has been considered in terms of (1) plant influent conditions, (2) the attenuation of these conditions through stormwater management, and (3) individual unit process conditions, with differing variability characteristics due to load modification by preceding unit process and recycle loads.

Influent Loading Variability. Wastewater treatment plants and their processes are commonly discussed and defined in terms of their average day capacity. As a practical matter, average day conditions are points on a curve of events that may not be observed on a daily basis. Sound design practice does not use average day conditions for anything except as a convenient point of reference for peaking factors that are actually of interest in the design. Conceptually, preferred practice applies two peaking factors: a hydraulic peak and a process peak.

Conduits, channels, and unit processes must be sized to pass the flow applied to them. The maximum hydraulic design of a plant should include capabilities throughout the plant to handle the capacity of the trunk sewer(s) or rated capacities of contributing pump stations. The hydraulic design of unit processes, using multiple treatment units, should also provide for applied flow to be accommodated even with one of the parallel unit processes removed from service for inspection or repairs.

Process design should be based on required performance attainment at maximum process loading conditions. Before Public Law 92-500 and its implementing regulations, performance and the loading basis on which performance was to be measured were considered in terms of seasonal or annual average conditions. Now, a minimum definition for this condition of design corresponds with the compliance interval included in the plant's NPDES permit. This interval typically represents the maximum month and week periods of compliance as noted in Chapter 2.

Effluent compliance standards, with pollutant limits for the maximum month and week per year, statistically dictate a design processing reliability up to the maximum month effluent restriction 92% of the time and a processing reliability up to the maximum allowable week effluent limitation approximately 98% of the time. Some regulatory agencies impose more stringent standards, defining effluent standards in terms of the maximum daily event per year (a design reliability of approximately 99.7%) or in terms of not exceeding or falling below certain criteria at any time (1 hour per year has a design or performance reliability of 99.99% of the time).

With the regulatory definition of the minimum compliance interval, design criteria may then be developed to establish the controlling condition(s) for the plant's most stressed month and week (or more stringent intervals, if appropriate). Usually, the controlling condition reflects one or more of the following constraints: maximum flow, maximum pollutant load or mass, most stringent effluent limitation, or most restrictive processing condition (for example, temperature). These constraints could occur simultaneously or at different times, depending on the particular circumstances of the plant.

The error most often encountered with maxima characterizations is the definition of the pollutant mass with a pollutant concentration that is independent of its flow regime. Then, the result of multiplying the pollutant concentrations by an independently developed or unlinked flow characterization may overstate the magnitude of the controlling design mass.

Minimum- and maximum-to-average day flow ratios or guidelines have been used frequently for design of sanitary sewers.[5,6] However, such information should be used with caution and only when actual plant information is unavailable because these ratios do not account for stormwater and infiltration introduced to the collection system from combined sewers and the leaking joints of an old collection system.

Plant designers lack generalized guidelines for the more reasonable process design criteria of maximum month and maximum week. Therefore, these criteria are now developed on a case-by-case basis.

Figure 3.1 shows the variation of influent flow regimes on an annual basis for two plants in the Northeast and the flow peaks derived from "Recommended Standards for Sewage Works."[5] Though influent to the smaller plant includes no significant industrial contribution, it does include flow from a seasonal influx of summer tourists; the tourist flow peak is not significant, however, because of concurrent seasonal declines of infiltration and inflow. The larger of the two plants has a dominant, mixed industrial

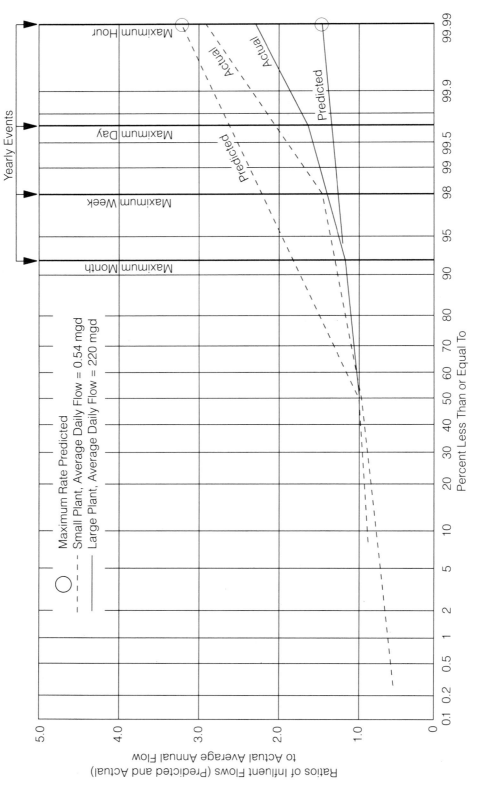

Figure 3.1 Variation of influent flow regimes for two plants (mgd × 3785 = m³/d).

base; its older service areas have combined sewers. Perhaps the key observation from these data is that any process design sized solely for acceptable performance at the maximum hour or day rate would be grossly oversized for a maximum week or month compliance interval.

Figure 3.2 presents a larger plant's influent BOD and suspended solids mass peaking factors, thus allowing further analysis of its wastewater characteristics. Note that all three influent design parameters exhibit similar peaking factors except for compliance periods more restrictive than the maximum week per year. Again, data strongly suggest that if a plant's entire process design were developed for meeting its permitted effluent limits during the maximum day, an overdesigned plant would likely result.

The designer should ensure that extreme influent and effluent pollutant observations or deviations from the norm noted at any plant are associated with real events and not sampling or analytical errors. The designer should guard against the inclusion of invalid data when defining maximum conditions.

In developing the flow and pollutant mass processing peaks for designing improvements of an existing plant, the engineer should use the most recent 3 full years of record (longer periods are appropriate if the service area is experiencing a prolonged drought) and compare data for each year. Such procedures will normally prevent a bias associated with an abnormally dry or wet year and disclose repeatable trends and possible abnormal peaks in any 1 year that should be further analyzed. The engineer should ensure that characterization of raw wastewater excludes recycle flows and that the plant's sampling program provides for representative samples.

Typically, regulatory sampling and monitoring requirements for smaller plants allow sampling programs that overstate the influent pollutant load (sampling restricted to 5 or fewer weekdays with samples collected for 6 or fewer hours) and potentially overstate the performance of the plant (limited sampling and collections, which fail to reflect the entire daily result).

The validity of the influent characterization can be checked by using typical values for per capita pollutant emissions. Typical BOD and SS values for municipal wastewater without significant industrial components approximate 0.1 kg/cap·d (0.2 lb/cap/d) ± 20%. Values outside this range should be regarded as suspect unless confirmed by detailed evaluation of the service area population (resident and transient) and significant industries. Data showing high, questionable peaks can be checked by examining reported data on waste sludge masses and performing mass balances.

Without a significant industrial contribution, pollutant mass peaking factors will generally be approximately 80 to 100% of the hydraulic peaking factor, although not necessarily concurrent, for a single day. This guideline can be used if other information is lacking. If the flow and mass loading data show peaks inconsistent with this relationship, this indicates possible data error and the need for review to verify the data.

Principles of Integrated Facility Design 77

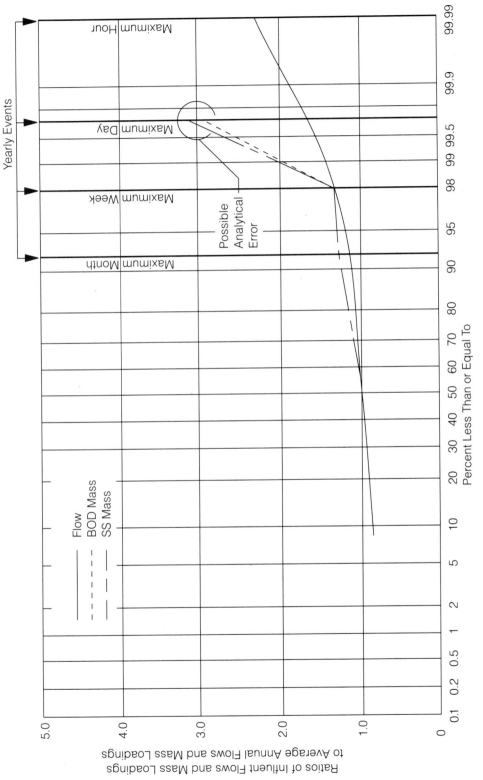

Figure 3.2 Flow and pollutant mass peaking factors for a single plant.

Table 3.4 offers further information regarding wastewater peaking characteristics at a major plant. As shown, the plant experiences most, if not all, of its peaks independent of each other. These data also suggest that the plant's BOD peaks in the maximum month and other times are due to extreme spike releases of soluble BOD (BOD and SS values are not correlated in the maximum BOD month) and some seasonal variation of industrial operations (BOD/SS minima in December and maxima in the summer). At this plant, because flow peaks are associated with random precipitation events, the conservative process design would be based on the assumption that the random rain event could be coincident with the plant's maximum month and maximum week pollutant masses.

Table 3.4 also defines the plant's minimum month pollutant loads. Minima determinations are significant to ensure that the process design can effectively handle the full expected range of pollutant loadings during the entire design period.

Based on the above considerations concerning influent peaking factors, two influent process peaks—a design hydraulic peak and a design process

Table 3.4 **Representative influent peaking factors for a major wastewater treatment plant.**

Parameter and peaking factor (PF)[a]	Minimum month	Average month	Maximum month	Maximum week	Maximum 3 consecutive days	Maximum day	Maximum rate
Influent flow basis (average daily flow = 220 mgd)[b]							
Month	October		April				
Flow PF	0.9	1.0	1.2	1.3	1.4	1.6	2.3+
BOD PF	1.0	1.0	1.1				
SS PF	1.0	1.0	1.0				
Influent BOD basis (average daily BOD = 300 mg/L)							
Month	December		August				
Flow PF	1.1	1.0	1.0	1.0	1.0	1.1	
BOD PF	0.9	1.0	1.1	1.3	2.0	2.9	
SS PF	0.9	1.0	1.1	1.2	1.3	1.8	
Influent SS basis (average daily SS = 450 mg/L)							
Month	December		September				
Flow PF	1.1	1.0	0.9	0.9	0.9	0.9	
BOD PF	0.9	1.0	1.1	1.0	1.2	0.8	
SS PF	0.9	1.0	1.3	1.3	1.7	3.0	

[a] All peaking factors (PF) relate to their respective average day values.
[b] mgd × 3785 = m^3/d.

peak—usually provide a sound basis for design. The design process peak should match the plant's compliance period interval for its effluent limitations (typically, the maximum month and week per year) under the most restrictive conditions for control of pollutants. Higher peaking factors, such as those for a maximum day, may be appropriate for hydraulic flows but not process performance. With few exceptions, flow peaks that cause difficulties at a treatment plant result from transient storm events and the nature of the plant's service area and collection system. Regulatory agencies and designers should exercise caution in using maximum effluent and process performance compliance standards based on the maximum day unless extra capacities are warranted because of the nature of the receiving water and its beneficial uses.

Because many liquid-processing facilities are sized, in part, through hydraulic criteria that apply to some maximum condition, strategic management of flow peaks in the plant, collection system, or both offers benefits of processing stability and cost savings. Plant design that couples flow peak mitigation opportunities found in the collection system with those of the treatment works design can proactively control and reactively respond to plant influent.

Collection system stormwater management technologies are growing in the U.S. and abroad.[7] Representative approaches for management of stormwater include

- Best management practices
  — Existing systems (I/I types of rehabilitation and regulator optimization) and
  — Newly developing areas (infiltration basins/trenches and extended detention/wet ponds).
- Stormwater management
  — Catch basin controls for shifting water away from combined area or storing surface flow,
  — Off-line storage tanks and surface impoundments for retention of surface flow, and
  — In-line/off-line underground storage detention tanks within collection systems.
- System optimization programs with variable static flow controllers to create and maximize the system flow control/storage capacity.
- Separation (selective or complete).

Some stormwater management options for collection systems are illustrated in Figure 3.3.[7] As shown, the many opportunities for flow peak mitigation in the collection system incorporate four simple principles:

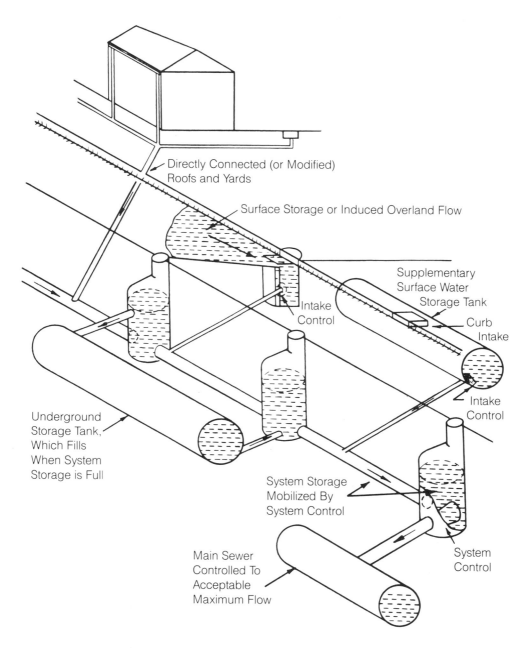

Directly Connected (or Modified) Roofs and Yards

Surface Storage or Induced Overland Flow

Supplementary Surface Water Storage Tank

Intake Control

Curb Intake

Intake Control

Underground Storage Tank, Which Fills When System Storage is Full

System Storage Mobilized By System Control

System Control

Main Sewer Controlled To Acceptable Maximum Flow

Figure 3.3   Stormwater management approaches.

1. Reduce the storm flow volume entering the collection system,
2. Optimize the peak flow attenuation capability and related storage capacity of the collection system,

3. Use off- and in-line surface and subsurface storage strategies to mitigate flow peaks, and
4. Selectively allow wet weather overflow after capture of the first flush to the extent permitted by water quality considerations.

The cost-effective approach to stormwater management at POTWs probably consists of some combination of upsystem (collection system) and downsystem (plant) control strategies. Upsystem or downsystem, the problem-solving principles that apply are the same: store flows and return them at a more favorable time or bypass flows and reintroduce them at a more favorable location.

Storage and bleedback systems are the most logical for plant-based flow peak control. They are most attractive for small plants at which flow peaks are the highest. The size of the storage tank depends on the cost-effectiveness analysis for the storm hydrograph of concern. As a preliminary planning guideline, a storage tank volume could range from 15 to 35% of the plant's average daily flow, with retention of a plant flow peaking factor of about 1.5 to 2 times the average daily flow.

Implementation of storage and bleedback systems has been hindered by problems with poststorm event cleanup. Flusher tanks included in German stormwater management practice are of particular interest because they appear to overcome the poststorm event cleanup problem.[7] Chapter 9 discusses equalization tanks.

Some have proposed an alternative approach that would replace the off-stream equalization tank and its cost with a capacity increase for a main stream treatment process such as the final sedimentation tank, a unit of continual use and value. Any advantage of this concept would be lost, however, if the design or operating MLSS level were too high or the design or operating return sludge rate were too low for the design storm event.

Another alternative, used most frequently at operating plants (whether or not included in the plant design), provides treatment equalization by bypassing the flow around one or more unit processes. Often, this bypass follows primary treatment, with the bypassed flow reentering the main flow stream just ahead of the disinfection unit. Another storm flow management option, if the soluble organic load is a concern, would provide for returning bypassed flow to the biological reactor effluent ahead of the final sedimentation tank where it dilutes and contacts organisms without imposing additional solids mass loading.

Plant designs frequently offer opportunities for treatment equalization through flow splitting and bypassing. The process designer may evaluate each intermediate pump station and process and then determine whether lower design peaks are appropriate. Examples for such evaluations include trickling filter pump stations and final effluent pump stations and filters. Additionally, where the plant's effluent limits are less stringent than the performance capability of the treatment technology (for example, an effluent SS limit of 12 mg/L, with an upstream effluent filter, or an effluent ammonium limit of 8 mg/L, with a nitrification system), the designer should provide the capability for low-flow and average-day flow splitting to allow savings of operating energy costs.

Finally, the plant designer should consider the opportunity to apply appropriate stormwater management control technologies in the plant or the collection system. One example of such a technology transfer is the vortex grit removal system (see Chapter 9). Similar approaches are also feasible for scum separation. Other stormwater technologies that may be especially appropriate in plant design include vortex valves and air-regulated siphons for flow regulation and diversion.

Unit Process Loading Variability. The development of appropriate unit process peaking factors for the plant represents the final decision before design begins. Usually these values are derived from the plant's operating records. They reflect not only the character of the wastewater but also the character of the upstream unit process and its proposed operating strategy, recycle returned ahead of the processing step, and the downstream processing step and its operating strategy.

Table 3.5 summarizes the considerations for developing peaking factors for unit process design. The table also presents representative peaking factors for those processing points where a case-specific analysis is unnecessary. The peaking factors presented in Table 3.5 may suffice if other information is unavailable. However, note that they were derived from a plant with a large northeast urban sewer service area generating an average daily flow of approximately 189 000 m^3/d (50 mgd) from its combined and separate sewers. The service area lacks unique industrial customers that would contribute significant floating scum. Sand and salt mixtures are used on the city and neighborhood streets during snowfall. Somewhat higher values for primary sludge and primary effluent organic mass may be encountered at smaller plants with flows of 3800 m^3/d (1 mgd) and less.

DYNAMIC FACTORS. Sound process design depends, in part, on understanding process dynamics, including the concept of equilibrium and the acclimation of organisms, each discussed below.

Equilibrium. Good design and the interpretation of investigative results depend on understanding the concept of equilibrium (or steady-state conditions). Equilibrium, in an applied sense, represents the inherent tolerance (or intolerance) of treatment processes for applied pollutant loads.

Figure 3.4 shows the response of a perfectly mixed system to a change in influent conditions. It applies equally well to a change in operating strategy. As shown for one theoretical detention time, if the new influent conditions

---

Table 3.5 Unit process peaking factors.

Substance	Comment	Peaking factors for consecutive days[a]			
		1	3	5	7
Screenings	Average day value will vary as function of screen size. Size container for maximum 3 consecutive days of screenings.	8	Yearly maximum 4	2.5	2
Grit	Average day value will vary as function of design mesh capture, service area, sewer age, locality, and snow removal practices. Size container for maximum 3 consecutive days of grit.	8	Yearly maximum 4	3	2
Raw wastewater scum	Convey in dilute slurry until ready for final disposal if at all possible. Average day value may vary with industrial base. Size for maximum week to maximum month at concentration or destruction step with excess return and easy ability to operate at significantly lower values.	8	Yearly maximum 4	Maximum month 1.5–2.0 3	2
Primary sludge	Average month peaking factors are usually adequate because of primary sedimentation tank solids storage capacity. This should be checked.	2–3 1.6–1.7	Yearly maximum 1.6–2.1 Average month 1.3–1.4	1.5–1.6 1.2–1.3	1.3–1.4 1.1–1.2
Primary effluent organic mass (excluding recycles)	Hourly interval values can be estimated form frequency distribution graphs (see Figure 2.2). Values highly sensitive to soluble industrial releases.	1.8–2.2 1.3–1.5	Yearly maximum 1.3–1.6 Average month 1.2–1.3	1.3–1.4 1.1–1.2	1.2–1.3 1.1–1.2
Nitrogen and phosphorus		1.8–2.2 1.3–1.5	Yearly maximum 1.3–1.6 Average month 1.2–1.3	1.3–1.4 1.1–1.2	1.2–1.3 1.1–1.2

Table 3.5   Unit process peaking factors (continued).

Substance	Comment	Peaking factors for consecutive days[a]			
		1	3	5	7
Oxygen demand and waste secondary sludge	Will vary as function of reactor configuration, cell residence time, operating solids concentration, secondary system hydraulic detention time and recycle rate, and the applied mass and form of oxygen-demanding materials. Actual waste secondary sludge mass will reflect solids storage reserve capacity (ability to operate at higher solids concentration) maintained in the reactor. Attached growth systems have no reserve.				
Recycles	Will vary as function of unit process and unit process operating strategy and, for solids processing, the operating strategies and processes for the upsystem reactors and separators and the main stream biological treatment system. Typically, granular media filters exert the greatest hydraulic stress (especially if backwash is discontinuous); the greatest biodegradable carbon recycle is associated with thermal sludge conditioning of high-rate biological sludge; the greatest nitrogen recycle is associated with anaerobic and composting digestion of high-rate secondary solids, with discontinuous supernating and dewatering more troublesome than digestion and dewatering and composting of raw solids; the greatest recycle of phosphorus occurs with anaerobic digestion of biologically enhanced phosphorus-laden secondary solids.				

[a] The peaking factor represents the result of dividing the average of maximum-consecutive day loadings within the tabulated time interval by the average daily loadings for the year.

Principles of Integrated Facility Design

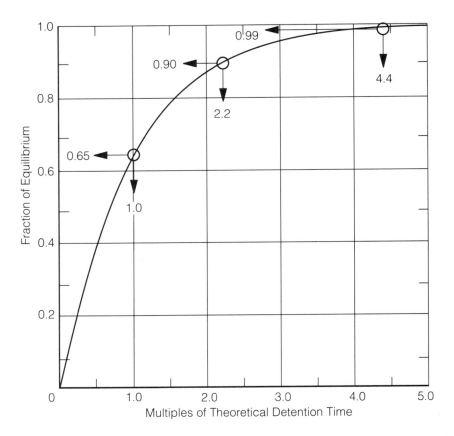

Figure 3.4 Response of perfectly mixed system to change in influent conditions.

were constant, the process would reach only 65% of steady-state conditions. Similarly, the process would not reach 90% of steady-state conditions until 2.2 times the theoretical detention time had elapsed; 4.4 times the theoretical detention time would be required to reach 99% of steady state. Such time lags for reaching equilibrium apply to observations of all plant processes depending on detention time, including the cell or solids residence times.

Even with a constant operating strategy, a wastewater treatment plant never fully reaches equilibrium, but rather continually adjusts to changes of flow and pollutant mass (and concentration), termed processing transients. The process designer influences the tolerance of the plant to transient conditions by selecting the unit processes. The process operator also influences the tolerance of the plant by choosing the operating strategy before and during the transient condition.

An example of an intolerant process selection would be a granular activated carbon filter with an empty bed contact time of several minutes for effluent from secondary treatment. At the other extreme, a highly tolerant process selection would be a stabilization pond with a detention time of several weeks. An example of an intolerant design and operating strategy for an activated sludge plant would be establishing a mixed liquor suspended solids operating level and return sludge rate based on average day conditions for operation under elevated flow conditions.

For process design and evaluations, Figure 3.4 suggests that all treatment processes represent imperfect attempts to achieve either plug flow or completely mixed conditions. Typically, a secondary treatment plant will have 12 or more hours of hydraulic retention in its main flow stream. This volume and the plant's internal recycle will dampen the extremes of daily pollutant loadings. If a biological stabilization process must nitrify in April, conditioning and equilibrium will require that conditions for nitrification be established in March to provide an adequate population of nitrifying microorganisms when needed in April.

The process designer must take into account and use the equilibrium concepts when developing responsive processing solutions. Sound sampling programs are based on and must consider concepts of processing and plant equilibrium.

Acclimation. Figure 3.5 shows application of the equilibrium concept to process design by presenting the relative masses of organisms acclimated (or more properly, exposed) to a 1-d/wk release of substrate for a range of commonly found design cell residence time (CRT) systems. As shown, the older system provides a better buffer and more resiliency for discontinuous substrate release. If this substrate is unique to the system, such as a spill or dump of a biologically resistant hazardous organic substrate or a readily biodegradable but foreign substrate, the material is more likely to pass through a young CRT system or accumulate in its sludge.

The theoretical capability of a biological treatment system to respond to a substrate depends solely on the intensity of immediate exposure (shock loading) of the organism and its history of exposure to the substrate. Sudden slugs of a unique substrate may pass through even with an acclimated culture if the culture was acclimated to a lower substrate mass or concentration.

The acclimation response presented in Figure 3.5 is at the heart of both municipal and industrial biochemical treatment strategies for continuous releases of biologically resistant organic compounds and discontinuous releases of uniquely different biodegradable materials. This response explains the greater soluble-phase variability and higher nominal values for low CRT systems compared with those observed with higher CRT systems. Such concepts are also used in the design of nitrification systems wherein older sludge cultures are selected to avoid the washout of slower replicating, nitrifying organisms.

The designer should recognize that, in many cases, the facility must serve the complex needs of compatible discharges from a variety of industrial customers and that the form and magnitude of these wastes may vary seasonally

---

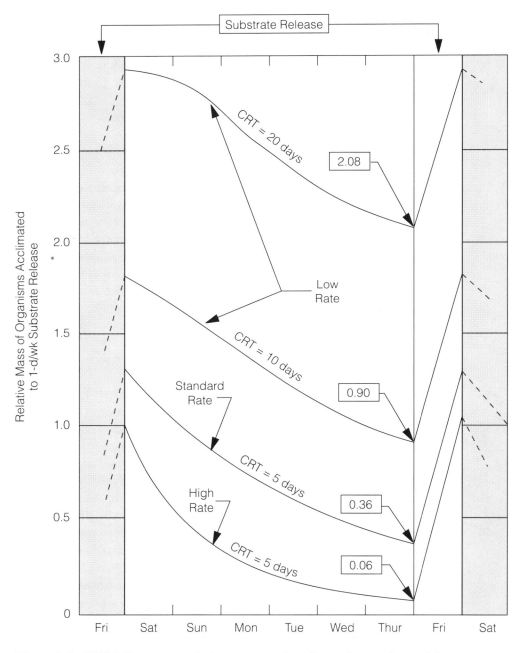

Figure 3.5   CRT influence on relative masses of acclimated organisms with discontinuous substrate release.

and daily. Besides the establishment of sewer use ordinances, manipulation of the plant's CRT provides the only natural means available for the plant to develop a compatibility for discontinuous release of foreign biodegradable material and continuous release of biologically resistant organic compounds. Plants that are designed to achieve effluent objectives over a reasonable CRT range can more readily respond to changing conditions.

Because BOD_5 is a major parameter of compliance for secondary treatment equivalency, plants that are designed or operated with a CRT of fewer than 5 days are progressively more likely to experience measurable soluble BOD passthrough problems as their departure from the 5-day interval increases. Such passthrough reflects the reduced acclimation opportunity resulting from variable industrial release practices and the longer CRT of the compliance test, though performed with unacclimated organisms, compared with the operating system.

PERFORMANCE EVALUATION AND INTERPRETATION. Recognizing the probabilities of sampling and analytical error, the designer must often interpret existing data from operating facilities or data derived from pilot-plant or prototype investigations. Also, the designer may be asked to assess the data interpretations and recommendations of others. This subsection discusses briefly opportunities for error in the interpretation of data and the usual limitations of pilot or prototype work.

Interpretation of Operating Records. The designer often reviews operating records to determine the performance of unit processes. In assessing the potential for error, this review should define the location, method, and frequency of sample collection; the point of introduction of all recycles; methods of flow measurement or estimation; instrument calibration frequency; and analytical methods.

Table 3.6 offers guidance for the interpretation of sampling results. Additionally, mass balances for conservative elements or pollutants that maintain their integrity through the processing step will allow the designer to gain a rapid understanding of the significant elements in terms of recycles, solids captures, and the validity of the sampling and measurement program. Inert solids measurements and balances can quickly provide an overall assessment of the validity of the plant's monitoring program and whether the performance of a solids destruction process is correctly defined. Percent accountable mass balance (out ÷ in) closures of 100 ± 10% are considered excellent; closures below 80% and above 120% reveal suspect results from one or more processing points.

Interpretation of Investigative Results. Results of special investigations, typically using pilot-plant or prototype equipment, are often incorporated in process design. Such investigations can provide valuable insights but do not

Table 3.6   Guidelines for interpretation of sampling results.

Waste stream	Guidelines
Liquid processing train	
Raw wastewater	Difficult to obtain representative sample, often includes recycles. Grit and floatable quantities typically not included in sample. Automatic samplers: clogging of sampling hose gives low suspended solids, overflowing of sampling bottle gives high suspended solids. Grab samples: suspect because around-the-clock diligence is difficult to enforce. Check validity with overall inert solids balance. SS error may be introduced by failure to fully filter sample. BOD_5 results may be understated with high SS because classical BOD test does not keep this material in suspension. False high-influent BODs may be encountered because of sulfides.
Primary effluent	Generally provides a good sample as variability associated with high suspended solids has been eliminated by the unit process. False, high BODs may be encountered because of sulfides and nitrifying flora recycles.
Aeration system	Dissolved oxygen level may not be reported at point of greatest exertion and time of greatest stress. If bulking sludges (for example, SVI > 150 mL/g) are encountered, look for DO stresses first. (Low DO-induced sludge bulking only requires the occurrence of occasionally low DO conditions sometime during the CRT history of the sludge.) Good measure of inert suspended solids. Metal salt addition phosphorus removal strategies will result in a false high measure of volatility due to the loss of the hydroxide water.
Return and waste sludge	Normally a grab sample has a sludge concentration reflecting the hydraulic conditions at the time of collection. Low total solids values (for example $\leq 0.5\%$) should be corrected by dissolved solids for valid SS estimate. The following calculated value is most valid: $$R, \text{mg/L} = (\text{MLSS, mg/L})\left[\frac{(Q+R)}{R, \text{mgd}}\right]$$
Secondary effluent	Unless CBOD tests are performed, BOD determinations are subject to false positives due to nitrification in the BOD bottle (likely if $NH_4\text{–}N \geq 1.0$ mg/L with $NO_3\text{–}N \geq 3.0$ mg/L, $BOD_5 >$ SS if SS > 10 mg/L; this suggests nitrification in BOD bottle, spring and fall changes yield transient false positives, or nonnitrifying plant design partially nitrifies in earlier years or in summer months). Grab-sampling programs may miss elevated effluent SS and diurnal stresses. Guard against samples after chlorination for BOD determinations without dechlorination and reseed.

Table 3.6   Guidelines for interpretation of sampling results (continued).

Waste stream	Guidelines
Solids-processing train	
General	Grab samples are typically in use. Sampling results are often higher than the real average because startup and shutdown conditions are not quantified and sampler may be interested in best rather than representative results (especially with product solids). Discontinuous operations likely to have more high-side bias than continuous operations. When the overflow sludge storage tanks are not filled, samples of gravity thickener overflow may not be representative. Scales, if calibrated, represent excellent point of mass measurement.
Stabilization process	Excellent point of inert solids measurement and long-term equilibrium with applied suspended solids characteristics where inert solids have to be equal to applied inert solids. Supernatant suspended solids recycles are often underestimated.

guarantee subsequent process success, particularly if they are improperly structured or interpreted. Similarly, caution should be used if investigative findings refute understandings gained from long experience.

Table 3.7 provides guidance for interpreting investigative results. The following paragraphs offer additional guidance for their use and interpretation.

WASTE TREATABILITY. Waste treatability studies are often conducted when the biodegradability or chemical treatability of industrial wastes are uncertain, unconventional processes are being explored, or process specific design parameters are uncertain. Equilibrium and acclimation considerations described in the previous subsection are vital to reaching valid conclusions and must be considered whenever a processing variable is changed. In conducting these studies, the investigator should remember that solids–liquid separation, oxygen transfer, and mixing considerations are likely to be significantly different (often worse) in a full-scale system and that the studies may only partially simulate the soluble effluent quality of the system.

With the 92% maximum month and 98% maximum week per year compliance standards previously described, the investigator should recognize that performance of any system must meet these reliability standards. If wastewater characteristics, loading stress, or both cannot correctly simulate these conditions, then average (or more correctly, median) results must be adjusted to account for the necessary reliability of the design. For soluble effluent qualities, maximum month and week per year values can be, respectively, about 1.5 to 2.5 and 2.5 to 3.5 times greater than median observations.

Table 3.7   Guidance for interpretation of investigative results.

Elements	Guidelines
Investigator	To achieve success, an investigator may underrate problems, overrate the universal applicability (significance) of the observations, and operate the pilot-plant unit with more diligence and tighter controls than likely to be realized with a real system.
Scale	Pilot-plant investigations are often conducted at a scale that fails to reveal practical design and operating problems with equipment and hydraulics. The mixing conditions found in pilot work are typically far greater than those found in real systems. This is especially important in the examination of results from pilot anaerobic digesters where performance may be enhanced by a brief period of vigorous, complete mixing.
Duration	Pilot-plant investigations are often conducted under limited time frames, which may fail to reflect adequate acclimation and equilibrium to one or more variables, such as season stresses, material considerations, yearly variations, or the effects of upsystem and auxiliary system, upsets. Equipment reliability and maintenance needs are rarely, if ever, adequately evaluated in any pilot system.
Auxiliaries	If success of the pilot-plant system under evaluation depends on the selection and performance of the auxiliaries that support it, use caution if the auxiliaries evaluations are not included in the investigation.
Methods	Pilot-plant performance should be evaluated under conditions that fairly represent the norm (batch, constant flow, and constant quality investigations are only of value for mechanism insight with municipal wastewaters). If possible abnormal conditions are not evaluated on a sustained basis, use caution.
Complexities	Solids separation, concentration, processing, or handling techniques are especially difficult to investigate and apply the results, especially where physical and processing considerations are time dependent on multiple-media regenerations (for example, filtration and activated carbon regeneration).
Constants	Most constants are not really constant or independent of the conditions from which they were derived. Some are linked to another, so that changing one may call for changing the other to ensure that the overall performance of the system is correctly described (for example, the cell yield and endogenous decay constants used to describe net oxygen requirements and solids production from an aerobic biological stabilization process). Biokinetic constants can be expected to vary as a function of soluble and particulate loads, recycle, and the reactor's overall age and environmental conditions.
Percent removals	Percent removals depend on the applied pollutant concentration. Influent and effluent pollutant levels are influenced by their soluble and suspended phases. Most if not all of the processes, for most if not all of the pollutants, perform to some limiting residual concentration rather than at a controlling percentage. Effluent pollutants in the soluble phase reflect influent levels, the nature of the upsystem reactor(s), and the performance of the upsystem separator(s). If any of the three depart from the conditions that resulted in the original percent removal characterization, the percent removal may change.

Finally, the investigator should remember that liquid-processing systems must not only successfully treat raw wastewater but also process recycles. Recycle high in suspended solids can cause a decline in the desired CRT. Solids-processing systems that solubilize liquid-processing pollutants of concern may impose transient or continuous liquid-processing stresses, which contribute to instability and noncompliance. Ideally, waste treatability studies should include consideration of these impacts and, if not, be appropriately qualified to avoid overly optimistic sizing and performance assumptions.

SOLIDS PROCESSING. The above considerations apply to solids-processing studies also. The achievable degree of solids stabilization depends on the feed, including the nature and mix of raw and secondary solids and the degree of stabilization of secondary solids. The sludge volume index (SVI) or densification characteristics of secondary solids often influence the performance of solids stabilization, thickening, and dewatering systems.

Investigations of solids thickening and dewatering require determining performance for a range of applied solids concentrations to assess the significance of this variable. The interpretation of an investigative run of limited duration should be tempered by the realization that feed sludge characteristics are influenced by changes in the liquid-processing train, changes caused by additives such as metal salts for phosphorus removal, and recycle impacts imposed by the downstream solids-processing train on the unit processes ahead of those being investigated. An often unevaluated but nonetheless frequently occurring impact results from excessive solids recycling associated with full and overflowing storage tanks, gravity thickeners, and poor secondary digester supernatant quality.

RANGE OF CONDITIONS. Sound design practice must also anticipate the range of conditions that the facility or process can reasonably be expected to encounter during the design period. The range of conditions for a plant typically varies from a reasonably certain minimum in its first year of operation to the maximum anticipated in the last year of the design period in a service area with growth of customers. The reverse applies for a service area with an anticipated decline of customers. Often the minimum is overlooked and the maximum is overstated, the result being that the plant or process lacks the capability for less costly operation if actual future conditions depart from those expected.

Table 3.8 provides guidance for the range of conditions that a process design should accommodate. As its main thrust, this guidance presents a range of operating options commensurate with the range of variability and reliability inherent in wastewater treatment. As an example, the guidance encourages a biological reactor design that would accommodate CRTs or F:M ratios varying 50% above or below the nominal design targets. Guidance also

---

suggests that the liquid-processing train be designed for higher CRT (lower F:M) and the solids-processing train can be designed for the waste solids generated from a lower CRT (higher F:M) condition.

Table 3.8   Design guidelines for the range of conditions.

Concerns	Guidelines[a]
Minimum values Hydraulics	Pipelines and channels must be carefully sized to avoid low velocity conditions (solids deposition). Typically encountered problem areas are conduits common to both existing facility and planned expansion configurations, headworks, and primaries. Design for solids suspension in the minimum month of the project life if possible. As troublesome deposits may be encountered even with careful design, means should be provided to facilitate cleaning. Access with a Sewer-Vac is an alternative to manual cleaning.
Reactor oxygen (air) supply and mixing.	Design for the minimum month in the project life. Special sensitivity to mixing needs is particularly appropriate and often limiting with fine bubble, high-transfer efficiency oxygen dissolution and mechanical aeration systems.
Integrated biological nutrient control systems	Check system performance for the minimum month (weakest) wastewater quality characteristics with the maximum and minimum recycle impacts and provide appropriate safeguards (backup addition of supplemental carbon source and/or metal salt) to assure compliance with effluent standards. The designer should remember that the success of enhanced biological phosphorus removal is strongly influenced by the solids-processing train. Resolubilization of a fraction of the removed phosphorus is likely in solids storage and under aerobic or anaerobic solids stabilization. Return of resolubilized phosphorus in sludge-processing recycles is likely until application of lime precipitation.
Alkalinity	Design for minimum monthly or weekly value (typically, high rainfall month) with maximum recycle.
Seasonal nitrification temperature	Design for average temperature of the month that precedes the most restrictive condition (effluent standard and/or CRT). Remember, single-sludge nitrification systems are CRT limited not nitrogen load limited in most municipal wastewater treatment applications.

Table 3.8   Design guidelines for the range of conditions (continued).

Concerns	Guidelines[a]
Unit equipment sizing (particularly pumps)	Design equipment to meet minimum processing needs where appropriate; for example, design pump station wet wells and pumps for reasonable operation during the minimum month to week condition and design minimum return sludge capacity (and if appropriate, waste sludge capacity) from minimum monthly flows (and loads) in design life with reasonably anticipated minimum MLSS concentrations and settled solids concentration of at least 10 000 mg/L or higher.
Maximum values Waste solids	Design for minimum CRT conditions under at least the maximum month loadings (may change with higher or lower storage) of the design life with anticipated average or lower solids concentration under anticipated operating schedule. Provide appropriate storage with minimum CRT and anticipated average or lower solids concentration.
Reactor sizing	Design for the controlling maximum month or maximum week condition of the design life with the maximum CRT (If the ratio of the maximum week to the maximum month condition is not greater than 1.5, the maximum month will typically control for process design).
Oxygen demand	Design for transient load of concern in above maximum period [applied load of concern will vary as a function of the actual reactor configuration (for example, plug flow versus complete mix) and other variables described in Table 3.5]. Re-member, oxygen requirements are less than proportional to the applied load under transient conditions with the degree of dampening (buffer) influenced by the period of the transient load of concern to the CRT and the form of the load (carbonaceous and nitrogenous) after transient synthesis removals.
Return sludge	Design for above maximum period so as to achieve balance with the applied MLSS with a return solids concentration of 7500 mg/L or less under the maximum day flow regime. (High-flow conditions may not coincide with high organic load conditions and the controlling condition should be checked and established by the designer.)

[a] Guidelines apply to a single-stage biological reactor and separator; when multiple stages of biological systems are used, guidelines would apply to the last stage in a treatment train. Less stringent guidelines would apply to the biological systems before the final stage because of equilibrium, buffering conditions.

PROCESS OPTIONS AND SELECTION CONSIDERATIONS

The plant designer should lean toward facilities with low maintenance, tolerance, and ample capacity to reflect the uncertainty of staffing, maintenance, and remedial action in a public marketplace and with political and public oversight where funding of major capital improvements depends on public indebtedness. The designer must carefully follow a narrow line between providing tolerant facilities that can respond to a multitude of future uncertainties and an overdesign, which results in the misuse of public monies for superfluous facilities and capacity.

This section discusses a designer's decisions at the onset of design. All can influence the reliability and cost effectiveness of new or improved treatment works. The designer may choose among a wide variety of unit processes and processing options. The following categorical discussion of these options offers information intended to result in a better understanding of the design technique applicable to each.

Procedurally, wastewater treatment consists of the serial application of reactors and separators. Reactors oxidize, reduce, solubilize, immobilize, or physically condition their contents and create gaseous products. Separators create two products depending on their separation objectives; for example, low and high suspended solids concentration product streams are derived from a solids separator. Reactor and separator processes can be passive or reactive; their performance (and design) is generally influenced by some key dependency on hydraulics, pollutant concentration, or pollutant mass.

When considering a passive unit process, the design engineer should recognize that its performance is not amenable to operational manipulation once the entire process becomes operational. For such a process, the designer should err on the conservative side. Well-designed, passive unit processes are the classical systems chosen for plants that will have little attention and a low personnel commitment. Generally, these processes require the highest capital investment and are the most unresponsive to new or unanticipated treatment requirements or inadequacies. Once upset, these processes may require the longest period to recover.

Active processes involve the opposite design considerations. The higher the activity of any process, the easier it is both to upset and to restore from an upset condition. The most active processes depend solely on adding some substance to a simple mix tank. Another way of making a process more active (or reactive) to the design or operating needs of the plant involves the use of recycle to manipulate the operating solids concentration and characteristics.

Active processes allow the plant and the processes to be optimized at the plant and usually offer opportunities to realize savings of both first costs and operating costs, especially early in the design period or when the facility must respond to seasonal treatment objectives. Although not the most demanding from either a design or operating standpoint, the most mechanistically complex active processes incorporate suspended growth biological treatment systems; their complexity makes them the most susceptible to either design or operating error.

A process's activity should not be confused with its reliability. The most demanding (and least reliable) unit processes for the designer and the operator are those that have a multitude of support systems and moving parts and operate under elevated temperatures, pressures, or both. Most of these processes exist in the solids-processing train of the plant, especially at large wastewater treatment plants.

When large volumes of sludge, in comparison to the average day sludge production volume (such as with a digestion process), are handled by the sludge-processing train, the designer should provide additional processing redundancy as an allowance for anticipated failures to realize "rated" capacity (with either new or old equipment), differing sludge characteristics, and maintenance or repairs of the principal unit process or one or more of the supporting unit processes or appurtenances. The success of the entire plant and all of its unit processes may ultimately depend on the ability to remove sludge from the plant. A single or undersized sludge-dewatering unit or residue management plan without a backup processing sequence or disposal outlet does not represent good practice.

Those processes that unavoidably solubilize pollutants create special design and operating issues that vary with technologies and operating strategies employed at the plant. Thermal processes, except for the thermal-conditioning system, achieve some beneficial destruction of solids. The engineer should use special care in evaluating and selecting processes that will reintroduce pollutants to the liquid-processing train, especially when the plant's design objective includes control of reintroduced pollutants.

MASS BALANCES

There is no better tool for understanding a plant and, in turn, designing it than the mass balance. In conceptual design, a mass balance should be prepared to yield preliminary guidance concerning design quantities and major differences between processing alternatives. In detailed design, the mass balance should be the first of the reference project documents created to ensure a commonality of understanding, consistent use of major design criteria and quantities, and a standard frame of reference and logic for the project team. The mass balance also provides the basis for the control logic of the subsequently prepared process and instrumentation diagram.

Mass flow diagrams can be created for any condition of interest by appropriate use of the factors that describe that condition. A mass balance is prepared through iterative calculations (readily amenable to computer solutions) that end when all recycle and conversion conditions balance (achieve equilibrium) within reasonable limits. The subsequent discussion explains some of the uses of mass balance information for process design.

SIMPLE SMALL PLANT. Figure 3.6 presents a processing schematic for a relatively simple and frequently encountered sequence of unit operations at small wastewater treatment plants. It notes the principal processes, processing streams, and points of air and chemical addition. Input systems noted as points of addition, reflecting the end result of a supply function often involving storage and mixing, represent smaller support unit processes.

In this example, the plant's treated effluent discharges continuously to a receiving stream with a 7-day (consecutive), 10-year low flow sufficient to allow a secondary treatment objective for the plants. The plant's sludges are used beneficially as a soil builder for seasonal farmland applications.

Table 3.9 presents the mass flow balance for the plant. The mass flow diagram excludes minor residue (screenings, grit, and scum) because of their relatively small, insignificant dry weight mass even though they are identified in the processing schematic. This exclusion may be inappropriate for some wastewaters. Principal assumptions associated with the performance of each unit process are summarized in Table 3.10.

The pollutant mass values are expressed in Table 3.9 as mg/L equivalents. This unifies the mass balance diagram in terms of any flow and allows a quick overview to check for reasonableness of the calculated or predicted pollutant values. Multiplication of the mg/L equivalents by the plant's average day design flow in mgd and the conversion factor of 8.34 lbs/mil. gal per mg/L yields the dry weight pollutant mass for the design flow.

Typically, a mass balance is created for each of the major pollutants with an NPDES limit or having significance for process control. Table 3.9 includes phosphorus (P) and nitrogen (N) because of their general interest; chemical oxygen demand (COD) is included because of its greater analytical reproducibility and its site-specific, correlative use to predict carbonaceous oxygen demand.

Inert suspended solids (ISS) are analyzed separately because this conservative material is neither created (except to a minor extent) nor destroyed during the sequence of unit operations. The conservative elements (ISS, P, and N in this balance) may be used to check the validity of calculations.

Metals and some low-volatile, biologically resistant organics are also conservative compounds. If these materials are fully insoluble, they will be concentrated in the plant's sludges (final effluent suspended solids and digested sludge cake), which total the 135 mg/L (20 + 115) equivalent in this example. This net sludge production provides a predictable concentration

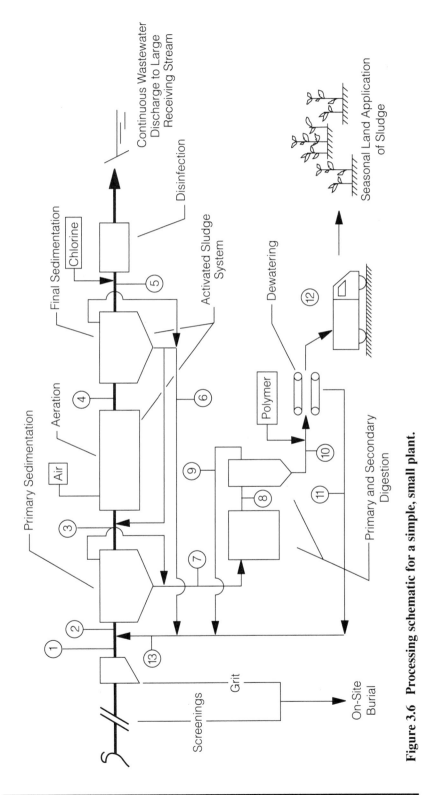

Figure 3.6 Processing schematic for a simple, small plant.

Table 3.9 Simple plant mass balance corresponding to schematic of Figure 3.6.

Processing point	SS			BOD			COD			Phosphorus as P			Nitrogen as N		
	ISS	VSS	TSS	SOL	PART	TOT	SOL	PART	TOT	SOL	PART	TOT	SOL	PART	TOT
1. Raw influent	50	150	200	65	135	200	130	270	400	5.5	1.5	7.0	22.5	1.5	30.0
13. Recycles[b]	26	67 + 7	93 + 7	5	34	39	12	96	108	0.7	1.3	2.0	3.9	6.4	10.3
2. Total influent	76	217 + 7	293 + 7	70	169	239	142	366	508	6.2	2.8	9.0	26.4	13.9	40.3
3. Primary effluent[c]	25	71 + 7	96 + 7	70	60	130	142	123	265	6.2	0.8	7.0	26.4	4.2	30.6
4. MLSS[d,e]	25	73	98	6	36	42	40	102	142	5.5	1.5	7.0	23.3	7.3	30.6
5. Final effluent	5	15	20	6	7	13	40	21	61	5.5	0.3	5.8	23.3	1.5	24.8
6. WAS	20	58	78	—	29	29	—	81	81	—	1.2	1.2	—	5.8	5.8
7. Primary sludge[f]	51	46	197	—	109	109	—	243	243	—	2.0	2.0	—	9.7	9.7
8. Digested sludge[f]	51	80 + 8	131 + 8	6	43	49	13	130	143	0.8	1.2	2.0	4.3	5.4	9.7
9. Supernatent	5	8 + 4	13 + 4	3	4	7	6	13	19	0.4	0.1	0.5	2.2	0.5	2.7
10. Sludge to dewatering	46	72 + 4	118 + 4	3	39	42	7	117	124	0.4	1.1	1.5	2.1	4.9	7.0
11. Filtrate	1	1 + 3	2 + 3	2	1	3	6	2	8	0.3	0	0.3	1.7	0.1	1.8
12. Sludge cake	45	71 + 1	115 + 1	1	38	39	1	115	116	0.1	1.1	1.2	0.4	4.8	5.2

[a] To find the actual mass, multiply the mg/L equivalent by 8.34 and the plant flow.

[b] To find the actual flow rates, multiply the sludge-processing point mg/L equivalents by plant flow and divide by expected actual sludge concentration and determine the liquid stream flows by difference.

[c] Recycles = sum of processing streams 6, 9, and 11.

[d] Average day oxygen demand can be calculated from COD mass balance results; for example, processing point 3 − 4 = 265 − 142 = 123 mg/L equivalents, which, with an

assumed nominal oxygen transfer efficiency of 5% and 0.017 5 lb of oxygen per cf of air, yields an average day air supply of $\dfrac{(123)(8.34)}{(0.05)(0.017\,5)(1440)}$ = 810 cfm/mgd.

[e] Actual MLSS concentration can be determined from the following relationship: (cited MLSS value from balance) (cell resident time, days, aerator only basis) ÷ (average day aerator detention time, days, influent flow basis). For example, with a CRT of 5 days and an aerator detention time of 0.25 days (6 hours), the actual MLSS would be [(98)(5)] ÷ (0.25) = 1960 mg/L.

[f] The (+) term introduced and carried through at this point, reflects the VSS solubilized but not destroyed in the digestion process.

Table 3.10   Assumptions for simple plant mass balance.

Plant components	Assumptions		
Primary treatment	Applied SS capture, %		
	Raw influent	60	
	WAS	90	
	Supernatant	50	
Mainstream biological treatment	No further destruction of recycled VSS associated with waste activated sludge, supernatant, and filtrate uncaptured by primary sedimentation tank.		
	Secondary treatment equivalency: soluble COD reduced to 40 mg/L and cell yield coefficient (Y, COD basis) and endogenous decay coefficient (k) of 0.5 and 0.1, respectively, with a cell residence time (CRT) of 5 days on the raw wastewater and soluble recycles. VSS produced characteristics: P:VSS = 0.02; N:VSS = 0.10; COD:VSS = 1.4; and BOD:VSS = 0.4.		
Anaerobic digestion	Volatile solids and COD, percent of applied		
		Destroyed	**Soluble residual**
	Raw wastewater solids	50	5
	Fresh biological solids	25	10
	Recycled solids	0	0
	Nitrogen and phosphorus solubilization equals the sum of the destroyed and soluble residual fractions for each respective solids stream.		
	The BOD entering and leaving the digestion process (or any reactor) is of different form and cannot be precisely balanced. BOD characterizations are most easily derived by BOD:COD or BOD:VS ratios. Supernating: for simplicity, assume recycles at least 50% of soluble components and 109% of applied particulate components.		
Dewatering	Fivefold reduction of volume and 98% capture of suspended solids.		

factor for the fully insoluble conservative compound (for example, metal) of 200/135 or 1.5 times or more in the net waste digested sludges and effluent suspended solids. For example, if 0.2 mg/L of an insoluble metal were measured in the influent SS of 200 mg/L, the dry weight metal concentration would increase from 1000 mg/kg in the raw sludge to 1500 mg/kg after treatment, resulting in a likely effluent metal concentration of 0.045 mg/L (0.001 5 × 30). Use of a balance in this manner allows rapid understanding and prediction of the magnitude of conservative compounds in product sludges from the plant and the significance of expected effluent suspended solids regarding compliance with effluent standards for pollutants that are present in low concentrations.

The assumptions of Table 3.10 reflect the important principle that unit processes behave differently for different materials. Note different digestibility assumptions for the recycled, raw influent, and biological or secondary suspended solids. The digestibility of secondary solids will vary, depending on the type of biological system that generated them (for example, suspended growth versus attached growth) and its CRT. Older CRT system solids will likely have lower relative BOD:VSS ratios and digestibility. The need for such a provision depends on the intended use of the information.

The following example demonstrates the strength of the mass balance in understanding processes and solving design problems. Example 3.1 evaluates the impacts throughout the plant of sludge wasting to the anaerobic digester. Discontinuous wastage is not only operationally convenient for a small plant but a practical necessity to maintain adequate velocities in pipelines.

Example 3.1. Calculate the impact on the simple plant (Table 3.9) caused by wasting primary sludge to a digester twice a day (1 hour in the morning and 1 hour at night).

Supernatant normally returns to the plant through simple hydraulic displacement of the secondary digester volume. As a worst case, this example assumes no concurrent sludge dewatering, typical of a small plant. Supernatant recycle is not assumed coincident with the plant's peak influent loading despite the likelihood that this would occur during the morning.

Table 3.11 presents a comparison of mainstream quality impacts of 24-hour recycle with those of wasting twice a day for a total of 2 hours based on the values of Table 3.9. The table clearly shows that the proposed wasting program would have significant impacts on the plant. A complete mix activated sludge system is a preferred means of attenuating the otherwise significant supernatant loading. If the process design lacks provisions for such loading, effluent deterioration would be expected. A small plant's sampling program (typically once or twice a week) may miss the effluent effects.

Inspection of the information presented in Tables 3.9 and 3.11 shows that the supernatant would be relatively benign if it were recycled continuously to the mainstream of the plant. However, returning the supernatant to the plant under the proposed sludge wastage program would cause an instantaneous near doubling of the plant's average daily concentration of carbonaceous oxygen demanding materials, a 150% concentration increase of suspended solids and phosphorus, and a concentration increase of 200% in the background influent level of nitrogen. Because of these impacts, many small plants avoid primary sedimentation and anaerobic sludge digestion. Example 3.1 demonstrates the need to use care with any discontinuous recycle and the value of a mass balance for characterizing recycle impacts.

Table 3.11 Comparison of effects on plant influent of two primary sludge wasting programs.

| Pollutant | Actual concentration, mg/L[a] | Main stream concentration and percent of average day raw influent characteristics | | | |
| | | With continuous 24-hour recycle | | With sludge wasting 2 hours/day | |
		mg/L[b]	%	mg/L[c]	%
VSS	3100	8	5	180	120
TSS	5000	13	6	290	145
SBOD	1200	3	5	70	110
TBOD	2700	7	4	160	80
SCOD	2300	6	5	140	110
TCOD	7300	19	5	430	110
SP	150	0.4	13	9	165
TP	190	0.5	7	11	155
SN	850	2.2	10	50	220
TN	1000	2.7	9	61	205

[a] Determine by the ratio of the total plant flow to the supernatant flow times the supernatant mg/L equivalents; factor used = (1/0.002 6) = 385.
[b] Values obtained directly from Table 3.9.
[c] Determined by multiplying flow ratio of primary sludge to supernatant by ratio of 24 to 2 hours by values obtained directly from Table 3.9; factor used = (0.004 9/0.002 6) (24/2) = 22.6.

High supernatant solids can be avoided by an operating strategy that dedicates the secondary digester to solids storage service instead of solids-concentrating functions. Such an operation, however, may reduce feed solids, thereby diminishing the yield and product solids concentration of the dewatering operation. In this mode of operation, the designer should consider thickening before digestion. These effects show the way in which a decision to avoid one problem (high supernatant solids) may create another (poor dewatering characteristics). The compensating decision (prior thickening) may result in a design and operation of increasing complexity and probable higher costs. Other options include nongravity thickening of the digested sludge or an expanded dewatering complex.

The second example examines the impacts of a discontinuous sludge management program on the required digested sludge storage volumes and dewatering facilities. The solids management recycles under this operating condition will impose a substantial load on the main process stream and the plant's design must allow for this load to avoid reducing the plant's performance. Without a compensating increase in the reactor size, such performance deterioration, resulting from a declining CRT and nutrient spikes, is especially likely in a plant designed for biological nitrification, nutrient control, or both.

Example 3.2. For the simple plant (Table 3.9), calculate the secondary sludge storage volumes and necessary sludge-dewatering capacity for 30-h/wk seasonal sludge-processing strategies.

Three optional seasonal strategies are 10, 8, or 6 access months per year. Table 3.12 compares sludge storage volumes and dewatering capacities for three seasonal options based on the values of Table 3.9. Table 3.12 shows expected increases of storage volume and dewatering needs as the number of access months per year is reduced. The impact of the additional dewatering needs depends partly on the plant size. For example, a 1-m belt press with two dewatering belts (one operating and another for standby) will suffice. If the plant size were 19 000 m^3/d (5 mgd), the belt selection choice would not be as simple because the yield of the belt and the actual time of operation would be significant. One choice is three 2-m belts (two operating and one standby).

Inspection of example 3.2 demonstrates how plant size influences the designer's decision making and that equipment selection is not a precise process. In dewatering and thickening applications, and throughout the plant, the engineer continually balances a reasonable range of expected results, anticipated quantities, and available equipment choices. With small plants, some equipment may be severely and unavoidably underused. Note that the backup or standby dewatering unit of example 3.2 is considered mandatory, regardless of equipment use, because this unit operation is critical to the performance of the plant.

Final points derived from the simple small plant examples are that costs per m^3/d (mgd) of these facilities are high because equipment needs cannot be closely matched with equipment availability. Also, discontinuous operations can disproportionately increase costs as plants become smaller. However, the operating convenience may outweigh the apparent savings of first

Table 3.12   Comparison of three seasonal sludge-processing options.

Access months per year	Storage volume needed[a]	lb sludge processed/mgd·hr at 30 h/wk[b]	No. operating 1-m belts at yield of		
			600 lb/hr[c]	1000 lb/hr[c]	1400 lb/hr[c]
10	0.14Q	270	0.45	0.27	0.19
8	0.28Q	340	0.57	0.34	0.24
6	0.42Q	490	0.82	0.49	0.35

[a] Determined by multiplying the sludge to dewater volume from Table 3.9 (0.002 3 Q) by the number of days that storage is required. (A standard, second-stage digester has a volume of 0.049 Q with a 10-day detention time.)

[b] Determined by multiplying sludge cake mg/L equivalent TSS from Table 3.9 by factor of (12/10) (7) (1/30) (8.34) = 2.34 for 10 access months per year; others calculated similarly.

[c] lb/hr × 0.453 6 = kg/h.

cost and operating cost, which might otherwise be attainable. As a result, many small wastewater treatment plants have excess processing reserves, with tolerance for a wide variety of conditions except precipitation-induced flow peaks.

COMPLEX LARGE PLANT. In contrast to a small plant, the large plant usually has continuous operations throughout the entire sludge management train. Depending on sludge-processing technologies, a continuous operational mode is typical for plant sizes of approximately 75 700 m^3/d (20 mgd) or larger. For these large plants, operation and maintenance responsibilities are usually not shared by the same group of plant employees, but rather assigned to specialized centers and groups. Also, larger plants have multishift staffing. These groups' loss of perspective of overall process objectives is another of the differences between large and smaller treatment facilities.

Based on the theoretical demands placed on the process engineer, the design of the large plant is actually simpler because peaking factors and recycles are less extreme. Practical demands, of course, impose more difficulties in the design of larger plants because they have more equipment, more unit operations, equipment operated closer to its rated capacities, more people with less tolerance for operational and maintenance procedures, and greater internal and external scrutiny.

Reflecting the foregoing perspective, Figure 3.7 shows processing concepts that may exist at a large, complex municipal wastewater treatment plant. For comparison with the small secondary treatment plant example, the assumed performance requirements for this large plant include nitrification and reductions of BOD, SS, and phosphorus to low levels. A need for post-aeration and dechlorination of chlorinated effluent was also assumed to be required by the plant's NPDES permit. For illustrative purposes only, this example plant comprises single-stage activated sludge, granular-media effluent filtration, metal salt addition to the activated sludge system for phosphorus removal, chlorination, and sulfur dioxide dechlorination. These processes are schematically illustrated in Figure 3.7. Figure 3.7 also shows unit process optimization strategies that may be employed at large plants. In this example, one of the principal strategies calls for separate processing of settled raw primary solids and secondary biological solids to the point of final preparation for ultimate disposal.

The ultimate disposal strategy for this hypothetical complex plant includes incineration of all waste residue, except screenings, and contractual arrangements for the subsequent transfer of the dry ash and compacted screenings to a dedicated landfill. Nearly all of the heavy metals in the plant's influent will appear in the plant's feed sludge to the incinerator and effluent suspended solids. Essentially all of the mercury and some lead, cadmium, and chromium will be contained in the incinerator's atmospheric emissions. The other metals will be largely retained in the ash. Metal salt addition strategies can

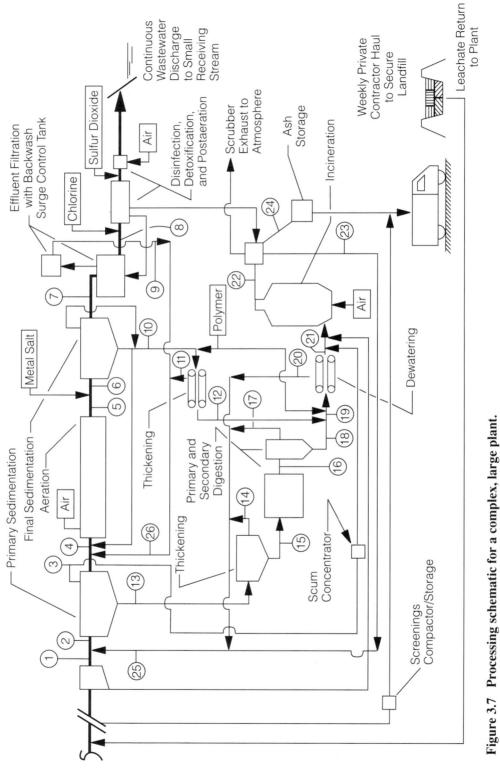

Figure 3.7 Processing schematic for a complex, large plant.

significantly increase the ash for disposal but will beneficially reduce the heavy metal concentration of the ash. (In this example, the insoluble, non-volatile metal content of 0.2 mg/L in the influent will concentrate from its dry weight concentration of 1000 mg/kg in the raw sludge to 2400 mg/kg in the incinerator ash; without the metal salt addition strategy, the metal content in the ash would have been approximately 4300 mg/kg.)

This example retains anaerobic digestion of the primary solids in the processing train only for comparative purposes. Although seldom used with an incineration operation, digestion may have limited merit derived from its load smoothing and generated methane gas that may be used as fuel for engines and as auxiliary fuel for the incinerator. Anaerobic digestion would also reduce the mass of emissions from the incineration, which may become more important in the future.

The second-stage digester in this example could also be used to store excess quantities of primary and secondary sludge during the maximum week condition. Typically, incineration systems operate with an applied peaking factor not more than 1.3, and perform efficiently and reliably with constant, sustained loading conditions.

In selecting the sizes of dewatering equipment, the designer should ensure that furnished dewatering equipment easily matches the rated capacity of the incinerator. Routinely, a given incinerator will have specifically dedicated dewatering equipment to facilitate incinerator feeding operations.

Results of the complex plant mass balance calculations are presented in Table 3.13 and attendant assumptions are presented in Table 3.14. Notes for Table 3.13 describe the way in which the mass balance may be used to determine requirements for oxygen transfer and metal salt addition and whether the wastewater has an adequate background alkalinity.

A comparison of simple and complex plant mass balances, Tables 3.9 and 3.13, discloses several other worthwhile points for evaluating and selecting processes for an integrated sequence of unit operations. Regarding recycle such as anaerobic digester supernatant, note the substantial improvement in quality when the waste biological sludge is not introduced to the digester. This is due to the lower net solubilization and nutrient content of the raw wastewater solids. Isolated processing and disposal of secondary solids may be preferred in some processing sequences, especially with old cultures, to mitigate impacts of digester supernatant. Anaerobic digestion of only secondary solids results in a product solids emulsion that is not amenable to gravity concentration.

As to the impact of nutrient control strategies, inspection of Tables 3.9 and 3.13 shows that the older nitrifying system of the complex plant has less nitrogenous oxygen demand and a similar concentration of phosphorus for immobilization. The higher nutrient removal potential of the younger secondary biological treatment system of the simple plant (Figure 3.6) is

Table 3.13   Complex plant mass balance corresponding to schematic of Figure 3.7.

Processing point	SS ISS	SS VSS	SS TSS	BOD SOL	BOD PART	BOD TOT	COD SOL	COD PART	COD TOT	P SOL	P PART	P TOT	N SOL	N PART	N TOT
1. Raw influent	50	150	200	65	135	200	130	270	400	5.5	1.5	7.0	22.5	1.5	30.0
25. Recycle 1[a]	12	16+4	28+4	3	13	16	6	29	35	0.5	0.4	0.9	2.3	0.8	3.1
2. Total influent	62	166+4	228+4	68	148	216	136	299	435	6.0	1.9	7.9	24.8	8.3	33.1
3. Primary effluent	22	62+4	84+4	68	56	124	136	113	249	6.0	0.7	6.7	24.8	3.1	27.9
26. Recycle 2[a]	10	10	20	—	2	2	—	14	14	—	1.0	—	1.0	1.0	—
4. Total primary effluent[b]	32	72+4	104+4	68	60	128	136	127	263	6.0	1.7	7.7	24.8	4.1	28.9
5. MLSS initial[b,c]	32	64	96	3	16	19	30	90	120	5.6	2.1	7.7	22.5	6.4	28.0
precipitate[c,d]	35	—	35	—	—	—	—	—	—	-5.0	+5.0	—	—	—	—
6. MLSS final[c]	67	64	131	3	16	19	30	90	120	0.6	7.1	7.7	22.5	6.4	28.9
7. Secondary effluent	10	10	20	3	2	5	30	14	44	0.6	1.0	1.6	22.5	1.0	23.5
8. Final effluent	3	3	6	3	1	4	30	4	34	0.6	0.3	0.9	22.5	0.3	22.8
9. Filtered backwash	7	7	14	—	1	1	—	10	10	—	0.7	0.7	—	0.7	0.7
10. WAS	57	54	111	—	14	14	—	76	76	—	6.1	6.1	—	5.4	5.4
11. WAS liquor	3	3	6	—	1	1	—	4	4	—	0.3	0.3	—	0.3	0.3
12. Thickened WAS	54	51	105	—	13	13	—	72	72	—	5.8	5.8	—	5.1	5.1
13. Primary sludge	40	104	144	—	92	92	—	186	186	—	1.2	1.2	—	5.2	5.2
14. Primary sludge overflow	4	10	14	—	9	9	—	18	18	—	0.1	0.1	—	0.5	0.5
15. Thickened primary sludge	36	94	130	—	83	83	—	68	168	—	1.1	1.1	—	4.7	4.7
16. Digested primary sludge	36	45+4	81+4	3	27	30	7	80	87	0.5	0.1	1.1	2.5	2.2	4.7
17. Supernatant	4	4+2	8+2	2	3	5	4	8	12	0.3	0.1	0.4	1.3	0.2	1.5
18. Thickened digested sludge	32	41+2	73+2	1	4	25	3	72	75	0.2	0.5	0.7	1.2	2.0	3.2
19. Sludge dewatering	86	92+2	178+2	1	37	38	3	144	147	0.2	6.3	6.5	1.2	7.1	8.3
20. Filtrate	2	2+2	4+2	1	1	2	2	3	5	0.2	0.1	0.3	1.0	0.1	1.1
21. Sludge cake	84	90	174	—	36	36	1	41	142	—	6.2	6.2	0.2	7.0	7.2

Equivalents, mg/L (column group heading over BOD, COD, Phosphorus as P, and Nitrogen as N).

Table 3.13 Complex plant mass balance corresponding to schematic of Figure 3.7 (continued).

Processing point	Equivalents, mg/L														
	SS			BOD			COD			Phosphorus as P			Nitrogen as N		
	ISS	VSS	TSS	SOL	PART	TOT	SOL	PART	TOT	SOL	PART	TOT	SOL	PART	TOT
22. Furnace ash	84	—	84	—	—	—	—	—	—	—	6.2	6.2	—	—	—
23. Scrubbing water	2	—	2	—	—	—	—	—	—	—	0.1	0.1	—	—	—
24. Net ash	82	—	82	—	—	—	—	—	—	—	6.1	6.1	—	—	—

a Recycle 1 = sum of processing streams 14, 17, 20, and 23; recycle 2 = sum of processing streams 9 and 11.

b Average day oxygen demand can be calculated from total COD and soluble nitrogen mass balance results from processing points 4 and 6, for example, carbonaceous oxygen demand = 263 − 120 = 143 mg/L equivalents; nitrogenous oxygen demand (assuming refractory N residual of 0.8 mg/L and 4.6 mg of oxidized) = (22.5 − 0.8) 4.6 = 100 mg/L equivalents, and total biological oxygen demand = 143 + 100 = 243 mg/L equivalents. This total oxygen demand with an oxygen transfer efficiency of 5% and 0.017 5 lb of oxygen per cf of air, yield an average day air supply of [(243) (8.34)] / [(0.05) (0.017 5) (1440)] = 1600 cfm/mgd.

c The nitrification reaction and the use of alum for phosphorus removal will cause a depletion of the wastewater alkalinity (7.1 mg $CaCO_3$/mg N oxidized and 5.6 mg $CaCO_3$/mg AL added, respectively). It is recommended that the alkalinity not drop below 50 mg/L $CaCO_3$ to protect the nitrification reaction and 30 mg/L $CaCO_3$ to protect the phosphorus precipitation reaction. In this example, the wastewater should have an alkalinity of 7.1 (21.7) + 5.6 (9.7) + 30 = 154 + 54 + 30 = 238 or 240 mg/L as $CaCO_3$ to avoid the possibility of pH depression problems on an average day basis.

d Selected dose of metal salts equals 2 M of metal ion per M of phosphorus remaining for precipitation. Phosphorus remaining for precipitation (that is, after natural background removals) taken to be 9.7 mg/L. Aluminum dose determined to be 5.6 mg/L, which yields a soluble phosphorus residual of 0.6 mg/L, 19 mg/L of aluminum phosphate, and 16 mg/L of aluminum hydroxide for a total precipitated solids mass of 35 mg/L.

Table 3.14   Assumptions for complex plant mass balance.

Plant components	Assumptions		
Primary treatment	Applied SS capture, %		
	Raw influent		60
	Thickener overflow, supernatant, and filtrate		50
	Scrubber water return		100
Mainstream biological treatment system performance	No further destruction of recycled VSS associated with VSS uncaptured by primary sedimentation tank and recycled with filter backwash and WAS-thickening liquors. Single-stage nitrification system: soluble COD reduced to 30 mg/L. Cell yield coefficient, Y, 0.5 (COD) and endogenous decay coefficient k, 0.1, with a cell resident time (CRT) of 10 days on the raw wastewater and soluble recycles. VSS produced characteristics: P:VSS = 0.02; N:VSS = 0.10; COD:VSS = 1.4; and BOD:VSS = 0.25.		
Phosphorus removal	Use single-stage metal salt (aluminum or iron) addition to end of activated sludge system and ignore hydroxide conversion to oxides during incineration in mass balance for convenience. (Multipoint addition, for example, to primary and secondary will yield lower overall chemical dose.)		
Anaerobic digestion	Volatile solids, percent of applied		
		Destroyed	**Soluble residual**
	Raw wastewater solids	50	5
	Recycled solids	0	0
	Nitrogen and phosphorus solubilization equal to the sum of the destroyed and soluble residual fractions for each respective solids stream. BOD entering and leaving the digestion process (or any reactor) is of different form and cannot be precisely balanced. BOD characterizations are most easily derived by BOD:COD or BOD:VS ratios. Supernating: for simplicity, assume recycles at least 50% of soluble components and 10% of applied particulate components.		
Thickening	Gravity thickening of primary solids = 90% solids capture.		
Dewatering	Digested primary sludge and thickened secondary sludge: fivefold reduction of volume and 98% solids capture.		

unrealized because its waste solids are anaerobically digested. This point highlights the importance of the solids-processing scheme when evaluating processing concepts to fully achieve the system potential for nutrient removal.

REFERENCES

1. "Analysis of Performance Limiting Factors (PLFs) at Small Wastewater Treatment Plants." EPA WH-546/OMPC 10-89, U.S. EPA, Office Munic. Pollut. Control, Office Water, Washington, D.C. (1989).

2. "Technology and Design Deficiencies at Publicly Owned Treatment Works." *Water Environ. Technol.,* 1, 4, 515 (1989).

3. D'Antoni, J.M., and Bahl, V., "Designs for the Future." Abstr. submitted to 63rd Annu. Conf. Water Pollut. Control Fed., Washington, D.C. (1990).

4. "Odor Control for Wastewater Facilities." Manual of Practice No. 22, Water Pollut. Control Fed., Alexandria, Va. (1979).

5. "Recommended Standards for Sewage Works." Great Lakes—Upper Miss. River Board State Sanit. Eng., Health Educ. Serv., Inc., Albany, N.Y. (1978).

6. "Gravity Sewer Design and Construction." Manual of Practice No. FD-5, Water Pollut. Control Fed., Alexandria, Va. (1982).

7. "Combined Sewer Overflow Pollution Abatement." Manual of Practice No. FD-17, Water Pollut. Control Fed., Alexandria, Va. (1989).

Chapter 4
Site Selection and Plant Layout

INTRODUCTION

Key elements to be considered early in the planning of a wastewater treatment project are the selection of an appropriate site, general plant facilities layout requirements, and the future needs regarding both site selection and plant layout. In this chapter, the section on site selection considerations identifies and, where appropriate, illustrates important considerations in the selection of a site for a new wastewater treatment facility. Because of increasing growth and development of areas surrounding major metropolitan centers, ideal sites may not be available. This chapter discusses innovation, such as multiple sites, that engineers are currently using to mitigate the shortcomings of less-than-ideal sites.

SITE SELECTION CONSIDERATIONS

The advantage of selecting a new site for a wastewater treatment plant is finding one that can be developed economically without unnecessarily stressing the environment. Although numerous considerations, such as social and environmental factors, are difficult to quantify, cost effectiveness and design and construction requirements can be quantified when choosing among potential sites.

It is important to involve the public early in the planning process. Public hearings held during the site selection process are recommended. Such considerations impact site and process selection and other design decisions. Both public participation and acceptance of the selected site are essential.

SURROUNDING LAND USE. Considering the effects of a wastewater treatment plant on the development of land in the area includes compliance with zoning regulations, impacts on adjacent property values, and compatibility with activities on neighboring properties. Generally, constructing a plant in an industrial neighborhood rather than in a residential area is more acceptable and less expensive. Wastewater treatment facilities located near airports may require Federal Aviation Administration (FAA) approval as airplane glide slopes may control the height of some structures. Lagoons and ponds attract birds, which also impact air traffic operations.

If the site selected is surrounded by residences, numerous measures must be taken to ensure that the plant does not irritate its neighbors. Such measures include minimizing noise; odors; aerosols, air particulates, chlorine, and other chemical hazards; insects; and intrusive lighting. Attractive architecture and landscaping and proper consideration of prevailing winds help make treatment plants more acceptable neighbors. Figure 4.1 illustrates a type of special architectural treatment required in some locations.

RECEIVING WATER CHARACTERISTICS. If treated effluent is to be discharged to a lake, river, or stream, receiving water characteristics and water quality standards impact the degree of treatment required. NPDES discharge permits and state and local health departments determine potential discharge requirements. Selecting a site that will require advanced treatment to comply with stringent water quality standards may not be the most cost-effective choice. The city of Sedona, Arizona, chose to pump raw wastewater 5 km (3 miles) beyond the city limits to a site that could support a secondary treatment plant, infiltration basins, and artificial wetlands rather than provide costly tertiary treatment processes required to discharge to a local waterway with stringent nutrient limitations.[1]

Downstream uses of receiving water may impact the cost of the treatment plant. Stream uses such as potable water supplies, shellfish waters, or water

---

Figure 4.1   Gainey Ranch plant in Scottsdale, Arizona, showing entrance area architectural design and landscaping that blends with the surrounding exclusive residential community.

contact sports require higher reliability standards than streams not used for such purposes.[2] Downstream uses also determine levels of redundancy in treatment units, process equipment, and power supply.

Plant personnel are better able to monitor effluent and its effects on water quality and aquatic life if a site is close to the receiving water; sampling upstream and downstream of the plant outfall is easier, and a convenient supply of upstream water is available for preparing various dilutions for bioassay testing.

SIZE REQUIREMENTS. The size of the site must accommodate present and anticipated future requirements. As the service area grows and treatment requirements increase, the plant will likely require additional space. Consider the potential for such demands when selecting a site.

The area required for a plant of a particular capacity depends on the following considerations:

- Degree of treatment required;
- Processes to be used;
- Degree of redundancy necessary;
- Space requirements of ancillary and support facilities; and
- Space required for access, circulation, and maintenance.

Layout and shape of process units can drastically impact the land area requirements. Stacking of process units is expensive and complicates operation and maintenance but has been done on occasion to accommodate restrictive sites. Likewise, square or rectangular tanks, when compared to circular tanks of like volume, save considerable space but may add to maintenance costs.

Space requirements of maintenance, administration, storage, laboratory, and staff services impact total space requirements. In some cases, related activities, such as maintenance crews and the equipment for the sewer collection system, may be housed at the treatment plant, and space requirements for these services must be met.

The amount of isolation and buffer area needed between plant processes and other property owners influences size requirements. In the absence of specified local regulations, the buffer in residential areas should be at least 46 to 76 m (150 to 250 ft). The distance required between residences and water supply wells is usually restrictive. If the land is available, consider purchasing as much land as is affordable to ensure that an adequate buffer can be maintained in the event of future plant or residential growth. Figure 4.2 presents the approximate amount of area required by activated sludge plants of various sizes. Area data, based on 53 recently designed treatment plants, include areas used for both process and administrative purposes. Dedicated sludge disposal sites and buffer areas are not included. As would be ex-

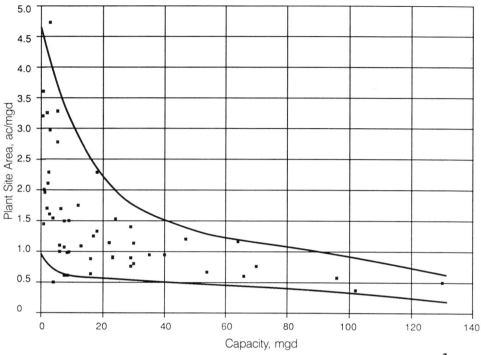

Figure 4.2 **Plant site area versus plant capacity (mgd × 3785 = m^3/d; ac × 0.404 7 = ha).**

pected, data vary considerably. Therefore, use the figure only as a preliminary guide to plant area requirements.

Because decisions regarding the land needed for treatment facilities will have long-term impacts, it is important to determine the requirements as accurately as possible. One of the best methods is to determine the "footprints" of various processes and appurtenant facilities using typical "textbook" values for detention time, overflow rate, and so on. Tentative layouts of process tankage, structures, and buildings can easily be made; square footage assigned to each use; and the total land area approximated, considering access, traffic, parking, landscaping, and buffer zones. To allow for unknowns in this analysis, the area required should be overestimated rather than underestimated.

MULTIPLE SITES. If one site cannot be found to satisfy the space requirements of the entire treatment facility, consider multiple sites. The liquid stream could be treated at one location while sludges are treated or disposed of at another.[3] Sludges from a number of plants could be trucked or pumped to one plant for treatment. This may be especially economical in the case of several small plants serving a community. Due to the need to handle recycle streams, it is best if the centralized sludge treatment facility also has liquid stream treatment processes.

Site Selection and Plant Layout 117

County Sanitation Districts of Los Angeles County and the City of Los Angeles have constructed a series of wastewater reclamation plants ranging in size from 57 000 to 378 000 m^3/d (15 to 100 mgd), virtually all of which are in populous urban residential or industrial areas. These projects provide hydraulic relief to overloaded trunk sewers while, at the same time, providing a source of tertiary treated reclaimed water for a wide variety of uses. Each of these facilities is located adjacent to a major trunk sewer. Wastewater is pumped from the trunk sewer, treated, and discharged to local watercourses or put to direct beneficial reuse. Solids are returned to the trunk sewer for processing at a central treatment facility located downstream. Because solids are processed elsewhere, upstream facilities occupy less land area, have less odor potential, and are more easily accommodated in the urban setting.

Different stages of liquid stream treatment could be accomplished at multiple sites. However, this arrangement usually increases total staffing, complicates handling of recycle streams, and increases overall treatment cost.

Alternatively, the intended service area could be separated into two or more plants, each containing full treatment capabilities. If each plant serves a distinct drainage area, this arrangement may reduce costs of construction and operation of the collection system. Due to economies of scale, however, total costs to construct and operate multiple treatment plants may be higher than if all treatment were accomplished at one larger facility. This may be a practical alternative where an existing treatment facility cannot be expanded farther due to site constraints and a second treatment plant could be constructed elsewhere to provide the needed additional capacity.

ELEVATION AND TOPOGRAPHY. A low-lying site facilitates the flow of wastewater from the service area by gravity and minimizes the number of pumping stations in the collection system, but such a site may also require flood protection. Adequate protection may be provided by building earthen dikes around the perimeter of the site. Constructing the tops of process tanks, buildings, and pipe gallery entrances and building finished floor elevations above the expected high water levels also provides flood protection but can be costly and may actually negate the advantages of selecting a low-lying site. Figures 4.3 and 4.4 illustrate two different methods of mitigating site flooding.

Protection against damage to plant structures and equipment from water elevation created by a flood or tidal surge with a 100-year recurrence interval (RI) is customarily provided. Protection against even greater floods may be warranted in some cases. As a minimum, the plant should remain operational during a 25-year RI flood; plants are usually designed to remain operational during larger RI floods. Effluent pumping or storage during episodes of high tide or high water may be required. When contemplating use of a site in a flood plain, contact the local flood plain management authority to learn of any restrictions on development of such property due to possible impacts on

(a)

(b)

Figure 4.3   **At the plant of the Carmel, California, Wastewater Agency, the buildings and structures are constructed with finished floors above the design flood level, an approach deemed more cost effective than filling the entire site or constructing a levee: (a) the chlorination building resting on concrete columns and (b) an elevated building entrance.**

(a)

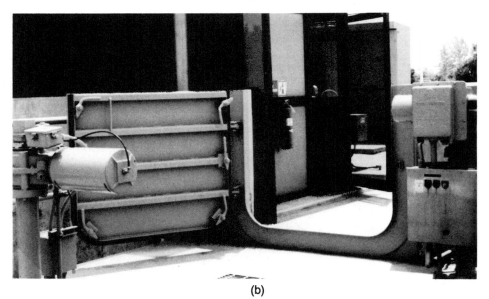

(b)

Figure 4.4 **(a) Gasketted waterproof door secures the tertiary filter pipe gallery and (b) gasketted half-height door protects another area, both at the Whittier Narrows, California Water Reclamation Facility, located upstream from a flood control dam and subject to flooding.**

upstream landowners. Further encroachment on the flood plain may be banned in some areas. Also, the location of the site within the flood plain impacts structural requirements of any flood protection provided. Design the structures within the floodway or the actual path of the rushing water to withstand resulting impacts and scouring.

A relatively flat site will generally facilitate construction activities. A naturally sloping site simplifies hydraulic design and facilitates gravity flow through the plant, avoiding the need for intermediate pumping.

A low-lying site surrounded by higher ground has some disadvantages. During certain atmospheric conditions, odors or other air emissions may easily be trapped and accumulate to uncomfortable and even unhealthy concentrations. Odor containment, collection, and control can mitigate this shortcoming. Also, noise may be a problem because surrounding hills reflect sound waves. The advantage of a low-lying site is that surrounding hills often hide the site.

GEOLOGY, HYDROGEOLOGY, AND SOILS. Site geology, hydrogeology, and soil types significantly impact construction costs. Design problems are more complex when a site is underlain by more than one type of soil. The allowable bearing capacity of the soils and the permissible differential settlement dictate the need to support tanks and structures on piles or caissons. Unfortunately, fine-grained alluvial soils and other unfavorable subsurface formations are often found near rivers and streams on otherwise ideal sites. Corrosive or "hot" soils, detrimental to buried metals, may dictate the use of more expensive coatings, nonmetallic substitutes, or a cathodic protection system. Bedrock depth may dictate the hydraulic profile of the plant and inflate construction costs if rock excavation is required.

A high groundwater condition also increases construction costs as dewatering requirements during construction become more extensive. Dewatering may cause subsidence and damage adjacent structures. It may also be more difficult to keep basements, deep dry-wells, and tunnels dry. Waterproofing membranes are recommended on the exterior (earth side) of these "dry" areas. Consider flotation of empty tanks and other buried structures but provide sufficient ballast or pressure relief drains to prevent flotation during the highest possible groundwater conditions. The use of pressure relief drains allows groundwater to enter any out-of-service tanks; however, use caution because relief valves can malfunction. Provisions may need to be made for the pumping of groundwater from tanks during maintenance work. Groundwater in a tank during cold weather may freeze and, when the tank is restarted, such equipment as fixed air diffusion systems and sludge collectors may be damaged.

Table 4.1 presents a number of measures that have been used at treatment facilities to mitigate site geologic and soil-related issues. Retaining a qualified geotechnical and soils engineering firm experienced in water-retaining

Table 4.1 Alternative methods for design mitigation of site geologic, soil, and seismic problems.

Site problem	Mitigation	Description or comment
Poorly densified granular materials,[a] seismic not a controlling factor	Overexcavate	Overexcavate poorly densified materials to a depth where stresses, due to surface and applied loads when appropriately distributed with depth, result in acceptable deformations. Replace with compacted material.
	Pile foundations	Drive or drill piles to firm material.
Poorly densified granular materials,[a] seismic considerations	Dynamic compaction	
	Vibro-flotation	
	Subsurface blasting	
Normally consolidated clays	Preloading	Preload site by placing a surcharge fill to increase stress in subsurface clays to a value exceeding the intended service stress. In wet or saturated clays, consider installing sand piles, stone columns, or wick drains to assist in drainage of water from clays.
	Pile foundations	Drive or drill piles to firm material.
Liquefaction	Pile foundations	Drive or drill piles to firm material not subject to liquefaction.
	Stone columns	Auger stone columns through liquefiable soils to assist in drainage.
	Sand piles	Auger sand piles through liquefiable soils to assist in drainage.
High groundwater	Pressure relief valves	Relief valves placed in invert of tank or reactor.
	Dewatering wells	Install dewatering wells with pumps actuated by high water level to lower groundwater table in the vicinity of the tanks.

Table 4.1 Alternative methods for design mitigation of site geologic, soil, and seismic problems (continued).

Site problem	Mitigation	Description or comment
High groundwater (continued)	Rupture slab	Design a thin slab section in center portion of the tank bottom slab designed to rupture in event of extremely high groundwater.
	Structurally reinforced, extra thick bottom slab	Tank bottom slab designed to withstand uplift stresses and sufficiently thick to withstand flotation. Extend bottom slab outward from walls, if necessary, to "pick up" weight of additional prism of soil. Note that the unit weight of soil is reduced where below the water table.
Potential differential settlement	Small, isolated structures with flexible pipe connections and adjustable weirs	Should settlement occur, the areas are isolated and little structural damage will occur. Weirs can be adjusted to maintain system hydraulics.
Potential landslide	Excavation	Remove loose soil material and/or flatten slope.
	Slough wall	Slough wall will retain minor land slippage and debris flow.
	Buttress fill	Construct engineered fill against landslide area as a buttress.
	Drainage	Install horizontal drains to remove groundwater from behind landslide.
	Retaining wall	Design retaining wall to resist loose soil material.
	Tie-backs	Anchor to firm, sound material.

[a] Poorly densified granular materials have standard penetration test (SPT) blow counts (N) less than 10.

structures during site selection and design of a project is essential. The American Concrete Institute (ACI) recommends a minimum of four soil borings for the first 930 m^2 (10 000 sq ft) of site construction area and a minimum of two additional borings for every additional 930 m^2 (10 000 sq ft).[4] Soil borings should extend to a depth well below the anticipated zone of influence of the structure and its contents. It is considered good practice to locate test borings directly under large foundations and major structures.

SEISMIC ACTIVITY. Site selection and plant layout for a wastewater treatment plant must consider the possibility of seismic activity in the area. In the past, earthquake-related damage has occurred at treatment plant sites located as far away as 80 km (50 miles) from the epicenter of an earthquake. Damage that occurred resulted from ground shaking, forces due to liquid movement in the tanks (sloshing), liquefaction, ground displacement, slope failure, and lateral soil spreading. Facilities adjacent to coastal areas have been damaged by tsunamis and seiches.

The severity of hazards varies throughout the world. For the U.S., Figures 4.5 and 4.6 show acceleration coefficients on firm ground with a 90% probability of not being exceeded in 50 years.[5] The information presented in these figures is a guide to the relative significance of seismic factors in design at various locations in the U.S., not a substitute for a complete soils and geotechnical evaluation of the site.

In most areas, detailed seismic and geotechnical analyses are necessary to determine the maximum expected intensity of the earthquake, probability of occurrence, sources of ground motion, and likely behavior of subsurface soil materials. Procedures for making such analyses are outlined by the Federal Emergency Management Agency (FEMA) and National Earthquake Hazard Response Program (NEHRP).[6] Procedures for design are presented in the American Concrete Institute's "Building Code Requirements for Reinforced Concrete"[7] and "Environmental Engineering Concrete Structures."[4] Internal forces on basins and reactor tanks from contained fluids caused by seismic activity are determined by the method presented in "Nuclear Reactors and Earthquakes" by the U.S. Atomic Energy Commission.[8]

Design for external earth loads and internal fluid loads caused by seismic activity is typically accommodated by performing a static analysis using equivalent static loads. Rarely is a dynamic analysis justified. A dynamic analysis may be warranted, however, where substantial structure irregularities exist. Installation of expansion joints and physically separated, adjacent structures having different loading and operating conditions is recommended.

Generally, the site selected for a treatment plant should not be located on or near an active fault. Movements of active faults can exceed several feet and are difficult to accommodate in design. An investigation of the presence of an active fault is necessary in areas of known seismic activity. At the

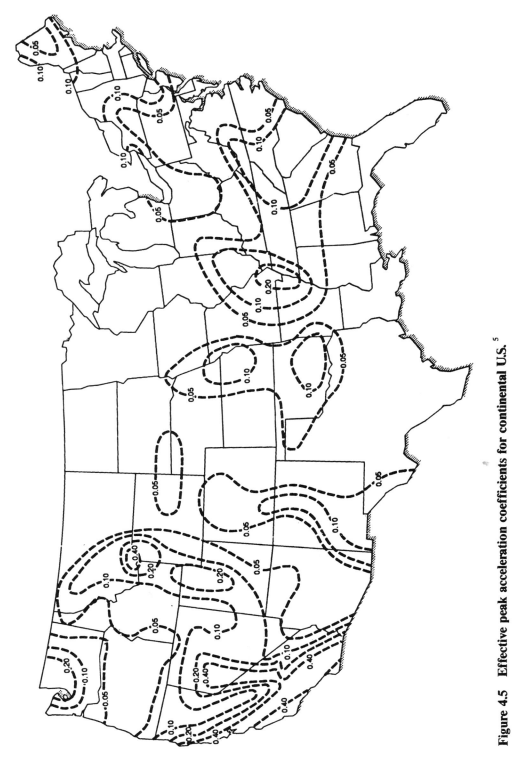

Figure 4.5 Effective peak acceleration coefficients for continental U.S. [5]

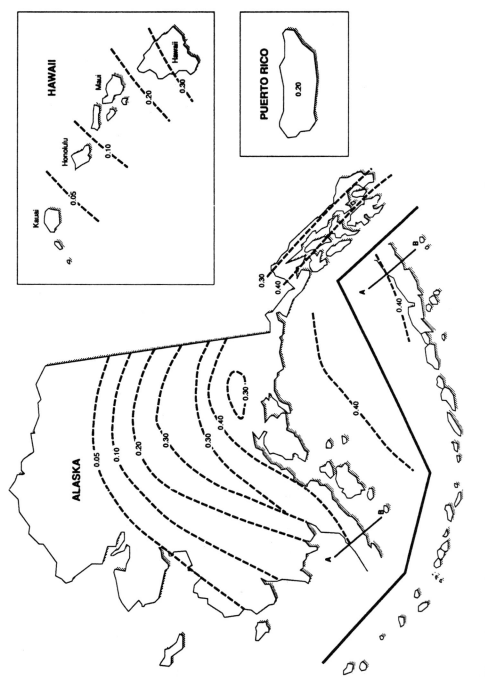

Figure 4.6   Effective peak acceleration coefficients for Alaska, Hawaii, and Puerto Rico. [5]

Malibu, California, wastewater reclamation plant site, deep trenching extending nearly to bedrock was performed along several transects. Soil strata were carefully logged and age dated. Some faulting in Preholocene materials was found, but because of its age and the lack of faulting in recent materials, the absence of active faults on the site was clearly demonstrated.

A seismic evaluation indicates the probability and consequences of fault displacement at the plant site. For essential structures, provide seismically stable foundations protected against excessive net differential settlement, lateral movements, and other earthquake-related failures. Pipelines, particularly joints and connections to structures, deserve similar consideration. Consider the potential for seismically induced slope movement. Do not place structures near the toe or at the top of a slope susceptible to failure during an earthquake.

Consider liquefaction of subsurface soil and cohesionless materials below the groundwater table. Liquefaction occurs when saturated loose sand and silt become fluidlike during ground shaking. The occurrence of liquefaction has resulted in the flotation of below-ground structures and the tilting and settling of above-ground structures. No distress to foundations has been reported due to liquefaction of sands and silts at depths greater than 9 m (30 ft) below the base of the structure.

TRANSPORTATION AND SITE ACCESS. It is essential that the plant site be accessible to personnel and delivery persons at all times. Access roads having the same degree of flood protection as the treatment plant ensure the safe ingress and egress of operating personnel. Plants located close to all-weather highways facilitate the delivery of equipment and chemicals and off-site disposal of grit, screenings, and sludge. Access roads having reasonable gradients, and wide curves allow the movement of large vehicles and heavy equipment; if possible, provide sight distances and curvature that permit comfortable vehicle operation at speeds of up to 56 km/h (35 mph). In large plants it may be advantageous to have two entrances—one for the staff and plant visitors and another for deliveries and the removal of residuals.

Service from a railroad spur offers the plant the option of accepting deliveries by rail, thus lowering the unit cost of bulk deliveries (for example, chlorine and other chemicals).

UTILITY SERVICES. A treatment plant must have sources of potable water, reliable electrical power, and telephone communication. Also, it is usually beneficial to locate the site near a supply of natural gas. The availability of such utilities in sufficient capacity is an important consideration when selecting a site.

NOISE CONTROL. Noise management is an important consideration in layout and design. Consider two aspects—first, the transmission of noise beyond plant boundaries and, second, the impact of excessive noise on the health and welfare of plant personnel. The latter is accommodated through proper specification of equipment and sound-adsorbing enclosures or isolation. Maximum noise levels for working areas are regulated under the federal Occupational Safety and Health Act (OSH Act).[9]

The effectiveness of noise control is determined by measuring the noise level and comparing it to a specified design standard. The noise or sound level is expressed as either sound pressure or sound power. Though they are different, both are measured in decibels (dB). (The decibel scale is logarithmic, beginning at 0 for the faintest audible sound and ending at 130 for the approximate threshold of pain.) Sound power level is related to the total acoustic power emitted by a source, expressed relative to a reference power quantity—typically one picoWatt (pW). Data on the sound power level of equipment are available from many manufacturers.

The sound pressure level, on the other hand, is a measure of the acoustic disturbance at some point and depends on the distance from the source, losses in the air, room effects, and other factors. The reference level for sound pressure is 20 micropascals (μPa). Sound power levels can be approximately converted to sound pressure levels.[10]

Sound-level meters used to measure sounds contain frequency-weighting networks (A-, B-, C-, and D-weighting scales) to emphasize sounds within a given frequency range. The A-weighted scale (dBA) is most commonly used in field measurements and is the scale often used in noise ordinances and codes. Table 4.2[11] presents typical background noise levels for various indoor and outdoor areas. This can be used as a frame of reference when

Table 4.2   Typical criteria for background noise.[11]

Space type	Noise level, dBA
Indoor	
Conference rooms, offices	42
Lobbies, laboratory, work areas	47–56
Light maintenance shops	52–61
Work spaces—communication required	56–66
Work spaces—no communication required, but with no risk of hearing damage	66–80
Outdoor	
Quiet residential	40–50
Average residential	50–60
Commercial	55–65
Industrial	60–70

analyzing the impacts of specific noise levels on the surrounding environment.

To reduce the sound level at the plant boundary, consider enclosing blowers, compressors, large pumps, incinerators, and other equipment that operate at high speeds in buildings of appropriate, sound-attenuating construction.[10] As a rule, sound levels are reduced as the distance from the source is increased. If possible, locate facilities that generate noise as far away as possible from potential receptors and erect sound walls, berms, and heavy landscaping in the surrounding area to minimize treatment plant noise.

A survey to determine ambient levels of noise is recommended at any proposed treatment facility site. A 3-dBA increase above ambient levels of noise by the treatment plant has little or no impact on surroundings; noise level increases of 3 to 15 dBA have a moderate impact; and noise level increases of more than 15 dBA severely impact the surroundings. To determine the impact of a wastewater treatment plant on the surrounding community, consider both hourly and seasonal fluctuations. In residential areas, noise is tolerated less at night and on summer evenings. Figure 4.7[10] is a guide to the acceptability of noise. Some sound levels are difficult to anticipate because the natural resonance of the completed structure can dramatically amplify noise levels. Inadequately isolated piping systems, equipment foundations, pressure control valves, and poorly designed equipment silencers also transmit and amplify noise.

Construction results in a temporary increase in the noise level,[10] which can be mitigated by restricting work hours.

ODOR CONTROL. Any treatment plant is a potential source of malodorous byproducts. Process upsets can occur that, if not properly handled, can be offensive. Generally, any odor traceable to the treatment plant is considered unacceptable. During site selection, consider the direction of the prevailing wind. In locations where wind shifts are frequent and public exposure is great (for example, adjacent to a busy highway, school, or residential development), odor containment and treatment may be the only option. Figure 4.8 shows what a large agency did to contain and mitigate odors.

AIR QUALITY CONTROL. Dust, smoke, and mists are aerosols and particulates common to treatment plants. Aeration tanks, cooling towers, spray irrigation, incinerators, and boilers are potential sources of such air emissions. Minimize their effects by providing scrubbers, filters, covers, electrostatic precipitators, afterburners, proper location of sources relative to prevailing wind direction, and isolation. Dust is usually not a major problem after construction activities are completed. Once paving is completed and lawns are established, dust problems are controllable by plant personnel.

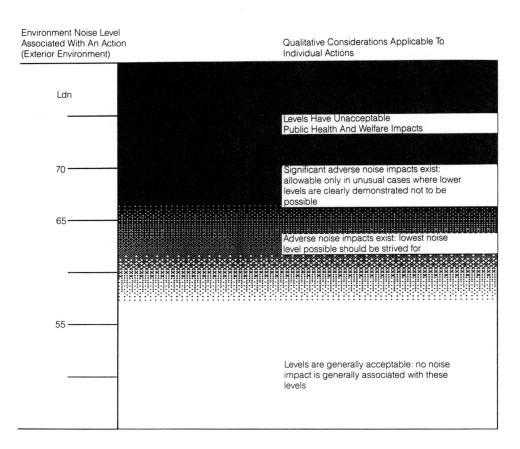

Environment Noise Level
Associated With An Action
(Exterior Environment)

Qualitative Considerations Applicable To
Individual Actions

Ldn

Levels Have Unacceptable
Public Health And Welfare Impacts

70

Significant adverse noise impacts exist:
allowable only in unusual cases where lower
levels are clearly demonstrated not to be
possible

65

Adverse noise impacts exist: lowest noise
level possible should be strived for

55

Levels are generally acceptable: no noise
impact is generally associated with these
levels

Figure 4.7 **Environmental noise levels for residential, hospital, and educational activities. Ldn equals combined day- and night-weighted sound pressure levels.** [10]

OTHER ENVIRONMENTAL IMPACTS. Consider also other impacts on the environment. Locating the plant in areas of special habitats, such as wetlands or other critical ecosystems, is often unacceptable. The Endangered Species Act protects threatened and endangered species and their habitats from adverse effects of proposed developments. Permits issued under section 404 of the Clean Water Act (CWA) by the U.S. Army Corps of Engineers are required for construction in wetlands.

Avoid sites with special natural features. The development of areas designated under the Wild and Scenic Rivers Act as wild, scenic, and recreational may be prohibited. Shorelines are often reserved for public use; this is especially important in urban areas where a shortage of open shoreline access exists. Facilities proposed along coasts should comply with the Coastal Zone Management Act.

Evaluate the presence of archaeological, historical, or other cultural resources of the site. If such resources are present, procedures prescribed by the

(a)

(b)

Figure 4.8 The 200-mgd plant of the County Sanitation Districts of Orange County, California, and its extensive odor containment and control system: (a) aerial view showing proximity of plant site to residential developments in Huntington Beach and (b) covered circular primary clarifiers, odor scrubbers, supporting chemical facilities, and the reinforced concrete chemical spill containment structure (mgd × 3785 = m^3/d).

Advisory Council on Historical Preservation and the State Historic Preservation Officer may be required to limit any adverse impacts.

Projects involving funding from the U.S. Government may require the preparation of an environmental assessment or environmental information documents to comply with the National Environmental Policy Act of 1969 (NEPA). The assessment describes anticipated significant effects of a project, including both primary (direct) and secondary (indirect) consequences of short-term and long-term duration. This may require the evaluation of alternative sites. If the assessment reveals that significant adverse environmental impacts are unavoidable, the preparation of an environmental impact statement may be required. Before choosing a site, investigate those located in or adjacent to existing or former industrial areas and waste disposal sites for soil and water contamination; site remediation is often costly.

POTENTIAL FOR EFFLUENT REUSE. Considering treated effluent as a resource may dictate that the plant be located near the reuse site. The reuse site may be an industrial plant, a groundwater recharge area, a golf course, a park, or a tract of agricultural land. The County Sanitation Districts of Los Angeles County and the City of Los Angeles, among others, constructed wastewater reclamation plants near the point of use. Experience shows that wastewater reclamation facilities located in close proximity to their principal point of use are cost effective.

COST EFFECTIVENESS ANALYSIS. Final site selection may ultimately be accomplished by performing a cost effectiveness analysis. For such an analysis, consider all life cycle costs, characteristics of the site, and their effects on construction and operating costs.

Also, evaluate other factors to which it is difficult to attach monetary worth such as public opinion, environmental impacts, and the effects on the community.

Establish a matrix to help select the best site among a number of alternatives. The matrix includes not only capital and operating costs, but also applicable "other" factors and considerations such as flood potential, public acceptance, soils, and seismic activity. Each factor is weighted on a scale of one to five, for example. A set of weighting criteria is established for each factor such that the final site "score" is the product of rating and weighting factors. It is essential to perform a sensitivity analysis to see what impact changing the weighting factors has on the final decision.

CASE STUDY. EPA presents other examples of how site environmental inventories were used in the site selection and evaluation process.[10] The following case study, prepared by a consulting firm, describes selecting a site for a new wastewater treatment plant.

Potential Plant Sites. In the facility plan, the recommended alternative calls for a new wastewater treatment facility to be located in the lower Dry Creek Basin. The general area for the plant was identified, but a specific site was not established. The scope of work for this study includes the evaluation of alternative plant sites and the recommendation of a site for approval by the city. Representatives from the city identified four sites within the proposed general vicinity for the plant that they considered to be suitable and acceptable for a wastewater treatment facility. The local airport authority was also asked to assess the feasibility of constructing a treatment facility close to the airport. Further, the city requested that an additional specific site be considered for the rapid infiltration treatment process.

Analysis Method. Several alternative sites were evaluated for the proposed Dry Creek Basin Wastewater Treatment Plant. Sites A, B, C, and D are each approximately 20 to 32 ha (50 to 80 ac) in size; each was analyzed for the siting of a traditional secondary wastewater treatment facility (with space available for advanced treatment, if required). Site E is approximately 200 ha (500 ac) in size and was analyzed for suitability as a rapid infiltration land treatment site. Figure 4.9 shows the relationship of the sites.

The analysis of the five alternative sites entailed input from the city planning office and specialists in vegetation, wildlife, and wetlands. Issues of major concern were the environment, odor impacts, zoning, land ownership and availability, flood plain designation, land costs, and engineering considerations. An environmental field reconnaissance of the five alternative sites was undertaken to provide comparative general information on wetlands and wildlife habitat. Additionally, an odor analysis based on wind frequency data from the airport was conducted to determine the potential odor impact on nearby areas. Information provided by the city pertained to land ownership, zoning, and estimates of land costs.

Description of Potential Sites. *SITE A.* Site A, approximately 32 ha (80 ac), is located east of Jones Road between First and Second Avenues and was recently annexed by the city. Of the total acreage, 6 ha (15 ac) are influenced by the airport "clear zone" and are not eligible for development. Approximately 90% of the remaining 26 ha (65 ac) is encumbered by flood plain or floodway restrictions for Dry Creek and Smith Gulch.

Of the five sites, site A is the farthest east and is located farthest upstream along the tributary drainage area. A plant on site A, therefore, would intercept less wastewater flow than would plants on the other sites.

Site A has a history of heavy grazing, which is indicated by weedy shrubs interspersed with nonnative forage grasses. Plains cottonwoods and peach-leaved willows grow scattered along the meanders of Smith Gulch. The herbaceous cover within the drainage channel reflects moist, but not persistently wet, conditions. No wetlands are present.

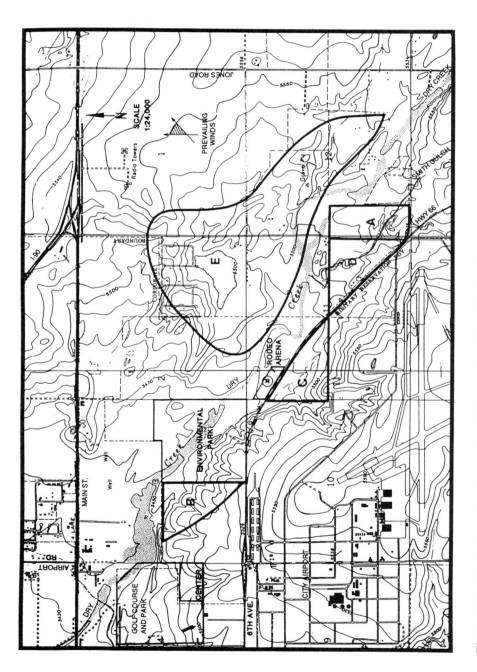

Figure 4.9 Alternative sites for a potential Lower Dry Creek Basin plant.

Wildlife use of the area appears to be minimal, with the exception of small cosmopolitan mammals and birds. There is no indication that the area is an important part of the wide-ranging habitat of any raptors, birds of prey of special concern in this region.

Prevailing winds in the vicinity are to the north and north–northeast. The distance to the nearest existing or proposed residential area is approximately 0.3 km (0.2 miles). The distance to the nearest residential receptor in the direction of the prevailing wind is approximately 0.6 km (0.4 miles). With possible residential development so close, winds could potentially carry plant odors toward residential receptors in site A. The wind frequency in the direction of the closest residentially zoned area is approximately 2.7%.

Development of site A could require extensive channelization of Dry Creek and Smith Gulch, the cost of which could exceed the value of the land. Land values, primarily influenced by floodway and flood plain designations, are estimated to range between $15 000 and $25 000/ha ($6000 and $10 000/ac).

SITE B. Site B, approximately 20 ha (50 ac) in size, is generally bounded by city parkland, including a golf course, an environmental park, and a recreation center. It is bounded on the southwest by the air installation compatible use zone (AICUZ) for the airport, with land use regulated in accordance with accident potential zone I-A (APZI-A) criteria. The property is zoned for light industrial use, and no flood plain or floodway designations appear to affect the area. Site B is the farthest west of the sites and is located farthest downstream along Dry Creek. A plant on this site, therefore, has the ability to intercept the maximum wastewater flow.

The site is currently in a cultivated and fallow condition, vegetated only by germinating weeds of various genera of the mustard family. The area, at present, appears to have minimal value to wildlife. No areas of wetland vegetation are present.

Prevailing winds are from the south and south–southwest, and the distance to the nearest residentially zoned area in the prevailing wind direction is 1.0 km (0.6 miles). Wind frequency in that direction is approximately 3.2%.

Land values appear to be primarily affected by the airport AICUZ and available utility service to Airport Road. Due to the proximity of site B to developed areas of the city, the land values may range from $220 000 to $270 000/ha ($90 000 to $110 000/ac).

SITE C. Site C, approximately 22 ha (55 ac) in size, is bounded on the west and south by the airport and on the northeast by State Highway 66. Site C is owned by three separate parties. The land is zoned planned community zone district (PCZD) industrial. No utilities currently serve the site.

Most of site C is in a cultivated and fallow state; however, the western edge and southwestern corner have not been cultivated because of steep slopes, a deep drainage channel, and the presence of some salty soils. On this uncultivated area, a perennial grass cover composed mainly of weedy annuals and nonnative forage grasses is found. The salty spot in the north central section of this unplowed strip has salt grass and alkali bluegrass as major species, while Spanish bayonet (*Yucca*) and buffalo grass dominate steep slopes in the southwest corner.

A prairie dog colony large enough to function as a significant prey base for the raptor has developed throughout much of the proposed site and the adjacent airport property. A ferruginous hawk was sited in the area. Additionally, this site is a potential habitat for the burrowing owl. The ferruginous hawk and the burrowing owl species are declining in the region due to loss of habitat. While there are no wetlands on the site, site C rates as a poor choice for development because of its habitat for wildlife and its value to the raptorial species.

As with the other parcels, prevailing winds are to the north and north-northeast. The distance to the nearest residentially zoned area is approximately 0.5 km (0.3 miles), while the distance to the nearest residential receptor in the direction of the prevailing wind is about 0.6 km (0.4 miles). Wind frequency in the direction of the closest residentially zoned area is approximately 2.7%.

It is estimated that the land value within site C may range between $160 000 and $220 000/ha ($65 000 and $90 000/ac). While access to the site is good, utility services are not currently available to the site.

SITE D. Site D, approximately 32 ha (80 ac) in size, is located west of Jones Road between State Highway 66 and First Avenue. The site comprises approximately 12 private lots and several residences along Highway 66. The confluence of Dry Creek and Smith Drainage lies directly north of the site, and approximately 70% of the area lies within the floodway designation for the creeks. A small portion of the property in the southern area lies within the airport "clear zone." Presently, the property does not lie within the city, but after annexation it will be zoned appropriately.

Site D, like site A (which is immediately adjacent to site D), reflects heavy agricultural use under dry conditions. Also similar is the presence of plains cottonwoods and willows along Smith Gulch. No wetland vegetation was observed, and no wildlife species of concern were identified. The site does not appear to be of value to wildlife species because several residences limit its usefulness to wide-ranging predatory species.

Prevailing winds are from the south and south-southwest, and the distance to the nearest residentially zoned area in the prevailing wind direction is 0.6 km (0.4 miles). Wind frequency in that direction is approximately 15.6%, indicating a greater potential for odor concerns from this site.

Land values for site D are estimated to range from $37 000 to $49 000/ha ($15 000 to $20 000/ac), primarily due to access along Highway 66 and

Jones Road. However, the flood plain designation may serve to devalue the property. No value has been assigned to residential improvements within this site, nor have relocation costs been considered.

SITE E. Site E contains approximately 200 ha (500 ac) and encompasses multiple ownerships with many large-lot residential and agricultural improvements. Approximately 35 of the 70 separate parcels have residential improvements on site. The property is currently in the unincorporated portion of the county.

This large site comprises residential areas in the northwest and north-central section, a cultivated area in the northeast portion east of Jones Road, and sand and gravel deposits and dry upland pasture in the southeast. Internal drainage located within the residential development is not persistently wet. Vegetation consists primarily of rank, weedy, annual forbs around the perimeter and foxtail barley in the center, which are not wetland indicators.

Site E presents the greatest potential for odor concerns. It is likely that the site will be bounded on several sides by residential development as close as 90 m (300 ft) from the treatment site.

The value of the land would probably range from $37 000 to $49 000/ha ($15 000 to $20 000/ac). No value has been considered for improvements or relocation costs.

Evaluation of Sites. Table 4.3 summarizes comparative evaluations of the parameters.

ENVIRONMENTAL ISSUES. No constraints due to wetlands are present on any of the five sites. Wetlands, in general, are relatively poorly developed in this area because of the arid climate and the tendency for topographic depressions to be well drained due to the sandy nature of the soils and substratum.

The only constraint regarding wildlife occurs on site C, where a prairie dog community forms a potentially important habitat for raptors.

POTENTIAL ODOR PROBLEMS. Potential odor problems associated with the proposed treatment facility are a function of wind direction and distance from the facility to the nearest receptor. Prevailing winds in the vicinity are to the north and north-northeast; therefore, areas in these directions have the greatest potential for odor problems. Distance to the nearest receptor in the direction of the prevailing wind ranges from 0.6 km (0.4 miles) for sites A, C, and D to 1.0 km (0.6 miles) for site B; the actual distance to the nearest residential receptor ranges from 0.2 km (0.1 miles) for site A to 0.8 km (0.5 miles) for site B. Of the proposed sites, the greatest potential for odor problems exists for site E, which may be bordered on several sides by residential areas, some as close as 90 m (300 ft) from the site. Sites A and D also present odor concerns due to the proximity of residentially zoned areas

Table 4.3   Analysis matrix for a site of a new WWTP located near residentially zoned areas (RZAs).

Site	Environmental issues	Odor considerations				Zoning/ownership of adjacent land
		Distance of site to nearest RZA, mile[a]	Wind direction to nearest RZA	Frequency of wind direction to nearest RZA, %	Distance of prevailing winds to nearest RZA, mile	
Site A	No wetlands No habitats for wildlife species of concern	0.20	ENE	2.7	0.40	Recently annexed—zoning to be established Many adjacent landowners
Site B	No wetlands No identified wildlife habitats	0.50	E	3.2	0.60	Light industrial zoning City owns most of adjacent land
Site C	No wetlands Hunting habitat for ferruginous hawk and burrowing owls	0.30	ENE	2.7	0.40	PCZD—industrial zoning Many adjacent landowners
Site D	No wetlands No habitats for wildlife species of concern	0.20	ENE	15.6	0.40	Proposed to be annexed—zoning to be established
Site E	No wetlands No identified habitats for wildlife	0.06	Several	NA	0.06	Primarily in unincorporated county with large-lot residential and agricultural use Many adjacent landowners

Design of Municipal Wastewater Treatment Plants

Table 4.3   Analysis matrix for a site of a new WWTP located near residentially zoned areas (RZAs) (continued).

Site	Land ownership/ availability	Flood plain designation	Land cost estimates	Engineering considerations	Comments
Site A	One landowner No residences on site	Flood plain/floodway restrictions 90% usable land in flood plain/floodway	$6 000–$10 000/ac[b]	50-acre plant size Farther upstream on Sand Creek Intercepts least flow	Floodway/flood plain designation restricts development Some residential odor potential
Site B	One landowner No residences on site	No flood plain/floodway restrictions	$2.00–$2.50/sq ft[c] ($90 000–$110 000/ac)	40-acre plant size Farther downstream on Sand Creek Intercepts most flow Available utility service in Tower Road	City owns most of adjacent land No floodway/flood plain restrictions Intercepts most flow Lowest residential odor potential
Site C	Three landowners No residences on site, many adjacent	40% of site in 500-year flood plain Small portion in 100-year flood plain	$1.50–$2.00/sq ft ($65 000–$90 000/ac)	50-acre plant size	Habitats for wildlife species of concern Parcel has 3 landowners
Site D	About 12 landowners Several residences on site	70% of site in flood plain/flood-way designation	$15 000–$20 000/ac	50-acre plant size restricts development	Floodway/flood plain designation Some residential odor potential Parcel has 12 landowners Several residences on site
Site E	Multiple landowners (70 parcels) Many residences on site	No flood plain/floodway restrictions	$15 000–$20 000/ac	500-acre plant size	Area 10 times larger than other sites Multiple landowners (70 parcels) Many adjacent landowners Many residences on site Prohibitive cost Some residential odor potential

[a] mile × 1.609 = km.
[b] acre × 0.404 69 = ha.
[c] sq ft × 0.092 9 = m^2.

and wind frequency, respectively. Site B appears to present the fewest odor concerns to residential areas but may present odor concerns to nearby city parkland.

ZONING. Zoning does not appear to present a major concern in the site selection process. The proposed sites are either zoned for industrial use or are presently located outside the city limits, with the appropriate zoning to be established after future annexation. Regarding land use, site B is entirely surrounded by city-owned or industrial land, while the other sites have residential or planned residential neighbors.

LAND OWNERSHIP AND AVAILABILITY. Sites C, D, and E present some concern with relation to land ownership and availability. The three sites have multiple owners, ranging from three for site C to 70 for site E. Site D has 12 owners, with several residences on site. Numerous residences have been constructed on site E. Therefore, land acquisition activities may be more difficult and time consuming for these sites.

FLOOD PLAIN DESIGNATION. Approximately 90% of the usable portion of site A and 70% of site D lie within the floodway designation for Dry Creek and Smith Gulch. Resultant increased costs of plant construction and operation and maintenance make these sites less attractive for development.

LAND COST. Land costs range from $15 000 to $25 000 /ha ($6 000 to $10 000/ac) for site A to $220 000 to $270 000/ha ($90 000 to $110 000/ac) for site B. Cost is particularly high for site E because of the large land area required for the rapid infiltration land treatment site.

ENGINEERING CONSIDERATIONS. From an engineering standpoint, including availability of future reuse supply, site B is the most suitable for the proposed Dry Creek Wastewater Treatment Plant. While higher in cost than the other sites, site B is situated to intercept the most flow, is surrounded by city and industrial land, and appears to present the fewest odor concerns to residential neighborhoods. Additionally, site B has no identified environmental constraints, has a single owner, and is not impacted by the 100-year RI flood plain designation.

While site B appears to be most suitable, a public awareness and participation program to seek public input and incorporate public concerns could be instituted to encourage community involvement into the final site acquisition and approval process.

*P*LANT LAYOUT CONSIDERATIONS

The layout of a wastewater treatment facility primarily considers the layout of treatment processes. Once the staff, administration, and ancillary facilities are sited, they are placed (based on function, hydraulics, and operation) for efficient, economical operation and the support and comfort of the staff.

TREATMENT FACILITIES. Arrangement of Treatment Processes. The arrangement of treatment processes on the plant site affects operating and construction costs. Before locating any facilities on the site, develop a rough, preliminary hydraulic profile to establish both tops and bottoms of principal structures. Using these key elevations maximizes the site's topographic features and relief. Locating raw wastewater pump stations near the point at which sanitary sewers enter the plant site minimizes the cost of building additional lengths of deep gravity sewers. Force mains can be relatively shallow and smaller in diameter than gravity interceptors.

Preliminary treatment units placed close to the pumping station simplify the control of a water hammer. A roadway, designed for heavy loads and located near treatment units, facilitates the loading and hauling of screenings and grit. Arrange the facilities to minimize the length of connecting and recycle piping. Locate chemical bulk storage facilities and chlorine storage areas along the main service road to make deliveries more convenient. Isolate potential sources of aerosols, particulate emissions, odors, and hazardous gases as much as possible and locate them downwind of the most critical, adjacent land. Locate sludge-dewatering facilities or liquid sludge loading stations along the main service road and away from sensitive odor receptors.

Where possible, group together similar unit processes to facilitate operation, minimize piping, and allow for expansion. Such groupings include sludge thickening, digestion, dewatering, and disposal; liquid stream and preliminary treatment units such as screening, grit removal, and grit-dewatering; and primary and secondary treatment units.

Arranging treatment processes to follow site contours helps maximize use of natural topography and reduces both pumping requirements and excavation costs for new structures.

Provisions for Future Expansion. Because most wastewater projects upgrade, refurbish, or expand sites already in existence, it is important during the layout of a wastewater treatment project to provide for future expansion or upgrading of treatment facilities. Provisions made to facilitate future expansion at the expense of short-term efficiency and convenience should be made only after due consideration. Maintaining adequate space during the arrangement of treatment units facilitates future construction. Providing adequate space where deep foundations or tanks are to be constructed adjacent to existing shallow foundations or tanks avoids undermining in the future. A good rule of thumb is to keep clear of a zone extending downward and outward 45 degrees from the bottom of a shallow foundation.

Consider the logical location of future tankage, the size of conduits connecting unit processes, and the way in which flows can be distributed among multiple tanks. Including future construction on building plans ensures that adequate space is reserved. It may be beneficial to establish a plant grid system that, for example, provides for capacity increases in the north and south directions and treatment upgrades in the east and west directions.

Determine the hydraulics of the plant with future expansion in mind. Provide sufficient fall between process stages to permit increases in flow rate and permit adequate distribution of flow among multiple tanks. It is not prudent design to construct a new plant that is hydraulically limited. Such plants are difficult and costly to expand or upgrade. Also, the existing plant should be able to operate while it is being expanded. This is facilitated by placing masonry plugs and stop plank grooves in chambers and channels to which future connections must be made. The installation of a valve upstream of a blind flange or other line plug enables the plugged line to be connected without shutting down the entire line.

Tank Geometry. Circular, square, and rectangular tanks are used for primary and secondary sedimentation. Circular tanks are commonly used for primary and secondary clarifiers for plants where land area is not restrictive. Tanks offer good performance with relatively simple mechanical equipment. Square tanks with center mechanisms require that corner sweeps be used on the rake arms and that more complex provisions be used for scum removal. These additions to the normal circular mechanism require more maintenance to achieve acceptable performance. Using common-wall construction for square and rectangular tanks minimizes space requirements and offers an opportunity to reduce construction costs. Figure 4.10 shows the way in which hexagonal-shaped biotowers constructed at the Monterey, California, Regional Wastewater Reclamation Facility take advantage of common-wall construction to save space.

Number of Treatment Units. Determine the number of treatment units necessary for each unit process carefully. Multiple units are required for all critical components of the treatment plant. Pumps needed to move flow through the plant and blowers providing air to an activated sludge system must have adequate capacity for peak conditions with the largest single unit out of service. Depending on the reliability classification of the plant, portions of the power distribution system must be duplicated.

Provide multiple bar screens or comminutors, grit tanks (especially in systems with existing combined sewers), primary and secondary settling tanks, aeration tanks (especially with fixed header air systems), sludge stabilization

Figure 4.10   Hexagonal biotowers of the wastewater reclamation plant at
Monterey, California.

units, and chlorinators. Provide a means to allow bypassing of each treatment
unit. Multiple units of the same size, make, and model number facilitate
maintenance and reduce inventories of spare parts.

When deciding the number of units to provide in each area of the plant,
consider the consequences of a unit being out of service. If the occurrence
would cause unacceptable results, then an additional level of redundancy is
necessary. If an alternate mode of operation can be used during an outage,
design necessary components to implement the alternate system. In developing the process design, recognize that, in some instances, facilities may
be allowed to function at a reduced level of service during an emergency
such as that following a damaging earthquake or other natural disaster.

In developing a risk evaluation of an outage, identify processes, functions,
structures, or facilities that can be dispensed with and the time that such an
outage can be tolerated. Such parameters would permit spending time, effort,
and money on mitigating damage from a disaster on essential rather than
nonessential portions of a facility. From parameters developed during risk
evaluation, identify weak system links in the flow diagram and process train
in terms of maintaining the minimum required serviceability after a disaster.
Where there is risk of an earthquake or other natural disaster, providing
redundancy and flexibility wherever practical or economical is desirable.
Evaluate treatment efficiency, assuming that the largest unit is out of service.

If the resulting treatment is unacceptable, consider arrangements using more but smaller units or slightly oversizing the units.

Flexibility of Operation. Flexibility in design to permit various operating modes allows the operator to accommodate changes in wastewater characteristics and other conditions. Routine maintenance, equipment breakdowns, loss of a utility (such as natural gas), or damage due to an earthquake or other natural disaster require flexibility to provide alternate modes of operation.

Increased flexibility also complicates the associated instrumentation and control system; for example, piping arrangements can become complicated and costly. This can lead to operator errors and, in the case of sludge piping, line plugging. The isolation of a line full of sludge can cause gas generation to develop, resulting in high pressures and possible rupture of the line. Consider provisions to flush and drain the lines.

Continuing Treatment During Construction. Planning the construction sequence carefully ensures that new facilities can be built without undue interruption of treatment. Include in the project specifications the number of process units that can be taken out of service at any one time, allowable electrical power shutdown periods, and temporary diversion schemes. State in the specifications whether the contractor or the owner of the treatment plant is responsible for draining and cleaning process tankage and piping.

Maintenance Considerations. To facilitate maintenance in cold weather areas, place the most critical equipment indoors. This is often not necessary in warmer areas where year-round outdoor maintenance is feasible. However, a roof over equipment even in warm or hot weather areas shields the equipment from direct solar radiation and provides a dry work area during rainy weather.

Proper maintenance requires convenient access to equipment. Size the buildings housing large process equipment and piping to allow sufficient space for repairs. Bridge cranes are appropriate in areas where heavy lifts are required. In other areas, the installation of monorails and hoists may be appropriate. Cast-in inserts capable of accepting a rig for lifting can be placed in concrete slabs above equipment and piping where infrequent use is anticipated. Size and locate floor doors and pass doors for the movement of components from one floor to another and to the exterior.

Place adequate electrical outlets throughout the plant. Specialty outlets such as those used for welding may be placed in some locations. Compressed air is often piped to various areas for equipment maintenance. Locate hose bibbs throughout the plant. Hot water or steam is useful in galleries containing scum or sludge piping or other solids-handling facilities.

Provide a means to adequately ventilate confined spaces. Such facilities may be permanent or may consist of portable ventilation equipment. Size the ventilation for adequate air changes based on the volume of the confined space in the absence of any liquid.

ADMINISTRATION, STAFF, AND SUPPORT FACILITIES. A useful tool in the layout of administration, staff, and support facilities is a functional diagram, or "bubble" diagram, which is shown in Figure 4.11. A functional diagram identifies major functional areas and interrelationships between these areas.

Operator Stations. Provide work stations for personnel in charge of operating the plant or recording data and making calculations. Locate such stations near an equipment-intensive area of the plant or within view of an alarm panel or major control panel. The area should be relatively quiet, clean, well lighted, and well ventilated. In warmer climates, the area should also be air conditioned.

Where distributed, computer-based process control systems are used, the operator station may be the appropriate location for video terminals. If this is the case, more stringent temperature control and air quality may be necessary for proper operation of this equipment.

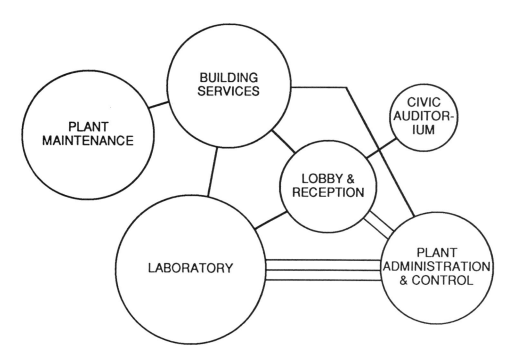

Figure 4.11 Functional diagram of administration building. Multiple connecting lines show areas of frequent communication and interrelationships.

If the plant laboratory is not always open, provide some laboratory bench space and a sink in an area adjacent to the laboratory or in some other area of the plant. For adequate process control, the operator needs to have access to basic laboratory equipment for timely analyses of process changes or plant upsets. A pH meter, dissolved oxygen meter, and 1-L graduated cylinder are examples of equipment that should be available to operators at all times. To ensure proper sample preservation, the operator needs 24-hour access to sample preparation and refrigerated storage areas.

Sample locations are necessary to satisfy NPDES monitoring requirements. Provide sample locations for evaluation of process performance and determination of recycle stream characteristics. To operate portable automatic samplers, it may be advantageous to extend a flow signal and electrical power to the sampling area. Permanent, automatic, refrigerated flow proportional samplers may be installed to meet permit requirements or provide more representative data. In other locations, providing access for taking grab samples should be adequate.

Locate sampling points in a well-mixed area. Consider the routing of recycle streams and their impact on sampling. Locate sampling of plant influent upstream of all recycle streams. Likewise, a secondary influent sample should not include return activated sludge or recirculation flows. Junction chambers, access manholes, or pipe taps are commonly provided for sampling.

Maintenance and Storage Facilities. Maintenance facilities are best located near the center of the plant or the most equipment-intensive area of the plant and should contain adequate, well-lighted bench space. The size of the maintenance staff and services to be performed in house determine the size of the facility and the tools to be provided. A shop located at the ground level best facilitates the handling of large equipment. If welding is to be done in the area, pay special attention to ventilation, isolation, and power supply. Plants with a high degree of instrumentation require a separate instrumentation shop and maintenance staff. Special static-protected test benches with various power supplies are useful for such work.

Locate maintenance records and documents, which may be extensive, near maintenance areas. Size the area to house, in addition to equipment records, maintenance equipment manuals and shop drawings. Provide facilities for personnel to review documentation and for computerization of maintenance records, inventory, and maintenance scheduling.

The storage area for the plant's spare part and maintenance supply inventory is best located near the shop area and should be large enough to accommodate an array of shelves, bins, and drawers. To facilitate deliveries, locate the storage area, normally secured, adjacent to the roadway of the main plant. Paints; lubricants; pesticides; herbicides; and similar toxic, flammable, and hazardous materials should be stored in an isolated, secure area.

Laboratory Facilities. A laboratory provides required influent monitoring and analyses of process control parameters. The size of the plant, the type of treatment provided, and the extent of laboratory analyses to be performed on site determine the size and layout of the laboratory. Many small plants find it cost effective to contract all but the simplest analyses to outside laboratories. During the design stage, identify the type and anticipated frequency of tests and the analytical equipment required to conduct them. Evaluate and factor future operating and monitoring requirements into the layout.[12]

Paint walls and ceilings a flat, light color. Use a moisture-resistant, easily cleaned paint. Finish the floors with a solid, light-colored, nonglare paint surface. Floors should be skid resistant; acid, base, and solvent resistant; easily cleaned; and comfortable to stand on.

Arrange benches to allow aisles that are at least 1.2 m (4 ft) wide; large laboratories can consider back-to-back benches. Use top surfaces that are chemical resistant and, as necessary, provide chemical-resistant sinks and cup sinks.

Store chemical reagents, in particular those chemicals considered dangerous, below bench tops. Arrange wall cabinets for storage of glassware, equipment, and safe chemicals. Provide a separate, isolated, specially rated floor cabinet to hold explosive and flammable chemicals. Anchor this cabinet securely to prevent tipping during an earthquake. Laboratory furniture manufacturers have standard bench and cabinet modules that can be used to simplify design and keep costs to a minimum.

Include fume hoods designed for fumes, gases, and vapors expected to be encountered. Dedicate an explosion-proof exhaust fan to each fume hood. Supply the fume hoods with alarms, light fixtures, cup sinks, and electrical outlets. Provide a specially designed unit if perchloric acid is to be used.

For an analytical balance, provide a separate balance table that is isolated as much as possible from vibrations.

Lighting should be provided such that technicians working at benches have adequate, shadow-free (preferably color-corrected) lighting. Design the electrical system to accommodate special power requirements of muffle furnaces, autoclaves, and stills, recognizing the need for an isolated, regulated electrical service for some sensitive analytical equipment.

Along benches, conveniently locate duplex electrical outlets, natural gas, vacuum, and compressed air fixtures. For back-to-back benches with center troughs, provide hot and cold water and, possibly, deionized water taps along the benches.

Some analytical equipment require that compressed gas cylinders be located close to the units. Provide sufficient floor space and a means to secure the cylinders. Storage arrangements for these gases must comply with local building and fire codes.

Safety is an important consideration in the laboratory. Provide space for a first aid kit, fire extinguishers, fire blanket, eye wash, and shower near the

---

main working area. Tempering tanks connected to all eye washes ensure that water supplied is close to room temperature. Provide at least two outward opening doors with panic hardware.

Laboratories should be well heated, air conditioned, and ventilated; to prevent the buildup of pollutants, do not provide for return air. Exhaust all supply air to the outside of the building. When designing the laboratory ventilation system, provide for air volume needs of the fume hoods. Prevent exhausted air from fume hoods from reentering the building through outside air intakes. To prevent false readings, it may be necessary to locate sensitive analytical instruments in separate rooms, perhaps with separate HVAC systems. Do not locate the laboratory in a building that also houses a treatment process. If this is unavoidable, take extreme care to provide completely separate heating and air-conditioning systems and prevent diffusion through the supply air system of corrosive gases from treatment areas to the laboratory.

The plumbing in the laboratory also requires special planning. Place backflow prevention devices on pipelines that supply potable water to the laboratory building and fixtures to which hoses may be attached. Make the drain piping of an acid-resistant material such as PVC or glass. Lines that may contain acids may have to discharge to a neutralization sump. To prevent plugging of the sump, a separate sink with a grinder connected to a drainage system that bypasses the neutralization sump may be desirable for dumping sludge samples and other neutral wastes. A centralized area for washing and cleanup is commonly provided to allow efficient laboratory operation.

A separate sample preparation area, beneficial in some situations, may contain refrigerated sample storage and sinks for cleanup of sampling containers. A large floor sink is useful in cleaning portable samplers. By providing a separate room, samples can be properly stored in refrigerators during hours when the laboratory is locked.

In areas vulnerable to earthquakes, anchor or restrain equipment such as refrigerators, incubators, stills, water heaters, analytical instruments, typewriters, computers, shelving, and cabinets. Keep chemicals and other stores in restrained cabinets with positive latch doors.

Administration Offices. Locate administrative offices near the front entrance of the plant so visitors may find them easily. A private office sized to hold meetings of three to six people provides the plant manager the privacy essential to handling personnel problems. Provide a manager easy access to plant operating records, personnel files, cost records, and operation and maintenance manuals.

If a plant has distinct organizational groups or departments, each manager may need a separate, private office. Grouping the individual offices of a management team promotes communication among the various groups. In larger plants, an assembly room large enough to accommodate meetings of plant personnel, on-site training, visitors, and public meetings is desirable.

Staff Services. For all operations and maintenance personnel throughout the plant, especially where employees are stationed and in locker rooms, provide restrooms containing toilets and lavatories. The Occupational Safety and Health Administration (OSHA) requires a minimum of one toilet for every 15 to 20 people the facility will serve. Enclose each water closet with partitions and doors to assure privacy.[13] If restrooms will be occupied by only one person and can be locked from the inside, separate toilet rooms for each sex need not be provided. In areas of the plant where handicapped workers or visitors could be present, provide stalls, urinals, and lavatories that meet access provisions for the handicapped. Unless local codes require otherwise, provide one shower for every 10 persons expected to complete work at the same time. Separate lockers for street clothes and work clothes may be required.

Providing lunchrooms and breakrooms for all operation, maintenance, and laboratory personnel discourages the consumption of food and beverages in unsanitary areas of the plant. Locate restrooms or separate washrooms close to the rooms where food or beverages will be consumed.

An essential part of a safe and efficient plant is a training room or area. Provide the training room with a large blackboard or other erasable board and a large wall suitable for pinning up diagrams and other training aids. Design the room for demonstrations and presentations that will use audio-visual aids such as slide projectors, overhead projectors, video monitors, and easels. An ideal arrangement for training personnel with "real life" situations is to place computers connected to the plant computer-based control system in the training area. Provide ample table space so that personnel may sit facing one direction.

Designate an emergency shelter for employees to use during hurricanes, tornadoes, earthquakes, and other natural disasters. Tunnels or basements may be the best locations for such shelters. Design underground structures used for shelters to resist dynamic soil pressures caused by seismic activity. Provide space for storage of emergency equipment, food, supplies, and cots in shelter areas. Large plants may consider having a separate first aid room.

WPCF Survey of Wastewater Treatment Plant Data. WPCF conducted a wastewater treatment plant survey in June 1989. Table 4.4 presents 119 responses to survey questions about plant layout. These responses represent existing conditions and not necessarily those recommended or even considered adequate.

OTHER LAYOUT CONSIDERATIONS. Site Access. If a plant is to run efficiently, the design of the plant's roadways is an important consideration.

Roadways must provide access to all points where deliveries will be made or where materials will be loaded onto trucks and transported off site. Where tractor-trailer rigs will be on site, plan pavement widths, curve radii, and grades accordingly. Pavement widths of 7.3 m (24 ft) provide comfortable two-way use. The pavement edge on right-angle curves follow a radius of at

Table 4.4   Employee and building space data from WPCF survey of treatment plants (June 1989).

Plant capacity	<1 mgd[a]	1–5 mgd	5–10 mgd	10–50 mgd	50–100 mgd	>100 mgd
Average capacity	0.38	2.48	7.51	18.29	68.14	181.67
No. of responses	23	55	13	18	7	3
Plant staffing						
Average	3.91	8.40	17.62	50.61	129.43	314.67
Lowest	1.50	1.00	5.00	17.00	30.00	154.00
25%	2.00	4.75	11.25	76.75	88.50	270.75
Median	3.00	6.00	15.00	136.50	147.00	387.50
75%	4.00	11.00	19.00	196.25	205.50	504.25
Highest	13.00	42.00	42.00	256.00	264.00	621.00
Employees per mg						
Average	15.93	3.51	2.27	2.43	1.82	1.59
Lowest	2.00	0.67	0.67	1.30	0.55	1.03
25%	7.25	2.45	1.64	1.43	0.55	1.31
Median	9.67	3.27	2.24	2.18	1.59	1.59
75%	25.42	4.34	2.53	2.79	2.39	1.86
Highest	50.00	9.00	4.20	6.83	3.30	2.14
Equipment maintenance, area in sq ft[b]						
Average	959.26	1 077.53	1 763.77	2 577.83	3 000.00	3 333.33
Lowest	0	0	0	0	0	0
25%	93	221	525	645	0	1 250
Median	280	613	860	1 500	2 000	2 500
75%	1 049	1 310	1 355	3 650	4 500	3 750
Highest	5 000	6 032	7 600	8 000	10 000	5 000

Table 4.4   Employee and building space data from WPCF survey of treatment plants (June 1989) (continued).

Plant capacity	<1 mgd[a]	1–5 mgd	5–10 mgd	10–50 mgd	50–100 mgd	>100 mgd
Electrical maintenance, area in sq ft						
Average	121.87	168.58	169.15	465.17	685.71	833.33
Lowest	0	0	0	0	0	0
25%	0	0	0	0	375	500
Median	0	0	0	153	750	1 000
75%	134	200	210	410	1 125	1 500
Highest	806	2 249	800	2 880	1 500	2 000
Laboratory & industrial pretreatment program, area in sq ft						
Average	160.22	449.15	769.69	1 310.78	942.86	4 166.67
Lowest	0	0	0	0	0	0
25%	0	0	275	358	675	2 000
Median	60	383	675	1 152	1 350	4 000
75%	234	600	1 037	1 650	2 025	6 000
Highest	675	4 534	2 175	4 800	2 700	8 000
Offices, area in sq ft						
Average	210.91	539.31	344.77	1 660.00	4 685.71	7 000.00
Lowest	0	0	0	0	0	0
25%	0	100	160	350	5 000	3 250
Median	96	278	297	620	10 000	6 500
75%	325	483	470	1 500	15 000	9 750
Highest	1 350	6 500	720	10 000	20 000	13 000
Meeting rooms, area in sq ft						
Average	82.52	282.25	343.77	772.33	585.71	3 933.33
Lowest	0	0	0	0	0	0
25%	0	0	0	0	575	2 500
Median	0	135	248	400	1 150	5 000
75%	15	347	436	1 170	1 725	7 500
Highest	540	3 000	1 250	3 500	2 300	10 000

[a] mgd × 3785 = m^3/d.
[b] sq ft × 0.092 9 = m^2.

least 12 m (38 ft). All grades, including that of ramps, should be no greater than 17%.

Consider emergency response to a chlorine leak or fire when planning roadways. The local fire department may have some specific requirements for equipment turnarounds and fire lanes. Outside emergency teams need a direct route to any critical areas. Provide for the storage of snow removed from roadways.

Provide parking for all personnel, including handicapped employees, and visitors. It is best to locate parking as close as possible to the area of the plant where personnel end their shift. For most of the staff, this might be at the locker rooms. Reserve a sufficient number of spaces for visitors to the plant. Large plants or plants anticipating visits from civic groups or schools can consider providing space for bus parking.

Provide paved sidewalks, at least 1.2 m (4 ft) wide, to all areas of the plant. Gravel-surface walkways, less expensive initially, require more frequent maintenance and are recommended only in areas where traffic is anticipated to be minimal. Connect sampling stations, buildings, tanks, and other areas requiring frequent monitoring with sidewalks. This provides personnel with safe footing under most conditions and helps minimize the impact of foot traffic on plant landscaping.

Plants located in harsh environments that extend over a large area might consider using tunnels. Tunnels that are designed to connect major process areas and, perhaps, double as pipe galleries facilitate the locating, repairing, or extending of piping, electrical conduits, or instrumentation cables. Providing tunnels and pipe galleries with large hatches, skylights, or removable top slab sections facilitates the installation of piping and equipment when the plant is expanded. Tunnels need to be well lighted, ventilated, and provided with additional exits for emergency use.

All treatment plant facilities located outside should be well lighted for access from one process area to another, providing approximately 54 lx (5 ft-c) throughout the area. To conserve energy, consider mercury vapor, high- or low-pressure sodium fixtures. Because these fixtures take some time to fully illuminate, connect fixtures in critical areas to a timer or photocell so they are on during the dark hours.

To control the points of entry, restrict access to the site. Provide a perimeter fence and lockable gates. In larger plants, closed-circuit television may be used to control plant access and maintain security. Security measures need to consider access to the site by outside emergency response teams.

Material Delivery, Handling, and Storage. A wastewater treatment plant is a diverse facility that receives a wide variety of materials. Chemicals, lubricants, disinfectants, spare parts, laboratory supplies, and various liquid wastes are among the materials frequently delivered to most plants. To handle and store such deliveries safely requires that proper provisions be made.

The delivery of liquid wastes such as septage and leachate presents a different set of problems. Because of the high organic and suspended solids concentrations, provide a below-ground holding tank. For leachate receiving, a holding tank is recommended to store the waste until it can be analyzed for material that is toxic to biological processes, harmful to sludge or effluent reuse, or otherwise detrimental to the treatment plant. Locate the unloading area close to the headworks area so wastes can drain by gravity from the holding tank to the influent raw wastewater pump station or be pumped to the screening, metering, and degritting facilities. Design the unloading area to contain and control odors and provide washdown facilities.

Liquid chemicals are frequently delivered in drums, carboys, or bulk. If either drums or carboys are received, a loading dock, monorail and hoist, or drum or cylinder dollies may be necessary to allow the containers to be properly unloaded and taken to a storage area. If a chemical is delivered in bulk, a chemical unloading area along a plant roadway is needed. Design the unloading area so that when deliveries are made, access to and circulation around the plant are not blocked. Install permanent piping between the unloading area and the bulk storage tanks. In some cases, a compressor available for air padding of the delivery tank is needed. In the interests of safety and handling efficiency, consider a central facility for chemical handling for the entire plant.

Lubricants are frequently delivered in drums. A loading dock or monorail and hoist facilitates the operation although other options are acceptable. Drums on pallets can be delivered to the proper level by way of a fork lift. If the delivery truck is equipped with a hydraulic tailgate, a drum hand truck can move the drum to a ground-level storage area.

Disinfectants such as chlorine, when delivered in 45- to 68-kg (100- to 150-lb) cylinders, can be moved laterally by hand trucks. For vertical movement, specially designed cradles or carrying platforms connected to a crane or hoist are used. Ton containers are typically removed from the delivery truck with a hoist and lifting beam and taken to storage. Empty containers are usually loaded onto the truck in the same way. The storage of ton containers can be outside if rails, stops, and a roof are provided. In colder climates, move ton containers into a heated room before they are put into service. Chapter 14 addresses provisions for the proper indoor storage of chlorine.

In areas subject to earthquakes, hurricanes, or tornadoes, secure chlorine containers, gas cylinders, and chemical tanks. Ton containers may be secured with heavy chains to rails. Effectively equipping the treatment plant with tie-downs, anchors, and flexible piping helps to prevent chlorine and chemical leaks to ensure that damage at the plant will not add to the impact of the natural disaster on a community.

Unloading large spare parts and replacement equipment and pallets and drums requires a forklift or large hoist. If the large-part storage area is on the ground floor, a forklift is satisfactory. A hoist on a monorail or bridge crane is more versatile; however, design the unloading area to allow the hoist to run out over the top of the truck.

Laboratory supplies are usually not as heavy, though cases of reagents and compressed gas cylinders cannot be easily taken up stairs. A laboratory not at ground level requires an elevator large enough for hand trucks. A loading dock for laboratory supplies is useful, although it is unnecessary if the delivery truck is equipped with a hydraulic tailgate.

Odor Control. The presence of odors is one of the most offensive conditions to people living near or working in a wastewater treatment plant. Do not underestimate the possible presence of odors and their effect on public relations. Even if odor control facilities are not installed at a plant, plan the layout and design to facilitate the addition of odor control systems should the need arise.

Odors in the wet stream usually are associated with characteristics of the wastewater received at the plant or the existence of an improper environment in biological treatment units. Aeration, chemical oxidation, or pH adjustment is used to reduce odors in plant influent. Grit and screenings in grit tanks and bar screens often are responsible for odors. Treating odors at sludge-handling facilities, which are often the source of odors, may require covering tanks or installing exhaust hoods. Foul air is then treated at a central location.

Table 4.5 rates the odor potential from various unit processes and operations. In site layout and design, locate areas that have a high potential for odor as far as possible from receptors. Particular attention to enclosing these areas and scrubbing odors is necessary. Table 4.6 lists odor control methods that have been used. Site layout and design should allow sufficient area to construct these facilities.

Noise Control. Some of the noisiest equipment typically found at a treatment plant are blowers, engines, engine-generator sets, and multiple-hearth incinerators, though silencers, mufflers, and acoustic architecture have successfully eliminated much of the noise that affects plant personnel. Separate enclosures or rooms are appropriate for noisy equipment. Whenever possible, select equipment that meets OSHA requirements for 8 hours of continuous exposure. Personal noise protection devices will have to be relied on in some cases.

Vehicle Storage. In cold weather areas, vehicles stored indoors start easier, last longer, and can be better maintained. Without heated indoor storage, some vehicles, especially those that are diesel powered, need engine block heaters to ensure that they will start. In warmer areas, vehicles can be stored outside.

Table 4.5   Odor potential for typical unit processes in a wastewater treatment plant.

Unit processes	Odor potential[a]
Liquid stream processes	
Flow equalization	H
Preaeration	H
Screening	H
Grit removal	H
Primary clarification	H
Stabilization	
Suspended growth	L
Fixed film	M
Chemical	H
Secondary clarification	L
Tertiary filtration	L
Disinfection	L
Sidestream returns	H
Sludge stream processes	
Thickening/holding	H
Aerobic digestion	M
Anaerobic digestion	M
Thermal conditioning	H
Storage lagoons	H
Dewatering	
Vacuum filter	H
Centrifuge	H
Belt filter	H
Filter press	H
Drying beds	H
Composting	H
Septage handling	H

[a] L = low, M = moderate, and H = high.

Cold Weather. Cold weather considerations affect process selection because certain processes will not operate properly in extremely cold weather, resulting in poor performance and an increase in operation and maintenance costs. Figure 4.12 is a cold weather zone map depicting areas in which problems may be experienced at wastewater treatment facilities (primarily the northern areas of the United States). Farther south, many areas experience freezing temperatures over extended periods of time. Freezing rain, another problem in these areas, can damage exposed equipment, especially electrical power lines. In warm weather areas, localized cold weather conditions also exist at higher elevations. Knowing the weather history of the site location helps the design engineer plan accordingly.

Table 4.6   Typical odor control methods for wastewater treatment plants.

Unit processes	Chemical, air, or O_2 addition upstream of plant	Aeration	Chemical addition	Covering with collection and treatment of foul air	Improved hydraulics to avoid turbulence	Improved O&M
Flow equalization	X	X	X			
Preliminary treatment						
Screening	X			X		
Grit removal	X			X		X
Preaeration	X			X		X
Liquid stream treatment						
Primary clarification	X		X	X		X
Suspended growth systems				X		X
Fixed film systems			X	X		X
Physical/chemical systems			X	X		X
Secondary clarification			X	X	X	X
Sidestream returns		X	X		X	
Sludge stream treatment						
Gravity thickening			X	X		X
DAF thickening				X		X
Blending and storage			X	X	X	X
Aerobic digestion						X
Anaerobic digestion						X
Chemical stabilization				X		X
Thermal conditioning				X		X
Mechanical dewatering			X	X		X
Drying beds			X			X
Composting				X		X
Septage receiving/holding		X	X	X	X	X

Design of Municipal Wastewater Treatment Plants

Cold weather causes[14]

- Ice formation on process equipment and freezing of process equipment, including viscosity changes in equipment lubricants;
- Snow and ice accumulation on structures, equipment, and roads; and
- Changes in the reaction rates of biological, physical, and chemical processes.

Designers of treatment facilities located in cold regions can anticipate, thereby reduce, potential problems of cold weather operation. Cold weather can result in damage to equipment because most exposed equipment drives are located near the wastewater; freezing of the condensation on the drive occurs frequently. In areas that experience more severe climates, weatherproof enclosures with heating are required to reduce weather-related problems. Review all exposed treatment plant equipment and consider the detrimental effects that freezing and icing have on them. For example, to prevent freezing, exposed piping may have to be insulated and heat traced. Exposed outdoor pipelines, when not in service, should be capable of being drained easily.

Consider the effects of cold weather and drifting snow in the design and layout of plant facilities and roadways. An inadequate plant layout can hinder

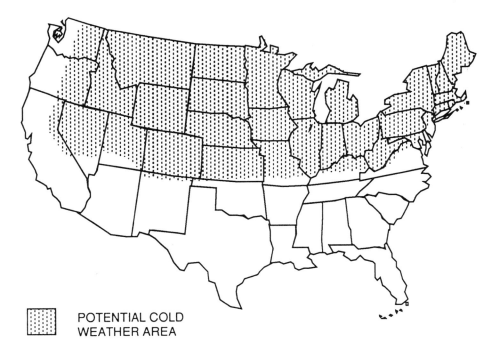

POTENTIAL COLD
WEATHER AREA

Figure 4.12   Cold weather zone map of continental U.S.

access needed to properly operate and maintain process equipment. The following general guidelines minimize the effects of drifting snow:[15]

- Use trees, shrubs, snowfences, or even structures to precipitate snow before it reaches the site proper. Where storms may occur from any direction, provide protection from other quadrants.
- Place major roads parallel to the wind.
- Do not locate roads directly upwind or downwind of large obstructions. Where possible, maintain a 30-m (100-ft) clearance upwind and a 60-m (200-ft) clearance downwind.
- Locate parking lots alongside roads to act as buffer zones. Do not place parking lots among buildings. Expect additional snow accumulation around parked vehicles, and provide ample room for snow storage on the downwind end of a lot away from the roads.
- Locate parking aprons alongside buildings and garages, not upwind or downwind of them.
- Orient surface structures with their longest dimension parallel to the wind. Doors are best located along the sides, toward the upwind end. Doors placed on the downwind end of the structure are subjected to suction forces during drift formation and are rapidly blocked with drifted snow. Those on the upwind face are difficult to seal.
- Orient large garage doors nearly parallel to the wind, even if this results in a building orientation perpendicular to the wind. Adjust this orientation slightly to ensure that the doors are not in the lee of the upwind corner of the building.
- Place structures in rows perpendicular to the wind, with enough space between them to permit effective snow removal. If a second row of structures is necessary, place them directly downwind of those in the first row.
- Locate priority buildings toward the downwind end of the facility where they are afforded protection by less important upwind structures.
- Provide snow-dumping areas to eliminate large piles of snow and windrows in the site area. Piles and windrows act as obstructions and increase the requirements of any future snow removal.

Always consider how the plant will operate in the winter months with freezing conditions and high winds and snow because cold weather affects all types and sizes of treatment processes in plants. The most frequent source of winter problems is settling tanks, particularly scum removal mechanisms. Whenever possible, avoid above-grade tanks, which maximize exposure to cold weather. If other conditions dictate the use of above-grade tanks, insulate tank walls and consider a cover. Also, earth mounding is an effective method of insulating exposed tank walls. Cold weather significantly affects sludge management facilities. Sludge drying beds are extremely susceptible to cold weather problems and are often covered to extend the drying season. Sludge disposal sites may not be available for extended periods of time due to weather conditions. Sludge may require storage for the winter.

In milder climates, temporary, preventive measures include using windscreens, flushing with warm water, and temporary heat. Consider changes in reaction rates in treatment processes. Slower reaction rates may result in increases in tank volumes and numbers of process units.

Hot Weather. Areas that normally experience extended periods of high temperatures and desertlike conditions also require precautions. Wind-blown sand can cause premature wear of mechanical and electrical equipment unless properly protected. Tank walls extended above grade (0.9 to 1.1 m [3 to 3.5 ft], if possible) minimize sand and dust deposition in treatment units. Building entrances can be arranged to minimize the entrance of blowing sand. Exposed window- and stairwells are not recommended.

Elevated ambient temperatures adversely impact electrical and instrumentation systems. Field components such as transmitters, analyzers, and panels need to be shaded from direct sunlight. Electrical controls may require environmentally controlled enclosures. Enclosures containing solid-state components may require an integral air-conditioning unit or be located in an air-conditioned building.

Desert areas and other areas of the United States are subject to severe thunder and lightning storms. Protecting electrical and instrumentation equipment and buildings from lightning is essential. Solid-state equipment is particularly sensitive to surges in the electrical system caused by lightning.

Architecture and Landscaping. Pleasing architecture and attractive landscaping greatly improve the facility's image within the community and provide a pleasant atmosphere for the staff. This will become especially important if the plant is located in a particularly scenic area or in a residential neighborhood. In such cases, additional costs of special architectural treatment and landscaping are justified. In residential areas, design the buildings and site to integrate with the surrounding neighborhood. In older industrial areas, a new, aesthetically pleasing treatment plant can form the nucleus for the revitalization of the surrounding area.

Avoid nonsymmetrical structures because earthquake forces can be amplified in undesirable directions in them, resulting in highly stressed structural members. Space multiple structures far enough apart to avoid contact during a seismic event. Design buildings and grounds to require minimal maintenance; minimize the area of glass to wash and surfaces to paint.

Treated, disinfected wastewater effluent can be used economically for watering vegetation that helps create a rich landscape. Clearly mark outlets and hose bibbs on the treated effluent and reclaimed water pipelines. Hardy, disease-tolerant plants that require minimum maintenance are the best choice for such sites. Scale plant groupings to the site and structures. Situate plantings to screen undesirable views and provide focal interest at appropriate locations. Provide landscaping that does not require extensive weeding or special care. Do not locate trees and shrubs that lose their leaves near open process tanks or outside air intake louvers.

Properly sloped lawn areas and a system of storm sewers prevent standing water. Flat areas of impervious soil may have to be tiled. Embankments normally have a maximum slope of three horizontal to one vertical (3:1) if the surface is to be mowed. Steeper slopes can be used if covered with a ground cover, which, when established, is sufficiently dense to inhibit weed growth. In all cases, facilities that use potable water for irrigation should consider native landscaping to minimize water requirements. A wastewater treatment plant, wherever it is located, can be a source of pride for the public and the employees.

*R*EFERENCES

1. "City of Sedona 1988 Wastewater Facility Plan and Environmental Assessment." Engineering-Science, Inc., Sedona, Ariz. (1988).
2. "Design Criteria for Mechanical, Electric, and Fluid System and Component Reliability." U.S. EPA, Office Water Program Oper., Washington, D.C. (1974).
3. Arora, M.L., *et al.*, "Steps to Create a Good Neighbor." *Public Works,* **121,** 12 (1990).
4. American Concrete Institute, "Environmental Engineering Concrete Structures." ACI-350R (1989).
5. "NEHRP Recommended Provisions for the Development of Seismic Regulations for New Buildings." Fed. Emergency Manage. Agency 96, Earthquake Hazards Reduction Series 18, Part 2 Commentary (1988).
6. "Recommended Provisions for the Development of Seismic Regulations for New Buildings, Parts 1 and 2." Fed. Emergency Manage. Agency 95 and 96, Natl. Earthquake Hazard Response Program (1988).
7. American Concrete Institute, "Building Code Requirements for Reinforced Concrete." ACI-318 (1989).
8. Thomas, T.H., Lockheed Aircraft Corporation, "Nuclear Reactors and Earthquakes." Pub. TID-7024, U.S. At. Energy Comm., Div. Reactors Dev. (1963).
9. Occupational Safety and Health Administration, "Occupational Noise Exposure." 36 FR, 105, 1910.95 (May 29, 1971).

10. "Direct Environmental Factors at Municipal Wastewater Treatment Works." U.S. EPA, Office Water Program Oper., Washington, D.C. (1976).

11. "Design of Wastewater Treatment Plants." Manual of Practice No. 8, Water Pollut. Control Fed., Washington, D.C. (1977).

12. "Estimating Laboratory Needs for Municipal Wastewater Treatment Facilities." U.S. EPA, Office Water Program Oper., Washington, D.C. (1973).

13. U.S. Code of Federal Regulations, 29 CFR 1910.141 (1987).

14. "Prevention of Freezing and Other Cold Weather Problems at Wastewater Treatment Facilities." U.S. Army Corps Eng., Rep. 85-11 (1985).

15. "Arctic and Subarctic Construction: Buildings." Dep. Army and Air Force, TM 5-852-9/AFM, 88-19 (1971).

Chapter 5
Plant Hydraulics and Wastewater Pumping

INTRODUCTION

Once the treatment concept has been selected, it is then necessary to consider how the liquid will flow through the unit processes and what is the most cost-effective method of doing so. Of primary concern is the cost of pumping the liquid versus the unit process structure depth and head loss through the plant.

The terrain of the treatment plant site and the influent sanitary sewer depth govern the need for pumping and the location of pumping facilities. Some systems have a uniform ground elevation reduction, which allows gravity flow to and through the main process train of the plant. In this case, only sludge pumping may be required. Other sites may have rugged, mountainous terrain or very minimal ground elevation differentials. In these cases, it may be necessary to collect the flow and pump through the plant or pump at other locations such as effluent pumping or intermediate pumping between primary and secondary treatment processes. The economics of building deeper plant structures should be considered in relation to building pump stations, which result in higher O&M costs and reduced reliability. This chapter reviews these considerations and addresses, in broad scope, the available pumping and flow-metering methods.

*H*YDRAULIC CONSIDERATIONS

HYDRAULIC PROFILE. The hydraulic profile is a pictorial summary of the hydraulic calculations. A profile is prepared for each of the major flow trains such as the wastewater process train and sludge treatment and disposal. The hydraulic profile is a necessity in establishing water surface elevations and the related optimum structure elevations. The hydraulic profile should present water surface elevations, hydraulic control devices such as control valves and weirs, structure profiles, and a vertical elevation profile. The profile may include ground surface and structure elevations. Figures 5.1 through 5.3 present example hydraulic profiles for several different wastewater processes.

The hydraulic calculations begin at a control point where the plant discharges to a receiving water course or body of water. The plant's water surface profile and pumping requirements are established from this control point. For instance, if the receiving body is a river, stream, or dry watercourse, the controlling elevation is the required flood elevation as calculated by accepted hydrological methods or as obtained from an agency such as the U.S. Army Corps of Engineers. The level of flood protection (for example, the 100-year flood elevation) will be established by the environmental regulatory agency that governs the respective area's plant design. If the receiving body is a storage basin or pond for 100% reuse applications, the controlling elevation is the pond overflow elevation. If the receiving body is a larger body of water such as a lake, wind setup and lake seiches must be considered in addition to the highest lake water surface elevation. If the

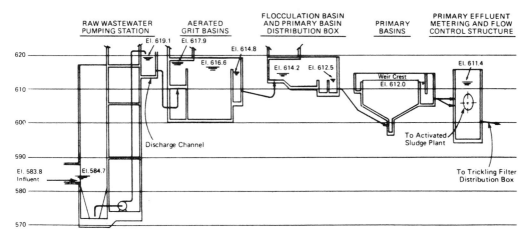

Figure 5.1 Typical hydraulic profile for influent pumping and primary treatment. Water surface elevations in feet above msl represent flow of 160 000 m³/d (42 mgd).

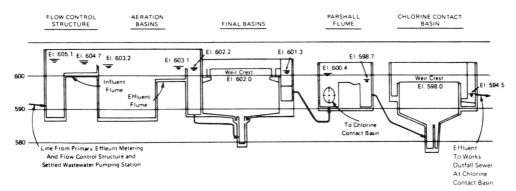

Figure 5.2 Typical hydraulic profile for an activated sludge plant. Water surface elevations in feet above msl represent flow of 160 000 m³/d (42 mgd).

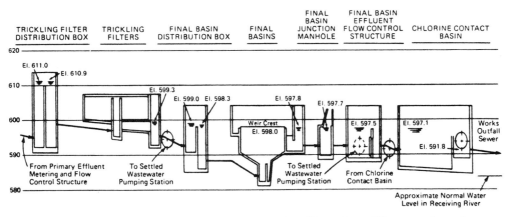

Figure 5.3 Typical hydraulic profile for a trickling filter plant. Water surface elevations represent flow of 85 000 m³/d (22 mgd).

receiving body is an ocean, the controlling elevation is high tide. Where discharging to an ocean, the designer should also consider the difference in density between seawater (1.025) and wastewater (1.00). For every 12.3 m (40 ft) of depth below the high tide, an additional 0.3 m (1 ft) of head must be provided to overcome the density differential.

The hydraulic profile is then calculated through the outfall line and each unit process up to the influent sewer by adding the calculated head loss to the receiving water elevation. If the influent sewer invert elevation and the required water surface elevation at the influent do not correspond, a cost-effectiveness analysis should be performed to determine the appropriate pumping location; that is, influent, effluent, or between processes. Head loss can be calculated using equations such as those discussed in the "Fundamentals of Hydraulic Engineering" section of this chapter.

FLOW RATES. A wastewater flow rate must be established as a basis for the hydraulic design. Peak flow is used for hydraulic design, whereas average flow is used for treatment process design. The problem is defining peak flow for the plant. Some plants may experience wide variations in diurnal and seasonal flow patterns; in this case the peak flow is a peak hourly flow rate. Some plants may experience infiltration/inflow problems and the peak flow may be a peak daily flow rate. In any case, the unit processes must convey the maximum flow unless this flow would cause a hydraulic washout of the treatment plant. In this situation, the designer should consider the use of equalization basins to minimize any negative impact on the treatment process. Unit process redundancy should be considered as discussed later in this chapter. Future plant expansion should be accounted for in sizing piping and channels. Minimum flow should be included in the hydraulic design to avoid low velocities and potential solids deposition.

UNIT PROCESS LIQUID LEVELS. Each unit process must be hydraulically designed to prevent the liquid from overtopping the walls of structures under all conditions. However, most designers will establish the elevations so that freeboard is maintained above the high water elevation. In addition to considering the water surface elevation in relation to structure walls, the designer must consider conditions such as

- Channel surface waves and foaming at higher velocities,
- Hydraulic jumps in channels,
- Aeration and the resulting surface disturbance and foaming,
- Redirection of flow from units temporarily out of service, and
- Water sloshing in regions subject to earthquakes.

Many different design philosophies exist concerning the optimum high water elevation in relation to the structure and the weir elevations. A designer with a conservative approach may establish a hydraulic profile for peak flow based on 8 to 15 cm (3 to 6 in.) of elevation difference between the weir elevation and the receiving weir trough water surface elevation. Another designer may establish a profile with the weirs flooded at peak flow. These philosophies are established based on the available head in the plant and the cost to pump the liquid. If the plant is such that sufficient head exists to allow a conservative design without pumping, the designer should allow as much head between unit processes as possible without interfering with the treatment design. This principle must be balanced against the cost of additional excavation or fill resulting from a steeper hydraulic profile.

UNIT PROCESS REDUNDANCY. The treatment design of a plant should include unit process redundancy. Redundancy generally means that each of the unit processes, depending on the mechanical equipment, can have one or more basins out of service at average flow without adversely affecting effluent quality. This may impact the hydraulic design. With one basin out of service, the flow to the unit process may exceed the peak flow for which it was designed and cause a hydraulic overload. In addition to the average flow conditions, the hydraulic designer must evaluate peak flow conditions with individual unit processes out of service. A hydraulic overload of a unit process under peak flow conditions may be acceptable temporarily or during an emergency but, in any case, the structure walls should not be overtopped.

FLOW DISTRIBUTION. To ensure proper treatment, it is essential to achieve equal flow distribution to each of the basins in a unit process. The optimum solution is to design the distribution to the basins for each unit process with an upstream control structure. Figure 5.4(a) illustrates this type of layout for two clarifiers. The use of symmetry alone will not ensure equal flow distribution. Any differential settling, particularly of basins such as clarifiers with long effluent weirs and minimal heads, will impair flow distribution. It is necessary to design flow equalization weirs in the upstream control structure. In many cases, the plant site or a complex treatment plant may not be suitable for a symmetrical layout of the unit processes. Figure 5.4(b) illustrates this type of plant layout. The use of flow equalization weirs in the upstream control structure, provided free fall over the weirs is maintained, will ensure equal flow distribution.

A common inlet or distribution channel or pipe may also be used to distribute flow to a unit process. However, in such cases the inlet ports or gates to each basin must be designed with significant head loss to ensure good distribution (as described in more detail later in this chapter).

PLANT HEAD LOSS. To demonstrate the variability of head loss in wastewater treatment plants due to different design philosophies, site conditions, or local factors such as power costs, WPCF conducted a design practice survey

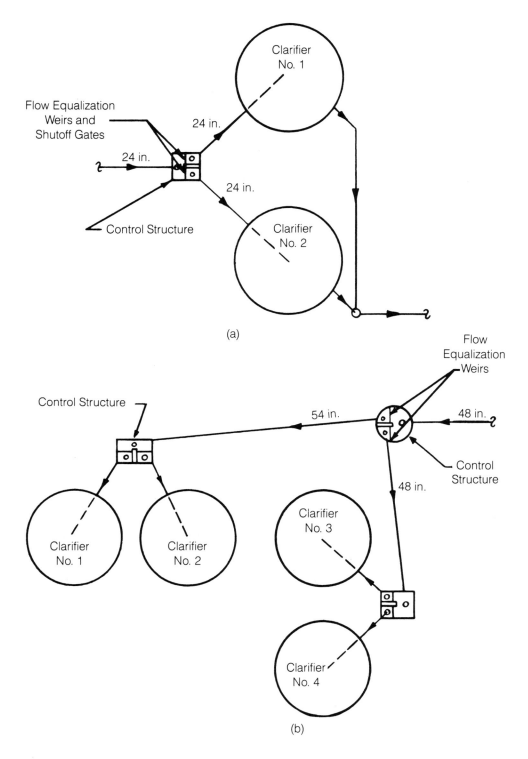

Figure 5.4 **Typical design examples of equal flow distribution among clarifiers: (a) symmetrical layout and (b) nonsymmetrical layout.**

in 1989. Thirteen private consultants in 10 states provided head loss information on 78 treatment plants ranging in peak flow capacities from 1510 to 2 158 000 m³/d (0.4 to 570 mgd), including processes ranging from secondary to sophisticated advanced treatment facilities. Figure 5.5 is a graph of head loss versus flow for secondary treatment facilities with and without disinfection, filtration, and reaeration. Figure 5.6 is a similar graph of tertiary treatment facilities with disinfection and with and without filtration and reaeration. The ranges are extremely variable; however, the graphs do indicate that smaller treatment plants [<114 000 m³/d (30 mgd)] tend to exhibit higher head loss through the plants. This may be a function of the cost of pumping large flows, limited site space for large structures, or other site-specific conditions.

Given the variability of this information, the designer should carefully evaluate the hydraulics of each plant in relation to site and configuration and avoid assumptions as to "typical" head loss through various types of unit processes.

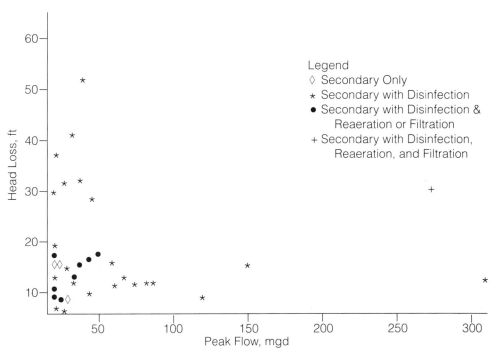

Figure 5.5 Plot of head loss at peak flow through secondary treatment plants
(ft × 0.304 8 = m; mgd × 3785 = m³/d).

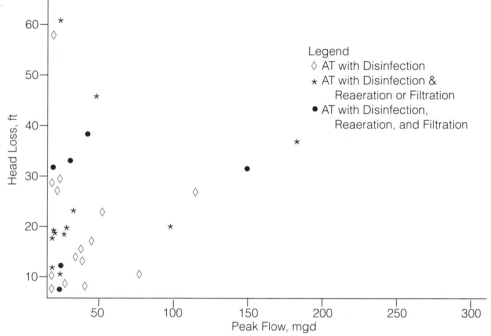

Figure 5.6 Plot of head loss at peak flow through advanced treatment (AT) plants. AT can include nitrification, denitrification, phosphorus removal, or a combination of these (ft × 0.304 8 = m; mgd × 3785 = m³/d).

*F*UNDAMENTALS OF HYDRAULIC ENGINEERING

HYDRAULIC HEAD. Hydraulic head calculations can be classified in three types: static head, velocity head, and friction (including turbulence) head. Static head or lift, considered in pump design, represents the potential energy measured by the difference in elevation between the liquid level in the wet well and the surface water elevation or the highest piping elevation, whichever is greater, to which the pump must deliver the liquid. Velocity head represents the dynamic or kinetic energy resulting from the movement of the liquid. When the flow velocity is reduced or stopped, this energy converts to static head and, in turn, when the flow velocity increases, a portion of the static head converts to velocity head. System head capacity design for pumps is discussed in the section, "Hydraulic Considerations for Pump Selection." Friction (and turbulence) head loss is defined as the loss in energy and the subsequent loss in hydraulic grade required to convey liquid through a pipe, channel, fittings, inlets, outlets, reducers/increasers, or other items that create resistance to flow.

PIPE FLOW. Two types of flow conditions exist in pipes: (1) gravity or free flow, referred to as open channel flow, and (2) pressure flow. Open channel flow between unit processes or within unit processes is a gravity flow condition. Generally, open channel flow in a pipe within a plant will occur only in the influent sanitary sewer or in the effluent outfall line. Pipes between unit processes within the plant are usually under a head or pressure flow condition. Open channels may also be used between unit processes. Head loss in a pipe for an open channel condition can be determined by using Manning's equation[1] for pipe flow. Open channel pipe flow is discussed in detail in WPCF's Manual of Practice No. FD-5, "Gravity Sanitary Sewer Design and Construction."[2]

Pressure pipe flow equations have been developed by Darcy-Weisbach and Hazen-Williams. Manning's equation, which was originally developed for open channel flow, can also apply to pressure flow. The theoretically correct Darcy-Weisbach[1] equation includes an f value related to the Reynolds number and is valid for any fluid in either laminar or turbulent flow. The Darcy-Weisbach equation is

$$h_L = f\,(L/d)(v^2/2g) \tag{1}$$

Where

h_L	=	friction head, m/lin m of pipe or ft/lin ft of pipe;
L	=	length of pipe, m or ft;
f	=	friction factor, dimensionless;
d	=	inside diameter of pipe, m or ft;
v	=	average velocity, m/s or ft/sec; and
g	=	acceleration due to gravity, 9.8 m/s^2 or 32.2 ft/sec^2.

In this equation, f varies with Reynolds number in the laminar flow regime. For transition flows (Reynolds numbers between 2000 and 3000), the flow regime is unstable; therefore, f is indeterminate. In the turbulent range (Reynolds numbers greater than 2000 to 3000), f varies with the surface roughness of the pipe as well as with viscosity and density. Values of f can be determined using the Moody diagram found in most hydraulic texts and handbooks.

The Hazen-Williams equation[2] is

$$h_L = [V/0.85C(R)^{0.63}]^{1.85} \times L \text{ (SI units)} \tag{2}$$

$$h_L = [V/1.318C(R)^{0.63}]^{1.85} \times L \text{ (English units)}$$

Where

h_L	=	friction head, m or ft;
V	=	mean velocity of flow, m/s or ft/sec;

R	=	hydraulic radius, m or ft ($d/4$ for circular pipes where $d =$ diameter);
C	=	surface roughness coefficient, dimensionless; and
L	=	pipe length, m or ft.

C varies with the pipe's type, age, and condition. Older ductile or cast iron pipe can have a C value of 100; new PVC pipe can have a C value of 140. When using new pipe and determining pump heads, the designer should consider two C values, one for the new condition of the pipe and the other for the design life of the pipe. Newer pipe will have less friction head and, if only the design life is considered, a pump might deliver more liquid at start-up than is anticipated. This could create problems for the pumping and the unit process receiving the pump's discharge.

The Hazen–Williams equation, developed empirically, has limitations. For most conditions encountered in treatment plants, this equation is satisfactory. However, for long piping systems, the error in head loss calculation can be significant. Friction factors depend on the Reynolds number, which is a function of the viscosity and density of the fluid. The C value does not account for the effect of the Reynolds number.

The accuracy of both pipe flow equations depends on selecting the correct surface roughness coefficient. The condition of the interior surface of the pipe varies by pipe material and over time with use, age, and the effects of corrosion. WPCF Manual of Practice No. FD-5[2] and the Hydraulic Institute Data Book[3] provide a listing of suggested roughness coefficient values.

Other equations exist for pressure pipe flow. However, the two equations discussed above are probably the most widely used. Books containing hydraulic tables, nomographs, and Moody diagrams based on these equations, programmable calculators, and computer spreadsheets can substantially reduce the time required to perform the hydraulic calculations.[1,4-6]

It should be noted that in both equations it is necessary to convert the energy losses in valves, fittings, entrances, exits, and sudden pipe enlargements or constrictions to equivalent lengths of pipe or to calculate the head losses in each. Since fitting losses are mostly due to turbulence, the equivalent length approach will be incorrect if the fluid viscosity differs from that for water. Head loss for pipe fittings such as elbows and bends may be calculated with the following equation:

$$h_L = K(v^2/2g) \tag{3}$$

Where

h_L	=	head loss, m or ft;
K	=	constant for type of fitting, dimensionless;
v	=	velocity in pipe, m/s or ft/sec; and
g	=	acceleration due to gravity, 9.8 m/s^2 or 32.2 ft/sec^2.

K values for commonly used fittings can be found in hydraulics handbooks. The K value accounts for head loss through the fitting. K is an empirical value; therefore, it accounts for the total head loss, including turbulence and friction losses. Turbulence composes the majority of head loss due to fittings.

OPEN CHANNEL FLOW. Open channels are used to convey liquid between unit processes or, as weir troughs and distribution devices, within unit processes. Manning's equation[1] is the most widely used formula for calculating head loss in open channels. However, its n value shares the same weakness as the Hazen–Williams C value. The Manning equation is

$$h_L = (Vn/R^{0.67})^2 \times L \text{ (SI units)} \tag{4}$$

$$h_L = (Vn/1.49R^{0.67})^2 \times L \text{ (English units)}$$

Where

h_L	=	head loss, m or ft;
R	=	hydraulic radius (cross-sectional area divided by wetted perimeter), m or ft;
n	=	roughness coefficient, dimensionless;
V	=	velocity, m/s or ft/sec; and
L	=	channel length, m or ft.

The n values depend on the roughness of the conveyance channel and range from that of glass (0.009) to more than 0.06 for natural channels. The most commonly used materials for channels in treatment plants are concrete and steel. New concrete and steel can have n values as low as 0.010; as the materials deteriorate, n values in the range of 0.013 to 0.016 can be expected. WPCF Manual of Practice No. FD-5 provides a listing of widely accepted n values.[2]

Minimum Velocity. When using any of the above equations previously described, the designer must consider maintaining the appropriate velocity in the pipe or channel. Generally, a minimum velocity of 0.62 m/s (2 ft/sec) for raw wastewater is required to prevent solids deposition in channels and pipelines. Typical flushing velocities for raw wastewater are 1.54 to 1.85 m/s (5 to 6 ft/sec). In some cases, the minimum velocity cannot be maintained due to specific process requirements. For example, the entrance velocity to a clarifier is reduced to 0.31 m/s (1 ft/sec) to improve the settling characteristics and prevent disturbance of the sludge blanket. The velocity required in aerated channels to prevent solids deposition is approximately double that for regular channels.

The design practice survey included questions regarding preferred design velocity, specific gravity, and viscosity values for raw wastewater and other internal plant flows, such as trickling filter recycle. Thirteen private consultants in 10 states responded to these questions. The preferred design velocities range from 0.23 to 1.83 m/s (0.75 to 6.0 ft/sec), with 64% preferring 0.62 to 0.93 m/s (2 to 3 ft/sec) and 18% preferring 0.31 to 0.59 m/s (1 to 1.9 ft/sec). All of the respondents preferred the same specific gravity and viscosity design values for wastewater as those for water.

In addition to head loss in pipes, fittings, and valves, head loss will also occur due to in-line equipment, flow meters and flumes, and weirs. In-line equipment and flow meter head loss can be obtained from equipment manufacturers. Flumes, such as Parshall or Palmer-Bowlus flumes, are used for open channel flow measurement. They are discussed in the flow measurement section of this chapter. Head loss from flow through weirs, open channels, ports, and headers is described in the following sections.

Weirs. Weirs are sometimes used for flow measurement in plants but more commonly serve as control devices to maintain a certain water surface elevation in a unit process. Weirs are classified in accordance with the shape of the notch. The weir types include rectangular, V-notch, trapezoidal, proportional, and parabolic. The upper edge of the weir plate or weir is the crest of the weir. The depth of the water over the crest is the head. The head, or depth over the weir, is measured as the difference between the upstream water surface elevation above the point where weir drawdown occurs and the weir crest elevation (see Figure 5.7).

Weirs are either sharp crested or broad crested as shown in Figure 5.7. The most commonly used weirs in treatment plants are rectangular and V-notch weirs. Both are generally sharp crested with weir plates bolted to a wall or trough so that they can be leveled independently of the structure. V-notch weirs are more accurate than rectangular weirs.

V-notch weir angles range from 22.5 to 120 deg. The 90-deg V-notch is the most commonly used notched weir. Under free flow conditions as shown in Figure 5.8(a), the head over a sharp-crested V-notch weir can be calculated with the following equation:

$$H = (Q/C \tan \phi/2)^{0.4} \qquad (5)$$

Where

Q =   flow, m³/s or cu ft/sec;
ϕ =   angle of the notch, deg;
H =   head over the crest, m or ft; and
C =   weir coefficient, 1.38 metric (2.5 SI) for a 90-deg weir with $H > 0.12$ m (0.4 ft).

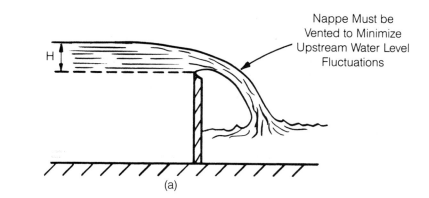

(a)

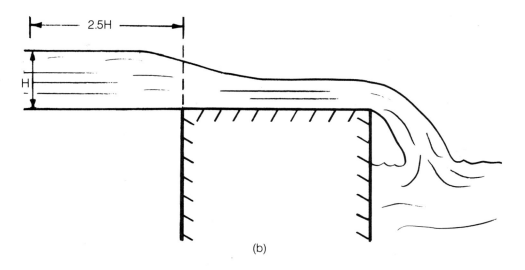

(b)

Figure 5.7   Weir sections: (a) sharp-crested weir and (b) broad-crested weir.

Head over a rectangular weir under free flow conditions can be calculated by the following equation:

$$H = (Q/C_w L)^{0.67} \qquad (6)$$

Where

Q =   flow, m^3/s or cu ft/sec;
L =   length of weir, m or ft;
H =   head over the crest, m or ft;

C_w = coefficient that accounts for the approach velocity in terms of the ratio of the weir plate depth to the head over the crest. The values range from 1.79 to 2.31 (SI) [3.24 to 4.18 (English)]. The most commonly used value, 1.82 (3.3), accurately represents deep water upstream with an extremely low approach velocity toward a sharp-crested weir.

Often, a rectangular weir has end contractions (that is, the ends of the weir project inward from the sides of the channel). These can be accounted for by subtracting $0.1H$ from L for each end contraction. The equation for a rectangular weir with a contraction on each end follows:

$$H = [Q/C_w(L - 0.2H)]^{0.67} \qquad (7)$$

Equations 5, 6, and 7 apply under free flow conditions. When the downstream water elevation rises above the weir crest, the weir becomes submerged as shown in Figure 5.8(b). Submerged weirs are not used for flow measurement because the free flow weir equations do not directly apply to submerged conditions, thus produce sizeable errors. H can be calculated from the free flow equations; however, the discharge flow must be corrected by use of curves developed experimentally.[1]

Other Open Channel Head Loss Equations. Three other basic formulas for calculating head loss in open channels are

- The formula for a channel water surface elevation from a side overflow weir, such as a clarifier weir effluent launder;
- The formula for head loss resulting from a partially open gate or port; and
- The formula for head loss resulting from a bar screen.

The side overflow formula, described in detail in "Open-Channel Hydraulics,"[7] within the section on spatially varied flow, is a differential equation for flow with increasing discharge. Most designers prefer to use a computer program to solve this equation.

Ports or gates are sometimes used to evenly distribute flow to unit processes. The sensitivity of the process dictates the need for even flow distribution to each parallel basin or across a basin. Head loss is created by these devices. The equation below represents the head loss relationship along a channel where the outflow is continuous or nearly so along the length of the channel:

$$h_p = \Delta h/(1 - m^2) \qquad (8)$$

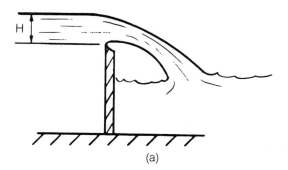

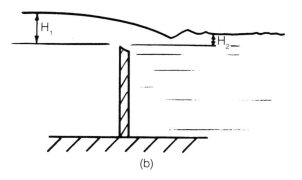

Figure 5.8   Weir flow conditions: (a) free flow and (b) submerged flow.

Where

h_p =   head loss through port, m or ft;

Δh =   hydraulic grade differential along the channel to the ports, m or ft; and

m =   ratio of the flow in the first port to that through the end port.

Equation 8 is used to determine the minimum head loss that must be allowed in the design of an inlet gate or port to ensure uniform distribution to all basins from a common inlet or distribution channel. The designer should recognize that to keep the flow to each basin within 5% of that to another basin, the head loss through the gate should be about 10 times the hydraulic grade line differential over the entire length of the distribution channel. To determine the actual head loss through a gate or port, the following equation applies:

$$H = Q^2/(CA)^2 2g \qquad (9)$$

Where

H = head loss through the gate, m or ft;

Q = flow, m³/s or cu ft/sec;

C = gate or orifice coefficient, dimensionless;

A = area of gate or port opening, m² or sq ft; and

g = acceleration due to gravity, 9.8 m/s² or 32.2 ft/sec².

Values of C for calculating head loss through ports under varying conditions may be found in the textbook, "Handbook of Applied Hydraulics."[5]

The hydraulic gradeline differential (Δh) is influenced by velocity head and head loss along the upstream header pipe or channel as shown by the following equation:

$$\Delta h = V_O^2/2g - h_L \qquad (10)$$

Where

Δh = hydraulic grade differential, m or ft;

V_O = header inlet velocity, m/s or ft/sec;

g = acceleration due to gravity, 9.8 m/s² or 32.2 ft/sec²; and

h_L = header head loss, m or ft.

When the outlet flow from a pressurized header approximates a uniform continuous outflow along the length of the header, the head loss along the header (h_L) can be estimated by calculating the head loss as if the inlet flow to the header were to be conveyed along the entire length of header and dividing the result by three. This estimation procedure is valid for m values greater than 0.9 in equation 8.

The influent screening system creates head loss due to the friction and turbulence of the wastewater flowing through the screen and screen blinding by debris. The equations and procedures for calculating head loss through bar screens are fully described in Chapter 9 as part of the procedures for designing bar screens.

*U*NIT PROCESS HYDRAULICS

The hydraulic calculations for each unit process may require the use of a combination of the equations discussed in the "Fundamentals of Hydraulic Engineering" section of this chapter. For instance, calculation of the difference in the water surface elevation between the effluent manhole and the clarifier shown in Figure 5.9 would require the use of the following equations:

- Pressure flow equation for calculating the head loss in the pipe between the manhole and the clarifier;
- Side overflow weir equation for calculating the highest water elevation in the clarifier effluent launder; and
- V-notch weir equation for calculating the head over the weir crest.

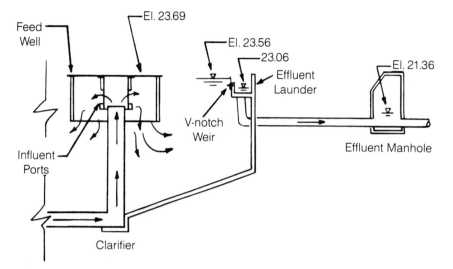

Figure 5.9   Unit process hydraulics for typical clarifier.

The hydraulic designer must understand how the liquid flows through each unit process and what water depths are required for the process. Within each unit process, devices are used to distribute flow, maintain a certain water depth, and control the flow. Some typical devices include shutoff gates, weir gates, valves, ports, weirs, baffles, orifices, launders, and underdrains. Each of these devices imposes a head upon the system and must be considered in the hydraulic calculations.[1]

Unit processes also impose heads on the system. Two examples of such processes are trickling filters and effluent gravity filtration. For trickling filters to operate, sufficient water surface elevation must exist to push the water through the distributor and keep it rotating over the media. For effluent gravity filters, sufficient water surface elevation must exist over the filter media to push the liquid through the media and the underdrains. Some processes can impose varying heads. One such process is the influent screening facility. The wastewater flowing through the bars of the screen creates head loss. Head loss increases if the screen becomes blinded or partially blinded by debris. As mechanical rakes clear the screen, head loss decreases. Typical head loss ranges for bar screens are 0.15 to 0.8 m (0.5 to 2.5 ft) and for comminutors and grinders are 0.05 to 0.3 m (0.17 to 1 ft). Chapter 9 covers design of screening devices.

Each unit process, its respective flow devices, and interconnecting piping must be carefully analyzed and the head loss added to the appropriate discharge elevation. In this manner, the water surface elevations can be calculated and the structure elevations and pumping needs can be established. This information is then summarized and presented in the form of the hydraulic profile.

In establishing the unit process hydraulic calculations, the designer must consider the need to control and equally distribute the flow into multiple tanks or within an individual tank. Distribution boxes, channels, and header pipes are used for this purpose. Although ports, gates, or valves can provide equal distribution, they create head loss, which raises upstream water surface elevations. "Treatment Plant Hydraulics for Environmental Engineers"[6] presents additional detailed hydraulic considerations.

WASTEWATER PUMPS

PUMP CLASSIFICATION. The design of wastewater treatment facilities confronts the engineer with a wide range of pumping applications, including raw wastewaters; treated wastewaters; mixtures of domestic and industrial wastes; raw and digested sludges; thickened sludges; grit; scum containing a mixture of grease, floating solids, and trash; return and waste activated sludges; chemical solutions; flushing water; spray water and pump seal water; tank drainage; and sump pump water. Figure 5.10 presents a summary of the major pump classifications commonly used in wastewater plants. Some typical wastewater treatment applications are shown in Table 5.1. References such as "Pump Application Engineering,"[8] "Hydraulic Institute Standards,"[9] "Pump Handbook,"[10] "Pump Selection: A Consulting Engineer's Manual,"[11] "Hydraulic Institute Engineering Data Book,"[3] "Centrifugal Pumps,"[12] and others will help in matching the demands of the design application with the characteristics of a particular pump.

CENTRIFUGAL PUMPS. The centrifugal pumps are classified into radial flow, mixed flow, and axial flow types according to the direction of flow in reference to the axis of rotation. These types can be further subdivided into single or double suction; and these subdivisions can be divided again into self-priming, single stage, and multiple stage. The mechanical construction of impellers provides a still further classification into open, semiopen, and closed types (Figure 5.11).[8] The axis of rotation of the pump shaft determines whether the pump is a horizontal or vertical unit. Vertical shaft pumps usually have submerged suctions and can be either dry pit or wet pit types. Figures 5.12 through 5.14 show typical pump sections of horizontal dry pit nonclog, vertical dry pit nonclog, and submersible types that are typically

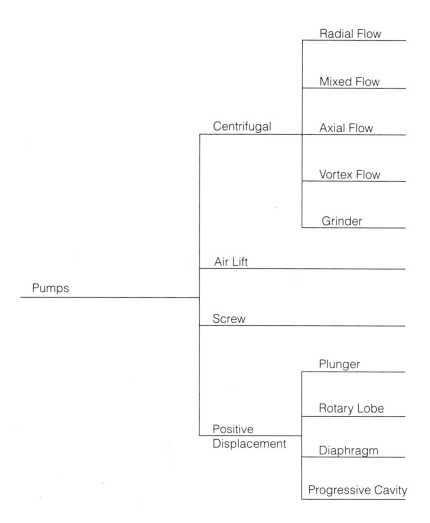

Figure 5.10 Pump classifications for wastewater treatment works.

used in wastewater treatment applications. The names recommended by the Hydraulic Institute for each of the pump parts are also presented in each of these figures. Table 5.2 compares operating characteristics of centrifugal pump types.

Two-port Nonclog Radial Type. These pumps derive their "nonclog" characteristics from the type of impeller used in their construction. (The term nonclog is a misnomer because material contained in most wastewater applications can clog any pump.) These pumps include types that will operate in wet and dry well installations. The impellers are well rounded and free from sharp corners and projections likely to catch and hold rags and stringy material. Although single-vane types are available, these pumps are most frequently of two-vane or two-port construction and of either the enclosed or semiopen type.

Table 5.1   Typical wastewater applications of some pump types.

Pump type	Usual maximum suction lift, m[a]	Maximum solids diameter generally handled, cm[b]	Some typical applications
Centrifugal, 2-port nonclog	4.6	>7.6	Raw wastewater. Primary sludge. Secondary sludge. Effluent.
Centrifugal, clear water	4.6	< 2.5	Flushing water for general cleaning. Spray water to disperse foam. Pre- and postchlorination. Solution water for chemicals. Seal water for pump stuffing boxes.
Centrifugal, vortex flow	4.6	>7.6	Sludge recirculation. Grit.
Mixed flow	4.6	>7.6	Raw wastewater. Sludge recirculation. Trickling filter recirculation.
Axial flow	4.6	<2.5	Effluent.
Air lift	3.0	>7.6	Raw wastewater. Return sludge.
Centrifugal grinder	3.0	Not applicable	Raw wastewater. Slurries.
Archimedean screw	0	>15	Raw wastewater. Return activated sludge.
Positive displacement, plunger, rotary, and diaphragm	6.7	Consult manufacturer	Primary sludge. Thickened sludge. Digested sludge. Slurries. Chemical feed.

[a] m × 3.281 = ft.
[b] cm × 2.54 = in.

Figure 5.11 Types of pump impellers: (a) open, (b) semiopen, (c) enclosed end suction, (d) enclosed double suction, (e) axial flow, and (f) mixed flow.[8]

The enclosed impeller consists of two or more vanes between two plates or shrouds. The lower, or outboard, shroud has a center opening into which the liquid to be pumped enters. The upper, or inboard, shroud contains the impeller hub, keyed or threaded to the pump shaft. The lower shroud has an extended lip around the circumference of the center opening that extends into the suction opening of the pump casing or volute. Both the impeller and the

---

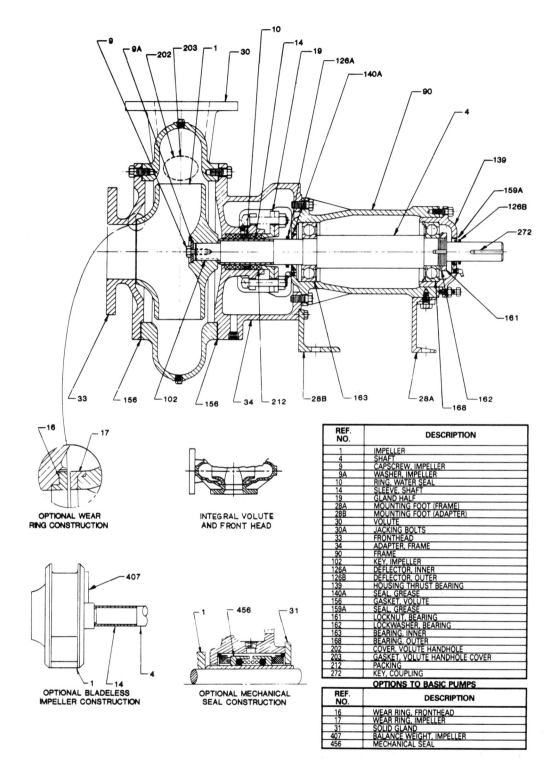

OPTIONAL WEAR
RING CONSTRUCTION

INTEGRAL VOLUTE
AND FRONT HEAD

OPTIONAL BLADELESS
IMPELLER CONSTRUCTION

OPTIONAL MECHANICAL
SEAL CONSTRUCTION

REF. NO.	DESCRIPTION
1	IMPELLER
4	SHAFT
9	CAPSCREW, IMPELLER
9A	WASHER, IMPELLER
10	RING, WATER SEAL
14	SLEEVE, SHAFT
19	GLAND HALF
28A	MOUNTING FOOT (FRAME)
28B	MOUNTING FOOT (ADAPTER)
30	VOLUTE
30A	JACKING BOLTS
33	FRONTHEAD
34	ADAPTER, FRAME
90	FRAME
102	KEY, IMPELLER
126A	DEFLECTOR, INNER
126B	DEFLECTOR, OUTER
139	HOUSING THRUST BEARING
140A	SEAL, GREASE
156	GASKET, VOLUTE
159A	SEAL, GREASE
161	LOCKNUT, BEARING
162	LOCKWASHER, BEARING
163	BEARING, INNER
168	BEARING, OUTER
202	COVER, VOLUTE HANDHOLE
203	GASKET, VOLUTE HANDHOLE COVER
212	PACKING
272	KEY, COUPLING

OPTIONS TO BASIC PUMPS	
REF. NO.	DESCRIPTION
16	WEAR RING, FRONTHEAD
17	WEAR RING, IMPELLER
31	SOLID GLAND
407	BALANCE WEIGHT, IMPELLER
456	MECHANICAL SEAL

Figure 5.12   Horizontal dry pit nonclog pump.

REF. NO.	DESCRIPTION
1	IMPELLER
4	SHAFT
9	CAPSCREW, IMPELLER
9A	WASHER, IMPELLER
10	RING, WATER SEAL
14	SLEEVE SHAFT
15	BASE
19	GLAND HALF
30	VOLUTE
30A	JACKING BOLT
33	FRONTHEAD
34	ADAPTER FRAME
44	SUCTION ELBOW
90	FRAME
102	KEY IMPELLER
126A	DEFLECTOR, INNER
126B	DEFLECTOR, OUTER
139	HOUSING THRUST BEARING
140A	SEAL, INNER GREASE
154	GASKET, ELBOW
156	GASKET, VOLUTE
159A	SEAL, OUTER CASE
161	LOCKNUT BEARING
162	LOCKWASHER, BEARING
163	BEARING, INNER
168	BEARING, OUTER
202	COVER, VOLUTE HANDHOLE
203	GASKET, VOLUTE HANDHOLE COVER
206A	RETAINER, INNER GREASE
206B	RETAINER, OUTER GREASE
212	PACKING
272	KEY COUPLING
290	COVER, HANDHOLE
291	GASKET, HANDHOLE
464	SUCTION ELBOW (BASE COMBINATION)

OPTIONS TO BASIC PUMPS	
REF. NO.	DESCRIPTION
16	WEAR RING, FRONTHEAD
17	WEAR RING, IMPELLER
31	SOLID GLAND
407	BALANCE WEIGHT, IMPELLER
456	MECHANICAL SEAL

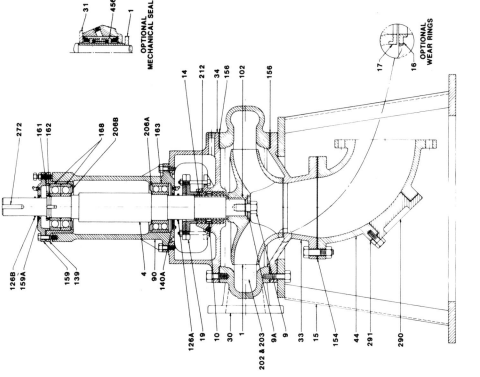

Figure 5.13   Vertical dry pit nonclog pump.

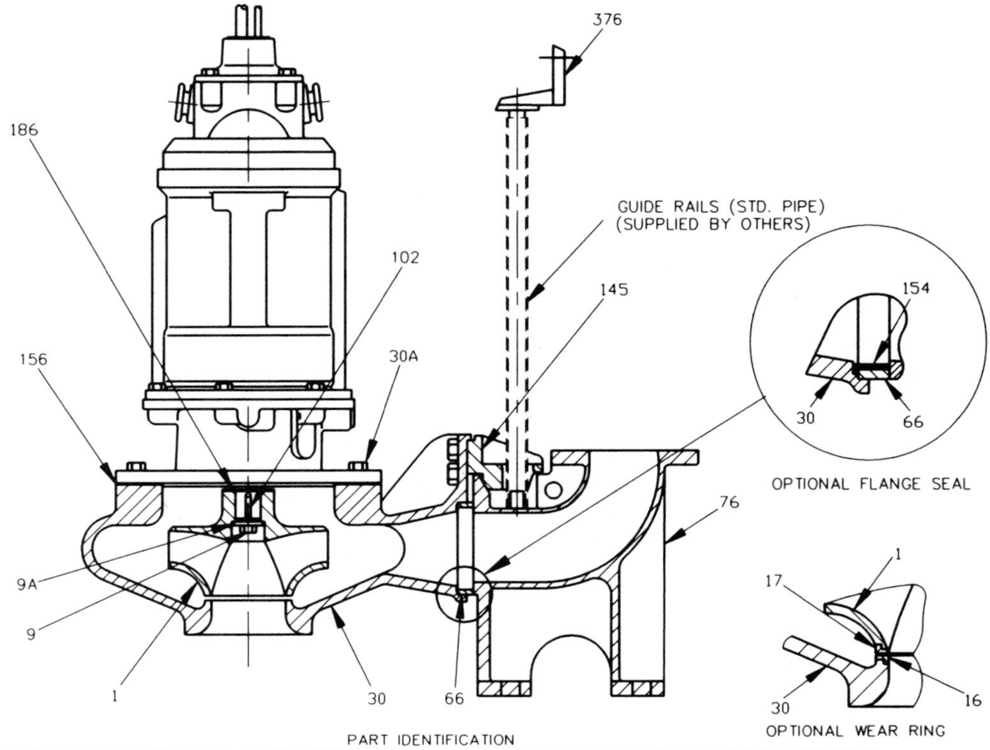

GUIDE RAILS (STD. PIPE)
(SUPPLIED BY OTHERS)

OPTIONAL FLANGE SEAL

OPTIONAL WEAR RING

PART IDENTIFICATION

NO	DESCRIPTION	MATERIAL	SPECIFICATION
1	IMPELLER	CAST IRON	A48 CL.30
9	IMPELLER CAPSCREW	STEEL	SAE BOLT STEEL GR.8
9A	IMPELLER WASHER	STEEL	A108 GRADE 12L14
30	VOLUTE	CAST IRON	A48 CL.30
30A	CAPSCREW, VOLUTE	STEEL	SAE BOLT STEEL GR.2
30B	GUIDE PINS (NOT SHOWN)	BRASS	COMMERCIAL
66	FLANGE RING	BRASS	B505 AL932
76	BASE ELBOW	CAST IRON	A48 CL.30
102	IMPELLER KEY	STEEL	A108 GR.1018
145	GUIDE BRACKET	BRASS	B584 AL 836
156	VOLUTE GASKET	TAG BOARD	D1170-G3111
186	IMPELLER SHIM	STAINLESS STEEL	A582 TYPE 303
376	RETAINER BRACKET	DUCTILE IRON	A532 65-45-12

OPTIONAL PARTS LIST

1	IMPELLER	BRONZE	B584 AL836
16	CASING WEARING RING	STAINLESS STEEL	A743-CA15 300-350 BHN.
17	IMPELLER WEARING RING	STAINLESS STEEL	A743-CA15 300-350 BHN.
30A	CAPSCREW, VOLUTE	STAINLESS STEEL	AISI TYPE 316
154	FLANGE SEAL	RUBBER	BUNA-N

NOTE: 1. ALL MATERIAL SPECIFICATIONS ARE ASTM UNLESS OTHERWISE NOTED
AND ARE FOR DESCRIPTION OF CHEMISTRY ONLY.

Figure 5.14 Submersible nonclog pump.

suction cover on an enclosed impeller pump can normally be furnished with renewable wearing rings.

The clearance, or spacing, between the extension of the lower shroud and the pump suction cover, relates directly to the pumping efficiency. The liquid recirculating from the higher pressure of the volute into the lower pressure area of the suction "eye" reduces the efficiency of the pump. The clearances

Table 5.2   Centrifugal pump characteristics.

Type	Capacity range, L/s[a]	Discharge head range, m[b]	Approximate optimum efficiency range, %
Nonclog radial	3–1300	8–60	60–85
Clean water radial	3–1300	8–150	80–90
Vortex	3–320	1–64	40–65
Mixed flow	63–5000	3–18	80–88
Axial flow	32–6300	0.3–12	75–85
Grinder	3.2–32	1–72	40–50

[a] L/s × 15.85 = gpm.
[b] m × 3.281 = ft.

normally allowed between the impeller and suction cover range from 0.250 to 0.640 mm (0.010 to 0.025 in.) depending on the size of the pump, the application, and other factors. After long periods of use or abnormal wear caused by abrasive materials, check the clearances, replace the wearing rings if necessary, and make other corrections to return the pump to its original efficiency.

The semiopen impeller, constructed with the vanes attached to an upper or inboard shroud, omits the lower shroud. The vanes are designed so that the contours of the open portion of the impeller follow the contours of the suction plate of the volute. The suction plate, therefore, serves as the lower shroud of the impeller. The entire lower edge of each impeller vane maintains a running clearance that prevents the higher pressure liquid from recirculating to the suction eye area of the impeller. The pump efficiency depends on maintenance of this constant clearance between the contour of the impeller and the suction cover.

Submersible pumps, typically used for wet well installations, generally are constructed as pumps vertically close-coupled to a submersible motor. The motor bearings serve as both thrust and radial bearings for the pump. In recent years, submersible pumps have gained much wider acceptance in both domestic and industrial wastewater treatment applications. They are available in motor sizes ranging from 2 to 450 kW (3 to 600 hp). The capital cost of submersible pump stations is generally less than that of installations with pumps in a separate dry well.[13] However, submersible pumps must be entirely removed from the wet well for servicing. Frequently, because of the special design of the motor, servicing can be accomplished only at the motor manufacturer's factory-certified service center.

Dry well installations may include either horizontal or vertical pumps. The horizontal pump and its flexible-coupled or close-coupled driver are normally mounted on a common base. The pump and the driver of large units may be mounted on adjacent bases. The pump construction positions the shaft horizontally with the thrust bearing mounted adjacent to the driver end of the shaft and the radial bearing mounted closest to the impeller.

The stuffing box, normally placed between the impeller and the radial bearing, prevents the flow of liquid along the shaft. The stuffing box houses four to six rings of packing and the water seal ring. For pumps with mechanical seals, the mechanical seal substitutes for the packing and the water seal ring. The portion of the shaft extending through the stuffing box is normally equipped with a replaceable sleeve, eliminating the need for replacing the entire shaft if it is scored by the packing. The stuffing box arrangement provides a lubricant access passage admitting grease or freshwater to lubricate the water seal ring. With mechanical seals, water for the seal is essential.

Vertical pumps normally have one of two configurations: one with the motor mounted on a pedestal directly connected to the pump (vertical pedestal mounted), and the other with the motor mounted at a level above the top of the wet well to protect the motor from flood damage. The motor is connected to the pump through one or more sections of extended shaft.

The horizontal pump offers the advantage of ease of maintenance and removal. The vertical pump, however, requires much less floor space and, with its motor mounted on a floor above, affords additional protection of the motor. Also, by merely disconnecting the pump end of the extended shaft, the pump can be serviced without removing the motor. When comparing the vertical pump with the motor mounted directly above (vertical pedestal mounted), the additional cost of the extended shaft must be balanced against the added protection for the motor and the additional ease of servicing. Extended vertical shafting should be equipped with safety cages and grease shields at universal joints. Extended shafts are designed to avoid rotational speeds that excite resonance frequencies of the shaft.

Two-port Radial Type for Clean Water. This centrifugal pump with a flexible coupling can be obtained as end-suction, single-suction, or double-suction units. They also include vertical split case or horizontal split case configurations, depending on the type of unit. These pumps also include single-stage units for lower heads and multistage units for higher heads.

While the double-suction horizontally split case pump generally costs more than the other flexible-coupled clean water centrifugals, the double-suction unit offers important advantages. Normally, the entire rotating assembly can be removed without disturbing either the piping or the driver to ease dismantling and replacement of worn bearings, impellers, or wearing rings. In addition, this type usually requires the least maintenance and outlasts the others.

Vortex Flow Type. The vortex, or torque flow, pump has an impeller completely recessed from the volute area. The multivaned, semiopen impeller

(Figure 5.15), recessed in the discharge casing and completely removed from the volute area, generates a whirling or circular motion that extends into the suction opening. The vortex created directs the flow of both the fluid and entrained solids entering the pump into the volute area, where centrifugal force pushes the flow out. The fluid and entrained solids do not enter the impeller but pass from the suction line through the volute into the discharge line. Thus, with a proper volute design, the pump will handle any solids that can pass the suction nozzle. Because of these design characteristics, vortex flow pumps are employed for sludges, grit, and other liquids containing large solids and abrasive grit. For such applications, vortex flow pumps are usually slow speed, belt driven, with "ni-hard" volutes and impellers.

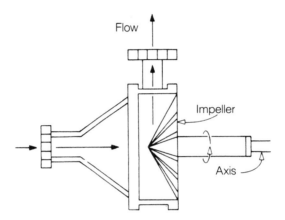

Figure 5.15   **Vortex or torque flow pumps have multiple vanes cut into the impeller.**

Available centrifugal vortex pumps will serve either wet well or dry well installations. Their construction is similar to that of the centrifugal radial non-clog pump. In general, vortex pump efficiencies are lower than those of centrifugal nonclog pumps.

Axial Flow Type. Axial or propeller flow pumps, commonly used in wet well installations, move liquid by the propelling or lifting action of the impeller vanes (Figure 5.11). This pump is primarily used for clean water or treated effluent, and should not be used for raw wastewater or sludge applications since large solids or stringy material can cause plugging.

The propeller is mounted near the bottom of the vertical, extended shaft and is enclosed in a housing or bowl that is totally submerged in the liquid to be pumped. A small housing inside the discharge pipe encloses the shaft, which is supported by sleeve bearings. The shaft may be driven by a vertical hollow shaft motor, vertical solid shaft motor, or a right-angle gear drive that can be powered by a horizontal electric motor or engine.

The propeller lifts and imparts velocity to the fluid, which flows vertically through the pump column and generally discharges at a right angle at the pump discharge flange. An adequate means of lubrication for the bearings in the line shaft must be ensured as well as the bearing normally provided in the suction bell.

Mixed Flow Type. The mixed flow impeller is an intermediate design between the radial and the axial flow propeller, with flow components of both. It is suitable for handling wastewater or stormwater because it is normally designed with wide unobstructed passages. For a comparable pump discharge size, the nonclog or radial pump will pass larger solids; however, the mixed flow pump is available in large discharge sizes that are suitable for many wastewater and stormwater applications.

The centrifugal mixed flow pump, available only for operation in a dry well installation, has both horizontal and vertical configurations. The construction is similar to that of the centrifugal radial nonclog pump. Mixed flow pumps, primarily available as packed pumps, usually have renewable wearing rings for impellers and casings to maintain the clearance necessary for efficient operation. Mechanical seals are available, but high costs usually prevent their use.

Grinder Pumps. Horizontal or vertical centrifugal grinder pumps are used primarily to handle raw domestic or industrial wastewaters containing large solids and break up digester scum. Their use is now increasing for pumping septage or holding tank waste. Grinder pumps may also be used to reduce screenings and primary sludge underflow solids to particle sizes of less than 10 mm (0.4 in.). Discharge heads are limited, and pumping efficiencies are low. Because of the abrasive materials that are pumped, these units require frequent maintenance. Cutters normally are rebuilt with hardened materials.

AIR LIFT PUMPS. Air lift pumps are commonly used for return and waste activated sludge applications in the smaller package treatment plants and for lifting grit from aerated grit chambers. The pump consists of a vertical pipe with its lower end submerged (Figure 5.16). Compressed air admitted to the bottom of the pipe reduces the average density of the mixture relative to the liquid outside of the pipe. Thus, at the proper air:liquid ratio, the liquid rises to the desired elevation. Air lift pumps can be used where a source of air pressure is available. They are relatively inexpensive to install, maintain, and operate. Although simple, air lift pumps offer the operator minimal turndown capacity. Generally, when the flow is reduced by throttling the air supply, pump operation becomes erratic, velocities decrease, and the pump may clog.

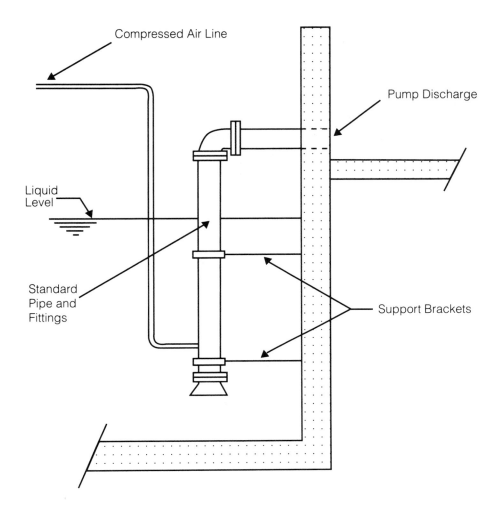

Figure 5.16 Air lift pump.

PNEUMATIC EJECTORS. Pneumatic ejectors can be used for pumping raw wastewater and scum, including solids that can enter the receiving container (Figure 5.17). Capacities of the units range from 1.9 to 9.5 L/s (30 to 150 gpm) at heads up to 18 m (60 ft).

A pneumatic ejector consists of a receiving container, inlet and outlet check valves, air supply, and liquid-level detector. When the wastewater or scum reaches a preset level, air forced into the container ejects the wastewater. Following the discharge cycle, the air supply is cut off and wastewater flows through the inlet into the receiver.

ARCHIMEDEAN SCREW PUMPS. Screw pumps range in size (rated by the diameter of the screw) from a minimum of 0.3 m (12 in.) to a maximum of 3.7 m (144 in.) in diameter. The normal rated capacities may range from 6.3 to 4400 L/s (100 to 70 000 gpm). Screw pumps, with both open and

---

closed configurations, are primarily used for applications with low lift, high capacity, and nonclog requirements. An open screw pump (Figure 5.18) has a lifting height limitation of 7.5 m (25 ft). Pump length is restricted due to

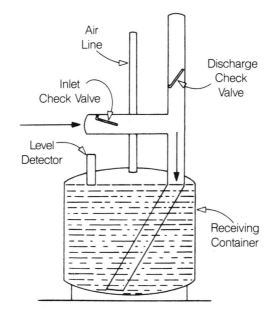

Figure 5.17 **Pneumatic ejector. As wastewater rises in receiving containers and contacts the electrode, air is forced into the container ejecting the wastewater.**

Figure 5.18 Open screw pumps.

deflection and strength considerations, and the incline angle is limited to 38 deg. The enclosed screw pump, installed at an angle up to 45 deg, can vertically lift more than 15 m (50 ft). In recent years, enclosed screw pumps have gained wider acceptance due to their higher pumping efficiencies and steeper incline angles (Figure 5.19).

The efficiency of the screw pump increases from its minimum capacity to its rated capacity, with near maximum efficiency realized within the top 70 to 80% of the pumping range. These units are well suited for variable capacity operation because the rate of discharge is controlled by the fluid level at the screw inlet. Therefore, a variable-speed drive is unnecessary. The horsepower varies almost directly with the pumping capacity, resulting in high efficiencies over a wide range of pumping capacities.

The revolving spiral limits the passage of solids to those smaller than the dimension of the opening between the flights. Any solids capable of passing through the flights are readily conveyed with the water through the pump.

In open screw pumps and enclosed screw pumps where the flights are not continuously welded to the exterior tube, water will flow back down the pump. If the backflow exceeds influent flow to the pump, the pump will cease to deliver liquid until the inlet level rises to a point where the pump again delivers flow. This surging condition may impair operation of downstream unit processes. Consult with equipment manufacturers to determine acceptable wet well water levels for optimum pump operation.

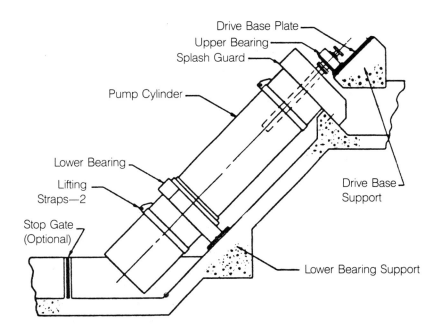

Figure 5.19   Enclosed screw pump.

Enclosed screw pumps with the flights continuously welded to the exterior tube will trap water in the flights when the pump is turned off. In cold weather, this trapped liquid can freeze and render the pump inoperable. Therefore, standby or inactive pumps require draining in cold weather to prevent freezing. Some pump designs require a special clutch to allow the pumps to be drained.

Because screw pumps expand and contract with changes in ambient temperature or exposure to direct sunlight, bearing designs must account for thermal expansion and contraction. Some installations include sun shades to minimize expansion and contraction.

POSITIVE DISPLACEMENT PUMPS. Positive displacement pumps are used because of their ability to move heavy and concentrated sludges containing entrained gases without losing prime. Efficiency is rarely a factor in pump selection.

Positive displacement pumps have one common characteristic: they lack a direct connection between inlet and outlet. Thus, these pumps usually do not significantly backflow. Generally, they operate with close clearances. Table 5.3 summarizes operating characteristics of positive displacement pumps, including plunger, diaphragm, progressing cavity, and rotary lobe. Chapter 17 describes these pumps and their sludge pumping applications.

Table 5.3   Positive displacement pump characteristics.

Type	Capacity range, L/s[a]	Discharge head, m[b]
Plunger	1–30	1–73
Diaphragm	1–10	1–115
Progressing cavity	1–25	1–73
Rotary lobe	1–106	1–73

[a] L/s × 15.85 = gpm.
[b] m × 3.281 = ft.

PUMP APPLICATIONS SURVEY. The WPCF Design Practice Survey and the Treatment Plant Data Survey, both conducted in 1989, collected information on the wastewater applications of various types of pumps. Table 5.4 summarizes the results of the Design Practice Survey and indicates designers' preferences for various pumping applications. Tables 5.5 through 5.7 summarize the results of the Treatment Plant Data Survey.

Type of pump	Raw waste-water	Grit	Grease	Primary sludge	Digested sludge	Return activated sludge	Waste activated sludge	Thickened activated sludge	Trickling filter recirc.	Clarified liquid	Final effluent
Centrifugal open impeller	9		1			8	7	1	6	4	2
Centrifugal recessed impeller	1	8		4	4	3	3	2	1		
Piston			3	5	4			3			
Diaphragm			3	4	2			1			
Air lift		3		1	1	2					
Propeller							3		2	4	5
Rotary lobe			1		2		2	5			
Progressing cavity			6	4	4		3	5			
Vertical turbine									3	5	7
Archimedes open	2					2	1		2		2
Archimedes enclosed	2					1	1			1	1
Other	1		1			1			1	2	1
Total responses	15	11	15	18	17	17	20	17	15	16	18

Table 5.5   **WPCF Treatment Plant Data Survey—pump type used by liquid stream.**

Type of pump	Raw wastewater	Primary effluent	Trickling filter recirculation	Effluent
Horizontal centrifugal	27	6	7	7
Vertical centrifugal	44	13	8	10
Enclosed archimedes screw	1			1
Open archimedes screw	5	5	2	1
Propeller	1	2	1	2
Vertical turbine		1	4	12
Other		3		1
Total plant responses	78	30	22	34

Table 5.6   **WPCF Treatment Plant Data Survey—pump type used by sludge stream.**

Type of pump	Response by sludge stream				
	Return activated	Waste activated	Primary	Digested	Digester recirculation
Centrifugal	67	62	28	29	42
Piston		2	30	20	6
Archimedes screw	3	2	1	2	
Air lift	16	10	5	2	3
Progressing cavity	3	13	15	23	8
Diaphragm			3		
Other	1		3	1	
Total plant responses	90	89	85	77	59

PUMP APPURTENANCES. Electric motors, pressure switches, mechanical seals, time clocks, pressure gauges, and priming equipment are briefly described below.

Electric Motors. The basic types of electric motors manufactured for pump drives are the squirrel cage induction motor, the wound rotor induction motor, the synchronous motor, and the shunt wound dc motor. Of these four types, the overwhelming favorite for wastewater pumping is the squirrel cage induction motor. Its advantages of maintenance simplicity, low initial cost,

Table 5.7   WPCF Treatment Plant Data Survey—pump type by scum and grit applications.

	Response	
Type of pump	Scum	Grit
Centrifugal open impeller	16	14
Centrifugal recessed impeller	10	16
Piston	23	
Progressing cavity	17	
Self-contained[a]	2	2
Air lift	10	8
Rotary lobe	2	
Other	5	5
Total responses	85	45

[a] In hauling vehicle.

and mounting versatility combine to make it the best choice for most installations. In older installations where large sizes [greater than 370 kW (500 hp)] and variable-speed control are necessary, the lower efficiency wound rotor induction or synchronous motor are generally used. However, recent advances in the development of higher power semiconductors have improved the power-handling capabilities of variable-frequency controllers. This has permitted use of the more efficient squirrel cage induction motor in variable-speed applications requiring up to 1500 kW (2000 hp). In rare instances, dc motors, with generally very large sizes, have been used where their superior efficiency offsets the high initial cost and high maintenance requirements. Manual of Practice No. OM-5, "Prime Movers," presents detailed discussions of motors and drives.[14]

Pressure Switches. Positive displacement pumps should be equipped with diaphragm-protected pressure switches to shut off the motor under excessive discharge pressures caused by a pipe plug or closed valve.

Mechanical Seals. Centrifugal pumps are often furnished with mechanical seals to minimize leakage around the pump shaft. Mechanical seals are generally water lubricated, although other lubricating fluids may be used with specially designed seals. Clean water is needed for shaft seals. Seal water pressure should generally equal 110% of maximum pump discharge pressure. Either treated effluent or potable water can be used for seal water. If potable water is used, the supply source should be protected by an air gap arrangement to prevent backflow of the pumped liquid. If treated effluent is used, it must be free from gritty material that could foul the seal or score the pump shaft. Generally, seal water is supplied only when the pump operates.

A solenoid valve, installed in the seal water supply line and wired across the pump starter, provides that control.

Packing. Packing may be used to minimize leakage around the pump shaft where it penetrates the volute casing. Packings are available in a wide variety of materials for specific applications. For wastewater pumps, packing is installed in the seal cage and held in place with a packing gland. Packing must be continuously cooled and lubricated while the pump is in operation. In wastewater and sludge applications, the lubricating fluid is treated effluent or potable water. Where potable water is used, the supply source should be protected with an air gap to prevent backflow of pumped liquid. All packings leak, and provisions should be made to convey leakage to appropriate drainage facilities.

Time Clocks. Time clocks serve best for intermittent operations that are not controlled satisfactorily by level controls. Time clocks are often used to pump sludge out of clarifiers intermittently, for example 10 minutes during every hour.

Where multiple units are to be pumped in sequence, time clocks with automatic programming of valves and pumps can be used. The sequence can be controlled with a secondary time clock or other devices such as those that measure density.

Pressure Gauges. Pressure gauges are useful for checking the pumping head to determine pressure buildups in the line with losses of pumping capacity. Pressure gauges need protection with a diaphragm that will keep solids from damaging the inner elements of the gauge. On pump discharge lines, the pressure gauges can be liquid filled to minimize vibration and wear. On pump suction lines, compound pressure gauges are necessary to indicate both pressure and vacuum.

Priming. Wastewater pumps have priming equipment with various designs. Some have automatic priming (self-priming) devices that become inactive after the priming is accomplished, and others incorporate a hydraulic device that is part of a self-priming pump. The more compact self-priming designs are usually preferred for portable use.

HYDRAULIC CONSIDERATIONS FOR PUMP SELECTION. When selecting pumps for a specific application, the design engineer must match the pump's performance with the head capacity curve for the system, taking into account the viscosity of the fluid.

System Head Capacity Curves. To determine the specific requirements of each pumping unit, it is first necessary to calculate and plot the head capacity curve for the system. This curve is plotted by summing the static lift and friction head loss in the system for various discharges. The static lift represents the elevation differential between the low water level in the wet well and the high point of the discharge force main or the high water level at the discharge, whichever is higher. The friction loss is the sum of friction losses through the suction pipe, suction fittings, discharge pipe, discharge fittings, and force main at various flow rates.

Figure 5.20 shows a typical system head curve, indicating portions of the total dynamic head (TDH) that comprise static head and friction head loss. Also shown are the head capacity curves of one pump, two pumps, and three pumps in parallel operation. Head capacity curves for multiple pumps operating in parallel are obtained by adding the capacity of each pump at each of several given heads and plotting the results for each pump grouping (that is, one, two, or three pumps, and so on).

In Figure 5.20, the design condition for one pump would be 38 L/s (600 gpm) at a TDH of 11 m (36 ft). If the station were designed initially with two pumps, this station would have a peak capacity of 63 L/s (1000 gpm). With the addition of a third pump, the station could ultimately have a capacity of 76 L/s (1200 gpm).

The maximum horsepower per pump would be required when one pump operates at its capacity of 38 L/s (600 gpm). With two pumps, the capacity of each would be 32 L/s (500 gpm), with the appropriate reduction in horsepower for each pump. Similarly, with three pumps operating, each would contribute 25 L/s (400 gpm), with a further reduced power requirement for each pump.

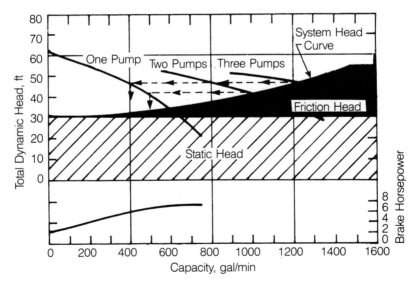

Figure 5.20   Typical system head curve showing static head and friction head loss (hp × 0.746 = kW; gal/min × 0.063 1 = L/s; ft × 0.305 = m).

System head and pump capacity curves can be combined to solve a number of complex pumping problems. In the case of two remote pumping stations discharging to a common force main, the system head curve would be a compound curve with a segment for each of the following conditions:

- Pump A on, Pump B off;
- Pump B on, Pump A off; and
- Pump A on, Pump B on.

When both pumps operate, the capacity contributed by each pump will vary depending on static head, friction losses, and individual pump head capacity curves. Therefore, pump selection for such a system is an iterative process.

Head capacity curves for centrifugal pumps operated in series are obtained by adding the operating head for each pump at a given capacity. Consult with pump manufacturers before designing series pump applications to ensure proper pump selection, thereby avoiding potential cavitation or motor overload conditions.

A family of curves represents variable-speed pump head capacities. Each individual curve corresponds to a discrete operating speed. Superimposing the system head curve on the family of head capacity curves identifies specific operating points for the system.

Where variable- and constant-speed pumps discharge to a common header, a complete hydraulic analysis for all operating speeds is required to prevent recirculation cavitation.

Effects of Viscosity. The liquid handled by a pump affects the unit's operating head and capacity, the required power input to the pump, and the construction materials.[8] Thus, pump selection, construction, and use will vary among liquids with differing viscosities. Although the effects of viscosity on pump performance have been tested, it is difficult to accurately predict the difference between a pump's performance when conveying a high- or low-viscosity liquid and its performance when conveying cold water.

High-viscosity liquids affect the performance of centrifugal pumps in three ways:

- The pump develops a lower head than when handling water,
- Pump capacity is reduced, and
- The power input required is higher.[8]

The Hydraulic Institute provides a procedure for analyzing viscosity effects for centrifugal pumps.[3]

For rotary and reciprocating pumps, the rated pump speed should be decreased for viscous liquids. Since pump capacity depends on the rated speed, capacity will decrease with an increase in viscosity. Pump manufacturers should be consulted to analyze viscosity effects.

The addition of polymers to wastewaters and sludges may improve flow conditions. Polymers reduce liquid viscosity by decreasing the cohesive forces between particles within the fluid. Chapter 17 presents viscosities for wastewater sludges.

Speed. Low capacity [63 L/s (1000 gpm) or less] and high-head nonclog wastewater pumps are currently designed for efficient operation at speeds as high as 1800 rpm. The maximum desirable speed depends on the head and the capacity required by the individual pumping unit. With lower speeds, less wear and lower maintenance costs can be expected. Lower speeds, however, generally result in higher capital equipment costs.

Constant-speed, multiple-pump stations are generally used where continuous discharge is unnecessary. A constant-speed drive is simpler, more reliable, and less costly than a variable-speed or multiple-speed drive. Constant-speed control systems are also simple, reliable, and economical. If discharge from the station must be continuous, a variable-speed drive is required to adjust the pumping speed to the influent flow rate. Pump stations with a large flow variation require more pumping units with a constant-speed system than with either a multiple-speed or variable-speed system.

*P*UMP CONTROL SYSTEMS

The selection of a control system and a specific control mode for pumping systems is at least as important as pump selection. To determine the proper type of control for any application, the engineer must first establish a set of parameters, including pressure and flow, to define the expected results. Then the engineer can select a control system that will allow the pumps to produce the established hydraulic effects. Considerations of efficiency or power factor should not supersede pump control's primary purpose of producing the necessary hydraulic effects.

Different processes or different pumping systems within a given process will require varying degrees of control of the two primary hydraulic parameters—pressure and flow. The simplest system that will reliably provide effective results will generally be the most satisfactory.

No rigid rules govern the weights assigned to any of the considerations for determining the type of control most suitable for any given application. Ultimately, the engineer must weigh such variables as capital and operating costs, efficiency, power factor, reliability, operational effects, and ease of operation and then choose the system best suited for the application. Such selections are not always obvious and require thoughtful consideration of pumping effects. For example, constant-speed pumps will cause hydraulic surges when influent flow is less than pump capacity. Because these surges can adversely impact some biological treatment processes, variable-speed pump control may be necessary in those cases.

The overall efficiency of a variable-speed system may exceed that of an on-off system, despite control losses. With the former system, the pump may operate against a lower average friction head, saving pump power to offset the power lost in the variable-speed control.

As another important consideration, the selected control system must be compatible with the training and experience of the operators or satisfactory operation will seldom be achieved.

MANUAL CONTROL. Manual control systems generally consist of push-button stations or selector switches that energize or deenergize the pump motor starter. Push-button stations (sometimes called "three-wire control") are electrically interlocked so that the units have to be restarted manually after a power outage, while a selector switch (sometimes called "two-wire control") remains in the "on" position and restarts automatically. Manual control is essential on all systems for maintenance but is rarely used on variable-speed pumps for operations.

AUTOMATIC ON-OFF OR SPEED CONTROL. Automatic control systems are commonly based on time, pressure, flow, or fluid level. Each of these is briefly described below.

Time. Pumps, started at regular intervals, operate for a preset length of time. Time-controlled systems are generally used for sludge pumping.

Pressure. Pressure drop, generally sensed by a standard pressure switch in a hydropneumatic tank, is used to start the pumps in plant water systems.

Flow. Wastewater pumps can be turned on when the required flow exceeds a certain limit or turned off when the required flow drops below a limit. Plant influent flow variations may also be used to start up or shut down return sludge or chemical feed pumps or to vary their speed. Influent flow signals generally come from a flow metering device with mA-dc control circuits. Flow meters are discussed in the "Flow Measurement" section of this chapter.

Fluid Level. Fluid level signals govern most of the automatically controlled constant- and variable-speed systems. Pumps are turned on or sped up as wet well levels rise and are turned off or slowed down as wet well levels fall. This method often controls influent and effluent pumps, sump pumps, and certain in-plant transfer pumps. The several level detection systems that are commonly used include float switches, enclosed floats, displacement switches, electrical conductance probes, electrical capacitance probes, captive fluid systems, bubbler systems, and ultrasonic level sensors. Each of these level detection systems is described below.

FLOAT SWITCHES. The simplest type of control is a float connected through a rod to a switch that opens and closes as the float moves with the fluid level in the wet well. These work well for "on–off" control but are not suitable for the adjustable speed type of control. The float switch can directly connect with the control circuit of the starter, thus starting and stopping the individual pump. In duplex or multipump installations, floats can be set to trip the switch at different fluid elevations in wet wells, thus operating the required number of pumps for each level. Alternation of the lead pump is often used in a duplex or multipump installation to even the wear among the pumps. This is generally accomplished by connecting the float switches to the individual pump starter through an electric alternator.

For variable-speed applications, the simplest float mechanism consists of a beaded cable strung over a notched pulley. The beaded cable has the float on one end and a counterweight on the other end. As the fluid level fluctuates, the beaded cable rotates the pulley, which can be linked to a current or pressure transmitter for adjusting the pump speed. Also, encapsulated mercury switches can be mounted on the pulley axle to tip and activate alarms or provide on–off control of pumps. The float is often installed in a stilling well that prevents lateral movement of the float in the flow stream. Figure 5.21 illustrates this type of float installation.[15]

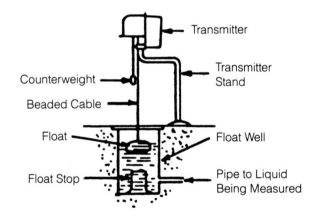

Figure 5.21   Simple float mechanism.

Float controls are economical, simple, and reliable when operated in effluent or clear water. However, when operated in raw wastewater or sludge, maintenance problems can develop from grease coating the floats, solids puncturing the floats, or corrosion of the floats or tapes. Corrosion-resistant material can be used for the floats or tapes, and flushing with clear water will help reduce grease buildup. Excessive turbulence in the wet well can result in cable fatigue and subsequent failure.

ENCLOSED FLOATS. Enclosed float switches consist of an encapsulated mercury switch inside a float, which is suspended into the wet well on a reinforced electrical cable. The switch may be either open or closed when the float is in the pendant position. As the liquid rises, the float inverts and changes the angle of the mercury switch, reversing its condition. These switches work well for on–off control but are not suitable for variable-speed control. These units are rugged, long-lived, and not subject to the corrosion problems that plague the rod- or tape-type float switch. Turbulence, however, can cause excessive wear and, in extreme cases, can cause the floats to bounce, resulting in false pump starts. Before selecting multiple enclosed floats, consider the potential tangling of long suspension cables.

DISPLACEMENT SWITCHES OR TRANSMITTERS. Displacement switches consist of one or more weights, usually made of porcelain, attached to a steel cable and suspended into the wet well. The cable hangs from a spring and is magnetically coupled to one or more mercury switches contained in the enclosure at the top of the assembly. When the water rises to the position of the weight and submerges it, the reduced force exerted by the submerged weight allows the spring to draw the weight-cable assembly up and activate a mercury switch. In addition to on–off control, these switches can be used for adjustable-speed control within small level ranges up to 0.6 m (2 ft). In this application they are referred to as displacement transmitters.

Displacement switches are more rugged than float switches and are better able to withstand turbulence. However, the spring will wear out and requires replacement periodically.

ELECTRICAL CONDUCTANCE PROBES. Electrical conductance probe controls are often used instead of floats or displacement switches because the probes have no moving parts. As limitations, however, they require a source of electrical power and cannot be used in hazardous locations or in nonconducting liquids without elaborate enhancements that reduce their reliability. These probes work well for on–off control but are not suitable for variable-speed type of control. With the use of relays, the probes can control a single pump, a duplex installation, or a multiple-pump installation. The number of probes depends on the number of pumps or functions needed. The use of conductance probe controls avoids the mechanical failures of the float-type controls.

Because grease can foul conductance probes, if grease may be present, the probe is enclosed in a sealed tube above a suspended bladder-type container with fluid that will conduct the slight current necessary for probe operation.

The fluid level in the sealed tube varies with the water level in the wet well. As each electrode or probe is immersed in the fluid, relays are energized, starting additional pumps or energizing an alarm circuit. The number of pumps to be operated or circuits to be energized is limited to the number of probes that can be suspended in the enclosed tube.

ELECTRICAL CAPACITANCE PROBES. Electrical capacitance probes have all the advantages of electrical conductance probes plus the ability to provide adjustable speed control. These probes consist of a conducting probe covered by a nonconducting material such as polytetrafluorethylene. This probe is immersed in the wet well along with another conducting probe that is uncovered. The two probes form an electrical capacitor with a capacitance that varies with the immersion length of the probes. This variation in capacitance, which can be linear or nonlinear with water level depending on the probe shape, is measured and converted to a current or voltage signal. Nonlinear probes are useful for applications involving wet wells with nonrectangular cross sections.

CAPTIVE FLUID SYSTEMS. Captive fluid systems have a diaphragm and small-diameter tubing for transmitting pressure signals to switches that turn pumps on and off. These systems, used with varying degrees of success for on–off control, can also provide adjustable speed control by transmitting the pressure signal to a transducer that converts the measured pressure to a current or pneumatic control signal. The disadvantages of these instruments are that the diaphragm is vulnerable and any leak in the transmission line prevents operation.

BUBBLER SYSTEMS. The bubbler system, a fairly maintenance-free approach to the measurement of liquid level, operates on the principle that a small, constantly regulated, and oilless air supply bubbling into a liquid at slightly higher pressure than the static head, will produce a back pressure equal to the static head. Because air is constantly bubbling out of the bubbler tube, the system is generally self-purging. However, grease or solids occasionally plug the bubbler pipe, causing erroneous readings. This problem can be largely mitigated by providing a minimum air supply of 0.5 L/s (1.0 cu ft/min) in a 13-mm (0.5-in.) diameter supply pipe. For maintenance, air supply pipe valving should be arranged to isolate the readout device while providing high purge flow through the bubbler tube for blow-down purposes if plugging does occur. In addition, a removable bubbler tube is needed for periodic inspection and cleaning. A removable cap will allow *in situ* rodding of the bubbler tube.

The bubbler system has no moving parts that contact the wastewater; therefore, they are generally reliable and easily serviced. Where the treatment process (such as anaerobic digesters) cannot tolerate air injection, water or

---

nitrogen can be used instead. Using water requires the pressure sensor's elevation to be below the minimum possible surface elevation of the measured liquid.

The range of a bubbler tube may be up to 56 m (185 ft) of head, requiring an air supply pressure of up to 550 kN/m^2 (80 psi). The tube's accuracy (±0.1% of the actual head except in low-head applications) is limited only by the accuracy and sensitivity of the device measuring the backpressure. The air signal generated by the bubbler tube, because it is proportional to the head, does not conform with the standard 20- to 100-kN/m^2 (3- to 15-psi) pneumatic signal for most receiving instruments. Therefore, the interface must be considered. Standard differential pressure cells, with one cell open to the atmosphere, may be used to standardize the signal. For very low heads [less than 76 cm (30 in.)], the response precision of a bubbler system is less than that of a float. Figure 5.22 illustrates a bubbler system schematic.[15]

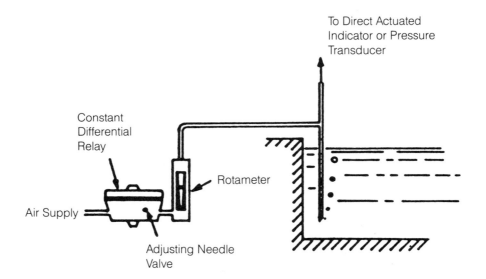

Figure 5.22   Schematic of bubbler system.

Recent developments in electronics have led to the increasing use of solid-state switching circuits. In this type of system, the bubbler pressure is changed to a voltage or current signal that is fed into an electronic network. Variations in signal level are used to turn the pumps on and off or adjust their speed.

ULTRASONIC LEVEL SENSORS. Ultrasonic sensors can be used to measure liquid levels. The sensor (transducer), mounted above the liquid level, transmits and then receives ultrasonic waves that bounce off the liquid surface.

The elapsed time for the signal to leave the sensor and return is a function of the distance to the liquid surface. Instrument accuracy is within ±1% of the distance to the liquid surface.

Mounting of the transducer above the liquid flow prevents fouling with rags, grease, solids, and ice. Ultrasonic level sensors should be separated from walls and any spray in the signal area. Readings can be affected by high winds, temperature differential, and foam or grease on the liquid surface. In the absence of such detrimental conditions, ultrasonic level sensors are capable of very high accuracy and can detect very small level changes.

VARIABLE-SPEED DRIVES. Variable-speed drives, the general class of equipment used to drive wastewater pumps at varying operational speeds, must be controlled by a signal based on level, pressure, or flow as described in the previous section. Most of the fully automatic control systems use a bubbler or ultrasonic level measurement system for primary level detection. The signal from one of these measurement systems controls a variable-speed drive that, in turn, controls the pump's speed. Variable-speed drives comprise two general equipment categories: variable-torque transmission systems and variable motor speed control systems. Each of these is described below.

Variable-torque Transmission Systems. This system, with a fixed-speed input shaft and a variable-speed output shaft, mechanically couples a fixed-speed motor to a pump. A distinct disadvantage of these systems is that they must dissipate energy in the form of heat whenever the pump operates at less than full speed. This results in poor efficiencies over their operating range. An advantage of these systems is they can be manufactured to accommodate very high power. The two most common types, eddy current clutches and liquid clutches, are described below.

EDDY CURRENT CLUTCHES. This clutch consists of a constant-speed input shaft directly close-coupled to a synchronous or induction motor and a variable-speed output shaft connected to the pump. An electromagnetic member is connected to the output shaft, and a field coil member is connected to the input shaft. The field coil member surrounds the electromagnetic member. As increased speed is required of the load, dc current enters the electromagnet, exciting it and causing eddy currents. The magnetic flux produced by the magnet develops torque that turns it and the connected load at an output speed proportional to the dc current. A current signal derived from any measurable variable controls the dc power applied to the field. Clutch efficiency is 2 to 3% below the efficiency of wound rotor motors with secondary resistance. The speed range is approximately 50 to 95% of motor rated speed. The speed of the selected motor should barely exceed the maximum speed of the pump. Otherwise, high slippage will result, with much of the input horsepower to the motor being wasted as heat rather than driving the pump.

Eddy current clutches are manufactured in a full range of sizes and speeds, with both horizontal and vertical configurations. These clutches provide a reliable method for remote manual or automatic speed control. However, this type of drive is not suited for applications where the motor drive is mounted above the driven load or directly to it (for example, directly to the pump frame of vertical dry pit pumps). For such installations, the weight of the combined motor and drive units could frequently exceed that of the pump, causing imbalance and requiring special bracing.

LIQUID CLUTCHES. Current models, called hydroviscous drives, are essentially wet clutches that transmit energy through the shear strength of the oil. The efficiency of these units, comparable with the eddy current clutch, is slightly higher at full speed and lower at reduced speeds. These clutches can have large horsepower ratings. The primary applications of these units include plant water or sludge handling pumps and very large influent and effluent pumps.

OTHER VARIABLE-TORQUE TRANSMISSION DEVICES. Other types of torque transmission equipment occasionally used include clutches using water as the hydraulic fluid and hydraulic pump–motor combinations. Most of these have either proven to be unsatisfactory or have offered no advantages.

Variable Motor Speed Control Systems. This category of variable-speed drives allows the motor to be close-coupled to the pump because these devices cause the motor speed to vary without mechanical couplings. The two major techniques are variable-frequency controllers with induction motors and secondary energy recovery units with wound rotor motors. These systems are fast becoming the most popular choices for variable-speed drives because their efficiencies exceed 85% throughout their operating ranges. These high efficiencies result from the avoidance of energy waste in the form of heat, as with variable-torque transmission systems and outdated non-regenerative wound rotor motor controllers. Because variable-frequency controllers rely on solid-state power semiconductors, their current handling capabilities are limited. Consequently, applications requiring motors rated above 746 kW (1000 hp) must use voltages over 600 V–ac. These higher voltages (for example, 2300 and 4160 V–ac) require expensive and large electronic equipment with high capital costs that must be weighed against the expected operational efficiency savings when selecting the equipment.

For present applications requiring over 746 kW (1000 hp), wound rotor secondary energy recovery systems are more practical than variable-frequency controllers. Nonetheless, the increased efficiency of variable-frequency controllers coupled with the steadily rising cost of energy is motivating industry to develop semiconductors with greater power-handling capacity that may, in the future, increase the feasible power capability of variable-frequency controllers above 746 kW (1000 hp).

VARIABLE-FREQUENCY CONTROLLERS. This type of drive is a true variable-speed drive because the motor will vary its output speed with or without the load connected. In operation, these supply systems use power semiconductors to change three-phase, 60-Hz power to dc (rectification); filter the dc; and then restructure (invert) the dc into an output of variable frequency and voltage. Two basic types of variable-frequency controllers are available: current sourced and voltage sourced. These terms refer to the technology used in the dc and invertor sections of the controller. Current sourced controllers are smaller and cheaper than voltage sourced controllers but must be matched to the reactance of the motor's windings. Furthermore, current sourced controllers can control only one motor at a time. Due to these limitations, voltage sourced controllers are preferred in the power range below 370 kW (500 hp) where the larger controller sizes and cost premiums are offset by the easier coordination of manufacturing and flexibility of application. Voltage sourced units are not available for motors operating above 600 V–ac.

An important consideration in the use of these variable-frequency controllers is the shape of the output wave form. Since it is produced by switching dc back and forth through the motor's windings, it releases significant energy in the harmonics of the output frequency.

A number of techniques are used to minimize harmonic output and improve power factor and efficiency. Some units have high efficiencies but have relatively poor power factors. Others have a constant high power factor but have a reduced efficiency. Some systems require motors with additional thermal capacity because of the motor heat generated by excessive harmonics. Even with good harmonic filtering, the motors should be rated for a 1.15 service factor and have high-temperature insulation.

The advantages of variable-frequency controllers include

- Allowing use of standard squirrel cage induction motors,
- Suitability for nearly any location or any mounting configuration,
- Limiting power draw to that necessary to drive the output, and
- Avoiding heat buildup caused by power wastage as with eddy current or hydraulic drives.

Efficiency of the controller over normal ranges will generally remain above 92%. The pumps can be operated at any speed up to 150% of the standard motor's full load speed, allowing normal pump impeller selection. However, operation above 100% of the motor's rated speed (associated with 60 Hz) will cause the motor to overheat. This practice is inadvisable as a normal operating strategy.

To prevent the harmonic feedback problems associated with variable-frequency drive equipment, variable-frequency drive control units need to include isolation transformers on their input circuits to isolate the control units from the plant's electrical system. To attain this objective, filters may also be required in some cases. In addition, each drive should have a separate circuit. The feedback can affect the electrical power bus protection relays, computerized control and data acquisition hardware, and radio communications. Generator manufacturers recommend derating the generators when a substantial portion of the electrical load supplied by the generators is for variable-frequency drive equipment. In these cases, the design engineer should consult with the generator manufacturer.

WOUND ROTOR MOTOR CONTROLS. When using wound rotor motors, speed control results from controlling voltage on either primary or secondary windings. Control of voltage on secondary windings is especially advantageous for motors over 75 kW (100 hp) or for voltages over 480 to reduce the amount of controlled power. Depending on the control system for these motors, restriction of full speed might occur. Advantages of these types of drives are that the inrush current draw of these motors during starting is greatly reduced, and the mounted weight of the motors (although generally slightly more than that of a normal squirrel cage induction motor) is considerably less than that of a motor–clutch combination.

The primary method of varying the rotor voltage is varying the impedance of the rotor windings. Older systems used such techniques as resistors, reactances, and liquid rheostats. More recently, solid-state electronic rheostats have been used to vary the rotor winding impedance. However, all of these methods required the dissipation of heat whenever the motor ran at less than full speed. Consequently, the efficiencies of these older designs were quite low. The advent of the solid-state electronic rheostat (Figure 5.23) led to regenerative controllers. Regenerative controllers convert the energy that would have been wasted as heat in R1 and R2 of Figure 5.23 into fixed-frequency ac power that feeds back into the plant's power grid. The average efficiency of regenerative controllers ranges from 85 to 90%. Since this method applies solid-state technology only to the energy drawn off the rotor at operational speeds below 100%, it is capable of handling larger motors and higher excitation voltages than variable-frequency controllers.

VARIABLE-SPEED APPLICATIONS SURVEY. A survey of design firms was conducted to characterize current design practice. Table 5.8 summarizes the results of the survey. Eleven responses were received. Every firm that responded recommended variable-frequency drives.

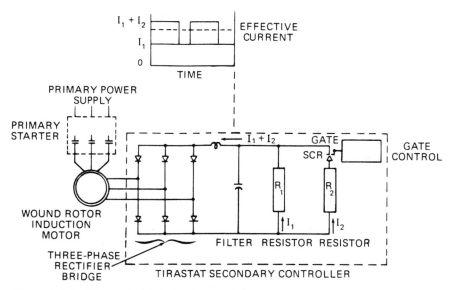

Figure 5.23 Schematic of wound rotor induction motor with electronic secondary controller.

Table 5.8   WPCF Design Practice Survey—preferences for variable-speed pump control systems.

System description	Number of design firms that would recommend
Eddy current clutches	4
Fixed stop resistors with wound rotor motor	1
Liquid rheostat with wound rotor motor	4
Reactance/resistance units with wound rotor motor	1
Electronic rheostat with wound rotor motor	2
Variable-voltage controllers	3
Liquid clutches	1
Variable-frequency drives	11
Slip recovery drives	1
Wound rotor with energy recovery	1
SCR with dc drive	1

PUMPING SYSTEMS

PUMPING STATION STRUCTURE. Since the pumping station structure imposes a major part of the station's capital cost, the structure must be

---

economically efficient. This calls for structural efficiency and spatial efficiency; that is, careful sizing of the wet well and dry well (if required) and efficient use of other space. Often, the pumping station structure houses, in addition to the pumping equipment, part or all of the other treatment process equipment. For this reason, the station's size and shape, particularly the superstructure, may be dictated by the space requirements of equipment other than the pumps.

The substructure, the below-ground part, of the influent pump station is often deep and expensive because of the low elevation of the influent sewer. Effluent and other treatment plant stations may have minimal substructure. Rectangular structures are more common than circular structures because the former offers more usable space. Nevertheless, round caisson-type structures may be economical for a deep substructure or poor foundation conditions, including high groundwater requiring use of the substructure as a sunk caisson during construction. In such situations, a concentric wet well–dry well arrangement deserves consideration. As circular wet wells may cause vortexing at the suction pipe, hydraulic modeling may be necessary to analyze this problem.

The floor of the superstructure and all openings to the wet well and dry well must be higher than the design flood elevation or be otherwise protected from flood waters and high water levels in the wet well. Where the substructure floor is beneath the groundwater table, the flotation forces must be offset by overall structure weight or by anchors. In earthquake-prone areas, the structural design should account for inertia and movement of the liquid in the wet well in addition to standard seismic considerations. Also, exterior wall pipe penetrations should be designed to withstand some settling of the structure, even in areas where seismic design is not required. The design engineer must ensure that the completed structure will conform with applicable safety and health provisions of all regulatory agencies and other local, state, and national codes.

WET WELL. No single method for sizing wet wells applies to all design situations. Proper wet well sizing considers three critical factors: detention time, pump cycle time, and turbulence at the pump intake.

As good practice, wet well detention times generally do not exceed 30 minutes for average flow to minimize generation of unpleasant odors. In colder climates, longer detention times may be acceptable. Where such detention time limitations would be impractical, odor mitigation must be accounted for in wet well design. Odor mitigation practice ranges from providing gastight covers to sophisticated chemical feed or off-gas scrubbing systems.

Pump cycle time refers to the elapsed time between successive motor starts. Excessive motor wear and shortened service life result from cycle times less than the manufacturer's recommendation. Minimum cycle times range from about 5 minutes for 4-kW (5-hp) motors to over 30 minutes for 150-kW (200-hp) motors. Consult with the motor manufacturer for minimum cycle time recommendations or special motor designs.

For constant-speed pumps, the minimum cycle time results if the influent flow equals 50% of the rated pump capacity. This limitation often determines wet well volumes for both single- and multiple-pump installations. For multiple-pump installations, alternating the lead pump after each pumping cycle effectively doubles cycle time and reduces wet well volumes accordingly. Wet well volumes can also be optimized with strategic pump "on" and "off" settings. In multiple constant-speed installations, the required wet well capacity represents the sum of the wet well capacities required for the individual pumps. Such an allowance will prevent cycling when lag or standby pump units enter service.

Some designers prefer to use detention capacity in the influent sewer line to minimize wet well volume by setting the pump "off" level above the invert elevation of the influent sewer. This practice may be acceptable in large installations where influent wastewater velocities are sufficient to minimize solids deposition in the influent sewer.

Wet wells for variable-speed pumping systems can be significantly smaller than for comparably sized constant-speed stations. However, cycle time will be limiting if the influent flow is 50% of pumping capacity at the minimum pump speed. In multiple variable-speed installations, the wet well capacity required is the sum of the wet well capacities required for the individual pumps. Such an allowance, similar to multiple constant-speed installations, will prevent cycling when lag or standby pump units enter service.

When determining the pump operating levels in the wet well, the design engineer needs to consider the net positive suction head (NPSH) requirements of the pump. Wet well designs should allow adequate submergence and clearance between pump intakes to prevent eddy currents and vortexes that could otherwise reduce pump efficiency or capacity. These requirements may dictate longer detention times than those necessary to meet pump cycling requirements. Consult pump manufacturer's recommendations and "Hydraulic Institute Standards"[9] for sizing sumps and wet wells. Figure 5.24 shows sump dimensions as a function of flow for axial flow and mixed flow pumps.[9] Figure 5.25 indicates arrangements suggested as good practice for axial flow and mixed flow pumps as well as those considered to be inadvisable. Adherence to these suggestions, based on testing by several pump manufacturers, will help ensure proper pumping station design and avoid costly future modifications. Figure 5.26 illustrates suggested positioning of suction inlets. For large complex wet well systems, where following established design practice would be impractical, hydraulic modeling will be necessary to ensure acceptable wet well performance.

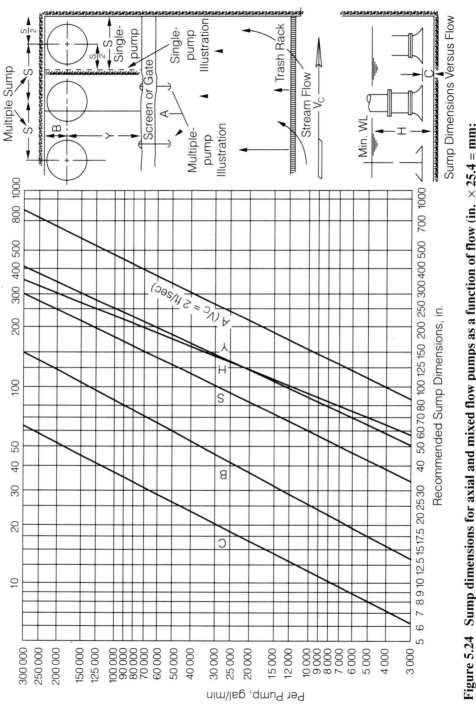

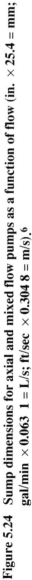

Figure 5.24 Sump dimensions for axial and mixed flow pumps as a function of flow (in. × 25.4 = mm; gal/min × 0.063 1 = L/s; ft/sec × 0.304 8 = m/s).[6]

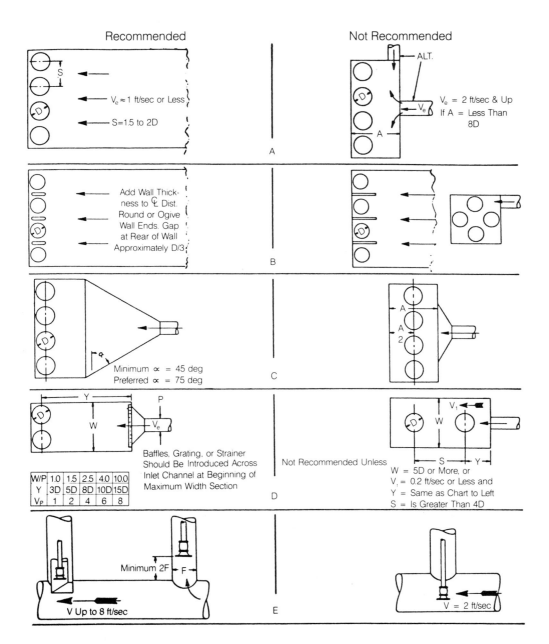

Figure 5.25 Multiple pump pit arrangements that are recommended and not recommended by Hydraulic Institute Standards. (Arrangements shown apply to sumps for clear liquid. For fluid-solid mixtures, consult the pump manufacturer.)[9]

Other wet well design considerations include
- Providing for sloped sidewalls or fillets designed to direct solids toward the pump intake, thus minimizing deposition;
- Arranging the influent sewer inlet to the wet well so that any splashing or eddy current is directed away from pump intakes to prevent air entrainment and pump air locks;
- Dividing the wet well into two sections that are properly interconnected to ease repairs, cleaning, and expansion if the pump station

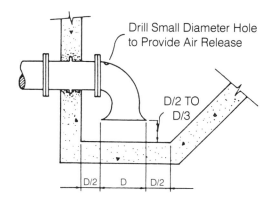

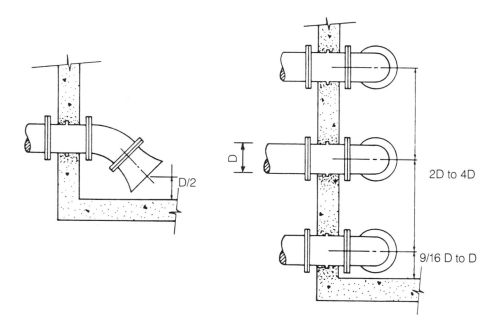

Figure 5.26 Positioning of submerged suction inlets. (Always consult with pump manufacturers for actual design recommendations.)

must operate continuously or future flows will substantially exceed existing flows; and

- Providing for a well-lighted wet well with vapor-proof and explosion-proof fixtures.

Wet well ventilation design considerations include

- Providing both supply and exhaust fans sized for 30 air changes per hour;
- Introducing supply air near the ceiling and, as a minimum, exhausting the air from near the maximum liquid level (some designers prefer exhausting from both the ceiling and near the maximum liquid level to quickly remove gases heavier and lighter than air); and
- Determining whether exhaust systems will operate only during personnel access or continuously.

DRY WELL PUMP ROOM. Submersible pump units do not require dry wells. If the installation requires a dry well, its size depends primarily on the number and type of pumps selected and on the piping arrangement. A good rule of thumb is to provide at least 0.9 m (3 ft) from each of the outboard pumps to the nearest side wall and at least 1.2 m (4 ft) between each pump discharge casing. Allow sufficient space between pumps to remove the pump from its base with ample remaining clearance between suction and discharge piping and room for on-site repairs, inspection, or removal from the pit to the surface for repairs.

Only the pump suction lines should penetrate from the dry well to the wet well. Bubbler level tubes, electrical conduit penetrations, and sump pump discharge lines must always exit the dry well above the top of the wet well and then enter the wet well from above. This practice will prevent inadvertent flooding of the dry well from a broken bubbler tube or poorly seated check valve on the sump pump. All electrical conduits from wet wells must be sealed.

Depending on the size of the pumping station, consider the installation of monorails, lifting eyes in the ceiling, and A-frames for the attachment of portable hoists, cranes, and other devices. The construction of such assemblies must allow safe, easy removal and disassembly of pumps, motors, control centers, piping, other heavy equipment, valves, and fittings that may be located in the dry well and the superstructure of the pump station. All doorways and openings likely to be used for installation and removal of the equipment need ample width and headroom.

The dry well must be well lighted and ventilated. Consider all applicable safety and other requirements to comply with local, state, and national safety codes and requirements of regulatory agencies.

The dry well design should include provisions for drainage from pump water seal connections. Stuffing boxes on pumps provided with water seals are equipped with drip pockets tapped generally for a 13-mm (0.5-in.) drain pipe connection. As the packing wears, the drippage may increase significantly. Unless these drain pipe connections are piped to a drainage gutter or drainage sump, the leakage from the stuffing box would likely make the equipment floor slippery and hazardous.

Dry wells need ventilation in accordance with local building and safety codes. Also, some designers prefer to dehumidify dry wells to prevent condensation on discharge piping and prolong the life of electrical equipment. Consult local building and safety codes for access and egress requirements.

STANDBY PUMPS. Reliable pumping performance must allow for adverse as well as normal circumstances. EPA has developed design guidelines[16,17] for reliability of treatment plant components, including pumps. Reliability considerations for pumping include the following:

- Generally, provide a standby pump for each set of pumps that performs the same function.
- Provide enough pumping capacity for a set of pumps to remove the largest pump from service and handle the peak flow with the remaining pumps.
- In some cases, one pump may serve as standby for more than one set of pumps.

FUTURE EXPANSION. When designing a pump station, allow for required future capacities. Three alternatives for future expansion are described below.

As one option, the original pumping units can be modified by installing a new impeller with a larger diameter. This option may require installing larger motors, starters, controls, or electrical systems. The least expensive modification would be increasing the impeller diameter if the motors, starters, standby generator, electrical system, and controls originally specified are large enough to power the pumps with larger impellers.

Future expansion can be achieved by adding pumping units if the structure and the piping are designed initially to accept additional units. As best practice, the added units should be identical to the original pumps. This will reduce maintenance costs since the parts will be interchangeable. If identical units are not practical, then the piping should be designed to handle the larger units when added.

Although capacity can be expanded by replacing pumps and motors with larger units, this would be a last-resort alternative, generally applied only to large expansions that preclude use of the original pumps. This type of expansion must account for pipe sizes, electrical capacity, and other elements.

PIPING AND VALVES. Piping material, sizes, and arrangements, particularly where high-viscosity sludges or chemicals are involved, are very important. Poor piping design alone could result in failure of the pump to perform as designed.

In recent years, the choice of materials for pressure piping has expanded greatly. Available materials include carbon steel, stainless steel, ductile iron, glass- or Teflon-coated pipe, concrete (with and without plastic liners), and plastic pipe. The material selected must be capable of withstanding the shutoff head of the pump as well as hydraulic transients such as water hammer. Preferably, the material should not abrade or corrode even after extended use.

Suction and discharge piping, valves, and fittings should be adequately supported to ensure that no weight is supported by the pump flanges. In larger installations, flexible suction and discharge connections to the pump are preferred.

Pipe sizes are selected on the basis of sustaining a reasonable head loss without substantially reducing velocity. In small plants, good design practice provides for suction pipes 100 mm (4 in.) in diameter or larger. For wastewater applications, the pipe diameter should be at least double the diameter of the largest sphere size that can pass the pump impeller. However, the designer must also consider the velocity required to flush the lines and prevent solids deposition.

Excessive velocities in piping systems can cause breakthrough in clarifiers and thickeners, resulting in low sludge solids concentrations. An improperly sized sludge pipe can encourage grease accumulation. Where low velocities are unavoidable, clean-outs and high-pressure flushing connections are necessary.

Suction Piping. Primary design considerations for suction piping include the following:

- Keep suction piping as short and straight as practical to minimize suction lift requirements,
- Ensure uniform flow distribution to the pump impeller, and
- Provide separate suction pipelines for each pump.

As good design practice, the suction piping should not include any elevated sections where air or gas could accumulate. Trapped air or gases can produce an air lock that restricts the flow to the pump. Also, avoid changes in size of horizontal suction piping. If reducers are required, the eccentric type should be selected and installed so the top of the suction pipe profile is level. Suction pipes should follow a slight upward slope from the pump to the wet well so that entrapped air will be released to the wet well. Some designers prefer a straight suction with a concentric flare enlarged at the point of

---

suction into the wet well. Other designers prefer a turned-down elbow with a flared fitting. However, if an elbow is used, a 12-mm (0.5-in.) diameter hole must be drilled in the top to allow trapped air and gases to escape. A shutoff valve, normally provided on the suction side of each pump, allows easy removal of the pump from service for inspection and repairs.

In wastewater applications, shutoff valves are normally gear-assisted plug valves or resilient-seated gate valves. Standard gate valves are not normally employed because the liquid contains solids, grit, and stringy materials that can prevent proper gate seating.

Suction piping should be sized to produce a velocity of at least 0.9 m/s (3 ft/sec) for all applications except services handling gritty materials and sludges. For these services, the velocity should be at least 1.2 m/s (4 ft/sec). The pump manufacturer will recommend maximum suction velocities. Suction pipelines conveying sludges and gritty materials normally have a 25- to 50-mm (1- to 2-in.) hose bibb connection so that the pipeline can be back-flushed periodically with water, steam, or compressed air.

Discharge Piping. A check valve followed by a shutoff valve is usually provided on the discharge side of each pump. The check valve will preferably be of the outside lever and weight type or outside lever and spring type as a means of preventing check valve slamming, visually checking pump operation, and moving the flapper to remove debris or rags. The check valve should always be mounted in a horizontal position. Check valves mounted in a vertical position accumulate debris on the top side of the flapper which restricts proper opening of the check valve.

Discharge piping systems are arranged to minimize dead zones where solids can accumulate (see Figure 5.27). Design velocities in discharge lines should range from 0.6 to 3 m/s (2.0 to 10.0 ft/sec).

Force Mains. The force main normally is sized for current and future needs so that it will convey wastewater or sludges without creating friction head loss beyond the head capacity capability of the selected pumps. In short-length systems where friction and fitting losses are a small percentage of the total head, higher velocities can be used than in force mains of considerable length. In a multiple centrifugal pump system, as each pump goes into service the friction head increases. Consequently, the discharge flow rate of each pump in service decreases. In certain applications, operating additional pumps may not result in significant increases in total discharge flow. This condition can result when the engineer does not fully analyze the sizing of the force main and the pumps selected for multiple-pump service.

The wastewater velocity in the force main should be at least 0.6 to 1.1 m/s (2 to 3.5 ft/sec). Long force mains, however, have been designed for lower velocities. Such applications need a means of periodically flushing the transmission pipe at a velocity of 1.1 m/s (3.5 ft/sec) by putting two or more

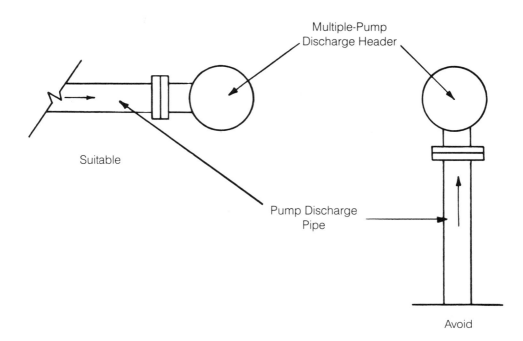

Figure 5.27 Horizontal pump connections with header pipes to minimize solids accumulations.

pumps in service at the same time or by using a pipe cleaner. Periodic flushing may require additional wet well volume to complete the flushing cycle. If the force main is to be sized for future flow, and current flows will not achieve flushing velocities, consider multiple force mains.

Good design practice consists of installing the force main on a gradual slope from the pump at the low end to the discharge point at the high end. This practice allows entrained air and gas to escape at the discharge point. Where a gradual slope would be impractical, air release valves are required at all high points in the force main. Where a high point in the force main is above the discharge elevation, combined air release and vacuum breaker valves are necessary.

The force main and piping on the discharge side of the pump need to be pressure rated to withstand the maximum hydraulic head on the system, including abnormal pressures that may be produced by hydraulic transients such as water hammer and surge pressures.

Surge Relief Protection. The sudden closing of a valve or hydrant, a power failure, or the starting and stopping of a water column as with reciprocating pumps can create enormous pressure variations in the pipeline. These pressure variations, called water hammer, are caused by any sudden change in flow inertia. This results in an oscillating pressure wave that travels at the speed of sound. The piping system design pressure must include an allowance for water hammer unless the system includes a measure for reducing the magnitude of the surge pressure that is likely to be created in the piping system. Because water hammer is a complex phenomenon, the assistance of an experienced hydraulics expert should be sought, particularly for complex pumping systems.

In small installations with low static head and short-length force mains, a standard check valve with external lever and weight may be adequate. By adjustment under actual operating conditions, the weight can be locked on the lever in a position such that the leverage causes the check valve to close quickly enough to prevent it from slamming. In some installations, it has been found necessary to add weights to those furnished with the check valve or to extend the length of the lever arm.

For larger installations with high static head, or low static head and force mains longer than 300 m (1000 ft), special valves, such as automatic cone or ball valves, may be necessary. These are constructed to close and open slowly as a means of reducing water hammer. Power-operated automatic cone or ball valves combine the functions of check and stop valves. For such installations, a gate valve on the discharge side of the cone or ball valve has often helped for inspection, repairs, and emergencies. Special design precautions, such as standby hydraulic systems, are necessary to control water hammer and prevent backflow through the pump when power failures occur.

Surge relief valves have been used to control water hammer. To relieve pressure surges, these valves are designed to open at a preset pressure and drain liquid to the wet well. Hydropneumatic pressure tanks can also be used to control water hammer. These devices consist of a pressure vessel containing air and water that act to dampen pressure surges.

Reciprocating pumps, such as plunger-type sludge pumps, should include air chambers on the suction and discharge side of each pump. An air connection for compressed air is necessary on the top of each vent to displace the water that gradually accumulates in the air chamber.

Air Vents and Blow Offs. All pumps have vent fittings. On vertical pumps, the vent opening is located on the top of the pump casing near the stuffing box. On horizontal pumps, the vent opening is on the top of the pump volute. A separate valved pipeline, preferably 13 mm (0.5 in.) in diameter, should extend from each pump vent to the wet well on a rising grade and without any low spots. Do not vent pumps to a common vent pipe.

F_{LOW} MEASUREMENT

Treatment plant process control decisions depend on measurements of a number of parameters, including flow rate. Plant streams typically measured include

---

Design of Municipal Wastewater Treatment Plants

- Influent,
- Primary clarifier sludge,
- Return activated sludge,
- Waste activated sludge,
- Digested sludge,
- Digester supernatant,
- Return and recycle flows (for example, trickling filter recycle),
- Disinfection process influent, and
- Effluent.

The above streams vary widely in physical characteristics that affect flow meter design. Devices used to measure influent, sludges, and certain recycle flows are designed to minimize problems associated with grease, grit, snails, rags, and solid materials. Final clarifier effluent and plant effluent contain much smaller quantities of these problem materials.

All flow meters perform two functions:

- Controlling the waste stream in a manner that causes a known relationship between flow rate and a measurable parameter such as depth, velocity, or pressure drop, and
- Measuring ("sensing") the parameter related to flow rate.

Devices used to control the waste stream are called primary devices. Devices used to measure depth, velocity, or pressure drop are called sensing devices.

For this discussion, flow meters are divided into two broad categories:

- Those that operate under open channel flow conditions, and
- Those that operate under pressure or full-pipe flows.

The following considerations apply to design of both primary and sensing devices for specific applications:

- Flow condition (that is, pressure or open channel);
- Available head loss;
- Physical characteristics of the measured stream;
- Required accuracy and repeatability;
- Spatial requirements and configuration;
- Peak to minimum flow rates (if the peak:minimum flow ratio is greater than 6:1 and the meter is sized for peak flow, the meter may not accurately measure low flows);
- Capital and operating costs; and
- Maintenance requirements.

ACCURACY. A note of caution applies to general accuracy of flow meters. Flow measurements are never exact and can rarely be precisely repeated. Inaccuracies are generally related to hydraulic variables such as velocity and pressure, which are, at best, difficult to control precisely. Additional inaccuracy related to calibration of sensing devices and recording equipment further compounds total error in recorded flow rate. The typically acceptable range for recorded flow versus actual flow is ±3 to 5%. With this range of acceptable accuracy, two acceptable, similar flow meters measuring the same flow rate can differ by as much as 10%.

PRIMARY DEVICES. Table 5.9 summarizes the advantages and limitations of some of the types of primary devices that are described below.

Flumes. Flumes, used to measure open channel flow conditions, consist of a channel constriction (throat) that causes the measured flow to achieve critical depth. Hydraulic jumps occur downstream where channel slopes permit subcritical flow. Flow rate through the flume depends on liquid depth upstream of the flume. Flumes are capable of accuracies in the 3 to 5% range. Their advantages and disadvantages are listed below.

- Advantages
  - Self-cleaning with respect to solids,
  - Moderate head loss, and
  - Calibration can be checked with manual measurements.
- Disadvantages
  - High cost,
  - Less accuracy when compared to pressure meters, and
  - Relatively long and straight approach channel required.

The list below includes general requirements for the design of flume installations:

- Locate a flume in a straight section of the open channel, without bends immediately upstream. As a general rule, avoid any bends or other flow disturbances within 10 throat widths of the entrance to the flume.
- Ensure a uniform approach velocity across the approach channel and avoid channel turbulence and waves.
- Avoid upstream solids deposition, which can occur if the flume approach velocity is less than 0.6 m/s (2 ft/sec).
- Account for head loss in overall plant hydraulic design.
- Consider the impact on accuracy of submergence due to downstream backwater.

Table 5.9   Primary devices summary.

Primary element	Advantages	Limitations
Open channel		
Parshall flume	Self-cleaning.	High cost.
Palmer–Bowlus flume	Self-cleaning. Suitable for installation in pipelines.	High cost.
Rectangular weir	Low cost. Measure wide flow range.	Upstream solids accumulation. May require ventilation.
V-notch weir	Low cost. High degree of accuracy.	Upstream solids accumulation. Narrow flow range.
Compound weir	Accurate over wide flow range.	Upstream solids accumulation.
Pressure		
Venturi tube	High degree of accuracy.	Subject to plugging. High cost. Pressure probes require periodic maintenance.
Velocity flow tubes	Some types are more economical than are Venturi tubes.	Lower accuracy than Venturi tubes.

Avoid submergence of the flume if possible. Where the flow difference between that for initial operation of the plant and the future design flow is large, consider temporary installation of a smaller flume liner in the throat of the larger flume. To preclude low velocities in the upstream and downstream channels, temporary brick fillers can be installed in the channels to reduce the cross section.

PARSHALL FLUME. The Parshall flume, originally developed for monitoring irrigation flow, can be formed from many different materials. It is produced commercially of metal and reinforced fiberglass materials, with throat widths ranging from 25 mm to 15 m (1 in. to 50 ft). For permanent installations, the manufactured flume can be used as the interior form for a cast-in-place concrete structure.

Figure 5.28 shows a Parshall flume.[18] The ratio H_b:H_a (expressed as a percentage) is called submergence and may be used to define free flow conditions (see Table 5.10). In cases where flumes are designed for free flow, the flow rate is proportional to the depth at a single point (H_a), which is inside the flume.

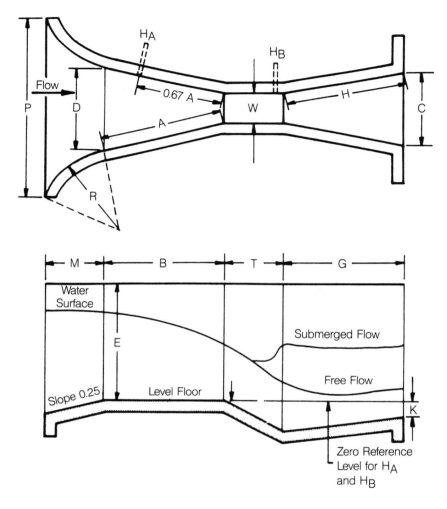

Figure 5.28 Parshall flume.

Table 5.10 Free flow conditions for Parshall flume.

Throat width, cm[a]	Maximum submergence $H_a/H_b \times 100$ for free flow condition, %
≤ 7.6	50
15.2	60
22.9	60
> 22.9 < 30.5	70
≥ 30.5	67

[a] cm × 2.54 = in.

Free flow designs are preferable to submerged designs since depth measurement is required at only one point. However, where available head loss is limited, the Parshall flume can be designed for a submerged flow condition. In submerged designs, depth must be measured at two locations, H_a and H_b. Rating curves and submergence correction factors may be found in "Stevens Water Resources Data Books"[19] and the handbook of the Soil Conservation Service.[20] Maximum:minimum flow ratios are generally limited to a range of 6:1 to 8:1.

PALMER–BOWLUS FLUME. The basic hydraulic principle for this type of flume is the same as that for a Parshall flume. The constriction of the sides and the step up in the flume bottom cause the flow to achieve critical depth. Critical depth can then be related to the flow rate.

The Palmer–Bowlus flume is available commercially in several materials, including fiberglass and steel. As its main advantage, the Palmer–Bowlus flume is easily adaptable to the circular cross section of pipelines.

Figure 5.29 shows the standard configuration of a Palmer–Bowlus flume.[18] Several manufacturers produce a Palmer–Bowlus prefabricated liner that can be grouted in circular pipelines at manholes or other structures. These liners can also be used as an interior form for a cast-in-place concrete

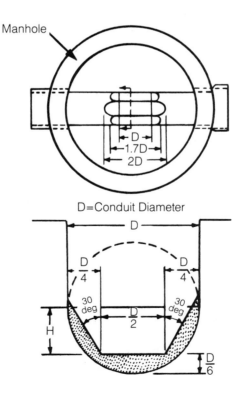

Figure 5.29 Standard Palmer–Bowlus flume dimensions.

structure. The designer should request rating curves from each manufacturer because the slightly different manufactured configurations can result in significant variations in rating curves. Figure 5.30 illustrates three different configurations.[2] Typical peak:minimum flow ratios for a Palmer–Bowlus flume range from 6:1 to 8:1.

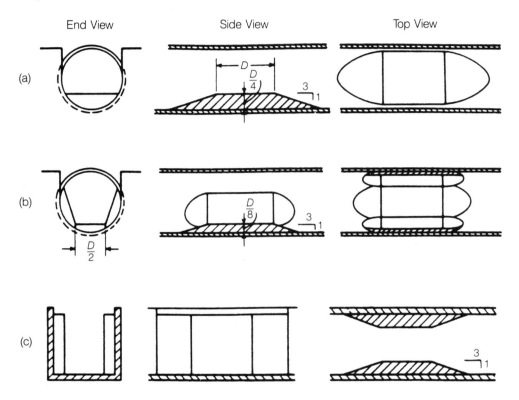

Figure 5.30   Three shapes of Palmer–Bowlus flumes.

Weirs. Weirs are commonly used as primary devices because of their high degree of accuracy, low capital and operating cost, and ease of installation. For flow measurement, sharp-crested weirs are the most common due to their greater accuracy. Weir design nomenclature is presented by Figure 5.31. Sharp-crested weir plates may be fabricated of materials with sufficient stiffness to withstand hydrostatic pressure requirements; such materials include steel, stainless steel, fiberglass, and aluminum.

For wastewater applications, solids and debris tend to settle just upstream from the weir. These deposits can create odor and corrosion problems as well as affect the accuracy of flow measurements. In waste streams carrying grit, solids, or other debris, weir designs should provide for solids removal by

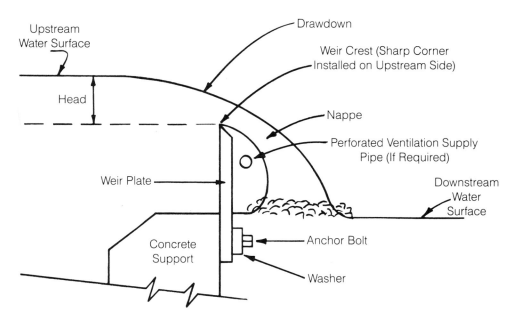

Figure 5.31   Weir nomenclature.

manual means or by periodic flushing. Advantages and disadvantages of weirs are listed below.

- Advantages
  - Low cost,
  - Easy to install, and
  - Easy to estimate flow from standard equations and nomographs.
- Disadvantages
  - High head loss,
  - Must be periodically cleaned (not suitable for channels carrying excessive solids), and
  - Accuracy affected by excessive approach velocities and debris.

A list of general design considerations follows:

- Construct the weir of a thin plate with a straight edge or a thick plate with a knife edge.
- To reduce approach velocity, the height of the weir (from the bottom of the channel to the crest) should equal at least twice the expected liquid depth above the weir crest.
- Provide for watertight connections between the weir plate and channel.
- To prevent a vacuum below the nappe, ventilate suppressed weirs with a perforated small-diameter pipe.

- The weir must be level.
- Provide for periodic cleaning of the weir crest and approach channel.
- Measure the head at a point upstream from the weir a distance of at least four times the expected liquid depth above the weir crest.
- Calibrate weirs under field conditions to ensure accurate measurements.
- Avoid use of weir equations and rating curves for submerged conditions.
- Provide for minimum flow depth over the weir crest of at least 0.06 m (0.2 ft) to ensure that a nappe will form.
- Use velocity head correction for approach velocities exceeding 0.15 m/s (0.5 ft/sec) for accurate measurement of flow. A velocity head correction adds $v^2/2g$ to the measured upstream flow depth to determine the total head.

Rectangular weirs are classified as contracted or suppressed. A contracted weir has end walls protruding from the channel sides to contract the nappe from the channel sides. A suppressed weir crest extends the full width of the channel. Suppressed designs usually require ventilation to prevent formation of a vacuum under the nappe, which will affect flow measurements. Figure 5.32 illustrates the difference between contracted and suppressed weir designs.

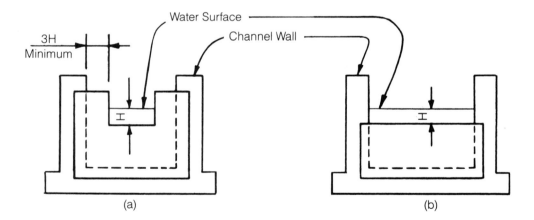

Figure 5.32   Rectangular weirs: (a) contracted and (b) suppressed.

V-notch or triangular weirs allow for more accurate flow measurements than rectangular weirs since small increases in flow over the V-notch weirs result in larger increases in measured head over the weir. V-notch weirs are thus preferable to rectangular weirs for controlling chemical feed rates. As an advantage offered by V-notch weir designs, they do not require ventilation.

Their disadvantage is that a given flow will require a greater head loss than that for rectangular weirs. Figure 5.33 shows a V-notch weir.

Compound weirs can measure low flows accurately, yet provide for measuring greater peak flows than a V-notch weir. Under low flows, the compound weir functions as a V-notch. To measure high flows, the compound weir may include either contracted or suppressed rectangular sections. Flow measurements in the transition range between the V-notch and the rectangle will be less accurate than either the V-notch or rectangular designs alone. Figure 5.34 illustrates a compound weir with a contracted rectangular configuration.

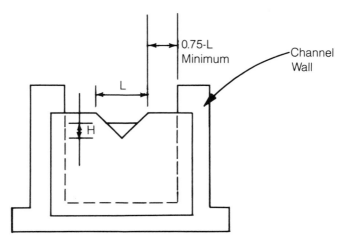

Figure 5.33 V-notch weir.

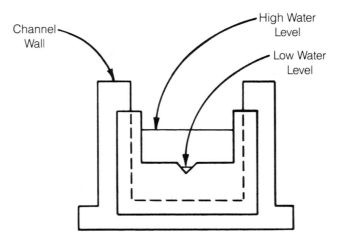

Figure 5.34   Compound weir.

Flow Tubes. In this chapter, the term flow tube refers generically to primary devices installed in pressurized pipelines. A flow tube may be as simple as a straight piece of pipe in a pressure system or as sophisticated as a venturi tube designed to relate flow rate to pressure drop across a constriction in the pipe.

Some general design considerations applicable to all flow tubes are listed below:

- Accurate measurements can only be obtained when the flow tube is full of liquid. Where possible, install flow tubes in vertical piping runs to ensure full flow conditions and to minimize buildup of entrained air.
- Install flow tubes to minimize inaccuracies due to hydraulic transients caused by bends or other fittings that create flow disturbances. As good practice, provide a straight run of pipe equal in length to at least 10 times the diameter of the flow tube immediately upstream of the flow tube. Provide a length of straight downstream piping equal to at least five diameters. Verify these distances with the flow meter manufacturer.
- Keep velocities through the flow tube in the range of 1.22 to 2.44 m/s (4 to 8 ft/sec) to minimize buildup of grease and solids.
- Provide for bypass piping to allow removal of the flow tube for maintenance, cleaning, or calibration.
- As a general rule, maintain velocities through the flow tube of over 0.7 m/s (2 ft/sec) for accurate results. Consult manufacturers for specific minimum velocity recommendations.

The venturi tube operates based on the principle that liquid flowing through a convergence and constriction of known shape will cause a pressure drop proportional to the flow rate squared. Pressure drop is measured by comparing pressure upstream from the convergence to pressure at the throat (constriction). Figure 5.35 shows a typical Venturi tube.[15]

Venturi tubes in clean water systems can measure flow rates with accuracies approaching ±0.75%. Accuracies for wastewater or sludges will generally depend on keeping the pressure connections clean. Venturi tubes for any wastewater applications should be equipped with water purge systems, manual tap cleaners, or pressure sensors isolated from the liquid flow.

Ultrasonic, Doppler, and magnetic sensing devices, described in the following section, rely on liquid velocity through flow tubes to measure flow rate. For magnetic flow meters, the flow tube is generally furnished with the sensor, ready for installation. Flow tubes for ultrasonic and Doppler sensors may be either a straight run of standard pipe or factory mounted with the sensor. Where standard pipes are used for flow tubes, accurate measurements

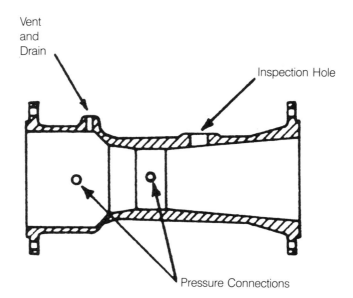

Figure 5.35 Venturi tube.

require the sensing device to be compatible with the exact interior dimension of the pipe and any lining material.

SENSING DEVICES. Sensing devices are required to sense the measurable parameter created by the primary device. Table 5.11 summarizes the advantages and limitations of different types of sensing devices.

Liquid-level Measurement Systems. Liquid-level measurement systems are used to measure the upstream head caused by primary devices mounted in open channels or in pipes flowing less than full. Common instruments include floats, displacement transmitters, capacitance probes, captive fluid systems, bubblers, and ultrasonic level transmitters. These devices are discussed in the previous "Pump Control Systems" section of this chapter.

Pressure Measurement Systems. Pressure measurement systems are used to measure the head loss caused by primary devices mounted in pipes that are flowing full. This application requires measurement of differential pressure to calculate the flow. Therefore, the pipe is tapped upstream of the constriction and at the constriction. Tubing connects these taps to opposite sides of a diaphragm that is mechanically connected to a transducer. The transducer converts the motion of the diaphragm [typically less than 0.1 mm (0.02 in.)] into either a pneumatic or electrical current signal.

Differential pressure transmitters use various technologies to convert the motion of the diaphragm into electrical or pneumatic signals. These technology types include resistance (strain gauges), capacitance, resonance,

---

Table 5.11   Sensing devices summary.

Device	Typical accuracy	Advantages	Limitations
Gravity			
Mechanical float	±1%	Simplicity of operation.	Requires frequent cleaning.
Bubbler	±0.5%	Relatively maintenance free. High degree of accuracy.	Less accurate in low head applications.
Capacitance probe	±2%	Low cost.	Requires frequent cleaning.
Ultrasonic sensor	±2%	Easy to calibrate. Nonfouling.	Affected by high winds, temperature, relative humidity, and foam.
Pressure			
Ultrasonic velocity	±1%	Simple to install. Low cost.	Requires periodic cleaning. Suspended solids, grease, or gas bubbles may affect readings.
Ultrasonic Doppler	±1%	Simple to install. Low cost. Available as clamp-on device for existing pipes.	Suspended solids must be present in flow. Pipe linings can affect results.
Magnetic	±1%	Available with wide range of lining materials. High accuracy over wide range.	Requires proper grounding. High cost. Requires periodic cleaning.

piezoelectric, and force balance. Typical accuracy of these sensing devices is 0.25% of measured span. Since the measured pressure drop varies with the square of the flow as discussed above, these meters must either use a square law scale on their output meter or incorporate a square root extractor to linearize the output signal with respect to flow.

Ultrasonic Transit Time Velocity Meter. The ultrasonic transit time velocity meter is an obstructionless flow-measuring device that can be installed in pipelines carrying liquids. This device usually consists of two ultrasonic transducers mounted in a diagonally opposed configuration to a precalibrated flow tube. The ultrasonic transducers are capable of sending and receiving ultrasonic pressure pulses. During operation, pulses are alternately transmitted through the fluid, against and then with the direction of flow. The pulse transit times, both upstream and downstream, are measured. The average fluid velocity is proportional to the difference between the two transit times. The meter calculates the time difference, and its output is, therefore, proportional to average fluid velocity.

The ultrasonic pulses can sometimes be blocked or attenuated by suspended solids or gas bubbles in the fluid. The manufacturers of modern meters claim that this problem has been overcome by the use of high-strength pulse trains, very high-gain preamplifiers that are sensitive to even severely attenuated signals, combined with proprietary circuitry that rejects pulses that do not conform to a particular pattern. Designers should research manufacturers' previous experience before specifying these flow meters for liquid streams containing significant quantities of grease, air, or suspended solids.

Accuracy at velocities of 0.3 m/s (1 ft/sec) or greater is specified to be within ±1% of reading for Newtonian liquids with a Reynolds number above 100 000. Most wastewater applications for flow measurement occur within this range.

Ultrasonic Doppler Velocity Meter. Ultrasonic Doppler meters are similar in configuration to the ultrasonic meters described above. However, the operating principle of Doppler meters differs as they can be used only for fluids with entrained solids or gas. They transmit a continuous beam of ultrasonic waves that are reflected to a receiver by suspended solids or gas bubbles in the liquid. The receiver measures the frequency of the reflections, which is Doppler-shifted in proportion to the velocity of the suspended solids or gas bubbles.

The transmitter and receiver may be a single ultrasonic transducer or they may be mounted separately. The transducer(s) may be factory mounted on a flow tube or may be supplied as clamp-on devices for use on standard lengths of pipe. Flow meter accuracies and design precautions are similar to those described for ultrasonic velocity meters.

Magnetic Flow Meter. The magnetic flow meter operates based on the principle that any conductor, whether a bar of steel or a column of conductive liquid, that passes through the lines of force of a fixed magnetic field will generate an electromotive force (dc voltage) directly proportional to the rate of the conductor's movement through the field. Figure 5.36 shows schematic operation of a magnetic flow meter.

The meter must always run full. It can be installed in any orientation but must be sized for a minimum velocity of 1.5 to 2.4 m/s (5 to 8 ft/sec) for

---

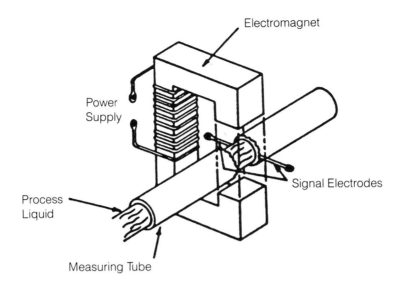

Electromagnet

Power
Supply

Process
Liquid

Signal Electrodes

Measuring Tube

Figure 5.36   Schematic of magnetic flow meter.

sludges to ensure scouring. The system accuracy is usually stated to be
±0.5% of meter scale for a velocity range of 0.9 to 9.1 m/s (3 to 30 ft/sec).
Below 0.9 m/s (3 ft/sec), the system accuracy deteriorates to ±3% of scale at
0.3 m/s (1 ft/sec).

To achieve reliable results, magnetic flow meters require proper electrical
grounding that includes the magnetic flow meter, piping system, and liquid
stream. The importance of proper grounding cannot be overemphasized. In
nonconductive piping systems or where the flow tube is lined, grounding
rings are needed at both ends of the flow tube, in addition to grounding the
meter and flow tube. Most manufacturers provide instructions on proper
grounding for specific applications. Good practice to ensure reliable measurements often requires exceeding a manufacturer's minimum grounding
requirements.

Most magnetic flow tubes in wastewater service are lined to protect the
tube from corrosion and abrasion. Typical lining materials for wastewater
service include polytetrafluorethylene, neoprene, rubber, and polyurethane.
Liner selection depends on the corrosive and abrasive properties of the liquid.

Grease and solids in the liquid flow frequently foul electrodes. Therefore,
as a means of mitigating fouling problems, most magnetic flow meters can be
furnished with ultrasonic cleaning devices. Electrode designs must always
provide for convenient removal and replacement for periodic inspection and
cleaning. Typical electrode materials include platinum, tantalum, stainless
steel, and Hastelloy C. Primary considerations in electrode material selection
are cost and service life.

Most manufacturers claim that magnetic flow meters can be installed in
any piping configuration, even near bends, contractions, and other arrange-

ments that create turbulence and hydraulic transients. Nevertheless, where reasonable accuracy is required over a wide range of flow, good practice dictates straight piping runs upstream and downstream of the magnetic flow meter.

SIGNALS. The most commonly used form of signal transmission in the waste treatment plant instrumentation is milliamp direct current (mA–dc). Although dc voltage transmission is still used, it has been displaced almost entirely by dc transmission in new installations. The major disadvantages of transmitted dc voltage signals are line resistance loss and errors induced by electromagnetic radiation. The major advantage of voltage signals is less power consumption than that of dc signals. The major advantages of mA–dc current systems are rapid transmission of information, relative immunity to environmental conditions, signal transmission over long distances without degradation, and, in most cases, no power source requirement at the transmitter. These advantages usually outweigh the disadvantage of the higher power consumption compared to voltage signals.

Instrumentation systems of most wastewater equipment manufacturers now use the 4- to 20-mA–dc signal level. This signal is now the standard although other ranges such as 10 to 50 and 1 to 5 mA have been available in the past. All of the above ranges provide a live zero, which allows easy detection of a loss of power or an open circuit. Although current is used for transmission, most receiving instruments convert the signal into a higher level voltage signal, in the range of 1 to 5 V, 0.5 to 2.5 V, or 0.25 to 1.25 V–dc.

The designer should recognize three common problems experienced in signal wiring:

- Some sensing devices can be located only a limited distance from recording equipment. Check with individual manufacturers for specific applications.
- If signal wires are installed in conduits alongside higher voltage power distribution wiring, the higher voltage lines can create an electromagnetic field that can disrupt the mA–dc signal and cause erroneous readings.
- Ground loops can shunt some (or all) of the signal current around the receiving device. These can be avoided by grounding each current loop in only one location, preferably at the source of the loop's electrical power.

RECORDING DEVICES. These devices include circular chart recorders, strip chart recorders, digital recorders, and flow totalizers, each described below.

Circular Chart Recorders. These devices, commonly used for trend monitoring, produce time-based records on flat circular charts. Their chief advantage is that an entire hour, shift, day, or other period of interest is displayed as a complete cycle. These recorders are frequently combined with controllers or cam programmers to provide control capabilities.

A chart radius usually ranges from 8 to 30 cm (3 to 12 in.). However, active graphing widths are usually 2 to 3 cm shorter than the radius to account for the driving hub and scale legend. The charts have concentric circular lines that form the scale for the indicated variable. Charts with equal intervals between concentric circles are most common. Straight or curved radial lines indicate the variable used as the base for rotation. Many circular recorders have a quick change feature that allows charts to be replaced rapidly. Some models feature automatic chart changers that provide continuous sets of records.

CHART DRIVES. Synchronous electrical drives and manually or solenoid-wound spring motors are used for chart rotation. In most cases, the chart is driven from a center spindle, but some edge-mounted friction drives also exist. Synchronous motors produce the most uniform rotation. Spring-driven systems are popular for areas that lack power or are hazardous. Electrically wound spring drives are particularly advantageous where temporary power interruptions are likely.

RECORDING MECHANISMS. Felt-tipped pens are most often used to produce the traces on chart paper, but capillary and pressurized ink systems also are used. Inkless methods are now widely used, especially where the recorders are unattended or are operated only after long intervals of nonuse. These methods solve the problem of wet ink, but their paper cost is high and their mechanisms are often expensive.

Thermal writing employs wire styli heated by electrical currents to mark specially coated chart paper. Electrosensitive recording uses a conductive paper that produces a trace on contact with the pen. Applying a potential of 200 to 300 V causes dielectric breakdown of the paper surface that results in a dark trace. Pressure sensitive or carbon transfer recording is also used.

Strip Chart Recorders. Strip chart recorders, widely used for monitoring and laboratory data presentation, combine the advantages of instantaneous graphic display with those of providing permanent records. The 10-cm (4-in.), 30-day strip chart recorder is used most often in areas where a large number of instruments are mounted and space is limited.

CHARTS. Certain chart parameters may determine the usefulness of the recorder information. For example, resolution can depend on chart width; the importance of wave shape determines the need for rectilinear or curvilinear coordinates. In addition, z-fold paper, with each page consecutively numbered, provides a convenient means of locating specific data with minimal search. However, if large volumes of information must be examined, roll papers may be scanned rapidly using motor-driven viewers.

TYPES OF DRIVES. The transport, an electrical or mechanical mechanism, moves paper through the recorder. Drive accuracy is necessary because the time axis is inferred from a distance along the chart by assuming a known uniform velocity. Input signals that vary at a slow rate can be recorded at low speeds to conserve paper. Rapidly varying signals require high paper rates to provide sufficient resolution.

RECORDER MECHANISMS. Most commercially available strip chart recorders, actuated electrically, transform a voltage or current signal into a displacement of the writing mechanism. However, mechanically and pneumatically actuated devices are also available.

WRITING TECHNIQUES. Several writing techniques are available for both continuous and dotted representation. Capillary pens supply a constant flow of ink at the paper surface, but at high writing speeds the line becomes thin and may even disappear. Therefore, pressurized systems are often used. They provide the additional advantage of reducing friction with the constant cushion of ink between paper and pen. The available systems offer a choice of felt or ball-point pens.

Inkless writing methods generally require special paper. Dot writing recorders commonly use ink or pressure writing techniques in which pins or styli are pressed on the paper.

Digital Recorders. Computer systems, now widely used for recording, offer the advantages of a much lower cost, where more than four variables must be recorded and of higher reliability than the mechanical mechanisms. Output can be viewed either on a color video display or a printed hard copy. Accessing historical data is much faster with computerized systems since the operator can simply request the dates of interest for nearly instantaneous computer retrieval of the data linked to those dates.

Flow Totalizers. In flow measuring systems, a device known as a totalizer is often used. The totalizer may be operated mechanically, either directly from the mechanical movement of the manometer transducer or from the movement of the chart pen arm. Such devices often include a cam, shaped to translate nonlinear measurements to a linear record scale.

Other devices receive the meter system signal—time pulse or pulse frequency (pneumatic or electric)—and convert it to a digital readout representing the accumulated flow. Industries use these devices with a calibration

---

constant to convert the digital number to engineering units. However, municipal wastewater treatment applications customarily employ units that readout directly in engineering units, sometimes using a multiple of 10 as a multiplier. When exceptional measurement accuracy is necessary, such as checking the delivery of a pump, the change in totalizer reading probably constitutes the most accurate method for measuring flow rate.

APPLICATIONS. The 1989 WPCF Design Practice Survey collected responses from 15 design firms as to their flow meter preferences for various wastewater applications. Table 5.12 summarizes the results of the survey.

Table 5.12   WPCF Design Practice Survey—flow meter preferences.

Type of meter	Raw wastewater	Primary sludge	Digested sludge	Return activated sludge	Waste activated sludge	Final effluent
Flumes						
Parshall	4			1		3
Other	3			2		2
Weir	1			2		6
Venturi	1	1				1
Ultrasonic	1	1	1	1		1
Doppler		2	2	2	1	
Propeller						1
Magnetic	5	11	11	9	11	4
Totals	15	15	14	17	12	18

REFERENCES

1. King, H.W., and Brater, E.F., "Handbook of Hydraulics." McGraw-Hill, Inc., New York, N.Y. (1963).
2. "Gravity Sanitary Sewer Design and Construction." Manual of Practice No. FD-5, Water Pollut. Control Fed., Alexandria, Va. (1982).
3. "Hydraulic Institute Engineering Data Book." 2nd Ed., Hydraulic Inst., Cleveland, Ohio (1991).
4. Shaw, G.V., and Loomis, A.W. (Eds.), "Cameron Hydraulic Data." Ingersoll–Rand Co., New York, N.Y. (1958).
5. Davis, C., and Sorensen, K., "Handbook of Applied Hydraulics." McGraw-Hill, Inc., New York, N.Y. (1969).

6. Benefield, L., *et al.*, "Treatment Plant Hydraulics for Environmental Engineers." Prentice–Hall, Englewood Cliffs, N.J. (1984).

7. Chow, V.T., "Open-Channel Hydraulics." McGraw-Hill, Inc., New York, N.Y. (1959).

8. Hicks, T.G., and Edwards, P.E., "Pump Application Engineering." McGraw-Hill Inc., New York, N.Y. (1971).

9. "Hydraulic Institute Standards." 14th Ed., Hydraulic Inst., Cleveland, Ohio (1983).

10. Karassik, I.J., *et al.*, "Pump Handbook." McGraw-Hill, Inc., New York, N.Y. (1976).

11. Walker, R., "Pump Selection: A Consulting Engineer's Manual." Ann Arbor Sci. Pub. Inc., Mich. (1972).

12. Karassik, I., and Carter, R., "Centrifugal Pumps." F.W. Dodge Corp., New York, N.Y. (1960).

13. "Pumping Station Design." Robert L. Sanks *et al.* (Eds.), Butterworth Pub., Stone, Mass. (1989).

14. "Prime Movers." Manual of Practice No. OM-5, Water Pollut. Control Fed., Alexandria, Va. (1984).

15. "Instrumentation in Wastewater Treatment Plants." Manual of Practice No. 21, Water Pollut. Control Fed., Alexandria, Va. (1978).

16. "Design Criteria for Mechanical, Electric, and Fluid System and Component Reliability." U.S. EPA, Office Water Program Oper., Washington, D.C. (1975).

17. "Municipal Wastewater Treatment Works Construction Grants Program." U.S. EPA, Washington, D.C. (1975).

18. "Existing Sewer Evaluation and Rehabilitation." Manual of Practice No. FD-6, Water Pollut. Control Fed., Alexandria, Va. (1982).

19. "Stevens Water Resources Data Books." 4th Ed., Leopold & Stevens, Inc., Beaverton, Oreg. (1987).

20. "SCS National Engineering Handbook." Soil Conserv. Serv., U.S. Dep. Agric., Washington, D.C. (1985).

Chapter 6
Occupational Safety and Health

INTRODUCTION

Designing an efficient, safe wastewater treatment plant poses a complex and difficult task. Designers face the formidable challenge of ensuring that the plant will meet all performance criteria while incorporating applicable safety and health considerations.

The wastewater treatment industry has historically experienced one of the highest rates of accident frequency and severity of all industries reporting to the National Safety Council. Results of the 1989 WPCF safety survey[1] indicate that, in 1988, the frequency of accidents for the wastewater treatment industry was about 4.5 times greater than the average for all industries in 1987. Approximately 20% of the accidents that are directly attributed to physical conditions call for special design emphasis because many could be prevented during design. The other 80%, caused by work methods, could be reduced indirectly by design changes that avoid hazardous work activities. Of the about 40 000 people that staff treatment plants in the U.S., about 4000, at current annual rates, will have accidents or work-related illnesses that will keep them away from work for about 12 days.

Because of the high injury rate in wastewater facilities, personal injury liability is an increasing concern for designers. This chapter provides design safety considerations for designers of wastewater treatment systems. Information applies to existing facilities that are being renovated as well as new facilities. During the design stage, old problems can be corrected and new ones can be prevented.

With proper design, accidents can be prevented or reduced to a lower level. This chapter identifies occupational hazards resulting from physical conditions commonly found in wastewater treatment systems and provides guidelines for eliminating them. It does not address equipment or machinery design, which is the responsibility of the manufacturer. Only basic design considerations for occupational health and safety concerns related to treatment plant personnel are introduced. Therefore, this chapter does not cover all situations, including public safety and safety program management.

WPCF sources of safety information include "Safety and Health in Wastewater Treatment Systems," Manual of Practice No.1;[2] "Guidelines to Developing a Wastewater Safety Program," Manual of Practice No. SM-2;[3] and "Operation of Municipal Wastewater Treatment Plants," Manual of Practice No. 11.[4]

This chapter deals predominantly with on-site hazards for wastewater treatment workers. However, wastewater facilities can create hazards for surrounding neighbors. Fires, explosions, chemical leaks, air pollution, sludge disposal, and wastewater reuse are potential hazards. In addition, air pollution and odors may emanate from liquid treatment or solids-processing activities such as drying and incineration. Only currently recognized hazards are described in this chapter. Because few in-depth occupational health studies have been performed in treatment facilities, particularly studies of the effects of long-term exposures to biological agents and air contaminants, these potential hazards are not described.

Several chapters in this manual include information on occupational health and safety because many of these concerns should be treated as integral parts of the design of each plant element, not as afterthoughts.

LAWS, REGULATIONS, AND GUIDANCE

Governments, at all levels, establish regulations, requirements, and codes that contain minimum occupational health and safety standards. Laws, regulations, codes, and agency guidelines require or recommend occupational health and safety provisions in wastewater treatment facilities. These provide the facility owner and designer with minimum requirements and guidance. Additional safety provisions may be warranted.

Most plants must be designed to meet Occupational Safety and Health Act (OSH Act) requirements.[5] Examples of fixed items covered by the OSH Act include machine guards, belt guards, handrails, floor openings, fixed ladders, stairs, working surfaces, and means of egress.[6] Several industrial associations provide information for designers, including the Chlorine Institute, Inc.,[7] the Manufacturing Chemists Association,[8] the National Fire Protection Association (NFPA),[9] and the National Safety Council.[10] In addition, safety organizations such as the American Society of Safety Engineers and the American Industrial Hygiene Association are active in many states and larger communities.

Codes contain many occupational health and safety requirements. Because keeping up with them is a challenge, a computerized service now summarizes building codes.[11] This database contains summaries for seven categories, including life safety and fire prevention. Designers must also consider which codes apply to wastewater treatment facilities. A variance, if justified, may be requested.

OSH ACT AND FEDERAL REGULATIONS. The most significant statute is the OSH Act of 1970.[5] Its goal is to eliminate unsafe working conditions in private industry and private businesses. Although this law does not directly apply to publicly owned treatment facilities, many states have adopted, by reference, standards set in regulations of the Occupational Safety and Health Administration (OSHA). It is common practice for public facilities to meet these regulatory requirements. They apply directly to privately owned treatment plants that treat municipal or industrial wastes. Because the status of state OSH Act requirements frequently changes, designers need to check state and local requirements.

The OSH Act, the basic guidance document for all general industry safety activities, establishes the standards for design. OSHA is responsible for enforcement of these regulations. Individual states that have federally approved programs may also enforce these standards. The regulations (copies are available at the nearest OSHA office) cover general safety concerns found in all industries and are not specific to wastewater facilities. They must be interpreted and applied with proper judgment. Topics covered in these documents include

---

- General safety and health provisions;
- Walking/working surfaces;
- Means of egress;
- Powered platforms, manlifts, and vehicle-mounted platforms;
- Occupational health and environmental controls;
- Hazardous materials;
- Personal protective equipment;
- General environmental control;
- Medical and first aid;
- Fire protection;
- Compressed gas and compressed gas equipment;
- Material handling and storage;
- Machinery and machine guarding;
- Hand and portable power tools and other hand-held equipment;
- Electrical systems; and
- Toxic and hazardous substances.

STATE REQUIREMENTS. Most state environmental agencies have developed manuals for staff to use when reviewing plans and specifications. These contain safety items that are to be incorporated into design and construction. The following partial list from Pennsylvania's guidelines[12] provides typical examples:

- Approval of the Department of Labor and Industry for features affecting the safety of employees;
- Approval of the Bureau of Occupational Health for confined-space entry;
- Site fence;
- Handrail and guards;
- First-aid equipment;
- "No Smoking" signs in hazardous areas;
- Protective clothing and equipment such as air packs, goggles, gloves, hard hats, and safety harnesses;
- Portable blower and sufficient hose;
- Portable lighting;
- Warning signs for slippery areas, nonpotable water fixtures, low head clearance, open service manholes, hazardous chemical storage areas, and flammable fuel storage areas;
- Ventilation; and
- Eyewash fountains and safety showers with a 122 to 130°C (50 to 90°F) water supply.

EPA GUIDANCE. Facilities that are constructed using EPA grants must consider EPA guidance. (Although the EPA construction grant program is no longer funded, some funded facilities are still under construction.) An EPA publication[13] contains health- and safety-related guidance for design of treatment works. This publication does not provide detailed design standards, but it does indicate many hazards that should be considered. Another EPA publication that describes how to prepare operating and maintenance manuals includes outlines for sections on safety and emergency operation and response.[14]

LOCAL CODES. Many requirements of existing and proposed fire codes are more detailed and stringent than OSHA regulatory requirements. In some instances fire codes dictate layouts, construction materials, and safety equipment. Local fire departments usually have the authority to establish requirements for site layouts to allow access for emergency response equipment and firefighters; to establish the number, location, and type of fire hydrants; and to set other requirements.

In most cases, local fire departments have been given the authority to establish requirements for prevention, control, and mitigation of dangerous conditions related to hazardous materials and to provide information needed by emergency response personnel. For example, the Uniform Fire Code[15] definition of hazardous materials is very broad, and it includes materials that are physical hazards (compressed gas) and health hazards (gasoline).

This code provides definitions of hazardous materials and describes storage, dispensation, use, and handling requirements. Examples of materials commonly stored and used in wastewater plants that are classified as hazardous include digester gas, chlorine, sulfur dioxide, ammonia, acid, ozone, oxygen, sodium hydroxide, fuels, and hydrogen peroxide. Portions of the code describe provisions for facilities to treat ventilation exhaust from areas containing hazardous materials, including alarms, multiple self-contained breathing apparatus, seismic analysis of several system components, signage, spill control, drainage containment, ventilation, explosion venting or suppression, protection from vehicles, fire access roads, water supply, gas detection units, smoke alarms, and a storage plan.

The Uniform Fire Code establishes requirements for features to be included by the designer and, in some cases, requires specific analyses by the designer, such as seismic analyses.

NFPA RECOMMENDATIONS. The NFPA committee, which includes several WPCF members, is finalizing a draft of recommended fire protection practices for wastewater treatment plants.[9] This publication will establish guidance for protection against fire and explosion hazards in wastewater treatment plants, including hazard classification of specific areas and processes. The document will call for a specific fire risk evaluation as part of the project design process. It will list locations, descriptions, electrical classifications, fire and explosion hazards, recommended materials of construction,

---

recommended ventilation practices, and suggested fire protection measures associated with liquid stream treatment and solids treatment processes from municipal collection systems. Detailed information on sources of ignition, sources of hazards, and mitigation measures will also be presented.

ACCIDENT CAUSES AND INJURIES

ACCIDENT CAUSES. The conditions or acts that caused previous accidents or injuries are some of the most important sources of information for preventing future accidents.

"Unsafe conditions" refer to physical conditions that cause an accident, while "unsafe acts" refer to personal actions that cause an accident. Any accident can result from both physical and personal causes. For example, an injury resulting from a fall on stairs leading to a pipe gallery could be attributed to a variety of causes. A physical cause (unsafe condition) could be inadequate lighting on the stairs. A personal cause (unsafe act) could be a person running down the stairs.

A proper investigation into an accident can identify physical hazards and lead to their correction. Conditions that might lead to an accident should be avoided in the design stage, rather than corrected after an accident occurs.

Unsafe conditions include workplace congestion, defective equipment or tools, excessive noise, fire and explosion hazards, hazardous atmospheres, inadequate support, inadequate guards, inadequate warning systems, poor illumination, poor housekeeping, and poor ventilation.

Most of these conditions are addressed by OSHA standards, and many can be reduced or eliminated through proper design.

INJURY FREQUENCIES. According to the 1989 WPCF safety survey,[1] the treatment plant injury rate in 1988 was 40.89 injuries per million hours worked (Figure 6.1), nearly 4.5 times the average reported for all industries for 1987.

INJURY TYPES. "Injury type" refers to the medical category of the injury, such as sprain, fracture, burn, and so on. This information may indicate the need for physical corrective actions. For example, back strains may indicate that material handling needs attention by designers. Sprained ankles may indicate that walking or working surfaces should be carefully evaluated. The 1989 WPCF safety survey[1] indicates that many injuries result from routine activities such as lifting, pushing, or pulling objects. Figures 6.2 and 6.3 show 1988 data on injury types.

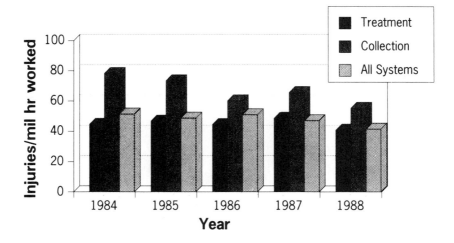

Figure 6.1   Five-year trend for treatment, collection, and all systems.

ACCIDENT LOCATION. This refers to the area in the plant where the injury occurred. A high injury rate at one location does not necessarily indicate a correctable condition, but it does call for special attention during design. Figure 6.4[1] shows 1988 data on plant locations where injuries occurred.

ERGONOMICS. An ideal work environment should complement worker expectations and minimize fatigue and stress without requiring performance of dangerous tasks or unusual physical or mental skills. Fitting the work space and machines to man is called ergonomics, human engineering, or human-factors engineering.[3,16-20]

Operation and maintenance requires a variety of physical and mental skills. Each designer makes assumptions about the physical characteristics and mental skills of the operating staff. If these assumptions differ greatly from actual staff characteristics or skills, the probable result will be more injuries. For example, if a design includes use of color-coded pipe or colored indicator lights, the designer has assumed that the staff have normal vision. As a result, color-blind staff are more likely to cause an accident (3.5% of the male population in the U.S. is, to some degree, color-blind). As a second example, a designer might assume that operators can readily lift and dump 23-kg (50-lb) bags of a chemical into a tank that is one meter (3 ft) above the floor. This imposes a risk of back injuries, especially for smaller, sedentary persons. A good reference for human physical characteristics and capabilities is available.[21]

The assumptions of designers and actual personnel practices (employee selection, training, job assignments, and so on) should be consistent. The owner is in the best position to ensure this consistency. There are many common expectations that should be noted by designers, or injuries may result. Examples include "red" signifies hot, an electrical switch in the "up" position signifies on or running, movement of a switch/dial to the right increases speed, a valve stem "up" is open, "green" is safe, clockwise rotation

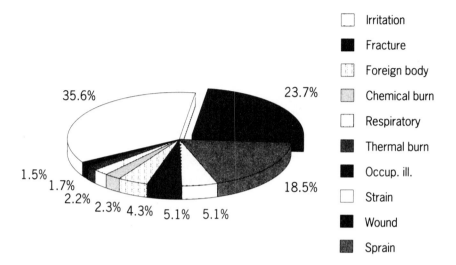

☐	Irritation
■	Fracture
▦	Foreign body
▨	Chemical burn
☐	Respiratory
▨	Thermal burn
■	Occup. ill.
☐	Strain
■	Wound
▦	Sprain

Figure 6.2   Percent of injuries by type of injury.

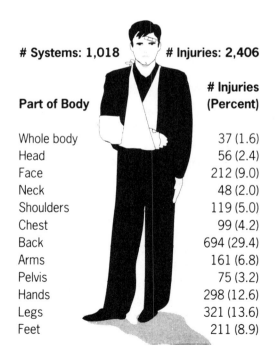

 # Systems: 1,018   # Injuries: 2,406

Part of Body	# Injuries (Percent)
Whole body	37 (1.6)
Head	56 (2.4)
Face	212 (9.0)
Neck	48 (2.0)
Shoulders	119 (5.0)
Chest	99 (4.2)
Back	694 (29.4)
Arms	161 (6.8)
Pelvis	75 (3.2)
Hands	298 (12.6)
Legs	321 (13.6)
Feet	211 (8.9)

Figure 6.3   Number of injuries per body part.

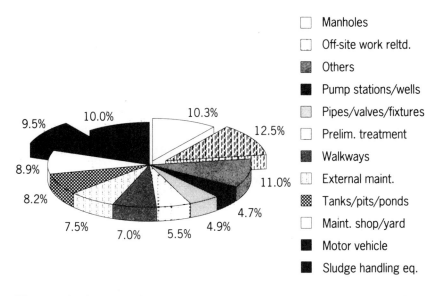

Figure 6.4   Percent of injuries reported by location of accident.

will open or increase speed, and so forth. These conventions are not consistent worldwide.

People have many natural reactions—particularly in emergencies—such as moving toward light in a suddenly darkened room, pulling away from heat, and running from fire. The natural reaction of moving toward light suggests that designers should place windows in or adjacent to doors and emergency lights near doors.

Fatigue and stress may lead to errors and accidents. Causes of fatigue and stress include noise, toxic substances, vibration, dust, humidity, light, temperature, and odor. Designers can partially control these causes by providing a comfortable environment, particularly at work stations.

PROJECT PHASES. Wastewater projects are usually accomplished in several phases to allow project review opportunities at the end of each phase. A number of agencies are typically involved in the review process, considering issues such as environmental impact, costs, schedules, coordination, technical content, completeness, regulation compliance, and so on. The designer is typically involved in project phases shown in Table 6.1, which also indicates potential health and safety tasks for each phase. During expansion of existing facilities, designers spend a great deal of time in the plant, and must, therefore, comply with the health and safety rules of the plant as well as those of their own employer.

OCCUPATIONAL SAFETY AND HEALTH SURVEYS. These surveys involve physical inspections of facilities to determine compliance with

---

Table 6.1   Health and safety tasks for the designer.

Phase	Tasks
Facilities planning	Consider safety risks as part of treatment process selection. Identify safeguards for chosen processes.
Preliminary design	Review existing accident reports. Identify codes and regulations that apply to the project. Conduct a study of code compliance. Conduct a safety and health survey.
Preparation of plans and specifications	Review plans and specifications for safety compliance. Prepare a list of needed tools, supplies, personal safety gear, safety equipment, signs, and first aid supplies. Prepare an emergency response plan. Prepare a safety section for the operations manual.
Construction	Require contractors to consider safety of plant staff during construction and to comply with plant health and safety rules. Conduct a health and safety survey of completed construction.
Start-up and performance testing	Provide safety training.

OSHA standards. While basic OSHA standards apply to the construction and use of any facility, technical expertise is required to interpret their application to processes and equipment found in wastewater treatment systems. Personnel performing this type of survey should have specific experience with wastewater treatment operations. Designers working on existing facilities should first determine if a survey has been conducted.

LIABILITY. Liability resulting from personal injuries is an increasing concern to designers, particularly in the U.S. The high injury rate in wastewater facilities exposes designers and owners to the potential for financial loss. Based on the limited information now available, few legal proceedings against designers have resulted from personal injuries incurred during operation of wastewater treatment facilities. However, high injury rate and frequent exposures to health and injury hazards in these facilities constitute a liability risk. A design firm can best protect itself against future accident litigation by strictly following applicable safety regulations and codes in the design of the wastewater treatment plant.

OCCUPATIONAL HEALTH

Occupational health hazards include harmful physical hazards, chemicals, biological agents, and environmental conditions. Designers can provide engineering solutions that will reduce exposure to these hazards.

Numerous chemicals are used for treating wastewater, conditioning sludges, and a variety of other operations and maintenance activities. Exposure to some of these chemicals is a hazard. In the industrial sector, the occupational safety and health standards in OSHA 29 CFR 1910.1200[5] require communication of chemical hazards to employees. These standards must also be complied with in the many states that have adopted OSHA. Workers may also be exposed to chemicals generated in the treatment system or discharged into the sewer. Although toxic gases generated in the system have received the most attention, chemical contamination of wastewater and sludges is also a potential hazard. Contaminants that have been found in wastewater and sludges include chlorinated hydrocarbons, pesticides, and polychlorinated biphenyls (PCBs).[22] The primary pathway of worker exposure to these chemicals is probably inhalation, although another potential route is direct contact with wastewater, scum, or sludges. Plant processes also represent a potential risk through air pollutants and chemical or biological agents that volatilize or are stripped from wastewater or sludges.

Wastewater workers are exposed to many disease-producing microbial agents. Most infections are caused by direct contact with wastewater and sludge.[23] For example, composting workers have experienced an excess of nasal, ear, and skin abnormalities and increased symptoms of burning eyes and skin diseases.[23] Although direct contacts can be controlled by appropriate work practices, designers may reduce the frequency of potential direct contact by selecting processes and equipment that minimize exposure.

In the past, hazardous materials such as mercury, asbestos, and PCBs were used in plant construction. Hazards associated with these materials have been extensively publicized, and their use is now prohibited.

Environmental conditions that cause fatigue and increase the probability of employee accidents include noise, humidity, odor, temperature, illumination, vibration, and chemical exposures. Occupational health issues are discussed fully elsewhere.[4]

SAFETY RESPONSIBILITIES

OWNER'S ROLES. The owner has prime responsibility to provide a workplace that is free from recognized hazards that cause or are likely to cause illness, death, or serious physical harm. The owner manages this responsibility

---

by delegating portions to others, including designers, suppliers, builders, managers, and employees. No standard exists for distributing responsibilities.

The work environment must have adequate light and ventilation, controlled noise levels, uncluttered work spaces, sanitation facilities, safe and properly guarded machinery, and appropriate tools. Sanitation facilities include a safe drinking water supply, toilets, wash facilities, locker rooms, and eating rooms. The owner is also responsible for protective clothing and equipment, adequate and safe tools and equipment for inspection and operation, first aid facilities, and rules and instructions on safe procedures for plant operation.

DESIGNER'S ROLES. The owner usually establishes the role of the designer in a services contract. Nonetheless, this contract often lacks specific delineation of designer responsibilities for occupational safety and health. Several organizations publish sample services contracts, including the Engineers Joint Contract Documents Committee and the American Institute of Architects. These contracts require the designer to provide drawings and specifications for construction that comply with applicable codes and regulations. Several of these codes and regulations address occupational health and safety requirements.

Many decisions that impact occupational health and safety, such as costs, operability, layout, operator skill, and convenience, are made by representatives of the owner. The designer is required by convention and, in many cases, law to comply with appropriate codes, regulations, and industrial standards. Registered professional engineers are required by registration laws to protect the public health and welfare.

Designers usually provide plans and specifications for the fixed facilities and some other facilities, such as portable fire extinguishers, equipment replacement parts, portable maintenance equipment, vehicles, laboratory equipment and supplies, and operating manuals. The designer is not likely to specify such items as maintenance tools, personal safety gear, protective clothing, or housekeeping chemicals. The owner and managers will often conduct safety training, emergency response planning, employee screening, or other tasks.

*D*ESIGNING A SAFE FACILITY

GENERAL PROCEDURES FOR ENHANCING SAFETY. In addition to compliance with codes and regulations, designers exercise judgment and use procedures that improve facility safety, including

- Reference to previously successful designs;

- Analysis of operating and maintenance tasks to develop layouts that will result in a simple and safe way to accomplish a potentially hazardous task;
- Operability reviews;
- Selection of equipment to reduce noise;
- Providing items to encourage safety, such as a safety library, training room, and safety signs;
- Building physical models; and
- Inspection of the constructed facility.

DESIGN PRACTICES FOR SAFETY. Many injuries and illnesses are caused by unsafe acts, rather than unsafe facilities. Consequently, owners must balance added training and supervision against the provision of less hazardous facilities. Costs are always a major consideration because they generally increase rapidly as risks diminish.

In addition to meeting applicable codes and regulations, implementation of measures to reduce added injury and occupational health risks is thus a compromise between costs for engineering solutions and costs for other approaches to reach an acceptable level of risk. An example would be a system for lifting pump components for maintenance. Options that might be considered are manual lifting, a tripod with chain hoist, a monorail with a manual trolley and hoist, a monorail with an electric trolley and hoist, and a robot. Assuming that these are listed in order of decreasing risk of injury and increasing cost, a selection must be based on tradeoffs and personal preference. As an increasingly common practice, the owner's staff may make these decisions with input from designers and others.

A second example is that of the owner providing clean work clothes. Options include designing a wash facility for the plant or providing an outside laundry service. The design of facilities for these two options would differ significantly. This choice usually depends on owner preference and would not likely be made by the designer. Another example is the ever-present pressure to cut costs by reducing building space. Reduced space often results in less than desirable clearances for operating and maintaining equipment and increases injury exposure.

The designer can reduce hazards by designing an inherently safer environment. Examples include use of less hazardous chemicals, use of more dilute forms of hazardous chemicals, design of processes to operate closer to ambient conditions, and reduction of inventories of hazardous chemicals.

*P*LANT HAZARDS AND SAFETY PROVISIONS

OCCUPATIONAL EXPOSURES. The environment of a wastewater treatment plant contains many potential hazards as a result of the nature of wastewater and its byproducts as well as the treatment processes and equipment.[2,4]

The following composite list of hazards and hazardous areas will help designers visualize potential risks:

- Abnormal atmospheres (ammonia, carbon dioxide, carbon monoxide, chlorine, ethane, gasoline, hydrogen chloride, hydrogen sulfide, methane, mixture of gases, natural gas, nitrogen, oxygen-deficient environments, oxygen-rich environments, ozone, sewer gas, sludge gas, sulfur dioxide, and temperature extremes);
- Airborne hazards (aerosols, biological agents, chemical dust, dust, toxic or explosive gases, and volatile solids);
- Burns;
- Chemicals (harmful);
- Confined spaces;
- Drowning;
- Earthquakes;
- Electrical shock;
- Elevated areas;
- Explosive gases or liquids;
- Falls;
- Fires;
- Flooding;
- Food contamination;
- Housekeeping (internal and external);
- Impact;
- Infections and diseases;
- Ingress and egress;
- Laboratory;
- Ladders, stairs, and ramps;
- Lifting;
- Materials handling;
- Moving machinery;
- Natural hazards;
- Night operations;
- Noise;
- Noxious gases and vapors;
- Openings;
- Open tanks;

- Overhead fixtures;
- Pinning and crushing;
- Radioactive material;
- Slips and falls;
- Spillage;
- Storms;
- Vapors and dust (gasoline, solvents, dried sludge, activated carbon, and so on);
- Vehicles;
- Ventilation;
- Vibration;
- Walkways;
- Weather (heat, cold, ice, and snow); and
- Yard work.

CHEMICAL HANDLING. Numerous chemicals and fuels stored at treatment plants for a variety of uses may pose a health hazard through normal use or an accident. Chapters 7 and 16 of this manual present design considerations for chemical handling systems, and their planning is described elsewhere.[24]

Common uses of chemicals at treatment plants include process, housekeeping, landscaping, agricultural, laboratory, maintenance, fuels, and odor control. Chapter 5 of MOP 11[4] presents safety considerations for many chemicals used in wastewater treatment facilities. In addition, material safety data sheets provided by chemical manufacturers describe proper handling of chemicals.

Commonly used chemicals include

- Alum,
- Ammonia,
- Caustic,
- Chlorine,
- Chlorine dioxide,
- Defoamers,
- Ferric chloride,
- Ferric sulfate,
- Hydrochloric acid,
- Hydrogen peroxide,
- Lime,
- Odor-masking agents,
- Oxygen,
- Ozone,
- Sodium bisulfite,
- Sodium hypochlorite,

- Sodium thiosulfate,
- Sulfuric acid, and
- Sulfur dioxide.

Flammables and explosive hazards include

- Activated carbon,
- Diesel fuel,
- Digester gas,
- Fuel oil,
- Gasoline,
- LP gas,
- Lubricating oils,
- Methanol,
- Paint and thinners, and
- Solvents.

CONFINED SPACES. Confined spaces are a major cause of death and serious injuries in the workplace. The National Institute of Occupational Safety and Health (NIOSH) publishes guides and criteria for working in confined spaces.[25-27] Confined spaces are defined in the NIOSH publications as "any space which by design has limited openings for entry and exit and unfavorable natural ventilation which could contain or produce dangerous air contaminants, and which is not intended for continuous employee occupancy." Specific examples of confined spaces in wastewater treatment facilities are

- Manholes (wastewater, stormwater, and so on),
- Large pipes and conduits,
- Channels,
- Tunnels,
- Digesters,
- Scum pits,
- Wet wells,
- Dry wells,
- Vaults (electrical, valve vaults, and so on),
- Grit chambers,
- Screening pits,
- Storage tanks and hoppers (chemicals, screenings, water, fuel, sludge, and so on),
- Septic tanks,
- Septage receiving tanks and pits,
- Gas holders,
- Excavated holes,

- Covered basins and channels, and
- Odor control systems.

The designer should avoid creating unnecessary confined spaces. If confined spaces are necessary, proper ventilation must be provided. (see Chapter 7).

ODOR CONTROL SYSTEMS. Odor control systems, increasingly common in treatment facilities, may present several hazards. Major elements of these systems are typically a collection structure (a cover over basins of wastewater or sludge), ducts, contact vessels, chemical makeup and feed systems, chemical piping, chemical solution recycle, blowers, and discharge stacks. Because these systems collect gases that could be explosive or toxic, they need to be carefully designed to avoid release of the collected gases into an operating space. The design of these systems should include monitoring for combustible or toxic gases.

Individual elements within odor control systems may be confined spaces (covered channels or basins, large ducts, contact vessels, and so on).

REPRESENTATIVE SAFETY PROVISIONS. The following safety provisions are intended to stimulate thinking rather than serve as a complete checklist. Many items will not apply to all facilities.

Building and Work Space.

- An open channel immediately ahead of where wastewater enters the influent structure to vent explosive gases and vapors;
- Wet wells located either in a separate structure or accessible only from the outside, properly ventilated;
- Monitored and alarmed screen room or shredder room, separated from other facilities with distinct outside access;
- Protection against flooding, including alarms as appropriate;
- Equipment, piping, valves, and other appurtenances within structures arranged for ease of access and ample space, including headroom and walk aisles;
- Work platforms for elevated equipment that may require adjustments, observations, or preventive maintenance;
- Access to windows, lights, and ceiling-mounted items that must be operated or maintained;
- Adequate space for equipment repair or removal;
- Dual entrances or accesses to potentially hazardous areas with tight-fitting, self-closing, open-out doors that are equipped with panic hardware;

- Panic hardware on exit doors and fusible links on doors in high-fire-risk areas, as appropriate;
- Potentially explosive areas provided with explosion venting, protective devices, suppression systems, or barricades;
- Restrooms convenient to work stations;
- Equipment maintenance shops with appropriate safety provisions for hazards associated with maintenance activities;
- Nonslip surfaces (such as broom-finished concrete or nonslip covering) for floors and ramps;
- Dust accumulation spots minimized (open truss members, ledges, light fixtures, and so on);
- Laboratories with two easily accessible exits that are reasonably remote from each other;
- Designed to withstand earthquake forces;
- Basement areas with two easily accessible exits that are remote from each other;
- Light interior colors in dark areas;
- Provisions for the safe collection of samples;
- Interior doors, where appropriate, that swing both ways and have see-through panels;
- Lightning protection;
- Adequate climate control (humidity, temperature, and so on) for comfort in offices, laboratories, eating areas, work stations, and selected work areas;
- Walking aisles and machine areas identified; and
- Allowable floor loadings posted.

Confined Spaces. These include manholes, valve vaults, electrical vaults, storage tanks, pits, and digesters. They should be identified with warning signs. Refer to the NIOSH guide to safety in confined spaces.[25]

Electrical.

- Medium- and high-voltage cables completely enclosed in either conduit or covered trays and adequately marked to warn personnel of contents;
- Switchboards with "dead front" and "dead rear;"
- Moisture-proof enclosures for switches, equipment, and lights in moist areas where there is no possibility of flammable gas accumulation;
- Ground fault circuit interrupters where required;
- Electrical equipment adequately grounded;
- Ground equipment to avoid static electricity sparks in explosive areas;
- Grounding straps for portable equipment;

- Wiring properly insulated, grounded, and nonexposed;
- Required clearances provided around electrical equipment;
- Electrical "lockout" facilities with padlocks and tags to prevent accidental starts when machinery and equipment are being worked on or otherwise taken out of service;
- Emergency shutoff switch, clearly labeled, at all machinery units;
- Oil-filled submersible motors equipped with thermal detectors to de-energize the motor before the ignition temperature of the oil is reached;
- Alternative power supply for critical lighting, ventilation, and sensory devices and alarms;
- Two separate power sources to the plant, or standby power to keep critical systems operational;
- Exterior floodlighting to provide for nighttime operation, maintenance, and inspection;
- Safe access for lamp replacement;
- Insulating floor mats at control centers and panels;
- Maintenance tools with insulated handles and flashlights with non-conductive cases;
- Electrical tools (drills, saws, and so on) grounded or double-insulated;
- Grounded extension cords; and
- For future construction, stub-outs designed so they are not a hazard.

Fire Control.

- Fire hydrants that meet local fire code for type and location;
- Landscaping that will not result in large quantities of combustible vegetation, particularly near structures;
- Smoke and fire alarms;
- External fire alarms as required by local fire code;
- Automatic fire suppression systems;
- Firefighting devices located in each separate structure at accessible points near the entrance to areas of likely conflagration;
- Fire extinguishers suitable for the area and the equipment to be protected;[9]
- Laboratory wall surfaces, ceilings, and furniture made of nonflammable or fire-resistant materials;
- Critical drains sized for fire flows;
- Containment for hazardous materials, fire flow, and precipitation;
- Provisions to allow use of adequately treated wastewater as a backup firefighting supply; and
- Equipment, buildings, and fire alarm systems in compliance with local, state, and national fire codes and OSHA and insurance company requirements.

First Aid.

- First aid supplies or kits (under some conditions, OSHA requires approval by a consulting physician. [OSHA standard, Part 1910, Subpart K—Medical and First Aid, 1910.151(b)]); and
- Posted instructions for calling a physician and hospital.

Fuel Storage.

- Separate storage for gasoline, diesel fuel, digester gas, liquid fuels, and propane;
- Containment for spills and overflows; and
- Floor drain traps for fuel spills.

Gas Collection, Piping, and Appurtenances.

- Gas protective devices in accordance with manufacturers' recommendations;
- Gas piping and pressure-vacuum relief valves on digesters with adequate flame traps;
- Drip traps designed to prevent release of gas;
- Waste burners and vents located a safe distance from buildings;
- Bypasses and valves to allow maintenance of gas equipment;
- Ventilated rooms for gas-burning equipment such as boilers and engines; and
- Automatic shutdown of gas systems at preset pressures.

Hazardous Materials. Refer to Chapter 14 for more information on chlorination and other disinfection systems, and to Chapters 7 and 16 for information on other chemical-handling systems.

- Compliance with fire codes;
- Unloading facilities well lighted and easily accessible by emergency response crews;
- Unloading station clearly marked;
- Separate receiving and storage areas for chemicals that react violently if mixed together;
- Temperature controlled;
- Ventilation provided;
- Containers shielded from heat sources;
- Leak detection provided;
- Leak repair kits provided;
- Vacuum relief devices on tanks;
- Tank liquid-level measuring devices and alarms provided;

- Pull-chain or pedal-operated deluge showers with pedal-operated, chest-level-high wash spouts and a floor drain adjacent to areas where hazardous chemicals are being handled or stored (alarm when used);
- Guard posts for equipment and storage tanks, including underground tanks to prevent damage by vehicles (fire codes often include specific requirements for post type and location);
- Restraints on gas cylinders;
- Fuel gas cylinders separated from oxygen cylinders;
- Ventilation exhaust ports adequately dispersed and located such that discharges will not contaminate air inlets in other areas;
- Treatment systems for hazardous gas releases;
- Repair and containment kits for cylinders and tanks;
- Light and ventilation switches located outside;
- Self-contained breathing apparatus provided;
- An automatic control to actuate forced ventilation and lighting when chemical rooms are occupied;
- Approved storage for flammables, thinners, solvents, and so on;
- Storage space adequate for peak demands;
- Dikes or curbs capable of holding the stored volume, plus a safety allowance in each liquid chemical storage area (designed to allow chemical to be recovered and reused);
- Health risks associated with chemicals considered (refer to chemical material safety data sheets);
- Piping minimized;
- Pumping and piping systems permanently installed for delivering liquid ferric chloride, sulfuric acid, and other corrosive liquid chemicals to the application point;
- Chemical pressure piping systems provided with pressure relief to storage areas;
- Chemical storage areas sited to eliminate the need to reach beyond safe handling limits;
- Nonslip floor surfaces in areas where polymers may be spilled;
- Dust collectors provided on chemical elevators;
- Materials and devices used for storing, transporting, or mixing hazardous chemicals compatible with the chemical(s) involved;
- Tanks, bins, and other containers labeled;
- Chemical material safety data sheets provided;
- Separate chlorinator/chlorine evaporator and chlorine storage rooms, each with aboveground ventilation only to outside air;
- Chlorination facilities with concrete floors and adequate, but separate, drainage from other facilities;
- View windows to the chlorinator/chlorine evaporator room and chlorine storage room for outside observation;
- Chlorine leak detection device provided;

- Chlorine leak containment system to capture and neutralize released chlorine (for large systems);
- Liquid chlorine containers stored in well-ventilated, fireproof structures with protection against direct exposure to the sun;
- Leak test kit; and
- Dry hypochlorite stored in a cool, dry area.

Housekeeping.

- Ample storage areas;
- Hose bibs, hoses, nozzles, and hose racks in spillage areas;
- Water-repellent wall surfaces for cleanup purposes;
- Sludge pumps with quick-closing sampling valves;
- Floors sloped and drained to facilitate cleaning;
- Cleaning equipment, including industrial vacuum cleaners, brooms, mops, high-pressure washers, steam cleaners, and so on;
- Splash guards and drip pans;
- Airtight, metal receptacles for solvent-soaked and combustible wastes; and
- Seal water discharged to hub drains adjacent to or integral to the equipment.

Hygiene Facilities.

- Walk-through shower facilities with hot and cold running water;
- Two lockers for each employee, one for work clothes and another for street clothes;
- Washing machine and dryer for work clothes;
- Pedal-operated laboratory sinks, toilets, and wash sinks; and
- Disinfectant dispensers, liquid soap dispensers, and towel dispensers.

Illumination.

- Adequate exterior and interior lighting throughout the plant, particularly in areas of operational activities such as repair and servicing of equipment, valves, and controls;
- Lights that promptly illuminate hazardous and interior areas;
- Emergency lighting (battery-operated lights) and exit lights for interior areas, particularly in the vicinity of stairways;
- Portable, explosion-proof lighting system;
- Emergency generator set; and
- Lighting of warning signs.

Incinerators.

- Dry sludge handling methods to preclude dust accumulation that results in potential dust explosion;
- Automatic signal for incinerator flame-out;
- Automatic shutdown controls in the event of incinerator flame-outs;
- Fully automatic ignition start controls;
- A proper safety train on the incoming fuel supply of the auxiliary fuel system;
- Burner system controls to ensure adequate purge time, including interrupted pilot, flame scanner, and safety controls to prevent the possible lighting or relighting of a burner in a potentially hazardous atmosphere;
- Adequate temperature controls; and
- Adequate ventilation.

Laboratories.

- Durable, nonslip floor material;
- Ventilation with adequate make-up air, explosion-proof motors, and laboratory hoods in special test areas;
- Eye wash and deluge shower;
- Clearly identified gas outlets equipped with substantial handles; and
- Lips on storage shelves.

Maintenance Shops.

- Provisions for protection against infrared radiation from combustion units, ultraviolet radiation from arc welding, and ionization radiation from radioactive substances;
- Exhaust facilities for welding and grinding;
- Enclosure and ventilation for sand-blasting, solvent-cleaning, and spray-painting areas; and
- Adequate materials handling equipment, including cranes and hoists.

Materials Handling and Storage.

- Chemical storage areas located so personnel do not have to stretch beyond safe handling limits;
- Provisions for minimizing manual lifting;
- Provisions for using hand trucks;
- Access to storage shelves for power lifting equipment;
- Well-planned, safe operations associated with railroad cars, including provision of derailers and wheel chocks;

- Fixed or portable electrical hoists with ceiling lifting devices for lifting heavy loads, including chemicals, pumps, motors, and equipment for repair or replacement;
- Hoists to remove and lower equipment into pit areas;
- Dust collectors on chemical elevators at loading points;
- Drum handling equipment;
- Rigging materials (ropes, chains, hooks, devices, pins, and so on) rated for intended service;
- Restraints on gas cylinders; and
- Provisions for earthquake forces, as necessary.

Noise Control.

- Equipment designed for noise reduction;
- Provisions for reducing noise from multiple equipment units;
- A maximum permissible noise level during operation, expressed in decibels of sound under standard test conditions; and
- Air compressors, vacuum pumps for filter units, centrifuges, blowers, standby power units, and other similar equipment producing high noise levels located either within isolated buildings or rooms or within acoustically sound-proofed structures for maximum sound reduction.

Openings and Hatchways.

- Railings with kickplates around openings and stairwells;
- Hatchway covers with springs or positive locking devices to hold the covers open (unless they swing free of opening and lie flat);
- Double handrails, fencing, or guards of proper height at floor and wall openings, pump wells, influent structures, open tanks, and above-ground ramps; and
- Portable handrail units or similar guard systems that can be placed around openings when gratings or covers are removed.

Personal Protective and Safety Equipment.

- Hard hats, work clothes, hearing protectors, face shields, safety goggles, rubber boots, hip boots, waders, safety shoes, rain gear, gloves, chemical aprons, and cold weather clothes;
- Laboratory protective gear;
- NIOSH-certified respiratory devices;
- Safety equipment located for easy access in an emergency;
- Safety equipment near location needed (separate storage); and
- Hard hats, safety glasses, and ear plugs for visitors.

Piping and Valves.

- Valves accessible and easily operated;
- Large, frequently operated valves power operated;
- Head clearance provided;
- Valves located above reach are chain or power operated;
- Influent and discharge pipes to pumps and other equipment valved so that dismantling them will not result in wastewater, sludge, gas, or chemicals entering the work area;
- Piping that will not block or restrict access for routine operation or maintenance;
- Selected valves provided with lock devices;
- Freeze protection;
- Supports required when systems dismantled for maintenance;
- Sludge pumps with pressure gauges to indicate buildup of gases when pumps are out of service;
- Safety and relief devices on heat exchangers;
- Cages or guards around accessible hot piping;
- Stubouts for future construction designed so they are not a hazard;
- Safety guards around check valve exterior levers; and
- Standardized color-coding of process piping and emergency equipment:
  - Orange: dangerous parts of machines or energized equipment and flammable gas lines;
  - Blue: potable water;
  - Yellow: chlorine;
  - Black: raw sludge;
  - Brown: treated sludge;
  - Purple: radiation hazards;
  - Green: compressed air;
  - Jade green: nonpotable process or flushing water;
  - Gray: wastewater;
  - Orange with blue letters: steam;
  - White: traffic and housekeeping operations; and
  - Red: fire protection equipment.

Rotating and Reciprocating Machinery.

- Caps or guards around exposed rotating shifts and all other moving parts (open-mesh type allows equipment viewing without removing guards);
- Guards that are easily replaced and fastened;
- Guards around long, exposed shafts to safeguard the worker from contact or injury from "whipping" if the shaft breaks;

- Shafts with painted spirals or other markings to indicate running conditions;
- Positive-displacement pumps with an air chamber and a pressure switch that will stop the pump at a preset pressure;
- Nonsparking pulleys, belts, and fan wheels used in explosive areas; and
- Warning signs on equipment that starts automatically or from a remote location.

Safety Equipment and Facilities.

- Safety equipment, including portable ventilation equipment such as air blowers and adequate lengths of noncollapsible ducting; indicators for hydrogen sulfide, combustible gases, methane, chlorine, carbon monoxide, and oxygen deficiency; proper self-contained air breathing apparatus; inhalators; resuscitators; decibel meter noise analyzers; explosion-proof flashlights; portable lifting equipment; first aid kits; safety tools (nonsparking); and nonconducting ladders with nonskid feet;
- Safety harnesses, ropes, tripods, and hoists for entering vaults or pits containing potentially harmful or explosive gases;
- Safety poles, life preservers, life jackets, or combinations of these at locations needed;
- Fire extinguishers;
- Barricades, traffic cones, warning signs, flashers, and reflective vests;
- Telephones, intercom systems, radios, and walkie-talkies for communication;
- Safety libraries; and
- Training rooms and training equipment.

Gas Monitoring Devices and Alarms.

- Alarm systems, both visual and audible, to detect explosive or combustible gases and vapors in screenings of shredder rooms, digester areas, flammables storage, and elsewhere, as needed;
- Sensing devices equipped with visual and audible alarms both nearby and at a centralized location, placed in all hazardous areas for combustible or explosive gases and vapors;
- Oxygen leakage detectors at appropriate points on oxygen supply tanks;
- Chlorine leak-detection devices to signal equipment failure in larger installations; and
- Visual and audible alarms.

Site Layout.

- Fencing around plant structures, railings, walls, locked doors, and so on where unauthorized entry could result in personal mishap or disruption of plant operations (avoid trapping personnel with these security measures);
- Provisions for emergency vehicles (work closely with the local fire department);
- Traffic control signs or signals;
- Sidewalks located for natural access routes;
- Delineated crosswalks and walkways visible to vehicle occupants and pedestrians;
- Landscaping that minimizes the need to use hand-operated mowers, hedge clippers, and so on;
- Safe landscaping maintenance equipment and associated personal safety equipment;
- Landscaping that avoids steep slopes that must be mowed;
- Landscaping that does not attract bees and dangerous pests;
- Layout that allows sun to melt ice and snow from walks and driveways;
- Areas for snow storage;
- Containers for storage of sand, salt, or other ice-melting chemicals;
- Signs to direct visitors to parking and reception areas to limit wandering of visitors;
- Designated parking for visitors and staff; and
- Provisions for safe transport of chemicals, fuel supplies, sludge, and so on.

Supplies Storage.

- Separate storage for dangerous chemicals, flammables, thinners, solvents, and so on;
- Storage space sufficient for peak demands;
- Storage for good housekeeping;
- Load ratings posted on storage shelves, floors, and so on; and
- Lips on shelf edges.

Ventilation. Note that ventilation adequate for control of fire and explosion might be insufficient for health protection.

- Separate, mechanical, forced ventilation for spaces such as influent channels, influent rooms, wet wells, dry wells, screen rooms, shredder rooms, grit chambers, disinfection areas, manholes, sumps, pits, sludge pump areas, sludge storage areas, sludge digestion areas, gas

---

control rooms, sludge storage and conditioning tanks, centrifuges, sludge-processing areas, digester buildings, boiler rooms, engine rooms, incinerator rooms, laboratories, garages, maintenance shops, laundry rooms, and shower rooms (even below-ground structures without a cover are hazardous; natural ventilation that is inadequate under some conditions has caused fatalities);

- Ventilation to force fresh air into wet wells so that the exhaust ventilator does not pull sewer gases from the influent sewer into the wet well;
- Forced mechanical ventilation automatically actuated when chlorination rooms, chemical handling rooms, and laboratories are occupied;
- Critical ventilation sustained during emergencies such as floods, fires, storms, or power failures (fire code may require break-glass-type emergency shutoff for hazardous materials locations);
- Ventilation exhaust ports adequately dispersed and located to discharge where there will be no contamination of air intakes;
- Adequate provision for makeup air for ventilators; and
- Treatment of hazardous materials in ventilation exhaust (required by some fire codes).

Walkways, Ladders, Stairways, and Ramps.

- In nonhazardous areas, manhole steps or permanently attached ladders inside of tanks, basins, or wet wells for entry or exit in case of emergency;
- Nonslip stair treads on landings and stairs;
- Stair risers of equal height and proper slope;
- Handrails of type and cross section such that they can be fully gripped with fingers and thumb;
- Separate handrail where entrance is provided by ship's ladders or manhole steps, and ladders or manhole steps extending above entrance level to provide a handhold;
- Fixed ladders more than 6.1 m (20 ft) long equipped with safety cages or ladder safety devices;
- Tall ladders provided with rest or offset landings;
- Rest landings on stairways;
- No manhole steps or fixed ladders to provide access to hazardous areas;
- Ramps with slopes commensurate with intended use;
- In climates with ice and snow, gratings on outside stairs and walkways on tanks wherever possible; and
- Lift-rings and grating locks flush-mounted to prevent tripping.

Water Supply.

- Potable water, when used for plant processes or other purposes such as washdown of equipment, protected by backflow preventers, vacuum breakers, or airbreak for all washdown hoses, pumpseals, and so on (backflow preventer provided in the plant supply);
- Warning signs near each nonpotable water outlet and color-coded, nonpotable water lines;
- Adequate supply for fire protection; and
- Adequate pressure to hoses for cleanup (excessive pressure can be a hazard).

Other Safety Considerations.

- Directive signs such as "No Smoking," "No Running," "Wear Life Vest," "Wear Hard Hat," or "Safety Glasses Required;"
- Hazard identification signs indicating dangers such as explosive gases, noise, chemicals, flammables, ice, slippery floors, high-pressure vessels, high-pressure pipes, overhead utilities, and underground utilities;
- Instructional signs to indicate correct procedures in critical locations and for critical operations;
- Signs to limit access;
- Special equipment bracing where required;
- Analysis of items such as piping and storage tanks for seismic loads; and
- Wind socks.

DESIGNING FOR CONSTRUCTION SAFETY

Construction safety is not an add-on, applied after construction begins or an accident has occurred. Designers include construction safety items in the contract documents, and often provide for construction safety monitoring. Communication with the construction contractor is an important part of ensuring proper safety considerations. Safety considerations are usually emphasized during bid specifications, prebid meetings, preconstruction meetings, and project safety monitoring.

FEDERAL REGULATIONS. Federal regulations cover all construction safety activities.[28] OSHA enforces these regulations. Individual states with federally approved programs may also enforce these standards. Copies of the regulations are available from the nearest OSHA office.

Hazard Communication Standard. The federal hazard communication standard has been extended to cover construction workers. Employers must provide training and information to construction workers concerning hazardous chemicals as required by this OSHA regulation.[28]

Excavation and Trenching Standard. The new OSHA trenching and excavation regulations set forth the procedures that must be followed for all trenching and excavation work. Most design engineers provide for compliance through project specifications and job site inspections.

PREBID SPECIFICATIONS. Prebid specifications often include relevant safety considerations and require the contractor to comply with all federal, state, and local safety regulations as well as any project-specific safety requirements. In addition, copies of the contractor's safety program should be part of the required submittals.

Specifications are often used to inform the contractor of safety hazards. For example, these might include

- Chemicals used, such as chlorine, sulfur dioxide, lime, ferric chloride, polymers, and so on;
- Confined spaces and associated hazards;
- Unusual process operations, such as the use of pure oxygen or ozone generation;
- Sludge-handling facilities;
- Rail or vehicular traffic;
- Availability of fire or rescue personnel; and
- Other hazards, as appropriate.

Informing the contractor of these exposures offers many advantages. Such information enables the contractor to protect employees, construction inspectors, and the public.

PRECONSTRUCTION MEETINGS. Preconstruction meetings offer an opportunity to reemphasize safety considerations necessary during the project. Emphasizing safety at preconstruction meetings demonstrates the concern for safety and provides documentation of the safety information given to the contractors. The preconstruction meetings offer the prospective bidders an opportunity to ask additional questions and identify other areas of safety concern that might have been overlooked. A walk-through of the proposed project area may also provide additional insights for the prospective bidders.

PROJECT SAFETY MONITORING. Monitoring of the conditions on the work site represents the final and most critical stage of the safety program for the construction project. Regardless of the comprehensiveness of the written safety program, if work-site inspections fail to ensure that the program is carried out, accidents will likely occur. Figure 6.5 shows a sample construction safety inspection report.

Inspector _____ Date _____

General Contractor _____

Project Name _____

Job Site Location _____

Contact _____

Item	Yes	No	Comments
Barricades/Tape			
Concrete Construction			
Forms Secure			
Guard Rails			
Proper Curing			
Electric			
Assured Grounding			
GFCI			
Ground Check			
Excavation			
Slope/Shore/Shield			
Guard Rails			
Barricades/Swing Radius			
Emergency Phone Numbers			
Fire Extinguishers			
First Aid Kits			
Hoisting			
Guys & Cables			
Signal Device			
Safety Bars/Barricades			
Side Protection			
Housekeeping			
Walkways Clear			
Work Area Clear			
Exposed Nails			

Figure 6.5   Sample form for construction safety report.

Item	Yes	No	Comments
Ladders			
Construction			
Length			
Training			
Openings			
Wall			
Floor			
OSHA Sign Posted			
PPE			
Eye			
Ear			
Foot			
Fall			
Hard Hat			
Respiratory			
Safety Meetings Held			
Scaffolding			
Construction			
Guard Rails			
Access			
Stairs			
Hand Rails			
Treads			
Clean			
Storage			
Gas Cylinders			
Fuel Containers			
Material			
Tools			
Storage			
Guards			
Grounds			

Figure 6.5 Sample form for construction safety report (continued).

REFERENCES

1. Hadeed, S. J., "1989 Safety Survey." *Oper. Forum,* **7,** 4, 24 (1990).
2. "Safety and Health in Wastewater Treatment Systems." Manual of Practice No. 1, Water Pollut. Control Fed., Washington, D.C. (1983).
3. "Guidelines to Developing a Wastewater Safety Program." Manual of Practice No. SM-2, Water Pollut. Control Fed., Washington, D.C. (1983).
4. "Operation of Municipal Wastewater Treatment Plants." Manual of Practice No. 11, Water Pollut. Control Fed., Alexandria, Va. (1990).
5. U.S. Code of Federal Regulations, Title 29, 29 CFR 1900-1910 (1990).
6. Hopf, P.S., "Designer's Guide to OSHA." McGraw-Hill, Inc., New York, N.Y. (1982).
7. Chlorine Institute, Inc., "Chlorine Manual." New York, N.Y. (1989).
8. Manuf. Chem. Assoc., "Chemical Safety Data Sheets." Washington, D.C. (1989).
9. National Fire Protection Association, "Recommended Practice for Fire Protection in Wastewater Treatment Plants." Publ. 820, Battery March Park, Quincy, Mass. (in press).
10. "Guide to Occupational Safety Literature." Natl. Safety Council, Chicago, Ill., 1 (1975).
11. "A Guide through the Code Maze." *Eng. News Rec.,* 26 (1988).
12. Pennsylvania Bureau of Water Quality Management, "A Guide for the Preparation of Applications, Reports and Plans." 6th Ed., Publ. No. 1 (1983).
13. "Construction Grants 1985." U.S. EPA, Washington, D.C. (1985).
14. "Considerations for Preparation of Operating and Maintenance Manuals." EPA-430/9-74-001, U.S. EPA., Washington, D.C. (1974).
15. "Uniform Fire Code." Int. Conf. Build. Off. and West. Fire Chiefs Assoc., Whittier, Calif. (1988).
16. Baker, C. A., and Grether, W. F., "Visual Presentation of Information." In "Human Engineering Guide to Equipment Design." McGraw-Hill, Inc., New York, N.Y. (1963).
17. Chapanis, A., "Human Factors—Engineering for Safety." *Professional Safety,* 16 (1980).
18. Sanders, M.S., and McCormack, E.J., "Human Factors in Engineering and Design." McGraw-Hill, Inc., New York, N.Y. (1987).
19. Woodson, W. E., and Conover, D. W., "Human Engineering Guide for Equipment Designers." 2nd Ed., Univ. of Calif. Press, Berkeley (1966).
20. Pailthorp, R. E., and Swanson, G. J., "Designing Wastewater Plants for Real People." *Water Eng. Manage.,* **November,** 35 (1986).
21. "Human Scale." MIT Press, Cambridge, Mass. (1981).

22. Elia, V. J., *et al.*, "Hazardous Chemical Exposure at a Municipal Wastewater Treatment Plant." *Environ. Res., 32,* 360 (1983).

23. Clark, C. Scott, "Potential and actual biological related health risks of wastewater industry employment." *J. Water Pollut. Control Fed., 59,* 999 (1987).

24. Mehta, H., *et al.*, "Planning a Chemical Handling System." *Oper. Forum, 4,* 15 (1987).

25. Pettit, T., and Linn, L., "A Guide to Safety in Confined Spaces." U.S. Dep. Health and Human Serv. Natl. Inst. Occupational Safety and Health, Cincinnati, Ohio (1987).

26. "Criteria for Recommended Standards for Working in Confined Spaces." U.S. Dep. Health, Educ. and Welfare Natl. Inst. Occupational Safety and Health, Cincinnati, Ohio (1979).

27. "American National Standard Safety Requirements for Confined Spaces." ANSI Z 117.1 Am. Natl. Stand. Inst., New York, N.Y. (1989).

28. U.S Code of Federal Regulations, Title 29, 29 CFR, Part 1926 (1987).

Chapter 7
Support Systems

INTRODUCTION

This chapter aims to review the state of the art and general design considerations for support systems of wastewater treatment facilities, thereby enabling the engineer to work effectively with other design disciplines when designing these systems. Numerous support systems are required to make a treatment facility fully functional and, in some cases, acceptable to the general public.

The selection, development, and design of individual wastewater treatment operations and processes requires significant effort. However, even a well-designed treatment system needs well-planned and developed support systems. As a result, the design of support systems can rival plant process design in complexity and potential for innovation and energy savings, especially at larger facilities. A well-coordinated process and support system design results in a treatment facility that effectively and efficiently provides the required degree of treatment.

RELIABILITY CRITERIA

In the early 1970s, when the emphasis was on building new wastewater treatment facilities, the U.S. Environmental Protection Agency (EPA) developed reliability design criteria for mechanical and electrical support systems.[1] The objective of the criteria was to ensure that new facilities maintained a high degree of effectiveness. In 1974, EPA published reliability design criteria,[1] which are still being followed. These reliability criteria apply not only to wastewater treatment processes, but also to mechanical and electrical support systems of the facility. The criteria, summarized below, define minimum reliability standards for three classes of wastewater treatment facilities.

- Class I: facilities that discharge to navigable waters that could be permanently or unacceptably damaged by effluent that was degraded in quality for only a few hours.
- Class II: facilities that discharge to navigable waters that would not be permanently or unacceptably damaged by short-term effluent quality degradations but could be damaged by continued effluent quality degradation.
- Class III: facilities not classified under class I or class II.

Once the facility classification is defined, specific design criteria to accomplish the desired reliability is found in the technical bulletin. Support systems that affect the criteria include electrical power systems, instrumentation and control systems, and other auxiliary systems such as plant water, plumbing, and chemical systems. For example, reliability provisions for an electrical power source and its distribution depend on the particular class (class I, II, or III) of treatment facility. In general, two separate and independent sources of electrical power should be provided to the facility either from two separate utility substations or from a single substation and a treatment facility based generator. If available from the electrical utility, at least one of the facility's power sources should be a preferred source, that is, a utility source that is one of the last to lose power from the utility grid if power-generating capacity is lost. Independent sources of power should be distributed to separate transformers at the facility to minimize common mode failures from affecting both sources. Power distribution reliability features within the facility should also include service to motor control centers, division of loads at motor control centers, power transfer, coordinated breaker settings or fuse ratings, equipment location, and emergency power generator starting.

Table 7.1 summarizes the results of an electrical power service reliability study. This table shows that two independent power sources provide a high level of reliability and that only under the most severe reliability restrictions would the expense for an additional source of power be justified.

Table 7.1   Power source reliability.

	Single 13.8-kV source	Single 115-kV source with two feeders to 13.8 kV	Two independent 13.8 kV
Simple radial system			
Failures/year	1.12	0.3	0.1
Hours down/year	7.0	5.5	5.0
Primary selective system			
Failures/year	1.0	0.3	0.06
Hours down/year	5.0	3.5	3.1
Secondary selective system			
Failures/year	1.0	0.2	0.03
Hours down/year	2.7	1.2	0.8

ELECTRICAL SYSTEMS

ELECTRICAL POWER DISTRIBUTION. The design of the electrical power distribution system for a wastewater treatment facility must conform to applicable codes and standards. In the absence of detailed local electrical code requirements, the design normally should be in substantial accordance with the National Electrical Code (NEC),[2] the National Electrical Safety Code,[3] and the requirements set forth in "Occupational Safety and Health Standards." [4,5] In addition, the design must comply with the operating rules of the local electrical utility. Therefore, it is important, early in the design process, to identify and thoroughly understand all of these requirements because, upon completion of construction, the contractor is often required to obtain and furnish certification of inspection and approval from all required authorities and underwriters. A good idea at the onset of the project is to meet with the local power utility to review their criteria.

In the final design of a power system, the objective is an adequately sized, reliable, safe, and economically sound system. Therefore, an accurate estimate of the electrical load and the associated characteristics is required. As soon as the general nature of the design has been established, a preliminary analysis of the project is usually made to determine the approximate size and location of major items of electrical equipment and furnish the local utility with the information required by its engineering department. The information generally desired by the utility includes plot plans showing the location of the plant and the various structures; the point of entry of the electrical service; total connected load for both lighting and power; maximum estimated demand; size and locations of motors and starting characteristics; and electrical load growth projection.

The physical layout of the plant affects the location of electrical equipment and the characteristics and configuration of the plant power system. A small or compact plant layout lends itself to centralized distribution and transformer location and radial low-voltage feeds to the individual loads. A larger and more expansive plant may be more suited to medium-voltage distribution, looped feeders, and the location of electrical equipment such as transformers, motor control centers, and power distribution centers near the load. Most cases are a combination of the two systems.

The nature of electrical loads in a wastewater treatment facility establishes the voltage levels required at the plant site. Lighting, convenience receptacles, and small motors [less than 0.4 kW (0.5 hp)] generally require a 120/240-V, single-phase, 60-Hz power supply. For lighting of both outdoor and large indoor areas, a more economical solution may be a three-phase, 60-Hz, 120/208- and 277/480-V power supply. Large motors [0.4 to more than 150 kW (0.5 to 200 hp)], which make up the bulk of loads in most wastewater treatment facilities, usually require a low-voltage (typically 480-V) three-phase, 60-Hz power supply. Larger motors (greater than 250 hp) normally require a medium-voltage (2400- or 4160-V, three-phase, 60-Hz) power supply. Loads requiring voltages in excess of 5000 V are rare.

The facility power load and characteristics are determined by preparing an equipment motor list and the nature of use, such as continuous, standby, or intermittent. Included with the preliminary design is a single-line diagram of the power distribution system showing all major components of the electrical system, such as power source transformers, switchgears, motor control centers, various power and lighting distribution panels, and all motors. Figures 7.1a and b present typical electrical single-line diagrams for a medium-sized wastewater treatment plant. The lighting load can be estimated from the areas of the various structures by using appropriate watt-per-area factors obtained from the National Electrical Code.[2]

It is important to examine the rate schedules of the power company when designing an electrical power system. In addition to charges for actual power consumed, most power companies add a maximum demand charge. Some power companies also impose a penalty charge if the plant's power factor is lower than a certain minimum value. The maximum demand charge, frequently based on the highest rate of usage in the preceding 12 months, penalizes the user for large short-time power usage. Uniform power usage is, therefore, economically important.

In addition to computing the connected load, the normal demand load, facility growth, and reliability should be considered. The total connected load is the sum of all electrical loads [expressed in kilovolt–amperes (kVA)] connected to the power system. Think of the demand load (in kVA) as the actual peak operating load of the plant. When calculating the demand load, assume maximum wastewater flow conditions. In designing for expansibility, select equipment and circuit ratings that are adequate for future load growth and allow space for additional electrical equipment and circuits. Reliability is generally maximized by providing alternate power supply sources and routes.

Supply and Distribution System Voltage. When power is obtained from the local electrical utility, the types of services available from the local utility limit the choice of power-supply voltage at the plant site for particular loads. As a result, the policy and main power distribution system of the local power utility affect the design of the power system. Therefore, to properly coordinate the facility design with the power utility, it is important to meet with representatives of the power company early in the design process. The most important electrical characteristics of the distribution system of a power utility are its power-producing capacity, distribution voltage, and available short circuit current.

POWER CAPACITY. Obviously, the power capacity of the distribution system should be at least equal to the load it is to supply. Generally, two separate and independent full-capacity sources of electric power are required. In this context, "independent" means that each power feeder comes from a separate substation of the power company; "full capacity" means that each feeder is capable of handling the total connected load of the power system. A less desirable, but acceptable, situation is to have two feeders, each capable of handling the total demand load and less than 50% of the total connected load. However, this may not be possible. It may be that only one substation is

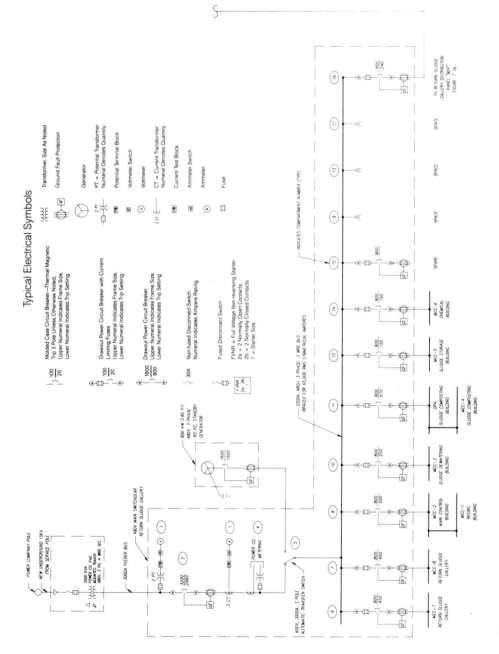

Figure 7.1a Typical electrical single-line diagram for a medium-sized wastewater treatment facility.

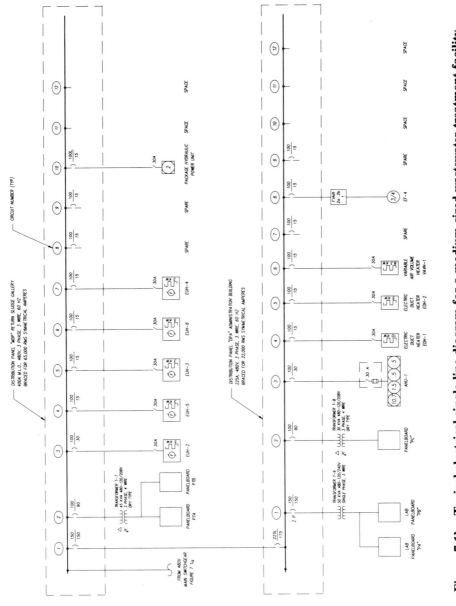

Figure 7.1b Typical electrical single-line diagram for a medium-sized wastewater treatment facility.

available or that the power company cannot provide a second full-capacity feeder. In this case, consider taking two feeders from a single substation and shedding all but the most essential load in case of power loss from the substation. On-site power generation or auxiliary engine generators can be used to supply the essential load.

The form of load transfer on power failure must also comply with power utility requirements. Consult the power company as to whether load transfers may be manual or automatic, in what form automatic load transfers will occur, and what provisions should be made to prevent paralleling two power utility lines by connecting them to each other.

DISTRIBUTION VOLTAGE. The distribution voltage of the power utility is normally much higher than voltages required for use on the plant site (usually in the 15-kV class), while plant loads are usually in the 5-kV or 600-V classes. A decision must be made either to distribute power on the plant site at the voltage of the power utility and transform voltage at the individual loads, provide initial transformation at the service entrance to the plant, or apply a combination of both. With primary service, the utility usually establishes the supply voltage. With secondary service, any of several standard distribution voltages can be made available. The terms primary and secondary feeder apply to feeders both to and from a transformer.

The use of primary versus secondary service requires careful consideration of

- The facility layout,
- Rate schedules for both services,
- Current and short circuit ratings of equipment,
- The cost of a substation,
- The cost of primary and secondary wiring and metering equipment to the plant for both services, and
- Plant personnel required to maintain the equipment.

The use of primary service generally achieves a lower electrical energy cost. With primary service, the utility provides and maintains feeders to the facility-owned and maintained substation. The amount of savings may not be sufficient to compensate for the cost and inconvenience of maintaining and replacing transformers and high-voltage equipment. Long delivery items are stocked by utility companies at all times. In addition, if the facility load is supplied by a secondary voltage of 480 V or less, the facility may not have qualified personnel to maintain high-voltage equipment and large transformers.

Metering of electrical energy can be installed on utility feeders before the transformers or on secondary feeders in front of the transformers. Final determination is usually made by the utility company. Energy loss on a transformer is about 1%.

When the facility load is large and expansive, secondary distribution is achieved by introducing higher voltages such as 2400 V, 4160 V, 6900 V, 7200 V, and 14 400 V. Commonly, the facility is responsible for secondary distribution of up to 4160 V. The maximum total voltage drop between the power source and final use for feeders and branch circuits should not exceed 5%.

SHORT CIRCUIT CAPACITY. The short circuit capacity of a power distribution system is the amount of power (or current) the system can supply to the point of short circuit. For most power utility distribution systems, the short circuit capacity at the distribution line is normally at least 50 000 kVA. At points in the facility power system, capacity will be reduced by the impedance of cables and transformers between the distribution line and the short circuit and will be increased by the effect of connected motors operating as generators at the time of short circuit.

All equipment in a power system must have a short circuit, or interrupting rating, greater than the short circuit capacity at that particular point in the system. If short circuit capacities are excessive, they can be reduced by increasing impedance of transformers or installing reactors in the system. An increase in short circuit equipment rating means a significant increase in equipment cost.

SELECTIVE DEVICE COORDINATION AND PROTECTION. A power distribution system, based on its significance and complexity, can be protected against short circuit current, overcurrent, overvoltage, undervoltage, grounding, differential current in the system or transformer, and loss of phase. Pay particular attention to adequately protecting transformers, motors, switchgears, and motor control centers. Fuses and circuit breakers protect against overcurrent, short circuit current, or both. Other protective devices are the various types of relays.

Selective device coordination isolates the problem spot of the power system, while the rest of the system operates without interruption. This is achieved by selective adjustment of the overcurrent and grounding protection devices. For proper protective device adjustment and selective coordination, it is necessary to perform short circuit calculations and load studies of the system. For proper motor protection, both the motor full load current and the starting current must be known. Large motors are additionally protected against overtemperature by resistive temperature detectors (RTDs).

Distribution System. The primary and secondary selective systems are the two basic configurations of dual feeders to electrical equipment such as the switchgear or motor control center. In the primary selective system, the equipment has a single bus to which the entire load is connected. In the event of a power failure, the feeder normally supplying the power is disconnected, and the second feeder is connected to the bus to supply the load. In the secondary system, the equipment has two buses; each bus supplies 50% of the load and is normally connected to a power feeder. An open bus tie is normally provided. If either feeder fails, the faulty feeder is disconnected, and the bus tie-breaker is closed so that the other feeder carries the entire load. In either case, the transfer may be manual or automatic. In most cases, positive means (such as key interlocking of switches or circuit breakers) must be taken to prevent paralleling the two feeders.

In a primary selective system, a transformer primary has a switch that selects one of two feeders and supplies a single load group, such as a motor control center. In a secondary selective system, the load group is designed to select one of two transformer secondaries. Both techniques can be combined to further increase reliability.

A radial feed system is one in which individual feeders from a central point serve individual load groups. A loop feed system is one in which a feeder from a central distribution point is taken to several load groups and returned to the central distribution point. The feeder may then be energized from either end. A third type of system, also referred to as a loop system, (although, strictly speaking, it is a looped radial system), is one in which each of two feeders is taken to the primaries of several transformers to make the transformers primary selective. Figure 7.2 shows a typical dual-feed power system.

LIGHTING SYSTEMS. Lighting requirements in general should follow procedures recommended by the Illuminating Engineering Society.[6] The three categories of lighting systems installed throughout wastewater treatment facilities—indoor, outdoor, and emergency lighting—serve various purposes. First and foremost, lighting systems are employed for the safe and effective operation and maintenance functions of the facility. In addition, adequate lighting provides a degree of site security at night. However, lighting near moving equipment or vehicular traffic must be selected to minimize glare.

Indoor Lighting Systems. During the daylight hours, properly used natural lighting minimizes the electrical lighting required indoors. Appropriately placed windows and skylights, coupled with proper surface coating systems, accomplish this. However, even with the best natural lighting, an electrical indoor lighting system is required. Table 7.2 summarizes suggested minimum indoor lighting requirements.

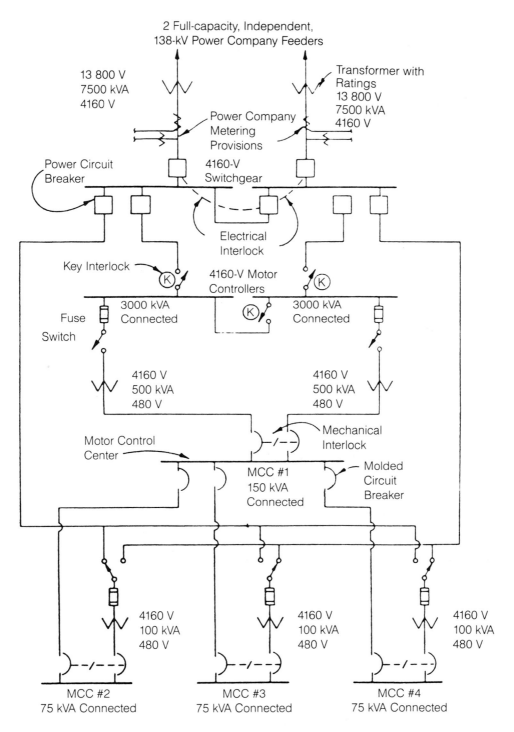

Figure 7.2 **Typical example of a dual-feed power system.**
(MCC refers to "motor control center.")

Table 7.2   Typical minimum indoor lighting requirements.

Location	Minimum illumination, ft–candle
Process areas	30
Pipe tunnels/galleries	15
Office areas/control rooms	100
Laboratories	100
Maintenance areas	30
Garages/storage areas	5

Typical indoor lighting sources include

- Incandescent,
- Fluorescent,
- High-pressure sodium,
- Metal halide, and
- Mercury vapor.

The most efficient lighting source is high-pressure sodium; the least efficient source is incandescent (See Table 7.3). Fluorescent fixtures are commonly used in offices, areas with low ceilings, and when the color of light is important. High-pressure sodium, mercury vapor, or metal halide fixtures are used in large indoor areas and areas with high ceilings 50 m (15 ft) and higher, and outdoors. Each of these three types of lighting sources—metal halide, mercury vapor, and high-pressure sodium—produces a different color of light.

Most fluorescent systems are 120-V, single-phase systems. High-pressure sodium systems operate at 277 or 480 V (single-phase voltages). When using high-pressure sodium or metal halide systems, consider the required warm-up time. As a result, they may not be appropriate for locations that require instantaneous lighting, such as passageways.

Lighting fixtures in areas defined as hazardous should comply with the NEC[3] requirements for class I, group D, and division 1. In areas where

Table 7.3   Efficiency of available lighting sources.

Lighting source	Efficiency, Lm/W	Average life, h
Incandescent	15–18	1 000
Mercury vapor	40–47	16 000–24 000
Fluorescent	40–70	6 000–20 000
Metal halide	60–72	10 000–20 000
High-pressure sodium	70–112	16 000–24 000

Design of Municipal Wastewater Treatment Plants

hazardous gases are normally contained in tanks and piping, the fixtures should comply with the NEC for class 1, group D, and division 2 requirements. To increase reliability and minimize maintenance, all lighting fixtures in these areas should be high quality and corrosion resistant.

Outdoor Lighting Systems. Outdoor lighting systems are generally used for the safety of the plant staff and overall site security. Outdoor lighting systems typically consist of pole-mounted, general area lighting with localized, supplementary illumination at exterior process units, building entrances, and plant entrance gates. Table 7.4 summarizes typical minimum outdoor lighting requirements. Current practice uses 480-V, single-phase high-pressure sodium fixtures and lamps. Outdoor lighting systems are controlled by photo cells, manual switches, or a combination of these. Depending on the size of the overall plant site, conduct an economic evaluation to determine the optimum selection of the number and height of pole-mounting luminaries needed to minimize glare and accomplish the general area lighting system goals.

Table 7.4   Typical minimum outdoor lighting requirements.

Location	Minimum illumination, ft–candle
General	0.5
Unoccupied site security	0.2
Outside entrances	5.0
Local task lighting	5.0

Emergency Lighting Systems. Specific requirements for emergency lighting systems are normally defined by the building code requirements that govern construction of the facility. Emergency lighting systems provide safe exit from the facility during a general power failure and are installed inside buildings and structures. In general these lighting systems are self-contained, battery-operated units. Luminaries are sealed-beam incandescent lamps, and the battery units are self-charging. Illuminated face exit lights are also typically installed to denote points of exit from a structure.

STANDBY POWER CONSIDERATIONS. The Institute of Electrical and Electronics Engineers (IEEE) Standard 446[7] presents recommended engineering practices for the selection and application of emergency and standby power systems. Standby power generating units are often used for increased reliability or for small critical loads within a facility where an affordable, additional independent power source is not available. Reciprocating engine-driven generators fueled with gasoline, diesel fuel, natural gas, or digester gas are most commonly used. For applications requiring more than 500 kW, gas turbine generators are sometimes preferred. Consideration must be given to providing adequate cooling and ventilation, fuel storage, starting readiness, and service availability. In addition, many areas have stringent air quality control ordinances that require permits to operate engines with certain types of fuel.

In large facilities with anaerobic sludge digestion, the digester gas often recovered is used as fuel for boilers and internal combustion engines that are, in turn, used for pumping wastewater, operating blowers, or generating electricity. A survey of 40 large wastewater treatment plants [of at least 3785 m^3/d (1 mgd) capacity] throughout the U.S. that have methane recovery facilities showed that many of the plants have also installed electric generating systems that use digester gas successfully. However, many questioned whether purchasing expensive equipment was justified. The most satisfied applications appeared to be those using gas for boiler (heating) applications and for sale to the area gas utility.

Generally, cogeneration in a combined facility is defined as a system for generating electricity and producing another energy at the same time (usually steam) that is more efficient than electrical generation would be alone. Thus defined, a cogeneration facility typically applies waste heat recovered from a gas engine to its heating needs for space and water.

When selecting a standby power system, consider the particular application needs. Some loads, such as computers, may not tolerate any outage and should be supplied by an uninterruptible power supply (UPS) system. Also, critical delays in switching to a standby utility line may affect its feasibility. In addition, the sizing of a standby power system must address voltage dip during the starting of a large motor because newer electronic systems and motor-starting coils are affected if voltage dips below 85 to 90% of nominal voltage.

OTHER CONSIDERATIONS. Grounding Systems. A properly designed grounding system provides for the safety of operating personnel against the possibility of injury caused by electrical shock in damp locations normally associated with wastewater treatment facilities. The system neutral of the low-voltage system (usually a 120-V to 208-V, three-phase, four-wire system) must be effectively grounded so that a fault in any phase will operate the protective device immediately. All electrical equipment and all conductive components of equipment enclosures or mounting electrical devices and building steel should be permanently and effectively grounded in accordance with section 250 of the National Electrical Safety Code.[3] Such equipment ensures that the voltage to ground of this equipment cannot reach harmful levels. All underground distribution system appurtenances (cable racks, manhole covers, and pull boxes) and aerial distribution system equipment (lightning arresters and transformer cases) are to be grounded also.

Load Transfer on Power Failure. A critical factor in the design of an alternate power source system is the means of switching to an alternate source or to the emergency power source. Both cost and complexity are functions of the transfer speed required. If an outage of 30 minutes can be tolerated, a manual transfer switch may be acceptable at a continuously manned facility. For an outage of several minutes, motorized primary system switching may be acceptable. To switch within 15 seconds, an automatic transfer switch is required. Critical equipment that cannot tolerate a momentary outage requires a solid-state transfer switch with instantaneous switching.

Hazardous Areas. Hazardous areas are classified by the National Fire Protection Association (NFPA)[8] according to the degree of risk of fire and explosions. Areas of prime concern in wastewater treatment facilities are areas at which methane gas is generated. Other areas of concern are where chlorine is handled and sulfur dioxide is produced and facilities where fuel is stored and handled. These areas, however, are mostly corrosive and need only be properly ventilated. In addition, electrical storage batteries generate hydrogen and must be considered.

The quantity and types of flammable liquids, gases, and dusts; type of ventilation; openness or confinement of the sources of hazard; and the distance of devices from the substances affect the need for and extent of the required explosion proof equipment. Due to the high cost of such equipment, the layout and design can minimize the amount of electrical equipment and devices required within the hazardous areas.

Seismic Protection. When electrical equipment is installed in regions prone to earthquakes, anchor heavy electrical equipment (transformers, motors, generators, and switchgear) to base pads or the building as required by most national building codes.[9] Resilient anchorage using either spring or rubber shock vibration isolation systems has performed well in earthquakes; snubbers or stops to limit movement must be provided. Conductors and connections must be sized and sagged to allow for expected movement during the earthquake, avoiding breakage of insulators and contact between adjacent conductors. This usually means wider spacing between conductors and greater sag than provided by normal nonseismic designs.

Electrical safeguards such as the following are often provided:

- Low-voltage and single-phase protection for motors;
- Pressure switches on the discharge piping to shut down pumps under low (ruptured pipelines) or high pressure (collapsed or blocked pipelines);
- Temperature switches to sense high discharge piping temperature, resulting from continuous water recirculation through a failed bypass pump control valve, a failed high discharge pressure switch, or a closed isolating valve; and
- Motor-winding temperature detectors to sense overloaded motor conditions or severely unbalanced electric power phase and voltage conditions.

Lightning and Surge Protection. Lightning and surge arresters limit the voltage impressed upon the winding of a transformer or motor due to lightning or switching surges. Surge arresters are connected between each phase and ground. The primary side of the main transformer is generally protected with lightning and surge arresters. The motors of 500 hp and greater should be protected by surge arresters and surge capacitors. The same is applicable for 200-hp motors and greater, supplied by overhead line. Locate surge arresters and surge capacitors as close to the motor as possible. Sometimes, for economical reasons, lightning and surge arresters are installed in the switchgear or motor control center and protect a group of motors tied to such a source.

"Clean" Power for Computer Systems. New, digital-based control systems are more susceptible to power line disturbances than older, analog-based systems. Power line disturbances fall into one or more of the following categories:

- Power outage (short-term to long-term),
- High and low voltage,
- Harmonic distortion,
- Frequency shifts,
- Short-term voltage fluctuations (surges, sags, and dips),
- Long-term voltage fluctuations (brown-out), and
- Noise impulses.

A range of techniques are available to condition power line disturbances. Analyze the power requirements of each control system. Then, use one of the following power-line-conditioning techniques to clean up the electrical power:

- Isolation transformer,
- Line voltage regulator,
- Line conditioners, or
- Uninterruptible power supplies (UPS).

INSTRUMENTATION AND CONTROL SYSTEMS

Instrumentation and control systems in wastewater treatment plants are as varied as the treatment plants.[10-16] This section is devoted to wastewater treatment plant control systems. Because digital systems, in the last few years, have become more economical than analog-based systems, analog control systems will not be addressed. However, analog controllers may be furnished as part of local control or digital control backup panels.

DESIGN CRITERIA. Design criteria for instrumentation and process control systems[10-17] are based primarily on guidelines established by the Instrumentation Society of America (ISA)[17] and manufacturers' recommendations. These include

- Standard S5.1—Instrumentation Symbols and Identification and
- Standard S20—Specification Forms for Process Measurement and Central Instruments, Primary Elements, and Central Valves.

PROJECT SCOPE DEVELOPMENT. In developing the scope of the instrumentation and control design, the following questions must be answered:

- Size of the plant,
- Type of treatment processes,
- Type of vendor-supplied controls,
- Amount of funds available,
- New plant or modification and expansion of existing plant,
- Current design standards,
- Special interfaces with other control systems, if any,
- Extent of automatic controls necessary, and
- Ability of the owner to maintain a control system.

Once the above questions have been answered, a control systems engineer determines the best approach for the control system and the amount of instrumentation required.

PROCESS AND INSTRUMENTATION DIAGRAMS (P&IDs). Symbols and identification codes used to identify instruments in wastewater plants are based on the ISA standards,[17] with modifications for wastewater treatment plants as detailed in Instrumentation in Wastewater Treatment Plants.[16] The basic use of a process and instrumentation diagram (P&ID) is to provide sufficient information for a knowledgeable person reading the P&ID to understand the means of measurement and control of the process without the need to go into the details of instrumentation that require the knowledge of an instrument specialist. The full details of the instrument are often left to be covered by a suitable specification, data sheet, or other document.

Types of P&IDs. Two basic types of P&IDs—the process flow diagram and the instrumentation system diagram—are discussed below. Because ISA serves different industries with different needs, its standards are flexible to allow for variations. The type used on a specific project will be a function of the owner's or design engineer's preference.

Process flow diagrams usually show valves, pipe size reduction, pipe size, flow stream, piping material identification and all mechanical equipment, instrumentation and control (I&C) panel designation, basic electrical requirements, numbers and types of signals for aid to electrical engineers in type, sizing and routing of conduit and wire along with the ISA bubbles, and tagging destinations (see Figure 7.3).

Instrumentation and control diagrams are usually simplified process flow diagrams, showing only major mechanical equipment and instrumentation and final control elements controlled by the control system. All other information is obtained from the process flow diagram, which shows only the in-line instrumentation (see Figure 7.4).

SELECTION AND SPECIFICATION OF FIELD INSTRUMENTS.
This section explains the process of properly selecting and specifying field instruments. The process starts with the development of the P&ID. Then, data sheets are developed for all instruments. Next, instruments are located on physical drawings. Lastly, special mounting details are developed as required.

P&IDs. As discussed in the previous section, P&IDs should be developed for all projects. Part of the process of developing the P&ID is to learn from the process engineer any special requirements that will affect instrument selection. Elements to be considered in development of P&IDs were defined above.

Instrumentation Data Sheets. The most important element in designing a control system is correct instrument specification. One key component of an instrumentation specification is the instruments data sheet. Because of the complexity of present day instruments and controls, ISA Standard S20,

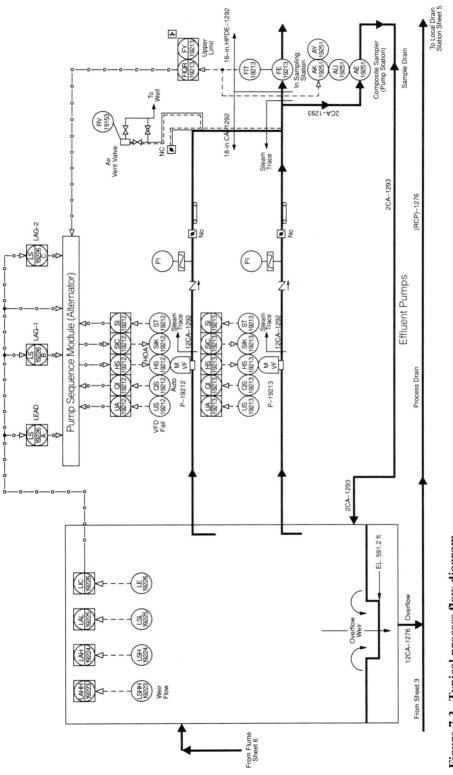

Figure 7.3 Typical process flow diagram.

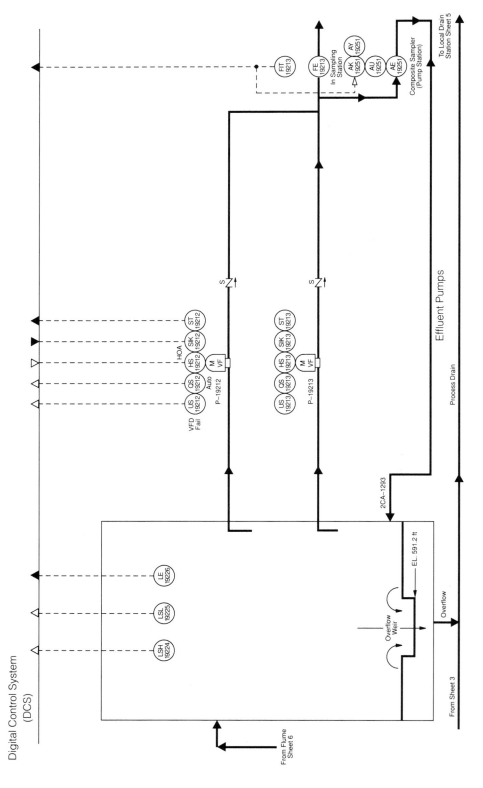

Figure 7.4 Typical instrumentation and control diagram.

"Specification Forms for Process Measurement and Control Instruments, Primary Elements and Control Valves,"[17] has standardized both the content and form of instrumentation data sheets. Figures 7.5 and 7.6 show a typical data sheet for magnetic flow meters.

©ISA S20

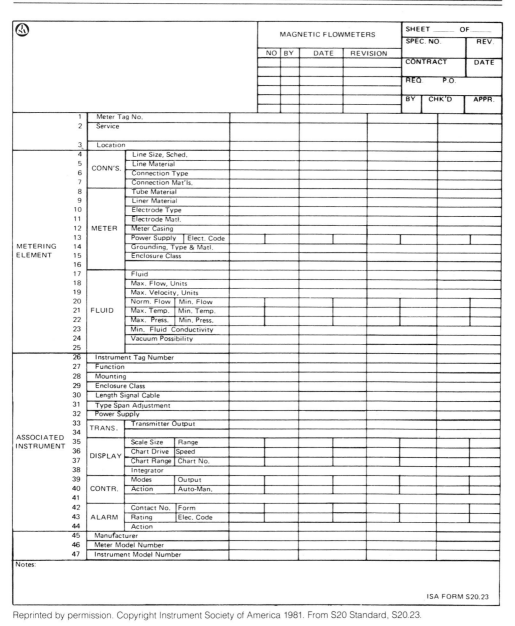

Figure 7.5   Typical ISA data sheet form.

MAGNETIC FLOWMETERS

Instructions for ISA Form S20.23

1. Tag number of meter only.

2. Refers to process application.

3. Show line number or identify associated vessel.

4. Give pipeline size and schedule. If reducers are used, so state.

5. Give material of pipe. If lined, plastic or otherwise non-conductive, so state.

6. Give connection type: FLANGED, DRESSER COUPLINGS, ETC.

7. Specify material of meter connections.

8. Select tube material. (Non-permeable material required if coils are outside tube).

9. Specify material of line.

10. Select electrode type: STD., BULLET NOSED, ULTRASONIC CLEANED, BURN OFF, etc.

11. Specify electrode material.

12. Describe casing: STD., SPASH PROOF, SUBMERSIBLE, SUBMERGED OPERATION, etc.

13. Give ac voltage and frequency, along with application NEMA identification of the electrical enclosure.

14. State means for grounding to fluid: GROUNDING RINGS, STRAPS, etc.

15. State power supply and enclosure class to meet area electrical requirements.

16.

17. State fluid by name or description.

18. Give maximum operating flow and units; usually same as maximum of instrument scale.

19. Give maximum operating velocity, usually in ft/s.

20. List normal and minimum flow rates.

21. List maximum and minimum fluid temperature °F.

22. List maximum and minimum fluid pressure.

23. List minimum (at lowest temp.) conductivity of fluid.

24. If a possibility of vacuum exists at meter, so state and give greatest value. (highest vacuum).

25.

26. List tag number of instrument used directly with meter.

27. Control loop function such as INDICATE, RECORD CONTROL, etc.

28. Mounting: FLUSH PANEL, SURFACE INTEGRAL WITH METER, etc.

29. Give NEMA identification of case type.

30. State cable length required between meter and instrument.

31. Span adjust: BLIND, ft/s DIAL, OTHER.

32. Give ac supply voltage and frequency.

33. If a transmitter, state analog output electrical or pneumatic range, or pulse train frequency for digital outputs, i.e., pulses per gallon.

34. List scale size and range.

35. List Scale Size and Range for indicating transmitter

36. Recorder chart drive — ELECT. HANDWIND, etc. and chart speed in time per revolution or inch per hour.

37. List chart range and number.

38. If integrator is used, state counts per hour, or value of smallest count; such as "10 GAL UNITS".

39. For control modes: (Per ANSI C85.1-1963, "Terminology for Automatic Control.") Write-in PI_f, I_f, PI_s, $PI_f D_f$, etc.

$$P = \text{proportional (gain)}$$
$$I = \text{integral (auto reset)}$$
$$D = \text{derivative (rate)}$$

Subscripts:

$$f = \text{fast}$$
$$s = \text{slow}$$
$$n = \text{narrow}$$

State output signal range, pneumatic or electronic.

40. Controller action in response to an increase in flowrate — INC. or DEC.

State auto-man. switch as NONE, SWITCH ONLY, BUMPLESS, etc.

42. Number of alarm lights in case. Give form of contacts; SPDT, SPST, etc.

43. Contact electrical load rating. Contact housing General Purpose, Class I, Group D, etc., if not in the same enclosure described in line 29.

44. Action of alarms: HIGH, LOW, DEVIATION, etc.

45. Fill in manufacturer and model numbers for meters
46. and
47. instrument after selection.

Reprinted by permission. Copyright Instrument Society of America 1981. From S20 Standard, S20.23.

Figure 7.6 Typical ISA data sheet form explanations.

Some designers use the forms directly from ISA, whereas other designers have modified the forms and entered them for use by automatic data-processing techniques. These standard forms also serve as a checklist for items to be considered when specifying a given instrument and have space for adding notes.

Some key information of the physical considerations of a flow meter include

- Pipe size (for in-line instruments),
- Minimum and maximum operating data (flow, pressure, temperature, and pH),
- Length between sensor and transmitter,
- Special installation hardware required,
- Environmental requirements (temperature, humidity, and corrosive or explosive atmosphere),
- Special tagging requirement, and
- Materials of construction for instrument and method of connection to physical system (flange).

Some specific instrument needs are

- Range of measurement;
- Accuracy;
- Repeatability;
- Vendor options that apply;
- Electrical requirements;
- Spare part requirements;
- Maintenance and calibration requirements;
- Output signal (linear or nonlinear);
- Input signals required;
- Type of setpoint adjustment (fixed or adjustable), if applicable; and
- Alarm switch along with limit settings and the amount of deadband required when switches return to normal.

Physical Drawings. It is a good idea to physically show field instruments on both mechanical and electrical drawings. Mechanical drawings show where in the process stream instruments are to be mounted by the mechanical contractor. For example, placing the instruments in the process stream allows a designer to check that the flow-measuring devices have their required up- and downstream straight run piping requirements to maintain the required accuracy.

Often, the mechanical contractor is responsible for installing the instruments requiring piping; therefore, instruments are shown on the mechanical drawings. By using the mechanical drawings as background, the electrical design engineer can locate the remaining instruments and control panels. The electrical engineer then uses the P&ID and physical drawings to develop the conduit and wire schedule. Some P&IDs have different wire types included as standard symbols, allowing for quick development of conduit and wire schedules. However, this may result in too much clutter on P&ID drawings.

The electrical engineer will also develop elementary wiring diagrams to show the control and signal-interlocking and interfacing logic requirements to be wired by electricians during construction (see Figure 7.7).

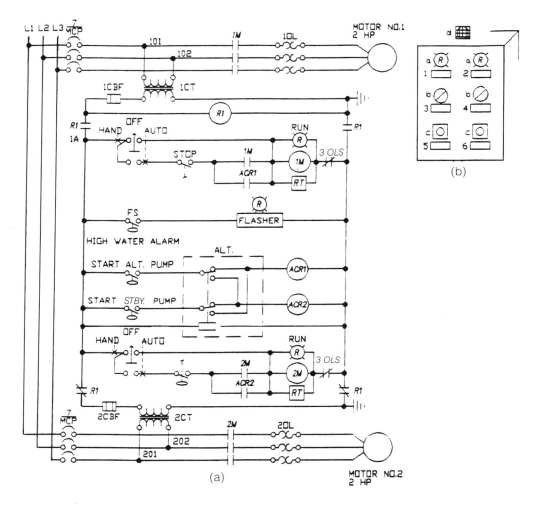

Figure 7.7   Typical elementary wiring diagram (a) duplex sump pump and (b) control panel.

Special Mounting Details. Special mounting details show the contractor what is required for proper installation (see Figure 7.8). The mounting details show all piping and sizing requirements, along with any special fittings or mounting material fabrication requirements for the instruments. A bill of materials may also be helpful and included.

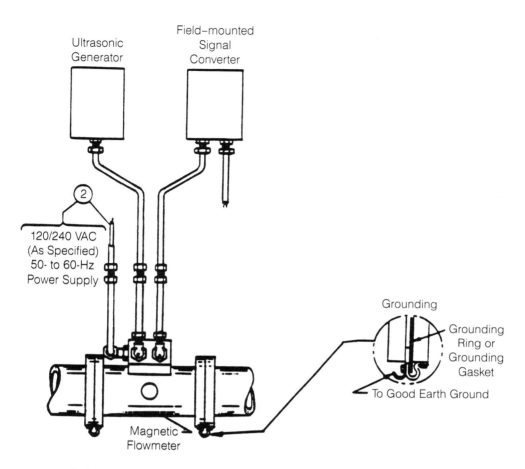

Ultrasonic
Generator

Field–mounted
Signal
Converter

2

120/240 VAC
(As Specified)
50- to 60-Hz
Power Supply

Grounding

Grounding
Ring or
Grounding
Gasket

To Good Earth Ground

Magnetic
Flowmeter

Figure 7.8   Typical mounting detail.

CONTROL PANELS. A typical wastewater treatment plant has many different kinds of control panels (such as vendor-supplied), custom panels (such as local control), main control and graphic panels, and termination panels. Given the widespread use of programmable logic controls and distributed computer control system technology, the current trend, however, is to reduce the analog-control-based panels installed in wastewater treatment plants.

Vendor Supplied. Vendor-supplied control panels are those panels supplied with major mechanical equipment (such as belt filter press, compressor, and centrifuges). The amount and type of control these panels furnish must be consistent with the overall, plant-wide control philosophy. However, from a practical point of view, enough manual control must be furnished so that vendors can prove that their mechanical equipment meets specifications independent of the plant-wide control system.

Custom Panels. Custom panels are installed to supplement vendor-supplied panels and integrate them to the control of multiple pieces of equipment or larger portions of the process stream. Typical custom panels include local control, area control, main control, and graphic panels.

LOCAL CONTROL PANELS (LCPs). Normally located on or near the associated piece of equipment, LCPs include on-off-auto switches. For maintenance, LCPs can also serve as termination panels for interfacing its process signals and control with a computer-based control system. Either "local/remote" or "local/computer" switches should be furnished with LCPs and the computer control system. These switches can switch the whole LCP or just a single piece of equipment. What is actually switched is based on the control design philosophy established for the plant.

AREA AND MAIN CONTROL PANELS. Area and main panels are designed control panels similar to LCPs. However, they generally integrate control of multiple treatment processes (area panels) or the entire plant (main panel) at one series of panels. This provides the plant operations staff with capabilities for controlling all or portions of the plant from designated control centers. The amount of information sent to area or main control panels is many times less than that available at the vendor-supplied or local control panels.

GRAPHIC PANELS. Graphic panels generally depict the process flow of the wastewater treatment plant in graphic form. Graphic panels vary in size and complexity depending on the information presented. They vary from simple strip graphics to more sophisticated, removable mosaic tiles with lights, instruments, switches, and annunciators integrated into the panel. Since the installation of the first color graphic video display terminal in a wastewater treatment facility, the need for graphic panels has decreased. They also have become relatively expensive compared to video display terminal-generated graphics. However, they still have their place in showing an overview of the treatment facility or wastewater collection system.

Termination Panels. Termination panels are specially designed panels for interfacing field signals with a computer-based control system. Normally no control is furnished with the termination cabinets. The panels are designed so each type of signal, analog or contact, is physically separated from the other to help eliminate signal interference. A 24-DVC redundant power supply is furnished for powering two-wire instrument loops. The panel may even contain a small uninterruptible power supply for critical loops that must remain powered during a power outage. Good design practice is to place an isolation switch in each contact (digital) loop and a fused switch in each analog loop. These fuses and switches can be separate or an integral part of the terminal block. They can protect the loop and allow for isolating failed loops.

For projects in which the computer-based control system and field instrumentation are furnished under separate contracts, termination cabinets define the contract boundaries for each contract. As a result, it is important that the designer clearly define which contractor is responsible for the electrical and instrumentation cable connections. In addition, the signal-numbering and identification system developed becomes critical for consistency between contracts and ensuring that contractors use the same identification for a given control signal.

The above describes termination cabinets for large projects where the cabinets are large due to the large amount of signals requiring terminations. On smaller or retrofit projects where space is a problem, the termination cabinet may be combined with LCPs.

TELEMETRY SYSTEMS. At wastewater treatment plants, telemetry systems are normally used to control and monitor wastewater collection systems. The type of telemetry system may vary for DC-powered phone lines with 3- to 15-second time pulse signals to personal computer (PC) digitally based supervisory control and data acquisition (SCADA) systems using modems over voice-grade phone lines, radio systems, direct burial cable, or cable TV. If radio-based telemetry systems are used, special attention should be given to the design and layout to eliminate any potential for radio frequency interference (RFI).

Interface to the plant control system ranges from a simple contact closure to a sophisticated digital data link with a special protocol requiring special software to be written and supported by the plant control system vendor. Try to avoid any special software because it may be difficult to get support from the vendor once the project has been accepted.

PROCESS CONTROL STRATEGIES. Process control strategies are a narrative of how the process is to be controlled. Process control strategies, normally developed for computer-based control systems, are useful for conventional analog systems as well. This narrative includes a listing of all input/output (I/O) signals used by the strategy, any required calculations, and interfaces with other strategies and defines color graphic video display requirements. These process control strategies must contain sufficient information to allow the control system vendor to implement this strategy using its own process control language software program.

Process Control Strategy Narrative. The process control strategy narrative contains sufficient detailed information to totally describe the control requirements in a functional format. Detailed information includes items such as

- Type of control
  — Manual,

- Local automatic,
- Direct digital control,
- Supervisory set point, or
- Batch (sequence);
- Hierarchy of control;
- Safety and process interlocks;
- Monitoring requirements;
- Operational requirements;
- Calculations required; and
- Associated I/O signals.

Interaction With Other Control Strategies. Interactions with other strategies are listed so that they are coordinated and so that the proper information and timing among the strategies are maintained.

Process Graphics. Graphics are laid out in a hierarchal manner, starting with an overview of the plant, then process areas, and finally the control elements of the specific process. Usually, process graphics are generated by using a simplified P&ID to represent the process. All live data points are then added, followed by the addition of the key alarm points and any fixed text that aids the operator in controlling the process.

Signal Listings. Signal listings must be included as part of the specifications and may be included as an appendix by multiplexer location or integrated as part of the control strategy.

DIGITAL CONTROL SYSTEMS. This section defines the two different types of digital control systems (DCS) most commonly used today and their application at typical wastewater treatment facilities. Also, the control hierarchy is discussed and compared with the different system architectures available. This knowledge will then be applied to three plants of different sizes.

Types of Process Controllers. The two basic types of DCS process controllers are programmable logic controllers (PLCs) and distributed process controllers (DPCs). PLCs were developed to replace relay ladder logic and scan their I/O in milliseconds. They are good at sequential control such as backwashing a filter. In the past, they have not had all of the analog control capabilities that DPCs have had. However, PLCs of today approach the versatility of microcomputer-based process controllers.

Distributed process controllers are process computers with control logic, memory, and I/O control, which are connected to other units or to a central control computer with a data or control highway. When selecting a DCS for a specific application, use a PLC-based system if discrete signals constitute a large portion of the I/O point count and the process control strategies require sequence control. Use a DPC-based system if the system contains a large portion of analog signals and the control strategies contain P&ID control. In the future, as PLCs acquire more analog control and DPCs acquire more sequencing capabilities, the differences between these two types may diminish. A design integrating PLCs to a DPC-based system could combine the advantages of both systems.

Building Blocks of a DCS. A distributor control system is composed of numerous basic building blocks, some of which are described below.

- Remote terminal units (RTUs) are microcomputer-based remote telemetry units that have the power to transmit data, accept commands, and replace controls normally performed with relay ladder logic. These units, normally remote from the treatment plant, are installed in wastewater lift stations in the collection system.
- Programmable logic controllers (PLCs) are relay ladder logic-based sequence controllers.
- Distributed process controllers (DPCs).
- Area control centers (ACCs) are control rooms in which an operator is on duty (on a continuous basis or for certain operating shifts only) and is responsible for the control of a certain area or group of plant processes. Equipment normally installed in this area includes DPCs, operator consoles, and certain other peripherals such as alarms and report printers.
- Central control centers (CCCs) are control rooms in which operator or supervisory personnel are on duty on a continuous basis and have control or oversight responsibility over the entire facility. Equipment normally installed in a CCC includes DPCs, operator consoles with full graphic displays, alarm and report printers, disk or tape drive units for data and software storage, and a programmer's console for making changes to the system.
- Data highways are defined as physical media-connecting devices in a DCS that provide communication among the different devices connected to data highways. Included are cables, modems, processors, and associated software that provides message handling, protocol, fault detection, time synchronization, and arbitration. Physical media used for a data highway can be coaxial cables, fiber optics, or other standard instrumentation cabling.
- Single loop controllers (SLCs) are single devices connected to the data highway that operate automatically to regulate a single controlled variable.
- Multiloop controllers (MLCs) are single devices connected to the data highway that operate multiple loops automatically to regulate multiple controlled variables.

DISTRIBUTED CONTROL SYSTEM APPROACHES. The three system sizes listed in Table 7.5 are general sizing guidelines only. Note that the size of the DCS does not have to follow the plant flow capacity. A 114-m^3/s (5 mgd) wastewater treatment plant may require a medium-sized DCS because the plant has sophisticated processes that require the computing power of a medium-sized system.

When actually determining the size of a system, the key items to use are the I/O signal point count, the size of the process computer hardware, and the software functions required. Other items in the guidelines further identify the capabilities of the different systems. Note that the items listed in Table 7.5 are general guidelines and can be changed to fit the requirements of a specific system.

Small Distributed Control Systems. A small DCS generally has a personal computer that functions as the process computer. As such, it performs all of the historical data functions, logs and reports alarms, does trending, and has a

Table 7.5   Distributed control system (DCS)—sizing guidelines.

Item	Small	Medium	Large
Input/output signal point count	<500	500 < 1500	>1500
Type host computer	Personal	Mini-computer	Large mini-computer
Other system interfaces available	Telemetry	Telemetry	Telemetry
Reports/historical data	Limited	Practical limit	Unlimited
Graphical screen representations	Limited	Practical limit	Unlimited
Control software	Resides in remote units	Resides in both remote units and hosts	Resides in both remote units and hosts
Trending	Limited	Practical limit	Unlimited
Local area network	Proprietary slow short	ANSI/IEEE carrier band fast long	ANSI/IEEE broadband/ carrier band fast longest

Design of Municipal Wastewater Treatment Plants

color video display terminal. The amount of logging and color graphic displays are limited by the size of the winchester disk. Figure 7.9 shows a typical block diagram for a small DCS.

The advantages of a small DCS are

- Low capital cost,
- Easily maintained locally,
- Ideal for small plants where control and monitoring are limited,
- Expandability of the system, and
- Menu-driven, easy to use, and supports commercially available software packages (word processing, database analysis, and spreadsheets).

Disadvantages of a small DCS are that the

- Maximum I/O point count is limited to 500;
- Numbers and types of log reports are limited;
- Numbers of color graphic CRT screens are limited;
- Plant optimization software is limited;
- Ability to interface with other computers is limited; and
- Local area network (LAN) is slow, short, and usually proprietary.

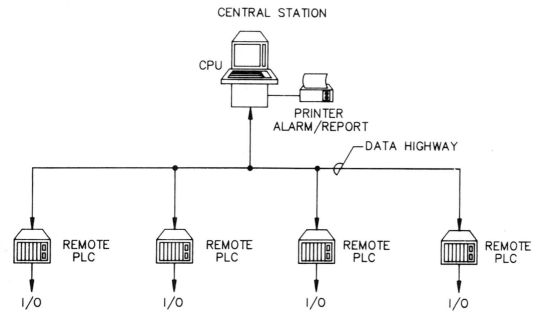

Figure 7.9   Small distributed control system (DCS) (I/O count- < 500 and CPU-personal computer).

Medium-sized Distribution Control Systems. A medium-sized DCS generally has a signal I/O point count ranging from 500 to 1500 and a medium-sized minicomputer that serves either as the host computer or as part of the operator's console. The base medium-sized DCS configuration contains at least one work station and can be expanded to two work stations. Thus, it serves as an area control center and a host computer, performing all of the global plant control strategies and all of the historical and data-reporting functions. Figure 7.10 shows a typical block diagram for a medium-sized DCS.

The advantages of a medium-sized DCS are

- The full control capabilities of the process control software,
- Work stations designed for plant control,
- The ability to interface easily with other computers,
- Reporting and logging capabilities,
- On-line diagnostics, and
- The ability to reconfigure control software from work stations and down-load changes.

The disadvantages of a medium-sized DCS are

- Capital costs are more than for a small DCS and
- A dedicated maintenance staff may be required.

Large Distributed Control Systems. The large DCS serves larger waste-water treatment plants, where the I/O point count is greater than 1500, and additional functions are required. The functions may include interfacing with off-site facilities, a maintenance management system (MMS), and an additional area control center and unlimited trending. Figure 7.11 shows a block diagram of a large DCS. The advantages of a large DCS are

- Ease of expanding the system,
- Ease of interface with other computer systems,
- Unlimited number of color graphic screens,
- Unlimited trending capability,
- Unlimited historical data,
- On- and off-line diagnostics,
- Maintenance management system,
- Plant optimization software, and
- Plant energy management software.

The disadvantages of a large DCS are

- Large capital investment and
- Dedicated maintenance staff required.

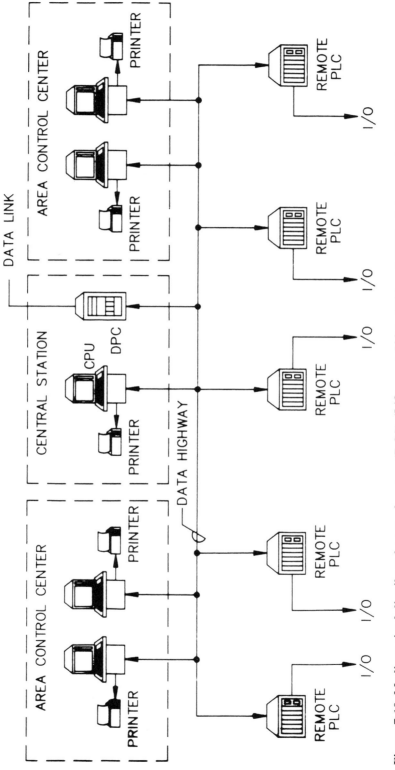

Figure 7.10 Medium-sized distributed control system (DCS) (I/O count-500 to 1500 and CPU-mini-computer, uVAX).

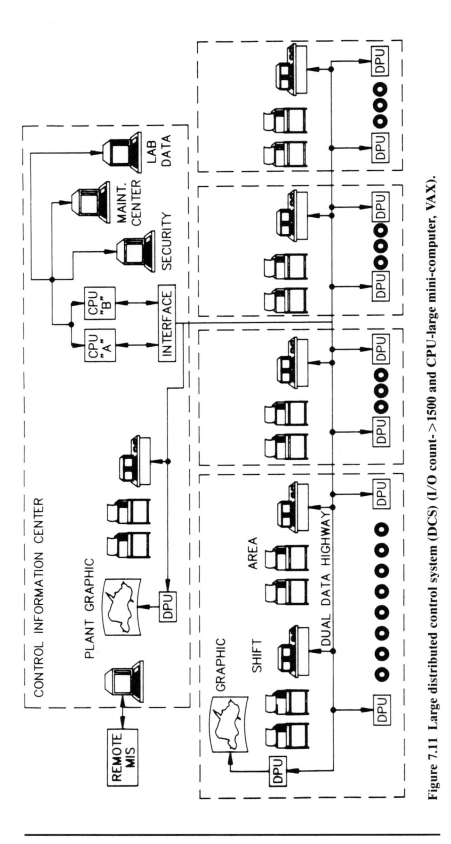

Figure 7.11 Large distributed control system (DCS) (I/O count->1500 and CPU-large mini-computer, VAX).

EARTHQUAKE PRECAUTIONS FOR INSTRUMENTATION AND CONTROL. Where a high probability for earthquakes exists, provide a means for the facility to continue operating if instrumentation and control systems fail. This applies particularly to pump stations. Providing for total redundancy of all instrumentation and control functions is impractical. Designing for damage to or failure of the central control system can be accomplished by providing separate switching at each pump motor site so that individual units can be manually operated, if necessary. This concept requires advance emergency training for operators.

Precision equipment such as residual analyzers, recorders, indicators, meter electronic instrumentation, electronic switching gear, equipment instrumentation, and communications systems should be mounted rigidly to avoid amplification of seismic accelerations. This type of equipment is a prime candidate for shake-table-testing qualification because analysis usually cannot demonstrate that operating capability will survive shaking. Use positive locking devices to hold circuit boards in place. Test all mechanical switching components (relays, for example) for their seismic response characteristics. Avoid mercury switches. Exercise caution in the use of gravity or light-duty, spring-controlled switches. Relays often respond adequately in the energized position but may fail in the nonenergized position. Exercise caution when using friction-restrained switches and components. Avoid the use of circuit board mounting on standoffs, as it may result in local resonance. Provide additional strengthening such as welded supports.

Provide communication equipment and critical instrumentation-controlling equipment with a dedicated emergency power supply (possibly batteries) and a station standby power supply. Provide manual overrides for all automatic control systems. Critical installations that cannot be designed to withstand seismic motion may be supported on an earthquake compatible floor vibration isolation system designed to attenuate motion.

*H*EATING, VENTILATING, AND AIR CONDITIONING (HVAC) SYSTEMS

Heating, ventilating, and air conditioning (HVAC) systems for buildings or other occupied areas are important support systems included in the design of wastewater treatment facilities. Not only are proper HVAC systems important to provide comfort for the plant's operations and maintenance staff, but they are also essential for the safety of the staff and the protection of the physical facilities. This section of the manual provides the process design engineer with some of the design criteria and equipment now in use for

HVAC systems. Topics covered include specific design criteria and HVAC equipment descriptions plus an introduction to design considerations associated with energy conservation and hazardous areas.

DESIGN CRITERIA. Heating, ventilating, and air conditioning (HVAC) systems in wastewater treatment plants are designed to provide a comfortable and safe atmosphere for all plant personnel and protect sensitive equipment. The HVAC engineer knows, understands, and applies

- The applicable local, state, and federal codes;
- Design criteria of special conditions prevailing in wastewater treatment facilities such as the minimum air changes per hour (ac/hr) required in hazardous areas; and
- Additional safety precautions such as monitoring systems for toxic and combustible gases and supplemental ventilation systems for the removal of such gases, alarms, and controls.

Heating Loads. Design criteria for heating loads are based on guidelines established by the American Society of Heating, Refrigerating, and Air Conditioning Engineers, Inc. (ASHRAE). Outside air temperature design parameters and heat loss calculations for structures are presented in the ASHRAE Handbook.[18] Design criteria for new construction and design parameters for occupied spaces are presented in ASHRAE's Standards 90A, 90B, and 90C[19] and Standard 55,[20] respectively. Table 7.6 presents some of the commonly used values for heating design temperatures for types of space normally found at wastewater treatment plants.

Ventilation Requirements. Ventilation requirements are presented in the ASHRAE Handbook[18] and other sources, including

- ASHRAE's "Ventilation for Acceptable Indoor Air Quality,"[21]
- "Recommended Standard for Sewage Works,"[22]
- "The Boca National Mechanical Code,"[23]
- "The Uniform Mechanical Code,"[24]
- "Industrial Ventilation,"[25]
- U.S. Department of Labor, Occupational, Safety and Health Administrator,[26]
- "National Fire Codes,"[8]
- "National Electrical Safety Code,"[3] and
- Local and state codes.

Table 7.6 presents some commonly used values for ventilation rates in spaces frequently found in wastewater treatment plants.

Table 7.6   Suggested HVAC design parameters.

Space	Ventilation rate, air changes/hr (ac/hr)	Heating design temperature, °F[a]	Remarks
Wet well	30 intermittent[b] 12 continuous[c]	50 minimum	Hazardous area, interlock fans with explosion proof equipment. Consider odor control. Step up to 24 ac/hr upon detection of hazardous fumes by air-monitoring equipment. Use 100% outside air.
Dry well	30 intermittent[b] 6 continuous	55–60	
Grit removal area	30 intermittent[b] 12 continuous[c]	55–60	Same as wet well.
Screening room	30 intermittent[b] 12 continuous[c]	55–60	Same as wet well.
Digester gas control rooms	30 intermittent[b] 12 continuous[c]	55–60	Same as wet well.
Sludge gas compressor rooms	30 intermittent[b] 12 continuous[c]	55–60	Same as wet well.
Enclosed grit truck loading areas	30 intermittent[b] 12 continuous[c]	55–60	Same as wet well.
Enclosed primary sedimentation tanks	30 intermittent[b] 12 continuous[c]	55–60	Same as wet well.

Table 7.6   Suggested HVAC design parameters (continued).

Space	Ventilation rate, air changes/hr (ac/hr)	Heating design temperature, °F[a]	Remarks
Scum concentration tanks	30 intermittent[b] 12 continuous[c]	55–60	Same as wet well.
Chlorine and sulfur dioxide rooms	60 occupied	55–60	Hazardous areas, toxic fumes, floor level exhaust required. Interlock fans with manual switches located at each entrance. Also interlock fans with chlorine and sulfur dioxide detection. Use 100% outside air.
Pump or blower room	d	55–60	Provide adequate ventilation to maintain summer temperatures below 104°F.
Filter room and dewatering area	12 continuous[c]	55–60	Consider odor control for exhaust air from dewatering area. Consider two-speed fans or multiple fans.
Garage, repair shops	d	60	Provide 100% outside air when occupied and 100% return air when unoccupied. For minimum ventilation, follow BOCA, UMC, ASHRAE, or local code.
Workshop	d	70	Satisfy exhaust requirements for equipment; comply with BOCA and UMC code and ASHRAE standards.
Locker room	d	70	Follow BOCA and UMC code, ASHRAE standards, or local codes.
Restrooms	d	70	Follow BOCA and UMC code, ASHRAE standards, or local codes.

Table 7.6   Suggested HVAC design parameters (continued).

Space	Ventilation rate, air changes/hr (ac/hr)	Heating design temperature, °F[a]	Remarks
Office, conference rooms, lunch rooms, control rooms	Air conditioning[d]	70	Follow BOCA, UMC, ASHRAE, or local codes. Use 78°F summer room temperature.
Laboratory and computer rooms	Air conditioning[d]	70	Follow BOCA, UMC, ASHRAE, or local codes. Use 75°F summer room temperature.
Paint rooms	[d]	60	Follow OSHA and local codes.
Battery rooms	[d]	60	Follow OSHA and local codes.
Welding areas	[d]	70	Follow OSHA and Industrial Ventilation, American Conference of Governmental Industrial Hygienists, or other local codes.

[a] °F × 0.555 (°F − 32) = °C.
[b] With intermittent design or start fans from automatic reset timers.
[c] Use two-speed fans—12 ac/hr at low speed and 24 ac/hr at high speed—or use multiple fans to achieve 12 and 24 ac/hr.
[d] See remarks column.

Air Conditioning Cooling Loads. Design criteria for air conditioning cooling loads and heating loads are found in the same references. Table 7.6 presents some of the spaces requiring air conditioning in wastewater treatment facilities.

HEATING, VENTILATING, AND AIR CONDITIONING SYSTEMS. The heating and air conditioning systems control the temperature, humidity, and quality and movement of air in a conditioned space or building, and provide comfort to the personnel and protection for the systems or equipment. The ventilation system provides the necessary environmental conditions for the safety of the workers, heat control in areas in which heat is dissipated from process equipment to the environment, spot cooling in working areas, and the means of preventing accumulation of combustible and toxic gases in hazardous areas. The final selection of the most appropriate system is based on an analysis of the life cycle cost for the system and owner preference.

Heating Systems. A heating system consists of an energy source (boiler), fuel use by the energy source (natural gas, fuel oil, digester gas, or electricity), the type of heat conveyance (steam, hot water, hot air, or a combination of the above), and the equipment required to distribute the heat to the spaces (air-handling unit with heating coils, unit heaters, baseboard, convectors, or cabinet fans).

Ventilating Systems. Two types of ventilating systems used in wastewater treatment plants are supply systems and exhaust systems. Supply systems are usually used for comfort or to replace air exhausted from the space. Exhaust systems can be used for heat control and removal of hazardous gases. When used as dilution systems, enough air is required to reduce the contaminated air to a safe level before it is discharged to the atmosphere. Local exhaust systems (hoods) capture the contaminated air close to the source. The local exhaust system requires less air; it is more effective and saves more energy because of lower flow rates.

Air Conditioning Systems. Air conditioning systems in wastewater treatment plants are generally all air systems using direct refrigerant coils (DX), electrically driven compressors, and air-cooled condensers. Air and refrigerant elements are located in air-handling units in mechanical rooms or on the roof, usually at some distance from the conditioned space. Treated air is brought through ducts to the conditioned space and distributed through air outlets (registers and gauges) within the space. Heating may either be separate or included in the air-handling unit. Self-contained or split refrigerant systems (DF) are used in small and intermediate tonnage air conditioning applications. Centrifugal and absorption chiller machines with chilled water-cooling coils in the air-handling units are used for higher tonnage systems.

Dehumidification. Dehumidification prevents rust, mold, and condensation. The use of dehumidifiers, another facet of conditioning air, lowers the relative humidity in enclosed spaces. Dehumidifiers used are normally electrically driven commercial units that operate independently of other heating and ventilating units. Two of the most common methods of dehumidification are refrigerant (DX) and desiccant (solid sorbents) methods.

Because of the high cost of using dehumidifiers, condensation on interior surfaces, walls, and piping can be prevented by providing adequate insulation in conjunction with heating and ventilating systems. In a humidity control analysis, consider moisture from the outside air, moisture in the process air leaving water surfaces, and evaporation from water surfaces. Adequate heating and ventilating in spaces with open water surfaces prevents fog formation. Experience to date indicates that a minimum of 4 to 5 air changes per hour (ac/hr) prevents condensation and fog formation.

ENERGY CONSERVATION. Energy conservation systems reduce energy consumption and operating costs by using or recovering waste heat before it is rejected from a facility. Sources of waste heat in wastewater treatment facilities include heated air exhausted from buildings as ventilation air, waste heat in the plant effluent, engine heat recovery, and large electric motors. When considering energy conservation, also consider the extra equipment costs required to conserve energy.

Air-to-Air Heat Recovery Systems. The three types of heat recovery systems often used are the run-around coil, heat pipe, and rotary heat-exchanger wheel. To prevent fouling of these heat recovery systems, air filtration on the exhaust and intake ducts is required. The heat recovery system of the run-around coil type transfers heat from the exhausted air to the supply air. The run-around coil system consists of a water system connected to the coil in the exhaust system that is piped to a coil in the supply system. This system operates using a circulating pump, control valve, and temperature control system. Heat pipe systems consist of a bank of refrigerant-filled tubes with half of each tube in the exhaust air stream and the other half in the incoming air stream. Warm exhaust air heats the refrigerant, which vaporizes and moves to the side of the tube located in the outside airstream. Heat transferred to the outside air condenses the refrigerant vapor. Liquid refrigerant flows back to the exhaust side of the tube. The heat recovery system of the heat pipe type consists of an air-to-air heat transfer device with no moving parts that transfers heat from exhaust air to the outside air being drawn in for the system. An optional tilt control offers summer and winter changeover, supply temperature regulation, and frost protection. If tilt control is not provided, face and bypass dampers on the exhaust side of the unit must be provided.

In HVAC systems, a rotary heat-exchanger wheel is installed between the exhaust duct and the supply air duct. The rotary heat-exchanger wheel

---

recovers energy from the exhaust air stream before it is vented to the atmosphere and transfers it to incoming fresh air. A built-in purge section reduces carry-over of the exhaust air to the fresh air.

Effluent Water Heat Pump. The heat pump extracts waste energy from the treatment plant effluent flow. This energy then becomes a source of hot water used to condition sludge and either heat or cool enclosed spaces. The heat pump operates on the principle of a closed loop refrigeration cycle, with major parts consisting of an electrically driven compressor, a condenser, and an evaporator.

The advantage of a heat pump is that the energy comes from residual heat in the treatment plant effluent. For heat pump applications, the quality of the effluent should be compatible for use in the evaporator. Consider using cupronickel tubes and carbon steel tube sheet, shell, and heads. To eliminate any plugging in the evaporator, use a strainer with a fine screen on the effluent supply to the evaporator. Primary treatment plant effluent is not recommended as a heat source; rather, use an effluent supply downstream of the final disinfection process.

Equipment Heat Exchangers. A good example of this type of heat recovery is the reciprocating engine. This system uses engine jacket water temperature, an exhaust heat recovery muffler, and a heat exchanger to transfer rejected heat to a secondary water supply.

Motors. Large electric motors, such as those used for pumps or blowers, are normally air cooled. This energy is recovered by ring water jacketed motors or ducted motor enclosures.

HAZARDOUS AREAS. Hazardous areas include those areas covered by the National Electrical Code (NEC)[3] and those areas not regulated by the NEC.

Hazardous locations covered by the NEC are class 1 locations (divisions 1 and 2) and are defined as those areas in which flammable gases or vapors are or may be present in the air in quantities sufficient to produce an explosion. Class 1 locations should use explosion-proof equipment.

Hazardous areas not covered by the NEC are those locations at which toxic gases or chemical liquids are or may be present in quantities sufficient to produce bodily injury to personnel.

Space Classification. Class 1, division 1:

- Wet wells,
- Screening rooms,
- Grit removal areas,

- Digester gas control rooms,
- Sludge gas compressor rooms,
- Grit truck-loading areas,
- Enclosed primary sedimentation tanks, and
- Scum concentration tanks;

Class 1, division 2: isolated operating galleries between the above areas; toxic gas areas:

- Chlorine rooms,
- Sulfur dioxide rooms,
- Chlorine and sulfur dioxide storage areas,
- Truck-loading areas, and
- Repair shops and vehicle areas; and

Explosion-proof areas and toxic gas areas:

- Battery rooms (sealed-type batteries),
- Welding areas, and
- Paint areas.

Minimum Ventilation Requirements. Ventilation for hazardous areas classified under class 1, division 1 shall be provided at 30 ac/hr intermittently or 12 ac/hr continuously. When high concentrations of combustible gases or toxic gases are detected in the continuous mode of operation, the ventilation rate shall be increased to 24 ac/hr. In the intermittent mode of operation, the rate remains the same upon detection of combustible gases or toxic gases.

Ventilation for all hazardous areas classified under class 1, division 2 shall be provided at 6 ac/hr continuously. An additional 6 ac/hr shall be provided when combustible gases or toxic gases are detected.

Ventilation for areas in which chlorine or sulfur dioxide is stored or used shall be provided at a minimum of 60 ac/hr intermittently.

Other areas, such as repair shops, vehicle areas, paint rooms, and battery rooms, are subject to additional requirements dictated by local or state code requirements. See Table 7.6 for additional ventilation requirements.

Design Recommendations. Ventilation systems in hazardous areas shall be designed for 100% outside air. "Air shall not be recirculated from any space in which flammable vapors, flyings, or dust are present in quantities and concentrations that would introduce a hazardous condition into the return-air system."[8] Electrical equipment in class 1, divisions 1 and 2 hazardous areas shall be rated explosion proof as per NEC.[3]

Fans and drives in class 1, divisions 1 and 2 areas shall be of spark-resistant construction in accordance with the Air Movement and Control Association (AMCA).[27]

Class 1, divisions 1 and 2 areas shall be monitored for combustible gases or toxic fumes. Ventilation rates shall be doubled if combustible gases or toxic fumes are detected. If gas concentration increases, local and remote alarms shall be energized and a flashing red light mounted outside the building shall be energized.

The ventilation system in chlorine and sulfur dioxide rooms shall be energized upon detection or upon entry to a room containing either gas. In addition, when gases are detected, a light outside the building and local and remote alarms shall be energized. Manual switches and pilot lights for fans serving such rooms shall be located outside each entrance door to the room.

Typical areas that may require supplementary ventilation for odor removal are wet wells, screen rooms, grit truck loading areas, raw sludge pumping stations, and sludge-processing areas.

ODOR CONTROL SYSTEMS

Odor associated with wastewater treatment is an area of growing concern. This section of the manual reviews the sources of odor from wastewater facilities, overviews the various odor control systems, and provides references for more information on odor control.

Wastewater treatment facilities, regardless of how well designed, at one time or another can generate odors as byproducts of the wastewater treatment process. As public awareness increases, odor complaints plague municipal officials. As urban development spreads, wastewater treatment facilities that were once isolated often become surrounded by residential developments. As a result, a treatment facility may quickly become an "unwanted" neighbor even though it may have been there long before the residential development. To address increasing public concerns, local governmental agencies are promulgating odor control ordinances with increasing frequency. Often, regulations that target the reduction of odors considered to be a nuisance are vague. Other regulations result in strict discharge limits imposed on point source discharges from odor control facilities. Therefore, early on in the design process, the design engineer needs to consult the local and state rules and regulations to ensure that odor control is properly addressed to meet local conditions.

SOURCES OF ODOR. Odors prevalent at wastewater facilities generally result from inorganic gases and vapors. The most common inorganic vapors are hydrogen sulfide and ammonia. Odorous compounds such as mercaptans, organic sulfides, and amines are common byproducts of decomposition of organic matter. Industrial waste discharged to the municipal wastewater system can also generate significant odorous vapors, some of which, if emitted at high enough concentrations, pose serious health and safety risks to plant personnel.

Common locations in wastewater facilities from which odors are released include collection systems, in-system pump stations, preliminary and primary treatment systems, and sludge-processing facilities. Of these locations, attention to odor control has primarily been focused at the pump station, headworks facilities (wet wells, screening facilities, and grit chambers) and, more recently, sludge storage, thickening, stabilization, and dewatering facilities.

During the early stages of the design of wastewater facilities, the design engineer should identify the potential sources of odor and define local regulatory requirements. Once this is accomplished, methods of odor abatement or control can be identified, evaluated, and selected for inclusion in the design of the support system.

APPROACHES TO ODOR CONTROL. Approaches to control odor emissions from wastewater facilities vary, including methods to inhibit the development of odorous vapor and the treatment of foul air streams.[28-31]

Upstream Controls. One approach to abate odors at a wastewater treatment facility or in-system pumping station is to control the introduction of an odorous vapor or its formation in the collection system. Controlling the introduction of wastes that may present special odor problems requires that sewer use and pretreatment ordinances be established and, more importantly, be enforced. In addition, the collection system should be designed to minimize the potential for the wastewater to become anaerobic. To accomplish this, construct the sewer to minimize travel time in the sewer and maintain velocity sufficient to scour deposits that may become a source of odor generation.

Because most collection systems are currently in place, it may not be possible to maintain a "fresh wastewater" in the collection system. In cases where odors are generated within collection systems, various attempts, including oxygen addition and chemical addition, have been made to inhibit or control the formation of the odorous vapors. Some chemicals used are chlorine, hydrogen peroxide, and metal salts.

Generally, these approaches have met with mixed results, but research is continuing. The design engineer should investigate the applicability of these methods if physical modifications to the collection system are not feasible or cost effective.[31]

Atmospheric Dispersion. At certain locations where odorous vapors can be collected for point source discharge, dilution by mixing and discharge to the ambient air may be acceptable. This approach to odor abatement depends on numerous factors, including odor levels, odor type, local meteorological and topographical conditions at the wastewater facility, and downwind land use.

To properly evaluate this approach, odor dispersion mathematical modeling may be appropriate. The modeling, besides predicting odor plume dispersion, is useful for proper siting and design of a discharge stack. Because regulatory agencies use various mathematical models for determining compliance with air quality regulations, the designer should consult with the local agencies regarding model selection.

Chemical Addition. Chemical addition has been used at wastewater treatment facilities to control odors. This approach, similar to that used for upstream odor control, generally has been more successful because it is done under more controlled conditions. Chemicals historically used to control common odors (hydrogen sulfide, for example) are strong oxidants such as chlorine and hydrogen peroxide. The chemical addition is typically performed at headworks facilities where hydrogen sulfide is a prevalent problem. Prechlorination has been the most common chemical addition approach used. Ozonation has been attempted with less success.

In addition to oxidants, iron salts have been used. Sulfides react with the iron salts to form precipitates, thus inhibiting the generation of hydrogen sulfide. Facilities associated with chemical addition are discussed in more detail elsewhere in this manual.

Odorous Air Treatment. Numerous technologies are available to treat the foul air where odorous vapors can be contained and collected. A brief description of each of these control methods follows. Table 7.7 summarizes advantages and disadvantages of each system.

ADSORPTION SYSTEMS. One approach to the treatment of odorous air is to pass the foul air through an adsorptive medium to which odorous compounds adhere. Adsorptive media include granular activated carbon (either virgin carbon or chemically impregnated), activated alumina, activated aluminum with potassium permanganate, and silica gel.

Granular activated carbon is the most widely used adsorption system today. Figure 7.12 presents a schematic diagram of an activated carbon adsorption type system. Facilities associated with this type of system include fans or blowers, pre- and postfilters, adsorptive media beds (one or more stages), corrosion-resistant ductwork, and necessary controls. Adsorptive media are regenerated by ambient temperature solubilization, in-place chemical regeneration, or using steam at high temperatures in a reducing atmosphere.

Table 7.7   Methods of odorous air treatment: advantages and disadvantages.

Treatment method	Advantages	Disadvantages
Adsorption systems	Simple operation Reliable and consistent Additives can enhance treatment Can accommodate varying gas flows	Costly media regeneration Adsorbent capacity used rapidly Additives can cause corrosive environment Particulate material can plug media
Biological systems	Simple operation No regenerant chemicals Treat variety of compounds Economical to treat high volume of gas	Require substantial space Reliability sometimes questionable Gas transfer limitations Process control limited Particulate material can plug media Require balanced environmental condition

Combustion systems	Reliable at high temperatures
	Oxidize compound not treated chemically
	  nor biologically
		Costly if dedicated system
		Can cause problems with incinerator
		Backup system required when incinerator
		  not in use

Ozonation	High-potency oxidant Simple operation	Costly Experienced significant operating problems Difficult to control dosage High maintenance costs Toxic off-gas if not properly destructed

Wet scrubbers
  Packed beds
	Wide-scale use
	Can economically treat high gas flows
	High mass-transfer efficiency
	Effectively handles changes in odorous
	  compound concentrations
		Recycle odorous compounds
		High chemical regenerant usage
		Can be high maintenance
		Chemical carry-over in treated gas
		Applicable to only certain compounds

Mist systems	Low-pressure drop Accommodate high flow rate High mass-transfer efficiency No chemical regenerant recycle	High energy requirements High maintenance Mist carry-over in treated gas Slow response to changing concentrations

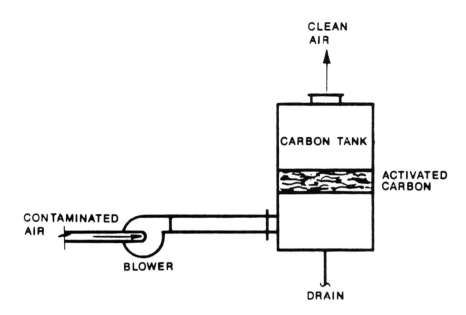

Figure 7.12   Activated carbon adsorption system.

BIOLOGICAL SYSTEMS. Treatment of odorous air using biological systems has been tried for many years. The basic treatment process involves biological oxidation of odorous compounds. Biological approaches attempted include using bulk media filters such as a soil filter (see Figure 7.13), using odorous air as feed air to an activated sludge treatment system, passing odorous air through a wastewater trickling filter, or using separate trickling filters dedicated to treat foul air. Few biological odorous air treatment systems are in use today. However, bulk media type filters are used in other parts of the world and are becoming more common in North America.

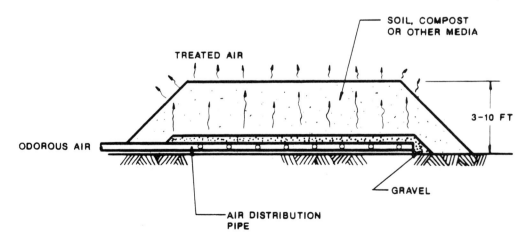

Figure 7.13   Biological filter odor control system.

COMBUSTION SYSTEMS. Combustion or high-temperature oxidation of odorous air has been an effective way to destroy odorous compounds. Odorous air may be used to meet a portion of the combustion air of a residuals incinerator. Another approach is to use fume burners to destroy odorous compounds. In most cases, separate combustion systems for foul air treatment have been too costly. Also, if foul air is used as combustion air to an incinerator, some type of backup system is required for periods when the incinerator is not in use.

OZONATION. Because ozone is a powerful oxidant, ozone treatment systems have been used to treat odorous air. Controlling the dosage of ozone to ensure that overdosing does not occur is critical and expensive. Overdosing results in an odorous ozone off-gas if an ozone destruct system is not employed. Associated facilities include ozone diffusion and ozone off-gas destruction systems. Ozonation systems that have been installed have experienced serious mechanical problems. As a result, most installed systems are no longer in use.

WET SCRUBBERS. In a treatment system that uses wet scrubber absorption, odorous air is brought into contact with a scrubbing liquid chemical. Odorous compounds in the foul air are absorbed in the scrubbing chemical and are thereby removed from the air. Strong oxidizing agents such as sodium hypochlorite and potassium permanganate are diluted and used as the scrubbing liquid chemical.

Two of several wet scrubber systems commonly used today are the wet packed bed scrubber system and the mist scrubber system. The packed bed scrubber, shown schematically in Figure 7.14, consists of a contact chamber with inert packing material (to increase the exposed surface), scrubbing liquid, a recirculating system, an air blower, ductwork, and controls. Odorous gases enter the scrubber and pass through the packed bed where the odorous gases come in contact with the liquid-scrubbing chemical. The scrubbing liquid collects at the bottom of the contact chamber and is recirculated through the unit as required along with a makeup chemical.

Mist scrubber systems are modified wet packed bed scrubber systems and are used to adsorb odorous gases to the liquid phase before oxidation. Mist systems depend on the contact between the odorous gas and small droplets or "fog" of a chemical-scrubbing liquid. As Figure 7.15 shows, the mist-type scrubber system consists of a contact chamber, chemical storage tanks, a chemical supply and dilution system, a dilute chemical distribution system, a compressed air system, ductwork, and controls. The scrubbing chemical, normally a solution of sodium hypochloride and sodium hydroxide, is converted to the mist or "fog" with compressed air before introduction to the contact chamber. Unlike the packed bed scrubber, the scrubbing chemical does not recirculate.

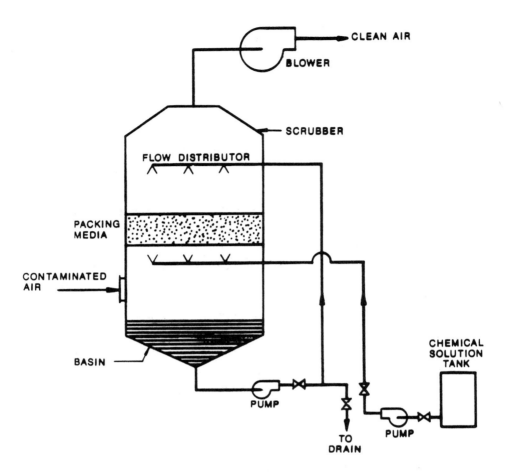

Figure 7.14 Wet packed bed scrubber odor control system.

PROCESS DESIGN. A final approach to odor control available to the designer is the proper hydraulic and process design of wastewater treatment facilities. This is especially true of any high-strength return or recycle streams from solids-processing facilities. For specific design methods to minimize odors from specific wastewater treatment operations or processes, see the chapters dealing with these systems.

CORROSION CONSIDERATIONS. When dealing with odor control, the designer must consider corrosion control, especially the approaches that employ collection, containment, and treatment of odorous vapors. To mitigate corrosion, consider providing adequate ventilation to reduce condensation and selecting proper equipment, materials, and coatings. For further information on the latter corrosion consideration, refer to Chapter 6.

Finally, outdoor air surrounding plant buildings may contain low levels of odors and concentrations of hydrogen sulfide, resulting in corrosion of electrical equipment, controls, and other system components the air contacts. For

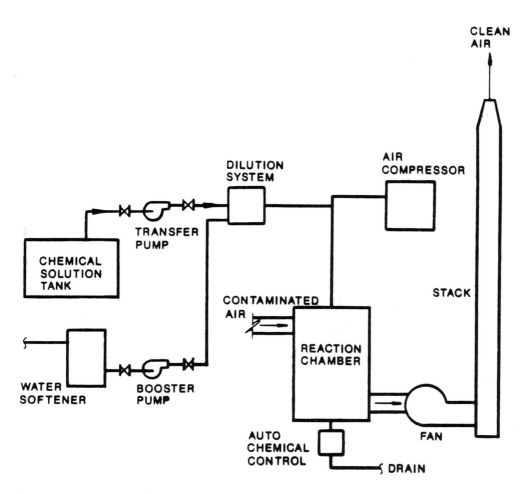

Figure 7.15 **Mist scrubber odor control system.**

removal of low levels of odor and hydrogen sulfide in outdoor air used for building ventilation, a bank of activated carbon panel filters or chemically impregnated absorbent is the most adaptable to normal building ventilation systems. A bank of panel filters requires less space, is readily maintainable, and requires minimal changes to the usual air-handling systems. Initial and maintenance costs of panel-type filters are relatively high, but the method remains the most practical means of reducing the odor and corrosiveness of plant ambient air used for building ventilation.

CHEMICAL SYSTEMS

Chemical systems are frequently used throughout a wastewater treatment facility to support the associated wastewater treatment process or serve as the process itself.

Procedures for choosing chemicals and chemical dosages for these applications vary and are discussed in chapters addressing specific chemical additions. Table 7.8 presents the most frequently used chemicals and their principal uses.

Table 7.8   Chemicals frequently used for wastewater treatment.

Chemical	Principal use
Activated carbon	Dechlorination
	Denitrification
	Odor adsorption
	Organics removal
	Sludge stabilization
Aluminum sulfate (Alum)	Suspended solids removal
	Phosphorus removal
	Sludge conditioning
Chlorine	Ammonia removal
	Disinfection
	Grease removal
	Prechlorination
	Odor control
	Sludge-bulking control
	Filter-fly control
	Sludge stabilization
Ferric chloride	Phosphorus removal
	Sludge conditioning
	Suspended solids removal
Ferric sulfate	Phosphorus removal
	Suspended solids removal
Ferrous sulfate	Odor control
	Phosphorus removal
	Sludge conditioning
	Suspended solids removal
Hydrogen peroxide	Odor control
	Sludge-bulking control

Table 7.8   **Chemicals frequently used for wastewater treatment (continued).**

Chemical	Principal use
Lime	Heavy metals removal
	Suspended solids removal
	Odor control
	Phosphorus removal
	pH adjustment
	Sludge conditioning
	Sludge stabilization
Methanol	Denitrification
Ozone	Disinfection
	Odor control
	Sludge-bulking control
Polymers	Suspended solids control
	Sludge conditioning
Potassium permanganate	Odor control
Sodium aluminate	Suspended solids removal
	Phosphorus removal
Sodium carbonate	pH adjustment
Sodium hydroxide	pH adjustment
Sodium hypochlorite	Disinfection
	Odor control
Sulfur dioxide	Dechlorination

CHEMICAL SELECTION. Many different chemicals are used in support of wastewater treatment.[32-42] Often, more than one chemical is suitable for a particular application. The selection of a chemical for a particular use includes both monetary and nonmonetary factors, which are discussed below.

Monetary Considerations. Cost is a significant consideration in the selection of a chemical to be used for a particular application. Capital, operation, and maintenance costs all need to be considered. A proper economic evaluation includes not only the direct cost of the chemical itself, but also costs related to the impacts of chemical application on the treatment facility such as changes to the quantity and quality of sludge generated and the pH impact on downstream treatment systems. Where more than one chemical is suited for a particular application, conduct an economic analysis that compares the present worth or annual equivalent costs of the possible chemicals. Note that potential cost savings may result if one chemical is used for multiple purposes, for example, chlorine, which is used for disinfection, odor control, or an aid in sludge-bulking control and oxidation of ammonia and organic substances.

Clearly the required chemical dosage affects the economic analysis. Therefore, accurate estimates of chemical dosages ensure the validity of the monetary analysis. Unfortunately, practice shows that theoretical stoichiometric relationships cannot always be used accurately to predict chemical dosages. As a result, laboratory tests such as jar tests, pilot plant studies, or on-line studies are recommended to more accurately determine optimum chemical dosages before recommending chemicals for a particular use.

Nonmonetary Considerations. When selecting the optimum chemical to be used for a particular application, consider certain nonmonetary factors along with economic factors. Several important nonmonetary factors are

- Effectiveness,
- Compatibility with other treatment processes,
- Reliability, and
- Environmental impact.

The effectiveness of using a chemical for a particular application varies from plant to plant and often depends on the specific waste or operating conditions. Use operating results from similar wastewater treatment facilities with caution. To most accurately determine the effectiveness of a chemical, laboratory or pilot tests using the proposed chemical on an equivalent waste sample may be needed. These tests also help to predict appropriate chemical dosages, thereby aiding the monetary analysis.

Compatibility with other treatment processes employed at the facility is an important nonmonetary factor to be considered when selecting the appropriate chemical to use. Again, pilot- or full-scale testing may be advisable to determine the impacts on other processes. This type of testing also helps to assess the impact of introducing the chemical to the process stream at various points.

The reliability of the supply of a chemical is also important. A chemical found to be the most economic and effective may not have a reliable supply source. This factor may negate the results of the other evaluations. It is important to include the projected quantity of chemical required in this assessment because, although the chemical may be reliably available, it may not be readily available at the required quantity.

Finally, assess the environmental impacts associated with the use of a particular chemical. Given current concerns about the safety of effluent toxicity, chemicals selected must be demonstrated to be environmentally safe upon disposal. An example is the trend requiring dechlorination of effluents where chlorine has been added to achieve disinfection. This is due in part to the determination that excessive chlorine residue in effluent can potentially link with other organic chemicals to form carcinogenic substances in the receiving body of water.

HANDLING, STORAGE, MIXING, AND FEED SYSTEMS. Physical facilities associated with the handling, storage, mixing, and feeding of a chemical are dictated by the form of the chemical used, its physical and chemical characteristics, the flow ranges of the wastestream, and the reliability of the equipment. Many chemicals used in wastewater treatment facilities are found in different forms (solid, liquid, or gas) at various stages in the chemical-handling system. For example, chlorine may be stored initially as a gas or liquid (depending on the quantity of chlorine used at a particular facility), transferred through a conveyance as a dry gas, and finally injected as a solution at the point of application.

The design of chemical operations involves not only the sizing of various unit operations and processes but also necessary appurtenances. Because of the corrosive nature of many of the chemicals used and the different forms in which they are available, give special attention to the design of chemical storage, feed piping, and mixing and control systems. This section includes brief discussions of these topics.

In domestic wastewater treatment systems, chemicals are generally in the solid or liquid form. Chemicals in the solid form generally convert to a solution or slurry form before introduction to the wastewater. Chemicals in the liquid form are usually delivered to the plant in a concentrated form and are diluted before introduction to the wastewater. Two types of chemical-feed systems, dry feed and liquid feed, are discussed below.

Dry Chemical-feed Systems. A dry chemical-feed system generally consists of a storage hopper, a dry chemical feeder, a dissolving tank, and a pumped or gravity distribution system (see Figure 7.16). Units are sized according to the volume of wastewater, the treatment rate, and the optimum length of time for chemical feeding and dissolving. Hoppers used with compressible and archable powder such as lime are equipped with positive hopper agitators and a dust collection system. Dry chemical feeders are either of the volumetric or gravimetric type. The volumetric type measures the volume of the dry chemical fed; the gravimetric type weighs the amount of chemical fed.

With a dry feed system, the dissolving operation is critical. The capacity of the dissolving tank is based on the detention time, which is directly related to the rate of chemical dissolution. When the water supply is controlled for

---

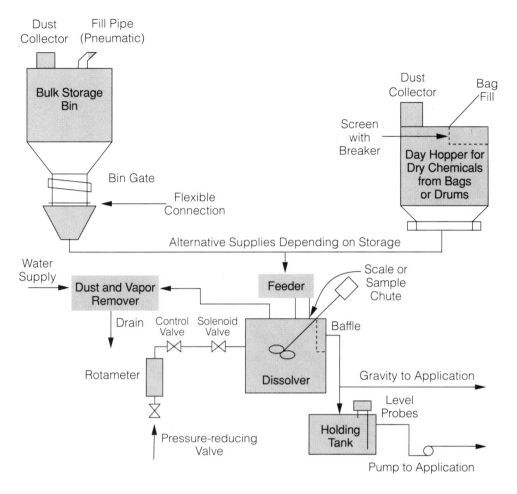

Figure 7.16 Typical dry chemical-feed system.

the purpose of forming a constant-strength solution, mechanical mixers are used. After dissolving, solutions or slurries are often stored and discharged by chemical-feed pumps to the application point at metered rates.

Liquid Chemical-feed Systems. Liquid chemical-feed systems typically include a solution storage tank, a transfer pump, a day tank for diluting the concentrated solution, and a chemical-feed pump for distribution to the application point (see Figure 7.17). In systems where the liquid chemical does not require dilution, the chemical-feed pumps draw liquid directly from the solution storage tank. Sizes of storage tanks are based on the stability of the chemical, feed-rate requirements, delivery constraints (cost and size of tank truck), and availability of the supply. For accurate metering of the chemical feed, solution-feed pumps usually are of the positive displacement type.

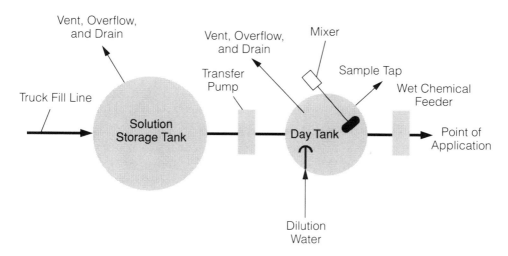

Figure 7.17 Typical liquid chemical-feed system.

While chemical-handling systems may appear simple, proper design of a complete and integrated system can be as complex as that of the associated wastewater treatment process. The design engineer, therefore, is cautioned to pay particular attention to the detailed design of all aspects of a chemical system.

APPLICATION POINTS. The optimum point for the application of a chemical to the wastestream ranges from somewhat obvious (as for chlorine) to more difficult to determine (as for chemical phosphorus removal). When selecting a location for introducing chemicals to the process stream, consider the following factors:

- Adequate mixing,
- Effects on subsequent treatment units, and
- Flexibility.

For any chemical addition to be effective, the chemical solution introduced to the process stream must adequately mix with the waste stream. Mixing is normally accomplished by hydraulic means, with static-type mixing, with mechanical mixers, or by diffused aeration. Based on past problems in achieving mixing in real-world situations, consider performing hydraulic modeling of critical application points to ensure that effective and desired mixing occurs. See Chapter 16 for additional information relating to rapid mixing, mixer types, fluid regimes, and design considerations.

The point of chemical application may, in some cases, adversely affect downstream treatment units. For example, if the waste becomes nutrient deficient for the biological microorganisms, phosphorus removal in advance

---

of a biological treatment system may decrease the efficiency of the biological system. Another example is if sulfur dioxide is added for dechlorination, overdosing of the sulfur dioxide can result in an oxygen depression, which may result in effluent limit violations.

On the other hand, some impacts are positive. For example, the addition of alkaline chemicals (lime and caustic soda) in the phosphorus precipitation process increases the alkalinity of the water and counteracts alkalinity destruction by nitrification in subsequent processes. Therefore, an analysis can assess potential impacts both adverse and positive on subsequent treatment processes and identify the mitigating measures to be incorporated to the design.

Wherever possible, include flexibility in the design of chemical application systems. In many cases, duplicate chemical storage and mixing tanks and the installation of several points for the introduction of a chemical to the process stream are included in a design for minor additional capital costs. Providing this type of flexibility enhances the effectiveness and perhaps reduces the required chemical dosage rates if minor changes in waste characteristics or other changes occur. For example, provisions to add polymers either upstream of sludge feed pumps or directly to chemical-conditioning tanks of a belt filter press add the flexibility to use different polymers that act differently when added to the sludge. One may require more reaction time than another, perhaps at a lower dose and unit cost. Therefore, the design engineer should consider providing the flexibility to a design as conditions warrant.

Finally, ensure that adequate flushing and clean-out connections are located throughout the system and that proper safety equipment is provided in chemical storage areas.

SLUDGE CONSIDERATIONS. Chemical addition for wastewater treatment can change the sludge characteristics. One frequent resulting characteristic is an increase in the inert fraction of the sludge. Also, the quantity of sludge to be processed and disposed of is increased. In some instances, lime addition for heavy metal removal generates a chemical sludge that is difficult to dewater and dispose of properly.

Given today's challenges in the disposal of wastewater sludge, the quantities of sludge generated must be carefully considered. Because it is difficult to obtain true estimates of sludge generated by chemical applications from strict chemical relationships, bench-scale testing has proved a valuable method to aid in these predictions. Several EPA publications present a methodology for calculating sludge quantities.[32,33,35-37]

O*THER SUPPORT SYSTEMS*

Along with major support systems that serve as the backbone to the wastewater treatment process, several minor systems warrant consideration during the design process. These minor systems include

- Fire protection,
- Site security,
- Plumbing (including water systems),
- Fuel,
- Compressed air, and
- Communication systems.

Design considerations associated with each of these systems are highlighted below.

FIRE PROTECTION. In general, the highest risk of fire and explosion is associated with wastewater collection, pumping operations, and the early stages of liquid and solids processing. Specific unit processes to consider include pumping station wet wells that handle raw wastewater or unstabilized solids, preliminary screening and grit removal processes, primary sedimentation, anaerobic digestion, scum-collecting processes, and fuel and chemical-handling and storage areas.

Principal control procedures used to minimize potential fire and explosion incidents at wastewater treatment plants include

- Risk evaluation,
- Process and equipment controls,
- Ventilation,
- Construction materials, and
- Education.

These control procedures also include proper electrical classification of hazardous locations and the selection, installation, and operation of electrical equipment, motors, and devices that are suitable for these locations. To implement and enforce the control procedures effectively requires an adequate safety program and the cooperation of the plant management and personnel and the public, private, and government sectors.

Specific design criteria for fire protection systems are based on guidelines established by the National Fire Protection Association (NFPA),[8,43] local codes and ordinances, and the owner's insurance carrier. To ensure compliance with codes and ordinances, the designer should review applicable requirements with the appropriate local officials early in the design process.

Building Classification. In the preliminary stages of any wastewater treatment plant design, the designer should specify which buildings or spaces within buildings require fire protection as per the applicable codes. Once this is determined, the designer selects the occupancy hazard of the buildings or structures or portions thereof and designs the appropriate fire protection system in accordance with the applicable fire prevention codes.

Most areas in wastewater treatment plants are classified as industrial use groups. In addition, most are unoccupied except for administration buildings, maintenance shops or areas, repair garage areas, and control rooms. Local fire codes and NFPA regulations apply mainly to occupied industrial, institutional, storage, and commercial buildings. Therefore, considerable judgment and local fire code interpretations determine which building or portion thereof requires fire protection and the type of occupancy hazard that applies.

Fire Protection Systems. The types of fire protection systems most often used in wastewater treatment plants are wet pipe sprinkler systems, dry pipe sprinkler systems, and halon fire-extinguishing systems. In addition, adequate portable fire extinguishers should be located throughout the facility for use by plant staff. Specific installation and safety requirements for these fire protection systems are normally covered by local codes or the NFPA standard.[8]

WET PIPE SPRINKLER SYSTEMS. The wet pipe sprinkler system is used in applications where the temperature is above freezing. Pressurized water is discharged from a sprinkler head immediately after heat actuation. Alarm check valves or water flow detectors are used to actuate local and remote alarms.

DRY PIPE SPRINKLER SYSTEMS. The dry pipe sprinkler system is used where a wet pipe system is impractical, such as rooms or buildings that are not heated. The dry pipe system uses automatic sprinklers connected to a piping system containing air under pressure. The release of air from the opening of a sprinkler permits water pressure to open a dry pipe valve. The water then flows into the piping system and out through the opened sprinklers. Dry pipe systems are less efficient than wet pipe systems because of the time needed to release the air.

HALON FIRE-EXTINGUISHING SYSTEMS. Halon fire-extinguishing systems are designed for use in applications where highly sophisticated electronic equipment, such as in computer rooms and electrical rooms, is to be protected or where water usage would cause severe damage. The system generally consists of fire detectors, high-pressure halon gas containers with a release mechanism, remote pipe-flooding nozzles, control panels, power supply, and auxiliary options such as audible and visual alarms, manual pull stations, and automatic door releases.

SITE SECURITY. In the past, site security systems consisted primarily of chain link security fencing to enclose the plant site. The purpose of the fencing was to protect the plant from theft and prohibit the general public from personal injury.

Given increasing concerns about the liability of a plant owner, site security systems are becoming more important. While site security fencing is still the mainstay of a plant security system, electronic systems are also becoming popular. This is especially true of unmanned remote facilities or a major plant with reduced staffing levels during the evening and night shifts. The most common electronic systems employed are intrusion alarms, which are installed at the primary structures of the plant. Intrusion alarms are connected either to the main control center of the plant, if manned during all shifts, or to other public facilities such as the local police station.

In addition to security fencing or electronic alarms, security guards control access to many plants. The design engineer, therefore, is required to provide proper facilities to house the security personnel, such as a guardhouse at the entrance of the main plant.

PLUMBING. Design Criteria. Design criteria for plumbing systems are frequently established by local codes of the municipality in which the facility is located. Many communities have adopted "The BOCA National Plumbing Code"[44] and "the Uniform Plumbing Code"[45] as their governing criteria. Most major cities have separate plumbing codes. When applying for a building permit, being aware of the local codes avoids conflicts with the code enforcement officer.

Equipment. To withstand the constant use and abuse of everyday service, all equipment for the plumbing system must be heavy duty industrial grade. To minimize operating costs and conserve resources, all fixtures should be of the water saver design and should be designed to accommodate handicapped persons, where required. Locate emergency shower and eyewash fixtures at all areas in which hazardous chemicals are handled. Eyewash fixtures shall be supplied with water temperatures between 9 and 32°C (50 and 90°F).[22] Laboratory drain piping from the sink outlet to the point of dilution in the main wastewater flow must be chemical resistant such as PVC, glass, or fiberglass. Garage drains must be provided with oil–water separators to prevent oil from entering plant treatment units.

Install backflow prevention at any connection between a potable water supply system and other source that is not potable. Break tank separation is the most secure form of backflow prevention. In such a system, potable water enters a tank through an air break to prevent the possibility of contamination. A float valve normally controls the level of the tank. Two pumps are recommended to provide standby reliability.

If a break tank system cannot be used, backflow preventers can be used. However, these devices require regular maintenance to ensure proper operation. Many local codes have strict requirements for the use of a backflow preventer.

Floor and equipment drain systems located in basements shall be drained by gravity to underfloor sumps where submersible wastewater-type pumps lift the wastewater up to the overhead sanitary drain. Sump pumps of the non-clog grinder type are highly recommended.

Water Supply Systems. The three types of water supply systems often used in wastewater treatment facilities are

- Potable water or city water,
- Service water or nonpotable water, and
- Effluent water.

Potable water is water that is satisfactory for drinking and is supplied to plumbing fixtures such as sinks, water closets, urinals, showers, lavatories, and water coolers. For other types of fixtures, refer to the applicable plumbing code.

Service water (nonpotable water) is water that is supplied to heat generation equipment, heat transfer equipment, and process equipment for cooling or makeup water purposes. Cross-connection between potable and nonpotable water is not allowed except where approved protective devices or means to prevent backflow into the potable water are installed.

In some applications, effluent water is used as service water. The amount of effluent water used depends on the availability and cost of potable water, water quality standards, and water treatment costs. Some of the most common uses of effluent water are irrigation systems, lawn watering, washdowns and flushing systems for process equipment, chlorine dilution, cooling systems, and heat recovery systems.

FUEL. Natural gas, propane gas, and fuel oil have numerous uses at wastewater treatment facilities. The type of gas and its specific use varies, depending on the treatment processes employed and the size and complexity of the facility. Typical fuel uses at a wastewater treatment facility may include the following:

- Building heating (central boiler or unit heaters),
- Hot water heaters,
- Laboratory fixtures,

- Sludge-heating systems,
- Sludge incinerators,
- Waste gas burner pilot lights,
- Emergency generators,
- Dual fuel engines, and
- Direct drive process equipment engines.

Digester gas, when available at a plant, can be used instead of other fuels for many of the above uses.

Service pressures for each of the above uses will likely vary and depend somewhat on specific requirements of the equipment manufacturers.

Early on in the design process, make preliminary estimates of natural gas requirements for the wastewater treatment plant. Estimates can then be used as bases of initial discussions with the local gas utility to determine service requirements and rates. Normally, each facility has only one connection to the local utility. However, in larger plants, it may be more appropriate and cost effective to have several connections, each served by an individual pressure regulator and meter.

Within the treatment plant site, it may be necessary to develop an internal natural gas distribution system that is then tapped to serve individual buildings or structures. If the internal distribution system is maintained at high pressure, it may be possible to reduce the pressure at each structure to meet the specific equipment requirements within that structure. It may also be appropriate to meter gas usage at each structure.

COMPRESSED AIR. Compressed air systems are used throughout a wastewater treatment facility. Some specific applications include

- Air operating pneumatic tools,
- Odor control systems,
- Sealed water systems (heating, ventilating, and air condition control systems), and
- Laboratory applications.

A clear understanding of the basic types of compressed air systems is helpful in all cases.

A power-driven device transforms air at some initial intake pressure (usually atmospheric) to a greater working pressure. The elements of a typical compressed air system are the source of air (the air compressor), which is connected to a storage tank, and the receiver in which the pressure is maintained between fixed limits. A pressure relief valve is required to prevent pressure from building up beyond a safe preset limit. Intake air filters remove dust and other particles from the air entering the compressor. A filter at the receiver discharge prevents foreign matter from causing malfunction of the regulator. The regulator, desirable in most compressed-air systems, maintains a constant pressure regardless of the rise and fall of line pressure at the compressor.

Some pneumatic systems require an air dryer to remove moisture. While a mechanical filter removes most of the particles in the air, it is not effective for removing water and oil vapors. An air dryer prevents condensation by reducing the humidity of the air and is normally installed after the receiver and upstream of the pressure-reducing valves.

COMMUNICATION. The owner of the wastewater treatment plant either owns or leases the plant communication system from one of the many communication companies currently offering such service.

Features provided by the communication system vary depending on the specific needs of the wastewater treatment facility. Along with standard incoming, outgoing, and internal voice communications, it is possible to provide intercommunication with on-site and off-site mobile vehicles, pocket personnel pagers, on-site fixed paging systems, and data communications.

Within the treatment plant site itself, current practice normally is to provide telephones in each major structure or building or at specific work stations within the buildings. In addition, paging speakers are appropriately located to provide full coverage of the treatment plant grounds and in unmanned areas such as service tunnels or at process tankage. While telephone conduits and low-voltage electrical conduits may be routed together through common electrical duct banks, it is generally not good practice to install telephone cable and electrical power or control wiring in the same conduit.

Review specific features of the communication system with the owner of the treatment plant during the design process. Once the required features of the communication system have been defined, solicit specific proposals from system vendors regardless of whether the communication system is to be owned or leased.

REFERENCES

1. U.S. EPA Design Criteria for Mechanical, Electrical, and Fluid System and Component Reliability (1974).
2. "National Electrical Code, An American National Standard." NFPA No. 70-1987, ANSI C1-1987, Natl. Fire Protection Assoc., Boston, Mass. (1986).
3. "National Electrical Safety Code." ANSI C2-1987, Am. Natl. Stand. Inst., New York, N.Y. (1986).

4. U.S. Dept. of Labor, Occupational Safety and Health Admin., "Occupational Safety and Health Standards." *Fed. Regist.*, **36,** 105 (1972).

5. U.S. Dept. of Labor, Occupational Safety and Health Admin., "Occupational Safety and Health Standards." *Fed. Regist.*, **37,** 202 (1972).

6. Illuminating Eng. Soc., "Lighting Handbook." Ref. Vol. (1981).

7. Inst. Electr. and Electronic Eng., Inc., "IEEE Recommended Practice for Emergency and Standby Power Systems for Industrial and Commercial Applications." IEEE Orange Book, ANSI/IEEE Stand. 446, New York, N.Y. (1987).

8. Natl. Fire Protection Assoc., "National Fire Codes." NFPA Codes, Stand. Recommended Practices, Man. and Guide, Quincy, Mass. (1990).

9. "The Uniform Building Code." Int. Conf. Build. Off., Whittier, Calif. (1988).

10. Skrenter, R.G., "Instrumentation Handbook for Water & Wastewater Treatment Plants." Lewis Publishers, Chelsea, Mich. (1988).

11. Shinskey, F.G., "Process Control Systems." McGraw-Hill, Inc., New York, N.Y. (1988).

12. Liptak, B.G., "Instrumentation Engineer's Handbook." Vol. II, Process Control, Chilton Book Company, Philadelphia, Pa. (1985).

13. Am. Water Works Assoc., "Automation and Instrumentation." AWWA Manual M2, Denver, Colo. (1983).

14. Stire, T.G., "Process Control Computer Systems Guide for Managers." Ann Arbor Sci. Publishers, Mich. (1983).

15. Lovuola, V.J., "Consider Your Control Choices." *Intech,* **36,** 3 (1989).

16. "Instrumentation in Wastewater Treatment Plants." Manual of Practice No. 21, Water Pollut. Control Fed., Washington, D.C. (1978).

17. Instr. Soc. of Am., "Standard and Recommended Practices for Instrumentation and Control." 10th Ed., Vol. I & II, Research Triangle Park, N.C. (1989).

18. Am. Soc. of Heating, Refrigerating and Air Conditioning Engineers, Inc. (ASHRAE), "Fundamentals." I-P Ed., Atlanta, Ga. (1989).

19. "Energy Conservation in New Building Design." ASHRAE Standard 90, Am. Soc. of Heating, Refrigerating and Air Conditioning Eng., Inc. (1980).

20. "Thermal Environmental Conditions for Human Occupancy." ASHRAE Standard 55, Am. Soc. of Heating, Refrigerating and Air Conditioning Eng., Inc. (1981).

21. "Ventilation for Acceptable Indoor Air Quality." ASHRAE Standard 62, Am. Soc. of Heating, Refrigerating and Air Conditioning Eng., Inc. (1989).

22. "Recommended Standard for Sewage Works." Great Lakes—Upper Miss. River Board of State Sanit. Eng. (1978).

23. "The Boca National Mechanical Code." Building Off. & Code Admin. Int., Inc. (1987).

24. "The Uniform Mechanical Code." Int. Conf. Building Off. (ICBO) and Int. Assoc. of Plumbing and Mech. Off. (IAPMO) (1988).

25. "Industrial Ventilation." 20th Ed., Manual of Recommended Practice, Am. Conf. Gov. Ind. Hygienists, Lansing, Mich. (1988).

26. U.S. Dept. of Labor, Occupational Safety and Health Admin., "Occupational Safety and Health Standards." *Fed. Regist.*, 20 CFR, 1900 (1987).

27. "Standards Handbook #99." Air Movement and Control Assoc., Inc. (AMCA), Arlington Heights, Ill. (1986).

28. "Odor Control for Wastewater Facilities." Manual of Practice No. 22, Water Pollut. Control Fed., Washington, D.C. (1979).

29. "Process Design Manual for Sulfide Control in Sanitary Sewerage Systems." EPA 625/1-74-005, U.S. EPA, Technol. Transfer (1974).

30. "Odor and Corrosion Control in Sanitary Sewerage Systems and Treatment Plants." EPA 625/1-85-015, U.S. EPA (1985).

31. Metcalf & Eddy, Inc., "Waste Engineering: Collection and Pumping of Wastewater." McGraw-Hill, Inc., New York, N.Y. (1981).

32. "Chemical Aids Manual for Wastewater Treatment Facilities." EPA 430/9-79-018, U.S. EPA (1979).

33. "Process Design Manual for Phosphorus Removal." EPA 625/1-76-001a, U.S. EPA (1976).

34. "Process Design Manual for Suspended Solids Removal." EPA 625/1-75-003a, U.S. EPA (1975).

35. "Process Design Manual for Upgrading Existing Wastewater Treatment Plants." EPA 625/1-71-004a, U.S. EPA (1974).

36. "Process Design Manual for Sludge Treatment and Disposal." EPA 625/1-79-011, U.S. EPA (1979).

37. "Physical-Chemical Wastewater Treatment Plant Design." EPA 625/4-73-002a, U.S. EPA (1973).

38. "Design Manual Phosphorus Removal." EPA 625/1-87-001, U.S. EPA (1987).

39. "Wastewater Treatment Plant Design." Manual of Practice No. 8, Water Pollut. Control Fed., Washington, D.C. (1977).

40. "Operation of Wastewater Treatment Plants." Manual of Practice No. 11, Water Pollut. Control Fed., Washington, D.C. (1990).

41. "Water Treatment Plant Design." Am. Soc. Civ. Eng., Am. Water Works Assoc., McGraw-Hill, Inc., New York, N.Y. (1990).

42. Metcalf & Eddy, Inc. "Wastewater Engineering: Treatment Disposal and Reuse." 3rd Ed., McGraw-Hill, Inc., New York, N.Y. (1991).

43. "Fire Protection in Wastewater Treatment Plants." Proposal NFPA 820, Natl. Fire Protection Assoc., Quincy, Mass. (1990).
44. "The Boca National Plumbing Code." Build. Off. and Code Admin. Int., Inc. (1987).
45. "The Uniform Plumbing Code." Int. Conf. Build. Off. (ICBO) and Int. Assoc. Plumbing and Mechan. Off. (IAPMO) (1988).

Chapter 8
Construction
Materials Selection

*I*NTRODUCTION

This chapter focuses on typical environmental conditions encountered in wastewater treatment plants and selection of construction materials that are compatible with these conditions. The first section discusses materials selection approaches and classifies environmental conditions typically existing in individual unit processes. The second section describes materials selection for various plant support systems, including structural; heating, ventilating, and air conditioning (HVAC); electrical; instrumentation; odor control; and chemical feed and distribution systems. The third section discusses materials selection for design of unit processes, generally in the order of the liquid and solids processes chapters. The fourth section contains a glossary of materials, along with simple descriptions of each material.

APPROACHES FOR MATERIAL SELECTION

PLANT ENVIRONMENTAL CONDITIONS. Wastewater treatment plants contain unique environments that challenge design engineers to select appropriate materials and equipment. Erosive and corrosive environments are created by the combined actions of acidic and basic wastewater constituents; high process temperatures, highly corrosive off-gases and byproducts; and weather extremes such as high heat, high humidity, cold temperatures, ice, snow, and, at plant sites near oceans, salt air.

There are three approaches to design equipment and structures to withstand a plant's aggressive environment. The first is to design equipment and structures to withstand the environment of a given unit process. The second is to alter the environment through the use of control equipment, and the third is a combination of the other two. Regardless of the selected approach, the engineer must select equipment and materials compatible with the expected environment for the unit process or other plant facility.

As a good design practice, the engineer must first identify the typical environment of the individual unit process. This will allow the designer, as a "first cut," to determine expected service conditions for the material or piece of equipment. This narrows the materials selection process to a few compatible types of materials.

Table 8.1 presents typical environmental conditions for a number of treatment unit processes. The sections of this chapter discuss selection of materials for unit processes and other plant facilities.

LIFE-CYCLE COST ANALYSIS. This analysis provides an objective basis for choosing a material for predetermined environmental conditions. It is important because selection of materials that are compatible with all of the environmental conditions often requires the greatest initial capital expenditure. With increasingly tight capital cost budget constraints, many designers select materials with a lower initial capital cost that do not fully meet all environmental conditions. This often shortsighted approach can be more expensive over the service life of the facility. It is important to remember that even

Table 8.1   Classification of typical environments.

Unit process	Submerged	Intermittently submerged	Nonsubmerged
Headworks	Wet/highly corrosive,[a] abrasive	Wet/highly corrosive, medium/high temperature/ corrosive	Wet/highly corrosive, medium/high temperature/ corrosive
Other preliminary treatment	Abrasive		
Primary	Wet/corrosive, abrasive	Wet/highly corrosive, abrasive	Wet/highly corrosive
Suspended growth	Wet/corrosive	Wet/corrosive	Wet/highly corrosive
Attached growth	Wet/corrosive	Wet/corrosive	Wet/corrosive
Disinfection	See text for individual chemicals		
Nutrient removal	See suspended growth		
Advanced treatment (chemical)	Wet/ highly corrosive[a]	Mildly humid/ corrosive[a]	Corrosive
Advanced treatment (filtration)	Wet/corrosive	Nonaggressive	Nonaggressive
Sludge handling	Corrosive,[a] wet/ highly corrosive,[a] wet/corrosive[a]	Wet/highly corrosive, wet/ corrosive	Wet/highly corrosive,[a] wet/corrosive
Sludge stabilization	Wet/corrosive	Wet/highly corrosive	Wet/highly corrosive, nonaggressive
Thermal	Corrosive, wet/ highly corrosive, abrasive	Corrosive, wet/ highly corrosive	Medium/high temperature/ corrosive, abrasive

[a] Addition of chemicals heightens corrosivity.

materials usually considered to be "corrosion resistant" may corrode or erode under certain combinations of environmental conditions and geometric configurations.

Materials Selection for Design of Plant Support Systems

STRUCTURAL SYSTEMS. Selection of materials for the principal components of plant buildings and structures are discussed in this section.

Concrete. Concrete should resist the action of chemicals, alternate wetting and drying, freezing and thawing cycles, and exposure to the elements. One of the principal causes of deterioration of concrete treatment plant structures is the corrosive action of sulfuric acid that forms above the water line when the wastewater contains a significant sulfide concentration.[1]

The resistance of concrete to acid attack can be increased by as much as five times by specifying concrete produced with limestone aggregates instead of granitic aggregates. The limestone aggregates should meet the requirements of ASTM C-33 for concrete aggregates. New or unproven limestone aggregate sources must undergo petrographic examination or other procedures to ensure that they do not have potential alkali reactivity. When limestone aggregates are used, type II low-alkali cement should be specified.[2]

If the concrete structure must be watertight and able to resist freeze–thaw cycles, then it should be air-entrained and should conform with the following specifications:

- Minimum 28-day compressive strength of 27 600 kPa (4000 psi);
- Type of cement as described above;
- Maximum water:cement ratio of 0.45;
- Fine and coarse aggregate conforms with requirements of ASTM C-33;
- Air content of 6% ± 1% for coarse aggregate size No. 57 [2.5 cm (1 in.) to No. 4] or No. 67 [2 cm (0.75 in.) to No. 4];
- Slump of 2.5-cm (1-in.) minimum and 10-cm (4-in.) maximum; and
- Admixtures as required to suit climatic conditions.

Use in concrete of pozzolanic admixtures, such as fly ash, generally increases resistance to aggressive attack by seawater, sulfate-bearing soil solutions, and naturally acidic waters.[3] To further increase resistance, reinforcement should be adequately covered to limit deflection cracking. Noncorrosive fittings and embedded items may be required. Structures that are in contact with moving liquids must resist abrasion. In some cases, sufficient durability can be attained only with use of special protective coatings or liners.

To ensure integrity of construction and expansion joints, they should be equipped with waterstops of rubber, vinyl, metal, or other acceptable material. Waterstops should be placed at joints that will be submerged. ASTM A 615, grade 60 steel should be used for reinforcement and, in highly corrosive areas, epoxy-coated reinforcement should be used. All reinforcement should be rust-free before concrete placement.

Grout. The following types of grout are commonly used in wastewater treatment facilities:

- Nonshrink, nonmetallic grout: this type, made of hard natural aggregates with expansive cement to overcome shrinkage, is used on column bases, railing posts, concrete saddles for steel tanks, and lighting standards.
- Nonshrink, epoxy grout: containing a resin, hardener, and aggregate, this type of nonshrink grout is resistant to impact and dynamic loads that could crack and disintegrate other grouts.
- Ordinary cement–sand grout: this type may shrink and settle unless admixtures are used to reduce shrinkage and water permeability.

Building Materials. Construction materials for wastewater treatment plants must withstand a variety of special conditions, including various degrees of corrosiveness, frequent water exposure, and the operations and maintenance activities performed by plant personnel. Selected materials should resist, and not nurture, bacteriological or mildew growth, and they should be amenable to cleaning by conventional methods. Ease of replacement is another materials selection consideration. These stringent requirements have historically resulted in less variation in the selection of architectural finishes for wastewater treatment plants than for applications with less demanding environments. For treatment plant applications, the final selection is often conservative because materials are usually expected to perform with little, if any, attention over the useful life of the facility.

Metals must be properly applied to ensure safe operation and minimize maintenance of plant facilities. The following guidelines represent good practice for selecting metals for key applications:

- Anchor bolts and expansion anchors are usually fabricated of carbon steel. Nevertheless, materials selection for anchor bolts and expansion anchors for equipment depends on the equipment's use and the environmental conditions it will encounter. Therefore, stainless steel is often used for equipment in corrosive areas.

- Nosings, thresholds, rungs, and steps are available in a variety of metals, but they are usually made of aluminum, frequently with an abrasive, non-slip-type finish.
- Stairs, platforms, and railings may be made of steel, aluminum, fiberglass-reinforced plastic (FRP), or stainless steel. These appurtenances are typically steel in administrative buildings, generating rooms, and boiler rooms; however, aluminum materials are usually chosen in corrosive process areas. Stainless steel and FRP deserve consideration for extremely corrosive conditions, including areas exposed to salt spray.
- Floor gratings, plates, and hatches are made of aluminum in most process areas. Aluminum that may contact other metals or concrete must be protected by back-painting to prevent galvanic currents that could otherwise corrode aluminum rapidly, particularly in moist areas. If steel is required for its superior strength in heavily loaded areas, then the steel should be galvanized after fabrication to limit corrosion. However, galvanizing of large, welded steel items should be avoided because warping may result. FRP grating and handrails may be required to resist corrosive conditions encountered in some chemical-handling areas.
- Weir plates, troughs, and appurtenances are usually of aluminum or FRP. Galvanized steel, precast concrete, or cast-in-place concrete is also satisfactory for most troughs.
- Connections between pipelines of dissimilar metals, including alloys of separate groups and distant from each other on the "Galvanic Series of Metals and Alloys," should be made with a dielectric to minimize galvanic corrosion.

Protective Coatings. To successfully apply paint, the surface of the material to be painted must be thoroughly cleaned to remove all rust, dirt, oil, grease, and other foreign material. The surface can be cleaned mechanically by a wire brush, abrasive or shot blasting, or sanding or by chemical conditioning using solvents, acids, or caustics.

Choosing and applying the prime coat is the next step. For metal surfaces, the primer, which forms a bond between the substrate and the second coat, should contain a corrosion inhibitor and should be suited to the expected service conditions. Some primers should not be used for submerged metals or damp conditions. For submerged steel work, coal tar coatings are often used, without a separate primer. However, coal tar epoxies are potentially hazardous and their use may be restricted in the future. Vinyl primers, which are also satisfactory for submerged surfaces, can tolerate slight condensation on the metal without any effect on their adhesive qualities.

Top coats protect the prime coat and decorate the surface. The top coats must be compatible with the prime coat and suitable for the service (for example, dry, damp, or submerged). Bituminous coatings are often used for submerged surfaces. Epoxy coatings, polyurethane, vinyl, and certain other paints have good corrosion resistance and provide excellent protection for submerged surfaces. A variety of paints are suitable for nonsubmerged metal surfaces. Chlorinated rubber paints resist moisture, condensation, and hydrogen sulfide gas. These paints may be used for concrete as well as metals. For severe exposure conditions, vinyl, polyurethane, or epoxy coatings should be considered. The "Steel Structures Painting Council" standards provide further information.

Galvanizing, a coating applied to steel by dipping it in a molten zinc bath, protects steel through preferential corrosion of the zinc coating. Steel surfaces can also be protected temporarily with a grease coating.

Rubber-based paints are superior to water-cement paints for concrete because they are easier to clean and more resistant to corrosive gases. Some rubber-based paints are affected by grease, but they are unaffected by lime or free alkali found in all masonry. As an alternative to painting, pigments may be added to concrete as it is being mixed. For concrete that will be submerged or will require a waterproof coating, bituminous or coal tar epoxy coatings are typically used. Where color is needed, vinyl coatings or epoxy amines can be used. In highly corrosive environments, concrete structures are commonly protected with a polyvinyl chloride (PVC) liner attached to the concrete surface with mechanical locking extensions that are cast into the concrete. Table 8.2[4] presents results of tests of concrete coatings reported by the Los Angeles County Sanitation Districts.

HEATING, VENTILATING, AND AIR CONDITIONING SYSTEMS.
Ventilation rates, placement of equipment, and selection of materials are primary variables for mitigating the effects of corrosion and providing a safer atmosphere for plant operators. Designers should refer to individual codes and state regulations because statutory requirements vary among states.

Air-handling Units. Air-handling units should not be located in spaces susceptible to corrosion, explosion, or fire. If this cannot be avoided, units will require special materials or ratings.

Fully insulated air-handling units with gaskets should have easily removable doors for access to various internal sections. External linkages should be fabricated of corrosion-resistant materials, and the exterior unit casing should have a weather- and corrosion-resistant finish such as baked-on epoxy.

Terminal Heating Equipment. Unit heaters are available for a variety of hazard classifications, ranging from general purpose to corrosion-resistant to explosion-proof classifications. Unit heaters may use one of several heating

Table 8.2   Summary of Los Angeles County Sanitation Districts coating study (updated September 1990).[4]

Coating	Test number	Performance[a] Application	Performance[a] Corrosion resistance	Performance[a] Bond
Senotex 3005	C-1 (urethane)	2	1	3
Zebron	C-3 and C-39 (urethane)	3	1	2
PR475	C-10 (urethane)	2	2	1
Quantum	C-17 (polyester)	1	1	1
Fosroc	C-22 (epoxy mortar)	3	3	1
Aquata-Poxy	C-25 (epoxy)	2	1	1
Vibrabond 500	C-26 (urethane)	4	1	2
Concresive 1305	C-28 (epoxy)	1	2	1
PVC Panels	C-29 (liner)	3	1	—
Acid Proof Cement No. 54	C-34 (liner)	2	2	—
Allied Urethylene	C-35 (liner)	1	2	3
GS 1490	C-36 (urethane)	4	—	3
Mainstay DS-4	C-37 (epoxy)	1	1	1
120 Vinester	C-38 (vinyl/ester; long cure time)	1	1	1
Allied Vinylthane	C-40 (liner)	1	1	—
Overkote V	C-42 (epoxy mortar)	2	2	1
IET System 3	C-44 (polyester)	2	1	1
Sancon 100	C-47 (urethane)	2	2	2
Semstone 140S	C-49 (epoxy)	1	1	1
Magma Quartz	C-50 (epoxy mortar)	2	1	1
IPI Crystal Quartz	C-53 (epoxy mortar)	1	1	1

[a] Each coating was ranked for application, corrosion resistance, and bonding: 1 = good; 2 = some problems but not significant; 3 = significant problems; and 4 = failure.

media, including electric resistance, hot water, steam, natural gas, and propane gas. Careful evaluation of environmental conditions governs the selection of heater type.

Fans and Blowers. Aluminum or corrosion-resistant, epoxy-coated steel should normally be specified for axial and tube axial fans. Some manufacturers, however, offer FRP tube axial fans for corrosive service.

Centrifugal blowers are used for extensive, duct-type exhaust systems in which higher static pressure is required to move larger quantities of air than ordinary fans can handle. These blowers are marketed with three material options, commonly selected for the particular application in the plant. Usually, select steel alloy, stainless steel, or FRP fans are used for odor control systems.

Ductwork. All ductwork should conform to the standards of the Sheet Metal and Air Conditioning Contractors National Association. Duct materials include galvanized steel, aluminum, type 304 stainless steel, type 316 stainless steel, and FRP. Galvanized duct work is used for air-conditioned spaces and clean, noncorrosive areas. Aluminum ductwork is used for toilets, lockers, shower areas, and other humid locations. In corrosive atmospheres, stainless steel ductwork is used, while in highly corrosive areas and odor control systems, FRP is used. Galvanized steel should not be used in corrosive areas.

Piping. Materials for heating system piping generally include

- Type "L" copper tubing, hard drawn, for piping diameters up to 64 mm (2.5 in.);
- Schedule 40 black steel pipe with welded fittings for piping diameters of 75 mm (3 in.) or larger; and
- Vapor barrier insulation for all piping, valves, and fittings.

Boilers. Because municipalities have strict ordinances regarding boilers and the installation of boilers in boiler rooms, boilers should not be located in process areas. All boilers must be constructed to the ASME Boiler and Pressure Vessel Code, Section I "Power Boilers" or Section IV "Heating Boilers." The three basic boiler classifications are

- Cast iron boilers: all pressure parts of these boilers are made of cast iron. The maximum operating pressure of these boilers is 100 kPa (15 psi) for steam and 1100 kPa (160 psi) for hot water.
- Fire-tube boilers: fire-tube boilers are usually welded steel, with fully immersed tubes that transport hot gases from the combustion chamber. Fire-tube boilers may provide both steam and hot water service. Steam pressures are generally limited to 2400 kPa (350 psi).
- Water tube boilers: water tube boilers are typically constructed of upper and lower drums that are connected by steel tubes. An exterior steel shell usually covers the entire assembly. The tubes and drums contain water, and hot furnace gases flow around the tubes. These boilers, furnished for both steam and hot water, have pressure ratings as high as 24 000 kPa (3500 psi).

Boilers may use one type of fuel or a combination of fuels such as natural gas and No. 2 fuel oil, digester gas and No. 2 fuel oil, or other combinations. Boiler selection should be based on the specific application, required operating pressures and temperatures, and fuel cost availability.

ELECTRICAL SYSTEMS. Like other systems in wastewater treatment plants, electrical equipment is often exposed to the corrosive conditions imposed by humidity associated with untreated wastewater and the corrosive gases produced as treatment byproducts. These conditions must be considered when locating electrical equipment, and should guide selection of appropriate corrosion-resistant materials and protective enclosures. Designers must maximize safety in hazardous or classified areas in treatment plants, where flammable gases or liquids may be present, and design systems that are non-spark-producing for these areas. In addition, appropriate explosion-proof or safety equipment may be necessary. The same general criteria apply to instrumentation selection, which is discussed in a subsequent section.

When designers select electrical materials and equipment for a wastewater treatment plant environment, the considerations described below are significant.

Conduits and Enclosures. The following guidelines generally apply:

- Conduits and electrical panels in outdoor areas, areas below grade, and areas subject to splashing of liquids or frequent washdown require a substantial level of waterproofing that should meet the National Electrical Manufacturer's Association (NEMA) 3R raintight requirements. Underground metallic conduit systems in corrosive soils should be protected by 8 cm (3 in.) of concrete all around, by PVC tape, or by a bituminous coating, depending on the degree of corrosiveness and loadings on the conduit.
- Indoor equipment in wet areas should meet NEMA 4 or 4X watertight requirements. In any wet area, watertight, threaded hubs should terminate the conduits.
- Corrosive areas, including wet wells and many chemical feed areas, need conduit systems of PVC-coated steel. In cases where the structural protection of steel is not required, Schedule 40 PVC may be appropriate. Enclosures should meet NEMA 4X watertight and corrosion-resistant requirements. Conduit and enclosure supporting systems should be PVC-coated or nonmetallic.
- Areas to be considered as hazardous locations are defined by the National Fire Protection Association (NFPA). These areas include headworks, wet wells, digesters, and other areas where methane and combustible mixtures of gases could accumulate. Materials and

---

equipment must be explosion-proof or otherwise suitable for installation in the specific class, division, and group area as defined by the National Electrical Code.

- All conduits and conductors entering an electrical enclosure should be sealed with expansion foam or an equivalent material at the enclosure entrance to keep out moisture, insects, and gases.
- Electrical conductors should have insulation that is compatible with the environment in which the conductors are located (for example, SJD cable for wet conditions).

Signal Cables. Because electromagnetic interference (EMI) may affect signal cables, they should be shielded, twisted pairs routed in metallic conduit.

INSTRUMENTATION. In general, considerations for selecting electrical equipment apply also to instrumentation. Materials for instrumentation and control equipment can be divided into three categories: panels/enclosures, panel mounting devices, and field instruments. Conservative materials selection is cost effective; for example, selection of a corrosion-resistant material costing 10% more than a less resistant material may more than double the instrument's service life and improve its reliability.

Panels/Enclosures. NEMA standards apply to differing environmental conditions for panel/enclosure rating selection. Criteria applied for selection of panel construction should be similar to that used for electrical switchgear, taking into consideration the dust sensitivity of the installed instruments. Use of NEMA-12 control enclosures adjacent to a NEMA-1 motor control center is appropriate. NEMA 4X panels are suitable for use in wet, corrosive, or outdoor environments. Designers must choose from epoxy-coated steel, stainless steel, or fiberglass materials. For outdoor locations, a stainless steel panel might be better than fiberglass because ultraviolet rays can break down the composition of fiberglass.

Conduits with sealed fittings must be used with explosion-proof panels. Where nonexplosive gases are present, a foam-type sealant at the conduit entry should be selected to prevent entry of gases. For all panel options, strength and rigidity depend on material thickness. The UL Standard (UL-50) should be used to determine appropriate material thickness as well as selecting the size of access doors.

Panel-mounted Devices. Any panel device must meet the NEMA rating of the panel. Wherever corrosive gas could migrate through conduits or doors, conformal coating of electronics should be applied.

Field Instruments. Generally, specification of the appropriate NEMA rating for the field device will afford the required degree of environmental protection for the installation. Wetted materials for devices should conform to the specified piping material. Because of the small material quantities for field instruments and the typically high cost of downtime, materials such as polytetrafluoroethylene (PTFE), ceramics, Alloy 20, and type 316 SS are normally cost effective. Because the instrument manufacturer usually furnishes mounting hardware, its material should be specified along with that of the instrument. Some instruments, such as magnetic flowmeters, include a liner that must also be selected. Aluminum oxide (available in limited sizes) and polyurethane liners afford resistance to abrasion and are suitable for untreated wastewater. As PTFE provides excellent chemical resistance, it may be used where abrasion is not significant.

ODOR CONTROL FACILITIES. Design and installation of odor control facilities has significantly increased at wastewater treatment plants recently as a result of heightened public awareness of environmental issues. Unit processes requiring odor control vary from plant to plant, depending on the release of odorous gases and the local perception of uncontrolled odors.

Odor control systems typically contain, transport, and treat the most corrosive gases found in wastewater treatment plant operations. Thus, selection of appropriate construction materials is critical for these systems. Foul air from wastewater treatment plants may contain a combination of water vapor, hydrogen sulfide, mercaptans, indoles, skatoles, volatile organic compounds (VOCs), ammonia, oil and grease, and carbon dioxide.

The key compound most commonly associated with plant corrosion is hydrogen sulfide. The proportion of hydrogen sulfide stripped out of solution depends on wastewater pH, turbulence, and temperature. Even though atmospheric hydrogen sulfide corrodes ferrous and cupric metals, corrosion of cementitious concrete occurs only after conversion of hydrogen sulfide to sulfuric acid. In the presence of oxygen and moisture in a contained space, gaseous hydrogen sulfide is converted to sulfuric acid by bacteria that inhabit nonwetted walls and covered undersides. Therefore, the presence of hydrogen sulfide in air does not always indicate a high potential for concrete corrosion.

Not only is the air stream corrosive, but odor control devices themselves can contain corrosive substances, chemicals, or both. For example, chemicals used in wet-scrubbing odor control systems are commonly oxidants (chlorine, sodium hypochlorite, hydrogen peroxide, or potassium permanganate), high- or low-pH solutions, or both. Dry odor control systems use abrasive and sometimes caustic oxidant-impregnated media. These chemical oxidants, pH adjusters, and dry media can themselves create corrosive conditions that require careful selection of construction materials. Construction materials for these chemical feed and distribution systems are discussed in the following section on chemical addition. In general, plastics and FRP chosen to resist specific chemicals have proven to be the most successful and cost-effective materials for odor control. Carbon steel, aluminum, copper, and copper alloys are subject to severe corrosion from chemicals and moisture conditions surrounding odor control facilities. Stainless steel may be satisfactory for certain equipment and piping if it is carefully selected for a compatible environment.

CHEMICAL FEED AND DISTRIBUTION FACILITIES. Some plant areas that are normally dry require corrosion-resistant materials because of storage or distribution of process chemicals. Materials capable of resisting a broad range of process chemicals in the gaseous phase include polyamide-cured epoxies, ceramic tile, structural glazed tile, and hard-coat anodized aluminum.

Glass, except for areas containing hydrofluorosilicic acid, is inert to most attacks from process chemicals likely to be in the gaseous phase. In areas where hydrofluorosilicic acid is present in the atmosphere, glass areas can be glazed with polycarbonate plastic; because of the combustible nature of polycarbonate, however, this material should conform with governing codes. Chemicals such as sodium bisulfide, ferric chloride, ozone, and hydrochloric acid can attack the type 300 series of stainless steel. Type 316 stainless steel, however, exhibits better resistance to sulfides and chlorides than type 304 and will demonstrate adequate resistance to corrosion from sulfuric acid. Stainless steel also resists bases such as lime and sodium hydroxide.

Care should be exercised in choosing the products discussed above. The brief information on materials for resisting various chemicals presented below is intended only as a general guide describing materials options for typical conditions. Chemical-resistance guide charts and material safety data sheets should be reviewed before selecting materials to resist any of the chemicals. In addition, manufacturers' recommendations in the selection of materials should be used. Chemicals and construction materials used to resist their attack are

- Sodium hypochlorite (NaOCl): pipes and containers for this chemical can be made from FRP with special fabrication and curing provisions, polyvinylidene fluoride (PVDF), PVC, or Hastelloy C. Because plastics have temperature limitations, they should not be used for temperatures above 60°C (140°F). Use of glass should be avoided.
- Ferrous sulfate ($FeSO_4$): polyethylene (PE), polypropylene (PP), FRP, and PVC are generally suitable materials of construction for storage tanks and process lines.
- Ferric chloride ($FeCl_3$): although available in both liquid and dry form, liquid ferric chloride is now used at most plants because of handling problems associated with the dry form. Extremely corrosive

liquid ferric chloride will attack most metals. Therefore, suitable handling materials are limited to most plastics, titanium, and, to a lesser extent, Hastelloy C. Storage tanks manufactured of fiberglass with vinyl or polyester resin materials are an economical alternative to rubber-lined steel tanks. Piping and equipment materials that safely handle liquid ferric chloride include chlorinated polyvinyl chloride (CPVC), PVC, PVDF, PTFE, PP, hypalon, and neoprene.

- Sulfuric acid (H_2SO_4): for fully concentrated acid, storage tanks may be constructed of unlined steel, with a suitable allowance for corrosion. Nevertheless, in the long run, type 316 stainless steel, PVDF-lined steel, or FRP may be better. Type 316(L) stainless steel may be used for piping, depending upon acid concentration.

- Hydrochloric acid (HCl): hydrochloric acid should be stored in tanks constructed of either rubber-lined steel or FRP. A vinyl ester resin is best. PVC is the best piping material if the piping will be adequately supported and temperatures will not exceed 60°C (140°F). For higher temperatures, CPVC, PVDF, or PTFE piping can be used.

- Nitric acid (HNO_3): materials used for hydrochloric acid are appropriate.

- Sodium hydroxide (NaOH): although this chemical is normally stored in tanks constructed of unlined steel, tanks fabricated of FRP are also suitable. Piping systems usually are steel or stainless steel.

- Sodium carbonate (Na_2CO_3): concrete or steel should be used for storage. If fed as a weak solution (usual procedure), this chemical will pose no handling problems. Best materials for handling soda ash solutions include stainless steel and many types of plastics.

- Alum ($Al_2(SO_4)_3-H_2O$): dry alum, noncorrosive and slightly hygroscopic, may be stored in concrete or mild steel bins. Alum is usually dissolved in water and fed as a corrosive liquid that will attack metals. Storage tanks can be fabricated of FRP, polyester materials, rubber-lined steel, or type 316 stainless steel. Materials of choice for piping systems include type 316 stainless steel, PVC [under 60°C (140°F)] and CPVC [up to 82°C (180°F)].

- Sodium aluminate ($Na_2Al_2O_3$): because neither the dry nor liquid form of this chemical is extremely corrosive, it can be shipped, stored, and fed in devices that are standard for the industry. Iron, steel, plastics, and rubber are suitable materials to handle either form.

- Sulfur dioxide (SO_2): sulfur dioxide, a colorless gas with a pungent, suffocating odor, is used as a dechlorination agent. The gas is normally stored in pressurized cylinders or containers similar to those used to store chlorine gas or liquid. Equipment and materials for feeding sulfur dioxide are identical to those for feeding chlorine, except that type 316 stainless steel may be also used for sulfur dioxide systems.

- Sodium bisulfite (NaHSO$_3$): sodium bisulfite, a commercially available solution delivered in strengths between 38 and 42%, serves as a dechlorination agent. The solution may be stored in stainless steel, PVC, or FRP tanks and is usually fed and metered by diaphragm-type metering pumps. A 38% solution contains the equivalent of 0.26 kg SO$_2$/L of solution (2.2 lb/gal).
- Hydrogen peroxide (H$_2$O$_2$): hydrogen peroxide, a commercially available clear and colorless liquid, is primarily used for odor control in collection systems and treatment plants. Other uses of the chemical include dechlorination and cyanide and phenol destruction by oxidation. Suitable materials are aluminum for piping and aluminum or aluminum alloys for storage tanks.
- Potassium permanganate (KMnO$_4$): potassium permanganate is a strong oxidizing agent used for odor control. As a fire and explosion hazard, it should be separated from combustible materials. The chemical is usually delivered to plants in dry form in drums; occasionally it is delivered in bulk-load quantities for safe storage in steel hoppers. The hygroscopic chemical is difficult to feed in dry form from storage hoppers during periods of high humidity. As the chemical is usually fed as a weak solution, most standard materials for the industry, such as iron, steel, and plastics, are suitable for handling the solution.
- Quicklime (CaO) and hydrated lime (Ca(OH)$_2$): both forms of lime are only available in dry form. Lime is used for pH adjustment, sludge conditioning, and precipitation of phosphates. Lime is usually fed as a slurry by adding water. Although this slurry is not corrosive, it is difficult to handle because of lime buildup in tanks, pipelines, and troughs. Most materials that are standard for the industry are suitable for handling the slurry.
- Polymers: polymers are available in both liquid and dry forms. In liquid form, polymers may be further subdivided into emulsions or solutions. For effective dispersion, polymers are usually diluted to concentrations of less than 0.25%. Polymers are not corrosive to stainless steel, PVC, FRP, and other polyolefin plastics.
- Methanol (CH$_3$OH): methanol, also known as wood alcohol, is used in a few plants as a carbon source in denitrification processes. Methanol may be fed with materials that are standard for the industry.
- Chlorine dioxide (ClO$_2$): as a powerful oxidizer occasionally used as a disinfectant, chlorine dioxide is unstable; it thus must be used as it is produced. Materials that normally are adequate to handle chlorine solutions will handle chlorine dioxide solutions as well (see discussion of chlorine in the subsequent disinfection section).

Construction Materials Selection   359

MATERIALS SELECTION FOR DESIGN OF UNIT PROCESSES

This section describes environmental conditions existing in the various unit process areas and offers information on materials typically used for these processes.

PRELIMINARY TREATMENT. Headworks and other preliminary treatment facilities inflict severe corrosive and destructive conditions on equipment and material. Typically, influent wastewater may be septic and contain dissolved gases, grit, rags, rocks, logs, grease and oils, waste chemicals, and constituents from internal recycle streams that together create a very corrosive and abrasive environment. Hydrogen sulfide and other gases such as ammonia are released at turbulent areas, including flow metering devices, screens, and weirs. Such gases, combined with high humidity, provide an aggressive environment that fosters corrosion.

In addition to corrosion and abrasion, a potential for fire and explosion exists whenever gasoline or other hydrocarbons spill into the collection system. At facilities where a high risk of spills exists, use of explosion-proof equipment and fire-proof materials is required. All preliminary treatment equipment must be carefully evaluated to ensure appropriate selection of materials.

Screens. Screens may be constructed of carbon steel, galvanized steel, stainless steel, aluminum, or plastic. Carbon steel is frequently selected because of its strength, hardness, ease of fabrication, and economy. However, corrosion resulting from moisture, sulfides, and other compounds may shorten the life and increase maintenance of steel screens. Galvanizing the steel will increase its corrosion resistance, but would be impractical on large welded racks or other welded pieces that could be warped by the heat of the galvanizing process.

Stainless steel screens have all the advantages of carbon steel, combined with corrosion resistance to most conditions existing in headworks. These screens cost substantially more than carbon steel screens, but they may be the most suitable choice for larger plants with more difficult corrosive conditions that would unduly increase maintenance costs and reduce life of carbon steel screens.

Aluminum screens may be used for smaller domestic wastewater facilities where increased corrosion resistance is needed and the abrasiveness of the wastewater is insufficient to cause serious wear on aluminum parts. Because aluminum is weaker and less resistant to abrasion than steel, its use must be carefully studied, but it may be an economical alternative to stainless steel.

Certain proprietary fine screen devices with plastic screens and stainless steel moving parts will successfully withstand severe environmental conditions.

Grit Removal. Several types of grit removal devices exist: velocity-controlled; aerated; constant-level, short-term sedimentation; and centrifugal. Abrasion is usually the most significant condition that must be resisted to attain long life from equipment and materials. Although subject to severe abrasion and corrosion, economical steel and hardened steel may provide a reasonable service life. Stainless steel is more corrosion resistant but may be too expensive for this application. Cast iron chains and sprockets are used to resist abrasion, but plastic and fiberglass-reinforced plastic chains, flights, sprockets, and wearing strips have proven increasingly economical and successful in resisting both abrasion and corrosion.

Flow Equalization. Service conditions encountered for flow equalization usually resemble those for primary treatment facilities discussed below. Materials and equipment selection should conform to the same criteria.

PRIMARY TREATMENT. Primary treatment facilities may have septic conditions and may release corrosive and odorous gases from liquid surfaces, particularly at inlet and outlet weirs and channels. Enclosed channels or tanks will likely have hydrogen sulfide corrosion. At wastewater treatment facilities located in populous areas, primary settling tanks are sometimes covered and provided with odor control equipment. This will control odors but accelerate corrosion. Designers must provide for the high humidity and corrosive gases that may occur in localized areas under the cover. Downtime of ventilation and odor control systems must be accounted for in the selection of protective coatings and materials.

Grit may enter primary settling tanks and cause rapid wear of sludge collection and, more importantly, sludge-pumping equipment. For example, "sugar-sand," a small-grain sand that passes through typical grit removal facilities, has been a widespread problem at plants in the Southeast.

Concrete is suitable for domestic wastewater tanks and channels. Submerged surfaces may be covered with asphaltic or epoxy coatings to resist corrosion, but this practice is usually unnecessary unless the wastewater contains corrosive industrial constituents or has a low pH. Carbon steel tanks with protective coatings can be used for small facilities, but they require substantial maintenance.

Collection equipment made of cast iron with wooden flights has been used successfully for years. However, grit abrasion and corrosion require frequent replacement of chains and appurtenances. Stainless steel chains and other alloys have been used successfully, but they will also experience wear. Plastic collection equipment resists corrosive, abrasive environments, but it may

---

not be suitable for the loadings of long tanks and heavy sludge because plastic lacks the strength of ferrous metals. FRP is usually selected in lieu of wood for sludge collection flights because it will not rot. Clear heart redwood, formerly used extensively, is increasingly expensive and sometimes unavailable. Other pressure-treated woods, however, may be used for this service. Weirs may be of FRP, aluminum, or stainless steel. Carbon steel weirs corrode quickly.

Tank launders and channel and tank covers may be of FRP or aluminum; scum removal equipment is usually made of ferrous metals. Sluice gates may be constructed of cast iron, although slide gates should be aluminum, stainless steel, or FRP. Carbon steel with a protective coating (usually epoxy) is used for flocculation and equipment for bridge-type and circular clarifiers. Railings, gratings, and plates may be of aluminum or stainless steel. Although carbon steel rails and gratings may be used, they require extensive protective coating, galvanizing, or both. Polymer handling equipment usually requires plastic or stainless steel material.

SUSPENDED-GROWTH BIOLOGICAL TREATMENT. Corrosive gases are not normally a problem for aerobic biological systems, although these gases may be intermittently present in relatively small quantities. High humidity and wet areas are characteristic of these units. At plants near populated areas, aeration tanks and settling basins may be enclosed, producing interior spaces that are vulnerable to corrosion. Concrete is the usual material for aeration and settling tanks and channels, although steel is used for small, package-type facilities.

Air piping for diffused air systems may be stainless steel, carbon steel, ductile iron, or FRP. Temperature, expansion, thrust, and noise limitations must be carefully reviewed, particularly if FRP is chosen. Air piping in tanks, including drop pipes and diffuser grids, may be of stainless steel or plastic. If plastic is used, the walls must be thick enough to withstand internal and external pressures, loads imposed by the diffusers, and physical damage. Plastic piping must be compounded to resist ultraviolet ray degradation. Stainless steel should not be used if an acid gas diffuser cleaning system is installed.

Aluminum is usually satisfactory for slide gates, weirs, railings, and ancillary items; stainless steel and FRP may be used at additional cost. Piping for sludge and effluent commonly consists of concrete or cement-lined ductile iron, although other materials such as cement-lined steel, plastic, and FRP may be selected. In earthquake-prone regions, "push-on" or rubber, ring-type joints with deep bells should be provided. Ductile iron fittings and valve bodies should be used with ductile iron pipe and steel valves with steel pipes to retain pipeline integrity when movement occurs. Sludge removal equipment in settling tanks is usually fabricated of carbon steel and coated to withstand humid and submerged conditions.

ATTACHED-GROWTH BIOLOGICAL TREATMENT. Attached-growth biological treatment processes include trickling filters and rotating biological contactors (RBCs). Materials selected to construct these facilities must be durable in highly humid and sometimes anaerobic atmospheres and must be strong enough to support the biological slimes characteristic of these process units. They also need to resist damage caused by sunlight or freezing conditions, depending on the location of the installation. Provisions must be made for differential expansion.

Trickling Filters (TFs). Although older TFs often have stone or slag media, modern designs use media of prefabricated plastic modules, specially designed plastic packing, or wooden slats. These media are often the single most expensive components in TF construction. In choosing among media, both installed cost and process performance should be considered. The following are typical materials for media:

- Blast furnace slag offers advantages over rock because slag has a higher proportion of surface area per unit volume because of its vesicular-type surface.
- Plastic media, usually constructed from PVC, is usually specified as insoluble, fire retardant, and not subject to deterioration by environmental conditions such as extreme temperature or sunlight. A wide variety of loose and modular plastic media having acceptable environmental and structural properties is available (see Chapter 12).
- Redwood slats are suitable as media. Treated lumber has not been widely used probably because chemicals used to preserve the wood are toxic to microorganisms.

Rotary distributors constructed of coated carbon steel, galvanized steel, stainless steel, or aluminum are used with most biological towers. Smaller towers may have fixed distribution grids constructed from pipe of these materials or from PVC or FRP pipe. Nozzles on rotary distributor arms and fixed pipes are fabricated of bronze or plastic.

Concrete is commonly used as the floor on biological towers. Precast filter blocks manufactured from vitrified clay or plastic are used to provide drainage for towers with rock media. Individual support systems will vary, depending on the recommendations of the media manufacturer.

Most rock-media biological towers have reinforced concrete walls to support the media and allow flooding of the towers for fly control. Walls have also been built of stone, brick, precast concrete, or steel plate. Plastic media allow use of economical alternative wall systems such as FRP panels or vertically aligned precast concrete T-beams because they require less structural support and do not need flooding to control flies. Designs containing walls with insulating properties or partially buried walls should be considered for installations in colder climates where heat loss may adversely affect process performance. Domes of fiberglass, aluminum, or even concrete may be used to cover towers where heat loss or odor control are design considerations.

RBCs. RBCs consist of plastic media supported by horizontal rotating shafts. RBC shafts are positioned over wastewater in a tank so that the media are always partially submerged. The tanks sometimes contain baffles or air diffuser systems. RBCs are usually covered or installed in a building.

Plastic RBC media consist of high-density PE sheets arranged in a variety of proprietary configurations. A small amount of carbon black is commonly added to the PE to reduce ultraviolet degradation.

RBC shafts are fabricated from carbon steel and covered with a protective coating suitable for water and high-humidity service. Shaft designs differ among manufacturers. Common to all designs is the need for carefully controlled welding procedures and proper application of protective coatings.[5] Manufacturers use a wide variety of techniques to attach media to the shafts. All of these techniques involve the use of corrosion-resistant hardware, fiberglass-reinforced plastics, stainless steel, or galvanized steel.

Concrete tanks are used for most RBCs except small, factory-assembled plants and some installations in buildings. Most of these exceptions are carbon steel tanks that require more maintenance than concrete tanks. Fiberglass tanks, which are more expensive, are available for very small plants. Corrosion-resistant coatings are usually used with concrete and steel tanks.

Depending on the configuration of the RBC shafts and process requirements, tanks may contain baffles or air diffusion systems. Baffles may be fabricated from treated lumber, aluminum, FRP, or stainless steel; the more economical materials are usually selected. PVC piping is usually used for air diffuser systems. Stainless steel is favored for fasteners, supports, anchor bolts, hinges, and other hardware. RBC covers are fabricated from FRP. Installation of RBCs in buildings creates additional requirements for corrosion-resistant materials and ventilation, heating, and humidity controls.[5]

NATURAL SYSTEMS. Materials used in natural systems are, by definition, available locally or on site; hence, they are not as readily definable as those for unit processes of typical wastewater treatment plants. Materials for natural systems require site-specific data because of myriad soil and plant types. Before a natural material is used in a treatment application, its compatibility with the wastewater and surrounding climate should be investigated.

Soil Absorption Systems. Materials selection for soil absorption systems is fairly straightforward because these systems are small and usually have relatively few corrosion problems. Cast iron pipe is typically used to convey the waste to a concrete septic tank. The tank may also be constructed of fiberglass, steel, or masonry, depending on local codes. Concrete tanks are most common because they offer better resistance to corrosive soils than steel tanks and are less susceptible to cracking than fiberglass tanks. If a dosing tank or a pump chamber is required, it is usually fabricated of the same material as the septic tank. PVC piping generally conveys the septic tank effluent to distribution or junction boxes that consist of concrete or fiberglass. Fiberglass is the least costly option if the boxes will not bear significant surface loads. Leaching lines (fields) are normally perforated PVC pipes in a washed gravel bed. If trigalleys or leaching tanks are specified, they are typically precast concrete in gravel.

Wastewater Ponds. Two considerations dominate materials selection for wastewater ponds: seepage must be prevented and the soil must not be reactive with wastewater. Therefore, the selection of a liner is critical. Although a native clay may be used, enhanced soils, geosynthetics, and bentonite geotextile mats usually have a much lower permeability. If spray-on liners are used, they pose a quality control problem. Designers should verify that the life expectancy of the liner will not be affected by reactions with the wastewater. If a berm is necessary for the existing site condition, it can be constructed of any soil, provided that the liner is properly integrated to block seepage. Chapter 13 discusses liner selection. If concrete, aerators, or other items are to be used in the pond system, materials selection should be based on criteria for exposure to weather and raw wastewater.

DISINFECTION. Materials selection for disinfection with chlorine, ozone, and ultraviolet radiation is discussed below.

Chlorine. Chlorine, which exists in either gas or liquid phases, is nonexplosive and nonflammable. Nevertheless, because of its strong oxidizing potential, chlorine can support combustion of some substances. Many organic chemicals react readily with chlorine, sometimes explosively.

Although water-operated injectors commonly feed chlorine at wastewater treatment plants, chlorine is only slightly soluble in water. When combined with water, chlorine reacts (hydrolysis) very rapidly to form a weak solution of hydrochloric (HCl) and hypochlorous (HOCl) acids. Because of these acids, chlorine in the presence of moisture, or as a water solution, is very corrosive to most common metals. Dry chlorine gas or chlorine liquid will not react with most common metals at ordinary temperatures. In the absence of water and at temperatures below 120°C (248°F), iron, copper, steel, lead, nickel, platinum, silver, and tantalum resist attack from chlorine.

These unique physical and chemical properties require special handling requirements and materials. Chlorine is supplied under pressure in various sized containers or tank cars. Piping systems for either dry gaseous or liquid chlorine under pressure are typically constructed of Schedule 80 steel pipe with 1360-kg (3000-lb) forged steel fittings. Gaseous chlorine vacuum systems and chlorine solution piping and accessories downstream of the water injector are typically Schedule 80 PVC with solvent-weld construction.

Ozone. Ozone, an unstable colorless to bluish gas, is one of the strongest oxidants used in wastewater treatment. Because of its short half-life in the gaseous phase or in water solution, ozone must be generated on site and immediately dispersed. Equipment required to successfully generate ozone is described in Chapter 14; generally, the ozone facility requires an air preparation system (unless high-purity oxygen is used) to produce dry air with a dewpoint temperature of less than $-50°C$ ($-58°F$), ozone generation equipment, ozone absorption and disposal equipment, and ozone destruction equipment for any and all exhaust gases.

Ozone is extremely toxic. Concentrations in the atmosphere of only 0.25 mg/L are considered unhealthy, and concentrations of 1.0 mg/L or more are hazardous to health and may even be fatal. Also a fire and explosion hazard, ozone must be separated from oil and other readily combustible materials.

Ozone is only slightly soluble in water (12 times less soluble than chlorine). However, when combined with pure water, ozone produces the free radicals HO_2 and OH, both with great oxidizing and disinfection potential. Because ozone is very corrosive, all materials of construction required to handle the dry gas must be carefully selected. Series 300 stainless steel is widely used in the U.S., while carbon steel is used extensively in Europe. Generators, piping, dispersion units, and off-gas destruct units all should be fabricated from Series 300 stainless steel, type 302 or 316. Use of PVC for piping systems has a deleterious effect on ozone use because of a side reaction that consumes ozone. Rubber products should not be used for handling ozone.

Ozone is normally fed as a gas under pressure that bubbles through a closed, vented tank or chamber. The residual feed pressure is a result of the air preparation pressure. Piping should be sized to minimize the pressure drop between the ozonator(s) and the point of application.

Ultraviolet (UV) Radiation. UV radiation is generated by low-pressure mercury vapor germicidal lamps for maximum energy output at a wavelength of approximately 253.7 nm. Efficiency of UV radiation depends mainly on effluent quality. According to EPA, effluents with over 20 mg/L of suspended solids severely affect the disinfection potential of UV radiation because the suspended solids act as a shield, reducing incineration of bacteria and viruses.

UV radiation performs differently as a disinfectant than chlorine and ozone. The two chemicals exert their toxic and oxidizing powers to destroy or inactivate bacteria and viruses; UV radiation uses radiation to biologically destroy or inactivate bacteria viruses.

UV systems typically have a series of UV lamps in a fabricated steel or concrete contact chamber. There are two commonly used types of UV systems: UV lamps housed in quartz tubes, with the plant effluent flowing around the tubes, and UV lamps arranged around PFTE tubes through which plant effluent flows.

INTEGRATED PROCESSES FOR NUTRIENT REMOVAL. Environmental conditions encountered in facilities with integrated processes for nutrient removal are essentially the same as those in suspended growth biological systems; therefore, materials selection criteria for these two types of processes are similar.

ADD-ON PROCESSES FOR ADVANCED WASTEWATER TREATMENT. Materials selection considerations for filtration, adsorption, and air stripping are discussed below.

Filtration. Filtration, a unit process for polishing wastewater after secondary or advanced (nutrient removal) treatment, produces a clear effluent containing little corrosive or abrasive material. The atmosphere at filtration facilities may be very humid, particularly when filters are covered. Because filter influent or backwash water is sometimes prechlorinated to prevent algal growth, some of this chlorine often escapes from solution into the atmosphere at the filter facility. To resist the high humidities and low concentrations of chlorine, good design practice for filtration units generally provides for the following corrosion-resistant materials:

- Inlet piping to large facilities is typically cement-lined fabricated carbon steel or, for smaller facilities, cement-lined ductile iron.
- Sluice gates are typically cast iron. When slide gates or stoplogs are used, they should be constructed from aluminum or stainless steel.
- Butterfly valves with cast iron bodies and alloy steel or stainless steel seats are generally used for isolation or control valves.
- Underdrain systems are constructed from noncorrosive materials because of the inaccessibility and resultant high replacement cost of the underdrains. Many systems are constructed with vitrified clay underdraining flow distribution blocks.
- Other systems use plates, nozzles, or both constructed from plastics or stainless steel to evenly distribute flow within filters. Porous plates constructed from stainless steel or aluminum oxide may also be used

for flow distribution. When inlet distribution systems are used for flow splitting to multiple cells within a single filter, stainless steel or other corrosion-resistant materials are used.

- Washwater collection troughs are commonly constructed of precast concrete, FRP, or stainless steel. Materials selection should account for the overall cost, which generally depends on the size of the facility. When air pulse or scour systems are used, piping, valves, and appurtenances are constructed of stainless steel because of the inaccessibility of the system. Surface wash systems and components are constructed from brass.

- The structural part of most filter facilities is concrete, but some smaller, package-type systems are constructed from carbon steel or stainless steel. A complete evaluation of the anticipated atmosphere at the filters will provide a sound basis for choosing the most appropriate materials for the system. Coating, where required, must be carefully selected to resist environmental conditions.

Adsorption. The adsorption process uses powdered or granulated activated carbon to remove organic compounds. Activated carbon removes organic material from wastewater by adsorption, filtration, and deposition processes. Adsorption basically results from forces at the surface of a particle; these forces cause adherence of soluble organic materials to the contact surfaces.

Activated carbon is corrosive, especially when mixed with or suspended in water. Materials of construction usually consist of concrete for gravity filter structures or a plastic-lined carbon steel vessel for pressure applications. Piping systems that convey the activated carbon slurry are constructed of rubber- or plastic-lined steel, type 316 stainless steel, or PVC.

Air Stripping. Air stripping facilities are used to remove ammonia–nitrogen from wastewater when the ammonia exists primarily in the un-ionized gaseous form at pH levels of approximately 10.5 to 11.5. Air usually enters countercurrent to wastewater flow. Materials of construction for the air stripping towers (packed columns) may be stainless steel, FRP, plastic-lined steel, or aluminum. Selection of materials for tower design depends on whether an acid spray or wash is needed to remove deposits of calcium carbonate and the actual size (area) necessary to achieve desired removals.

Packing or internals for air stripping towers may be constructed of wood slats, plastic pipe, a PE grid, or a combination of plastic pipe and a PE grid. Spacing of packing or internals is proportional to the degree of ammonia removal required; tighter spacing achieves higher removals.

SLUDGE HANDLING AND CONCENTRATION. Sludge characteristics vary over a wide range, from a water-like fluid to a slurry to a semisolid cake. Sludge contains materials such as gravel, sand, rags, grease, oil, and plastics. Sludge can be corrosive and abrasive; areas for treating sludge are often dirty and damp.

Equipment used for sludge handling and concentration is complex and costly. Although exclusive use of corrosion-resistant materials for sludge handling equipment would prolong system service life and might ease maintenance, costs would likely be prohibitive. Therefore, design of facilities for sludge handling and concentration must use a systematic approach for selecting the most cost-effective, operable system. The following material selection considerations apply:

- Corrosion- and abrasion-resistant materials should be chosen selectively to meet service conditions. For example, all wetted parts that directly contact sludge should be composed of corrosion-resistant material such as stainless steel or plastic.
- Parts subject to rapid wear or deterioration should be readily accessible and easily replaced.
- Prolonged contact with sludge may accelerate corrosion. Therefore, provisions for intermittent or continuous cleaning should be included and facilities should be designed for ease of cleanup.
- Good ventilation should be provided to minimize accumulation of corrosive gases such as hydrogen sulfide.
- Critical or vulnerable components should be protected from sludge contact or corrosive environments.

Because of its low cost and many desirable properties, carbon steel is frequently used. However, if carbon steel will contact sludges, it should be sandblasted to near white metal in accordance with SSPC-10 standards, and it should be coated with high-build epoxy suitable for the service.

Sludge Storage. Large, liquid-sludge storage tanks are generally constructed of concrete. Smaller tanks are often constructed of carbon steel with a suitable coating system. Tank equipment often includes an aeration system, mechanical mixers, or a recycle system for mixing. All equipment within the tank should be constructed of a corrosion-resistant material such as PVC, PE, or stainless steel. Sludge cake is typically stored in hoppers constructed of carbon steel with a suitable coating.

Sludge Processing. Sludge-processing systems improve sludge quality before concentration. They include grinders to reduce particle size, grit separation devices, or screening equipment to remove large particles.

SLUDGE GRINDERS. Sludge grinders cut or shear large sludge particles to smaller, treatable particles, reducing plugging of pumps, lines, centrifuges, and other downstream units that may clog. Sludge grinders must resist abrasion—the most severe problem encountered. Grit, gravel, rags, rocks, oversized particles, and even metal tools may inflict extensive abrasive damage to wetted components. Abrasion-resistant, high-impact materials such as heat-treated alloy steel (300 bhn to 500 bhn) should be specified for impellers and cutter bars.

The grinder cutters are extremely critical because they must resist high and potentially damaging impacts of incoming abrasive materials. Therefore, cutters are fabricated of special high-grade, highly abrasion-resistant stainless steel. Usually 304 and 316 stainless steels meet the service requirements of other wetted components. Higher grades, however, are available at higher costs. For longer service lives, all parts exposed to abrasion are hard-faced, generally with tungsten carbide or chrome boride. Seals must often bear high pressures. As excessive wear and deterioration of seals may occur, stellite or tungsten carbide may be used to lengthen seal service.

SLUDGE-DEGRITTING EQUIPMENT (CYCLONE DEGRITTERS).
Cyclone degritters consist of a volute feed chamber, vortex finder, cylindrical and conical sections, apex valve, and accessories. Degritters are normally designed with a replaceable liner to protect from grit abrasion in high-velocity sludges. Vortex finder materials deserve careful consideration because of direct contact with incoming sludge. The major pieces of equipment and their suggested materials are presented in Table 8.3.

Overflow launders collect degritted sludge, and underflow launders collect separated grit and sludge. Both types are normally constructed of welded steel plates equipped with neoprene or rubber protective liners. All bolting and fasteners are type 316 stainless steel. Launders normally include removable, fiberglass-reinforced polyester covers.

Table 8.3   Materials for cyclone degritters.

Item	Material
Distribution box	Steel
Box liner	Neoprene
Cover, cylinder housing, and cover housing	Steel
Liners for above items	Neoprene
Vortex finder	Ni-hard steel
Apex housing	Aluminum
Apex liner	Neoprene

Grit classifiers consist of an inlet box, settling compartment, and screw- or rake-type conveyor. Screws and rakes are equipped with replaceable-wear shoes, and submerged bearings are watertight to prevent grit abrasion. Table 8.4 lists the major equipment items and suggested materials.

Table 8.4   Materials for grit classifiers.

Item	Material
Conveyor	Heavy steel
Wearing shoes	Abrasion-resistant steel
Settling compartment and inlet box	Stainless steel
Inlet box liner	Neoprene or rubber [13 mm (0.5 in.) minimum thickness]
Bolting	Stainless steel, type 316, or cadmium-plated steel

SLUDGE SCREENING. Sludge screens remove coarse material, rags, and other items from waste sludge before sludge treatment. Preferred models of the many types of sludge screens do not use moving parts permanently located below water. Front return-rake bar rack designs have been used successfully. This screen's major components and suggested materials are presented in Table 8.5.

Table 8.5   Materials for sludge screens.

Item	Material
Bar rack and mainframe	Steel
Dead plate and rake	Stainless steel, type 304
Pin rack and log wheels	Hardened steel

Sludge Conditioning. Sludge conditioning refers to physical and chemical methods used to alter sludge properties to enhance dewatering characteristics. The most common methods are heat (thermal) conditioning and chemical addition. Materials of construction for thermal conditioning are discussed later in this section. The most common chemicals used for chemical conditioning are lime, ferric chloride, and polymers. These chemicals and resistant materials for each were previously described.

Sludge Pumping. The transport of sludge by pump and pipeline offers several advantages over conveyor systems. These include improved odor control, minimal space requirements, and no spillage of material. Three types of pumps are commonly used for sludge transport: positive-displacement piston, positive-displacement rotary, and centrifugal. For all sludge-pumping equipment, any external carbon steel material should be protected with an epoxy coating system. All pump hardware should be stainless steel. Materials selection for individual sludge pump types is discussed below.

- Piston pumps: positive-displacement piston pumps are used to transport sludges with high solids concentration. Piping distances over 120 m (400 ft) are possible for dewatered sludges. Cylinders constructed of hard-chrome-plated material with pump pistons constructed of Buna-n are normally used for sludge; poppet valves are normally constructed of hardened steel. The pump frame is generally constructed of structural steel and protected with a coating system.
- Rotary pumps: positive-displacement rotary pumps are used in situations where a uniform, pulsation-free flow of sludge is required. The pump body is commonly constructed of cast iron; however, type 316 stainless steel can be used if severe service warrants the extra costs. Pump internal parts, including rotors, are constructed of alloy steel or stainless steel. The pump stator is usually constructed of ethylene propylene diene monomer (EPDM).
- Centrifugal pumps: these pumps are generally used to transfer primary, return activated, and waste activated sludges. As a result of the high grit content of primary sludge, impellers, the pump body, wear rings, and volutes should consist of a hardened alloy. For return activated and waste activated sludge, either cast iron or ductile iron impellers, pump body, wear rings, and volutes are normally used. For all applications, a type 316 stainless steel shaft and shaft sleeve should be used.

Sludge Conveyance. Sludge conveyors are normally used to transport sludges that are not easily pumped. The conveyor belt materials must resist the corrosive effects of dilute sulfuric acid (formed by the reaction of hydrogen sulfide and moisture), deterioration from oils and grease, and the abrasive action of grit. Selection of materials for conveyors is discussed below.

- Belt conveyors: belts are normally made of neoprene or PVC, depending on the application. To minimize belt failures, an "endless" belt is preferred. If this type of belt is impractical, stainless steel fasteners should be used. Idlers are generally constructed of carbon steel, stainless steel, or lubricated thermoplastic material. The supporting framework and drainage system is normally fabricated of carbon steel or hot dipped galvanized steel.
- Screw conveyors: screw conveyors may be constructed of either carbon steel or stainless steel. Depending on the material conveyed, screw conveyors may be designed with either a partial or full cover to prevent odors. As a result, the cover may enclose a highly corrosive space. If so, either hard-facing of the internal parts with tungsten carbon or complete stainless steel is necessary.

Sludge Thickening. Sludge thickeners are normally used to increase solids concentration and decrease water concentration to levels appropriate for the handling requirements of the dewatering equipment. Generally, either gravity or flotation thickeners are used. Both harbor a corrosive environment.

GRAVITY THICKENERS. Gravity-induced settling, one of the more common methods of sludge thickening, typically includes a circular tank equipped with a slow-moving mechanism that helps remove water from the sludge. Septic conditions that often develop in gravity thickeners result in release of corrosive and odorous gases. Also, the large surface area and warm sludge tends to produce a high humidity environment. This combination of conditions makes the enclosed area particularly vulnerable to corrosion. In addition, grit, gravel, and other particles in the sludge abrade equipment components.

Most gravity thickener tanks are constructed of concrete, although small tanks are sometimes constructed of carbon steel. The standard material of construction is carbon steel for most components, such as collector arms, center support columns, center cages, and feed wells. All carbon steel components must be protected with a suitable vinyl or epoxy. Scraper blades are typically equipped with adjustable spring brass or stainless steel squeegees to scrape settled sludge along the tank bottom to the center trench for removal. Center drive housings, gears, shafts, and so on are typically constructed of cast iron or steel designed to prevent entry of moisture and dirt. Above the liquid surface, other items such as walkways, structural support, handrails, grating, and weirs should be constructed of aluminum, stainless steel, or FRP.

FLOTATION THICKENERS. Flotation thickeners may be housed in carbon steel tanks or in cast-in-place concrete tanks. Many of the support systems associated with flotation thickening are enclosed (that is, the sludge is contained within pumps, pipes, and tanks). Within the tanks, components that contact the sludge, such as skimmers and bottom sludge collectors, should be constructed of corrosion-resistant materials such as plastic or stainless steel.

OTHER THICKENING METHODS. Centrifuges and hybrid versions of belt filters, discussed below, have been used for sludge thickening. Although basket-type and disk-nozzle centrifuges have been used for thickening, most of this equipment has been replaced with solid-bowl centrifuges. Materials of construction for centrifuges and belt filters are discussed below.

Sludge Dewatering. Although thickeners reduce the water content of sludge by some degree, further water reduction is generally necessary to reduce weight and improve handling characteristics. The abrasive and wet nature of sludge, combined with the great physical forces exerted by the dewatering device to drive the water from the sludge, creates a severe environment that requires strong, corrosion-resistant materials.

BELT FILTER PRESSES. This type of equipment is complex, and its materials vary widely among manufacturers. As good practice, all wetted parts are constructed of corrosion-resistant material, and items requiring maintenance or adjustment are easily accessible. Stainless steel is preferred for areas subject to corrosion, abrasion, or strong physical forces.

Filter belts are composed of corrosion-resistant materials suitable for the sludge being dewatered, and they are designed to withstand a tensile force of at least three times the maximum expected tension of the belt. For most municipal wastewater sludges, monofilament polyester should be specified. Structural frame members should be galvanized or epoxy-coated carbon steel. Perforated rollers should be constructed of type 304 or 316 stainless steel. Nonperforated rollers should consist of carbon steel with a rubber, nylon, or FRP coating designed for abrasion and corrosion resistance.

CENTRIFUGES. Centrifuges commonly have wetted parts of alloy steel; nevertheless, manufacturers may use stainless steel for especially corrosive applications. High rotational speeds require the use of special materials to minimize wear from abrasive materials in the sludge. Tungsten carbide hard surfacing or another suitable liner is required on the feed compartment walls, feed ports, leading surfaces and tips of the scroll conveyor flights, solids discharge ports, and solids discharge compartments.

VACUUM FILTERS. The standard vacuum filter materials for most sludge applications are type 304 stainless steel for the drum and painted mild steel for the drumheads and vat. Stainless steel heads and vats are also available. Internal filtrate pipes are type 304 stainless steel. The choice among the several types of available drum deck material available depends on the type of filter material selected.

SLUDGE STABILIZATION. All carbon steel items used in digester tanks require protection with epoxy coatings. Stainless steel, usually type 304 or type 316, may be used instead of carbon steel for many items if costs are not excessive. Type 316 is superior to type 304 because of better resistance to oxidizing and reducing agents. For welded construction, type 304L or type 316L stainless steels are best because of their low carbon content.

Aerobic Digesters. The oxygen-rich liquids inside aerobic digestion tanks may rapidly corrode bare carbon steel. Generally, the following materials are used successfully in aerobic digesters:

- Tanks: both steel and concrete tanks are used for smaller installations, while concrete tanks are usually used for larger installations.
- Mechanical aeration: the shafts and propellers of mechanical aerators consist of stainless steel or carbon steel.
- Diffused aeration: for diffused aeration systems (usually the coarse-bubble type), air piping may be stainless steel, carbon steel, ductile iron, or PVC. Diffusers are either stainless steel or plastic.
- Lift-out header systems: both the drop pipes and headers should be constructed of metal. In selecting the metal for piping, the weight of the assembly should be considered; stainless steel weighs less than carbon steel or ductile iron.
- Swing-type headers and hanger pipes: these should be of stainless steel.
- Floor-mounted fixed header systems: drop pipes are constructed of metal because of high air temperatures. For plastic diffusers, fixed header pipes are PVC or metal; for stainless steel diffusers, the fixed headers are metal.

Anaerobic Digesters. In anaerobic digestion, acid-forming bacteria convert organic materials to volatile organic acids; therefore, all materials below the liquid surface must resist organic acids. These acids, in turn, are changed into methane and carbon dioxide by methane-forming bacteria. If the collection system retains raw wastewater for an extended period, the raw sludge fed into the anaerobic digesters will likely generate hydrogen sulfide gas, which is very corrosive in the presence of moisture. The ambient temperature variation can affect the design of fixed steel and concrete domes as well as selection of materials for the domes. In some parts of the U.S., the temperature in the domes may vary during a summer day from 21°C (70°F) at night to 60°C (140°F) during the afternoon.

DIGESTER TANKS. Digester tanks may be steel or concrete with either floating or fixed covers. Floating covers are normally of steel construction with steel roofs. Infrequently, wood construction with composition roofing or concrete with composition roofing is used. Steel fixed covers should be used for steel tanks; for concrete tanks the covers may be either steel or concrete. If the tanks are likely to contain a significant concentration of hydrogen sulfide gas, the concrete domes and a portion of the internal sidewalls should be protected with a PVC liner or suitable coating system. Flexible membrane covers of thermoplastic elastomer are also available.

MIXING SYSTEMS. External recirculation pumping should consist of recirculation pumps, piping, and nozzles. Pumps are generally built of standard cast iron with discharge nozzles of abrasion-resistant steel. Suction columns are steel or stainless steel; piping is typically ductile iron with glass or cement lining.

For confined draft-tube mechanical mixing, external draft tubes, because of their large size, are normally constructed of carbon steel. The internal draft tube is also of carbon steel, except for the top, exposed portion, which is of either steel or stainless steel. Mixer propellers are of stainless steel, bronze, or cast iron. For unconfined mechanical mixing, the turbine propeller and shaft normally are stainless steel.

GAS MIXING SYSTEMS. In an unconfined gas mixing system, gas diffusers are generally of cast iron, and piping is either carbon steel or stainless steel. Because of the small size of the diffuser piping, stainless steel pipes are best. Compressors are usually of cast iron construction.

In a gas injection mixing system, the lances are carbon steel or stainless steel pipe. The gas compressors are similar to those in an unconfined gas mixing system.

A confined gas lifting system will include draft tubes, gas piping, and gas compressors. Materials for piping and compressors are the same as those for unconfined gas mixing systems. Draft tubes are normally carbon steel.

Gas Collection, Storage, and Distribution. Steel pipe with flanged and welded joints usually collects and conveys gas from digesters, but stainless steel pipe may also be used. Cast and ductile iron pipe are sometimes used for low-pressure lines. Gas piping accessories, including valves, sediment traps, drip traps, and flame traps, are usually constructed of standard materials. The flare may be carbon steel, stainless steel, or a combination of the two metals. Carbon steel is used for floating gas holder covers or spheres.

THERMAL PROCESSING OF SLUDGE. This section discusses materials for thermal processing of sludge, including thermal conditioning and wet air oxidation, thermal drying, and thermal destruction.

Thermal Conditioning and Wet Air Oxidation. These processes have many similarities. They both use similar types of equipment and equipment configurations and subject sludge to simultaneous application of heat and pressure. Thermal conditioning is used to improve the dewaterability of sludge without the addition of chemicals, while the goal of wet air oxidation is to reduce sludge volumes through oxidation of volatile solids.

The two basic types of thermal-conditioning systems are heat treatment (HT) and low-pressure oxidation (LPO). These systems have similar configurations, except that LPO includes the addition of air to the sludge. Wet air oxidation (WAO) is similar to LPO, but WAO requires more air and operates at considerably higher temperatures and pressures. When selecting materials for corrosion resistance, criteria for LPO and WAO systems are similar.

Both abrasion and corrosion must be considered. Grit removal must be maximized before thermal conditioning or WAO to reduce abrasion and plugging of heat exchanger tubes, pumps, and valves. Also, materials selection for piping and equipment must avoid the use of dissimilar metals that provide the potential for galvanic corrosion. In terms of corrosion potential, processed sludge is slightly more corrosive than influent or raw sludge. For example, a reduction in pH can be observed between the influent and effluent sludge. Also, the hot side of the process (closest to the reactor) is the most corrosive.

Major equipment components include sludge grinders, positive-displacement pumps, heat exchangers, and reactors. A positive-displacement air compressor is also required for systems using air. Sludge grinders, pumps, and compressors are discussed elsewhere in this chapter.

Heat exchanger tubes, the reactor, and piping for a heat treatment system are usually constructed of carbon steel. However, when an acid-washing system is used to remove and prevent scale formation, materials similar to those used for LPO and WAO systems deserve consideration. Because both LPO and WAO add air to sludge, carbon steel if used in these systems would be subject to severe corrosion. Materials selection must recognize, as a significant concern, the potential of stress corrosion or stress cracking from a combination of high temperature and chlorides. For lower chloride concentrations, stainless steel is the most common material used for the heat exchanger tubes, reactors, and piping. Although either type 304 or 316 stainless steel is used, type 316 is more resistant to stress corrosion. Low-carbon stainless steel provides even more resistance to stress corrosion. For high chloride concentrations, use of stainless steel should be avoided; materials such as titanium, Inconel 625, and Hastelloy C are required. Heat exchanger tubes and piping are generally thin walled and constructed of a single material. The reactor can also be constructed of a single material, or, alternatively, constructed of carbon steel with an appropriate liner.

Thermal Drying. Sludge drying involves reducing a sludge's water content by vaporization of water into air. Mechanical processes used to dry sludge include flash dryers, spray dryers, rotary dryers, multiple-hearth dryers, fluid bed dryers, and the Carver–Greenfield process. The thermal drying process typically follows dewatering and produces a dried sludge with a low moisture content (typically 60 to 70% solids). This process, which often precedes ultimate disposal of sludge as processed fertilizer or in a landfill, is used less often than thermal destruction of sludge. These two processes are not used in combination because it is not cost effective to thermally dry and then thermally destroy most sludges. Each type of thermal dryer described above has its own unique environment; accordingly, materials compatible with these environments should be selected.

FLASH DRYERS. Carbon steel is typically used for the drying column, cage mill, cyclone separator, ducts, and fans. In certain cases where specific corrosives are anticipated, stainless steel or ANSI 600 series iron-based superalloys may be used.

Erosion control and strength must be considered in accordance with the manufacturer's practice. At locations such as elbows or the cyclone, where changes in direction or high velocities occur, special abrasion-resistant ceramic liners deserve consideration. Dryer components and commonly used materials are listed below:

- Preheat chamber: carbon steel plate is commonly used for the outer shell construction of the preheat air chamber, with a refractory lining material in the flame area.
- Mixers, paddle breakers, and fans: these components represent auxiliaries to the system. Wear-resistant surfaces or replaceable tips should be considered for abrasive materials. The blade tips of mixers or paddle breakers may be hardened, and rotors may be coated with FRP.

ROTARY DRYERS. Typically, rotary chambers, including the rotating drum and flights, are constructed of carbon steel. Alloys should be selected when high temperatures or corrosive environments are anticipated. A carbon steel plate outer shell is commonly used for preheat air chambers, with a refractory lining material in the flame area.

FLUID BED DRYERS. The fluid bed dryer resembles the fluid bed furnace, but lining materials differ because the typical operating temperatures of a dryer are much lower than those of a furnace. Although refractory and insulating materials for dryers may be selected to withstand temperatures that are lower than those in thermal destruction furnaces, these dryer materials must nonetheless withstand the intrinsically abrasive environment.

INDIRECT DRYERS. Rotating components of paddle, hollow-flight, or disc dryers are typically carbon steel; nevertheless, the use of alloy steels is required if corrosive conditions are anticipated, as in some industrial applications. Rotors may be surface hardened for handling abrasive materials, but the extra cost may not be warranted. Instead, the entire assembly may be considered as a wear item and replaced when necessary.

The shell and stator area of these dryers may be of carbon steel if the heating medium is circulated for drying. In cases where such drying is not practiced, the material should be an alloy steel. The assistance of the manufacturer should be sought for selection of the optimum materials for both the rotor and the stator assemblies.

Thermal Destruction. This section describes selection of materials commonly used for multiple-hearth furnaces, fluid bed furnaces, breeching devices, heat recuperator equipment, and air pollution control devices.

MULTIPLE-HEARTH FURNACE. Designers must take special care in selecting refractory and insulating materials systems, using appropriate temperature ratings and reference to ASTM standards (where applicable). A properly designed furnace refractory and insulation system is essential to reduce failures of insulation and the outer shell. Designers are urged to consult furnace manufacturers for their recommendations before finalizing insulation design. Requirements for municipal wastewater sludge incinerators may differ from those for industrial or hazardous waste sludge incinerators. Furnace components and their typical materials are listed below:

- Furnace shell: carbon steel plate is commonly used for outer shell construction of the furnace; steel structural members are used for the top frame. Shell lining material is usually a combination of high duty firebrick and insulating block, with mineral fiber fill adjacent to the steel shell. As an alternative, castable insulation with insulating block may be used.
- Furnace hearths: hearths are generally constructed of high-duty or super-duty firebrick.
- Furnace roof: the furnace roof may be constructed of poured castable or arched firebrick. The roof topping is usually of block insulation finished with hard insulation topping.

- Furnace bottom: the furnace bottom is commonly made of carbon steel plate supported on steel structural members, and the lining material is usually insulating block or brick.
- Center shaft: the air-cooled center shaft is usually made of cast iron insulated with lightweight, reinforced castable refractory insulation. The bottom air housing may be constructed of either cast iron or fabricated steel. Internal seals should be of heat-resistant, chromium–nickel alloy material.
- Rabble arms and teeth: chromium–nickel alloy steel is the usual choice for air-cooled rabble arms. Although the top and bottom hearths are not exposed to the intense heat used in burning hearths, all furnace rabble arms are usually constructed of the same material. Designers should carefully choose the alloy to suit operating conditions and temperature peaks occurring within the furnace. For example, 25% chrome and 12% nickel, ASTM A447, type II alloy has been used for rabble arms in a number of sludge incinerators. Rabble teeth, like rabble arms, must consist of heat-resistant chrome–nickel steel.

FLUID BED FURNACE. As with the multiple-hearth furnace, materials must be carefully selected with regard to anticipated temperatures and environmental conditions such as highly corrosive sulfur gases and other products of high-heat processes. Abrasion is much more severe in fluid bed furnaces than in multiple-hearth furnaces because of the turbulent superheated sand in the fluid bed. As another significant difference between the two furnace types, no moving mechanical parts exist in a fluid bed furnace. Furnace components and their materials are listed below:

- Outer shell: carbon steel plate with all-welded construction is typically used in the roof, freeboard, bed, and windbox.
- Refractory and insulation: the roof includes high-duty or super-duty firebrick using insprung arch construction. Lightweight castable insulation is also used. The freeboard and bed section contains high-duty or super-duty firebrick with double-wall construction and, typically, block insulation. The windbox contains high-duty or super-duty firebrick with high-strength castable insulation. The roof of the windbox is typically sprung-arch type using super-duty firebrick. Fluidizing nozzles penetrate this roof or dome, and they are mounted using high-temperature cement.

BREECHING. Breeching, used to convey gases for furnaces, is exposed to temperature extremes. Thus, in materials selection, designers must consider longitudinal and lateral thermal expansion and contraction. Components and their materials are briefly described below:

- Outer shell: carbon steel plate of all-welded construction is typical.
- Refractory and insulation: inner lining is high-strength castable, with lightweight castable insulation.
- Expansion joint: a breeching expansion joint, typically of the multi-convolution bellows-type, is usually composed of stainless steel. Highly corrosion-resistant, chromium–nickel–molybdenum–carbon alloys prevent failures under extreme conditions.

HEAT RECUPERATOR. These flue-gas-to-air heat exchangers are used to preheat combustion air in all recent installations. The shell is of welded carbon steel plate. Heat exchanger tubes are traditionally made of type 304 stainless steel, but the use of better heat- and corrosion-resistant alloys is increasing. Components and their materials are listed below:

- Expansion joints: joints for both tubes and tube sheets are typically type 321 stainless steel or another alloy with increased heat and corrosion resistance.
- Tube sheets: these are typically type 304, 20cb-3 stainless steel or hot tube sheet.

AIR POLLUTION CONTROL DEVICES. Selection of materials for Venturi scrubbers and electrostatic precipitators is described below:

- Venturi scrubber impingement tray cooler: the shell may consist of type 304 stainless steel in cases where chloride concentrations are limited; otherwise, type 316 is preferred. In some cases, FRP has been used to reduce cost. The internal trays and de-mister are of type 316L stainless steel, type 321, or another alloy with increased heat and corrosion resistance if excessive corrosion is anticipated. Fiberglass internals are available.
- Electrostatic precipitator: materials requirements for this component are greatly influenced by stringent regulatory requirements that are now being promulgated for air pollutants. Manufacturers should be consulted to determine optimum materials.

*M*ATERIALS GLOSSARY

Various materials typically used in the equipment and structures of a wastewater treatment facility are described below. This generic information serves only as a general reference. Manufacturers and other sources should be consulted for use in specific applications.

MASONRY. Brick Masonry. This material is no longer used in applications where it may come in contact with wastewater, except for acid-resistant brick used in special applications. Brick masonry is used as facing for aesthetic or architectural purposes.

Concrete Unit Masonry. This masonry is more vulnerable to moisture penetration than ordinary concrete. Thus, should not be used in high-moisture environments where freezing may occur. If it is used, appropriate coatings should be employed to minimize moisture penetration.

METALS. Aluminum. Aluminum is a high-ductility and relatively low-strength metal that nonetheless has a high strength:weight ratio. Typically, aluminum alloys (aluminum combined with other metals of higher strength and hardness) are used. Aluminum materials resist corrosion because a protective aluminum oxide film forms quickly. However, aluminum should not directly contact concrete because the alkali material in the concrete may attack the aluminum.[6]

Aluminum Oxide (Al_2O_3). A common abrasive compound used to impart antislip properties to nosings, sills, and floor plates; to act as a grinding compound; and to serve as the principal compound in anodized aluminum coatings.

Brass. A copper–zinc alloy often used as facing material.

Bronze. A copper–tin alloy, generally stronger and harder than brass, that resists corrosion that would attack the two lone elements. It is frequently used for small fittings and nozzles for long service with a mildly corrosive wastewater, steam, or fresh water.

Cadmium. Often used with zinc to provide a corrosion-resistant coating for ferrous metal items and manufactured items, cadmium is highly resistant to marine conditions and alkalis.

Carbon Steel. A hot-rolled, all-purpose material used in structural and miscellaneous steel members, carbon steel is the most common, economical, and versatile metal used in industry. Because unprotected steel will rust, it is typically coated. Carbon steel may become embrittled if it remains in contact with alkaline or strong caustic fluids. Direct reaction of hydrogen sulfide gas with unprotected steel may occur. Contact with acid accelerates corrosion.

Cast Iron (Gray). Primarily iron, with carbon as the main alloying element, it offers excellent abrasion resistance, economy, and strength, but it is quite brittle.

Chromium. A hard metal used to plate pump impellers and other moving parts that contact gritty material, chromium also provides a low-friction surface. It is the essential addition to stainless steel (12% or more chromium by weight) that gives stainless steel excellent resistance to corrosion and oxidation.[6] Its primary treatment applications are for abrasion resistance and hard facing.

Copper. Copper is a ductile, malleable metal that does not corrode easily but oxidizes in continuous contact with the atmosphere.[7] It corrodes rapidly when exposed to oxidizing agents such as chlorine, ozone, and hydrogen sulfide. It is used primarily in electrical wiring and potable water piping.

Ductile Iron. Ductile iron has better strength, toughness, and wear resistance than gray cast iron. It is a hard, nonmalleable ferrous metal that must be poured in molds to form patterns or shapes. It is used for piping, pumps, gears, shafts, and other applications requiring strength, shock resistance, and machinability. It has good resistance to corrosion, but it will react directly with hydrogen sulfide to form ferrous sulfide.

Inconel 625. A nickel-based superalloy, Inconel 625 has excellent high-temperature strength and good corrosion resistance.

Hastelloy C. A nickel-based superalloy, it offers good resistance to wet chlorine, hypochlorite bleach, ferric chloride, and nitric acid.

Nickel. Nickel offers excellent resistance to certain corrosive chemicals. It is most often used in very tough, corrosion-resistant alloys with high-temperature strength.

Ni-Hard. A hard, abrasion-resistant cast iron containing varying amounts of nickel, chromium, carbon, manganese, and smaller amounts of other elements, it is used frequently in pump casings and impellers for pumping abrasive liquids.[8]

Lead. A heavy metal that melts at a low temperatures, making it easy to cast and form, lead resists attack from most corrosive chemicals and provides excellent service in vibration- and sound-dampening applications.

Platinum. An expensive, malleable, ductile, virtually inert metal that offers excellent corrosion resistance, platinum is used to convey high-purity chemicals.[9]

Silver. The least expensive of the precious metals, silver is very malleable, ductile, and corrosion resistant. It is particularly suitable for use with alkaline solutions such as caustic soda and potash.

Stainless Steel. Steel, when alloyed with chromium; nickel; and also possibly molybdenum, copper, manganese, silicon, and other elements, provides a broad spectrum of corrosion-resistant materials. Chromium is the element that makes stainless steel "stainless." Of the hundreds of stainless steel types, the more commonly used types for wastewater treatment applications are type 304 and 316. Stainless steel may corrode and crack under stress when exposed to chlorides. With molybdenum addition, type 316 can withstand the corrosive attack of many chemicals, including sodium and calcium brines, phosphoric acid, sulfite liquors, and sulfurous acids.

Stellite. A nonferrous alloy of varying amounts of chromium and cobalt and small amounts of molybdenum or tungsten, it is extremely hard and is used for severe abrasive service.

Tantalum. A refractory metal characterized by a high melting point, high-temperature strength, and excellent corrosion resistance, it is subject to accelerated oxidation at temperatures above 195°C (383°F) and requires a protective coating. It is used in chemical process equipment.

Titanium. Titanium is a corrosion-resistant, lightweight metal that has a high strength:weight ratio. In corrosive environments, either pure titanium or a titanium-0.2% palladium alloy is used.[9]

PLASTICS. Chlorinated Polyvinyl Chloride (CPVC). A chlorinated form of PVC that provides increased resistance to heat, its use parallels that for PVC up to a temperature of 104°C (219°F).

Fiberglass-reinforced Plastic (FRP). FRP consists of glass fibers imbedded in a variety of thermosetting resins (chiefly vinyl esters, polyesters, and epoxies). It can be formed into a wide variety of structural shapes, paneling, grating, tanks, pipe, and fasteners. FRP has excellent resistance to a broad spectrum of corrosive materials. However, it is relatively costly, lower in strength than metals, and has a high coefficient of thermal expansion. It may be made fire resistant by adding certain compounds to the resin.

Pultruded Fiberglass. This material is formed by pulling fiberglass rovings and continuous-strand mats or other reinforcing materials through tanks of thermosetting resin and then through a curing and forming die to form a completed composite shape.

Polyethylene (PE). A thermoplastic material with excellent resistance to a broad spectrum of corrosive materials, it is used in pipes, tubing, tanks, sheets, and also as RBC media.

Polypropylene (PP). A thermoplastic material with excellent resistance to chemical attack, it is virtually unaffected by aqueous solutions of inorganic salts, mineral acids and bases, and most organic chemicals.

Polyurethane. This group of polymers is classified as either plastic or elastomer materials, depending on the form. In thermoset form, it offers maximum wear resistance. The thermoplastic form offers good toughness and durability. In general, polyurethane offers excellent abrasion and cut resistance and good chemical resistance.

Polytetrafluoroethylene (PTFE). A fluoroplastic characterized by excellent chemical and electrical resistance, it is used for slide bearings, gaskets, and liners for pipes and tanks.

Polyvinyl Chloride (PVC). A thermoplastic with excellent resistance to a broad spectrum of corrosive materials, it is available in a wide variety of pipe sizes and classes and in sheet and molded form as biological filter media.

Polyvinylidene Fluoride (PVDF). The toughest of the thermoplastics, it offers good resistance to chemicals such as halogens, acids, bases, and strong oxidizing agents. It has excellent abrasion resistance, and it can withstand a wide temperature range.

Carbon Black. An additive to thermoplastics, carbon black inhibits degradation by ultraviolet light.

RUBBER AND ELASTOMERS. Buna-N (Nitrile Butadiene). Buna-N offers excellent resistance to petroleum oils, aromatic hydrocarbons, and gasoline. It has good abrasion resistance and better resistance to heat aging than natural rubber.[9]

Hypalon (Chlorosulphonated Polyethylene). It has a service range of 4 to 110°C (39 to 230°F). As its principal advantage, it can handle certain strong oxidizing chemicals and certain mineral acids. It is excellent for use with ozone.[8]

Ethylene Propylene Diene Monomer (EPDM). It offers broad resistance to chemical attack. Its resistance to oil oxidation and ozone exposure is better than natural rubber.[9]

Construction Materials Selection   385

Neoprene (Chloroprene). One of the most rubberlike of the synthetic rubbers, it has excellent resistance to oils, ozone, oxidation, and flame. It does not have the flexibility of natural rubber at low temperatures, however.[9]

Rubber (Natural). The best of the general purpose rubbers, it is excellent for flexure, cut resistance, abrasion resistance, and general endurance because of its low heat buildup. It offers poor resistance to oil, ozone, and oxidation.[8]

OTHER MATERIALS. Bentonite Geotextile Mat. These mats, which consist of sodium bentonite between geotextile layers, have expected permeabilities in the range of 10^{-9} cm/s.

Enhanced Soil. This soil is created by tilling natural soils and mixing them with sodium bentonite or, less commonly, cement. Permeabilities of enhanced soils are typically less than 10^{-7} cm/s.

Firebrick. A refractory material usually made of a blended mixture of flint clay and plastic clay, mixed with water, it is formed to shape, dried, pressed, and burned in a kiln at a temperature between 1200 and 1480°C (2200 and 2700°F).

Geosynthetics. These are synthetic materials, including geomembranes, geonets, geotextiles, and others, manufactured for geotechnical applications such as erosion control, channel stabilization, membrane liners for lagoons, soil filters, landfill leachate control, and roadway improvement.

Glass. A silicate substance that resists most alkaline and acid solutions and abrasion, it has been used for lining piping to decrease friction, prevent corrosion, and decrease effort required to remove oil and grease. Glass lining is susceptible to mechanical damage.

Vitrified Clay. A fired clay product used for pipe and filter blocks, its inert, impervious surface resists corrosion. However, this material is brittle.

Redwood. A naturally rot-resistant wood formerly used for clarifier flights and trickling filter media, its relatively high cost now restricts its application.

Treated Lumber. Wood treated with chemicals that preserve it from rot and insect attack, it is used for supports, scraper flights, baffles, and other uses.

REFERENCES

1. "Sulfide in Wastewater Collection and Treatment Systems." Manual of Practice No. 69, Am. Soc. Civ. Eng., New York, N.Y. (1989).
2. "Sulfide in Sanitary Sewerage Systems." EPA 625/1-74-005, U.S. EPA, Cincinnati, Ohio (1974).
3. "Admixtures for Concrete." Pub. No. 212.1R, Am. Concrete Inst., Detroit, Mich. (1986).
4. Redner, J.A., "Evaluation of Protective Coatings for Concrete." Paper presented at 59th Annu. Water Pollut. Control Fed. Conf., Los Angeles, Calif. (1986).
5. "Summary of Design Information, Rotating Biological Contactors." EPA 430/9-84-008, U.S. EPA, Cincinnati, Ohio (1984).
6. Lipsett, C.H., "Metals Reference and Encyclopedia." Atlas Pub. Company, Inc., New York, N.Y. (1968).
7. "Metal Product Outline—Division 5 Metals." Natl. Assoc. Architectural Metal Manufacturers, Oak Park, Ill., 15 (1986).
8. "Perry's Chemical Engineers' Handbook." 6th Ed., McGraw-Hill, Inc., New York, N.Y., 23 (1986).
9. "Machine Design." Ref. Issue Series, Penton Publishing, Inc., Cleveland, Ohio, 223 (1988).

Chapter 9
Preliminary Treatment

INTRODUCTION

Preliminary treatment prepares wastewater influent for further treatment by reducing or removing problem wastewater characteristics that could otherwise impede operation or unduly increase maintenance of downstream processes and equipment. Typical problem characteristics include large solids and rags; abrasive grit; odors; and, in some cases, unacceptably high peak hydraulic or organic loadings. This chapter presents descriptions of and design considerations for the preliminary treatment processes—that is, screening, grit removal, septage handling, odor control methods, and flow equalization.

SCREENING

The screening of wastewater, one of the oldest treatment processes, removes gross pollutants from the waste stream to protect downstream operations and equipment from damage. Screening that removes oversize solids is distinguished from comminution and grinding, which reduce the sizes of solids. Comminution and grinding of solids, discussed in this section, are used for preliminary treatment where handling of screenings would be impractical.

Some modern wastewater treatment facilities employ both coarse and fine screens. Coarse screens, with 6-mm (0.25-in.) and larger openings, remove large solids, rags, and debris from the wastewater. Fine screens, with 1.5- to 6-mm (0.059- to 0.25-in.) openings, can be used to remove material that may significantly increase operation and maintenance in downstream liquid and sludge processes, particularly in systems without primary treatment. Very fine screens, with openings of 0.2 to 1.5 mm (0.008 to 0.059 in.), following coarse or fine screens can reduce suspended solids to near the primary treatment level. These very fine screens plus microscreens, with openings of 1 μm to 0.3 mm (3.9×10^{-5} to 1.2×10^{-2} in.), can be used for "effluent polishing" to upgrade secondary effluent to tertiary standards. This chapter covers coarse screens and comminution or grinding; fine, very fine, and microscreening are discussed in Chapter 10, "Primary Treatment."

Coarse screens, normally employed as the first unit treatment process, remove solids and trash that could otherwise damage or interfere with the downstream operations of treatment plant equipment such as pumps, valves, mechanical aerators, and biological filters. Included in the coarse screen category are mechanically and manually cleaned bar screens, including trash racks. Bar screens consist of vertical or inclined steel bars spaced at equal intervals across a channel through which wastewater flows. Criteria used in their design include bar size, spacing, and angle from the vertical, as well as channel width and wastewater approach velocity. The most commonly used coarse screens are mechanically cleaned bar screens. This section also includes comminutors and grinders because of their purpose and position ahead of primary treatment. Table 9.1 summarizes typical size ranges and design criteria for coarse screening equipment.

BAR SCREENS. The various types of bar screens are described below.

Trash Racks. Trash racks are bar screens with large openings, 38 to 150 mm (1.5 to 6 in.), that are designed to prevent logs, timbers, stumps, and other large, heavy debris from entering the treatment processes. Trash racks, normally followed by bar screens with smaller openings, are principally used in combined systems that carry high amounts of large debris, especially during storms. Trash racks may be mechanically or manually cleaned. Where space

Table 9.1   Typical design criteria for coarse screening equipment.

Item	Range[a]	Comment
Trash rack		
Openings	38–150 mm	Commonly used on combined systems—opening size depends on equipment being protected.
Manual screen		
Openings	25–50 mm	Used in small plants or in bypass channels.
Approach velocity	0.3–0.6 m/s	
Mechanically cleaned bar screen		
Openings	6–38 mm	18-mm (0.75-in.) opening considered satisfactory for protection of downstream equipment.
Approach velocity (max.)	0.6–1.2 m/s	
Minimum velocity	0.3–0.6 m/s	Necessary to prevent grit accumulation.
Continuous screen		
Openings	6–38 mm	This type of screen effective in the 6- to 18-mm range.
Approach velocity (max.)	0.6–1.2 m/s	
Minimum velocity	0.3–0.6 m/s	
Allowable head loss	0.15–0.6 m	
Comminutor (size reduction only)		
Openings	6–13 mm	Opening a function of the hydraulic capacity of unit.
Grinder (size reduction only)		
Openings	6–13 mm	
Typical head loss	300–450 mm	In open channel.

[a] mm × 0.039 4 = in.; m/s × 3.281 = ft/sec; m × 3.281 = ft.

is limited, plants sometimes have basket-type trash racks that are manually hoisted and cleaned.

Manually Cleaned Screens. Manually cleaned bar screens typically have 25- to 50-mm (1- to 2-in.) openings with the bars set 30 to 45 deg from the vertical to facilitate cleaning. The screenings are manually raked from the screen onto a perforated plate where they drain before removal for disposal. If the screens are cleaned infrequently, the backwater caused by the buildup of a solids mat between cleanings, when released by cleaning, may cause flow surges. These high-velocity surges can reduce the solids-capture efficiency of downstream units.

Although manually cleaned screens require little or no equipment maintenance, they demand frequent raking to avoid clogging. Manually cleaned screens exist primarily in older, small (<1 mgd) treatment facilities and in bypasses of comminutors and mechanically cleaned screens.

Mechanically Cleaned Screens. Mechanically cleaned screens have openings typically ranging from 6 to 38 mm (0.25 to 1.5 in.), with bars set from 0 to 30 deg from the vertical. Mechanical cleaning, compared with manual cleaning, tends to reduce labor cost; improve flow conditions and screening capture; reduce nuisances; and, in combined systems, better handle large quantities of stormwater debris and screenings. A mechanically cleaned screen is, therefore, almost always specified for new plants of all sizes. Many types of mechanically cleaned bar screens are manufactured, including, but not limited to, chain or cable driven with front or back cleaning, reciprocating rake, catenary, and continuous.

CHAIN- OR CABLE-DRIVEN SCREENS. Mechanical bar screens with a chain or cable mechanism to move the rake teeth through the screen openings are the oldest mechanized screening devices. These types of screens are manufactured in several configurations: front clean/front return, front clean/rear return, and back (or through) clean/rear return. The front clean/front return type most efficiently retains captured screenings by minimizing carryover. Such screens are used extensively in ordinary municipal wastewater applications.[1,2] Front clean/rear return screens are used for heavy-duty applications.[1] A disadvantage of a rear return configuration is the possible return to the screened wastestream of debris that is not dislodged by the rake-cleaning mechanism. For any of the front-cleaned configurations, the bars should be trapezoidal in cross section, thus providing a tapered opening to prevent wedging and trapping of solids between the bars. Disadvantages of front-cleaned screens include the possibility of bottom jamming by unusual deposits of trash, particularly if no trash racks are employed. Also, the chain- or cable-driven raking mechanism consisting of submerged sprockets or other mechanical devices is subject to fouling by grit and rags. Frequent inspection and maintenance of the drive mechanism are typically required and, depending on the screen design, channel dewatering may be necessary.

With the back-cleaned, or through-cleaned screening configuration, the rake and operating mechanisms are protected by the bars from damage due to large objects in the wastewater stream. The back-cleaning action pushes solids away from the bars, thus preventing solids from being compressed and jammed between the bars. However, the longer rake teeth protruding through the bars are more susceptible to bending and breakage than are the teeth of front-cleaned screens.

RECIPROCATING RAKE SCREENS. The reciprocating rake screen can be equipped with a back clean/back return mechanism (Figure 9.1) or with a front clean/front return mechanism that minimizes solids carry-over (Figure 9.2). The up-and-down reciprocating motion of the rake, similar to that of a person raking a manual bar screen, minimizes the possibility of jamming. These screens remain popular because their lack of submerged moving parts allows easy, trouble-free inspection and maintenance without dewatering the channel.

Headroom requirements for reciprocating rake screens are greater than those for other types of screens. The estimated headroom requirement can be determined by adding the vertical depth of the screen to the screenings discharge height above the floor.[1] Headroom deserves special attention from the design engineer.

Although many drive mechanisms are available—chain and cable, hydraulic, and screw operated—the most successful designs include a

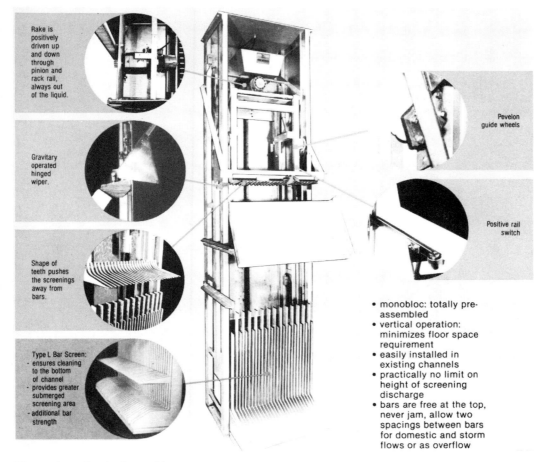

Rake is positively driven up and down through pinion and rack rail, always out of the liquid.

Gravitary operated hinged wiper.

Shape of teeth pushes the screenings away from bars.

Type L Bar Screen:
- ensures cleaning to the bottom of channel
- provides greater submerged screening area
- additional bar strength

Pevelon guide wheels

Positive rail switch

- monobloc: totally pre-assembled
- vertical operation: minimizes floor space requirement
- easily installed in existing channels
- practically no limit on height of screening discharge
- bars are free at the top, never jam, allow two spacings between bars for domestic and storm flows or as overflow

Figure 9.1   Back cleaned bar screen.

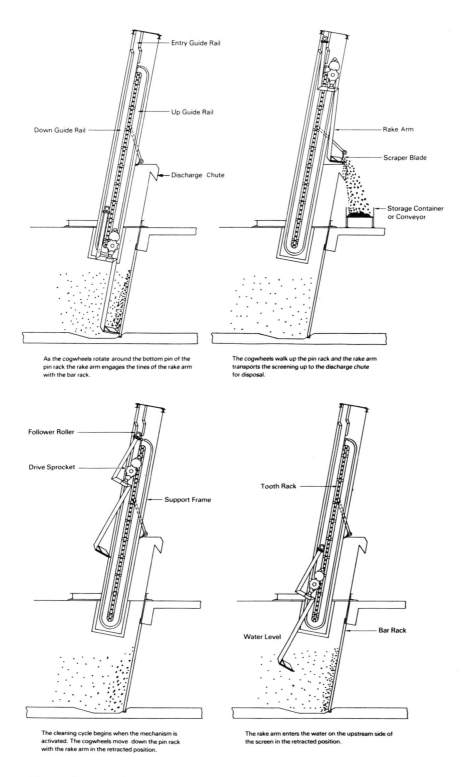

Entry Guide Rail

Up Guide Rail

Down Guide Rail

Discharge Chute

Rake Arm

Scraper Blade

Storage Container
or Conveyor

As the cogwheels rotate around the bottom pin of the
pin rack the rake arm engages the tines of the rake arm
with the bar rack.

The cogwheels walk up the pin rack and the rake arm
transports the screening up to the discharge chute
for disposal.

Follower Roller

Drive Sprocket

Support Frame

Tooth Rack

Water Level

Bar Rack

The cleaning cycle begins when the mechanism is
activated. The cogwheels move down the pin rack
with the rake arm in the retracted position.

The rake arm enters the water on the upstream side of
the screen in the retracted position.

Figure 9.2   Reciprocating rake bar screen.

cogwheel-type drive. For these designs, the entire cleaning rake assembly, including the gear motor, is carriage mounted on cog wheels that travel on a fixed pin or gear rack. The drive mechanism is typically designed to allow the rake to ride over obstructions encountered during the cleaning stroke. In the unlikely event that the rake becomes completely jammed, a limit switch is activated to turn off the drive motor.

A disadvantage of reciprocating screens is the single rake, which limits their capacity to handle extreme loads, especially for deep applications where cycle times are long. Also, these systems require high overhead clearance, which might limit their use in retrofits.

Reciprocating rake screen installations include

- Huntington Beach Plant No. 2, Orange County, CA—five, 2.4 m (8 ft) wide by 4.5 m (14 ft, 8 in.) deep with 25.4-mm (1-in.) openings;
- Metropolitan Water Reclamation District of Greater Chicago, Calumet City WWTP, IL—four, 0.8 m (2.5 ft) wide by 1.5 m (5 ft) deep with 9.5-mm (0.375-in.) openings; and
- Patapsco WWTP, Baltimore, MD—two, 1.7 m (5 ft, 8 in.) wide by 7.3 m (24 ft) deep with 19.1-mm (0.75-in.) openings.

CATENARY SCREENS. The cleaning mechanism of a catenary screen consists of heavy tooth rakes, held against the screen by the weight of its chain (Figure 9.3). The screen's name, "catenary," stems from the catenary loop formed by the operating chain ahead of the screen. A curved transition piece at the base of the screen allows for efficient removal of solids captured at the bottom. Like reciprocating rake screens, all sprockets, shafts, and bearings are located out of the flow stream to reduce wear and corrosion and to ease the required maintenance. Because the cleaning rake is held against the bars primarily by just the weight of the chains, the rake can be pulled over large rags or solids, which may be tightly held against the screen. Examples of successful catenary screens include those at the 19 000-m^3/d (5-mgd) Victor Valley Wastewater Reclamation Authority treatment plant, Victorville, CA; the Salado Creek WWTP, San Antonio, TX; and the Cro WWTP, Lexington, KY.

CONTINUOUS SELF-CLEANING SCREENS. Continuous self-cleaning screens consist of a continuous "belt" of plastic or stainless steel elements that are pulled through the wastewater to provide screening along the entire length of the screen (Figures 9.4 and 9.5). The screen openings are designed with both horizontal and vertical limiting dimensions; the vertical spacing is slightly larger than the horizontal spacing. Continuous screens may have openings as small as 1 mm (0.039 in.) ranging up to more than 76 mm (3 in.). The greater solids-handling capacity of these screens allows smaller openings to be used with resultant greater capture of solids from the waste stream. Also, the continuous screening action of these units allows efficient removal

---

Figure 9.3 Catenary bar screen.

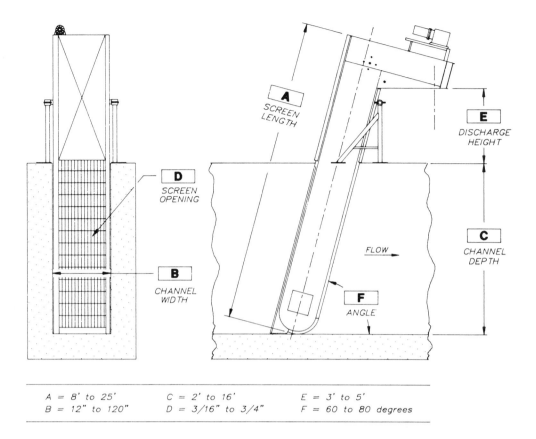

A = 8' to 25'   C = 2' to 16'   E = 3' to 5'
B = 12" to 120"   D = 3/16" to 3/4"   F = 60 to 80 degrees

Figure 9.4   Continuous self-cleaning screen (in. $\times$ 25.4 = mm; ft $\times$ 0.304 8 = m).

of large quantities of solids. Continuous screens have either a lower gear sprocket or a guide rail at the channel bottom to support the screen elements. Construction of a recessed notch or step in the channel at the screen bottom is a good practice that can help prevent buildup of grit and debris ahead of the unit. Continuous screens are designed to be pivoted up and out of the channel for maintenance. Some screen systems include spray bars and brushes to improve cleaning of screen elements. A disadvantage of continuous screens is possible solids carryover resulting from the front clean/back return design.

Examples of successful continuous-screen installations include the 984 000-m^3/d (260-mgd) Point Loma WWTP, San Diego, CA; the 114 000-m^3/d (30-mgd) Davenport, IA, WWTP; and a 91 000-m^3/d (24-mgd) facility located in Manatee County, FL (Southwest region). A continuous screen [13-mm (0.5-in.) opening size], recently installed at a 19 000-m^3/d (5.0-mgd) treatment facility in Burlingame, CA, was found to effectively remove high screening loads from numerous hotels and motels in the service area. The unanticipated increased organic solids removal was found to improve overall

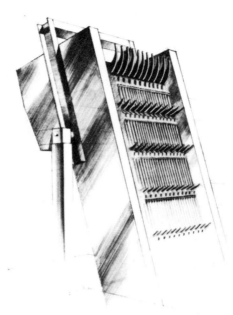

Figure 9.5   Aquaguard® continuous self-cleaning screen.

plant operation, principally because of reduced organic solids loadings on the anaerobic digesters and the activated sludge system.

COMMINUTORS AND GRINDERS. Comminuting devices are installed in the wastewater flow channel to screen and shred material into sizes from 6 to 19 mm (0.25 to 0.75 in.), without removing the particles from the flow. The 1989 WPCF Treatment Plant Data Survey[3] indicates that many small plants [<19 000 m³/d (<5 mgd)] continue to use comminutors. The use of such a device is intended to reduce odors, flies, and unsightliness often associated with screenings that are handled by other means and to essentially eliminate the steps of screenings removal and handling. Unfortunately, solids from comminutors and screenings grinders have caused downstream problems, including deposits of plastics in digestion tanks and rag accumulations on air diffusers. Pulverized synthetic materials will not decompose in the sludge digestion process and, if allowed to remain in the sludge, they could bar public acceptance of the sludge for reuse as a soil amendment. Comminutors may create "ropes" or "balls" of material (particularly rags) that can clog treatment equipment (for example, mechanical aerators, mixers, pump impellers, sludge pipelines, and heat exchangers). As experience with comminutors in the U.S. and elsewhere has been generally unsatisfactory, they are avoided in new designs and are being removed from many existing plants.

In-line, slow-speed wastewater grinding devices, used for over 10 years, are typically located in piping or in channels to avoid removal and handling of screenings. Such grinders are sometimes used at smaller plants where screening handling is not practical and on primary sludge lines (see Chapter 17). The grinder, shown in Figure 9.6, consists of two sets of counterrotating, intermeshing cutters that trap and shear wastewater solids into a consistent particle size, typically 6 mm (0.25 in.). The cutters are stacked on two steel or stainless steel drive shafts with intermediate spacers. The shafts counterrotate at different speeds to produce a self-cleaning action of the cutters.

Because of their design, the grinders aggressively chop, rather than shred, wastewater solids into particle sizes of 6 to 9 mm (0.25 to 0.38 in.). The chopping action helps reduce, but not eliminate, the formation of rag "balls" and rag "ropes" produced by comminutors. Where wastewater contains large quantities of solids and rags, such as prison wastewater, grinders are sometimes installed downstream of coarse screens to help prevent frequent jamming and excessive wear. The grinders' restricted opening area, compared to

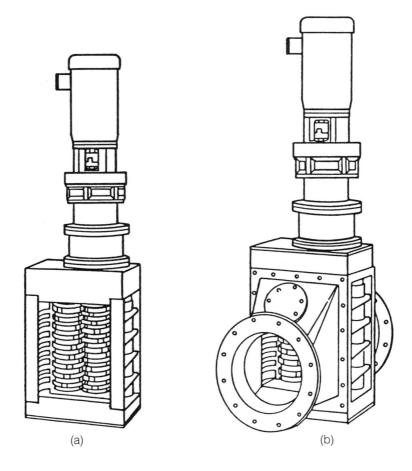

(a)   (b)

Figure 9.6   Wastewater grinders: (a) channel unit and (b) in-line unit.

screens, tends to cause greater head losses, which must be accounted for in the hydraulic design. Grit and other solids impose severe wear on grinders; they therefore require routine inspection every 6 months and replacement of bearings and cutter teeth every 1 to 3 years.

SCREENINGS QUANTITIES AND CHARACTERISTICS. Both the quantities and characteristics of screenings to be collected and removed require careful consideration prior to design of screening systems.

Quantities of Screenings. The quantity of coarse screenings to be removed can vary significantly depending on the bar screen opening, wastewater flow, characteristics of the municipality, type of collection system, and type of screen. These quantities are difficult to estimate without actual operating data, which should be obtained whenever possible. Table 9.2[4,5] shows quantities of screenings collected for each of 39 municipal WWTPs.

For separate sewer systems, the most important variable affecting screenings quantity is usually the clear opening between bars. Published information such as that shown in Figure 9.7 provides average and

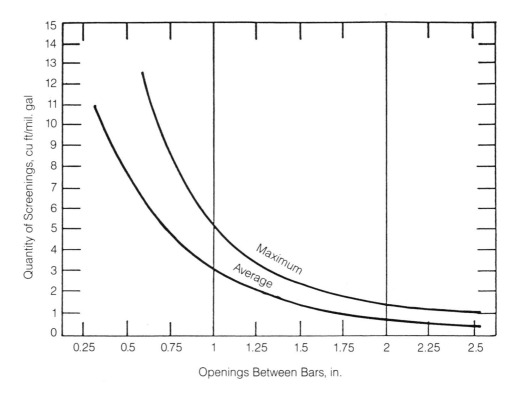

Figure 9.7 **Average and maximum volumes of coarse screenings as a function of openings between bars (cu ft/mil. gal $\times$ 0.007 5 = L/m^3; in. $\times$ 25.4 = mm).**

Table 9.2   Grit and screenings quantities.[4,5]

Plant location	Flow, mgd[a]	Grit, cu ft/mil. gal[b]	Screenings cu ft/mil. gal[b]
Norwalk, CT	11.75	3.3	0.17
Portsmouth, VA	9.7	0.39	0.82
East Hartford, CT	4.0	2.4	1.33
Oklahoma City, OK (Southside)	25.0	1.95	2.1
Taunton, MA	3.5	1.11	1.0
Uniontown, PA	3.0[c]	10.5	0.9
Fargo, ND	2.7[c]	1.0	4.55
Minneapolis–St. Paul, MN	134.0[c]	5.2	0.9
Waterbury, CT	15.0	4.15	2.35
Bridgeport, CT (Eastside)	14.0[c]	1.25	2.04
Duluth, MN	12.0	0.8	0.56
Marshalltown, IA	4.0	3.4	0.25
Richmond, IN	6.2[c]	2.0	3.44
Detroit, MI	450.0[c]	4.0	0.47
E. Bay Mun. Utility, S.D. No. 1 WPCP	128.0	1.26	0.83
Chicago, IL (Northside)	333.0	0.41	0.83
New York, NY, Jamaica WPC	100.0	2.24	0.42
New York, NY, Port Richmond WPC	60.0	0.50	0.17
New York, NY, North River WPC	220.0	1.50	1.0
San Francisco, CA, Southeast	30.0	—	11.7
Boston, MA, MDC, Nut Island	112.0	0.68	0.41
St. Louis, MO, Lemay	167.0	2.69	0.06
Passaic Valley Treatment Plant	225.0	3.82	0.76
Allegheny Co., PA, Alcosan WTP	200.0	3.32	0.38
Fort Worth, TX, Village Ck. WTP	45.0	1.29	0.72
Hampton Roads Sanitary District, Chesapeake-Elizabeth WPCP	12.0	2.17	1.17
Lamberts Point WPCP	20.0	4.85	1.20
County of Milwaukee, South Shore WTP	120.0	0.48	0.60
Twin Cities Metro, WTP	218.0	4.82	1.15
Santa Rosa, CA, College Avenue	—	0.88	—
San Jose, CA	143.0	2.5	—
Manteca, CA	—	5.2	—
Santa Rosa, CA, Laguna	12.0	5.0	—
Seattle, WA, West Point	125.0[c]	2.6	—
Dublin–San Ramon, CA	—	7.0	—
Los Angeles, CA, Hyperion	420.0	2.0	—
Livermore, CA	6.25	1.0	—
Gary, IN	—	8.6	—
Renton, WA	—	4.1	—

[a] mgd × 3785 = m^3/d.
[b] cu ft/mil. gal × 0.007 45 = m^3/ML.
[c] Combined sewer area is large share of total service area.

maximum volumes of screenings per unit of wastewater flow for various bar openings. The design engineer can adjust such basic information according to the type of collection system and community characteristics. Figure 9.7 represents data obtained from 133 installations of manually and mechanically cleaned bar screens from a general cross section of separate collection systems in the U.S.

For bar openings between 25 and 50 mm (1 and 2 in.), the volume of screenings removed per unit of flow is approximately proportional to the clear opening dimension. In this range, for each 13-mm (0.5-in.) reduction of clear opening size, the volume of screenings will approximately double. For screen openings less than 25 mm (1 in.), the volume of screenings removed will increase more rapidly as the clear opening decreases to about 6 mm (0.25 in.). Removals by trash racks have little relation to the volume of flow.

The quantity of screenings removed will partly depend on the length and slope of the collection system and whether pump stations exist. Screenings quantities may be greater with a short, gently sloping collection system with low turbulence than with lengthy, steep interceptor systems or systems with pump stations. This difference stems from disintegration of solids exposed to long-duration turbulence. The impact of turbulence will be more significant for smaller bar openings of less than 12 mm (0.5 in.), which capture more organic solids.

Since combined systems may produce several times the coarse screenings produced by separate systems, the data presented in Figure 9.7 do not apply to the former systems. In addition, the quantity of screenings removed in a combined system during wet periods will greatly exceed that during dry periods. The quantity of screenings has been observed to increase greatly during storm flows. Industrial and commercial wastes, such as those from textile mills and institutions, can affect the volume of screenings removed. In addition, peak daily removals may vary considerably (by as much as a 20:1 ratio on a hourly basis) from average conditions. Table 9.3[3-5] presents typical screening quantity information for both separate and combined sewer systems.

Characteristics of Screenings. Composition, as well as volume, affects the disposal of screenings. Coarse screenings consist of rags, sticks, leaves, food particles, bones, plastics, bottle caps, and rocks. With smaller openings near 6 mm (0.25 in.), cigarette butts, fecal matter, and other organic matter are included in the screenings. The normally large organic portion of screenings readily decomposes; plastics and inorganics are generally not biodegradable. In screenings, some of the material is inherently odorous and other material absorbs odorous compounds from the wastewater. Some of the pathogenic organisms from the wastewater will also likely enter the screenings.

Screenings, normally containing about 10 to 20% dry solids, will have a bulk density ranging from 640 to 1100 kg/m^3 (40 to 70 lb/cu ft). The volatile

Table 9.3   Typical design properties of coarse screenings.[3-5]

Item	Range[a]	Comment
Quantities		
Separate sewer system		
Average	3.5–35 L/1000 m^3	Function of screen opening size and system characteristics
Peaking factor (hourly flows)	1:1–5:1	
Combined sewer system		
Average	3.5–84 L/1000 m^3	
Peaking factor (hourly flows)	2:1–>20:1	
Solids content	10–20%	
Bulk density	640–1100 kg/m^3	
Volatile content of solids	70–95%	
Fuel value	12 600 kJ/kg	

[a] L/1000 m^3 × 0.134 = cu ft/mil. gal; kg/m^3 × 0.062 4 = lb/cu ft; kJ/kg × 0.43 = Btu/lb.

content of the solids ranges from 70 to 95%. Table 9.3 summarizes information pertaining to screenings quantities and composition.

DESIGN PRACTICE. Design considerations for mechanical bar screens include

- Bar spacing, materials of construction, and dimensions;
- Channel depth, width, and approach velocity;
- Discharge height to accommodate screenings conveying equipment;
- Angle of screen;
- Screen cover to block wind and improve aesthetics;
- Materials of construction and coatings for overall unit;
- Drive unit service factor;
- Drive motor size and enclosure (for example, totally enclosed fan cooled);
- Spare parts;
- Provision of redundant screen or bypass manual screen; and
- Head loss through unit.

A bar opening of 19 mm (0.75 in.) or smaller will adequately protect downstream equipment. The smaller the opening, the greater the removal of solids and degree of protection. However, exercise caution when selecting screen openings smaller than 13 mm (0.5 in.) for plants served by gently sloping gravity collection systems due to the potential increased capture of fecal and other organic matter. Excessive capture of such solids should be avoided

---

because they are better handled by downstream primary and secondary treatment processes. In general, selection of smaller screen openings down to 6 mm (0.25 in.) applies best where steeply sloping sewers or pumping stations provide sufficient turbulence in the collection system to fragment organic matter ahead of the screen.

The 1989 WPCF Design Practice Survey[6] included the following design practice preferences for wastewater screening:

- Trash racks are now seldom used ahead of coarse screens or grinders, except for combined systems with significant quantities of stormwater and at prisons with heavy screening quantities. Trash rack bar openings generally range from 38 to 76 mm (1.5 to 3 in.).
- Coarse screen openings now selected range from 6 to 25 mm (0.25 to 1 in.); about half the respondents favor openings of 19 to 25 mm (0.75 to 1 in.). A 19-mm (0.75-in.) bar spacing is sometimes too large because twigs, ball-point pens, and other plastics can pass sideways through the bars. Hence, an increasing number of plants are switching to continuous-type screens with smaller openings in the 6- to 13-mm (0.25- to 0.5-in.) range. Alternatively, the primary sludge from plants with larger screen openings passes through grinders to prevent problems with downstream sludge-handling processes.

Location Considerations. In determining the location of a coarse screening device, the major concern is the protection of equipment. The largest particle allowed to pass may be governed, for example, by the minimum opening in a pump, pipe, fitting, or piece of equipment following the screening device. If the sewer system is very deep at the treatment plant site or if a pumping station exists at some distance from the plant processes, provision of only a trash rack ahead of the station to protect the pumps may be the best approach. In this case, a coarse screen with smaller clear openings could follow the pumping station at a location better suited for maintenance of the screen and handling of screenings. For combined sewer systems, trash racks are normally placed ahead of the pumps to protect them from debris.

Should placement of a screening device in a deep pit or channel be necessary, the overhead approaches to the screen must be broad enough for comfortable access and suitable for the elevation of collected materials. In addition, the screening system design should allow ample space for maintenance work and provide for proper ventilation in accordance with applicable safety standards (see Chapter 6).

When locating a screen, the design engineer needs to consider the effects of the backwater caused by the head loss through the screen. Many installations include an overflow weir to a bypass channel to prevent upstream surcharging if the screen becomes blinded due to power failure or mechanical problems. Screens normally precede flow metering devices to prevent flow backup from affecting the accuracy of the flow measurement.

Contemporary design practice varies as to the location of screening devices at the treatment plant. Present practice calls for coarse bar screens to be located upstream from the grit chamber and the raw wastewater pumps. In the past, screening devices were often located downstream from grit chambers to prevent gritty material from causing premature failure of comminutor teeth and combs. However, operational experience has shown that an upstream location resulting in more frequent replacement of teeth and combs costs less and causes fewer problems than rags in the grit chamber. However, whether the screening device precedes or follows the grit chamber, the screen should be readily accessible because of the nature of the materials handled and the need for frequent inspection and maintenance.

The question of whether screening equipment should be located in an enclosed structure depends on two conditions: the equipment design and the climate. In climates with freezing temperatures, a heated enclosure is necessary. A structure will not only protect the equipment, but also ease maintenance and improve aesthetics. Regardless of whether the screening equipment will be housed, the drive mechanisms of a mechanically cleaned bar screen should be enclosed. In windy areas, the screen rake area and the discharge chute area need covers to prevent screenings from blowing away.

Any structure that contains a screening device, particularly mechanically cleaned bar screens, requires good ventilation to reduce the accumulation of moisture and obnoxious odors. This will provide a safer and better working environment. It will also prolong the life of the equipment by retarding formation of a corrosive atmosphere. In areas of excessive humidity, combining good ventilation with air drying or heating units may be warranted.

Hydraulic Considerations. The hydraulic considerations described herein apply to design of a bar screen installation. For design of comminuting and grinding devices or other coarse screens, follow the manufacturer's data and ratings of these units for channel dimensions, capacity ranges, upstream and downstream submergence, and horsepower requirements.

The velocity distribution in the approach channel has an important influence on bar screen operation. A straight channel ahead of the screen ensures good velocity distribution across the screen and maximum effectiveness of the device. Use of other than a straight approach channel has often resulted in diverting much of the flow along one side of the channel and the resultant accumulation of debris on one side of the screen.

The design engineer must ensure that the wastewater's approach velocity to the screen does not fall below a self-cleaning value or rise enough to dislodge screenings. The lower the velocity, the more material that will be removed from a given waste and the more solids that will be deposited in the channel. Ideally, the velocity in the rack chamber of bar screens should exceed 0.4 m/s (1.3 fps) at minimum flows to avoid grit deposition if grit chambers follow bar screens. However, this is not always possible with the typical diurnal fluctuation in wastewater flows. As a reasonable compromise, the channel could be designed so that a velocity of at least 0.8 m/s (2.5 fps) for resuspending solids is attained during the peak flow periods of the day. Where significant stormwater must be handled, approach velocities of approximately 0.9 m/s (3 fps) are needed to avoid grit deposition at the bottom of a mechanically cleaned screen, which might otherwise become inoperative during storms when most needed.

The velocity of flow ahead of and through a bar screen affects its operation substantially. Satisfactory designs have provided for velocities of 0.6 to 1.2 m/s (2 to 4 fps) through the openings of mechanically cleaned screens and velocities of 0.3 to 0.6 m/s (1 to 2 fps) through the openings of manually cleaned screens. Presently, the lower velocities are preferred for low flows, and a maximum velocity of 0.9 m/s (3 fps) for peak instantaneous flows in common design practice.

Usually accepted practice calls for a minimum head loss allowance through a manually cleaned bar screen of 150 mm (6 in.), assuming frequent screen inspection. The maximum head loss allowance through clogged racks and bar screens is generally limited to 0.8 m (2.5 ft). A nearly constant head loss for continuously cleaned mechanical screens of any type can be maintained with a constant flow. Curves and tables for head loss through the screening device are usually available from the equipment manufacturer.

The head loss for comminutors usually ranges from 50 to 310 mm (2 to 12 in.), depending on the flow and screen opening size; nevertheless, the head loss can reach 1 m (3 ft) or more in large units at maximum flow. The design engineer must take steps to ensure that the head loss through the screening device at maximum flows will not surcharge the influent sewers enough to impair upstream services. To prevent flooding of the screening area caused by blinding of the screen due to a power failure or another problem with the screen, the design should provide for an overflow weir or gate and a parallel bypass channel allowing overflows to go around the screen.

Based on an assumption of steady, two-dimensional flow, the clean water head loss through a bar screen can be calculated by using the Bernoulli equation. This equation states that the depth of flow plus the velocity head before the screen equals the depth of flow plus the velocity head after the screen plus friction losses. The actual head loss necessarily varies with the quantity and nature of the screenings allowed to accumulate between cleanings. The head loss created by a clean screen may be calculated based on the flow and the effective area of the screen openings. The effective area of the screen openings equals the sum of the vertical projections of the screen openings.

The head loss through a clean or partially clogged bar screen can be represented by the following equation:

$$h = \frac{k(V^2 - v^2)}{2g}$$

(1)

Where

h	=	head loss, ft;
V	=	velocity through bar screen, ft/sec;
v	=	velocity upstream of bar screen, ft/sec;
g	=	acceleration due to gravity, 32.2 ft/sec^2; and
k	=	friction coefficient (typically 1.43).

The head loss through only a clean bar screen is indicated by Kirschmer's equality:

$$H_L = \beta \, (w/b) \, 1.33h \sin \varnothing$$

(2)

Where

H_L	=	head loss, ft;
β	=	a bar shape factor (Table 9.4);
w	=	maximum cross-sectional width of bars facing upstream, ft;
b	=	minimum clear spacing of bars, ft;
h	=	upstream velocity head, ft; and
$\varnothing$	=	angle of bar screen with horizontal.

Table 9.4   Kirshmer's values of β.

Bar type	β
Sharp-edged rectangular	2.42
Rectangular with semicircular upstream face	1.83
Circular	1.79
Rectangular with semicircular upstream and downstream faces	1.67

Handling of Screenings. The method of cleaning the screens, manual or mechanical, relates to the method of removing and transporting screenings from the screens to a disposal site. When designing a manually cleaned screen, relatively shallow screening channels [<2 m (<6.6 ft)] are needed, along with a drainage plate to allow drainage of the screenings prior to shoveling (Figure 9.8). The container used to carry the screenings to a truck or other transport may range from a wheelbarrow to a bin carried by an

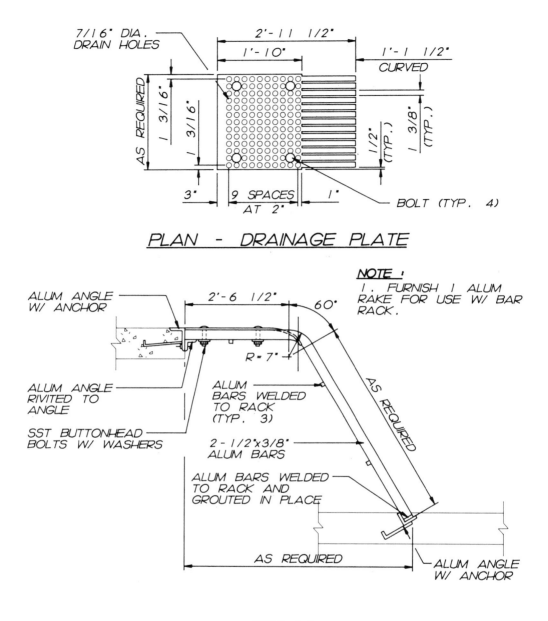

PLAN - DRAINAGE PLATE

7/16" DIA. DRAIN HOLES

2'-11 1/2"
1'-10"
1'-1 1/2"
CURVED

AS REQUIRED

1 3/16"
1 3/16"

1/2" (TYP.)
1 3/8" (TYP.)

3"
9 SPACES AT 2"
1"

BOLT (TYP. 4)

NOTE:
1. FURNISH 1 ALUM RAKE FOR USE W/ BAR RACK.

ALUM ANGLE W/ ANCHOR

2'-6 1/2"
60°

R = 7"

ALUM ANGLE RIVITED TO ANGLE

ALUM BARS WELDED TO RACK (TYP. 3)

SST BUTTONHEAD BOLTS W/ WASHERS

2-1/2"x3/8" ALUM BARS

AS REQUIRED

ALUM BARS WELDED TO RACK AND GROUTED IN PLACE

AS REQUIRED

ALUM ANGLE W/ ANCHOR

SECTION

Figure 9.8   Manual bar screen detail (in. × 25.4 = mm; ft × 0.304 8 = m).

overhead crane or monorail. Whatever the means of conveyance, operator safety, including nonslip platforms and railings, deserves special attention during design.

In mechanically cleaned units, the rakes move the screenings up the screen to above the deck level where they discharge to a conveyor, pneumatic ejector, compactor, or removable containers. The drainage plates for a manually cleaned screen are unnecessary because the screenings drain as they are lifted from the wastewater. Allow sufficient clearance [typically 1.3 to 1.5 m (4.3 to 4.9 ft)] under the discharge chute for easy placement and removal of the container. Clearance for the container should be allowed for, even for units with conveyors or ejectors as a safeguard in case of mechanical breakdown.

Conveyors and pneumatic ejectors are the prime means of mechanically transporting screenings; grinders followed by solids-handling pumps are occasionally used. Compactors can transport screenings a short distance into a receiving container. Although generally reliable, conveyors, unless covered, generate odors and attract insects. In the past few years, conveyors that curve around corners and up inclines have been used more. If direction changes are needed, these conveyors eliminate the need for multiple-conveyor units. An advantage of conveyors is that the belt is the only portion of the system that normally contacts the screenings, thus minimizing plugging or jamming of mechanical components.

When designing a conveyor system, provide drainage for the screenings. One method maintains at least a slight incline throughout the conveyor length. Another method uses a perforated conveyor to allow water to drip into a trough beneath the conveyor; the trough drains back to the wastewater flow stream. The holes should be located to prevent dripping on the carriage assembly. The belt material is either rubber or another material that will not rust. Typically, a spray wash cleans the belt and prevents the buildup of sticky solids. The best belt will be concave across its width to prevent spillage and will have small ribs across the belt to prevent screenings from sliding back if the conveyor is steeply inclined.

Pneumatic ejectors offer several advantages over conveyors because the ejectors are less odorous and usually require less overall space. However, conveying sticks, rags, stones, and other debris in screenings through pipelines can clog and excessively wear the bends. Therefore, the system design needs to include cleanouts at bends and a high-pressure water service for flushing the discharge piping.

Compaction of the removed screening materials is an option for dewatering and reducing the volume of screenings. Such devices, including hydraulic ram and screw compactors, receive screenings directly from the screens and discharge the compacted screening "sausages" a short distance to portable bins or a conveyor. Compactors can reduce the water content of the screenings by up to 50%, with a volume reduction of 6.5 to 1. However, sticks and large objects in the screenings may cause mechanical breakdowns. As a good practice, controls should sense jams, automatically reverse the mechanism, and actuate an alarm when a motor overloads. Compactors handle well the increased quantity of solids removed by the smaller openings of continuous screens.

Grinders disintegrate the removed solids materials and then return the fine particles to the wastewater flow. In some cases, the fine particles are processed further. The problems associated with returning ground screenings to the wastewater were described previously in the section on comminutors and grinders. Ground screenings are easily pumped to a convenient point for dewatering, usually by static screens. The high maintenance associated with grinding, pumping, and dewatering usually eliminates grinding as an economical option unless the screenings are to be incinerated. For ultimate disposal in a sanitary landfill, grinding of screenings is generally unwarranted.

In large treatment plants, the high volume of screenings requires careful consideration of conveyance, storage, and transport to point of disposal. For example, the County of Sacramento (CA) Regional treatment plant [757 000-m^3/d (200-mgd) peak capacity] grinds the screenings from mechanical bar screens with a 25-mm (1-in.) opening, conveys the ground screenings slurry by pipeline using grinder pumps to covered elevated storage hoppers, and hauls the screenings and grit for disposal to an on-site landfill. Plant staff are currently planning to dewater the ground screenings on fine drum screens and transport the dewatered screenings by pneumatic ejector to the elevated hoppers. Static screens used in the past for dewatering the screenings have not worked satisfactorily. As another example, the City of Roseville (CA) treatment plant [151 000-m^3/d (40-mgd) peak capacity] discharges screenings from the mechanical bar screen chutes into wheeled containers on tracks leading to a loading area where the full containers are hoisted by an overhead monorail and discharged into a large dumpster. The dumpster is transported to an off-site landfill. A well-ventilated building with exhaust air treated by the plant odor control system houses the entire screening system.

To complete the handling of screenings, the material must be transported to a disposal site for disposal in accordance with local, state, and federal regulations. Lime is sometimes added to the screenings to control odors and insects. Any transportation or disposal alternative must be sensitive to traffic, air emission, water quality impacts, and environmental regulations. The 1989 WPCF Treatment Plant Data Survey[3] shows that most plants dispose of screenings at landfills, while a few bury or incinerate the screenings on site. The Sacramento County (CA) Regional facility is designed with the capabilities of either incinerating ground screenings or disposing of the screenings at an on-site landfill. The New York City Department of Environmental Protection, Bureau of Wastewater Treatment, disposes of screenings from its 14 plants at landfills.

On-site incineration, another method of screenings disposal, usually involves mixing the screenings with other treatment plant solids. Nevertheless, a few large plants incinerate screenings separately because of burning problems otherwise stemming from incomplete mixing of these thermodynamically different solids. Screenings will tend to clog feed mechanisms and should be ground or shredded and dewatered before being fed to the incinerator.[4] If the screening:sludge volume ratio is small, the material should be mixed with the sludge feed at a relatively constant ratio. The batch addition of screenings to an incinerator can cause uneven burning and hot spots.

Sometimes screenings hauled from the plant site are mixed with municipal refuse to be burned at the municipal incinerator. Although screenings contain more water and have a higher organic content than does municipal garbage, the relatively small volume of screenings has negligible impact on incineration.[4]

Instrumentation. For mechanically cleaned bar screens the following types of controls can be used alone or in combination:

- Manual start and stop,
- Automatic start and stop by timer control,
- High-level switch,
- Differential head-level actuated starting switch on cleaning mechanism, or
- Variable-speed operation with one of the following equipment items:
  — Two-speed motors,
  — Variable-frequency drive ac motors, or
  — Mechanical speed reduction devices.

Mechanical bar screen installations should include the following alarms:

- High upstream level and
- Screen start-fail.

The upstream-level alarm can be a bubbler system, submerged transducer, or float device. Exercise care in locating a float to avoid areas of excessive turbulence or velocity in the main influent channel.

A well-designed screening system will interlock auxiliary equipment such as spray wash, conveyors, or screenings compactors with the operation of the screen. This will ensure coordination among the process operations.

GRIT REMOVAL

Grit removal, an important part of wastewater treatment, prevents unnecessary abrasion and wear of mechanical equipment, grit deposition in pipelines and channels, and accumulation of grit in anaerobic digesters and aeration basins. Plant operational experience has shown that much grit is present in wastewater conveyed by either separate or combined sewer systems, with far more in the latter. Without grit removal, grit would be removed in the primary clarifiers or, if the plant lacks primary treatment, in aeration basins

---

and secondary clarifiers. Provision for grit removal is now common practice for any treatment plants with mechanical equipment and sludge-handling processes that grit could otherwise impair.

Depending on the type of grit removal process used, the removed grit is often further concentrated in a cyclone, classified, and washed to remove lighter organic material captured with the grit. The washed grit is more readily stored and disposed of than unwashed grit.

Wastewater grit materials, generally characterized as nonputrescible, have a subsidence velocity greater than that of organic putrescible solids; the grit particles are generally discrete rather than flocculent. Grit materials include particles of sand, gravel, other mineral matter, and minimally putrescible organics such as coffee grounds, egg shells, fruit rinds, and seeds.

The earliest use of grit removal systems was for plant influents from combined sewers. Use of these systems has since shown that whether a sewer system is separate or combined, large or small, operation of the treatment facility can benefit from removal of grit from the incoming flow. Grit removal is critical for protection of sludge dewatering centrifuges and high-pressure progressing cavity and diaphragm pumps; all are easily damaged by grit. Grit removal is normally omitted in natural treatment systems where grit and sludge are applied to land or allowed to accumulate on the bottoms of lagoons.

GRIT REMOVAL PROCESSES. The quantity and characteristics of grit and its potential adverse impacts on downstream processes are important considerations in selecting a grit removal process. Other considerations include head loss requirements, space requirements, grit removal efficiency, organic content, and economics. A variety of grit removal devices have been applied successfully over the years. The basic categories of grit removal processes include aerated grit chambers, vortex-type (paddle or jet-induced vortex), detritus tank (short-term sedimentation basins), horizontal flow type (velocity-controlled channel), and hydrocyclone (cyclonic inertial separation).

Aerated Grit Chamber. In aerated grit chamber systems, air introduced along one side near the bottom causes a spiral roll velocity pattern perpendicular to the flow through the tank. The heavier particles with their correspondingly higher settling velocities drop to the bottom, while the roll suspends the lighter organic particles, which are eventually carried out of the tank. The rolling action induced by the air diffusers is independent of the flow through the tank. The nonflow-dependent rolling action allows the aerated grit chamber to operate effectively with a wide range of flows. The heavier particles that settle on the bottom of the tank are moved by the spiral flow of the water across the tank bottom and then into a grit trough or hopper. Chain and bucket collectors, screw augers, clamshell buckets, or recessed impeller or air lift pumps remove the collected grit from the trough or hopper. The grit collection trough, depending on the chamber design, requires special design attention. A properly designed aerated grit chamber will produce a relatively clean grit.

When the wastewater flows into the aerated grit chamber, the grit particles will settle to the bottom at rates dependent on the size and specific gravity of the particles and on the velocity of roll in the tank. The rate of air diffusion and the tank shape govern the rate of roll and thereby the size of the particle with a given specific gravity that will be removed. The diffused air serves essentially as a velocity-control method with sufficient flexibility to accommodate varying conditions. The diffused air also requires only a minimal head loss through the unit. Proper design depends on an understanding of the variables affecting the air lift pumping energy and its impact on the roll pattern in the tank. Some empirical expressions include the variables influencing the roll pattern.[7] Table 9.5 summarizes some of the advantages and disadvantages of aerated grit chambers.

Poor performance of some full-scale aerated grit chamber installations has resulted in high percentages of organic matter in the removed grit, low grit removal efficiencies, and periodic abnormal grit loads on the collection

Table 9.5   Advantages and disadvantages of aerated grit chambers.

Advantages	Disadvantages
The same efficiency of grit removal is possible over a wide flow range.	Power consumption is higher than other grit removal processes.
Head loss through the grit chamber is minimal.	Additional labor is required for maintenance and control of the aeration system.
By controlling the rate of aeration, a grit of relatively low putrescible organic content can be removed.	Some confusion exists about design criteria necessary to achieve a good spiral roll pattern and proper hopper and grit removal system.
Preaeration may alleviate septic conditions in the incoming wastewater to improve performance of downstream treatment units.	Significant quantities of potentially harmful volatile organics and odors may be released from wastewaters containing these constituents.
Flexibility to remove grit can adapt to varying field conditions.	
Aerated grit chambers can also be used for chemical addition, mixing, preaeration, and flocculation ahead of primary treatment.	

equipment. The following causes of poor performance in aerated grit chambers have been observed:

- Improper baffle location resulting in poor hydraulic control and short circuiting,
- Unsteady and uncontrolled air supply to diffusers resulting in erratic roll patterns,
- Improper diffuser location, and
- Less than ideal tank geometry causing inefficient operation.

A typical layout of an aerated grit chamber is shown in Figure 9.9. Typical aerated grit chamber design criteria that will prevent operation problems are described below and summarized in Table 9.6.[3,7,8]

- Air rates typically range from 4.6 to 12.4 L/s·m of tank length (3 to 8 cu ft/min/ft). Rates as low as 1.5 L/s·m (1 cu ft/min/ft) have been used for shallow, narrow tanks. Rates above 7.7 L/s·m (5 cu ft/min/ft) are often used for deep, wide tanks. Also providing valves and flow meters is good practice for monitoring and controlling the air flow rate to each bank of diffusers. Aeration, preferably with coarse-bubble wide-band diffusers, is tapered to allow even removal of grit along the length of the chamber.[9] The plant process air supply may provide the air for the grit chamber, but separate dedicated blowers are preferred. See Chapter 11 for design of an air supply system.
- As good practice, a minimum hydraulic detention time of 3 minutes at maximum instantaneous flow rates will reliably capture 95% of the 0.21-mm (0.008-in.) grit (specific gravity of 2.6). Longer detention times improve grit removal and may be necessary to provide additional preaeration or to capture smaller grit particles. Detention time, however, is a less important criterion than baffle and diffuser location, air flow rates, and system designs.[8]
- A chamber length:width ratio ranging from 2.5:1 to 5:1 is appropriate. Nevertheless, square tanks have been used successfully with proper air diffuser location (perpendicular to the flow through the tank) and baffles to prevent short circuiting.[8]
- The tank inlet and outlet are positioned so that the flow through the tank is perpendicular to the spiral roll pattern. Inlet and outlet baffles will serve to dissipate energy and minimize short circuiting. A good design should include intermediate baffles across the width of the tank to prevent short circuiting through the center of the roll pattern.[10]
- A longitudinal baffle is positioned approximately 1 m (3.0 ft) from the wall along the air diffusers to help control the roll pattern.

- With proper adjustment the aerated grit chamber, at a level 150 mm (6 in.) below the top of the water, should produce a roll velocity of 0.6 m/s (2.0 ft/sec) near the tank entrance and 0.4 m/s (1.5 ft/sec) at the tank exit.[11]

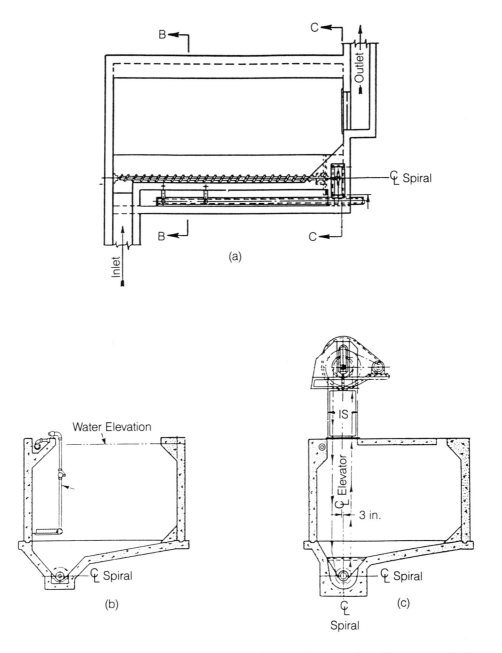

Figure 9.9 **Typical layout of an aerated grit chamber: (a) plan, (b) section B–B, and (c) section C-C (in. × 25.4 = mm).**

Table 9.6   Typical design criteria for aerated grit chambers.[3,7,8]

Item	Range[a]	Comment
Dimensions		
Depth, m	2–5	Varies widely
L:W ratio	2.5:1–5:1	
W:D ratio	1:1–5:1	2:1 typical
Minimum detention time (at peak flow), min	2–5	3 typical
Air supply		
$m^3/min \cdot m$	0.27–0.74	0.45 typical
Type of diffuser	Medium to coarse bubble	
Distance from bottom, m	0.6–1.0	
Transverse roll velocity, m/s	0.6–0.75	Provide valves and flow meters to allow proper adjustment

[a] m × 3.281 = ft; $m^3/min \cdot m$ × 10.76 = scfm/ft; m/s × 3.281 = ft/sec.

Vortex Grit Removal Systems. The vortex grit removal system relies on a mechanically induced vortex to capture grit solids in the center hopper of a circular tank. The incoming flow straightens in the inlet flume to minimize turbulence at the inlet of the chamber. At the end of the inlet flume, a ramp causes grit that may already be on the flume bottom to slide downward along the ramp until reaching the chamber floor where the grit is captured. At the end of the flume an inlet baffle, positioned so that the flow entering the chamber and the flow inside the chamber impinge on its sloped surface, deflects the flow downward. At the center of the chamber, adjustable rotating paddles maintain the proper circulation within the chamber for all flows. This combination—paddles, inlet baffle, and inlet flow—produces a spiraling, doughnut-shaped flow pattern that tends to lift lighter organic particles and settle the grit. Grit solids, removed from the center hopper by air lift or recessed impeller pumps, are further concentrated and washed.

Vortex grit systems include two basic designs: chambers with flat bottoms and a small opening to collect grit and chambers with a sloping bottom and a large opening to the grit hopper. Figure 9.10 shows a typical plan, section, and dimensions for a flat-bottom type. As the vortex directs solids toward the center, rotating paddles increase the velocity enough to lift the lighter organic materials and return them to the flow passing through the grit chamber. All grit passes under the paddles for removal of organic materials before the grit falls into the storage chamber.

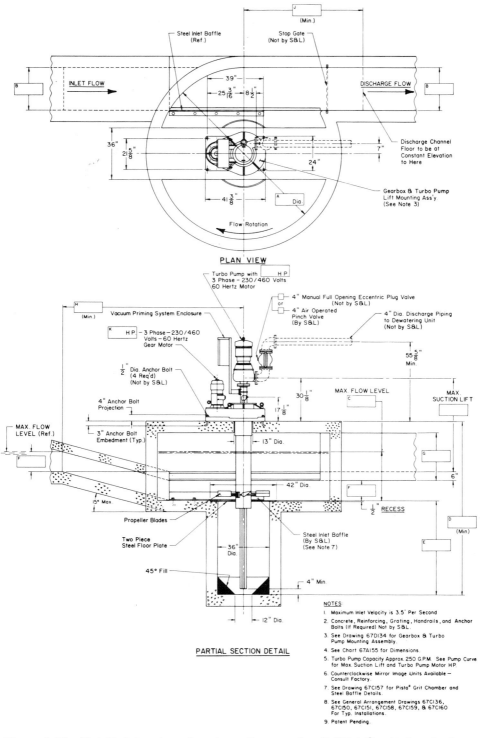

Figure 9.10 Detailed drawing of vortex grit removal unit (Pista® grit chamber)
(in. × 25.4 = mm; gpm × 0.063 08 = L/s).

NOTES

1. Maximum Inlet Velocity is 3.5' Per Second
2. Concrete, Reinforcing, Grating, Handrails, and Anchor Bolts (If Required) Not by S&L.
3. See Drawing 67D134 for Gearbox & Turbo Pump Mounting Assembly.
4. See Chart 67A155 for Dimensions.
5. Turbo Pump Capacity Approx. 250 G.P.M. See Pump Curve for Max. Suction Lift and Turbo Pump Motor H.P.
6. Counterclockwise Mirror Image Units Available — Consult Factory.
7. See Drawing 67C157 for Pista® Grit Chamber and Steel Baffle Details.
8. See General Arrangement Drawings 67C136, 67C150, 67C151, 67C158, 67C159, & 67C160 For Typ. Installations.
9. Patent Pending.

When sufficient grit has accumulated in the storage chamber, the grit may be air scoured to remove additional organics. It is then conveyed to the dewatering device by means of an air lift or recessed impeller pump. Typically, the system includes air or water scour to loosen the grit in the hopper just before its removal from the storage chamber. Table 9.7 summarizes the advantages and disadvantages of the vortex grit removal system.

Ideally, the flow into a vortex grit chamber should be straight, smooth, and streamlined. As good practice, the straight inlet channel length should be seven times the width of the inlet channel or 15 ft, whichever is greater. The ideal velocity in the influent channel ranges from 0.6 to 0.9 m/s (2 to 3 ft/sec). This ideal range should approximate flows between 40 and 80% of the peak flow. The minimum acceptable velocity for low flow is 0.15 m/s (0.5 ft/sec) because lower velocities will not carry grit into the grit chamber. If velocities as low as 0.15 m/s (0.5 ft/sec) will be experienced, provisions for flushing are necessary to move the settled grit into the tank. The flushing system must avoid washing grit through the grit chamber.

A baffle, located at the chamber's entrance, helps control the flow system in the chamber and also forces the grit downward as it enters the chamber. Some larger models of vortex grit systems, however, do not require the inlet baffle. The grit chamber's effluent outlet, with twice the width of the influent flume, results in a lower effluent velocity than that of the influent, thereby preventing grit below the opening level from being drawn into the effluent flow.

Sizing of the proprietary vortex grit chambers is based on recommended dimensions provided by the equipment manufacturers.[12-14] The units, usually marketed in standard nominal sizes, are rated on a peak flow basis.

Table 9.7   Advantages and disadvantages of vortex grit removal system.

Advantages	Disadvantages
Effective over a wide flow variation.	Proprietary design.
No submerged bearings or parts that require maintenance.	Paddles may collect rags.
Requires a minimum of space, thus reducing construction costs.	Grit sump may become compacted and clog. Requires high-pressure agitation water or air. Air lift pumps are often not effective in removing grit from the sump.
Minimal head loss (typical 6 mm).	
Energy efficiency.	
Removes high percentage of fine grit (up to 73% of 140 mesh size).	

Typical detention times for these units at peak design flows are short—20 to 30 seconds. Selecting a unit larger than necessary may result in improved performance, but operating costs will be higher. The manufacturers should verify that the given unit dimensions have been field tested to determine performance parameters. Deviation from the recommended dimensions without the manufacturer's prior approval could void any performance guarantees.

After selecting the vortex grit unit, additional information must be obtained from the manufacturer's drawings and design data to provide the appropriate entrance and exit channels and the concrete chamber in which to install the grit removal equipment. The design engineer then evaluates downstream flow control devices, such as a Parshall flume or weirs, to ensure that water levels will remain within the proper operating range of the unit.

Vortex grit removal systems are becoming increasingly popular because of their efficiency in removing fine grit at a reasonable cost. The following large-scale installations have reported successful operation:

- Virginia Beach, VA,
- Milwaukee, WI, and
- Mill Valley, CA.

Detritus Tank. One of the earliest grit chambers was a constant-level short-detention settling tank called a detritus tank (square tank degritter) (Figure 9.11). Because these tanks settle heavy organics as well as grit, they require a grit washing step as part of the process to remove organic material. Some designs incorporate a grit auger and a rake that removes and classifies grit from the grit sump. Advantages and disadvantages of detritus tanks are summarized in Table 9.8.

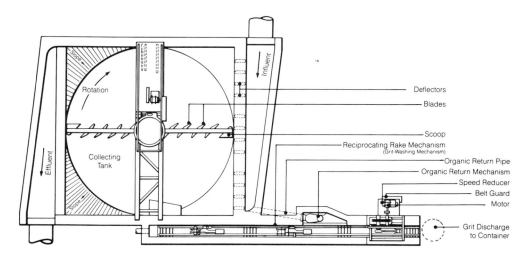

Figure 9.11   Plan view of detritus tank installation.

Table 9.8   Advantages and disadvantages of a detritus tank.

Advantages	Disadvantages
Flow control is not required.	Inlet baffles cannot be adjusted to achieve a uniform flow distribution over a wide range of flows.
Bearings and moving mechanical parts are above the water line.	
Units are sized on an area basis; thus, all grit is removed, washed, and classified up to the design flow.	Tanks will remove significant quantities of organic material, especially at low flows requiring grit washing and classifying.
Minimal head loss across the unit.	In shallow tanks [<0.9 m (3 ft)], grit can be lost due to agitation created by the rake arm.

Detritus tanks are sized on a overflow rate basis governed by the grit particle sizes (see Table 9.9). Design considerations for tank depth selection include minimizing the horizontal velocity and turbulence while, at the same time, maintaining a short detention time (typically less than 1 minute). An additional 150 to 250 mm (6 to 10 in.) of depth is provided for the raking mechanism. The detritus tank relies on well-distributed flow into the settling basin. Allowances for inlet and outlet turbulence as well as short circuiting are necessary to determine the total area required. Thus, good design practice generally applies a safety factor of 2.0 to the calculated overflow rate as an offset for all of these hydraulic inefficiencies.

Horizontal Flow Grit Chamber. One of the earliest types of grit removal systems, velocity-controlled systems (horizontal flow grit chambers), use

Table 9.9   Theoretical maximum overflow rates for detritus tanks.

Particle size		V_s,[a] cm/min (ft/sec)	Theoretical required overflow rate[b]	
Diameter, mm	Approx. mesh		m^2/1000 m^3/d	gpd/sq ft
0.83	20	494 (0.27)	7 120	174 500
0.59	28	363 (0.20)	5 200	128 000
0.46	35	247 (0.14)	3 550	87 000
0.33	48	186 (0.10)	2 670	65 500
0.25	60	165 (0.09)	2 370	58 000
0.21	65	131 (0.072)	1 890	46 300
0.18	80	116 (0.063)	1 670	40 900
0.15	100	91 (0.050)	1 320	32 300

[a] Based on liquid temperature of 15.5°C (60°F) and specific gravity of 2.65.
[b] A 2.0 safety factor for the overflow rate accounts for inlet and outlet turbulence and hydraulic inefficiency.

proportional weirs or rectangular control sections (such as a Parshall flume) to vary the depth of flow and keep the velocity of the flow stream at a constant 0.3 m/s (1 ft/sec). Operational experience has shown that this velocity allows the heavier grit particles to settle while lighter organic particles remain suspended, or become resuspended, and are carried out of the channel. Chain and flights are used in larger installations to scrape the grit to a hopper at the chamber's inlet end, where screw augers or chain and bucket elevators remove the collected grit. In some small plants, grit is shoveled manually from the channel. Table 9.10 lists advantages and disadvantages of horizontal flow grit chambers.

In designing a horizontal flow grit chamber, the settling velocity of the target grit particle (see Table 9.7) and the flow control section/depth relationship govern the length of the channel. For example, to capture 0.21-mm (0.008-in.) grit particles at a depth of 1 m (3 ft) with a horizontal flow velocity of 0.3 m/s (1 ft/sec), the theoretical length would be 12.7 m (41.7 ft):

$$12.7 \text{ m} = 0.30 \text{ m/s} (0.91 \text{ m}/0.022 \text{ m/s})$$

To determine the actual length of the channel an allowance for inlet and outlet turbulence must be added. The cross-sectional area will be governed by the rate of flow and the number of channels. Allowances for grit storage and removal equipment are included in determining the channel depth.

Table 9.10   Advantages and disadvantages of horizontal flow grit chambers.

Advantages	Disadvantages
Flexibility to alter performance is possible by adjusting the outlet flow control device.	Difficulty in maintaining a 0.3-m/s (1-ft/sec) velocity over a wide range of flows.
No unusual construction is required.	Excessive wear on submerged chain and flight equipment and bearings.
With effective flow control, removal of grit not requiring further classification is possible.	Where effective flow control is not achieved, channels will remove significant quantities of organic material requiring grit washing and classifying.
	Flow control weirs typically require free discharge and hence a relatively high head loss (typically 30 to 40% of flow depth).
	Proportional weirs may cause higher velocities at the channel bottom leading to bottom scour.

Table 9.11 presents a summary of typical design criteria for horizontal flow grit chambers.

Table 9.11   Typical design criteria for horizontal flow grit chambers.

Item	Range[a]	Comment
Dimensions		
Water depth, m	0.6–1.5	Depends on channel area and flow rate.
Length, m	3–25	Function of channel depth and grit settling velocity.
Allowance for inlet and outlet turbulence, %	25–50	Based on theoretical length.
Detention time (at peak flow), sec.	15–90	Function of velocity and channel length.
Horizontal velocity, m/s	0.15–0.4	Optimum velocity is 0.3 m/s.

[a] m × 3.281 = ft/sec; m/s × 3.281 = ft/sec.

Hydrocyclone. The hydrocyclone type system is typically used for separating grit from organics in grit slurries (see the following grit washing section) or for removing grit from primary sludge (see Chapter 17 for further information about primary sludge degritting).

Hydrocyclone systems are sometimes used to remove grit and suspended solids directly from the wastewater flow by pumping at a head of 4.2 to 9.1 m (12 to 30 ft). Coarse screening ahead of these units is needed to prevent their clogging with sticks, rags, and plastics. The centrifugal forces developed in the cyclone cause the heavier grit and suspended solids particles to concentrate along the sides and on the bottom, while the lighter solids, including scum, are removed from the center through the top of the cyclone. Theoretically, cyclones could remove as much solids as a primary clarifier, but this would entail the problematic disposal of combined primary sludge and grit. Cyclones operate best at constant flow and pressure. If flows depart from design flows, solids will be lost to the centrate stream.

A relatively new process known as the "tea cup"[TM] generates a sub-cyclonic vortex with a tangentially entering flow. Centrifugal and gravitational forces within the cylindrical unit remove grit particles with densities greater than water by forcing the particles to the wall where they fall by gravity to the bottom; the lighter organics exit with the effluent through the top. Organic materials entrapped with the grit are partially removed by scour at the bottom of the unit.

Like a hydrocyclone, the "tea cup" relies on hydraulic head, requiring as much as 0.6 to 1.5 m (2 to 5 ft) of head loss to generate a sufficient vortex for removal of fine [0.1-mm (0.004-in.)] grit. The units operate most efficiently at the peak design flow rate; their efficiency is reduced with the less intense vortex generated at lower flows. The capacity of individual units is 7600 m^3/d (2.0 mgd) or less, thus limiting their use for larger installations.

GRIT QUANTITIES AND CHARACTERISTICS. The quantity and characteristics of grit removed from wastewater will vary over a wide range. The variables influencing grit quantity include the type of collection system (combined or separate), characteristics of the drainage area, use of household garbage grinders, condition of sewer system, sewer grades, types of industrial wastes, and efficiency of grit removal equipment. If available, actual plant data should be used for design. In lieu of data, grit removal quantities may be estimated based on the data in Table 9.12.[3-5] For comparison, a listing of grit quantities from municipalities in the U.S. is presented in Table 9.1.

Grit solids characteristics will vary with a solids content from 35 to 80% and a volatile content from 1 to 55%.[4] A well-washed grit should achieve a solids content of 70 to 80% with a minimum of putrescible solids. The moisture and volatile content will be influenced by the degree of washing the grit receives. The bulk density of dewatered grit will range from 90 to 110 lb/cu ft. Dewatered grit can contain pathogens unless it is incinerated.

Table 9.12   **Estimated grit quantities.**[3-5]

Type of system	Average grit quantity, m^3/1000m^3[a]—typical range	Ratio of maximum day to average day
Separate	0.004–0.037	1.5 to 3.0:1
Combined	0.004–0.18	3.0 to 15.0:1

[a] m^3/1000 m^3 × 133.7 = cu ft/mil. gal.

DESIGN PRACTICE. Grit particle sizes, for design purposes, have traditionally included particles larger than 0.21 mm (0.008 in.) (65 mesh) with a specific gravity of 2.65.[5] Removal of 95% of these particles has traditionally been the target of grit removal equipment design. Modern grit removal designs are now capable of removing up to 75% of 0.15-mm (0.006-in.) (100-mesh) material because of recent recognition that plants often need to remove particles that are small to avoid their adverse effects on downstream processes.

Grit removal normally follows bar screening or comminution that prevents large solids from interfering with the grit-handling equipment. The grit removal process should precede primary clarification; in secondary treatment plants without primary clarification, grit removal should precede aeration. In some cases, grit is allowed to settle in the primary clarifiers and is removed from the primary sludge (see Chapter 17). Grit removal ahead of raw wastewater pumping equipment usually requires deep placement of the grit removal processes with the associated high construction and O&M expenses. Normally, as a more economical approach, the raw wastewater is pumped with the grit included, despite the increased wear of the pumps. This allows the grit chamber to be located conveniently just ahead of the other treatment processes.

Generally, a single grit removal unit with a bypass channel around the unit will suffice for small installations [<15 000 m³/d (< 4 mgd)] or for other plants where infrequent flows of wastewater containing grit can be tolerated in downstream processes. For large plants, plants served by combined sewers, or plants with grit-sensitive unit processes such as centrifuges, multiple grit removal units are necessary to allow periodic removal of units from service for cleaning, maintenance, and repair.

Similar to screen design, wastewater flow extremes must be known so that the grit chambers can be designed to efficiently remove grit from all flows. The quantity of grit entering the treatment plant is usually greatest during peak flows when scour velocities and grit transport rates are highest in the collection system. Grit chambers are sized to remove grit effectively at peak flows but to avoid removing excessive organic material at lesser flows.

The limited survey of design professionals by WPCF[5] suggests that no single grit removal process is favored exclusively over the other processes. In general, horizontal flow systems are being used less and vortex systems are increasingly being specified. Both aerated grit chambers and detritus tanks are still popular and used frequently.

The grit removal processes described earlier ultimately depend in some manner on gravity to collect removed grit solids at the bottom of a tank, trough, or hopper. The collected grit must then be removed from the grit chamber, dewatered, washed to remove organic solids (if necessary), and conveyed to a final disposal site. Each of these grit-handling steps can be accomplished by well-established techniques, which are described below. Manual of Practice No. 11, "Operation of Municipal Wastewater Treatment Plants,"[11] describes the operation and maintenance of grit-handling facilities.

Methods of Removing Grit from Grit Chambers. Removal of grit from each of the grit chambers can be accomplished with varying degrees of success by a number of different methods, primarily automatic. Some small plants use manual methods. Manual grit removal (shoveling) requires at least one redundant tank, capable of handling peak flows, to allow isolation and dewatering of the tank to be shoveled.

Large treatment plants or plants that accept wastewater from combined sewer systems require some type of automatic grit removal equipment. There are four methods that automatically or semiautomatically remove grit from the chamber hoppers:

- Inclined screw or tubular conveyors,
- Chain and bucket elevators,
- Clamshell buckets, and
- Pumping.

Grit removal is sometimes accomplished in two steps: horizontal conveyance of grit in a trough or channel to a hopper and then elevation of the grit from the hopper. Most aerated grit chambers have a tank bottom sloped downward to one side. The sweeping action of the water induced by the air roll pattern moves grit along the bottom of the chamber to a trough along the low side of the chamber. For most designs, the collection mechanism in the trough is a horizontal screw conveyor that pulls the settled grit to a hopper located at the head of the tank. In many longer tanks, however, two screw conveyors pull grit to a hopper located at the midpoint. Chain and flight mechanisms are sometimes used instead of the screw conveyors, especially in horizontal flow grit chambers. Any method for conveying grit to the hopper and lifting grit out of the hopper must reliably move heavy and abrasive grit—a severe, heavy-duty operation.

INCLINED SCREW. Inclined screw or tubular conveyors not only lift the grit out of the chamber, but also provide, as ancillary benefits, some washing and dewatering of the grit. Such washing and dewatering may suffice for the selected mode of transportation to the disposal site as well as for the planned disposal practice. A specific design concern for inclined conveyors is their inefficient use of plant area because of slope limitations. For the screw conveyor, long screws may require intermediate support and intermediate couplings; both are sensitive to wear. When selecting the screw conveyor system, ensure that the motor has sufficient horsepower to drive the selected screw conveyor when it is fully loaded with grit. The motor size selected for a combined sewer system must handle the sudden, high peak loads.

CHAIN AND BUCKET ELEVATORS. Chain and bucket elevators are used primarily with horizontal flow and aerated grit chambers. The design considerations related to peak loading and motor sizing are similar to those for the inclined screw collector mechanisms. The major additional concern with the chain and bucket system is the wear and potential jamming of the submerged sprockets. Because of grit's abrasive nature, this system should

---

Preliminary Treatment 425

always include metallic chain. A disadvantage of the elevator system is that at least one additional system, mechanical or manual, is required to move the grit to the transport vehicle. Also, this system does not provide effective grit washing and dewatering.

When using either screw conveyors or chain and bucket mechanisms, the chambers must be dewatered for periodic repair or routine maintenance. Therefore, at least one additional or redundant chamber is required beyond those needed to handle peak flows.

CLAMSHELL. A clamshell bucket arrangement moved by overhead monorail track affords yet another means of removing settled grit from grit chambers. This method has provided reliable service at the large Blue Plains WWTP in Washington, DC [1 400 000-m^3/d (370-mgd) average flow] and at the Chicago (IL) Northside Sewage Treatment Works [1 022 000-m^3/d (270-mgd) average flow]. The method, however, provides inconsistent grit removal, requires discontinuing flow to the chamber during grit removal, lacks effective dewatering and washing, and may generate odors. Nonetheless, most plant managers using this method seem to be satisfied with its performance.

PUMPING. Pumping grit from hoppers in the form of a slurry offers distinct advantages over other methods of grit removal but also has some disadvantages. Pumping grit from hoppers makes economical use of space because the pumps are located in a dry well adjacent to the chambers. This flexible arrangement allows each pump to serve any of the multiple chambers.

In addition to saving space, the use of common grit pumps does not necessarily limit the operation of individual grit chambers to one grit-handling system. If one grit pump fails, then another pump in the pumping station can easily enter service to replace it. As a disadvantage, this system includes a piping header and several valves, which require intensive maintenance due to the abrasive grit.

The grit can be pumped directly to washing and dewatering equipment. Design of the piping layout for grit pumping must address several critical considerations:

- Minimize the horizontal and vertical bends to reduce plugging by sticks and rags.
- Provide cleanouts at the bends to readily clear any blockages.
- Consider providing a redundant piping system because of severe piping wear and likely frequent plugging. This would allow maintenance without interrupting grit removal.
- Maintain velocity of 1 to 2 m/s (3 to 6 ft/sec) to keep the grit and other solids moving while minimizing pipe abrasion.

- Use discharge piping with nominal diameters of at least 100 mm (4 in.) to avoid high scouring pressures and velocities that would cause excessive wear.

Vortex or recessed impeller pumps and air lift pumps normally handle grit slurries. Both types of pumps can handle abrasive grit slurries with less wear than a centrifugal pump. Although the vortex pump is not as efficient as the air lift pump, the vortex pump is very reliable and operates with a higher head than does an air lift pump. Waterjets or compressed air will be needed for pumping grit that has become compacted in a hopper. Frequent pumping and introduction of an air stream to the hopper to loosen the grit before air lift pumping have solved the compaction problem in vortex grit chambers of the plant in Tucson, AZ.

Grit Washing. After removing the collected grit from the grit hopper, the grit is normally washed to ease handling. Grit removal methods have historically attempted to minimize the organic material removed with the grit while, at the same time, efficiently removing the grit particles. Facilities have met this dual objective with varying degrees of success. Thus, when choosing a possible washing system, the design engineer should judge the nature of the grit to be removed from the grit chamber. This assessment is easily accomplished for facility renovations, where grit chambers are already in place.

In most cases, a reduction in grit volume by removing the water contained in the grit saves transportation costs and eases transport and handling during disposal. Also, washing the grit (accomplished by some removal techniques) to remove putrescible organic material makes grit handling and disposal more manageable. Removing putrescible organic material prevents the odors and nuisance conditions caused by decomposition of the material as well as the added difficulty of disposing of sloppy material. Perhaps of most importance, removing the organic material from the grit will reduce the number of odor complaints that facility operators would otherwise receive from citizens near grit storage facilities, transportation routes, and disposal sites.

For any grit slurry pumping system, concentrating and washing the grit removed from the grit chamber are required. These operations are often conducted together in series, likely with compatible hydrocyclone and grit-classifying equipment provided by a single manufacturer operated intermittently. A hydrocyclone separator concentrates the grit centrifugally, requiring a steady feed of the grit slurry at an inlet pressure of 34 to 138 kPa (5 to 20 psi). The constant feed rate will typically be within the range of 13 to 32 L/s (200 to 500 gpm), depending on the size of the cyclone. The cycle times for intermittent operation can vary widely from 5 minutes to 8 hours; peak grit loadings may require continuous operation. Frequent grit removal cycles will tend to reduce grit accumulation in the hopper and its associated compaction and plugging and will dilute the grit slurry. Removal of excessively diluted grit slurry from the hoppers causes inefficiencies, including increased energy costs for recycling reject water through the headworks.

Hydrocyclone separator sizing is based on the cycled feed flow rate and the grit slurry solids concentrations. Hydrocyclones work best at feed concentrations of less than 1% solids. The centrifugal action created in the hydrocyclone separators increases the solids content to an average of 5 to 15%.[15] Approximately 90 to 95% of the feed flow rate discharges through the vortex finder at the top of the hydrocyclone. This flow volume reduction saves transportation and storage and reduces the required classifier size.

Grit classifiers, either the inclined screw or reciprocating rake type, wash the grit by separating the putrescible organics from the grit. Classifiers are sized based on the settling velocity of the particles to be settled, the feed flow capacity, and the grit raking capacity. For a target particle size and flow rate, the design engineer selects a minimum pool area and overflow weir length. The design engineer checks the classifier slope to ensure removal of the desired marginal particle size. Flatter slopes will remove finer grit particles.

Classifiers offered by major manufacturers are inclined from 15 to 30 deg from horizontal. In addition to slope, proper flight tip speed (rpm) and pitch (usually half or double pitch) assist in particle removal. Sectional flight construction may perform better than helicoid flights. Hardened flight edges should be used to resist the abrasive action of the grit. The screw or rake is sized to convey anticipated peak grit mass loading. An example of a typical grit cyclone and classifier is shown in Figure 9.12.

Pumps may be needed to boost the influent wastewater to the required pressure and flow rate. Where the water surface elevation in the grit tank exceeds the elevation of the cyclone, the piping design must avoid continuous siphoning from the grit hopper back through the pumps and the cyclone. Grit slurry pumps for removing grit from the chambers are generally sized to meet the high head requirements of the cyclone separators, static head, and pipe and fitting friction losses. Since head loss through the cyclone is a function of flow rate and size, consult the manufacturer's pressure and flow rating information.

The influent piping to the cyclone separators or classifiers is designed to ensure an even flow distribution in each unit. Isolation valves are necessary to allow removal of units from service for repair. Screening of the cyclone separator overflow has been found to reduce maintenance requirements by removing plastics and rags that accumulate in the system. Also, placement of the cyclone separators and classifiers above and near the discharge to the disposal truck or hopper reduces the need for conveyance.

Grit Disposal. Grit can be conveyed directly to trucks, dumpsters, or storage hoppers. Containers need covers to prevent odors during storage and hauling. Conveyors are frequently used for transporting grit from the handling facilities to the containers. If grit is transported directly to trucks or

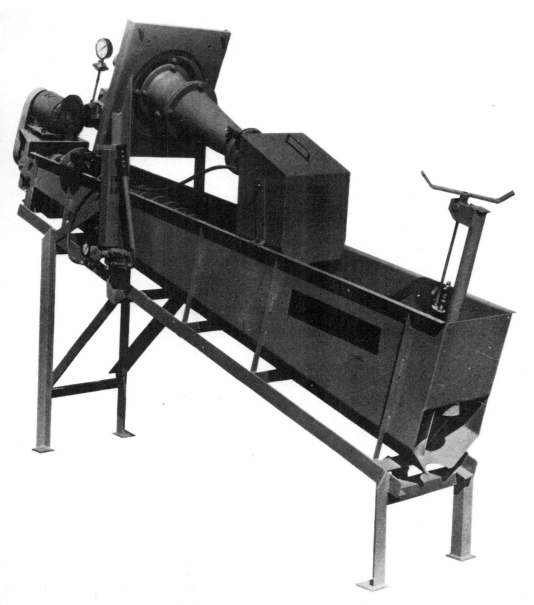

Figure 9.12 Grit cyclone and classifier installation.

dumpsters, the container should have enough capacity to handle daily peak
grit loads because inclement weather or other reasons might prevent replacement trucks from arriving or the loaded truck from leaving. Also, two bays
are advisable to ensure continued loading if the loaded container must remain
because of a truck breakdown.

Overhead storage hoppers that discharge into truck containers avoid the
need for keeping a truck always at the facility. Such overhead storage hoppers should be equipped to prevent grit bridging in the hopper and to bypass
the hopper if it fails to open. As a good practice to prevent bridging,

---

provide minimum hopper side slopes of 60 deg from horizontal. In northern climates, locate the storage hoppers in areas that can be heated to prevent freezing of the grit.

SEPTAGE

Separate pretreatment of septage may not be required if a limited amount of septage is discharged to an interceptor upstream of a plant or if a relatively small volume is discharged to the headworks of a large existing treatment facility with adequate preliminary processes. However, preliminary treatment facilities often must be designed to accept septage loads from septic tank pumpers. The following considerations apply to design of septage-receiving facilities:[16,17]

- Septage contains hair, grit, rags, stringy material, plastics, and is highly odorous.
- Septage is difficult to feed at controlled rates unless a receiving station exists.
- Septage receiving and storage facilities with separate screening and grit removal constitute the best arrangement.
- Provide for 150-mm (6-in.) diameter line sizes as a minimum.[16]
- Provide for a covered aerated grit chamber with air scrubbing for preliminary treatment of septage.

Septage characteristics can vary widely for many reasons, including

- User habits;
- Septic tank size and design;
- Pumping frequency;
- Climate;
- Types of in-home appliances used, such as garbage grinders or washing machines; and
- Difficulties in sampling septage.

Characteristics of septage are presented in Table 9.13. In general, septage is much stronger than typical domestic wastewater and has much higher concentrations of all constituents. Table 9.14 shows the ratios of typical concentrations of constituents in septage to those in domestic wastewater.

RECEIVING STATION DESIGN. Receiving facility design must account for the anticipated volume of septage, impacts of septage on plant processes, and odor control.

Table 9.13 Septage characteristics[4,16,18–29] with suggested design values.[a,b]

Parameter	U.S.				Europe/Canada				EPA mean	Suggested design value
	Average	Minimum	Maximum	Variance	Average	Minimum	Maximum	Variance		
TS	34 106	1 132	130 475	115	33 800	200	123 860	619	38 800	40 000
TVS	23 100	353	71 402	202	31 600	160	67 570	422	25 260	25 000
TSS	12 862	310	93 378	301	45 000	5 000	70 920	14	13 000	15 000
VSS	9 027	95	51 500	542	29 900	4 000	52 370	13	8 720	10 000
BOD$_5$	6 480	440	78 600	179	8 343	700	25 000	36	5 000	7 000
COD	31 900	1 500	703 000	469	28 975	1 300	114 870	88	42 850	15 000
TKN	588	66	1 060	16	1 067	150	2 570	17	677	700
NH$_3$-N	97	3	116	39	—	—	—	—	157	150
Total P	210	20	760	38	155	20	636	32	253	250
Alkalinity	970	522	4 190	8	—	—	—	—	—	1 000
Grease	5 600	208	23 368	112	—	—	—	—	9 090	8 000
pH	—	1.5	12.6	8.0	—	5.2	9.0	—	6.9	6.0
LAS	—	110	200	2	—	—	—	—	157	150

[a] Values expressed as mg/L, except for pH.
[b] The data presented in this table were compiled from many sources. The inconsistency of individual data sets results in some skewing of the data and discrepancies when individual parameters are compared. This is taken into account in offering suggested design values.

Table 9.14   Comparison of septage to municipal wastewater.[a,16]

Parameter	Septage[b]	Wastewater[c]	Septage:wastewater ratio
TS	40 000	720	55:1
TVS	25 000	365	68:1
TSS	15 000	220	68:1
VSS	10 000	165	61:1
BOD_5	7 000	220	32:1
COD	15 000	500	30:1
TKN	700	40	17:1
NH_3-N	150	25	6:1
Total P	250	8	31:1
Alkalinity	1 000	100	10:1
Grease	8 000	100	80:1
pH	6.0	—	—
LAS	150	—	—

[a] Values expressed as mg/L, except for pH.
[b] Based on suggested design values in Table 9.13.
[c] From Metcalf and Eddy, 2nd Ed., "medium strength sewage."[2]

When discharging septage directly to the head of a treatment plant, equalization facilities are necessary to control the flow of septage proportionately to the wastewater flow. Equalization is not generally necessary where septage is discharged to an interceptor at a point far enough upstream from the plant to allow complete mixing with the wastewater, provided that the total quantity of septage discharged represents less than 1% of the wastewater flow at that time and location. This can often be achieved by avoiding septage discharge during daily low-flow periods.

Dumping Station. The dumping station, the initial point where septage enters a receiving facility, needs a slightly sloped ramp to tilt the truck for complete drainage and accommodate washed spillage to a central drain. Figure 9.13 shows a basic layout of a dumping station.[16] The dumping station design should prevent tank trucks from releasing septage without any hose connection. Otherwise, spillage and release of odors will inevitably result.

Design considerations for dumping station equipment include

- Provide hoses and other washdown equipment, including steam equipment in colder climate for thawing lines and valves, at a convenient location to aid cleanup by individual haulers.
- Ensure septage can discharge only through a hose from the truck to the station; provide watertight connections.
- Provide a quick-release discharge tube for hose connections in the dumping station.

- For colder climates, include heater cable installation in the dump chamber bottom to prevent freezing.
- Extend the discharge tube below the liquid level in receiving chambers to prevent release of odorous gases; the tube diameter is generally 10 cm (4 in.).[16]

The amount of septage entering a dumping station and the rate at which it passes through the pretreatment facility must be accurately estimated during design. The design of the receiving facilities requires an accurate assessment of both the septage volumes and the range of daily septic flows expected. The critical limitation on a dumping station's peak flow capacity may be the number of discharge points (that is, unloading docks and hose connections). Therefore, multiple discharge points deserve consideration if heavy traffic is expected during peak hauling periods. Similarly, the access arrangement should permit efficient queuing of several pumper trucks in the dumping station area.

Storage and Equalization. Septage holding basins can be used for storage, equalization, mixing, and aeration of the septage prior to further treatment. Such holding facilities allow a controlled outflow of septage to downstream treatment processes to prevent hydraulic and organic shock loadings.

Holding tanks (where septage is handled independently or at existing wastewater treatment facilities) function mainly to equalize flow and attenuate variations in septage characteristics among loads. In cotreatment applications, a holding facility is necessary to allow proper metering of septage addition as a proportion of plant flow.

Pumping stored septage into the treatment plant has been accomplished successfully with air-operated diaphragm pumps. If the septage is to be added directly to a sewer or to a primary treatment unit, use of mechanical or diffused-air aeration and mixing in the holding tank is a good approach for improving treatability and preventing settling of organic solids. However, this tends to aggravate the odor problem (due to the air stripping effect) and therefore requires the use of enclosed tanks to control odors.

The major design criterion for a holding tank is detention time. As a rule, the holding tank storage capacity will equal at least a 1-day maximum expected volume of septage. However, storage of several days peak flow may be necessary, depending on the sensitivity of downstream treatment processes and the expected variation among daily septage volumes. The design of the equalization basin requires a site-specific approach that depends on the type and magnitude of the input flow variations and the facility configuration. If other preliminary treatment functions, such as preaeration, are to be combined with flow equalization, then the equalization basin should provide adequate detention times for these functions. Studies have shown negligible changes in the characteristics of finely screened septage after 24 hours of

---

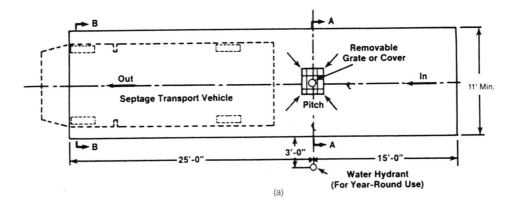

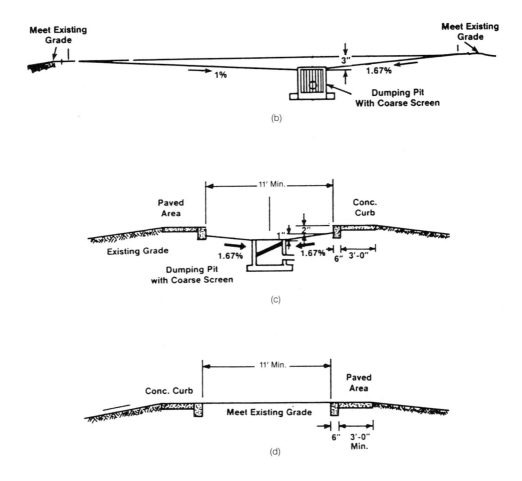

Figure 9.13 Typical layout of septage dumping station: (a) plan, (b) profile at
pavement centerline, (c) section A-A, and (d) section B-B
(in. × 25.4 = mm; ft × 0.304 8 = m).

aeration.[30] Detention periods of less than 48 hours are considered good practice.

ODOR CONTROL FOR SEPTAGE. Designers of septage-receiving facilities must address odor control during the design process rather than as a retrofit measure in response to pressure from nearby residents. Odor problems at septage-receiving facilities can be solved in advance of construction by proper siting and use of existing technologies, including chemical scrubbers, filters, combustion, biological processes, and so on. The principal sources of odor at the facility must be identified and only the odorous gases treated. As a simple approach for isolating the odorous gases, the component of the facility generating the odors could be enclosed. The gases confined in this housing structure would then be isolated from nonodorous air, thereby reducing the volume of air to be treated and thus the overall cost. Designers must recognize the dangers of closed spaces to operating personnel (see Chapter 6). See Chapter 7 for further information on air scrubbing methods.

ODOR CONTROL CONSIDERATIONS

Wastewaters entering treatment plants often contain odorous compounds that can escape from open channels and tanks in the preliminary treatment system, particularly at points of turbulence. In warm weather, screenings and incompletely washed grit can develop obnoxious odors. Prevention and control of these odors are required to prevent complaints from nearby residents, provide a reasonable working environment for plant operators, and reduce the corrosive effects of sulfide-bearing gases on equipment.

Malodorous gases emanating from municipal wastewater may contain numerous components, including hydrogen sulfide (H_2S), indoles, skatoles, amines, ammonia (NH_3), carbon dioxide (CO_2), methane (CH_4), and others.[19] These compounds may form in wastewater collection and preliminary treatment systems where anaerobic conditions are allowed to develop. Odors are commonly generated in long force mains where anaerobic slime layers can grow and in slow-moving sewers and stagnant channels where anaerobic deposits and scum layers can develop. Most odor control methods focus on the removal and control of hydrogen sulfide.

ODOR CONTROL METHODS. A brief description of the odor control methods used in the preliminary treatment area includes prevention, chemical treatment, preaeration, and air scrubbing. Each of these is described below. For more information on odor control techniques see Chapter 7.

Prevention. This entails minimizing the formation of anaerobic conditions, largely through good housekeeping at the headworks.

Chemical Treatment. Chemical addition into the wastewater ahead of the headworks can help to minimize formation and release of odorous substances. Some of the common chemicals used to control odors in wastewater are described below.

CHLORINE AND CHLORINE COMPOUNDS. Since most treatment plants use chlorine for disinfection within the plant, it is also normally available for controlling odors. Chlorination, one of the oldest odor control practices, is effective because chlorine is a strong oxidizing agent. Chlorine will oxidize hydrogen sulfide to elemental sulfur and to sulfate based on the following reactions:

$$HOCl + H_2S \rightarrow S + HCl + H_2O$$

$$4HOCl + H_2S \leftrightarrow SO_4^{-2} + 4Cl^- + 6H^+$$

Approximately two parts of chlorine (as Cl_2) by weight are theoretically required for each part of sulfide (as H_2S) based on the first reaction; 8.3 parts on the basis of the second reaction. Since chlorine reacts preferentially with sulfide, all of the chlorine demand of the wastewater need not be satisfied. As a rule of thumb, use 8 to 10 parts by weight of chlorine for each part of hydrogen sulfide present. Typical chlorination doses for odor control of fresh wastewater range from 2 to 12 mg/L. Septic wastewaters or those containing sulfite wastes from tanneries or food processors can require as much as 40 to 50 mg/L of chlorine.

HYDROGEN PEROXIDE. Hydrogen peroxide (H_2O_2) is a weak disinfectant but a practical oxidant for odor control. The chemical reaction for hydrogen peroxide and hydrogen sulfide is as follows:

$$H_2O_2 + H_2S \rightarrow 2H_2O + S$$

Theoretically, one part of H_2O_2 will react with one part of H_2S. In practice, one to four parts are required.

METAL SALTS. Many plants add ferric and ferrous iron salts to improve suspended solids removal in the primary clarifiers and realize ancillary benefits of phosphorus removal and control of odors as well. Dissolved iron will combine with sulfide to form highly insoluble iron salts such as smythite (Fe_3S_4), pyrite (FeS), and marcasite (FeS_2). Also formed are partly soluble compounds such as pyrrhotite (Fe_5S_6) and ferric sulfide (Fe_2S_3).[19] By precipitating insoluble forms of sulfide, hydrogen sulfide is also removed due to a shift in its dissociation equilibrium. An optimum mixture of one part ferrous to two parts ferric salts (as Fe) has been used to effectively precipitate sulfides.[19,31]

Ferric chloride is added upstream of the Point Loma advanced primary treatment plant in San Diego, CA, to both control sulfide concentrations in the incoming wastewater and provide coagulation to achieve advanced primary treatment. The ferric chloride is added at two places: upstream from the plant at a pumping station where approximately 30 mg/L is added and at the headworks where 10 mg/L is added. For more information regarding feeding and storage of ferric salts, see Chapters 7 and 16.

Zinc salts have been successfully used by the city of Los Angeles to decrease sulfide concentrations and control odor. As a general rule, two parts of zinc will be required for each part of hydrogen sulfide to effectively remove the sulfide.[19]

pH ADJUSTMENT. Caustics or alkalies can be added to wastewater to adjust pH and decrease the release of hydrogen sulfide by shifting the predominant species to hydrosulfide and sulfide ions. Chemicals commonly used to raise pH include caustic soda, soda ash, and lime slurry. Addition of alkalies can have the added benefit of increasing the alkalinity of the water to counteract alkalinity destruction by nitrification or by addition of primary coagulants. Typically, addition of alkaline agents is not cost effective unless they would be necessary, either alone or with other primary coagulants, for phosphorus removal.

Preaeration. Preaeration replenishes DO in the wastewater, thereby reducing septicity and odor. Preaeration will strip hydrogen sulfide and volatile organics from the wastewater and, hence, may require a cover and an air scrubbing system. Normal practice meets both preaeration and grit removal objectives by providing an aerated grit chamber system or accomplishes preaeration in an aerated channel or in a flow equalization basin, which requires aeration and mixing. The detention time in a aerated grit chamber must exceed the normal minimum for grit removal to ensure sufficient preaeration. In addition to odor control, other preaeration benefits include

- Promoting uniform distribution of solids to clarifiers,
- Providing grease separation,
- Preventing septicity, and
- Flocculating solids.

Typical design criteria for preaeration to control odors are listed in Table 9.15.

Table 9.15   Typical design criteria for preaeration.

Item	Value
Detention time	10–15 min
Air requirement	0.82 L/L[a] or 90–350 L/min·m[a] of channel

[a] L/L × 0.134 = cu ft/gal; L/min · m × 0.011 = cfm/ft.

Air Scrubbing. Scrubbing of gas streams collected from covered preliminary treatment processes can provide effective and economical odor control. Odor scrubbers (described in Chapter 7) remove odorous compounds from the air stream by many methods, including activated carbon adsorption, biological towers, adsorption on soil mounds, dissolution into liquids containing oxidants using packed towers or aerosol contact vessels, ozonation, and combustion. Ozonation of off-gas from wastewater pumping stations has worked very successfully in the Metropolitan Water Reclamation District of Greater Chicago. Typical chemicals used for liquid absorption systems include buffered potassium permanganate ($KMnO_4$), sodium hypochlorite (Na_2ClO_4), caustic ($NaOH$), hydrogen peroxide (H_2O_2), chlorine (Cl), and chlorine dioxide (ClO_2). If possible, the optimal solution for scrubbing should be based on a pilot study.

Air-moving equipment is sized to provide air changes frequently enough to prevent accumulation of odorous gases in the enclosed space. Air-moving equipment must tolerate the corrosive conditions often caused by the conversion of hydrogen sulfide to sulfuric acid in the presence of oxygen, humidity, and bacteria. For further information regarding off-gas treatment, see Chapter 7.

DESIGN PRACTICE. Design considerations for control of odors from preliminary treatment facilities include

- Proximity of headworks to residences or commercial areas,
- Prevailing wind patterns in the area,
- Characteristics of the sewer system (for example, long force mains and long gravity sewers with gentle slopes tend to develop odors), and
- Degree of turbulence created in the preliminary treatment system.

In most installations, designs that provide good housekeeping around the headworks help to prevent development of odors. Common design defects that can increase odor generation include

- Stagnant or backwater areas in open or closed conduits that allow solids to accumulate and decompose with attendant odors,

Design of Municipal Wastewater Treatment Plants

- Square corners that provide pockets where solids can easily accumulate,
- Difficult-to-clean rough surfaces on which organic matter and slime can adhere and build up,
- Points of turbulence such as a flumes, drops, or aerated chambers without covers and ventilation systems to collect and treat odorous gas,
- Flow equalization tanks where septic conditions can develop due to insufficient mixing or aeration to keep solids in suspension,
- Lack of covered airtight containers to store grit and screenings, especially in warm climates, and
- Insufficient nearby hose bibbs to allow convenient cleanup around the headworks area.

*F*LOW EQUALIZATION

Accommodating wide variations in flow rates and organic mass loadings is one of the major challenges faced in the design of wastewater treatment facilities. Because of the naturally occurring variations in the generation of wastewater and the related effects of infiltration and inflow, all municipal wastewater treatment facilities must process unsteady wastewater flows. Efficiency, reliability, and control of the unit process operations within the plant can be adversely affected by the cyclic nature of the waste generation, resulting in possible violations of effluent standards. Equalization of influent flow can dampen the diurnal variations and the variations caused by infiltration and inflow to achieve a relatively constant loading of downstream treatment processes. Equalization of effluent discharge is used occasionally before subsurface discharge or other wastewater reuse applications.

Two types of equalization of influent flows may be used: flow equalization and waste strength equalization. The primary objective of flow equalization basins for municipal wastewater plants is simply to dampen the diurnal flow variation and thus achieve a constant or nearly constant flow rate through the downstream treatment processes. An additional benefit is a reduction in the variability of the concentration and mass flow of wastewater constituents by blending in the equalization basin. This more uniformly loads downstream processes with organics, nutrients, and other suspended and dissolved constituents.

Waste strength equalization, commonly used in industrial applications, dampens the variability of the strength of the waste by blending the wastewater in the equalization basin. For this purpose the volume of wastewater in the equalization basin normally remains a constant. With this type of equalization, the flow remains variable.

APPLICATIONS. Equalization may be used to minimize the effects of loading variations from dry weather flows, wet weather flows from separate sanitary sewers, or storm-related flows from combined sewers. In some cases, providing equalization storage for wet weather flows in sanitary sewers may be feasible, depending on the magnitude of the infiltration and inflow component. For combined sewer systems, the design engineer should consider temporary storage of peak flows either within the collection system sewers or off line, for later regulated release back to the plant.

Flow equalization, a useful upgrading technique, can significantly improve the performance of an existing treatment facility. In the case of new plant design, flow equalization can reduce the required size of downstream facilities. However, economic and operations analyses should determine the feasibility of equalization versus increasing the size of the required unit operations based on the loading variations anticipated. Flow equalization basins are generally not cost effective for new plants with separate sanitary sewer systems, nonexcessive infiltration and inflow, and small contributions of industrial wastes.

Equalization basins are commonly located in the plant following the grit and screening units. However, based on the individual circumstances, equalization storage is sometimes located in the collection system to relieve overloaded trunk lines or pump station force main systems. Also, equalization of solids-handling sidestreams is used frequently to regulate and reduce the impact of sidestream returns on the liquid treatment system.

In some instances, large interceptor sewers entering the treatment plant can be effectively used as storage basins to dampen peak diurnal dry weather flow variations. In such cases, nightly or weekly drawdown of the interceptor system is necessary to increase flow velocity, thereby flushing solids that have accumulated during the storage period.

Equalization basins, located at the treatment plant site, may be designed as either in-line or side-line units. For the in-line design shown by Figure 9.14, all the flow passes through the equalization basin.[32] This results in significant concentration and mass flow dampening. The side-line design, also shown in Figure 9.14, diverts only the flow portion exceeding the daily average flow through the equalization basin.[32] This design minimizes pumping requirements but less effectively dampens variations of pollutant concentrations.

An equalization basin was added to the Naval Station Treasure Island WWTP in San Francisco Bay (CA) to dampen flows from more than 30 pumping stations that discharged into a common force main entering the plant. The basin also affords emergency storage for plant malfunctions. The equalization basin comprises rectangular concrete tanks with jet aeration equipment for mixing and aeration. Screw lift pumps regulate flow from the basin by discharging a constant flow regardless of the depth of water in the

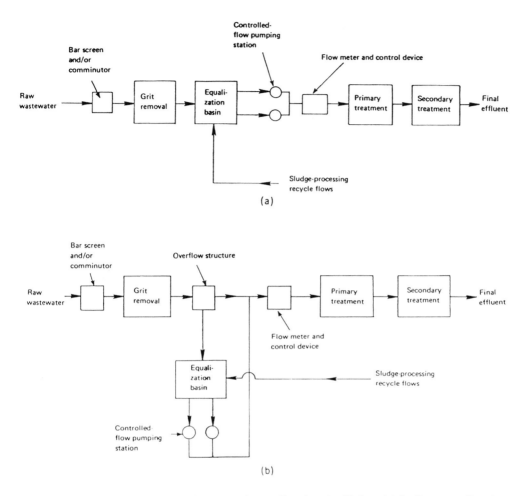

Figure 9.14 Schematic flow diagram of equalization facilities: (a) in-line equalization and (b) side-line equalization.

basin. To ease cleaning of the basins, the basin floor slopes downward to sumps that drain to the plant drain.

BENEFITS. Equalization will reduce the size of unit processes in a new facility or relieve overloaded unit processes in an existing facility. By providing a relatively constant loading, equalization can improve the efficiency, reliability, and operability of some types of facilities. For example, activated sludge and trickling filter plants that are sized for peak organic loadings could be downsized. Also, the protection from shock loading provided by equalization adds reliability to any biological process.

    The use of equalization to either reduce the size of, or relieve an overloaded condition on, an individual treatment process would seldom be economically justified. Nevertheless, if the cumulative effects of equalization throughout the plant are accounted for, equalization may be more cost effective than modifying existing unit processes.

Primary Treatment. The benefits will not likely match the costs of equalization for plants with primary treatment unless the influent peaking factor exceeds 2:1. For such systems, equalization benefits related to primary clarification include

- A reduction in the area required for primary clarification since the units will be designed to accept the equalized flow rather than the normal peak flow, and
- For an overloaded primary treatment system, a reduction of the peak flow, thereby relieving the stress on the units.

Another possible benefit related to primary clarification is improvement in performance due to preaeration in the equalization basin. Studies have shown that prolonged preaeration will improve primary clarifier performance up to a maximum of about 15% because preaeration preflocculates the suspended solids to improve their settling characteristics.[22,23] Unless designed specifically for preaeration, flow equalization basins may not perform efficiently as preaeration systems (see previous discussion of preaeration). Often the flow equalization unit process is optimally located after the primary clarifier to reduce O&M requirements.[24]

Secondary Treatment. Equalization benefits the secondary treatment units by dampening both the flow and the waste strength, allowing the process to operate more nearly at steady-state conditions. Enhanced primary treatment efficiency resulting from equalization also reduces loads entering the secondary units.

Dampening the flow by equalization can reduce the size of new secondary clarifiers and the solids loading on existing clarifiers, thereby improving suspended solids and associated BOD_5 removals. In existing plants, the reduced peak flows from equalization may offer an opportunity to increase the MLSS concentration in the aeration system and still maintain an acceptable solids loading on the secondary clarifiers. An increase in MLSS concentration, in turn, will increase the SRT and decrease the F:M ratio. This may improve the reliability of nitrification and reduce biological sludge production. Conversely, equalization may allow reducing the MLSS in response to lower primary effluent BOD_5, thus reducing solids loadings on secondary clarifiers and improving treatment performance.

Dampening of the wastewater strength variations may reduce aeration requirements and associated power requirements. In addition, prolonged aeration in an equalization basin is likely to somewhat reduce BOD. A 10 to 20% BOD reduction would likely be an upper limit for an in-line basin equalization of raw wastewater. Nonetheless, studies of such BOD reduction should precede detailed design.

Advanced Wastewater Treatment System. The reliability and efficiency of sensitive biological nutrient removal systems may benefit from equalization through dampening variations of both flows and mass constituents. Dampening of the mass loading to chemical coagulation and precipitation systems will improve chemical feed control and process reliability. As a result, this may reduce instrumentation complexity and cost. With biological phosphorus removal, stabilization of the BOD:P ratio, a key to performance, may enhance reliability of the process. A constant flow rate to filters will lead to more uniform solids loading and a higher level of performance.

DESIGN PRACTICE. The design methodology for equalization involves determining the necessary volume, the mixing and aeration requirements, and how to control flows leaving the basin.

Volume Determination. The first step in estimating the volume necessary for equalization involves determining the diurnal variations of the wastewater. Whenever possible, this step should be based on actual operating data. Diurnal flow patterns will vary from day to day, and even seasonally, depending on the nature of the community (for example, tourist related, winter residences, agricultural food processing). It is therefore important to select a pattern that will ensure a large enough volume for equalization, taking into account conditions such as infiltration and storm-related inflow that influence variability of the influent flow.

Several methods proposed for estimating the equalization volume required for normally varying wastewater flows and strengths are[32]

- Flow balance ("mass diagram"),
- Concentration balance,
- Combined flow and concentration balance,
- Sine wave method, and
- Rectangular wave method.

These methods, sufficient for most municipal wastewater treatment applications, offer relatively simple procedures for estimating equalization volume requirements. If needed, other more sophisticated methods of equalization analysis are available.[21-29]

Mixing Requirements. The successful operation of equalization basins requires proper mixing and aeration. Design of mixing equipment provides for blending the contents of the tank and preventing deposition of solids in the basin. To minimize mixing requirements, grit removal facilities will normally precede equalization basins. Aeration is necessary to prevent the wastewater from becoming septic. Mixing requirements for blending a municipal wastewater having a suspended solids concentration of approximately 200 mg/L range from 3.9 to 7.9 W/1000 L (0.02 to 0.04 hp/1000 gal) of storage.[32] To maintain aerobic conditions, air should be supplied at a rate of 0.16 to 0.25 L/s·1000 L (1.25 to 2.0 cu ft/min/1000 gal) of storage.

Mechanical aerators, one method of providing both mixing and aeration, have oxygen transfer capabilities varying from 4 to 6 kgO_2/MJ (3 to 4 lb O_2/hp-hr) in tap water under standard conditions, but the oxygen transfer efficiency (OTE) in wastewater is much lower. A reasonable OTE value for design would approximate 1.3 to 2.3 kgO_2/MJ (1.0 to 1.5 lb O_2/hp-hr). Minimum operating levels for floating aerators generally exceed 1.5 m (5 ft) and vary with the horsepower and design of the unit. Low-level shutoff controls are needed to protect the unit. Baffling may be necessary to ensure proper mixing, particularly with a circular tank configuration.

The horsepower requirements to prevent deposition of solids in the basin may greatly exceed that needed for blending and oxygen transfer. In such cases, the most economical approach might be to provide mixing equipment to keep the solids in suspension and a diffused air system to supply the air requirements. As another option, a surface aerator blade could be mounted on the mixer.

Diffused aeration systems may use coarse- or intermediate-bubble diffusers. Ceramic fine-bubble diffusers are unsuitable because of possible clogging from biological slime growths and inorganic deposits. With variable volume systems (and therefore variable water depths), ensure that the blowers have pressure regulation controls. Mixing requirements are 0.5 to 0.8 L/s·1000 L (30 to 50 cu ft/min/1000 cu ft) of basin volume.

Flow Control Methods. The use of flow equalization will, in most cases, involve pumping either before or after the equalization basins. If flow is pumped to the equalization basin, the basin effluent must be controlled by a flow-regulating device. The design engineer must recognize the acceptable flow range of the regulating device and the range of basin operating water levels that must be throttled. For pumped flow from the equalization basin, a variable-speed pumping system will be required to carefully regulate the hydraulic loadings, providing a near-constant flow to the downstream units. Influent pumping to the equalization basin requires a pump of sufficient capacity to handle diurnal peaks.

A flow-measuring device downstream of the basin must monitor the equalized flow. Instrumentation should be provided to control the preselected equalization rate by automatic adjustment of the basin effluent pumps or flow-regulating device.

The use of waste strength equalization, which generally requires a constant volume, may require pumping into the equalization basin with a variable outflow equal to the input flow.

*R*EFERENCES

1. Pankratz, T., "Screening Equipment Handbook for Industrial and Municipal Water and Wastewater Treatment." Technomic Pub. Co. (1988).
2. Metcalf & Eddy, "Wastewater Engineering: Treatment, Disposal, Reuse." 2nd Ed., McGraw-Hill, Inc., New York, N.Y. (1979).
3. "Wastewater Treatment Plant Data Survey." Water Pollut. Control Fed., Alexandria, Va. (1989).
4. "Process Design Manual for Sludge Treatment and Disposal." EPA-625/1-79-011, U.S. EPA, Washington, D.C. (1979).
5. "Preliminary Treatment Facilities—Design and Operational Considerations." EPA-430/09-87-007, U.S. EPA, Washington, D.C. (1987).
6. "Design Practice Survey." Water Pollut. Control Fed., Alexandria, Va. (1989).
7. Albrecht, A.E., "Aerated Grit Chamber Operation and Design." *Water Sew. Works,* **114,** 9, 331 (1967).
8. Morales, L., and Reinhart, D., "Full-scale evaluation of aerated grit chambers." *J. Water Pollut. Control Fed.,* **56,** 337 (1984).
9. Finger, R.E., and Patrick, J., "Optimization of grit removal at a WWTP." *J. Water Pollut. Control Fed.,* **52,** 2106 (1980).
10. Anderson, M.M., *et al.*, "Designing to improve grit removal at the Point Loma wastewater treatment plant, San Diego, CA." Paper presented at the 63rd Annu. Water Pollut. Control Fed. Conf., Washington, D.C. (1990).
11. "Operation of Municipal Wastewater Treatment Plants." Manual of Practice No. 11, Water Pollut. Control Fed., Alexandria, Va. (1990).
12. Smith and Loveless, Pista Grit Chamber Tech. Info., Appl. Data, Lenexa, Kans. (1988).
13. Jones and Atwood, Grit Removal Systems, Libertyville, Ill. (1988).
14. John Meunier, Inc., MecTan Grit Chamber, Montreal, Quebec, Can. (1989).
15. WEMCO, "Hydrogritter Separator." Bulletin No. 11-86, Sacramento, Calif. (1990).
16. "Handbook of Septage Treatment and Disposal." EPA-625/6-84-009, U.S. EPA, Washington, D.C. (1984).
17. Segall, B.A., and Ott, C.R., "Septage treatment." *J. Water Pollut. Control Fed.,* **52,** 2145 (1980).

18. Neighbor, J.B., and Cooper, T.W., "Design and Operation Criteria for Aerated Grit Chambers." *Water Sew. Works,* **112,** 12, 448 (1965).

19. "Odor Control for Wastewater Facilities." Manual of Practice No. 22, Water Pollut. Control Fed., Alexandria, Va. (1979).

20. "Process Design Manual for Upgrading Wastewater Treatment Plants." EPA Technol. Transfer, U.S. EPA, Washington, D.C. (1974).

21. LaGrega, M.D., and Keenan, J.D., "Effects of equalizing wastewater flows." *J. Water Pollut. Control Fed.,* **46,** 123 (1974).

22. Roe, F.C., "Pre-aeration and Air Flocculation." *Sew. Works J.,* **23,** 127 (1951).

23. Seidel, H.F., and Bauman, E.R., "Effect of preaeration on the primary treatment of sewage." *J. Water Pollut. Control Fed.,* **33,** 339 (1961).

24. Ongerth, J.E., "Evaluation of Flow Equalization in Municipal Wastewater Treatment." EPA-600/2-79-096, U.S. EPA, Munic. Res. Lab., Cincinnati, Ohio (1979).

25. Wallace, A.T., "Analysis of Equalization Basins." *J. Sanit. Eng. Div., Proc. Am. Soc. Civ. Eng.,* **94,** 1161 (1968).

26. "Process Design Techniques for Industrial Waste Treatment." C.E. Adams, Jr., and W.W. Eckenfelder, Jr. (Eds.), Enviro Press, Nashville, Tenn., 37 (1974).

27. DiToro, D.M., "Statistical Design of Equalization Basins." *J. Environ. Eng.,* **101,** 917 (1975).

28. Novotny, V., and Stein, R.M., "Equalization of Time Variable Waste Loads." *J. Environ. Eng.,* **102,** 613 (1976).

29. McInnes, C.D., *et al.,* "Stochastic Design of Flow Equalization Basins." *J. Environ. Eng.,* **104,** 1277 (1978).

30. Condren, A.J., "Pilot-Scale Evaluations of Septage Treatment Alternatives." EPA-600/2-78-164, U.S. EPA, Washington, D.C. (1978).

31. "Process Design Manual for Sulfide Control in Sanitary Sewerage Systems." U.S. EPA, Washington, D.C. (1974).

32. "Flow Equalization." EPA Technol. Transfer, U.S. EPA, Washington, D.C. (1974).

Chapter 10
Primary Treatment

*I*NTRODUCTION

Primary sedimentation, the principal form of primary treatment, is the oldest and most widely used unit operation in wastewater treatment. Fine screens, instead of primary sedimentation, are occasionally used for primary treatment in the U.S. According to the U.S. EPA, in the U.S. in 1988, at least 5341 treatment facilities included some form of primary treatment.[1] The objectives of primary treatment are to produce a liquid effluent suitable for downstream biological treatment and to achieve solids separation, which results in a sludge that can be conveniently and economically treated before ultimate disposal.

Historically, primary treatment has often been overlooked in the research and design of wastewater treatment facilities, particularly where downstream secondary treatment processes could compensate for the poor performance of the primary treatment facilities. Primary treatment is the first "line of defense" in wastewater treatment and reduces suspended solids and BOD₅ loading on downstream treatment processes. Loading reductions minimize operational problems in downstream biological treatment processes, lower the oxygen demand, and decrease the rate of energy consumption for oxidation of particulate matter.[2,3] These effects enhance soluble substrate removal during aeration and reduce the volume of waste activated sludge that is generated.[2,3] Primary treatment also removes floating material, thereby minimizing operational problems in downstream treatment processes (that is, buildup of scum in secondary treatment processes) and improving the plant's overall aesthetics (that is, visual blights and odors).

Primary treatment equalizes raw wastewater quality and flow to a limited degree, thereby protecting downstream unit processes. The design of the primary treatment process should assess the economic impact of primary treatment on the operation and maintenance of downstream unit processes. Money spent on primary treatment often provides the greatest return on the investment in terms of dollars per kg of pollutant removed. For instance, primary treatment with anaerobic sludge stabilization may be more economical than incremental capacities of downstream processes necessary to handle the increased load resulting from the omission of primary treatment.[4] The most common form of primary treatment is quiescent sedimentation with skimming; collection; and removal of settled sludge, floating debris, and grease. Preaeration or mechanical flocculation, often with chemical addition, can be used to enhance primary treatment.

*S*EDIMENTATION

Raw wastewater contains suspended particulates heavier than water; these particles tend to settle by gravity under quiescent conditions. Sedimentation tank design has historically been based on empirically derived relationships or design criteria. These criteria, when coupled with a theoretical understanding of the sedimentation process, may be used for the design of reliable and efficient sedimentation basins.

SEDIMENTATION THEORY. Gravity sedimentation is an effective removal method for raw wastewater suspensions, which range from a low concentration of nearly discrete particles to a high concentration of flocculent solids. In discrete particle settling, the settling velocity of a particle is a function only of the fluid properties of the wastewater and the characteristics of the solid particles. The terminal settling velocity of a discrete particle is represented by Stokes' law:

$$V_s = gd^2(p_s\,p_f)/(18u) \qquad (1)$$

Where

V_s = terminal settling velocity, m/s;
d = spherical particle diameter, m;
g = gravity acceleration, 9.8 m/s^2;
p_s = particle density, kg/m^3;
p_f = water density, kg/m^3; and
u = absolute dynamic viscosity of water, kg/m^3.

Camp divided the ideal sedimentation basin into four zones: the inlet, settling, outlet, and sludge zones (Figure 10.1).[5,6] In the inlet zone, wastewater solids and flows are uniformly distributed over the cross-sectional area of the sedimentation basin. In the settling zone, a uniform concentration of particles settles at the terminal settling velocity to the sludge zone at the bottom of the basin where sludge is collected and removed. Effluent discharges over a weir in the outlet zone. Camp determined the removal efficiency of discrete particles to be a function of only the overflow rate. The overflow rate of the sedimentation basin can be expressed by the following equation:

$$V_o = Q/A \qquad (2)$$

Where

V_o = ideal basin overflow rate, m/s (ft/sec);
Q = flow to sedimentation basin, m^3/s (cu ft/sec); and
A = surface area of sedimentation basin, m^2 (sq ft).

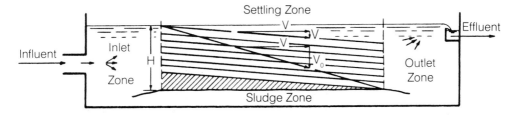

Figure 10.1 Zones of an ideal sedimentation basin.[5,6]

The overflow rate represents the minimum settling velocity necessary for sedimentation. Hence, all particles with a terminal-settling velocity (V_s)

Primary Treatment   **449** equal to or greater than the overflow rate (V_o) will settle in the basin; only the fraction (V_s/V_o) of the particles with a velocity less than the overflow rate will settle in the basin. Other references provide information on performing and interpreting settling column tests for determining discrete particle removal efficiencies.[7-9]

Ideal settling tank theory presents a methodology for sedimentation tank design but, in reality, the actual settling performance cannot be adequately predicted because of the unrealistic assumptions regarding discrete particle settling. Departures from ideal discrete particle settling include particle inter-action (flocculation) and currents in the settling zone. Suspended solids in wastewater are not discrete particles and vary in size and other charac-teristics. Under quiescent settling conditions, large and heavy particles settle faster than small and light particles. As these two types of particles pass each other and make contact, they agglomerate and grow in size in a process known as flocculation. The flocculation process increases removal efficiency but cannot be adequately represented by equations. Settling column tests can provide information for sizing sedimentation tanks. Other references offer in-formation on procedures for conducting and interpreting settling column tests.[7-9]

TYPES OF SEDIMENTATION TANKS. Rectangular (Figure 10.2[7]), circular (Figure 10.3[9]), square, and stacked (Figures 10.4[10] and 10.5[10]) are four types of sedimentation tanks. The selection of the type of sedimentation tank for a given application can be governed by the size of the plant, local regulatory authorities, local site conditions, the experience and judgment of the design engineer, and the economics involved. Rectangular and circular sedimentation tanks are commonly used for wastewater treatment. Of the 13 design firms responding to a national survey, 9 firms indicated a preference for circular tanks, 1 firm indicated a preference for rectangular tanks, and 3 firms indicated that it was a site-specific decision.[11]

Rectangular Sedimentation Tanks. Typically, these tanks range from 15 to 90 m (50 to 300 ft) in length and 3 to 24 m (10 to 80 ft) in width.[9] Depths typically should exceed 2 m (7 ft).[2] Large rectangular tanks seem to achieve better removal efficiencies than do comparably sized circular tanks, espe-cially if diameters exceed 38 m (125 ft).[12] Operation of rectangular and cir-cular sedimentation tanks with identical parameters in Winnipeg, Canada, resulted in no significant differences in performance.[13] Rectangular tanks with common-wall construction are advantageous for sites with space con-straints. Rectangular tanks are discussed in more detail later in this chapter.

Circular Sedimentation Tanks. Diameters of these tanks vary from small, 3 m (10 ft), to over 90 m (300 ft). Depths typically range between 2.4 and 4.0 m (8 to 13 ft).[2,7] Circular sedimentation tanks can use relatively

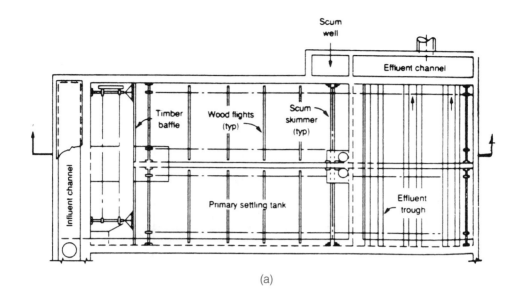

(a)

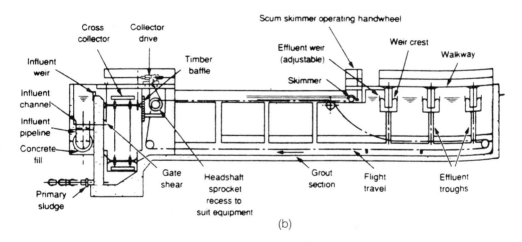

(b)

Figure 10.2 Typical rectangular primary sedimentation tank: (a) plan and (b) section.[7]

trouble-free circular sludge removal equipment (drive bearings are not under water).[8] Walls of circular tanks act as tension rings, which permit thinner walls than those for rectangular tanks. As a result of such advantages, circular tanks have a lower capital cost per unit surface area than that for rectangular tanks.[8] For small facilities, circular tanks generally require more yard piping than rectangular tanks do. Circular tanks are discussed in more detail later in this chapter.

Square Sedimentation Tanks. These tanks are used occasionally for primary sedimentation (for example, in Delavan, Wisconsin). Square sedimentation tanks are hydraulically similar to circular sedimentation tanks. Because square tanks typically have the same sludge equipment as that of

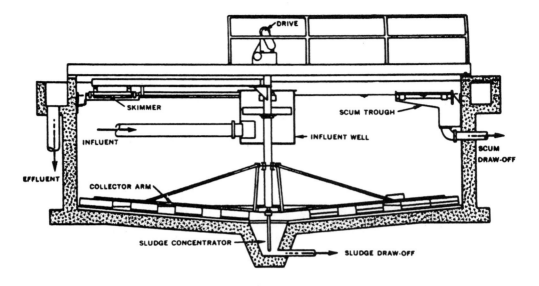

Figure 10.3   Typical circular primary sedimentation tank.[9]

circular tanks, sludge removal from the corners can cause problems. Square tanks may use common wall construction but require thicker walls than circular units. Because square tanks are rarely used, they are not discussed any further in this chapter.

Stacked Sedimentation Tanks. This concept was originally proposed by Camp in 1946.[14] In areas where land for treatment facilities is not available or is extremely expensive, stacked sedimentation tanks (tray clarifiers) can be used. Use of stacked sedimentation tanks for both primary and secondary facilities has resulted in plant area requirements ranging from 0.13 to 0.5 m^2/m^3·d (0.12 to 0.47 ac/mgd).[15] Series flow and parallel flow are two types of stacked sedimentation tanks.

In the series flow unit, wastewater enters the lower tray, flows to the opposite end, reverses direction in the upper tray, and exits the effluent channel (Figure 10.4).[10] Baffles straighten the flow paths and minimize turbulence at the influent point in the lower tray and at the turnaround on the top tray.

In the parallel flow unit, pipes convey wastewater from the influent channel to both the upper and lower trays (Figure 10.5).[10] Influent baffles on each tray straighten the flow path and minimize turbulence. Effluent is removed from both trays by longitudinal launders along the top tray. The parallel tray unit is the most common stacked configuration that has been used to date for primary sedimentation.

Chain and flight collectors are used for sludge collection and removal from stacked tanks. Scum is removed only from the top tank. The inlet and outlet design is considered a weakness of stacked tanks because the wastewater flow patterns might possibly intersect with those of sludge.[16] The

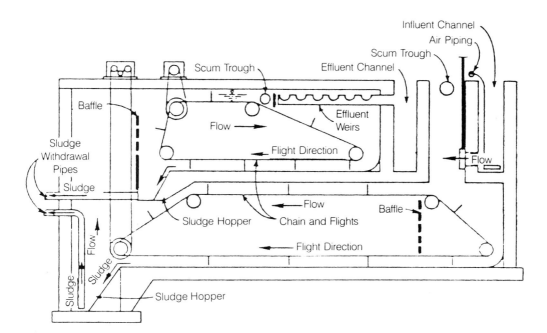

Figure 10.4   Stacked sedimentation tank: series flow type.[10]

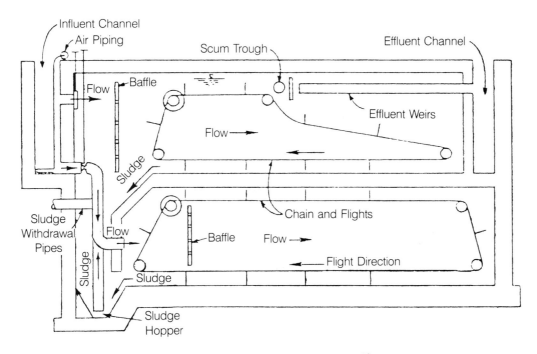

Figure 10.5   Stacked sedimentation tank: parallel flow type.[10]

lower trays of stacked sedimentation tanks are confined spaces subject to confined space entry requirements. The underlying position of the lower tray makes its maintenance and operational observation particularly difficult.

Rectangular sedimentation tanks have been stacked two or three deep at 37 wastewater treatment plants in Japan to increase the treatment capacity per unit surface area.[10,17,18] Treatment plants range in size from 95 000 to 380 000 m^3/d (25 to 100 mgd), with average overflow rates between 18 and 43 $m^3/m^2 \cdot d$ (370 and 1060 gpd/sq ft) and weir loading rates between 84 and 174 $m^3/m \cdot d$ (6800 and 14 000 gpd/ft).[10]

At Osaka City, Japan, wastewater is distributed by outlet control rather than inlet control between the upper and lower tiers of parallel-tray-type sedimentation tanks.[18] Outlet troughs with V-notch weirs are used because density currents are minimal. Osaka City has operated stacked facilities with satisfactory performance records for 10 to 20 years.[18]

Four batteries of stacked primary sedimentation tanks are under construction at the 4 900 000-m^3/d (1.3-mgd) Deer Island treatment facility in Boston, Massachusetts.[16] The design of stacked primary and secondary sedimentation tanks has reportedly reduced the "footprint" of the processes by 40%.[16]

Stacked primary sedimentation tanks under construction in Mamaroneck, New York, are designed for peak influent flows of 350 000 m^3/d (92 mgd).[10] The design overflow rates are 22.3 $m^3/m^2 \cdot d$ (547 gpd/sq ft) at average flow and 44.6 $m^3/m^2 \cdot d$ (1095 gpd/sq ft) at peak flow; the design detention times are 4.52 hours at average flow and 2.26 hours at peak flow.[10]

DESIGN CONSIDERATIONS. Historically, sedimentation basin design has relied on empirically derived criteria such as basin overflow rate, depth, surface geometry, hydraulic detention time, and weir rate. These criteria are helpful for design but are not accurate enough to permit prediction of actual sedimentation performance.

The effect of anticipated frequency and duration of extreme conditions (high and low flows) on sedimentation tank performance should be evaluated during design. In a survey of wastewater treatment plants in the U.S., 8 of 69 (12%) respondents indicated that influent flow variations are a significant problem.[19] Sedimentation tanks sized at average flow conditions should be checked at extreme flow conditions (such as peak storm flows with recycle flows and tanks out of service) to verify that operating parameters are acceptable. Degraded performance at peak flows should be considered in the design of secondary treatment processes. Design of sedimentation tanks must identify and take into account the flow characteristics of the flow stream. Recycle and WAS flows should be considered in sizing primary sedimentation tanks. Combined sewer systems are subject to wider flow variations than are separate sanitary systems.

Typically, dry weather raw influent to wastewater treatment plants varies throughout the day as shown by the curve presented in Figure 2.2. Organic loading to the plant, also shown in Chapter 2, follows a similar curve but with a 2- to 4-hour lag behind flow.[20,21] The ratio of peak dry weather flow to average flow is typically 1.5:1 to 3:1 or more.[7] Peak to low flow ratios may be 5:1 to 10:1 in extreme cases. The design of primary sedimentation tanks should be flexible enough to allow successful operation during low flow start-up conditions.

Although dry weather flow conditions prevail throughout most of the U.S., peak storm or wet weather flows must also be considered. Wet weather flows depend on the location, intensity, and duration of rainfall plus characteristics of the sanitary sewer system. Therefore, wet weather flows are more difficult to predict than are dry weather flows. Substantial infiltration or inflow from sanitary sewers or the existence of combined storm and sanitary sewers might result in wet weather flows that are several times higher than normal dry weather flows.

Recycle streams, such as waste activated sludge or trickling filter underflow, may cause surges in flow. These surges should be avoided if possible or returned to the plant influent stream during low flow periods. In a survey of wastewater treatment plants in the U.S., 13 of 72 (18%) respondents indicated plant recycle streams to be a significant problem.[19] Table 10.1 lists the type of recycle streams returned by 81 treatment plants responding to a national survey.[19] Influent pumps, typically variable speed or multiple constant speed, are normally designed to provide a smooth gradual transition of flow to primary sedimentation tanks.

Table 10.1   Types of recycle streams for 81 plants included in WPCF Survey.[19]

Recycle stream	Percentage of treatment plants (81) reintroducing recycle stream ahead of primary treatment
Sludge-thickening supernatant	58
Aerobic digester supernatant	43
Anaerobic digester supernatant	73
Heat treatment liquor	83
Sludge-dewatering filtrate	66
Tertiary filter backwashing	63

Less conservative design criteria may be considered in those cases where secondary treatment follows primary treatment. However, the economic and operational impacts of less conservative design criteria on downstream treatment processes require consideration.

Overflow Rate. Overflow rates (surface loading rates) recommended by various entities for primary sedimentation tank design are summarized in Table 10.2.[2] Table 10.3 lists the design criteria for overflow rates used by 12 design firms in the U.S.[11] State regulatory agencies typically review and approve overflow rates for plant design and many have adopted standards for overflow rates governing design. These criteria, generally guided by the so-called idealized settling theory, rely on many favorable conditions that are not attainable in practice. A graphical representation of the relationship between overflow rate and performance on an idealized basis is shown by the curve in Figure 10.6.[2,19] Such performance is not always achieved because many conditions not accounted for by theory affect the performance of operating facilities. These include inappropriate application of design details, loading variability, the soluble-to-insoluble ratio of organic content, volatile/inert proportions, and recycle flow proportions. These limitations and their contributions to variability often distort the expectation of the overflow rate design criteria as illustrated by the WPCF survey (1989) of primary sedimentation tank performance graphically depicted in Figure 10.6. Therefore, use of overflow/performance relationships calls for caution, recognizing that more favorable ratios require careful and prudent consideration of the many other factors that affect performance. Use of high overflow rates may

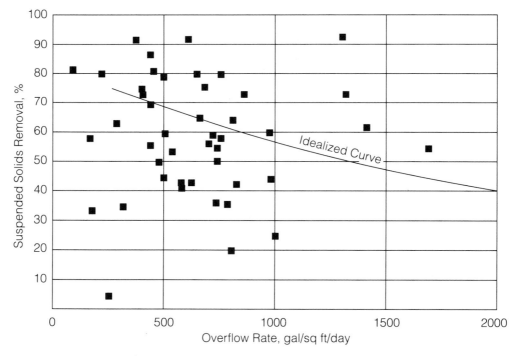

Figure 10.6 Primary clarifier suspended solids removals versus overflow rates showing idealized curve[2] with data.[19]

Design of Municipal Wastewater Treatment Plants

Table 10.2   Summary of overflow rates and side water depths recommended by various entities for primary sedimentation tanks.[a]

Source	Recommendations and remarks
Metcalf and Eddy, Inc.[7]	For primary settling followed by secondary treatment—32 to 48 $m^3/m^2{\cdot}d$ at average flow, 80 to 120 $m^3/m^2{\cdot}d$ at peak flow. For primary settling with waste activated sludge—24 to 32 $m^3/m^2{\cdot}d$ at average flow, 48 to 70 $m^3/m^2{\cdot}d$ at peak flow. Recommended side water depth: 3 to 5 m for rectangular clarifiers, 3.6 m typical. 3 to 5 m for circular clarifiers, 4.5 m typical.
"Naval Facilities Design Manual"[22]	49 $m^3/m^2{\cdot}d$ at maximum 24-hour flow with all units in service. 81 $m^3/m^2{\cdot}d$ at peak flow with all units in service. 163 $m^3/m^2{\cdot}d$ at peak flow with one unit out of service. Side water depth = 3 m.
"Ten State Standards"[23]	For primary clarifiers: 41 $m^3/m^2{\cdot}d$ at average design flow, minimum side water depth = 2.1 m. 61 $m^3/m^2{\cdot}d$ at peak hourly flow, minimum side water depth = 2.1 m. Area used is the larger of the two areas calculated using the criteria above. For intermediate tanks following fixed film processes: 61 $m^3/m^2{\cdot}d$ at peak hourly flow, minimum side water depth = 2.1 m.
EPA Process Design Manual "Suspended Solids Removal"[4]	For primary settling followed by secondary treatment—33 to 49 $m^3/m^2{\cdot}d$ at average flow, 81 to 122 $m^3/m^2{\cdot}d$ at peak flow, side water depth = 3 to 4 m. For primary settling with waste sludge—24 to 33 $m^3/m^2{\cdot}d$ at average flow, 49 to 61 $m^3/m^2{\cdot}d$ at peak flow, side water depth = 4 to 5 m.
U.S. Army[24]	Allowable overflow rate depends on plant design flow. Varies from 12 $m^3/m^2{\cdot}d$ for design flow not exceeding 38 m^3/d to 41 $m^3/m^2{\cdot}d$ for design flow above 37 850 m^3/d. Side water depth is dependent on clarifier dimensions, between 2.5 and 4.5 m.
Steel and McGhee[25]	24 to 60 $m^3/m^2{\cdot}d$, side water depth = 1 to 5 m.
"Guidelines for the Design of Wastewater Treatment Works" (1980 edition)[26]	"Depth: The liquid depth of mechanically cleaned settling tanks shall be as shallow as practical, but not less than 7 ft (2.1 m). Where activated sludge is returned to the settling tanks, the liquid depth of the tank shall not be less than 8 ft (2.4 m)."

[a] $m^3/m^2{\cdot}d$ × 24.55 = gpd/sq ft; m × 3.281 = ft.

Primary Treatment   457

Table 10.3   Design criteria for overflow rates used by 12 design firms.[11]

Overflow rate, gpd/sq ft[a]	Number of firms
800 (average design flow)	1
800 –1000 (average design flow)	1
1000 (average design flow)	2
2500 (peak flow)	1
2000 (peak flow)	1
"Ten State Standards"	4
WPCF MOP FD-8, "Clarifier Design"	1
Site specific	1
Total	12

[a] gpd/sq ft × 0.040 74 = $m^3/m^2 \cdot d$.

be appropriate, in some cases, where well-designed secondary treatment will follow primary treatment. For instance, removal efficiencies for 65% SS and 44% BOD5 were achieved with overflow rates between 54.5 and 66.4 $m^3/m^2 \cdot d$ (1340 and 1630 gpd/sq ft) at the Encina treatment plant in Carlsbad, California, in 1987.[22] During this period, primary sludge concentrations were maintained at 5.76%.

Depth. The opportunity for contact between particles and flocculation increases with depth. Hence, theoretically, the removal efficiency should increase with depth. In actual practice, it is uncertain whether better removals can be obtained or higher overflow rates can be applied with deeper sedimentation tanks. Sedimentation tanks must be deep enough to accommodate mechanical sludge removal equipment, store settled solids, prevent scour and resuspension of settled solids, and avoid washout or carry-over of solids with the effluent. Shallower depths may be acceptable with continuous sludge removal. Excessive depth is to be avoided if the solids detention time could cause anaerobic conditions. Rectangular sedimentation tanks in Louisville, Kentucky, have a depth of 4.4 m (14.5 ft) and the tanks at St. Paul, Minnesota, are 4.7 m (15.5 ft) deep. Designers' opinions differ as to the depth required for optimum sedimentation tank performance. Table 10.2 lists a range of depths recommended by several entities.

Hydraulic Detention Time. Sufficient time for contact between solids particles is necessary for flocculation and effective sedimentation. Typical hydraulic detention times for primary sedimentation tank design are summarized in Table 10.4.[2] Some states have set limits (high and low) for detention times. Design considerations should include effects of low flow periods to ensure that longer detention times will not cause septic conditions. Septic conditions increase potential odors, solubilization, and loading to

Table 10.4 **Summary of hydraulic detention times recommended by various entities for primary sedimentation tanks.[2]**

Source	Recommendations and remarks
Metcalf and Eddy, Inc.[7]	Primary settling followed by secondary treatment—range 1.5 to 2.5 hours, 2.0 hours typical.
	Primary settling receiving waste activated sludge—range 1.5 to 2.5 hours, 2.0 hours typical.
Steel and McGhee[23]	1 to 2 hours based on peak flow.
Fair *et al.*[24]	Minimum detention period of 2.0 hours in 3-m (10-ft) side water depth.
Sunstrom and Klei[25]	1 to 4 hours.
U.S. Army[26]	2.5 hours except where clarifier precedes an activated sludge system. 1.5 hours where clarifier precedes an activated sludge system. Selection of optimum detention time depends on side water depth and overflow rate.

downstream processes. Detention times of more than 1.5 hours without continuous sludge withdrawal may result in resolubilization of organic matter. Consider detention times increased by the following multiplier for cold climates with wastewater temperatures below 20°C (68°F).[2]

$$M = 1.82\, e^{-0.03t} \qquad (3)$$

Where

M = detention time multiplier and
t = temperature of wastewater, °C.

Weir Rate. Weir rates have little effect on the performance of primary sedimentation tanks, especially with side wall depths in excess of 3.7 m (12 ft).[7,27] Typical weir rates for primary sedimentation tanks are summarized in Table 10.5.[2] In practice, weir loadings often do not exceed 120 m³/m (10 000 gal/ft) in treatment plants handling 3785 m³/d (1.0 mgd) or less and 190 m³/m (15 000 gal/ft) in treatment plants handling more than 3785 m³/d (1.0 mgd).[30] Regulations of some states govern weir loadings.

Linear Flow-through Velocity. In practice, the linear flow-through velocity (scour velocity) has been limited to 1.2 to 1.5 m/min (4 to 5 ft/min) to avoid resuspension of settled solids.[31] The critical scour velocity may be calculated from the following equation:[14]

$$V_H = \{[8k(s-1)gd]/f\}^{0.5} \qquad (4)$$

Primary Treatment 459

Where

V_H = critical scour velocity, m/s;

k = constant for type of scoured particles;

s = specific gravity of scoured particles;

g = acceleration due to gravity, 9.8 m/s^2;

d = diameter of scoured particles, m; and

f = Darcy–Weisbach friction factor.

Typical k values are 0.04 for unigranular material and 0.06 for sticky, interlocking material. Typically, f values range between 0.02 and 0.03. f values are a function of the Reynold's number and the characteristics of the settled solids surface.

Table 10.5 Summary of weir loading rates recommended by various entities for primary sedimentation tanks.[2,a]

Source	Recommendations and remarks
Metcalf and Eddy, Inc.[7]	Primary settling followed by activated sludge—125 to 500 m^3/m·d at average flow, 250 m^3/m·d typical. Primary settling receiving waste activated sludge—125 to 500 m^3/m·d at average flow, 250 m^3/m·d typical.
"Ten States Standards"[28]	Not to exceed 124 m^3/m·d for plant flows of 3785 m^3/d or less. Higher loadings may be used for higher flows but should not exceed 186 m^3/m·d. If pumping is required, weir loadings should be related to pump delivery rates to avoid short circuiting.
"Naval Facilities Design Manual"[29]	For primary clarification—1240 m^3/m·d at maximum 24-hour flow. For intermediate clarification—372 m^3/m·d at maximum 24-hour flow.
U.S. Army[26]	Not to exceed 63 m^3/m·d for plants designed for less than 379 m^3/d. Not to exceed 126 m^3/m·d for plants designed for between 379 and 3785 m^3/d. Weir loadings for plants designed for in excess of 3785 m^3/d may be higher but must not exceed 149 m^3/m·d. When pumping is required, the pump capacity shall be related to tank design to avoid excessive weir loadings.

^{a} m^3/m·d × 80.52 = gpd/ft.

Surface Geometry. As indicated above, currents in the settling zone may hinder settling or result in scouring of settled solids. To minimize these effects, surface geometry is another design variable that has been used in attempting to control scouring of solids from high linear flow-through velocities or wind. Although the length-to-width ratio of rectangular tanks has historically been used as such a design tool, it is not considered to be reliable. Common length-to-width ratios employed for design range from 3:1 to 5:1. Length-to-width ratios for existing rectangular tanks range between 1.5:1 and 15:1.[2] Width is often controlled by the availability of sludge collection and removal equipment.

Inlet Conditions. Inlets should be designed to dissipate the inlet port velocity, distribute flow and solids equally across the cross-sectional area of the tank, and prevent short circuiting in the settling tank. Concentration and density differences between the influent and the tank contents significantly affect the hydraulic performance of the tank.[21] Inertial currents and wind direction may also impact the hydraulic performance. Horizontal velocity variations across the width of rectangular tanks can adversely affect sedimentation efficiency.[21] Vertical variations are thought to have little effect on sedimentation if scour is avoided.[21] The minimum distance between the inlet and outlet should be 3 m (10 ft) unless the tank includes special provisions to prevent short circuiting. Wherever possible, wastewater should enter the tank parallel and symmetric with the tank center line. Because this configuration is often not possible, wastewater must then approach the tank from a right angle channel.

Inlet channel velocities should be high enough to prevent solids deposition. The inlet channel design typically allows a minimum velocity of 0.3 m/s (1 ft/sec) at 50% of design flow. Other alternatives to high inlet velocities for prevention of solids deposition are inlet channel aeration or water jet nozzles.[32]

Velocities are typically dissipated through some type of inlet baffle. Baffles are usually installed 0.6 to 0.9 m (2 to 3 ft) ahead of the inlets and submerged 46 to 61 cm (18 to 24 in), depending on tank depth. The top of the baffle should be far enough below the water surface to allow scum to pass over the top.

Influent flow can be distributed by

- Inlet weirs,
- Submerged ports or orifices with velocities between 3 and 9 m/min (10 and 30 ft/min), and
- Gate valves and perforated baffles.

One of several possible combinations will describe the flow pattern in a circular sedimentation tank. The flow pattern can be center-feed with

---

center-withdrawal, center-feed with peripheral-withdrawal, peripheral-feed with center-withdrawal, or peripheral-feed with peripheral-withdrawal. Center-feed with peripheral-withdrawal is the most common type of flow pattern. Peripheral-feed configurations require horizontal pipe supports that impair flow distribution and scum collection. Center-feed configurations eliminate these disadvantages.

Circular sedimentation tanks typically have a feedwell with a diameter 15 to 20% of the tank diameter. Manufacturers' recommendations for submergence vary significantly. In practice, the feedwell has typically been extended at least half of the tank depth.

Many designs of inlet baffling devices have been used with varying degrees of success.[2] Figure 10.7 illustrates the inlet diffuser for rectangular sedimentation tanks in Long Beach, California. Figures 10.8 and 10.9 show inlet diffusers used for primary sedimentation tanks in Valencia, California. Figure 10.10 illustrates various inlet configurations for circular sedimentation tanks.

Practical considerations usually govern the size of openings, but principles of jet diffusion may serve as a design guide.[21] The prime causes of skewed flow distributions in a tank are, typically, uneven flow distribution in the inlet channel and deflection by the baffle in the inlet zone.[32] More uniform flow can be achieved by locating inlet ports away from tank sides, adding partitions or baffles in the inlet zone to redirect the influent, and creating a higher head loss in the inlet ports relative to that in the inlet channel.[32] The higher head loss may break up flocs. In plants with multiple sedimentation tanks, general practice has provided for equal distribution of flow and solids between the tanks. Splitter boxes and common channels are sometimes used for this purpose. The flow distribution system should include devices for measuring flow to each tank so that adjustments can be made to equally split flow.

Various devices and procedures can influence flow distribution.[2] Figures 10.11 and 10.12 show the configuration of inlet diffusers and finger baffles at the Encina treatment plant in Carlsbad, California. Inlet baffles with finger baffles are used at Sunnyvale, California, for influent distribution (Figure 10.13). Target and finger baffles are used at the Renton treatment plant in Seattle, Washington, for influent distribution (Figure 10.14).

Outlet Conditions. Proper sedimentation tank operation depends on outlet conditions. Effluent should be uniformly withdrawn to prevent localized, high-velocity gradients and short circuiting. Figure 10.15 illustrates the prevailing velocity gradients (drift) in a rectangular sedimentation tank. If these velocity gradients reach the scour velocity, settled particles can be swept into the tank effluent. Density flow, rather than a high approach velocity, often causes sludge carry-over of the effluent weirs. Therefore, effluent should be withdrawn from the tank in a manner that minimizes these currents. Typically, effluent is withdrawn from a sedimentation tank by an

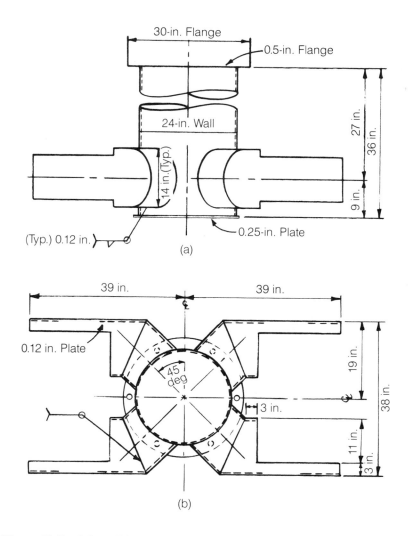

Figure 10.7 Inlet diffuser for primary sedimentation tanks at Long Beach, California: (a) plan and (b) elevation.

overflow weir into a launder or effluent channel. The overflow weir must be level to control the water surface elevation in the sedimentation tank and promote uniform effluent withdrawal. Weirs may be either straight edged (Figure 10.16) or V-notched (Figure 10.17). V-notched weirs provide better lateral distribution of outlet flows than straight-edged weirs that are imperfectly leveled.[8]

Submerged launders have also been used for effluent withdrawal (Figure 10.18). Collection pipes or launders with submerged orifices are two types of submerged launders. Orifices should be sized for uniform flow distribution. Compared with overflow weirs, submerged launders offer some advantages. The submerged launders avoid free fall of wastewater with the

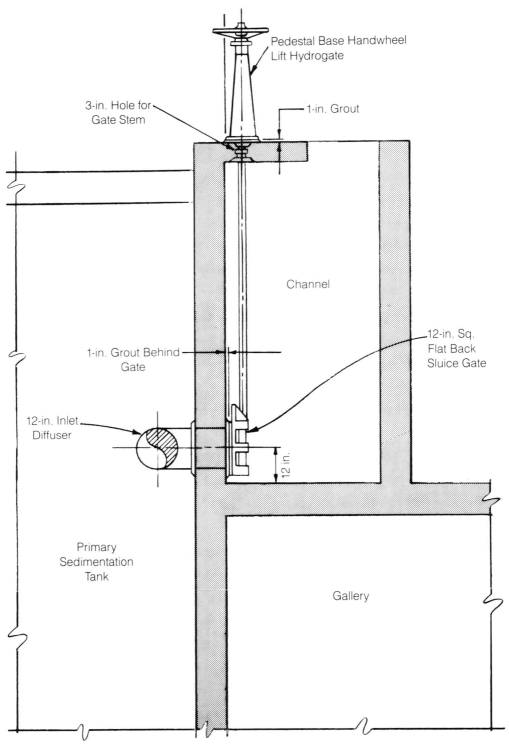

Pedestal Base Handwheel
Lift Hydrogate

3-in. Hole for
Gate Stem

1-in. Grout

Channel

1-in. Grout Behind
Gate

12-in. Sq.
Flat Back
Sluice Gate

12-in. Inlet
Diffuser

12 in.

Primary
Sedimentation
Tank

Gallery

Figure 10.8   Inlet configuration of primary sedimentation tanks at Valencia, California.

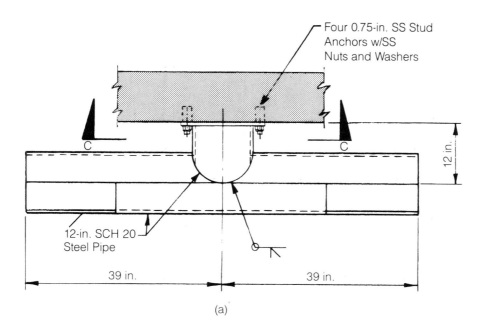

Four 0.75-in. SS Stud Anchors w/SS Nuts and Washers

12 in.

C   C

12-in. SCH 20 Steel Pipe

39 in.   39 in.

(a)

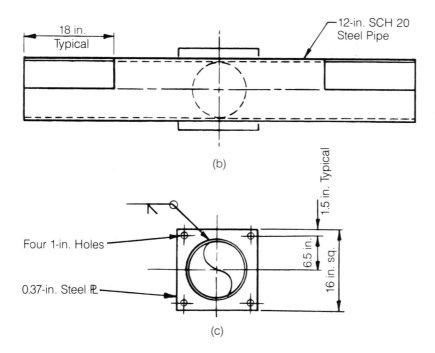

18 in. Typical

12-in. SCH 20 Steel Pipe

(b)

1.5 in. Typical

Four 1-in. Holes

6.5 in.

16 in. sq.

0.37-in. Steel PL

(c)

Figure 10.9   Inlet diffusers for primary sedimentation tanks at Valencia, California: (a) plan, (b) elevation, and (c) section C–C.

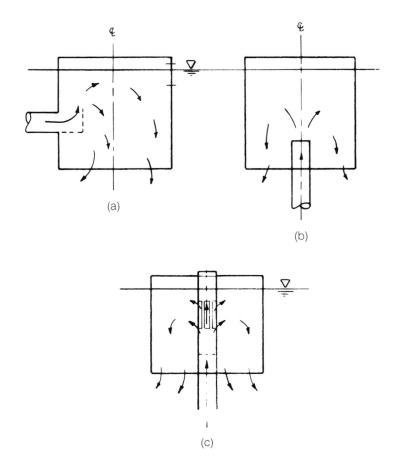

Figure 10.10   Various designs of conventional center feed inlets: (a) side feed, (b) vertical pipe feed, and (c) slotted, vertical pipe feed.

consequent release of entrained odorous gases and allow surface skimming at the end of the tank. A disadvantage of submerged launders is that orifices sized for uniform flow distribution at average flows will not be effective at peak flows. Thus, a separate modulating flow control device or primary effluent pumps are required with submerged launders. These devices should be located and sized properly for effective scum removal. Unique launders with both submerged orifices for average flows and V-notch weirs for overflow capability are used at the Renton treatment plant in Seattle, Washington.[22]

The two principal approaches to weir and launder design are the long-launder and short-launder options. The long-launder approach assumes weir placement and length to be as important as that for secondary clarifier design. Using this approach, weirs and launders for rectangular sedimentation basins would be designed to cover from 33 to 50% of the basin length.[12,33] Long launders control the water elevation in the sedimentation basin within a narrow range but are ineffective when bottom-flowing density currents exist in the basin.[33] In cold regions, long launders might not be best because fluc-

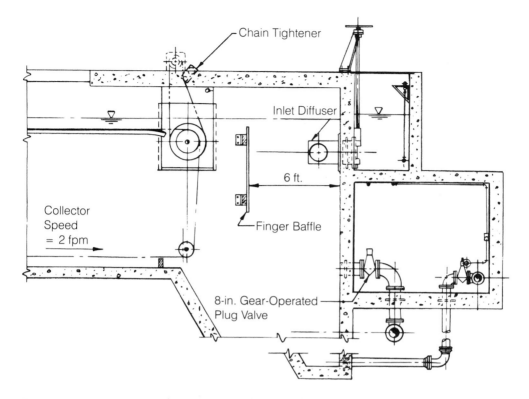

Chain Tightener

Inlet Diffuser

6 ft.

Collector
Speed
= 2 fpm

Finger Baffle

8-in. Gear-Operated
Plug Valve

Figure 10.11   **Inlet configuration of primary sedimentation tanks at Encina plant in Carlsbad, California.**

tuating water levels with short launders would minimize ice attachment to launders and basin walls.[33] The short-launder approach assumes weir length to be unimportant. A simple tank-width weir is used at the end of the tank for outlet control at Valencia, California (Figure 10.16).

Designers' opinions regarding launder spacing also differ. Some designers believe that launders should be spaced between 5 and 6 m (16 to 20 ft) apart.[12] A launder spacing of only 2.4 m (8 ft) exists at the San Jose Creek treatment plant in Whittier, California. Launders are typically arranged transversely across the basin with chain and flight sludge collection equipment. With traveling bridge sludge collection equipment, launders must be arranged longitudinally as parallel finger weirs supported on piers. Parallel finger weirs are used at Sunnyvale, California, with flight and chain collection equipment. Weirs and launders for circular sedimentation basins are typically mounted on the peripheral wall of the tank. Experience at some plants has demonstrated that weirs and launders should be placed at least 15% of the basin radius inboard from the periphery of the tank.[12] Such placement minimizes wall flow disturbance and draws effluent from a broader area.[12]

Launder stability is an important design consideration. Launders should be designed with provisions to relieve loadings during basin draining and prevent buoyancy uplift during basin filling. A 5-cm (2-in.) diameter hole in the bottom of each launder section is one possible method.[12]

Wave harmonics from wind, earthquakes, or fluid flow may cause launders to oscillate or vibrate, thereby possibly deflecting or deforming the

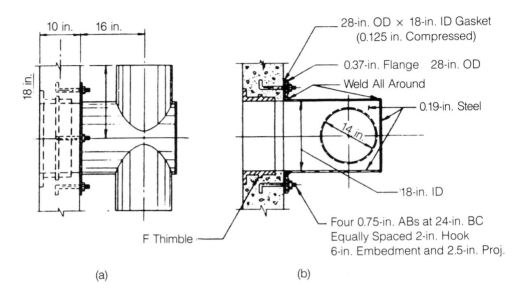

Figure 10.12 **Inlet diffusers for primary sedimentation tanks at Encina plant in Carlsbad, California: (a) plan and (b) section.**

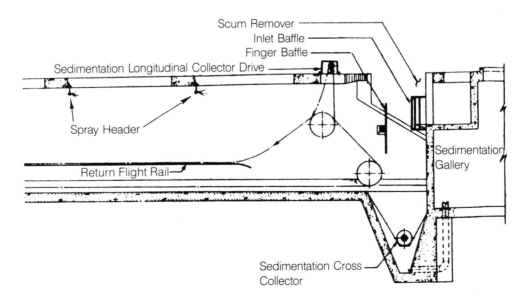

Figure 10.13 **Inlet configuration of primary sedimentation tanks at Sunnyvale, California.**

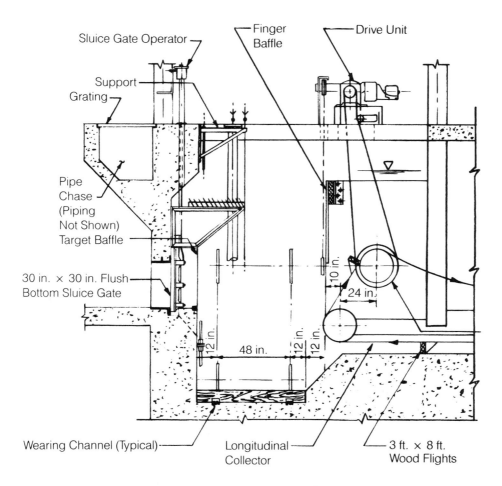

Figure 10.14   Inlet configuration of primary sedimentation tanks at Renton plant in Seattle, Washington.

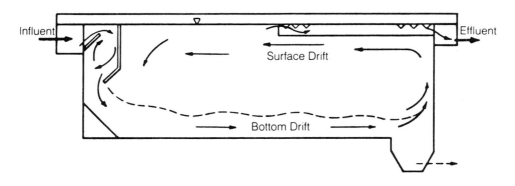

Figure 10.15   Displacement vectors in a real basin showing the prevailing drifts.

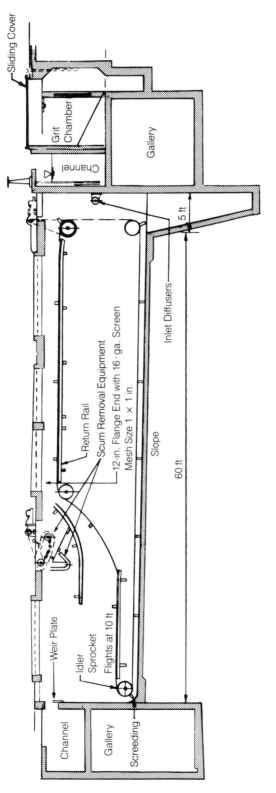

Figure 10.16 Typical cross section of primary sedimentation tanks at Valencia, California.

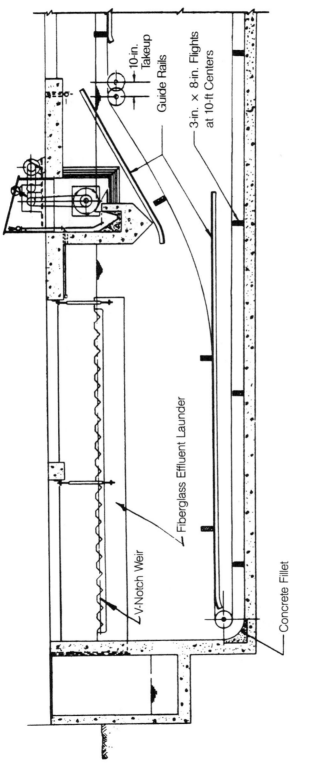

Figure 10.17 Outlet configuration of primary sedimentation tanks at the Encina plant in Carlsbad, California.

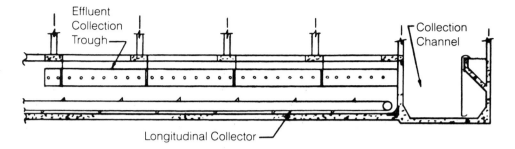

Effluent
Collection
Trough

Collection
Channel

Longitudinal Collector

Figure 10.18 Outlet configuration of primary sedimentation tanks at the Renton plant in Seattle, Washington.

launders and damaging the structural support system. New light materials and long launders aggravate this problem. Launders and weirs should be anchored to resist seismic forces because of wastewater sloshing in the basin. The large launders in center-feed, circular sedimentation tanks are particularly vulnerable to this type of damage. In some cases, designing a breakaway launder support system that would allow for easy replacement might be more economical and practical than designing a structure to withstand earthquake-induced loadings from sloshing wastewater. Break-away designs must prevent sections from falling to the bottom of the tanks and potentially damaging the sludge collection and removal equipment.

Weather Conditions. Weather conditions can affect the performance of sedimentation tanks and must be considered in their design. Rectangular sedimentation tanks should be oriented, if possible, with the length perpendicular to prevailing winds to shelter tanks from the wind. However, some believe that the width should be perpendicular to the predominant wind direction because the water surface would tend to remain level at the outflow weirs. Wind shelter is especially needed for circular sedimentation tanks to avoid nonuniform withdrawal rates and short circuiting caused by wind created turbulence.[4] At Milford, Iowa, 1.1-m (3.5-ft) walls provide wind protection for the 17-m (56-ft) diameter primary sedimentation tanks (Figure 10.19).[34]

Wind may cause the water surface on the leeward side to be higher than on the windward side. This may lead to unbalanced weir rates, especially for large circular sedimentation tanks. Surface skimmers should be oriented so that prevailing winds will push scum towards the collector. Design considerations for wind mitigation include orientation of tanks, installation of windbreaks or covers, increase of tank freeboard, and reduction of circular tank diameters to 37 m (120 ft) or less.

Cold weather conditions that also deserve consideration include freeze protection of surface sprays, insulation of piping, installation of underground piping at greater depths (below the freeze line), and provisions for drainage

Figure 10.19 Concrete wall windbreak for 17.1-m (56-ft) diameter primary clarifier.

piping that conveys intermittent flow. Because scum collection equipment is especially prone to freezing, it needs adequate protection. Occasionally, these steps are insufficient in areas of severe cold and freezing; then sedimentation tanks require covers to avoid operational problems.

Ventilation of Enclosed Areas. The presence of hydrogen sulfide gas and oxygen depletion are extremely hazardous and common occurrences in enclosed areas. Enclosed areas of primary treatment facilities need to be well ventilated (with appropriate provisions for odor release and control) to provide a safe, nuisance-free atmosphere for operation and maintenance of facilities. Local regulatory codes (NFPA) typically require minimum ventilation rates for enclosed areas. The design should include alarms to alert personnel, before entering an enclosed area, of malfunction of the ventilation system. Sensors for a hazardous atmosphere (low oxygen, hydrocarbons, and hydrogen sulfide) should also be considered.

Maintenance Provisions. Two or more sedimentation tanks will allow the process to remain in operation while a tank is out of service for maintenance or repair.[7] Primary treatment facilities should include provisions for necessary maintenance. For example, pumps or other equipment located in buildings or vaults require access for maintenance and repair. Access measures include lifting eyes in the overhead, traveling bridge cranes, access openings

---

Primary Treatment   473

for use of a crane outside the structure, and adequate lighting. Sufficient clear space should be provided around pumps, meters, valves, and other equipment to accommodate maintenance and repair. This is particularly important when replacing the rotor and stator on progressing cavity pumps. Valves should be operable from the floor level.

A good design will allow dewatering of sedimentation tanks for servicing sludge collection equipment or removing an obstruction from an inlet baffle. Dewatering measures can include sloped bottoms for draining, permanent pumps, and piping or connections for temporary equipment. Also necessary are provisions to isolate a tank that is out of service from the remainder of the plant, which must remain in service. Sludge collection and removal mechanisms with their critical components remaining above the water line merit consideration. Ample flushing ports and cleanouts are needed at critical tees, elbows, and ends.

Routine maintenance needs should be taken into account. For example, the plant influent wet well should be pumped down periodically to control the buildup of grease and other floating material. To relieve the heavy load of floating material on flow distribution devices it must be possible to remove scum in inlet distribution boxes or channels with submerged orifices when necessary. Sluice gates should be downward opening wherever possible to avoid the buildup of scum on the water surface and prevent deposition of solids in the track, which impedes full closure of the gate. Eliminate corner pockets and dead ends to minimize the possibility for septic conditions and use corner fillets and channeling where necessary. In practice, the tops of sub-merged troughs, beams, and other construction features have often been sloped 1.4:1 and their bottoms have been sloped 1:1; this will reduce or prevent the accumulation of solids and scum.[28] Provisions should also be made for cleanup after maintenance (for example, frequent use of hose bibs and sump pumps). Hose bibs should be provided at each tank, scum trough, sump, and pump station.

E*NHANCED SEDIMENTATION*

Primary sedimentation can be enhanced by preaeration or chemical coagulation, each discussed below.

PREAERATION. Preaeration of raw wastewater before sedimentation promotes flocculation of finely divided solids into more readily settleable flocs, thereby increasing suspended solids and BOD_5 removal efficiencies. Preaeration also improves scum flotation and removal. Other benefits include scrubbing of VOC odor components from raw wastewater, the addition of dissolved oxygen, and the prevention of septicity during primary sedimentation. A grit chamber with preaeration before each sedimentation tank will also promote uniform distribution of flow to sedimentation tanks and improve grit separation (for example, Sunnyvale, California; New Dublin–San Ramon, California; and Renton treatment plant in Seattle, Washington). Of 1003 wastewater treatment facilities responding to a national survey, 29% use preaeration.[35]

Design Considerations. Detention times of 20 to 30 minutes are necessary for floc formation and improved suspended solids and BOD_5 removals. This range exceeds the 10 to 15 minutes suggested for odor control. The exact quantity of air required is a function of wastewater characteristics and tank geometry. The minimum air rate typically provided is 0.82 L/L (0.11 cu ft/gal). Chapter 9 presents additional information on preaeration.

Plant Information. Preaeration for 45 minutes improved suspended solids and BOD_5 removals by 7 to 8% during research at Ames, Iowa.[36] Data from the city of Los Angeles' Hyperion wastewater treatment plant indicates that preaeration does not improve settling characteristics of an anaerobic raw wastewater with high oxygen demands.[37]

CHEMICAL COAGULATION. Chemical coagulation of raw wastewater before sedimentation promotes flocculation of finely divided solids into more readily settleable flocs, thereby increasing suspended solids, BOD_5, and phosphorus removal efficiencies. Sedimentation with coagulation may remove 60 to 90% of the TSS, 40 to 70% of the BOD_5, 30 to 60% of the COD, 70 to 90% of the phosphorus, and 80 to 90% of the bacteria loadings.[7,38] In comparison, sedimentation without coagulation may remove only 40 to 70% of the TSS, 25 to 40% of the BOD_5, 5 to 10% of the phosphorus loadings, and 50 to 60% of the bacteria loading.[7,23,38] Chapter 16 contains additional information on the selection and application of chemicals for phosphorus removal.

Advantages of coagulation include greater removal efficiencies, the ability to use higher overflow rates, and more consistent performance. Disadvantages of coagulation include an increased mass of primary sludge, production of a sludge that is often more difficult to thicken and dewater, and an increase in operational cost and operator attention. The designer of chemical coagulation facilities should consider the impact of enhanced primary sedimentation on downstream sludge-processing facilities.

Chemical Coagulants. Historically, iron salts, aluminum salts, and lime have been the chemical coagulants used for wastewater treatment. Iron salts have typically been the most common coagulant used for primary treatment (for example, the cities of Ludington, Escanaba, Greenville, and West Branch, Michigan). Only a few plants use lime as a coagulant for primary treatment (for example, Martinez, California) because lime addition produces more sludge than do metals salts and is more difficult to store, handle, and feed. To enhance sedimentation, Maysville, Ohio, and Milan, Michigan, use aluminum salts (alum).

Coagulant selection for enhanced sedimentation should be based on performance, reliability, and cost. Performance evaluation should use jar tests of the actual wastewater to determine dosages and effectiveness. Table 10.6[39] lists coagulant dosages for various plants with chemical coagulation. Operating experience, cost, and other relevant information drawn from other plants should be considered during selection. Organic polymers are sometimes used as flocculation aids.

Rapid Mix. During rapid mix, the first step of the coagulation process, chemical coagulants are mixed with the raw wastewater. The coagulants destabilize the colloidal particles by reducing the forces (zeta potential), keeping the particles apart, which allows their agglomeration. The destabilization process occurs within seconds of coagulant addition. At the point of chemical addition, intense mixing will ensure uniform dispersion of the coagulant throughout the raw wastewater. The intensity and duration of mixing must be controlled, however, to avoid overmixing or undermixing. Overmixing may reduce the removal efficiency by breaking up existing wastewater solids and newly formed floc. Undermixing inadequately disperses the chemical, increases chemical usage, and reduces the removal efficiency.

The velocity gradient, G, is a measure of mixing intensity. Velocity gradients of 300 m/m·s are generally sufficient for rapid mix, but some designers have recommended velocity gradients as high as 1000 m/m·s[40-42] Formulas for calculating the velocity gradient for various mixer configurations are presented in other references.[4,38,43]

Mechanical mixers, in-line blenders, pumps, baffled compartments, baffled pipes, or air mixers can accomplish rapid mix.[44] The mixing intensity of mechanical mixers and in-line blenders is independent of flow rate, but these mixers cost considerably more than other types and might become clogged or entangled with debris. Air mixing eliminates the problem of debris and can offer advantages for primary sedimentation, especially if aerated channels or grit chambers already exist. Pumps, Parshall flumes, flow distribution structures, baffled compartments, or baffled pipes—methods often used for upgrading existing facilities—offer a lower cost but less efficient alternative to separate mixers for new construction. The methods listed above are less efficient than separate mixers because, unlike separate mixing, the mix intensity depends on the flow rate. Chapters 7 and 16 contain additional information on the design of rapid mix facilities.

Flocculation. During the flocculation step of the coagulation process, the destabilized particles grow and agglomerate to form large, settleable flocs. Through gentle prolonged mixing, chemical bridging, physical enmeshment

Table 10.6 Performance and coagulant dosages for enhanced sedimentation.[39]

Location	Flow, mgd	Advanced primary performance						Chemical addition		
		BOD			TSS				Concentration, ppm	
		Inf., mg/L	Eff., mg/L	Removed, %	Inf., mg/L	Eff., mg/L	Removed, %	Type		Duration
Point Loma City of San Diego	191	276	119	56.9	305	60	80.3	$FeCl_3$ Anionic polymer	35 0.26	Continuous
Orange County Plant No. 1	60	263	162	38.4	229	81	64.6	$FeCl_3$ Anionic polymer	20 0.25	8 hours peak flow
Orange County Plant No. 2	184	248	134	46.0	232	71	69.4	$FeCl_3$ Anionic polymer	30 0.14	12 hours peak flow
JWPCP Los Angeles County	380	365	210	42.5	475	105	77.9	Anionic polymer	0.15	Continuous
Hyperion City of Los Angeles	370	300	145	51.7	270	45	83.3	$FeCl_3$ Anionic polymer	20 0.25	Continuous
Sarnia Ontario, Canada	10	98	49	50.0	124	25	79.8	$FeCl_3$ Anionic polymer	17 0.30	Continuous

of particles, or both occur. Flocculation is slower and more dependent on time and agitation than is the rapid mix step. Typical detention times for flocculation range between 20 and 30 minutes. Increasing the detention time beyond this range offers only marginal benefits.[45] Detention times as short as 5 minutes have been reported. Flocculation can occur in separate structures or in baffled areas of channels, tanks, or existing structures serving other purposes. Aerated and mechanical grit chambers, flow distribution structures, and influent wells are areas that promote flocculation upstream of primary sedimentation. Advantages and disadvantages of different configurations resemble those for rapid mix facilities.

Like rapid mix, the velocity gradient, G, achieved with each configuration should be checked. Velocity gradients should be maintained from 50 to 80 m/m·s.[38] Formulas for calculating the velocity gradient for various configurations are presented in other references.[4,38,43,46,47]

Polymers are sometimes added during the flocculation step to promote floc formation. Polymers should enter as dilute solution to ensure thorough dispersion of polymers throughout the wastewater. Polymers may provide a good floc with only turbulence and detention in the sedimentation tank inlet distribution.[4] Chapter 16 contains additional information on the design of flocculation facilities.

Coagulant Addition. Supplementing the conventional primary sedimentation with chemical coagulation requires minimal additional construction. The optimal point for coagulant addition is as far upstream as possible from primary sedimentation tanks. Figure 10.20 shows a possible flow diagram for coagulant addition.[38] The optimum feed point for coagulant addition often varies from plant to plant. If possible, several different feed points should be considered for additional flexibility. Dispersing the coagulant throughout the wastewater is essential to minimize coagulant dosage and concrete and metal corrosion associated with coagulant addition.[48] To promote dispersion, multiple injection points or a chemical solution header

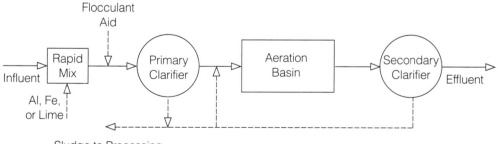

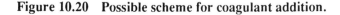

Figure 10.20 Possible scheme for coagulant addition.

(Figure 10.21) can be used.[48] Flow-metering devices should be installed on chemical feed lines for dosage control.

Plant Information. All raw wastewater at the Los Angeles, California, Hyperion wastewater treatment plant 1 438 300 m^3/d (380 mgd) receives enhanced primary treatment by chemical coagulation. The addition of approximately 20 mg/L of ferric chloride and 0.2 mg/L of polymers to the headworks before primary sedimentation increased sludge production by approximately 45%.[49] Approximately 30% of this increase results from improved suspended solids removal and the remaining 65% stems from chemical precipitation and removal of colloidal material.[49] Additional sludge production is partially offset by a decrease in waste activated sludge production. Chemical coagulation increased suspended solids removals from 65 to 75% and BOD_5 removals from 30 to 45%. Primary sludge has an approximate solids content of 4.5%.

Evaluation of 30 different polymers at the Joint Water Pollution Control Plant (JWPCP) at Carson, California,[50] found anionic polymers to be the most effective type for enhancing primary sedimentation. Polymer addition (0.15 mg/L) at JWPCP increased suspended solids capture from 66 to 83%,

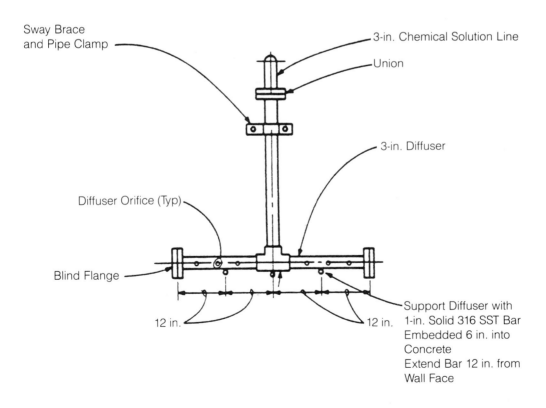

Figure 10.21 Chemical solution diffuser.

but the settled sludge concentration decreased from 5.3% in 1976 to 4.5% in 1984.[20] The decrease is likely attributable to a combination of reasons, including polymer addition, simultaneous increase in the contribution of waste activated sludge from upstream plants, and implementation of pretreatment regulations for industrial discharges. Turbulence in the aerated headworks inlet channel provides the necessary mixing for flocculation.

Removal efficiencies and coagulant doses for various plants with enhanced primary sedimentation by chemical coagulation are listed in Table 10.6.[39] Sedimentation tanks with chemical coagulation operate with overflow rates between 69 and 81 $m^3/m^2 \cdot d$ (1700 and 2000 gpd/sq ft) in Los Angeles and San Diego.[34] These rates are twice the conventional overflow rate of 33 $m^3/m^2 \cdot d$ (800 gpd/sq ft).[34] Research in Sarnia and Windsor, Ontario, on the effects of chemical coagulation on primary sedimentation tanks indicated that overflow rates up to 98 $m^3/m^2 \cdot d$ (2400 gpd/sq ft) do not significantly affect effluent quality.[51] Hence, the size and number of sedimentation tanks required for primary treatment can sometimes be reduced substantially by chemical coagulation.[39] The average removal efficiencies of 56 Norwegian treatment plants employing chemical coagulation are listed in Table 10.7.[52]

Table 10.7   Results of direct precipitation representing the average of 56 Norwegian treatment plants in 1985.[52]

Pollutant	In, mg/L	Out, mg/L	Reduction, %
SS	387	24	94
BOD	229	38	83
P	6.2	0.5	92

S*LUDGE COLLECTION AND REMOVAL*

SLUDGE COLLECTION. Settled sludge is generally scraped into a hopper where it is removed by gravity or pumping. The sludge hopper for rectangular tanks is usually located at the inlet end of the tank to minimize the travel time of sludge particles to the hopper. For circular tanks, the hopper is usually located in the center of the tank. The hopper, up to 3 m (10 ft) deep, generally has steep sides with a minimum slope of 1.7:1.[28] Hopper wall surfaces should be smooth with rounded corners to avoid any solids buildup.[28] In practice, the hopper bottom has a maximum dimension of 0.6 m (2 ft).[28] Sedimentation tanks with steep sides and widths over 3 m (10 ft) often need more than one hopper to reduce its depth.

Common sludge withdrawal pipes from two or more hoppers often remove sludge unequally from the hoppers. Therefore, multiple tanks and hoppers need separate pipes and pumps or valves on each outlet.

Sludge collection equipment for rectangular tanks generally includes chain and flights (for example, in Valencia, California; St. Paul, Minnesota; and Greensboro, North Carolina) or traveling bridge (for example, in Janesville, Wisconsin). Chain and flights (Figure 10.16) consist of two endless loops of chains with cross scrapers (flights) attached at approximately 3-m (10-ft) intervals. Revolving flights push the settled sludge to the sludge hopper at the end of the tank. Chain and flight sludge collectors are limited in width to approximately 6 m (20 ft); some installations, however, have used side-by-side collectors without common walls in tanks as wide as 24 m (80 ft) or more.[7] Historically, cast iron chains and wood flights were used. Designers now select almost exclusively nonmetallic (plastic) chains and fiberglass flights (for example, in Valencia, California). Chain and flights for a pair of tanks usually move about 0.6 m/min (2 ft/min), driven by a single-drive unit located on the wall between the two tanks.[12] A flight speed of 0.9 m/min (3 ft/min) is used at Long Beach, California, and Pomona, California.

Flights travel along the long axis of the tank and, as the upper flights move away from the sludge hopper, can skim the surface, pushing floating material toward the scum removal mechanism. At the end of the tank, the flights drop to the floor and drag heavy, settled material to a hopper for removal. Single tanks can have either a single or double hopper. Multiple tanks or tanks with more than one sludge collection assembly [tank widths in excess of 6 m (20 ft) generally require more than one assembly] often use a cross collector in a transverse trough. The cross collector, typically 1.2 m (4 ft) wide and 0.9 to 1.2 m (3 to 4 ft) deep, runs the width of one or more tanks as it conveys sludge to a hopper.[12] Cross collectors are typically flight and chains on 1.5-m (5-ft) centers, which travel along the transverse trough at between 0.6 and 1.2 m/min (2 and 4 ft/min).[12] Screw-type cross collectors, sometimes used, usually rotate at 10 rpm (Figure 10.22).[12]

A traveling bridge (Figure 10.23) consists of a sludge scraper blade mechanism mounted on a bridge or carriage that travels approximately 1.8 m/min (6 ft/min) toward the sludge hopper on tracks or rails mounted on top of the tank. As it travels away from the sludge hopper at approximately 3.7 m/min (12 ft/min), the mechanism, largely out of the water, acts as a skimmer, pushing floating material toward the scum removal mechanism. As it reaches the end of the tank, the mechanism drops to the floor of the tank, reverses direction, and travels toward the hopper end of the tank, pushing settled sludge to a hopper.

Traveling bridges are generally easier to maintain than chain and flight collection equipment. Traveling bridge collectors cannot be used if covers are required on primary sedimentation tanks. Traveling bridge collectors may

---

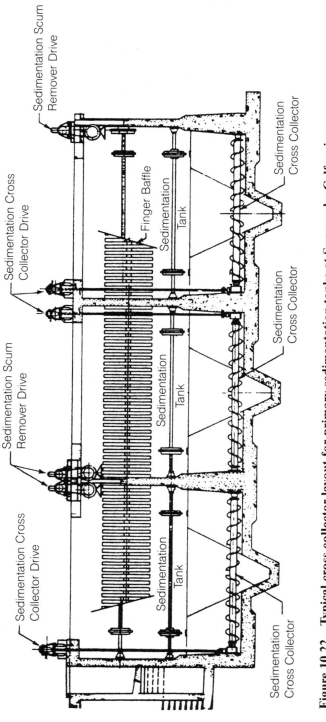

Figure 10.22 Typical cross collector layout for primary sedimentation tanks at Sunnyvale, Calfornia.

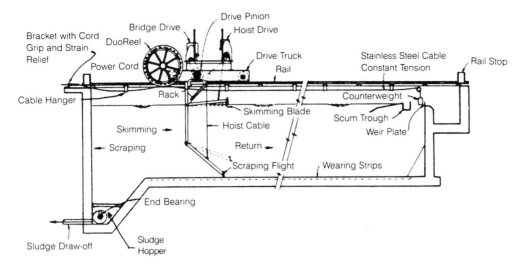

Figure 10.23 **Typical section of a traveling bridge sludge collector.**

span tank widths up to 30 m (100 ft). In earthquake zones, a restraining mechanism should be considered to resist derailing from seismic ground motion. In cold climates, design of traveling bridge collectors should provide for control of snow and ice buildup on rails or tracks. Otherwise, this buildup could derail the bridge or reduce traction between the wheel and rail.

Circular primary sedimentation tanks typically have plow-type sludge collection equipment (for example, in Bismark, North Dakota, and Midland, Texas). The plow type (Figure 10.3) consists of scrapers that drag the tank floor at a tip speed of approximately 1.8 to 3.7 m/min (6 to 12 ft/min).[12] Plows are located at an angle to the radial axis to force sludge towards the hopper, normally at the center of the tank, as the device rotates. The center hopper is normally a vertical-sided sump where the sludge is removed by pumping. The rotating element of the device can be driven from either the center or the outside tank wall. Torque must be sufficient to move the densest sludge expected.

During testing, plow-type collectors at Kansas City, Missouri, did not adequately remove sludge from circular sedimentation tanks.[53] Collectors left a sludge blanket of up to 0.91 m (3 ft) thick. This contributed to septic conditions, increased the fraction of soluble BOD_5 in the raw wastewater, and decreased the BOD_5 removal efficiencies to only 1 to 17%.[53] Modification of the plows with a picket fence (Figure 10.24[53]) reduced the sludge blanket to less than 130 mm (5 in.), minimized septic conditions, lowered the fraction of soluble BOD_5, and increased the BOD_5 removal efficiency to approximately 30%.[53]

The suction-type sludge collector should not be used for primary sedimentation because of the high sludge density and the risk of clogging the suction

Figure 10.24 Temporary picket fence modification of plow-type collectors at Kansas City, Missouri.[53]

arm orifices with items such as rags, though suction-type sludge collectors are used in Westville, Indiana.

SLUDGE QUANTITIES AND PROPERTIES. Primary sludge production can be estimated from the following equation:

$$S_M = \frac{Q \times TSS \times E}{1000} \text{ (metric)} \tag{5}$$

$$S_M = Q \times TSS \times E \times 8.34 \text{ lb/gal (English)}$$

Where

S_M	=	mass of sludge, kg/d (lb/d);
Q	=	primary influent flow, m^3/d (mgd);
TSS	=	primary influent total suspended solids, mg/L; and
E	=	removal efficiency, fraction.

TSS removal efficiencies in primary sedimentation tanks usually range between 50 and 65%. Many designers assume a removal efficiency of 60% for estimating purposes.[7]

Enhanced primary sedimentation with chemical coagulation can increase primary sludge mass by 50 to 100%.[48] Chemical sludge quantities can be estimated by the stoichiometric relationship between raw wastewater and coagulants. The stoichiometric quantity should be increased by approximately 35% for aluminum and iron salts to account for increased BOD_5,

COD, and TSS removal.[38,54] Table 10.8[55] lists typical sludge quantities for estimating where site-specific data are not available. Sludge volumes may be estimated by assuming a solids concentration for removed sludge. In a survey of wastewater treatment plants in the U.S., 11 of 71 (15%) respondents identified measurement of the sludge flow rate to be a significant problem at the plant.[19]

Table 10.8 Predicted quantities of suspended solids and chemical solids removed in a hypothetical primary sedimentation tank.[55,a]

Sludge type	No chemical addition[c]	Chemical addition[b]		
		Lime[d]	Alum[e]	Iron[f]
Suspended solids, lb/mg	1041	1562	1562	1562
Chemical solids, lb/mg	—	2082	362	462
Total sludge production, lb/mg (kg/m^3)	1041 (0.13)	3644 (0.44)	1924 (0.23)	2024 (0.24)

[a] Assumes no recycle streams (for example, recycle of waste activated sludge to primary sedimentation and digester supernatant). Secondary solids production would be cut from 833 lb/mg without chemical addition to 312 lb/mg with chemical addition in this hypothetical plant. lb/mg × 0.000 12 = kg/m^3.
[b] Assumes 10 mg/L influent phosphorus concentration (as P), with 80% removed by chemical precipitation.
[c] Assumes 50% removal of 250 mg/L influent TSS in primary sedimentation.
[d] 125 mg/L Ca (OH)$_2$ added to raise pH to 9.5.
[e] 154 mg/L Al$_2$ (SO$_4$)$_3$·14 H$_2$O added.
[f] 84 mg/L FeCl$_3$ added.

Composition of primary sludge is variable and depends on the nature and degree of industrial development in the collection area. Table 10.9[55] lists typical characteristics for primary sludge. Chemical sludges are gelatinous, with a high water content, low suspended solids content, and high resistance to mechanical or gravitational dewatering. This sludge composition merits careful consideration in design of sludge-handling and processing units.

SLUDGE THICKENING. Primary sludge is thickened in primary sedimentation tanks, sludge stabilization facilities, or separate thickening units. Primary sedimentation tanks can be operated to produce a thickened sludge of 6% solids or more by allowing a blanket of solids to build up and compact the sludge.[56] Typical solids concentrations for primary sedimentation sludge range from 4 to 12%.[7] Higher concentrations can be achieved but often cause problems in the conveyance system.[56]

Table 10.9   Primary sludge characteristics.[55]

Characteristic	Range of values	Typical value	Comments
pH	5–8	6	—
Volatile acids, mg/L as acetic acid	200–2000	500	—
Heating value, Btu/lb[a]	6 800–10 000	—	Depends on volatile content and sludge composition; reported values are on a dry-weight basis.
		10 285	Sludge 74% volatile.
		7 600	Sludge 65% volatile.
Specific gravity of individual solid particles	—	1.4	Increases with increased grit and silt.
Bulk specific gravity (wet)	—	1.02	Increases with sludge thickness and with specific gravity of solids.
		1.07	Strong wastewater from a system of combined storm and sanitary sewers.
BOD$_5$:VSS ratio	0.5–1.1	—	—
COD:VSS ratio	1.2–1.6	—	—
Organic N:VSS ratio	0.05–0.06	—	—
Volatile content, percent by weight of dry solids	64–93	77	Value obtained with no sludge recycle, good degritting; 42 samples, standard deviation 5.
	60–80	65	
	—	40	Low value caused by severe storm inflow.
	—	40	Low value caused by industrial waste.
Cellulose, percent by weight of dry solids	8–15	10	—
	—	3.8	—
Hemicellulose, percent by weight of dry solids	—	3.2	—
Lignin, percent by weight of dry solids	—	5.8	—
Grease and fat, percent by weight of dry solids	6–30	—	Ether soluble.
	7–35	—	Ether extract.
Protein, percent by weight of dry solids	20–30	25	
	22–28	—	—
Nitrogen, percent by weight of dry solids	1.5–4	2.5	Expressed as N.
Phosphorus, percent by weight of dry solids	0.8–2.8	1.6	Expressed as P$_2$O$_5$. Divide values as P$_2$O$_5$ by 2.29 to obtain vlaues as P.
Potash, percent by weight of dry solids	0–1	0.4	Expressed as K$_2$O. Divide values as K$_2$O by 1.20 to obtain values as K.

[a] Btu/lb × 2.326 = kJ/kg.

Sedimentation tanks are sometimes operated with continuous withdrawal of dilute primary sludge to minimize thickening, maximize removal, and prevent anaerobic decomposition of settled sludge. Anaerobic or septic conditions will result in the resolubilization of BOD_5. Removal efficiencies for raw wastewater with a large fraction of soluble BOD_5 will be considerably lower than those for the same wastewater with a smaller fraction of soluble BOD_5. Solubilization and septicity are especially troublesome in hot climates (Southwest U.S. and Hawaii) and where collection systems have long detention times. The Sand Island treatment plant in Honolulu, Hawaii, has difficulty meeting the primary treatment requirements imposed by EPA because the large percentage of soluble BOD_5 present in the wastewater reduces sedimentation removal efficiencies.

Soluble BOD_5 accumulated in the sedimentation tank during attempts to thicken sludge at the Renton Plant in Seattle, Washington.[22] This was attributed to the development of septic conditions in the tank and scouring of the sludge blanket at peak flows. The BOD_5 loading to the aeration tanks was increased by approximately 20%.[22] In general, thickening of sludge should not be attempted with overflow rates greater than 100 $m^3/m^2 \cdot d$ (2500 gpd/sq ft).[22] Such rates call for separate thickener facilities. Chapter 17 includes additional information on separate thickening methods.

SLUDGE TRANSPORT AND HANDLING. The sludge drawoff system should be designed with the capacity to allow either continuous withdrawal or intermittent withdrawal at a rate that will control the sludge blanket depth. If primary sedimentation tanks are to be operated to achieve additional primary sludge thickening, drawoff piping and pumps must be designed to handle the more concentrated sludge. Sludge withdrawal lines should be at least 100 mm (4 in.) in diameter. As the sludge solids content increases to more than 6%, risk of plugging increases due to the increased viscosity of the thickened sludge and its tendency to clog the piping. For this reason, the shortest and straightest possible lengths of suction piping should be provided along with access to the suction piping for rodding, pigging, or flushing to clear obstructions. A sight glass or sludge density meter is necessary on the suction side of the primary sludge pump. The primary sludge line should include a sampling port and flowmeter. Primary sludge pumps should be positioned to maintain a net positive suction head. Time clocks on sludge pumps should be capable of being set at 30-minute increments. Where practical, standby pumps should be provided for sludge pumping instead of interconnected piping that could become clogged with debris and grease and thus fail to function properly when needed.

Scum Management

Removal of floating materials, or scum, is an important function of primary treatment. Oil, grease, plastics, and other floating materials increase the organic load to downstream treatment processes and might cause various operational troubles, including visual blights, odors, and the buildup of scum in downstream treatment processes. In a survey of wastewater treatment plants in the U.S., 19 of 74 (26%) respondents identified scum collection and pumping as a major problem at the plant.[19]

SCUM COLLECTION. Scum collection has typically been located on the effluent end of the primary sedimentation tank (Figure 10.2). Some plants, however, have located scum collection on the influent end of the sedimentation tank to decrease the travel distance of scum to the collection point and ensure rapid removal of all flotage (Figure 10.25).[57] The Renton treatment plant in Seattle, Washington, and the plant at Sunnyvale, California, employ this approach. Manual scum collection is employed at some plants (for example, in Olcott, New York; West Branch, Michigan; and Durham, North Carolina). Automated scum removal mechanisms may be operated by the sludge collection mechanism or a separate operating device. The scum removal mechanism should extend the full width of the tank to bar floating material from reaching the effluent weir. The inlet design should allow scum to freely enter sedimentation tanks without being trapped in inlet channels or behind baffles. Slots should be provided on the circular feed well for this purpose. In some cases (for example, influent high in oil and grease content), a separate skimmer mechanism or water spray system is necessary for the center feed well. A nearly constant water surface elevation in sedimentation tanks should be maintained for proper operation of scum collection equipment.

Generally, two types of mechanisms—tilting trough (Figure 10.26[2]) and sloping beach (Figure 10.27[2])—are used for removing scum from primary sedimentation tanks. A tripping device on the sludge collector activates the tilting trough (for example, in Pasco, Washington, and Tampa, Florida). The trough tilts to allow collected material on the surface to flow into the trough and then to a wet well. Wet well scum is pumped to another treatment process. The sloping beach is a stationary device with a collector trough. Floating material may be directed to the sloping beach by air jet sprays (for example, in Sunnyvale, California), water jet sprays (for example, in Martinez, California, and Sunnyvale, California), the sludge collection mechanism, or a separate, blade-type scraper. A separate, blade-type scraper moves the floating material up the beach and into a trough. Carrier water (primary effluent) flushes scum through the trough into a wet well where it is pumped to another treatment process.

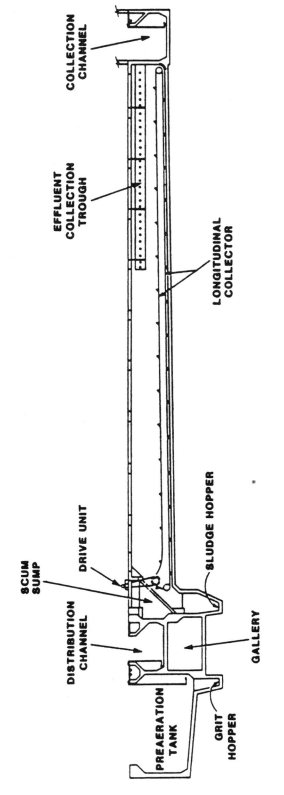

Figure 10.25 Primary sedimentation tank with scum collection at influent end of tank.

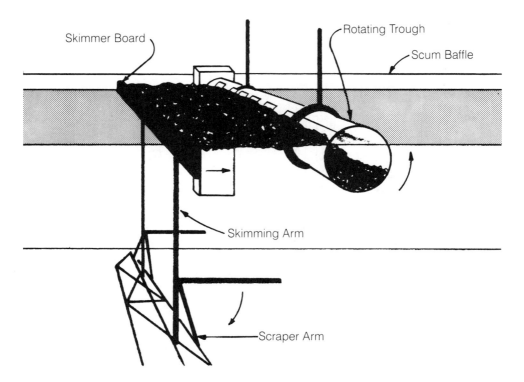

Skimmer Board

Rotating Trough

Scum Baffle

Skimming Arm

Scraper Arm

Figure 10.26   Tilting trough scum collector.[2]

Collectors for rectangular tanks are manually tilted with a lever, rack or pinion, worm gears, or motor-operated devices. The rotating device acts as a collector for surface floating material, which is pushed to either a sloping beach or a tilting trough skimmer. A spring-loaded section of the rotating arm rides up the beach, wipes material into a trough, and drops back into the water on the far side. The tilting trough usually extends into the tank just short of the rotating arm. As the arm or surface collector passes, it physically tilts (rotates) the trough, allowing it to skim the surface.

SCUM QUANTITIES AND CHEMICAL COMPOSITION. Scum quantities and chemical composition for 19 facilities have been summarized in Table 10.10.[55,58,59] The dry-weight scum quantities ranged between 0.1 and 19 mg/L, with a median value of approximately 5 mg/L. Quantity and chemical composition of scum are both highly variable and depend on several factors, including the degree and type of industrial development in the collection area, recycled plant sidestreams, and scum removal efficiency of upstream processes and scum removal equipment.

SCUM TRANSPORT AND HANDLING. Progressing cavity pumps, pneumatic ejectors, and recessed impeller centrifugal pumps, both with and without cutting bar attachments, have been used to pump scum. The design

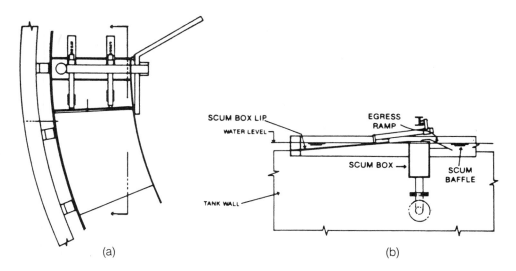

Figure 10.27   Sloping beach scum collector: (a) partial plan and (b) section.[2]

of scum removal equipment includes measures to keep the scum tank or hopper contents mixed during pumping to prevent scum from crusting or coning. The bottom of the scum tank should be sloped. Some designers use glass-lined pipe, which is kept reasonably warm (15°C or higher) to minimize blockages.[55] Flushing connections, pigging stations, and cleanouts are provided where blockage could occur. Whenever practical, standby pumps are provided for scum pumping rather than interconnected piping to avoid clogging of the piping with grease and debris that would impair its use when needed.

SCUM CONCENTRATION, TREATMENT, AND ULTIMATE DISPOSAL. Methods that have been used for scum concentration, treatment, and disposal are listed in Table 10.11.[55,58] Historically, scum has usually been landfilled (for example, in Escanaba, Michigan; Vermillion, South Dakota; Rockford, Illinois; and Bismark, North Dakota), coprocessed with wastewater treatment sludges, and digested (for example, in Palm Springs, California; Tampa, Florida; JWPCP in Carson, California; Yonkers, New York; Midland, Texas; Marshfield, Wisconsin; Downers Grove, Illinois; and Oneonta, New York) or incinerated (for example, in Martinez, California; Fort Edward, New York; and St. Paul, Minnesota). Adequate digester mixing must be provided when scum is discharged to a digester to ensure complete digestion and minimize scum blanket formation. Scum digestion, often based on convenience instead of value, sometimes deferred the scum disposal problem or shifted it to another process. Scum and sludge grinding in the 1970s and 1980s have eliminated aesthetically offensive floating plastic and rubber articles. Scum may be concentrated by flotation

---

Table 10.10 Raw wastewater scum characterizations (from primary sedimentation facilities).[55,58,59]

Treatment plants	Quantity (dry weight), mg/L	Percent volatile	Percent oil & grease	Fuel value Btu/lb	Fuel value kJ/kg
Northwest Bergen County, N.J.	2.3				
Minneapolis–St. Paul, Minn.		98		5 600	13 000
East Bay, Oakland, Calif.	9.8	96	91	6 000	14 000
West Point, Seattle, Wash.	2.9				
Not stated		89	80	7 200	16 800
Three New York City Plants, N.Y.	0.1–2.0				
Los Angeles County Sanitation Districts, Calif.	10				
Albany, Ga.	17				
Milwaukee, Wisc.	3.1				
Central Contra Costa County, Calif.	5.6				
Sacramento, Calif.	14.4				
Passaic Valley Sewerage Commissioners, N.J.	6				
Detroit, Mich.	3				
Wards Island, New York City, N.Y.	4.8				
Bissell Pt., St. Louis, Mo.	10.5				
Calumet, Gt. Chicago WRD, Ill.	1.8	Average of all four is	Average of all four is	Average of all four is	Average of all four is
Southwest, Gt. Chicago WRD, Ill.	5.3				
Westside, Gt. Chicago WRD, Ill.	4.8				
Northside, Gt. Chicago WRD, Ill.	0.5	91	73	6 600	15 300
Eimco (BSP)		96	77	6 500	15 100
Number of plants	19	8	7	8	
Maximum	17	98	91[a]	7 200	16 800
Median	4.8				
Average	5.3	93	77[a]	6 500	15 000
Minimum	0.1	89	73[a]	5 600	13 000

[a] Likely saponifiable (biodegradable) content is 60 to 70%.

Table 10.11   Methods of handling raw wastewater treatment scum.[55,58]

Method	Advantages	Disadvantages
Coprocessing with wastewater treatment sludges and biological stabilization		
• Aerobic	— Partial decomposition occurs — Avoids complexity of separate handling — Widely used	— May cause grease balls to form — May cause petroleum contamination of sludge, impacting reuse — May degrade appearance of sludge if to be reused — May cause scum buildup if remaining residue is not completely removed
• Anaerobic	— See above	— Same as above — If digester not strongly mixed, scum layer is certain to form. Some scum layer is likely with any digester; digester should be sized for this layer — Removal of scum layer is difficult except on cleaning; disposal of layer on cleaning is still difficult — Requires good decanting to avoid the introduction of significant quantities of water with the scum when it is applied to the digester — Avoidance of significant water may lead to line clogging unless special precautions are taken

Table 10.11 Methods of handling raw wastewater treatment scum (continued). [55,58]

Method	Advantages	Disadvantages
Coprocessing with wastewater treatment sludges and incineration	— Low theoretical incremental cost	— Requires good decanting — Can tax furnace if not introduced in a nearly continuous manner — Special air pollution concerns with particulates and volatiles
Individual processing without wastewater sludges and incineration	— Little residual ash	— See above — High first and maintenance costs — Unacceptable at small plants
Individual processing without wastewater sludges and landfill	— Low capital cost	— May have high operating cost — Requires good decanting to minimize volume and fluidity — Product may be unacceptable due to odors and infectious character
Individual processing without wastewater sludges and reuse as animal feed or low grade soap manufacture	— Low capital cost	— Doubtful attractiveness because of risks (most toxic organics tend to concentrate in grease) and superior, low-cost alternatives

(Scranton, Pennsylvania; Willoughby, Ohio; Rockford, Illinois; and JWPCP in Carson, California) and self-cleaning, rotating screens. In Boston, Massachusetts,[58] chemical fixation of scum with lime, Portland cement, soluble silicates, and cement kiln dust yielded an easily handled product with low permeability and no indicator organisms. Additional study of chemical fixation of scum with cement kiln dust is underway at Deer Island Treatment Plant in Boston, Massachusetts.[58] Chemically fixed scum facilitates handling and ultimate disposal and offers possible beneficial reuse as a structural fill or interim or final landfill cover.[58]

COTHICKENING OF WASTE ACTIVATED SLUDGE

Waste biological sludge is sometimes discharged to the influent end of primary sedimentation tanks for cothickening (settlement and consolidation) of primary sludge. Separate WAS thickening has replaced this practice in most cities because of its detrimental effect on primary sedimentation. Cothickening will typically result in a combined sludge concentration of 2 to 6% solids instead of the 4 to 12% solids achievable with primary sludge settling and thickening only.[7] In practice, high overflow rates are avoided with cothickening and sludge is removed rapidly to minimize solubilization of BOD_5 and resuspension and carry-over of sludge solids.[57] Typical design criteria for overflow rates at peak flows are presented in Table 10.12.

In a number of trickling filter and solid contact facilities, cothickening of waste solids with raw wastewater has not adversely affected primary sludge settling or thickening.[57] These plants, which exist in areas with moderate climates, do not have excessive overflow rates. Total solids concentrations were restricted to 3.5% in warmer climates, with water temperatures well above 20°C instead of the 4 to 6% associated with temperate climates. This may reflect the development of biological activity in the thickening zone of

Table 10.12 Typical primary clarifier design overflow rates at peak flow for cothickening of primary sludge.

Reference	Without waste activated sludge, $m^3/m^2 \cdot d^a$	Cosettling, $m^3/m^2 \cdot d^a$
Metcalf and Eddy, Inc.[7]	80–120	48–70
EPA "Process Design Manual"[55]	81–122	49–61
Ontario guidelines[60]	80–120	50–60

[a] $m^3/m^2 \cdot d \times 24.55$ = gpd/sq ft.

primary sedimentation tanks.[57] Rapid sludge removal eliminated the increase in soluble BOD_5 across the sedimentation tank caused by biological activity at one facility.[57] Cothickening of waste activated sludge and raw wastewater during primary sedimentation in Winnipeg, Canada, reduced the sludge concentration from 7.5 to 5.4% but did not otherwise adversely impact the settling characteristics of the raw wastewater.[13,60]

IMHOFF TANKS

Imhoff tanks are sometimes used for small communities with raw wastewater flows in the order of 945 m^3/d (250 000 gpd).[61] A 1988 needs survey identified 415 facilities with Imhoff tanks.[1] Regulations in some states (for example, in Wisconsin) prohibit Imhoff tanks for new construction except for special circumstances. As shown in Figure 10.28, the Imhoff tank consists of a top compartment, which serves as a settling basin, and an unheated lower compartment in which the settled solids are anaerobically stabilized.[7,62] Effluent flows over the sidewall weir to a channel and gas vents to the atmosphere through openings along the side. Digested sludge is periodically removed by gravity from the tank. Imhoff tanks have no mechanical equipment and typically have low maintenance requirements; they nonetheless have operational problems, including the periodic production of an odorous foam, excessive accumulation of scum in the gas vents, and production of an offensive sludge.[3] Heating the lower compartment of the Imhoff tank is not economical because heat will dissipate through the gas vents to the sedimentation compartment. Hence, the tank volume required exceeds that for separate, heated tanks. Imhoff tanks have a limited future, but might be an inexpensive alternative for adding primary treatment to existing community septic tank or stabilization pond systems.

Table 10.13[7] includes design parameters for an Imhoff Tank and Figure 10.28 shows a typical design.[7,62] Conventional Imhoff tanks are usually rectangular, although circular tanks are sometimes used.[7] Imhoff tanks should not be located near residential areas because an odorous gas vents to the atmosphere. A multiport distribution channel that discharges behind a baffle should initiate an even flow pattern across the compartment and limit short circuiting.[3] Outlet weirs should be V-notched and extend across the width of the settling compartment.[3] Like other unit processes, two units in parallel can provide additional backup capacity. All walls below the water line should be insulated by earth embankments to keep the wastewater temperature higher than 15°C for prevention of foaming during spring warm-up.[3] Sludge withdrawal should be limited to twice a year to ensure adequate time for sludge stabilization.

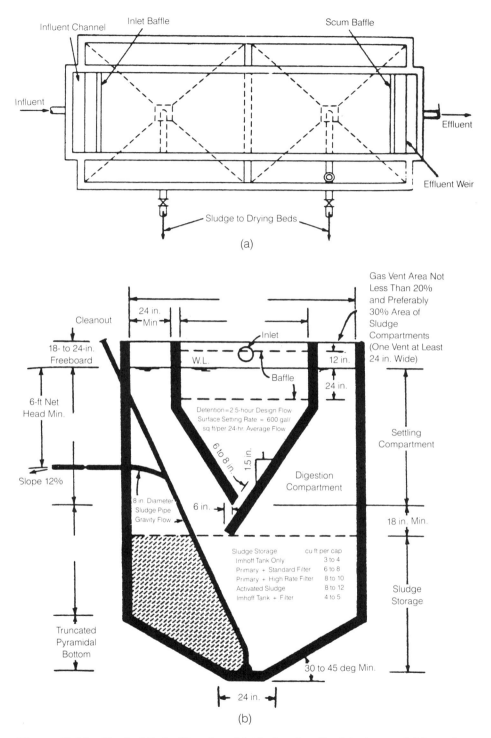

Figure 10.28 Typical Imhoff tanks with design details: (a) plan and (b) section. Use larger sludge storage allowances for garbage solids and for communities with less than 5000 population (in. × 2.54 = cm; cu ft × 0.028 3 = m³) [7, 62]

Table 10.13 Typical design criteria for unheated Imhoff tanks.[7,a]

Design parameter	Value	
	Range	Typical
Settling compartment		
Overflow rate peak hour, gpd/sq ft	600–1000	800
Detention time, hours	2–4	3
Length-to-width ratio	2:1–5:1	3:1
Slope of settling compartment ratio	1.25:1 to 1.75:1	1.5:1
Slot opening, in.	6–12	10
Slot overhang, in.	6–12	10
Scum baffle		
Below surface, in.	10–16	12
Above surface, in.	12	12
Freeboard, in.	18–24	24
Gas vent area		
Surface area, percent of total surface area	15–30	20
Width of opening,[b] in.	18–30	24
Digestion section		
Volume (unheated), storage capacity—		
cu ft/capita		6 months of sludge
Volume[c]	2–3.5	2.5
Sludge withdrawal pipe, in.	8–12	10
Depth below slot to top of sludge, ft	1–3	2
Tank depth		
Water surface to tank bottom, ft	24–32	30

[a] gpd/sq ft × 0.040 7 = $m^3/m^2 \cdot d$; in. × 25.4 = mm; cu ft × 2.831 7 × 10^{-2} = m^3; and ft × 0.304 8 = m.
[b] Minimum width of opening must be 18 in. to allow a person to enter for cleaning.
[c] Based on a 6-month digestion period.

*F*INE SCREENS

Fine screens can be used in lieu of sedimentation for primary treatment but will not achieve removal efficiencies of sedimentation (for example, in Aberdeen, Idaho; Milwaukee, Wisconsin; and Huber Heights, Ohio).[4,7] As another application, fine screens can upgrade existing primary sedimentation facilities. Of 1003 wastewater treatment facilities responding to a national survey, 6% reported using fine screens.[35] Fine screens with openings from 1 to 6 mm (0.04 to 0.25 in.) typically only achieve removal efficiencies of 15 to 30% for suspended solids, 15 to 25% for BOD5, and 10 to 20% for bacteria loadings.[7,23] Hence, downstream treatment processes should be designed on the basis of the anticipated screen performance.

DESIGN CONSIDERATIONS. The most common fine screens used for primary treatment are the inclined, self-cleaning (static) screen (Figure 10.29), the rotary drum screen (Figure 10.30), and the rotary disk screen (Figure 10.31).[7] Table 10.14[4,7] lists typical design information for fine screens. The clear-water head loss through fine screens may be obtained from the manufacturer or calculated by the orifice formula:[7]

$$h_L = [1/c(2g)](Q/A)^2 \qquad (6)$$

Where

h_L	=	head loss, m (ft);
c	=	coefficient of discharge for screen;
g	=	gravity acceleration, 9.8 m/s^2 (32.2 ft/sec^2);
Q	=	discharge through screen, m^3/s (cu ft/sec); and
A	=	effective open area of submerged screen, m^2 (sq ft).

Values of c and A may be obtained from the screen manufacturer. A typical value of c for a clean screen is 0.60.[7] Although the clear-water head loss may be useful for preliminary assessment of different fine screens, determination of the head loss during screen operation with wastewater is more

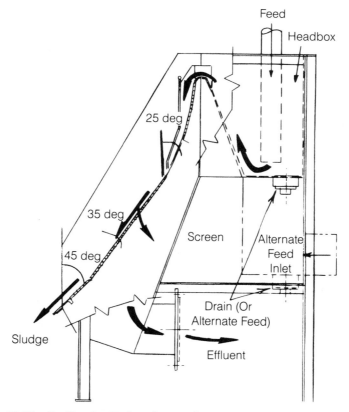

Figure 10.29   Inclined self-cleaning static screen.

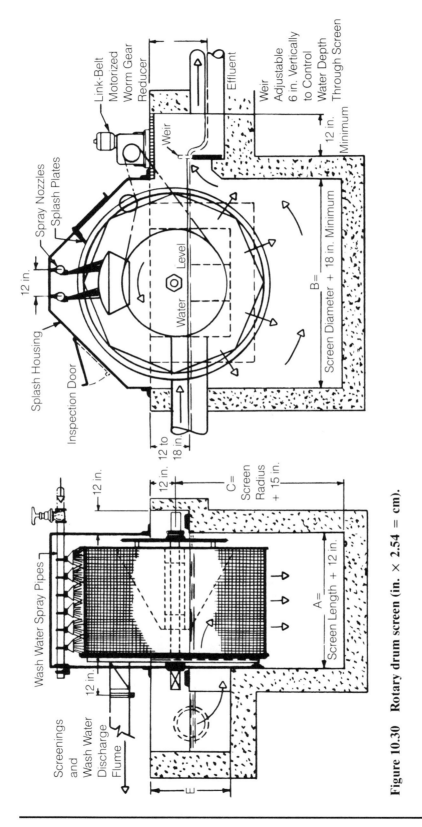

Figure 10.30 Rotary drum screen (in. × 2.54 = cm).

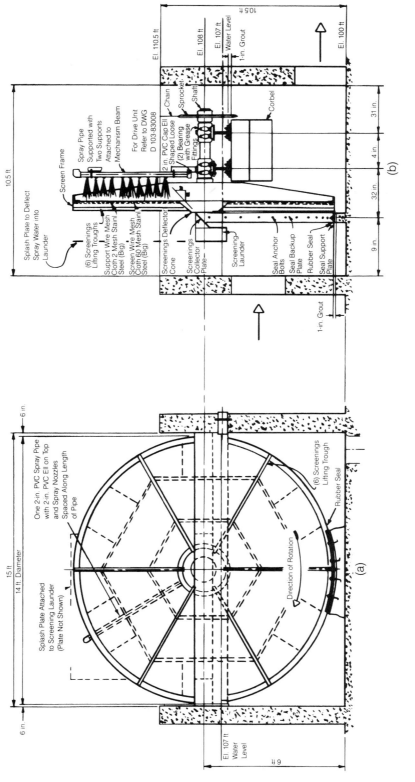

Figure 10.31 Rotary disc screen: (a) screen front (inlet side) view and (b) screen section (in. × 2.54 = cm; ft × 0.304 8 = m).

Table 10.14 Typical design information on screening devices used for the primary treatment of wastewater.[4,7,a]

	Type of screen		
Item	**Inclined**	**Rotary drum**	**Rotary disk**
Screening surface			
Size classification	Medium	Medium	Fine
Size range, μm	0.01–0.06	0.01–0.06	0.001–0.01
Screen material	Stainless steel wedge wire	Stainless steel wedge wire	Stainless steel woven wire
Hydraulic capacity, gpm/sq ft	15–60	0.12–1.0	0.10–1.0
Composition of waste solids— solids by weight, %	10–15	10–15	6–12
Suspended solids removal, %	15–30	15–30	40–50

[a] gpm/sq ft × 0.040 7 = $m^3/m^2 \cdot$min.

important for design. Head loss during operation depends on the quantity and type of solids in the wastewater, the size of screen openings, and the frequency of screen cleaning.

Mechanically cleaned coarse screens, racks, or other protective devices should precede fine screens. Comminutors should not be used upstream of fine screens because lower removal efficiency for the fine screens would result. An installation should have a minimum of two fine screens, with each unit capable of independent operation at peak capacity while the other unit remains out of service. The buildup of grease on screens, especially in colder climates, requires periodic cleaning. Hosing equipment should be included in the design to facilitate cleaning.

HANDLING, TRANSPORT, AND ULTIMATE DISPOSAL OF SCREENINGS. Tables 10.15[55] and 10.16[55] show some of the properties of fine screenings and handling methods. Screenings–water mixtures can be pumped with positive displacement or centrifugal pumps. Fine screenings are odorous and will attract rodents and insects that must be controlled.[55] As a rule, screenings should never be returned to the treatment process after removal.

PLANT INFORMATION. Eight rotary fine screens are used for primary treatment at the Jones Island (Milwaukee, Wisconsin) wastewater treatment facility, with a peak flow of 1 140 000 m^3/d (300 mgd). Average monthly discharge standards for the plant are 30 mg/L TBOD5, 30 mg/L TSS, and 1 mg/L TP. Each drum has a 2.4-m (8-ft) diameter by 2.4-m (8-ft) long stainless steel screen equipped with a brush mechanism. The screen drum has

Table 10.15 Analyses of screenings.[55]

Solids content, % dry solids	Volatile content, %	Fuel value, Btu/lb dry solids[a]	Bulk wet weight, lb/cu ft[b]	Comments[c]
20	—	5400[d]	60	Coarse screenings. Fine screenings may have lower solids content
10–20	80–90	—	40–60	Common values
8–23	68–94	—	53–67	Various plants, fine screens, 0.03- to 0.12-in. openings
6.1	96	—	—	Thickened ground screenings from 0.75-in. racks; after grinding, screenings were thickened on a static screen with 0.06-in. openings
17	96	—	—	Dewatered ground screenings from 0.75-in. racks; after grinding, screenings were dewatered on a rotating drum screen with 0.03-in. openings
—	86	7820	—	Fine screenings

[a] Btu/lb × 2.326 = kJ/kg.
[b] lb/cu ft × 16.02 = kg/m^3.
[c] in. × 25.4 = mm.
[d] Computed.

2.4-mm (0.1-in.) wide by 51-mm (2-in.) long slots. Bar screens and grit removal for preliminary treatment precede the rotary fine screens. Figure 10.32 shows the flow sheet for the facility. The screenings are trucked to a landfill for disposal.

Inclined static screens with slot openings of 1.5 mm (0.06 in.) are used in lieu of primary sedimentation tanks at Huber Heights, Ohio. The screens removed an average of 25% of the influent suspended solids.

ODOR CONTROL

Odor control is a major design consideration for primary and preliminary treatment. In a survey of wastewater treatment plants in the U.S., 14 of 74 (19%) respondents identified odor releases from primary sedimentation tanks as a major problem at the plant.[19] Odors emanate from raw wastewater with high hydrogen sulfide concentrations and anoxic conditions. Release of hydrogen sulfide and other similarly noxious odors occurs in areas of high turbulence such as sedimentation tank inlets, effluent launders, and weirs. Increased turbulence increases the water surface area for gas transfer with H_2S

Table 10.16   Methods of handling screenings.[55]

Method	Advantages	Disadvantages
Removal from main stream, draining or dewatering, landfill	Keeps screenings out of other sludges. Can be fairly well mechanized.	Transport of screenings may be difficult. Unless carefully designed and operated, causes fly and odor nuisances and health hazards. Regulations for landfill disposal may strongly affect operations.
Removal from main stream, dewatering, incineration, landfill of ash	Keeps screenings out of other sludges. Ash is small in volume and easy to transport and dispose of. If incineration is used for other sludges and/or grit, then screenings can be added at modest cost. Pathogen kill.	High cost if an incinerator is required for screenings alone. Unless incinerator is properly designed and operated, air pollution (odor and particulates) will be serious. Not well adapted to wide fluctuations in screenings quantities unless screenings are only a small part of the total incinerator load.
Anaerobic digestion of fine screenings alone (not mixed with other solids)	—	Digestion was tested at large scale at Milwaukee, Wisconsin, but found to be impractical.
Anaerobic digestion of screenings together with scum but separate from other sludges	—	Tested at Malabar plant, Sydney, Australia, but found to be inoperable. Material handling was the chief difficulty.

release to the atmosphere. Scum and settled sludge-handling systems can also be significant sources of odors.

Volatile organic compounds (VOCs) are released to overlying air by volatilization or air stripping of organic compounds from the surface of the sedimentation basin and inlet and outlet structures. Concern over the release of VOCs has led to stringent control requirements in some parts of the U.S. Odor and VOC control programs usually consist of one or more of the methods described below.

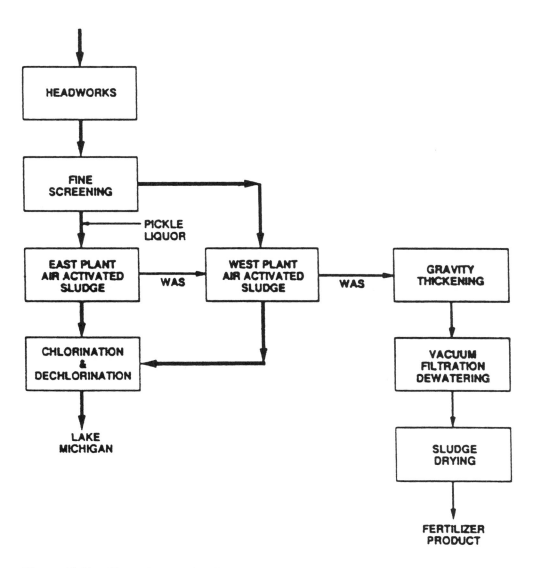

Figure 10.32   Flow diagram for Jones Island plant.

SOURCE CONTROL. Improvement of operating conditions or source control should be the first method considered for odor control. Low dissolved oxygen concentrations, high temperatures, and a pH range of 1 to 6 support biological growth and lower the solubility of H_2S in water. These conditions will increase the generation and release of H_2S to the atmosphere. Turbulent flow conditions should be minimized and sludge withdrawal should be frequent to limit sludge residence times to less than 1 hour under average flow conditions, thereby preventing septic conditions.[63]

The odor potential of sludge process sidestreams (often septic) should be considered during design. Sidestreams should be returned to a carefully chosen unit process in a manner that limits subsequent odor and VOC

Primary Treatment   505

emissions. The return of recycle flows to primary sedimentation tanks at a point below the water line may minimize odors. Scum collection and fine screen hoppers should be covered, emptied, and cleaned frequently. Provisions for flushing, degreasing, and disinfecting scum collection and screening equipment should be included during design. Chapter 7 includes additional information on the design of odor control facilities.

CHEMICAL TREATMENT. Chemical treatment methods include the addition of chlorine, hydrogen peroxide, nitrates, metallic salts, and lime. Chlorine, hydrogen peroxide, and nitrates oxidize sulfide. Hydrogen peroxide reacts with hydrogen sulfide as follows:

$$H_2O_2 + H_2S \rightarrow S + 2H_2O$$

A secondary reaction that increases the theoretical H_2O_2 demand has also been postulated:[64]

$$4H_2O_2 + H_2S \rightarrow SO_4^{2-} + 2H^+ + 4H_2O$$

During a demonstration in Baltimore, Maryland, maximum dissolved H_2S levels at the primary effluent weirs were reduced from 2.5 to 1.0 mg/L by H_2O_2 dosing 150 (490 ft) upstream of the headworks. H_2O_2 dosing also reduced gaseous emissions at the primary effluent weir from an average of 0.32 to 0.11 mg/L. Full-scale evaluation of H_2O_2 dosing of a 76 000-m^3/d (20-mgd) plant influent by the Hampton Roads Sanitation District, Virginia, indicated H_2O_2 addition to be effective but expensive (Table 10.17[64]).

Table 10.17 Hydrogen peroxide dosing to reduce sulfide emissions from plant influent to the air.[64]

H_2O_2 dose, mg/L	Dissolved H_2S concentration of influent, mg/L as S^{2-}	Sulfide level in air, ppm
0	15	>250
7	8	>250
15	7	—
50	2	<10

Chlorine reacts with hydrogen sulfide as follows:

$$H_2S + 4Cl_2 + 4H_2O \rightarrow SO_4^{2-} + 8Cl^- + 10H^+$$

Full-scale evaluation of prechlorination by the Hampton Roads Sanitation District at a 91 000-m^3/d (24-mgd) plant indicated that prechlorination was an effective H_2S removal method (Table 10.18[64]). In the study, combined hydrogen peroxide and chlorine addition was also shown to be an effective H_2S removal method.[64]

Table 10.18   Prechlorination to reduce H_2S concentrations of plant influent.[64]

Influent flow, mgd[a]	Chlorine dose, kg/d	Dissolved H_2S concentration of influent, mg/L
24	0	18
24	7258	9
11	7258	0

[a] mgd × 3785 = m^3/d.

The precipitation of sulfides by the addition of metallic salts such as ferric chloride ($FeCl_3$), ferrous sulfide ($FeSO_4$), or lime will also reduce emissions of odors. Ferrous sulfide reacts with hydrogen sulfide as follows:

$$FeSO_4 + H_2S \rightarrow FeS + 2H^+ + SO_4^{2-}$$

Plant testing by Hampton Roads Sanitation District, Virginia, indicated that ferrous sulfide addition resulted in incomplete H_2S removal.[64] Addition of $FeCl_3$ in the headworks before primary treatment reduced H_2S emissions, thereby reducing the odor and corrosion potential at the city of Los Angeles' Hyperion treatment plant.[49]

PREAERATION. Preaeration of raw wastewater will reduce odor emissions while improving settling performance. Full-scale evaluation of preaeration by the Hampton Roads Sanitation District, Virginia, indicated that preaeration could oxidize 80% of the dissolved influent sulfides.[64] Table 10.19[64] lists the oxidation potential and associated anaerobic bacteria activity. Chapter 9 contains additional information on the use of preaeration for odor control.

CONTAINMENT, COLLECTION, AND TREATMENT OF GASES. Containment of odorous gases and VOCs within a cover with collection and treatment of exhaust gases should be considered only where source control, chemical treatment, or preaeration would be insufficient or impractical. The use of covers inhibits access to and visual observation of the sedimentation

---

Primary Treatment   507

Table 10.19   Anaerobic bacteria activity associated with oxidation–reduction potential (ORP).[64]

ORP, mV	Condition
+50	No action by anaerobic bacteria
0	Poor anaerobic activity
−100 to −200	Maximum efficiency for anaerobic activity
−50 to −300	Favored by sulfate-reducing bacteria for production of sulfides

process and equipment. Exhaust gases are collected and treated by odor removal devices such as chemical scrubbers, activated carbon filters, or both. Off-gas piping should be corrosion-resistant to prevent attack by H_2S gas and sulfuric acid. Fiberglass pipe is sometimes used for this purpose.

Caustic scrubbers and activated carbon filters treat off-gases from the covered primary sedimentation tanks at JWPCP in Carson, California. Covers (used at the Hyperion Los Angeles, California, and Boston, Massachusetts, plants) control VOCs and odors. The on-going plant upgrade and expansion at Hyperion includes chemical scrubbers to remove H_2S from gaseous emissions collected from the primary sedimentation tanks.[49] Primary sedimentation tanks have fiberglass covers at Sacramento, California; exhaust gases vent to a series of granular activated carbon beds. Primary sedimentation tanks at Rochester, Minnesota, and Beloit, Wisconsin, are enclosed within a building for odor control. Exhaust gases from the enclosure at Rochester are treated by either a chlorine scrubber or activated carbon bed. Exhaust gases from primary sedimentation tanks and other unit processes at Kalamazoo, Michigan, are withdrawn by the centrifugal blowers and forced into the powdered activated carbon secondary aeration basins. Also, exhaust gases from covered primary sedimentation tanks enter the aeration tanks at the San Jose Creek treatment plant in Whittier, California. This approach has been effective for odor control and has only slightly increased the maintenance of the activated sludge process air compressors. This approach option has been eliminated for new plant construction in Southern California by the local air quality management district (SCAQMD).

ODOR MASKING. In some areas where odors are infrequent, control of odors by the methods described above might not be feasible. Masking agents—stronger, less offensive odorants—have served as short-term mitigative measures, with varying degrees of success. Nonetheless, this approach should not be used to mask toxic gases such as H_2S. The design of primary treatment facilities should anticipate locations of potential odor release and include permanent corrective measures instead of masking techniques.

Corrosion control

Important design considerations include corrosion control of concrete and metallic surfaces such as those of inlet and outlet structures, structural members, gratings, covers, walkways, and equipment. Severe corrosion of concrete and metal work has occurred at numerous plants (for example, in Kansas City, Missouri; JWPCP in Carson, California; and Hyperion in Los Angeles, California). Figure 10.33[53] illustrates concrete corrosion damage of a circular primary sedimentation tank at the North Kansas City, Missouri, treatment plant.

Figure 10.33 Corrosion of circular primary sedimentation tanks at Kansas City, Missouri.[53]

Most corrosion problems are associated with hydrogen sulfide (H_2S) gas. Anaerobic bacteria oxidize the elemental sulfur contained in municipal wastewater into dissolved sulfides during collection and transport to the treatment facility. As illustrated in Figure 10.34,[7,63] H_2S is the significant form of dissolved sulfide within the normal pH range of 6.0 to 8.0 for municipal wastewater because H_2S is the only form of sulfide that can be released as a gas to the atmosphere. The concentration of H_2S gas varies according to Henry's law.[63] Figure 10.35 illustrates the equilibrium concentration of H_2S gas as a function of temperature for a range of concentrations of dissolved H_2S.[63]

Odor containment may result in high hydrogen sulfide (H_2S) concentrations and high humidity conditions in contained areas. This moist, warm environment supports the biological oxidation of H_2S to sulfuric acid by

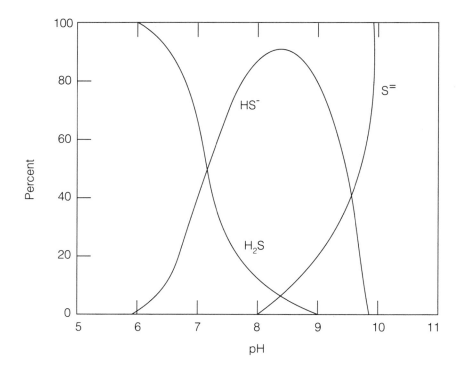

Figure 10.34 Effect of pH on hydrogen sulfide equilibrium.[7]

sulfate-reducing bacteria in the anaerobic zone of a slime layer on surfaces above the water level. This reaction can be represented by the following equation.[63]

$$H_2S + 2O_2 \rightarrow (bacteria) \rightarrow H_2SO_4$$

Sulfuric acid with a pH below 2 rapidly corrodes concrete and metal surfaces.[63] Fluctuations in the daily raw wastewater flow causes intermittent wet–dry and freeze–thaw conditions, which can accelerate corrosion. Addition of chlorine or ferric chloride may also produce a local corrosive environment at the point of chemical addition.

Corrosion of primary treatment tanks, screens, and equipment can be minimized by reducing the production or release of H_2S gas from wastewater. This can be accomplished by source control, chemical treatment, and pre-aeration, discussed in the section on odor control. Dehumidification, ventilation, or both of enclosed areas will also minimize corrosion potential. Unsubmerged equipment should be used infrequently, protected, or both from corrosion in contained areas. Concrete and steel surfaces should also be protected with an acid-resistant coating. The designer should consider cast-in-place plastic lining or epoxy coatings for new construction. Epoxy coatings have not demonstrated a long life in corrosive conditions

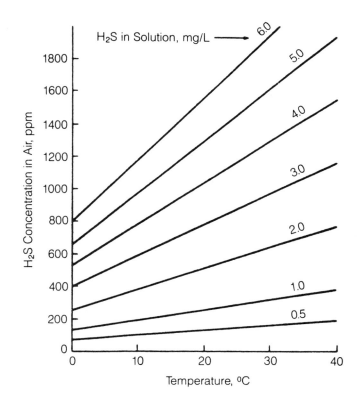

Figure 10.35 Equilibrium concentration of H₂S in air at 1 atm.

encountered at the Long Beach, California, plant. Sand-extended, coal-tar epoxy coatings, polyurethane mastic with PVC liner, polyester resin coatings, and 100% solid epoxy–amine resin coatings on existing concrete structures have demonstrated resistance to corrosive conditions at numerous plants in Southern California (for example, at the JWPCP in Carson, California). The effectiveness of the corrosion control method depends also on preparation of surfaces to be protected and application of the protective coating. Chapter 8 and other references present additional information on construction materials and corrosion control.[63]

*R*EFERENCES

1. EPA Needs Survey (1988).
2. "Clarifier Design." MOP FD-8, Water Pollut. Control Fed., Alexandria, Va. (1985).
3. Linvil, R.G., "Low-Maintenance Mechanically Simple Wastewater Treatment Systems." McGraw-Hill, Inc., New York, N.Y. (1980).
4. "Process Design Manual for Suspended Solids Removal." EPA 625/1-75-003a, U.S. EPA, Washington, D.C. (1975).

5. Camp, T.R., "A Study of the Rational Design of Settling Tanks." *Sew. Works J.,* **8,** 742 (1936).

6. Camp, T.R., "Studies of Sedimentation Basin Design." *Sew. Ind. Wastes,* **25,** 1 (1953).

7. Metcalf and Eddy, Inc., "Wastewater Engineering: Treatment, Disposal, Reuse." 3rd Ed., McGraw-Hill, Inc., New York, N.Y. (1991).

8. "Water Treatment Principles and Design." James M. Montgomery Consult. Eng. Inc., John Wiley and Sons, Inc., New York, N.Y. (1985).

9. Weber, W.J., "Physiochemical Processes for Water Quality Control." John Wiley and Sons, Inc., New York, N.Y. (1972).

10. Kelly, K., "New Clarifiers Help Save History." *Civ. Eng.,* **58,** 10 (1988).

11. "Design Practice Survey." Water Pollut. Control Fed., Alexandria, Va. (1989).

12. "Water Treatment Plant Design." Am. Soc. Civ. Eng./Am. Water Works Assoc., McGraw-Hill, Inc., New York, N.Y. (1990).

13. Ross, R.D., and Crawford, G.V., "The influence of waste activated sludge on primary clarifier operation." *J. Water Pollut. Control Fed.,* **57,** 1022 (1985).

14. Camp, T.R., "Sedimentation and the Design of Settling Tanks." *Trans. Am. Soc. Civ. Eng.,* **3,** 2285, 895 (1946).

15. Lagnese, J. F., Jr., "Report on Visit to Japan Concerning Use of Stacked Clarifiers for EPA Region 1." Unpublished (1988).

16. Lager, J.A., and Locke, E.R., "Design Management Keeps Boston's Wastewater Program on Track." *Public Works,* **121,** 13 (1990).

17. Matsunaga, K., "Design of multistory settling tanks in Osaka." *J. Water Pollut. Control Fed.,* **52,** 950 (1980).

18. Yuki, Y., "Design of Multi-Story Sewage Treatment Facilities in Osaka City." *Sew. Treat. Works Jap.,* 110 (1990).

19. "Survey of Wastewater Treatment Plant Data." Water Pollut. Control Fed., Alexandria, Va. (1989).

20. Carry, C. W., and Moshiri, M., "Wastewater Treatment for Ocean Disposal." Presented at Joint Tech. Seminar on Waste Treat. Technol. Jap. (1984).

21. Hamlin, M.J., "Paper 2. Preliminary Treatment and Sedimentation." In "Advances in Sewage Treatment, Proceedings of the Conference of the Institution of Civil Engineers." London, England (1972).

22. Uhte, W.R., Personal Communication (1990).

23. Steel, W. E., "Water Supply and Sewerage." McGraw-Hill, Inc., New York, N.Y. (1979).

24. Fair, G.N., *et al.,* "Water and Wastewater Engineering (Vol. 2)." John Wiley and Sons, New York, N.Y. (1968).

25. Sunstrom, D.W., and Klei, H.E., " Wastewater Treatment." Prentice-Hall, Englewood Cliffs, N.J. (1979).

26. "Domestic Wastewater Treatment." U.S. Army Tech. Manual No. 5-814-3, St. Louis, Mo. (1978).

27. Graber, S.D., "Outlet weir loading and settling tanks." *J. Water Pollut. Control Fed.,* **46,** 2355 (1974).

28. "Recommended Standards for Sewage Works." Great Lakes–Upper Miss. River Board of State Sanit. Eng., Health Educ. Serv., Inc., Albany, N.Y. (1978).

29. "Civil Engineering Pollution Control Systems." U.S. Naval Facilities Design Manual 5.8, Philadelphia, Pa. (1979).

30. Merritt, F.S., "Standard Handbook for Civil Engineers." 3rd Ed., McGraw-Hill, Inc., New York, N.Y. (1983).

31. Theroux, R. J., and Betz, J. M., "Sedimentation and Preaeration Experiments at Los Angeles." *Sew. Ind. Wastes,* **31,** 1259 (1959).

32. Yee, L.Y., and Babb, A.F., "Inlet design for rectangular settling tanks by physical modeling." *J. Water Pollut. Control Fed.,* **57,** 12 (1985).

33. Kawamura, S., and Lang, J., "Re-evaluation of launders in rectangular sedimentation basins." *J. Water Pollut. Control Fed.,* **58,** 12 (1986).

34. Wall, D.J., and Petersen, G., "Model for Winter Loss in Uncovered Clarifiers." *J. Environ. Eng.,* **12,** 1 (1986).

35. "A Survey of Technology and Design Deficiencies at Publicly-owned Treatment Works: Correlations of Problems with Plant Parameters." Special Publ., Water Pollut. Control Fed., Alexandria, Va. (1990).

36. Seidel, H.F., and Baumann, E.R., "The effects of preaeration on primary treatment of sewage." *J. Water Pollut. Control Fed.,* **33,** 4, 339 (1961).

37. Bargman, R.D., *et al.,* "Aeration Requirements of High Oxygen Demand Sewage." *Sew. and Ind. Wastes,* **23,** 2, 127 (1951).

38. "Design Manual for Phosphorus Removal." EPA/625/1-87/001, U.S. EPA, Cincinnati, Ohio (1987).

39. Harleman, D.R.F., and Morrissey, S.P., "Chemically-enhanced Treatment: An Alternative to Biological Secondary Treatment for Ocean Outfalls." *Proc. Hydraul. Eng.,* HYDIV/ASCE Natl. Conf., San Diego, Calif. (1990).

40. Hudson, H.E., "Water Clarification Process: Practical Design and Evaluation." Von Nostrand Reinhold Co. (1981).

41. Sanks, R.L., "Water Treatment Plant Design." Ann Arbor Sci., Mich. (1981).

42. Kawamura, S., "Considerations on Improving Flocculation." *J. Am. Water Works Assoc.,* **65,** 6, 320 (1976).

43. Camp, T.R., "Flocculation and Flocculation Basins." *Trans. Am. Soc. Civ. Eng.,* **120,** 1 (1955).

44. Klute, R., "Rapid Mixing in Coagulation/Flocculation Processes—Design Criteria." Chem. Water and Wastewater Treatment Schr.-Reine Verein Wa Bo Lu 62 G. Fischer, Verlag, New York, N.Y. (1985).

45. Andreu–Villegas, R., and Letterman, R.D. "Optimizing Flocculation Power Input." *J. Environ. Eng.*, **102,** 251 (1976).

46. Moll, H.G., "Fluid Mechanical Principles of Flocculation in Pipes." Verlag, New York, N.Y. (1985).

47. Grohmann, A., "Flocculation in Pipes: Design and Operation." Verlag, New York, N.Y. (1985).

48. Soap and Detergent Assoc., "Principles and Practices of Nutrient Removal from Municipal Wastewater." New York, N.Y. (1989).

49. Chaudhary, R., *et al.*, "Evaluation of Chemical Addition in the Primary Plant at Los Angeles' Hyperion Treatment Plant." Paper presented at Water Pollut. Control Fed. Annu. Conf., San Francisco, Calif. (1989).

50. Parkhurst, J.D., *et al.*, "Wastewater Treatment for Ocean Disposal." Paper presented at Am. Soc. Civ. Eng. Natl. Conf. Environ. Eng. Res. Dev. Design, Univ. Wash., Seattle (1976).

51. Heinke, G.W., *et al.*, "Effects of chemical addition on the performance of settling tanks." *J. Water Pollut. Control Fed.*, **52,** 12 (1980).

52. "Sewage Treatment the Scandinavian Way." Kemira Water Treatment, Inc., Kemira Kemi AB, Helsingborg, Swed. (1989).

53. Schmidt, O.J., "Wastewater treatment problems at North Kansas City Missouri." *J. Water Pollut. Control Fed.*, **50** (1978).

54. Mertsch, V., "Characteristics of Sludge from Wastewater Flocculation/Precipitation." New York, N.Y. (1985).

55. "Process Design Manual for Sludge Treatment and Disposal." EPA 625/1-79-011, U.S. EPA (1979).

56. "Operation of Municipal Wastewater Treatment Plants." Manual of Practice No. 11, Water Pollut. Control Fed., Alexandria, Va. (1990).

57. Kemp, F.D., and MacBride, B.D., "Rational Selection of Design Criteria for Winnipeg South End WPCC Primary Clarifiers." Paper presented at Water Pollut. Control Fed. Annu. Conf., Washington, D.C. (1990).

58. Mulbarger, M.C., *et al.*, "Scum Management: Past Practices/New Approaches." Paper presented at Water Pollut. Control Fed. Annu. Conf., San Francisco, Calif. (1989).

59. Feja, F., Personal Communication (1986).

60. Parker, D.S., "The TF/SC Process at Eight Years Old: Past, Present and Future." Paper presented at 59th Annu. Conf. Calif. Water Pollut. Control Assoc. (1987).

61. Hardenbergh, W. A., and Rodie, E. R., "Water Supply and Waste Water Disposal." Int. Text Book Co., Scranton, Pa. (1961).

62. Salvato, J.A., "Environmental Engineering and Sanitation." John Wiley and Sons, Inc., New York, N.Y. (1972).

63. "Odor and Corrosion Control in Sanitary Sewerage Systems and Treatment Plants—Design Manual." EPA/625/1-85/018, U.S. EPA, Cincinnati, Ohio (1985).

64. Waltrip, G.D., and Snyder, E.G., "Elimination of odor at six major wastewater treatment plants." *J. Water Pollut. Control Fed.,* **57,** 10 (1985).

Chapter 11

Suspended-growth Biological Treatment

INTRODUCTION

This chapter defines suspended-growth systems as those aerobic processes that achieve relatively high microorganism (biomass) concentrations through the recycle of biological sludge solids. These microorganisms convert biodegradable, organic wastewater constituents and certain inorganic fractions into new cell mass and byproducts, both of which can subsequently be removed from the system by gaseous stripping, settling, and other physical means. Suspended-growth systems are already the most common of the secondary treatment processes in use today and will remain so as new facilities are built to comply with the Clean Water Act requirements. Table 11.1, developed from data in the 1984 Needs Survey of the U.S. EPA,[1] shows the number of existing plants and those to be built. These projections show that an approximate 50% increase in the number of plants is needed, but only a 10% increase in overall volumetric capacity is needed. The difference reflects the large number of small facilities that remain to be constructed.

The design engineer can choose among a number of available suspended-growth systems, system modifications, and processes. Considered in this chapter are conventional, complete-mix, plug, and series flow systems; high-rate, extended aeration, contact stabilization, and tapered aeration modifications; and deep shaft, pure oxygen, and sequencing batch reactor processes.

Table 11.1   Summary of biological unit processes.

	Now in use		To be built	
Process	**Number**	**Flow, mgd[a]**	**Number**	**Flow, mgd[a]**
Stabilization ponds	5 298	3 138	2 783	118
Aerated lagoons	1 368	1 516	1 494	148
Containment ponds	834	252	433	30
Aquaculture	2	2	3	2
Trickling filters	2 463	6 345	107	408
Activated biofilter	8	21	5	8
RBCs	347	940	276	433
Activated sludge	5 690	27 302	2 585	2 713
Oxidation ditch	741	500	474	131
Biological removal				
Nitrification	860	6 303	1 533	2 533
Denitrification	40	226	42	373
Phosphorus removal	18	222	9	55
High-purity oxygen				
Activated sludge	240	5 800	20	1 500

[a] mgd × 3785 = m^3/d.

PROCESS DESCRIPTION. Figure 11.1 presents a general schematic of the systems, modifications, and processes listed above. Wastewater and sludge solids are first combined, mixed, and aerated in a reactor. Typically, the process operates in a continuous-flow mode, but can also be operated as a batch process. Contents of the reactor, referred to as mixed liquor, consist of wastewater; microorganisms (living as well as dead); and inert, biodegradable, and nonbiodegradable suspended and colloidal matter. The particulate fraction of the mixed liquor is termed mixed liquor suspended solids (MLSS).

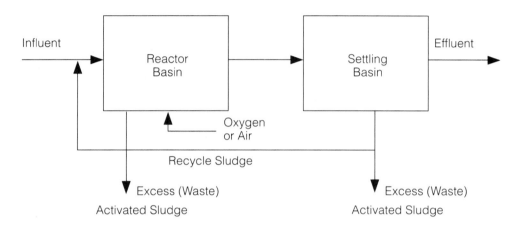

Figure 11.1 Schematic diagram of a typical activated sludge process.

After sufficient time for the biological reactions, the mixed liquor is transferred to a separate settling basin or clarifier to allow gravity separation of the MLSS from the treated wastewater. The settled MLSS are then recycled to the aeration basin to maintain a concentrated microbial population for degradation of influent wastewater constituents. Because microorganisms are continuously synthesized in this process, a means must be provided for wasting some of the MLSS from the system. Wasting generally is from the clarifier, although removal from the aeration basin is an alternative. Depending on the design and operation of the process, either maximizing or minimizing production of biological sludge solids is possible.

BASIC SYSTEM COMPONENTS. A basic suspended-growth system consists of a number of interrelated components:

- A single aeration basin or multiple basins designed for completely mixed flow, plug flow, or intermediate patterns and sized to provide a hydraulic retention time (HRT) in the range of 0.5 to 24 hours or more.

- An oxygen source and equipment to disperse atmospheric or pressurized air or oxygen-enriched air into the aeration basin at a rate sufficient to keep the system aerobic.
- A means of mixing the aeration basin contents to keep the MLSS in suspension.
- A clarifier to separate the MLSS from the treated wastewater. [In a sequencing batch reactor (SBR), mixing and aeration are stopped for a time interval to permit MLSS settling and treated wastewater decanting, thereby eliminating the need for a separate clarifier.]
- A method of collecting the settled MLSS in the clarifier and recycling them to the aeration basin. (This is not required in an SBR system.)
- A means of wasting excess MLSS from the system.

The basic activated sludge system, commonly used for carbonaceous BOD_5 removal, can be designed to achieve nitrification as well. Both carbonaceous BOD_5 removal and nitrification systems are discussed in this chapter. Basic system adjuncts can include those to effect phosphorus removal and denitrification. Chapter 15 describes nutrient removal process considerations and design.

HISTORICAL OVERVIEW. The activated sludge process was initially developed by Fowler, Ardern, Mumford, and Lockett at the Manchester, England, wastewater treatment plant in 1914.[2] In the 1920s several installations started operation in the U.S. However, widespread use of the process did not begin until the 1940s. Early investigators noted that the amount of biodegradable organics applied to a system affected microorganism metabolic rate. Initial design methods were entirely empirical in nature and aeration basin detention time was one of the first parameters used. Generally, short HRTs were chosen for what were considered to be "weak" wastewaters and long HRTs were chosen for "strong" wastewaters.

Various design loading criteria eventually were developed, usually relating to the mass of BOD_5 applied per day per mass of microbial solids present in the aeration basin. Only within the last 40 years have design equations been developed based on the concepts of microbial growth kinetics and mass balances. The different design approaches offered by Eckenfelder,[3] McKinney,[4] Lawrence and McCarty,[5] and Ramanathan and Gaudy[6] have been shown to yield similar results.[7] In recent years, more sophisticated, computerized models, such as those of Bidstrup and Grady[8] and the IAWPRC,[9] have evolved.

Solution of the more sophisticated design equations and computer models requires knowledge of microbial metabolism and kinetics. Although laboratory or pilot-plant studies can be used to determine reaction rates for a particular wastewater, such studies are generally beyond the capabilities of most design firms. Alternatives to such studies are (1) to assume certain wastewater characteristics and embark on a semi-empirical design or (2) to use an entirely empirical approach relying on some state standards. This chapter follows the first alternative by offering the design engineer values of coefficients based on numerous observations from municipal treatment plants and pilot-plant and laboratory studies. The first approach allows design of a process on the basis of the amount of time biomass spends in a system [generally termed mean cell residence time (MCRT)] or on a food:microorganism ratio (F:M). Wastewaters with a significant industrial fraction (more than 10% of the organic load) warrant special attention, such as pilot-plant studies, to establish values of coefficients. Such wastewaters may also warrant use of more sophisticated models such as those listed above.[8,9]

BIOLOGICAL TREATMENT FUNDAMENTALS

THE ACTIVATED SLUDGE ENVIRONMENT. An activated sludge process employs a suspension of flocculent microorganisms comprised of bacteria, fungi, protozoa, and rotifers to treat wastewater. The dry weight of these microorganisms typically comprises 70 to 90% organic and 10 to 30% inorganic substances. Composition of the organic fraction is generally represented by the empirical formula, $C_5H_7O_2NP_{0.2}$. Inorganics include K, Na, Mg, S, Ca, Fe, and other trace elements. The characteristics of the wastewater, environmental conditions, process design, and its mode of operation determine the predominant microorganism group that will develop. Successful plant performance will depend on the development of a microbial community that will assimilate target waste materials and form a flocculent biomass that is readily removed by gravity separation.

The microbial population is normally dominated by heterotrophic organisms that require biodegradable organic matter for energy and new cell synthesis. These microorganisms include bacteria, fungi, and some protozoa. Autotrophic bacteria, including those that oxidize ammonia to nitrite and nitrate, have the ability to use inorganic materials for energy and cell synthesis. Such autotrophs are usually present in varying concentrations. Their relative presence depends on many conditions, including the mode of operation of the facility and reactor concentrations of biomass and organic and inorganic materials. Many protozoa and all rotifers in suspended-growth systems are predators, feeding on bacteria and likely enhancing both flocculation and clarification. Protozoa and rotifers can constitute as much as 5% of the mass of organisms in a system.

Nuisance organisms might occur sometimes and can interfere with successful plant operation. Most nuisance organism problems are associated with sludge settling and the formation of heavy scum-like growths. Control

---

of these organisms is the subject of considerable research.[10] Additional information concerning the effects of microbial population distributions on system performance can be found in the literature.[11,12]

BIOLOGICAL GROWTH AND SUBSTRATE OXIDATION. An empirical relationship depicting the stabilization of biodegradable organic matter in aerobic suspended-growth systems can be expressed by the following equation:

$$COHNS + O_2 + Nutrients \xrightarrow{Microbes} New\ microbes + CO_2 + H_2O$$

where COHNS represents the organic matter.

The equation summarizes a complex series of biochemical reactions that can be simplified into three fundamental activities: oxidation, synthesis, and autoxidation. Oxidation is the coupled release of energy through the conversion of organic matter to lower-energy products (CO_2 and H_2O). Synthesis is the conversion of a portion of the organic matter, assisted by the energy released during oxidation, into new biomass. Autoxidation is the conversion of some of the cell constituents to low-energy products, with the release of additional energy. Details of these biochemical reactions and their stoichiometry can be found elsewhere.[13-15]

An equation similar to that above represents the oxidation of ammonia to nitrate (nitrification) by select autotrophs:

$$NH_4^+ + O_2 + CO_2 + HCO_3^- \xrightarrow{Microbes} New\ microbes + H_2O + NO_3^- + H^+$$

This equation represents the net result of both oxidation and synthesis reactions for conversion of ammonium to nitrite and then to nitrate. The stoichiometry of this reaction is better understood than that for the oxidation of organic matter because the substrates are well defined and only a few species of microorganisms can carry out nitrification. Details of nitrification biochemistry and stoichiometry can be found in the literature[12-17] and in Chapter 15.

PROCESS DESIGN FOR CARBON OXIDATION. Figure 11.2 illustrates a typical suspended-growth system flowchart. A reactor with a volume (V) receives a forward flow (Q) plus a sludge recycle flow (Q_r). Forward flow contains a soluble, biodegradable substrate at concentration S_o and suspended solids at concentration M_o. The suspended solids comprise microorganisms (X_o) and other particulates (X_{io}). These "other particulates" include nonvolatile solids (Z_{io}), nonbiodegradable volatile solids (Z_{no}), and biodegradable volatile solids (Z_{bo}). Recycle flow contains soluble

Design of Municipal Wastewater Treatment Plants

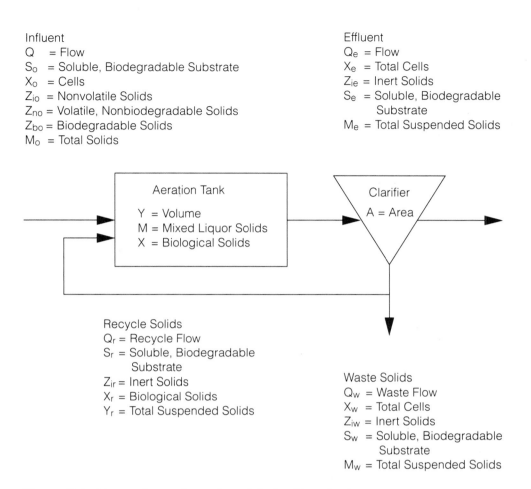

Influent
Q   = Flow
S_o = Soluble, Biodegradable Substrate
X_o = Cells
Z_{io} = Nonvolatile Solids
Z_{no} = Volatile, Nonbiodegradable Solids
Z_{bo} = Biodegradable Solids
M_o = Total Solids

Effluent
Q_e = Flow
X_e = Total Cells
Z_{ie} = Inert Solids
S_e = Soluble, Biodegradable Substrate
M_e = Total Suspended Solids

Aeration Tank
Y = Volume
M = Mixed Liquor Solids
X = Biological Solids

Clarifier
A = Area

Recycle Solids
Q_r = Recycle Flow
S_r = Soluble, Biodegradable Substrate
Z_{ir} = Inert Solids
X_r = Biological Solids
Y_r = Total Suspended Solids

Waste Solids
Q_w = Waste Flow
X_w = Total Cells
Z_{iw} = Inert Solids
S_w = Soluble, Biodegradable Substrate
M_w = Total Suspended Solids

Figure 11.2 Nomenclature for activated sludge flow sheet.

biodegradable substrate at concentration S_r, biological solids at concentration X_r, and inert suspended solids at concentration Z_{ir}.

In design, it is important to evaluate all components of the incoming wastewater that will influence calculations of sludge production and oxygen demand. Many simplistic models do not properly account for the incoming suspended solids and, therefore, underestimate solids generation. Bench-scale and pilot-plant investigations are valuable in providing useful estimates of these factors.

When pilot-plant studies are not possible, wastewater components must be estimated. For typical domestic wastewater, influent microorganisms (X_o) are normally assumed to be negligible relative to those in the reactor. However, acknowledging their presence is critical during system start-up. They can also influence oxygen demand patterns in a system's aeration tank (see Grady and Lim, Chapters 13 and 14).[13] Nonvolatile suspended solids (Z_{io}) can be calculated as the difference between influent TSS and VSS.

Nonbiodegradable volatile suspended solids (Z_{no}) are normally presumed to approximate 40% of the influent VSS.[18] Biodegradable volatile suspended solids (Z_{bo}) are often presumed to be rapidly absorbed onto the biomass and subsequently hydrolyzed (solubilized). Therefore, they are often ignored in calculations. However, slowly hydrolyzable biodegradable VSS can significantly affect system kinetics and mass balances.

Numerous models have been developed for suspended-growth systems. The model used herein was originally developed by Lawrence and McCarty,[5] but it has been modified and expanded over the years. A survey of major consulting firms in the U.S. [WPCF Design Practice Survey (1989)] suggests that the basic Lawrence and McCarty approach is most widely applied. This approach also provides the basis for the IAWPRC model currently being used in a computerized form to automate design calculations.[7,9]

Basic Lawrence and McCarty design equations used for sizing systems are presented below. Details of their development can be found in the references.[5,13,19]

$$\Theta = V/Q = \text{hydraulic residence time} \tag{1}$$

$$\Theta_c = VX/Q_wX_r = VM/Q_wM_r = \text{mean cell residence time} \tag{2}$$

$$1/\Theta_c = [YkS_e/(K_s + S_e)] - b \tag{3}$$
$$\text{(completely mixed aeration basin)}$$

$$XV = \Theta_cQY(S_o - S_e)/(1 + b\Theta_c) \tag{4}$$

$$Y_{OBS} = Y/(1 - b\Theta_c) \tag{5}$$

$$M = \Theta_c/\Theta\{[Y(S_o - S_e)/(1 + b\Theta_c)] + Z_{io} + Z_{no}\} \tag{6}$$

$$P_x = Q\{[Y(S_o - S_e)/(1 + b\Theta_c)] + Z_{io} + Z_{no}\} \tag{7}$$

$$\alpha = Q_r/Q = X/(X_r - X) = M/(M_r - M) \tag{8}$$

$$R_c = Q(S_o - S_e)(1 + b\Theta_c - BY)/(1 + b\Theta_c) \tag{9}$$

$$R_n = 4.57\, Q\,(N_o - N) - 2.86\, Q(N_o - N - NO_3) \tag{10}$$

Where

V	=	aeration tank volume, L^3;
Q	=	wastewater forward flow, L^3/t;

Design of Municipal Wastewater Treatment Plants

Q_r	=	sludge recycle flow, L^3/t;
X	=	reactor biological solids, m/L^3;
Y	=	true cell yield, m/m;
S_o	=	influent soluble biodegradable substrate, mg/L;
S_e	=	effluent soluble substrate, m/L^3;
Q_w	=	sludge waste flow, L^3/t;
K_s	=	half-velocity coefficient, m/L^3;
k	=	maximum rate of substrate utilization per unit weight of biomass, $1/t$;
b	=	endogenous decay coefficient based on biomass in aerated zone, $1/t$;
Y_{OBS}	=	observed cell yield, m/m;
M	=	total mixed liquor suspended solids, m/L^3;
Z_{io}	=	influent nonvolatile suspended solids, m/L^3;
Z_{no}	=	influent volatile nonbiodegradable solids, m/L^3;
R_c	=	mass of oxygen required per unit time to satisfy carbonaceous oxidation, m/t;
R_n	=	mass of oxygen required per unit time to satisfy nitrification oxygen demand, m/t;
N_o	=	influent oxidizable nitrogen, m/L^3;
N	=	effluent oxidizable nitrogen, m/L^3;
NO_3	=	effluent nitrate–nitrogen, m/L^3;
P_x	=	mass of total activated sludge solids generated or wasted per day, m/t; and
B	=	oxygen equivalent of cell mass, often calculated as 1.42 mass O_2/mass VSS, mg/mg.

In the above definitions, the units for each variable are expressed in terms of the fundamental dimensions of mass (m), length (L), and time (t).

Design of a system requires determination of the following items:

- Volume of the aeration basins, V;
- Quantity of sludge wasted, P_x;
- Total oxygen demand, $R_c + R_n$;
- Sludge recycle requirements; and
- Size of clarifiers.

The mathematical development represented by equations 1 through 10 assumes that the loss of TSS in the clarifier effluent is negligible. Because up to 30% of the TSS that should be removed from a system to maintain equilibrium can be in the clarifier effluent, the designer may add an additional term, $Q_e X_e$ (where Q_e = effluent flow rate and X_e = effluent TSS), to the denominator of equation 2, resulting in:

$$\Theta_c = VX/(Q_w X_r + Q_e X_e) \qquad (11)$$

With the use of this revised equation for MCRT, subsequent equations of the development must also be modified. Note also that the development assumes a constant ratio of VSS to TSS in the mixed liquor, recycle flow, and effluent. As a further caution, such constancy does not necessarily occur.

Volume of Aeration Basins. Sizing of aeration basins is based on two important factors. The first is providing sufficient time to remove soluble BOD_5 (and oxidize ammonia nitrogen, if required) and to allow biomass activity, as estimated by oxygen uptake rate (OUR) measurements, to return to an endogenous level. The second is maintenance of flocculent, well-settling MLSS that can be effectively removed by gravity settling.

For municipal systems, the value of S_e can be as low as only a few mg/L. Experience has shown its exact value to be somewhat unpredictable. Therefore, it would be inadvisable to reduce aeration tank volume by assuming a higher S_e value, such as in the discharge permit of a facility.

Mean cell residence time (Θ_c) is often used to relate substrate removal time requirements to biological kinetics and stoichiometry as indicated in equation 3. For nitrification, these well-documented kinetic and stoichiometric parameters[5,13,16-19] may be applied for estimating Θ_c. As good practice, this estimate is multiplied by a safety factor.[5,16] Alternatively, information from the literature or pilot-plant studies may be useful for estimating Θ_c and other kinetic parameters.

From a practical point of view, selection of values of Θ_c for carbonaceous BOD_5 removal systems usually is not based on kinetics considerations, but rather on experience. Typically the design is based on providing a high enough value of Θ_c for the system to yield a well-flocculated sludge that settles well. Figure 11.3,[20] which represents a nonfilamentous sludge grown on a soluble waste (glucose plus yeast extract), shows that a minimum Θ_c value of about 3 days is required. In practice, a value of Θ_c from 3 to 5 days is typically used during warm weather and up to 15 days during cold weather. Values of Θ_c outside this range are selected in situations where environmental conditions warrant lower or higher values. In very warm climates where nitrification is not desirable, Θ_c values of 1 to 2 days are employed. Also, very long Θ_c values are often used in extended aeration systems where secondary goals require minimization and stabilization of the excess sludge solids generated. Figure 11.4 presents suggested ranges of Θ_c (SRT) for BOD_5 removal and nitrification at various temperatures.[20]

Once a design value of Θ_c has been selected, equation 4 can be used to estimate the required aeration tank volume, V. Calculation of V requires an estimation of the mixed liquor biomass concentration (X) and the stoichiometric coefficients (Y and b). Selection of X is not a trivial exercise. It depends on the oxygen transfer capabilities and mixing properties of the aeration system and on the allowable solids loading rate for the secondary clarifiers. Conventional air activated sludge system MLSS concentrations

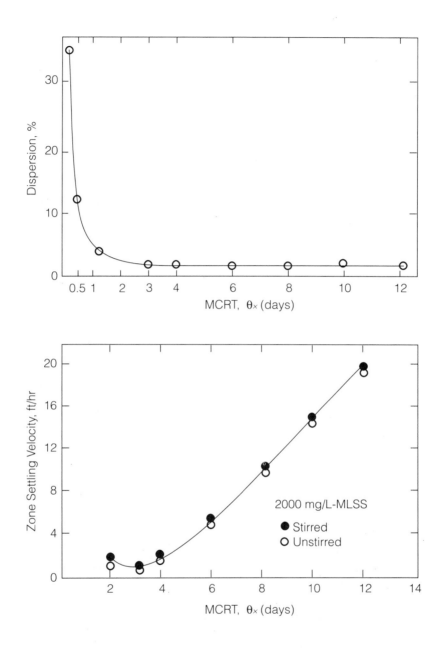

Figure 11.3 Effects of MCRT on the amount of dispersed growth in activated sludge effluent and the settling velocity (ft/hr × 0.304 8 = m/h) of activated sludge mixed liquor.

ranging from 1500 to 3000 mg/L are often used. Values of 2000 mg/L MLVSS and 2500 mg/L MLSS are most common. It is possible, however, for these systems to accommodate higher concentrations. Pure oxygen systems can be operated at MLSS concentrations in excess of 10 000 mg/L.

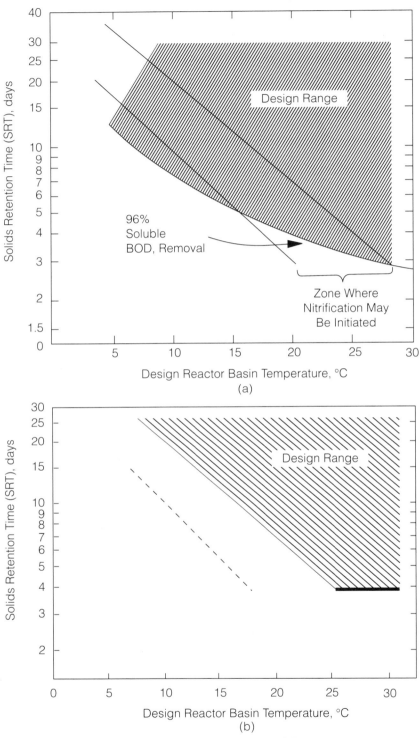

Figure 11.4 Design SRT for (a) carbonaceous BOD$_5$ removal and (b) single-stage nitrification (toxicity not present, MLSS washout controlled at pH 7.5–9.0).

Solids settling and thickening properties often dictate final selection of the MLSS concentration. For air activated sludge systems, design for concentrations over about 5000 mg/L is seldom economical.[21] Figures 11.5 and 11.6 show suggested values as functions of SVI and temperature. In pure oxygen systems, the upper boundary may be higher because better-settling sludges can be generated; however, most operating pure oxygen plants in the U.S. do not exceed 5000 mg/L MLSS; the MLSS levels for some plants are as low as 1500 mg/L.

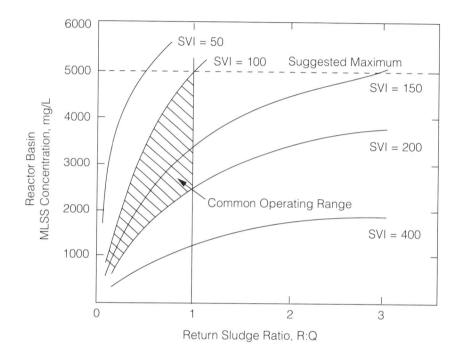

Figure 11.5 **Design MLSS versus SVI and return sludge ratio (high-rate sludge removal mechanism) at a reactor basin temperature of 20°C.**

Selection of the stoichiometric coefficients (Y and b) can be based on designer experience, pilot-plant studies, or values found in the literature. Tables 11.2 and 11.3 present typical values. The term b decreases with increasing SRT because the active biomass fraction of the MLSS decreases and the inert organic and inorganic fractions increase, especially for plants with little or no primary treatment.

Design of completely mixed aerated lagoons with sludge recycle can be based on the design equations presented above since they function as a system operated in the extended aeration mode. Aerated lagoons without recycle can also be designed using the above equations, with MCRT (Θ_c) set equal to

---

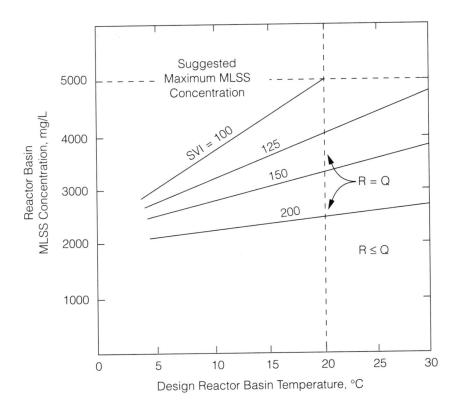

Figure 11.6 **Suggested maximum MLSS design versus temperature
and SVI (SVI at aerator temperature, not ambient
temperature).**

Table 11.2   Coefficients for the activated sludge process.[19,20]

Coefficient	Basis	Range[b]	Typical[b]
Y[a]	g VSS/g BOD$_5$	0.4 to 0.8 (0.6 to 0.84)	0.6 (0.71)
Y[a]	g VSS/g COD	0.25 to 0.4	0.4
b	day^{-1}	0.004 to 0.075 (0.02 to 0.10)	0.06 (0.064)

[a] The yield coefficients represent only the conversion of biodegradable
primary effluent substrate to cell matter. Total sludge yield must also include nonbiodegradable and fixed suspended solids that leave the system.
[b] Numbers in parentheses represent values obtained from 15 major practicing
consulting engineering firms in the U.S. in 1989. It is assumed that domestic wastewater primary effluent is being treated.

Table 11.3 Typical kinetic coefficients for suspended growth nitrification process.[16-19]

Coefficient	Basis	Range	Typical
k_n	days^{-1}	0.3–3.0	1.0
K_{sn}	$NH_4 - N$, mg/L	0.2–5.0	1.4
Y_n	$NH_4 - N$, mg VSS/mg $NH_3 - N$	0.04–0.29	0.15
b_n	days^{-1}	0.03–0.06	0.05

the HRT (Θ). Another approach to designing aerated lagoons without recycle is to assume that the observed BOD5 removal (either total or soluble BOD5) can be described by first-order kinetics. For a single, completely mixed lagoon, the first-order equation is[19]

$$S_e/S_o = 1/(1 + k_1\Theta) \tag{12}$$

Where

k_1 = observed BOD5 removal rate constant (base e) and
Θ = HRT.

Reported values of k_1 have ranged from 0.25 to 1.0 day^{-1} for overall BOD5 removal.[19] Additional details on aerated lagoon design are presented elsewhere.[22,23]

Sludge Generated and Wasted. The amount of sludge generated can be estimated using equation 7. Note that the total sludge mass includes non-volatile, biodegradable volatile, and nonbiodegradable volatile suspended solids. Any precipitates that form from the addition of iron or aluminum salts for phosphorus removal should also be included in this calculation. Figure 11.7 illustrates net secondary treatment system sludge production for the stated waste characteristics, both with and without primary sedimentation. This figure assumes typical values for the coefficients Y and b and assumes primary sedimentation removes 60% of the influent TSS.

Total Oxygen Demand. Oxygen demand in an aeration tank can be calculated using equations 9 and 10. Additional oxygen demand can also result from the presence of certain readily oxidizable components in the waste-water[24] such as the sulfide ion. Oxygen demand usually varies both spatially and temporally in a suspended-growth system. Temporal variations can be estimated from statistical analyses of data collected for influent loadings (carbonaceous BOD5 and nitrogenous oxygen demand) of the process.

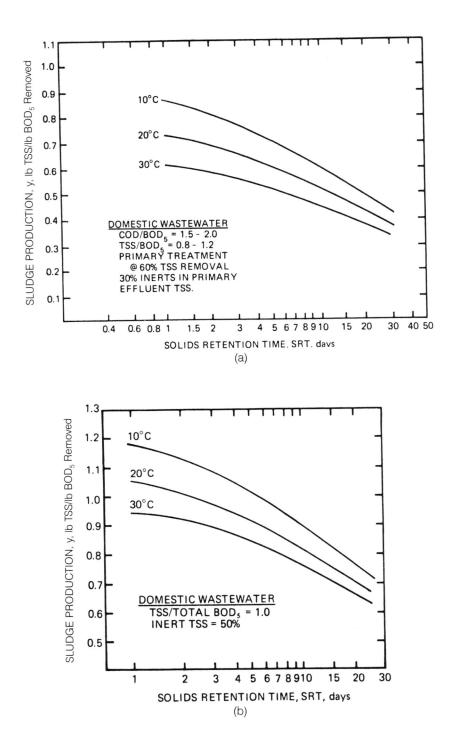

Figure 11.7 Net sludge production versus SRT and temperature (a) with primary treatment and (b) without primary treatment.[81]

Design of Municipal Wastewater Treatment Plants

Spatial variations depend on the kinetic relationships between growth rates of the biomass and substrate removal rates and the dissolved oxygen (DO) concentrations; they also depend on the flow regime and HRT of the process. Good estimates of this latter variation can be obtained from process design tools such as the IAWPRC model described earlier.[7,9] A more empirical approach to estimating spatial oxygen demand is based on designer experience or data in the literature. Table 11.4 presents such data collected in the United Kingdom[25] for long, narrow aeration tanks ($L/W > 20$). Estimates of nitrogenous demand assumed that nitrification progressed uniformly along the entire tank length.[25] Typical spatial distributions observed in the U.S. may differ from Table 11.4 values, depending on tank geometry, recirculation rates, and mixing intensities. More details on estimation of oxygen demand can be found in the references.[24,26]

Table 11.4 Variation in proportion of oxygen demand along the length of a plug flow aeration tank ($L/W > 20$).[25]

Proportion of aeration tank volume, %	Carbonaceous demand		Carbonaceous + nitrogenous demand	
	Proportion of demand, %	Diurnal range, %	Proportion of demand, %	Diurnal range, %
20	60	40–85	46	33–62
20	15	5–20	17	10–20
20	10	5–15	14	10–17
20	10	5–15	13	10–16
20	5	<1–10	10	7–13

Total oxygen demand for design should be based on the peak loadings anticipated during the design year. As a minimum, the design requirement for a conventional system should be based on the 24-hour demand of the average day of the peak month. Some designers prefer to use the peak 4-hour demand of the average day of the peak month. Basing the requirement on the peak day demand plus 50% of the peak 4-hour rate for the peak day has also been suggested.[27]

Sludge Recycle Requirements. Requirements for sludge recycle pumping capacity can be estimated from equation 8. The recycle ratio (α) deserves careful consideration since it affects the size of the final clarifiers without influencing the size of the aeration tanks. No generalizations can be offered as to the best ratio because that will depend on sludge settling and thickening characteristics. As a rough guide, the design value of α should range from 20 to 100% of the facility design flow. More discussion of recycle ratios follows in a subsequent section of this chapter.

Clarifier Sizing. The sizing of the secondary clarifier is an important function, integral with the design of other components of a suspended-growth system. Details of clarifier sizing appear later in this chapter.

PROCESS DESIGN FOR NITRIFICATION. Nitrogen contained in municipal wastewater occurs predominantly in the organic and ammonium nitrogen forms. Values reported in the literature for total nitrogen range from 20 to 85 mg/L (as N), with a median value of 40 mg/L.[19] About 40% of the total occurs in the organic form and 60% in the ammonium form. Typically, less than 1% is present as nitrate or nitrite. Additional information on influent quality characteristics is presented in earlier chapters.

The growth of new cells will remove some of the influent nitrogen. This nitrogen removal will approximate 12% of the mass (on a dry weight basis) of the net biomass accumulation. For given conditions of SRT and carbonaceous BOD_5 removal, nitrogen removal by assimilation can be estimated from equation 3 and will normally range from 8 to 20% of the total influent nitrogen. This range indicates the interdependence of MCRT and endogenous decay on net biomass retention of nitrogen. Nitrogen assimilation depends on the ratio of BOD_5 to N in the influent, and thus can be significant in systems treating wastewaters with high concentrations of BOD_5.

Ammonium nitrogen is oxidized to nitrate by the staged activities of the autotrophic species *Nitrosomonas* and *Nitrobacter*. Each gram of ammonia oxidized to nitrate (both expressed as N) will result in

- 4.18 g of O_2 consumed,
- 7.14 g of alkalinity (as $CaCO_3$) destroyed,
- 0.15 g of new cells produced, and
- 0.09 g of inorganic carbon consumed.

The degree of biological nitrification will depend on the mass of nitrifying organisms allowed to remain in the system. Their presence depends on the relative growth rates of the autotrophic and heterotrophic populations involved, system MCRT, and other conditions such as temperature, ammonium ion, organic substrate, and DO concentrations. Thus, mechanisms exist to either limit or promote nitrification.

Biological oxidation of ammonia to nitrate can be achieved in combined carbonaceous BOD_5 removal–nitrification (single-stage) systems or in separate nitrification (two-stage) systems. The degree of nitrification in a combined, single-stage process depends on the ratio of carbonaceous to nitrogenous oxygen demand and on the MCRT of the system, provided that a population of nitrifying autotrophs can be maintained. The degree of nitrification, therefore, is governed to a large extent by the design parameters (Θ and Θ_c for a nitrification system). Two-stage systems allow some separation of the carbonaceous and nitrogenous oxidation processes. In the first stage (aeration basin with clarification and solids recycle), most of the carbonaceous BOD_5 removal occurs and nitrification is limited. The second stage (separate aeration basin and clarifier with solids recycle) can, therefore, maintain more favorable conditions for nitrification of the wastewater.

The basic design equations for nitrification in suspended-growth systems are presented below. Details of their development can be found elsewhere.[16]

$$f_N = Y_N(N_o - N)/[Y(S_o - S_e) + Y_N (N_o - N)] \tag{13}$$

$$\Theta_{c,\,min} = 1/\mu_{N,\,max} \tag{14}$$

$$\Theta_{c,\,design} = \Theta_{c,\,min} (SF) \tag{15}$$

$$\mu_{N,\,max}(T) = \mu_{N,\,max}[e^{0.098\,(T-15)}] \tag{16}$$

$$\mu_N = \mu_{N,\,max}(N/K_N + N)[DO/(K_o + DO)] \tag{17}$$

$$q_N = \mu_N/Y_N = (N_o - N)/f_N X\Theta \tag{18}$$

Where

N_o, N	=	influent and effluent nitrogen concentrations, respectively, mg/L;
f_N	=	nitrifier fraction, g/g;
Y_N	=	nitrification yield coefficient, g/g;
$\Theta_{c,\,min}$	=	minimum solids retention time for nitrification;
$\Theta_{c,\,design}$	=	design solids retention time;
$\mu_{N,\,max}$	=	maximum nitrifier specific growth rate at 15°C, 1/t;
μ_N	=	nitrifier specific growth rate, 1/t;
K_N, K_o	=	half-velocity constants for ammonia and oxygen, respectively, mg/L;
q_N	=	specific rate of nitrification, 1/t (mass of N oxidized/mass MLNVSS/time); and
SF	=	safety factor.

Single-stage Nitrification. The fraction of nitrifying organisms required for (or resulting from) a given degree of ammonia oxidation can be estimated by equation 13. This approximation neglects the endogenous decay rate of the heterotrophs and autotrophs involved and is a function of the influent substrate concentrations (N_o and S_o) and the nitrogenous and carbonaceous yield coefficients of the biomass. Reported individual cell yield coefficients in terms of nitrifying volatile suspended solids (NVSS) per mass of ammonia nitrogen oxidized are as follows:[16]

- *Nitrosomonas*—0.04 to 0.29 g NVSS/g NH_3-N oxidized to NO_2-N;
- *Nitrobacter*—0.02 to 0.08 g NVSS/g NO_2-N oxidized to NO_3-N; and
- Design value—0.15 g NVSS/g NH_3-N oxidized to NO_3-N.

To maintain a population of nitrifiers in a combined carbonaceous BOD_5 removal–nitrification process, the system MCRT must be greater than the minimum value required for nitrification given by equation 14. When this condition is not met, nitrifier washout will occur. As a good practice, the minimum estimated MCRT for nitrification is multiplied by a safety factor (SF) of 1.5 to 2.5. Such an SF will normally accommodate the impact of load peaking on system performance.

Equation 18 provides the specific rate of nitrification as a function of the combined carbonaceous and nitrogenous biomass fractions and actual ammonia removal. Maximum nitrifier specific growth rate ($\mu_{N, max}$) at 15°C is approximately 0.45 day^{-1}. Typical values for the half-velocity constants (K_N and K_o) are 0.5 and 1.0 mg/L, respectively.[16]

Two-stage Nitrification. Use of a two-stage suspended-growth system is also effective. Some engineers believe it offers certain advantages over a single-stage system. The first stage (aeration tank plus clarifier) is designed to remove a substantial amount of carbonaceous BOD_5. Equations 1 through 10 can be used to establish the requirements of the first stage. Typically, relatively high-rate systems are used for the first stage since the second stage can achieve a minor level of additional carbonaceous BOD_5 removal if periodic bleed-through occurs. Removal of essentially all of the carbonaceous BOD_5 in the first stage allows the second stage (aeration basin plus clarifier) to more specifically function as a system to convert ammonium ions to nitrate ions. Equations 1 through 10 and 13 through 18 can be used in the design of the second stage. A plug flow aeration tank with tapered aeration is suggested to ensure proper oxygen distribution, including the maintenance of at least 2.0 mg/L of DO throughout the entire tank at all times. This geometry will also minimize ammonium ion bleed-through into the effluent. Clarifiers in both stages might receive mixed liquors that have a slightly poorer settling characteristic than that typical of a single-stage nitrification system. This uncertainty dictates use of extra caution in the system design.

OTHER ENVIRONMENTAL EFFECTS. Temperature, dissolved oxygen, nutrients, toxic and inhibitory wastes, pH, and variability affect the performance and thus the design of activated sludge systems. Each of these effects is described below.

Temperature. Temperature will affect reaction kinetics, stoichiometric constants, and oxygen transfer rates. Most temperature corrections used in biological treatment designs follow the modified van't Hoff–Arrhenius equation:

$$K_{T_2} = K_{T_1} \Theta^{(T_2 - T_1)} \qquad (19)$$

Where

K_{T_1}	=	a specific kinetic, stoichiometric, or mass transfer coefficient at temperature T_1;
K_{T_2}	=	a specific kinetic, stoichiometric, or mass transfer coefficient at temperature T_2;
Θ	=	1.00 to 1.04 for carbonaceous BOD removal systems except for aerated lagoons; or
	=	1.06 to 1.12 for aerated lagoons.

K_{T_1} represents a specific kinetic, stoichiometric, or mass transfer coefficient at temperature T_1. Temperature corrections have been made for k, K_s, Y, and b.[13,16,17,19] Both k and b will increase with increasing temperature; however, the effect on K_s and Y also depends on the type of system. Values of Θ for k in carbonaceous BOD_5 removal systems range from 1.00 to 1.04, typically about 1.02. The range of values of Θ for k in aerated lagoons is somewhat higher at 1.06 to 1.12. Temperature corrections for b in carbonaceous BOD_5 oxidation systems typically use a Θ value of 1.04. Note that all of these values are approximate and thus need careful review. Nitrification systems are also sensitive to temperature changes. The dependence of nitrifier specific growth rate (μ_N) on temperature is presented in equation 16.

Dissolved Oxygen. DO concentration is another important control parameter. In systems designed for carbonaceous BOD_5 removal, a minimum average tank DO concentration of 0.5 mg/L is acceptable under peak loading conditions and 2.0 mg/L under average conditions. In nitrifying systems, engineers normally use a minimum average tank DO concentration of 2.0 mg/L under all conditions.

Nutrients. An adequate nutrient balance is necessary to ensure an active biomass that settles well. Typical nutrient requirements are based on a ratio of BOD_5 to N to P of about 100:5:1. Systems with higher values of Θ_c generally require fewer nutrients because of their recycle resulting from biomass autooxidation. Normal domestic wastewater typically contains ample nutrients.

Toxic and Inhibitory Wastes. The presence of certain inorganic and organic constituents can inhibit or destroy suspended-growth system microorganisms. An excellent listing of many of these is presented elsewhere.[13] Nitrification processes are particularly sensitive to toxic inhibition.[16,17]

pH. The pH of mixed liquor should range from 6.5 to 7.5 for optimum cell growth in carbonaceous BOD_5 removal systems. Nitrifying systems are more sensitive to system pH.[16] Pure oxygen systems often attenuate pH more than air systems because the former lacks nitrogen gas flow to help strip dissolved CO_2 from the mixed liquor. To avoid pH attenuation, the designer should strive for a residual alkalinity of at least 50 mg/L (as $CaCO_3$) with classical aeration systems.

Variability. Effluent quality from most nitrification plants varies substantially from day to day.[28] Figure 11.8 shows that the average effluent total nitrogen concentration for the maximum month typically will approximate 140% of the annual average effluent concentration for total nitrogen. About the same variability would be expected for effluent ammonia concentrations.

*P*OPULATION DYNAMICS AND CONTROL

A well-designed biological treatment reactor provides an environment promoting the growth of desired microorganisms. In modern designs, this can include favoring the growth of nutrient-removing species and discouraging the growth of nuisance types. The environment must always favor the growth of microorganisms that settle and thicken well. Design elements promoting the growth of microorganisms that settle and thicken well, as well as discouraging the growth of nuisance species, are discussed below.

MICROORGANISM CLASSIFICATION. Microorganisms that settle and thicken well are typically referred to as floc forming. Such biomass includes a mixture of bacteria, protozoa, and some metazoa. Some of the more common types are listed in Table 11.5. Those that do not settle and thicken well are typically referred to as bulking microorganisms. Another nuisance type is associated with the formation of dark viscous foams on aeration tanks and clarifiers.

Most bulking microorganisms are filamentous bacteria. When viewed under a microscope, they are typically long and stringy in appearance. Such filaments protruding from flocs are believed to prevent biomass compaction. The difference between a nonbulking and bulking biomass may be envisioned as the difference between a container of smooth balls (nonbulking) and a similar container containing balls with spikes protruding, porcupine-

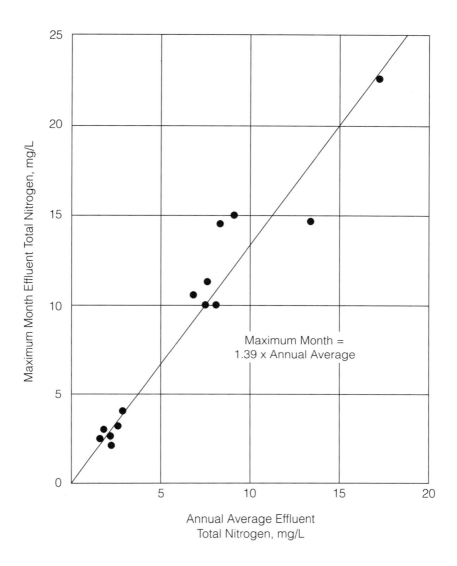

Figure 11.8 Variability of effluent total nitrogen for seven biological nutrient removal facilities.

like, from them (bulking). A much rarer and less well-understood type of bulking is referred to as viscous bulking[29] or zoogleal bulking.[30] This type of bulking is believed to be caused by secretions of exocellular material from the microorganisms, causing them to be dispersed and highly water retentive.[31] Viscous "cushions" between these flocs prevent normal compaction. To depict this latter type of bulking, the balls in the previous analogy may be envisioned as being suspended in, or thickly coated with, petroleum jelly.

Some researchers[31,32] contend that an ideal floc contains just the right mixture of filamentous microorganisms and floc formers, with the filaments forming the backbone of the floc (Figure 11.9a). They contend that if the

Table 11.5   Common floc-forming organisms found in activated sludge.

Bacteria	Protozoa
Pseudomonas	Paramecium
Achromobacter	Aspidisca
Flavobacterium	Vorticella
Alcaligenes	
Arthrobacter	
Citromonas	
Zooglea	
Acinetobacter	

floc lacks enough filaments, the floc falls apart (Figure 11.9b) and effluent quality deteriorates. If too many filaments exist, the sludge becomes bulky (Figure 11.9c).

Several major recognized causes of filamentous bulking are

- Low DO,
- Low F:M,
- Septic wastewater,
- Nutrient deficiency, and
- Low pH.

Filamentous organisms typically associated with each of these causes are summarized in Table 11.6. The designer's task is to design a system that discourages the growth or accumulation of bulking and nuisance

Table 11.6   Dominant filament types as indicators of conditions causing activated sludge bulking.[31]

Suggested causative conditions	Indicative filament types
Low DO	Type 1701, *S. natans, H. hydrossis*
Low F:M	*M. parvicella, H. hydrossis, Nocardia* sp. types 021N, 0041, 0675, 0092, 0581, 0961, 0803
Septic wastewater/sulfide	*Thiothrix* sp., *Bergiatoa* and type 021N
Nutrient deficiency	*Thiothrix* sp., *S. natans* type 021N, and possibly *H. hydrossis* and types 0041 and 0675.
Low pH	Fungi

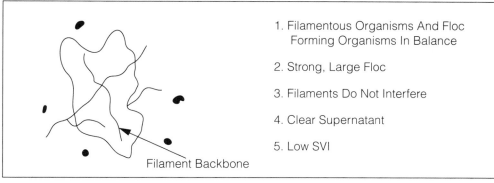

(a)

1. Filamentous Organisms And Floc Forming Organisms In Balance

2. Strong, Large Floc

3. Filaments Do Not Interfere

4. Clear Supernatant

5. Low SVI

Filament Backbone

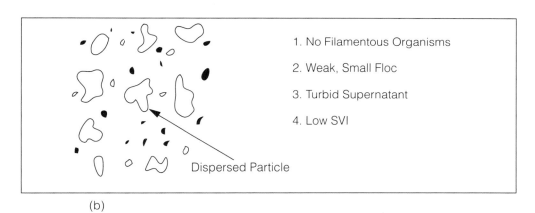

(b)

1. No Filamentous Organisms

2. Weak, Small Floc

3. Turbid Supernatant

4. Low SVI

Dispersed Particle

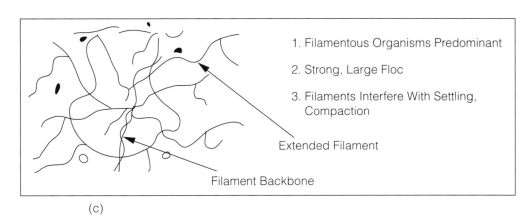

(c)

1. Filamentous Organisms Predominant

2. Strong, Large Floc

3. Filaments Interfere With Settling, Compaction

Extended Filament

Filament Backbone

Figure 11.9 Effect of filamentous organisms on activated sludge floc structure: (a) ideal, nonbulking activated sludge floc, (b) pin-point floc, and (c) filamentous, bulking activated sludge.

microorganisms. The designer also must provide, in the design, flexibility that allows the operator to control any nuisance organisms that may appear in the system.

PROCESS DESIGN. Design elements that can be used to control nuisance microorganisms include reactor geometry, proper distribution of air or oxygen, return activated sludge (RAS) pumping rate, RAS or sidestream chlorination, chemical addition, and specific *Nocardia* control procedures. Each is discussed below.

Process Configuration. In general, and particularly in low F:M systems, a good approach is to design at least the first portion of the aeration tank with a plug flow configuration. This configuration will effectively minimize the growth of low F:M types of organisms that result in a bulking sludge. Lightly loaded, completely mixed activated sludge (CMAS) systems are more prone to this type of growth than are plug flow reactors.

Some cases merit providing a step feed capability. This consists of designing the aeration tank so that some or all of the influent flow can be added at each of several points along the length of the aeration tank. Typically, two to four addition points are used. Return sludge is added only to the first portion of the aeration tank (Figure 11.10).[33] This type of design, for a given tank volume and F:M ratio (or MCRT), allows lower solids loading rates on the

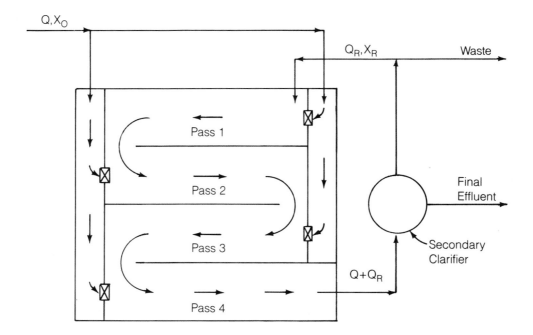

Figure 11.10 Typical configuration of step feed activated sludge plant.

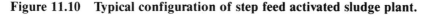

final clarifiers, thus allowing required treatment levels to be attained despite a relatively bulky sludge.[33] Step feed also allows the oxygen demand to be more evenly distributed along the length of an aeration tank.

Selector Design. To promote the growth of floc-forming microorganisms while suppressing filamentous growth, special reactors called selector tanks can be provided ahead of conventional aeration basins. Typical selector tank configurations are illustrated in Figure 11.11. Designed as either a separate reaction stage for a CMAS process or as individual compartments in a plug flow system (to prevent backmixing), selector tanks are characterized by short HRTs and high F:Ms. These tanks can be aerobic, anoxic (absence of free oxygen), or anaerobic; some configurations are patented.[34,35] Table 11.7 compares the three types of systems.

Mechanisms for the control of microbial population dynamics to achieve selection of a nonbulking, low-SVI mixed liquor have not been entirely delineated. As a result, no definitive criteria or methods are available for the design of selectors. Table 11.8 presents a summary of selector design parameters from the literature.[36-45] Design parameters for aerobic selectors include a minimum DO of 2.0 mg/L[38] with an oxygenation capacity of about

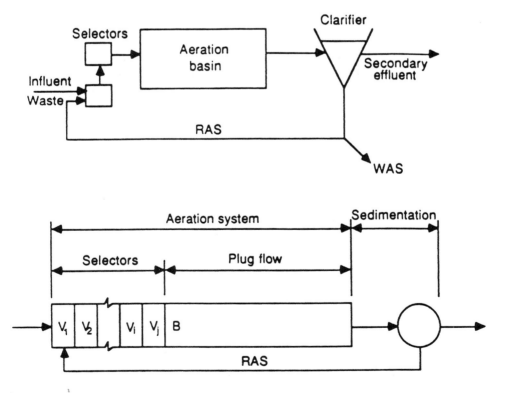

Figure 11.11 Typical selector configurations.

4 kg/m^3·d (0.25 lb/cu ft/day).[43] Anoxic and anaerobic selectors can also be applied for *Nocardia* control in nitrifying systems.

As good design practice, pilot studies of proposed selector systems and evaluation of all potential causes of excessive filamentous microorganism

Table 11.7   Comparison of biological selectors.[34,35]

Selector type	Advantages	Disadvantages
Aerobic	Simple process, no additional internal recycle streams other than RAS. Relies on basin geometry, not nitrification.	Does not reduce O_2 requirements. Requires more complex aeration system design to meet maximum O_2 uptake rate in the initial high F:M zone. May require patent fee if operated within a certain range of DO conditions.
Anoxic	Tends to buffer nitrification (recovers approximately 3.5 lb alkalinity as $CaCO_3$/lb of NO_3^- –N denitrified. Lowers O_2 demand in a nitrification process (recovers approximately 2.86 lb O_2/lb of NO_3^- reduced). The initial high F:M region occurs in the anoxic zone, with the high O_2 demand met by NO_3^- instead of O_2.	Cannot be used with a process that does not nitrify. Uses an additional recycle stream. Requires care in design and operation to minimize the introduction of O_2 in the anoxic zone. Poor system design can induce low-DO bulking.
Anaerobic	Simple design, no internal recycle other than RAS. The simplest of selector systems to operate. Can be used for biological phosphorus removal.	A patented process requiring a licensing fee. Does not reduce O_2 requirements. May not be compatible with long SRTs. Requires care in design and operation to minimize the introduction of NO_3^- and O_2 in the anaerobic zone. Poor system design can induce low-DO bulking.

growth should precede retrofitting existing plants. Examples of pilot- and full-scale investigations of selector system performance have been reported.[46,47]

Application of Air or O₂. The kinetics of BOD₅ removal in a plug flow system require supply of most of the air to the first portion of the aeration tank. If air addition does not match the oxygen demand profile, the DO concentration may drop below a critical value and bulking organisms may form. Figure 11.12 shows the importance of DO as a function of loading on a CMAS system for controlling the growth of filaments.[48] Two means of matching oxygen demand or oxygen uptake rate (OUR) profiles are tapered aeration and the addition of pure oxygen.

Table 11.8   Design parameters found in the literature.

Parameter	Suggested value	Reference
Initial substrate concentration	>80 mg COD/L[a]	36
Initial F:M	20–25 g COD/g MLVSS·d[b]	36
	20 g COD/g MLVSS·d	37
	>4 g BODᵤ/g MLVSS·d[c]	38
	8–12 g COD/g MLVSS·d	39
Initial floc loading	50–150 g COD/g MLSS[d]	40, 41
	No significance	36
Fractional substrate removal in all selectors (soluble and degradable)	50–70%	40, 41
	80%	42
Substrate gradient between first selector and main aeration basin	25 mg/L	43
Aeration basin dispersion number	<0.2	44
Soluble substrate leaving selector	<60 mg COD/L	39
Number of compartments	1–5	45
Hydraulic retention time	12–25 minutes	45
	5–20 minutes	38

[a] In first compartment.
[b] F:M = (Influent COD − soluble effluent COD)/total mass MLVSS, including clarifier.
[c] The F:M BODᵤ (ultimate BOD) recommended by Casey *et al.*[38] can be estimated by 1.5 × soluble BOD₅, and the MLVSS is multiplied by an activity coefficient according to Casey *et al.*[38]
[d] Floc loading = (influent COD − return sludge COD) × influent flow/(recycle flow × recycle MLSS).

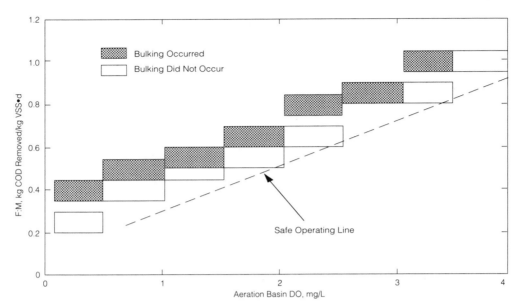

Figure 11.12   Combinations of F:M and aeration basin configurations related to bulking and nonbulking sludges.

Return Activated Sludge Considerations. Because bulking sludges typically require high RAS rates,[49] this must be accounted for in the design of return sludge systems. Current practice generally allows a maximum RAS pumping rate capacity of 100% of design average flow. For systems that carry high MLSS concentrations (for example, oxidation ditches), a 150% allowance may be justified. Turndown to as low as 20% is also common. Such flexibility requires special consideration of variable-speed RAS pumps as well as RAS concentration measurements. Mass balance calculations around the final clarifiers can help to estimate the actual RAS pumping rate required once the concentrations of the MLSS and RAS are known.

Provision for RAS or Sidestream Chlorination. Bulking sludges can be partially controlled by RAS or sidestream chlorination. A typical design for a low (5- to 10-hour) HRT system uses 5 to 15 kg Cl_2/kg MLSS·d, with the chlorine added to the RAS system. Longer-HRT systems might need chlorine added to a sidestream or to multiple points in the aeration tanks (Figure 11.13). Hydrogen peroxide can be substituted for chlorine in many cases. Further design and sizing details can be found elsewhere.[31] Note that RAS chlorination can interfere with nitrification.

Chemical Addition. Most municipal nonnitrifying systems do not have nutrient deficiencies, low pH, or both. If, however, even an occasional deficiency is expected, providing the capability to add nutrients (nitrogen and phosphorus) would be prudent. Similarly, for wastewaters with pHs outside

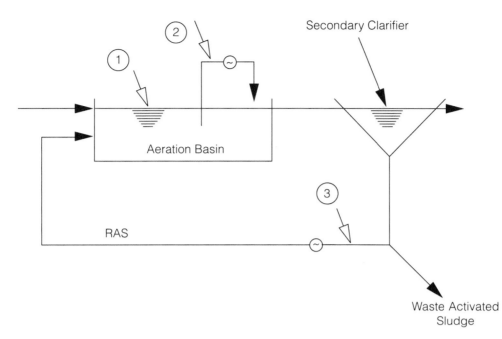

Figure 11.13 Chlorine-dosing points for bulking control.

of the normal 6 to 8 range, provisions must be made to adjust the pH by addition of acids or bases. Since pretreatment of highly septic wastewater with chlorine can improve overall process control, this capability should be included at facilities receiving such wastewater. However, trihalomethane (THM) formation might be a problem, depending on receiving water use or air emission standards. Addition of cationic polymers at concentrations of typically less than 1 mg/L has been shown to be effective in improving mixed liquor settleability.

Nocardia **Control.** Enough *Nocardia* bacteria can develop to create a nuisance foam condition. Systems with high MCRTs or those using pure oxygen seem to be prone to this type of growth. Low influent BOD_5 concentrations and grease are also believed to stimulate such growth. In pure oxygen systems with low MCRTs, a *Nocardia* accumulation problem might be due, in part, to baffling that prevents scum from passing downstream to clarifier skimming mechanisms that would otherwise remove it from the secondary treatment system.

No single approach or device has been found totally successful in controlling *Nocardia* foam. In the 1989 WPCF survey of 15 U.S. consulting firms, the numbers of firms using various approaches were compiled:

- Water spray—10;
- Selectors—5;

- Chlorinating the RAS—11;
- All of the above—2;
- Mechanical skimming of mixed liquor in the aeration tank or aerated channels—1;
- No devices—1; and
- Other
  - Full radius skimming of final clarifiers—2,
  - Free surface discharge to final clarifiers—1, and
  - Lower the MCRT—2.

For nonnitrifying systems, *Nocardia* can often be controlled by reducing the MCRT. In some wastewaters, particularly those with high temperatures, *Nocardia* control was found to require operation at MCRTs below 1 day.[50] Because low-MCRT operation is almost always accompanied by dilute RAS and WAS concentrations (sometimes as low as 1500 mg/L), such operation will require a relatively large capability for sludge wasting.

Designers might consider maintaining an open water surface on the aeration tanks and continuing this open surface to that of the secondary clarifiers. This will prevent accumulation of *Nocardia* foam in the aeration basins and allow its collection on the surface of the secondary clarifiers. Avoiding return of scum from secondary clarifier skimmers to the aeration basin merits consideration, although such return is desirable during plant start-up. In some cases, spraying of *Nocardia* foam with chlorinated water has been helpful, but few operating facilities have found this to be satisfactory by itself. One design innovation at a plant having *Nocardia* problems consisted of locating a surface skimmer just downstream of the aeration basins, in the channel transporting mixed liquor to the clarifiers.[51] Foam and all of the WAS were removed from this channel. The best method of *Nocardia* control appears to be prevention by careful reactor design and adjustment of sludge age.

INSTRUMENTATION. Most plants are equipped with built-in or portable DO meters. Some also monitor TSS or collect samples frequently enough to know their approximate concentration at all times. The trend to increase the level of instrumentation has resulted in use of on-line systems in many of the larger plants. Instantaneous readings of DO, MLSS, OUR, flow rates, F:M, and sludge age can be obtained. Such data, coupled with computers, can assist in optimizing control of effluent quality and energy use. The concept is generally more attractive to larger plants with sufficient staff to keep the instrumentation operational and calibrated. Savings in energy costs at large facilities can more than offset those associated with installing and maintaining instrumentation.

OXYGEN TRANSFER AND MIXING SYSTEMS

The supply of oxygen to suspended biomass represents the largest single energy consumer in an activated sludge facility. Recent studies have indicated that the aeration system accounts for 50 to 90% of the total power demand.[52] A general survey in 1982 of municipal and industrial wastewater treatment facilities suggested that approximately 1.3 million kW of aeration equipment with a value of $0.6 to $0.8 billion had been installed in North America.[53] Annual operating costs for these systems were estimated to be about $0.6 billion.

Over the years, oxygen transfer equipment has evolved enough to give an engineer a wide selection of devices to meet the specific needs of a facility. Table 11.9[54] summarizes the general characteristics of major types of equipment currently available. More detailed information is contained in WPCF Manual of Practice No. FD-13.[54] The 1989 WPCF Design Practice survey of 15 major consulting engineering firms in the U.S. revealed that more than 50% preferred fine-pore diffusers, 25% semifine pore equipment, and the remainder divided their preferences between coarse-bubble and mechanical aeration systems.

Transfer rates for diffused air systems are typically reported as oxygen transfer efficiency (OTE), expressed as a percentage, or as oxygen transfer rate (OTR) in units of mass/time. Mechanical devices are normally rated on the basis of OTR or aeration efficiency (AE), the latter expressed in units of mass/time/unit of power. Testing procedures and translation from clean to process water conditions are subsequently discussed in this section on oxygen transfer and mixing systems.

As a secondary function, aeration devices furnish sufficient energy for mixing. Ideally, mixing energy should be sufficient to thoroughly disperse dissolved substrate and oxygen throughout a given segment of an aeration tank and keep the MLSS suspended. This does not necessarily mean that both soluble and suspended material should be uniformly mixed throughout the entire aeration tank. For example, plug flow tanks and reactors with point source oxygen addition do not rely on uniformity for proper operation.

Note that the power required to satisfy oxygen demand depends on substrate and biomass concentrations, flow rate, and reactor volume; the power for mixing depends on aeration tank volume and, to a lesser degree, on MLSS concentration. For certain combinations of biomass and substrate concentrations and other variables listed above, power requirements for mixing may exceed those for oxygen transfer. In systems with high biomass concentrations, oxygen demand will typically control power requirements; power requirements for aerated lagoons are often dictated by mixing. The

---

Table 11.9   Characteristics of aeration equipment.[54]

Equipment type	Equipment characteristics	Processes where used	Advantages	Disadvantages	Reported clean water performance[a]
Diffused air					
Porous diffusers	Ceramic, plastic, flexible membranes; dome, disc, tube, plate configurations; total floor grids, single or dual roll; fine bubble.	High-rate, conventional, extended, step, contact stabilization activated sludge systems.	High efficiency; good operational flexibility; turndown about 5:1.	Potential for air- or water-side clogging; usually require air filtration; high initial cost; low alpha.	SOTE: 13–40%
Nonporous diffusers	Fixed orifice, perforated pipe, sparger, slotted tube, valved orifice, static tube; coarse bubble; typically single or dual roll, some total floor grids.	Same as for porous diffusers.	Do not usually clog; easy maintenance; high alpha.	Low oxygen transfer efficiency; high initial cost.	SOTE: 9–13%
Others					
Jets	Compressed air and pumped liquid mixed in nozzle and discharged; fine bubble.	Same as for porous diffusers.	Good mixing properties; high SOTE.	Limited geometry; clogging of nozzles; requires blowers and pumps; primary treatment required; low SAE.	SOTE: 15–24%
Porous rotating disc	Porous top plate secured to rotating hollow shaft; air blower through shaft; fine bubble; rotates at 80 rpm.	Same as for porous diffusers.	Good mixing properties; high SOTE.	Limited geometry; requires blowers and pumps; low SAE.	N/A
U-tube	30- to 300-ft shaft; air blown into inlet of down leg.	Activated sludge with limited geometry.	High efficiency because driving force is increased.	Limited geometry; typically effective for strong waste.	N/A

Table 11.9   Characteristics of aeration equipment (continued).[54]

Equipment type	Equipment characteristics	Processes where used	Advantages	Disadvantages	Reported clean water performance [a]
Mechanical surface					
Radial flow, low speed (20–60 rpm)	Low output speed; large diameter turbine; floating, fixed-bridge, or platform mounted; used with gear reducer.	Same as for porous diffuser.	Tank design flexibility; high pumping capacity.	Some icing in cold climates; initial cost higher than axial flow aerators; gear reducer may cause maintenance problems.	SAE: 2.0–4.5 lb/hp-hr [b]
Axial flow, high speed (300–1200 rpm)	High output speed; small diameter propeller; direct, motor-driven units mounted on floating structure.	Aerated lagoons and reaeration.	Low initial cost; easy to adjust to varying water level; flexible operation.	Some icing in cold climates; poor maintenance accessibility; mixing capacity may be inadequate.	SAE: 2.0–2.5 lb/hp-hr
Horizontal rotor	Low output speed; used with gear reducer; steel or plastic bars, plastic discs.	Oxidation ditch, applied either as an aerated lagoon or as an activated sludge.	Moderate initial cost; good maintenance accessibility.	Subject to operational variable, which may affect efficiency; tank geometry is limited.	SAE: 2.5–3.5 lb/hp-hr
Submerged turbine	Units contain a low-speed turbine and provide compressed air to diffuser rings, open pipe, or air draft; fixed-bridge application, may employ draft tube.	Same as for porous diffusers, oxidation ditches.	Good mixing; high capacity input per unit volume; deep tank application; operational flexibility; no icing or splash.	Require both gear reducer and blower; high total power requirements; high cost.	SAE: 1.7–3 lb/hp-hr

[a] Manufacturers data in clean water at standard conditions; diffused air units expressed as SOTE, mechanical devices as SAE. Range of values accounts for different equipment, geometry, gas flow, power input, and other factors.
[b] lb/hp-hr × 0.170 = kg/MJ; pso × 6.895 = kN/m^2.

following sections and Manual of Practice No. FD-13 discuss mixing in greater detail.[54]

DIFFUSED AERATION. Diffused aeration has been employed since the turn of the century.[55] Early applications introduced air through open tubes or perforated pipes located at the bottom of aeration tanks. The desire for greater efficiency led to the development of porous plate diffusers that produce small bubbles. These diffusers, used as early as 1916, became the most popular method of aeration by the 1930s. Unfortunately, serious fouling problems occurred at a number of installations, which gradually discouraged their use. Systems requiring lower maintenance gained dominance during the period of relatively inexpensive energy (prior to 1972). Typically, these low-maintenance devices used fixed orifices [6 mm (0.25 in.) or more in diameter] to produce relatively large bubbles. Rapid escalation of power costs commencing in the early 1970s rekindled interest in porous media devices and triggered vigorous efforts to optimize the oxygen transfer efficiency of all types of aeration systems.

Diffused aeration, defined as the injection of a gas (air or oxygen) below a liquid surface, covers all equipment described in this section. However, hybrid equipment that combines gas injection with mechanical pumping or mixing equipment is arbitrarily classified herein as diffused aeration equipment. These hybrid devices include jet aerators, porous rotating discs, and U-tube aerators. Another hybrid device, the combination turbine–sparger aerator, is arbitrarily classified as a mechanical device. These will be discussed later.

The wastewater treatment industry has witnessed the introduction of a wide variety of air diffusion equipment. In the past, the various devices commonly were classified as either fine bubble or coarse bubble, designations that supposedly reflected oxygen transfer efficiency. Unfortunately, the demarcation between coarse and fine bubbles is difficult to define.[55] Also, applying this classification to specific equipment generated confusion and controversy. For these reasons, the industry now prefers to categorize air diffusion systems by the physical characteristics of the equipment. In the following discussion, the various devices have been divided into three categories: porous diffusers, nonporous diffusers, and others. No one should generalize about performance based solely on these classifications, which relate more to organization than performance.

Porous Diffuser Systems. Use of porous diffusers has gained renewed popularity because of their relatively high OTE. An excellent reference on this subject was published in 1989 by the U.S. EPA in cooperation with the ASCE Committee on Oxygen Transfer.[56] Much of the information presented in this section was derived from that source. The reader is encouraged to review the report for further information on specific topics.

Numerous materials have been used in the manufacture of porous diffusers. They generally can be divided into two categories: rigid materials of ceramic or plastic and perforated membranes of flexible plastic or rubber.

The oldest and most common rigid porous diffuser is of the ceramic type. It consists of rounded or irregular-shaped mineral particles bonded together to produce a network of interconnected passageways through which compressed air flows. As the air emerges from the diffuser surface, the pore size, surface tension, and air flow rate interact to produce a characteristic bubble size. Currently, the most common rigid porous diffusers are manufactured from aluminum oxide. A recent development is the use of porous plastic materials. Claimed advantages of the plastic material over aluminum oxide are its lighter weight; inert composition; and, depending on the actual material, greater resistance to breakage. Disadvantages include the brittleness of some plastics and their susceptibility to creep and pore or orifice collapse over time. Porous plastics are made from a number of thermosetting polymers. The two most common are high-density polyethylene (HDPE) and styrene-acrylonitrile (SAN). Also available is a nonrigid porous plastic material that is extruded from a combination of rubber and HDPE.

Membrane diffusers differ from rigid diffusers because the former does not contain a network of interconnected passageways. Instead, mechanical methods create preselected patterns of small, individual orifices (perforations) in the membrane to allow gas passage. These perforated diffusers consist of a thin, flexible membrane made from either a thermoplastic or an elastomer. Like porous plastics, perforated membranes offer the advantage of being lightweight. As one disadvantage, their properties can change with time, depending on materials of construction, shape and dimensions, and certain environmental conditions.[56]

Porous diffusers are available in plates, domes, discs, and tubes; some are shown in Figures 11.14 and 11.15. Although plates were once the most popular, since 1970 their use has declined as the use of dome and disc configurations has increased. The dome, developed in 1954, is typically an 18-cm (7-in.) diameter disc with downturned edges. The dome, consisting of a ceramic material, is mounted on a PVC saddle attached by a center bolt. The disc, a newer development, varies in diameter from approximately 18 to 24 cm (7.0 to 9.5 in.) for ceramic and porous plastic materials and from 20 to 51 cm (8 to 20 in.) for perforated membranes. Like the dome, the disc is usually mounted on a PVC saddle but may be fastened with either a center bolt or a peripheral clamping ring.

Most tubular diffusers have the same general shape, typically 51 to 61 cm (20 to 24 in.) long with an outside diameter of 6 to 8 cm (2.5 to 3 in.). Materials used include ceramics, porous plastics, and perforated membranes.

With the exception of the old plate designs, each porous diffuser is equipped with a flow control orifice to ensure uniform air distribution to each diffuser. Typical air flow rates for domes range from 0.01 to 0.07 m^3/min

---

Suspended-growth Biological Treatment   553

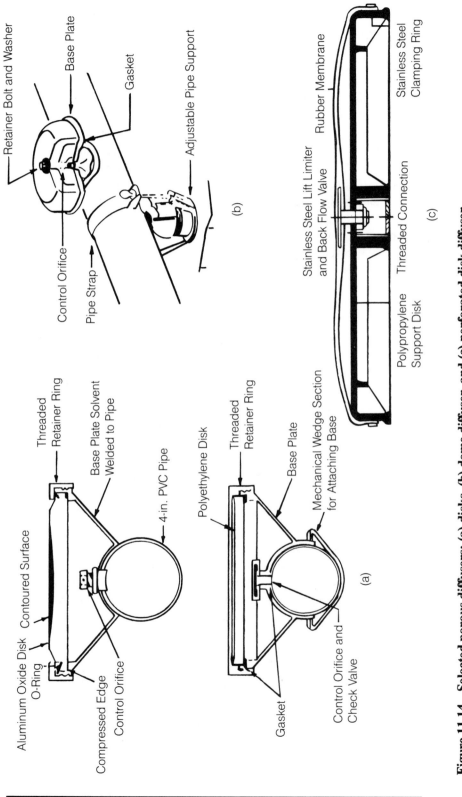

Figure 11.14 Selected porous diffusers: (a) disks, (b) dome diffuser, and (c) perforated disk diffuser.

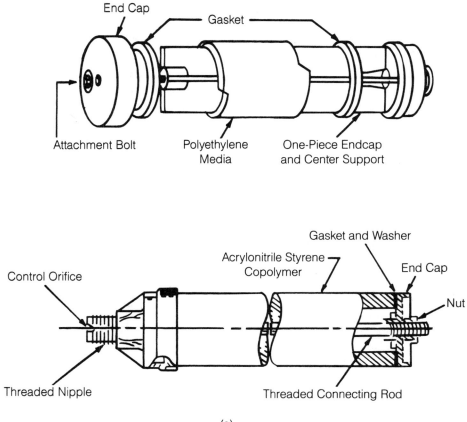

End Cap

Gasket

Attachment Bolt

Polyethylene
Media

One-Piece Endcap
and Center Support

Gasket and Washer

Acrylonitrile Styrene
Copolymer

End Cap

Nut

Control Orifice

Threaded Nipple

Threaded Connecting Rod

(a)

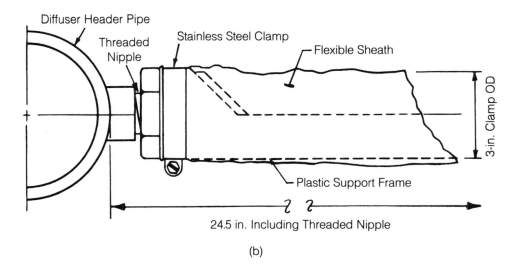

Diffuser Header Pipe

Threaded
Nipple

Stainless Steel Clamp

Flexible Sheath

3-in. Clamp OD

Plastic Support Frame

24.5 in. Including Threaded Nipple

(b)

Figure 11.15   Selected porous tubes: (a) rigid plastic and (b) flexible perforated membrane.

Suspended-growth Biological Treatment   555

(0.5 to 2.5 cu ft/min). For discs, the range is 0.01 to 0.08 m^3/min (0.5 to 3.0 cu ft/min) for ceramics and 0.03 to 0.57 m^3/min (1 to 20 cu ft/min) for perforated membranes, depending on their size. For tubes, the range is 0.03 to 0.14 m^3/min (1 to 5 cu ft/min).

Plates, domes, and discs are normally installed in a total floor coverage configuration, but plates also have been placed along the sides of aeration tanks to generate single- or dual-roll spiral mixing patterns. Discs and domes are normally arranged in a grid pattern with variable spacing. Tube diffusers are typically installed from swing arms along one or both sides of an aeration tank. They can also be placed in a more efficient full-floor coverage configuration.

Nonporous Diffusers. Nonporous diffusers, available in a wide variety of shapes and materials, have larger orifices than porous devices (Figure 11.16). The fixed orifices vary from simple holes drilled in piping to specially designed openings in metal or plastic fabrications. Perforated piping, spargers, and slotted tubes are typical nonporous diffuser designs.

Valved orifice diffusers include a check valve to prevent backflow when the air is shut off. Several also allow adjustment of air flow rate by changing either the number or size of the orifices through which air passes.

Typical system layouts for fixed and valved nonporous diffusers closely parallel those for porous diffuser systems. The most prevalent configurations are the single- and dual-roll spiral patterns using either narrow- or wide-band diffuser placement. Mechanical swing-type headers, which allow removal of the diffusers for cleaning, are common. Cross-roll and full-floor coverage patterns may also be employed. Fixed and valved orifice diffusers are commonly employed where mixing is more important than oxygen transfer. These applications include aerated grit chambers, channel aeration, sludge and septage storage tank aeration, flocculation basin mixing, and so on.

The static tube, another type of nonporous diffuser, resembles an air-lift pump, except that the tube has interference baffles placed within the riser. These baffles are intended to mix the liquid and air, shear coarse bubbles, and increase contact time. With this type of system, the tubes, typically about 1.0 m (3 ft) in length, are anchored to the basin floor in a total floor coverage pattern.

Other Diffused Aeration Systems. Other diffused aeration devices include jet and U-tube systems which are described below.

JET AERATION. Jet systems, which are placed on the floor of aeration tanks, combine liquid pumping with air diffusion. The pumping system circulates mixed liquor in the aeration basin, ejecting it through a nozzle assembly. Air, usually supplied from a blower, is introduced into the mixed liquor prior to

---

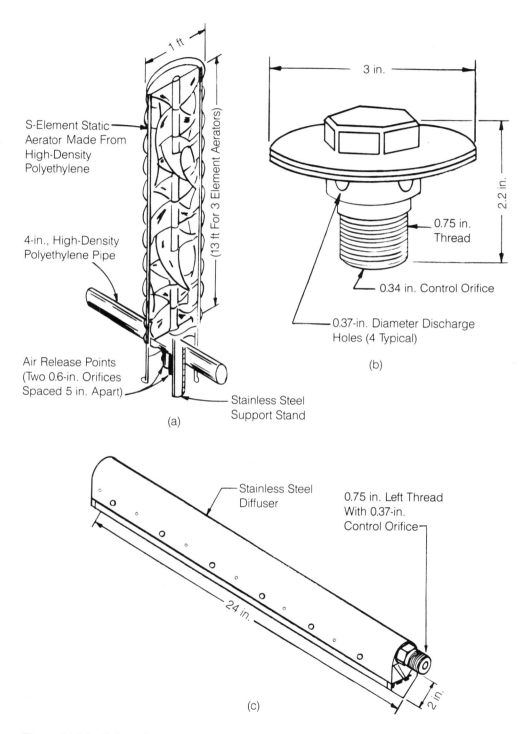

S-Element Static Aerator Made From High-Density Polyethylene

(13 ft For 3 Element Aerators)

1 ft

4-in., High-Density Polyethylene Pipe

Air Release Points (Two 0.6-in. Orifices Spaced 5 in. Apart)

Stainless Steel Support Stand

(a)

3 in.

2.2 in.

0.75 in. Thread

0.34 in. Control Orifice

0.37-in. Diameter Discharge Holes (4 Typical)

(b)

Stainless Steel Diffuser

0.75 in. Left Thread With 0.37-in. Control Orifice

24 in.

2 in.

(c)

Figure 11.16   Selected nonporous diffusers: (a) static tube, (b) orifice, and (c) tube.

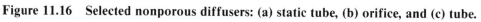

its discharge through the system's nozzles. Typically, the jets are configured in either cluster or directional arrangements as shown in Figure 11.17. Distribution piping and nozzles are usually of fiberglass construction.

Normally, the recirculation pump is a constant-capacity device. Turndown for the aerator is accomplished by varying the air supply rate from the associated blower. A typical nozzle has a 2.5-cm (1-in.) throat through which the air and mixed liquor pass. To overcome potential clogging problems, some systems are equipped with "self-cleaning" features.

U-TUBE AERATION. A U-tube system consists of a 9- to 150-m (30- to 500-ft) deep shaft that is divided into an inner and an outer zone. Air is added to the influent mixed liquor in the downcomer zone. The mixture travels to the bottom of the tube and then back to the surface through the return zone where the effluent is removed. The great depths of the mixed liquor result in high pressure that enhances oxygen transfer efficiencies.

The cost-effectiveness of U-tube aeration can be related to waste strength. For normal-strength wastewater (100 to 200 mg/L BOD_5), the air needed to circulate the mixed liquor through the shaft governs the amount of air added, not the oxygen demand. For high-strength wastes (above 500 to 600 mg/L BOD_5), the oxygen demand of the waste governs the amount of air added. Under these conditions, all oxygen forced into solution is likely to be consumed. With this efficient power use, the U-tube process is most cost effective for treating stronger wastewaters.

MECHANICAL SURFACE AERATORS. Surface aerators can be grouped into four general categories: radial flow low speed, axial flow high speed, aspirating devices, and horizontal rotors. Each is used widely and has distinct advantages and disadvantages, depending on the application.

Mechanical aerators are generally not well suited to plug flow geometry (with the exception of basins in series). They are better suited to other geometries, particularly to large basins where high power input is required. With this situation, substantial savings in capital costs might be realized because of the larger tankage and fewer aeration devices needed compared with other aeration systems.

Some surface aerators are equipped with submerged draft tubes that ensure mixing by bringing liquid from the bottom of the basin up through the tube and into the impeller. Draft tubes are normally used where basins are deeper than 4.5 m (15 ft) and where mechanical aerators alone might not provide enough mixing throughout the entire basin.

As an alternative to the draft tube, an auxiliary submerged mixing impeller can be provided. This submerged impeller will increase the amount of liquid pumped from the bottom of the basin. The impeller is usually of an axial flow design to maximize pumping efficiency. Note that such submerged impellers increase system power requirements. The optimum location of the

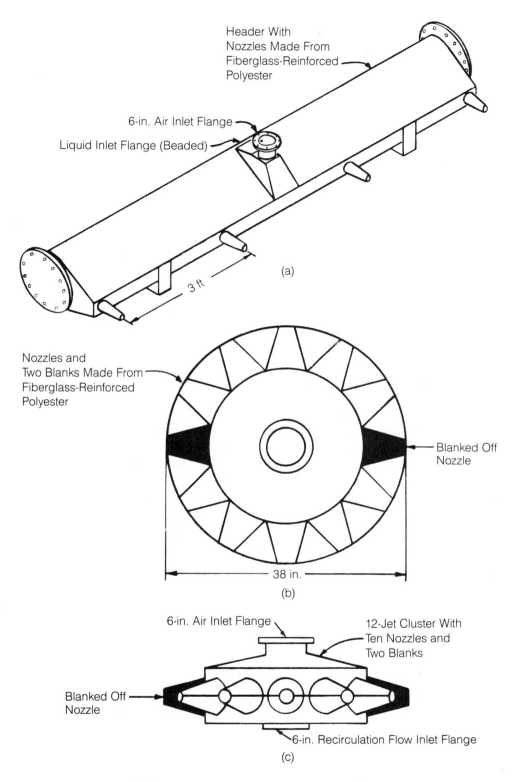

Header With
Nozzles Made From
Fiberglass-Reinforced
Polyester

6-in. Air Inlet Flange

Liquid Inlet Flange (Beaded)

3 ft

(a)

Nozzles and
Two Blanks Made From
Fiberglass-Reinforced
Polyester

Blanked Off
Nozzle

38 in.

(b)

6-in. Air Inlet Flange

12-Jet Cluster With
Ten Nozzles and
Two Blanks

Blanked Off
Nozzle

6-in. Recirculation Flow Inlet Flange

(c)

**Figure 11.17   Jet diffusers: (a) header-type, (b) plan of radial-type, and (c) elevation
of radial-type.**

impeller depends on its configuration. Radial flow impellers are generally located 0.5 to 0.7 times the impeller diameter above the tank bottom; axial flow impellers are located at 60 to 65% of the tank depth, measured from the water surface. Water depths using unsupported shafts (no bottom bearings) range up to 9 m (30 ft). With this unsupported length, shafts can transmit high side loads that the gear box must be designed to withstand.

All of the mixed liquor pumped through a draft tube is dispersed into the air. Without draft tubes, a portion of the pumped fluid flows beneath the liquid surface and is not aerated. Mixed liquor pumped in either way creates liquid momentum that tends to circulate the mixed liquid directly around the aerator. This momentum keeps the MLSS in suspension and the circulation does not impair oxygen transfer efficiency. Designers should recognize that mechanical aerators provide point source oxygen input. Pumped fluid flows radially outward from the aerator with decreasing velocity. Dissolved oxygen reaches its maximum near the impeller blade where surface turbulence is greatest and decreases as the fluid flows back below the surface of the aeration tank toward the aerator.

The action of surface aeration devices, particularly splashing from high-speed units, can generate mists with attendant nuisance odors. The odors can result from insufficient oxygen supplied by the aerator or from influent waste-water containing sulfides or other volatiles. Mists can freeze in cold climates, coating equipment with ice. Such freezing can cause hazards to both the facility staff and the equipment. Splashing effects can be minimized with proper geometric design of the aeration tank and use of deflector plates. Another cold-weather problem is basin heat loss induced by surface aerators.

Surface aerators are usually either bridge or platform mounted. Platform and bridge designs should address torque and vibration. Bridges should be designed for at least four times the maximum moment (torque and impeller side load) anticipated. The aerator manufacturer can provide the magnitude of this moment. Platform- and bridge-mounted aerators are quite sensitive to changes in the depth of impeller submergence. An increase in submergence results in increased fluid pumpage and can decrease gear box life expectancy. Some mechanical aerators tend to increase pumpage in the main body of the tank rather than lifting more mixed liquor for aeration when the impeller submergence increases.

Radial Flow Low Speed. Low-speed mechanical aerators have gained in popularity for several reasons. They can provide a higher oxygen transfer efficiency than coarse-bubble aeration equipment and they are good mixing devices.

These aerators generally operate in the range of 20 to 100 rpm and include a gear box to reduce the impeller speed below that of the motor. The gear box, a critical item, needs a service factor rating of 2.0 or higher to ensure mechanical reliability. Without a service factor, adjacent aerators operating at different speeds or units placed close to tank walls can cause drive overloads.

Manufacturers produce this type of aerator in several configurations (Figure 11.18). The simplest use is an impeller that operates at the water surface. For another configuration, submergence of the impeller can be adjusted to control power draw and oxygen transfer. Movable weirs (manual or automatic), usually part of such systems, allow this control. These aerators are available in normal power increments up to 110 kW (150 hp), with either floating or fixed mounting structures. They are also available as two-speed units, with turndown ratios of approximately 50% at the lower speed. Impellers range up to 3.6 m (12 ft) in diameter and operate at top peripheral velocities of 4.5 to 6 m/s (15 to 20 ft/sec).

Clean water standard aeration efficiencies (SAEs) of these units typically range from 1.2 to 2.7 kg O_2/kWh (2.0 to 4.5 lb O_2/hp-hr). Efficiency depends on many variables, including the design of the impeller itself, tank geometry, effects of adjacent walls, input power/tank volume, impeller size and speed, and other factors that are less well understood.

Axial Flow High Speed. High-speed aerators are normally used for stabilization lagoons where dispersed organism growth or other sources such as

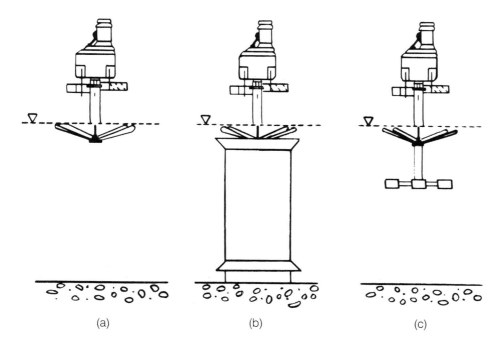

(a)   (b)   (c)

Figure 11.18   **Low-speed surface aerators: (a) surface aerator, (b) with draft tube, and (c) lowering impeller.**

benthic deposits exert oxygen demands. This limited primary application stems from concerns regarding disruption of sludge floc, possibly impairing settling.

High-speed aerators have limited depth of mixing and oxygen transfer capabilities. Neither of these limitations generally impedes aeration of lagoons, but they can restrict performance of aerators in conventional plants. Typically, high-speed units exhibit lower SAEs than radial flow devices. Like radial flow devices, performance of high-speed units is affected by basin geometry and other factors. High-speed units, available in standard motor increments up to 95 kW (125 hp), are most often mounted on floating structures.

Aspirating Devices. Another aeration device is the motor-driven propeller aspirator. One such device, shown in Figure 11.19, consists of a 1.2-m (4-ft) long hollow shaft with an electric motor at one end and a propeller at the other. Propeller rotation draws air from the atmosphere through the shaft. Air velocity and propeller action create turbulence, forming small bubbles. These devices can be positioned at various angles to reach different levels for aeration, mixing, or circulation. The portable units can be mounted on booms or floats in aeration tanks and oxidation ditches. An aspirator with a disk rather than a propeller disperses bubbles at a 90-deg angle to the shaft. Operation during very cold weather has been reported to occasionally cause icing of the aspirator pipe at the air inlet end, shutting off the air supply.

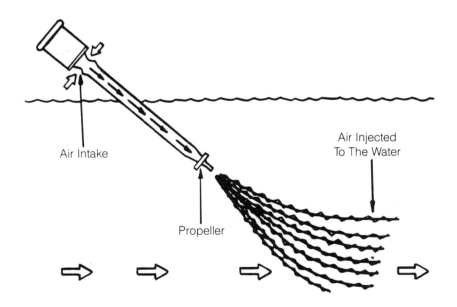

Air Intake

Air Injected To The Water

Propeller

Figure 11.19   Aspirating device.

Horizontal Rotors. This type of unit, available in several configurations, has a horizontal impeller (rotor). The impeller agitates the surface of the basin, transferring oxygen and concurrently moving the liquid in a horizontal direction.

Clean water SAE of these rotors approximates that of low-speed surface aerators; 1.2 to 2.7 kg O_2/kWh (2.5 to 3.5 lb O_2/hp-hr). Small changes in rotor submergence do not affect transfer efficiency. Lesser submergence will decrease total oxygen transfer; nonetheless, the mass of oxygen transferred per unit of power input remains about the same. The units are made in various sizes up to a maximum length of approximately 7.5 m (25 ft). Two rotors with one centrally located drive can aerate a 15-m (50-ft) wide, 3.5-m (12-ft) deep basin or ditch.

SUBMERGED TURBINE AERATORS. A submerged turbine consists of a motor and gear box drive, one or more submerged impellers, and piped air from a blower to a point below the impeller. Impeller designs vary, but generally are either the axial or radial flow type (Figure 11.20). With an axial flow turbine, the pumped liquid has sufficient velocity to drive the released air downward and disperse it across the bottom of the tank. In the radial flow design, the air flows into the impeller, mixes violently with the liquid, and disperses outward, driven by the impeller blades. With either

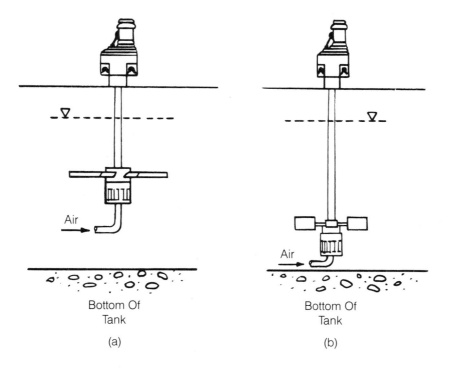

Figure 11.20   Submerged turbine aerators: (a) axial flow and (b) radial flow.

type, the operator must carefully control the air flow rate. Too much air will cause the turbine to backflood, reducing the amount of oxygen transferred and the pumping capacity of the impeller.

Air enters the turbine through either an open pipe or a diffuser ring. Although an axial flow unit can transfer a higher percentage of oxygen than can a radial flow unit, a radial flow design can handle a higher volume of air per kilowatt. Both units can be designed to transfer oxygen at rates of up to several hundred milligrams per liter per hour. This maximum rate far exceeds that typically required because even industrial applications with high-BOD_5 wastewaters seldom use more than 200 to 300 mg/L·h, and municipal plants are normally designed for demands of 30 to 100 mg/L·h.

The area of influence of submerged turbines is somewhat smaller than that of surface aerators. Low-speed axial flow units achieve maximum influence. The area of influence will vary from 4 to 13 m^2/kW (30 to 100 sq ft/hp), depending on reactor geometry and size. Turbine speeds employed are a function of impeller design and power input. Most operate in the range of 37 to 100 rpm. The gear box service factor should be 2.0 or higher to accommodate hydraulic loads imposed by adjacent walls and other aerators.

Submerged turbines transfer varying amounts of oxygen, depending on the air:mixer kW ratio, mixer design, and basin geometry. Clean water SAEs of these turbines range from 1.2 to 2.0 kg O_2/kWh (2.0 to 3.3 lb O_2/hp-hr), including both air blower and mixer power requirements. These SAEs are slightly lower than those of slow-speed surface aerators, but a submerged turbine offers the advantage of an adjustable gross oxygen input by modulating input air flow rate. In addition, where operation as either an aerator or a mixer (as in a combined nitrification/ denitrification reactor) would be useful, the submerged unit offers a significant advantage.

In those areas where basin cooling is a concern, the submerged turbine agitates the surface only slightly, with little consequent loss of basin temperature. Use of a submerged turbine in circular or square tanks requires baffling to prevent rotation of the entire tank contents. Rectangular-tank baffling requirements might be less than those for circular tanks. Regardless, the turbine manufacturer should specify the baffling requirements for any tank shape.

Available submerged turbine aerators match common motor sizes up to 110 kW (150 hp). Special designs include motors up to and exceeding 260 kW (350 hp). Air flow rates vary from 4 to 8 L/s (8 to 20 scfm) per aerator, and oxygen transfer rates range from 15 to 35%, depending on impeller configuration and depth.

Another submerged turbine aerator consists of a downward-pumping, air-foil-type impeller in a draft tube, with an air sparge ring mounted directly below the impeller (Figure 11.21). Coarse bubbles are sheared into smaller ones by the flow energy from the high pumping rate impeller. The bubbles are forced downward through the draft tube and baffles before rising to the surface. A major element of this design is the highly efficient axial flow impeller. Introduction of air disrupts the axial flow pattern, resulting in some reduction in pumping efficiency. Such aerators, normally deck mounted above the tank, can be installed in a conventional CMAS tank or in a total barrier oxidation ditch with a J-tube. Tank depths ranging from 7.6 to 9.1 m (25 to 30 ft) are used in CMAS applications. Clean water SAEs are reported to range from 1.0 to 1.8 kg O_2/kWh (1.7 to 3.0 lb O_2/hp-hr).

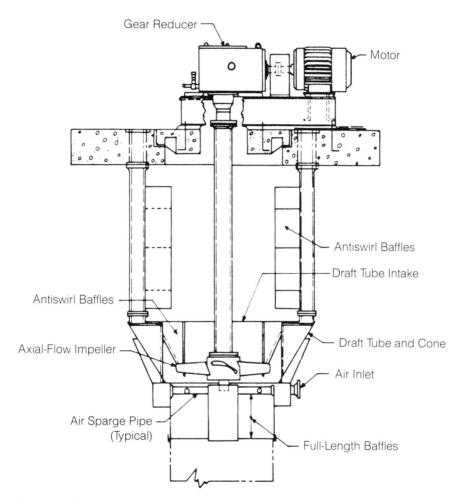

Figure 11.21   Draft tube submerged turbine aerator.

AIR SUPPLY SYSTEM. Although selection and design of air diffusion systems typically receive the most attention, design of a supply system demands attention to ensure that overall process goals are met. An air supply system consists of three basic components: air filters and other conditioning equipment (including diffuser cleaning systems), blowers, and air piping. Air filters remove particulates such as dust and dirt from the inlet air to protect the blowers and diffusers from mechanical damage and clogging. Blowers are designed to develop sufficient pressure to overcome static head and line losses while delivering gas at the required flow rate to the diffusion system. Piping conveys air or oxygen to the diffusers.

Air Filtration. The degree of air cleaning required depends on supply air quality, blowers installed, and diffusers employed. Diffuser systems traditionally have been designed with integral air filtration equipment. Particulate matter, rust and scale, construction debris from poor site cleanup, and oil in air supply lines have all caused air-side fouling of diffusers.

Cleaning efficiency is the primary filter design parameter. For protecting blowers, a typical minimum requirement for inlet air filtration is 95% removal of particles with diameters of 10 µm and larger. Standard practice in design for fine pore aeration devices is to remove 90% of all particles greater than 1 µm in diameter.[54,56] Manufacturers of perforated membrane diffusers indicate that filters designed to protect blowers will suffice for their equipment.[56] Some evidence, based on recent studies, indicates that this same degree of filtration might suffice for ceramic diffusers in new installations.[56] Designers should be cautious, however, in providing less-efficient air filtration than is currently considered acceptable.

TYPES OF AIR CLEANING SYSTEMS. Three basic types of air cleaning systems are viscous impingement, dry barrier, and electrostatic precipitation. These systems, manufactured in a variety of forms and sizes, can be either manually or automatically operated.

Viscous filters remove dust by impingement and retention of particles on a labyrinth of oil-coated surfaces through which air passes. These units will handle much dust and effectively remove large particles. However, a high percentage of small, low specific gravity particles will pass through such units. Consequently, viscous filters serve best as a preliminary filtration device. A coarse viscous filter ahead of an electrostatic precipitator can perform well in dusty areas and thus would be a good investment.

Dry barrier systems use a very fine filter material, such as paper, cloth, or felt, to entrap particles. These systems typically comprise a coarse prefilter followed by a fine filter. The prefilter normally consists of a sheet of fiberglass cloth mounted on a frame. Fine filters are housed in racks behind the prefilter. Such replaceable systems occupy little space and offer easy maintenance.

Designers of large plants may wish to investigate bag house collectors. These dry barrier units are constructed as steel enclosures that house sets of cloth stocking tubes. The tubes are precoated with filter aid before being placed in service and after each cleaning. Efficiency increases during a filter run because retained particles increase the effectiveness of the filtering medium. A bag house collector, properly installed and maintained, will protect up to recommended standards if the atmosphere is not too smokey. Size, expense, and precoat requirements of bag house filter systems have reduced their selection for new installations.

Electrostatic precipitators impart an electric charge to particles so that they can be removed by attraction to elements of opposite polarity. These units require 30 to 50% the area of bag houses and have relatively simple maintenance needs. This type of device will remove very fine dust particles and will protect up to recommended standards when operated at velocities of less than 2 m/s (400 ft/min). These devices are especially useful in areas with smokey atmospheres.

FILTER SELECTION. Of the air cleaning systems available, replaceable filter units are the simplest to construct and operate. Their capital costs are about 12% those of electrostatic precipitators. Bag house dust collectors are bulky and expensive, though relatively maintenance free. Replaceable air filters are a good selection except where poor air quality would require replacement of the fine filtration elements more frequently than once per year. In such cases, electrostatic precipitators might be cost effective.

DESIGN CONSIDERATIONS. In addition to the design recommendations of the filter manufacturers, other needs require special attention. Since a treatment plant must operate continuously, facilities for equipment maintenance are needed. Continuous operation calls for weather protection of the air intake structure. Good louvers and an ample chamber between the louvers and filters are essential. In freezing climates, preheating the air might be necessary to prevent snow or water vapor from freezing onto the filters. A simple method of preheating relies on ducts and dampers to direct part of the air flow inside the blower building. Designers need to exercise care in locating the filter inlet to prevent drawing excessively moist air into the filters. This could soak the filter medium, reducing its performance and throughput capacity. Housing for air filters consists of corrosion-resistant materials.

Blowers. The term blower typically applies to air-conveying equipment that generates pressures up to about 103 kPa (15 psi). As shown in Figure 11.22, many different types of blowers are available. Two types of blowers used for aeration are rotary positive displacement (PD) and centrifugal units.

As a major advantage, PD blowers are capable of operating over a wide range of discharge pressures. Some disadvantages are that they are difficult to operate at variable air flow rates, they require more maintenance, and they are noisy. Advantages of centrifugal blowers include less noisy operation and smaller foundations. Their disadvantages include a limited operating pressure range and reduced volumes of air delivered with any increase in backpressure due to diffuser clogging.

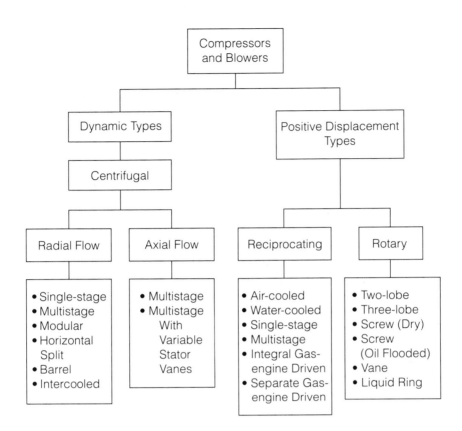

Figure 11.22 Blower selection.

Blower capacity can be specified in several ways. The most useful method is the actual volume of air per unit of time. PD blowers have capacities from 2 to 23 600 L/s (5 to 50 000 acfm). Centrifugal blower capacity typically ranges from 240 to 70 800 L/s (500 to 150 000 acfm). Impellers of a centrifugal unit can be arranged in stages for higher discharge pressures.

TURNDOWN. The selected blower system must be capable of supplying the volumes of air necessary to meet varying oxygen demands over the design life of the facility. Therefore, blower selection should take into account the minimum air requirement for plant start-up (normally dictated by mixing requirements) and the maximum requirement for design conditions. Integral with this range is the air requirement variation associated with diurnal loading conditions. Such conditions can impose a substantial peaking factor for actual air requirements; this peaking factor might correlate somewhat with the HRT of the aeration basins selected. For example, a basin with a 24-hour HRT will have substantially less variation in required air supply rate than that for a basin with an HRT of 6 hours.

SELECTION OF BLOWERS. As previously discussed, each type of blower has distinct operating characteristics. A blower must be compatible with the normal operating mode of the treatment system. Other factors, such as noise, maintenance, and operator preference are also considered in the selection process. If the aeration system is designed for operation with a fairly constant water depth, typical of most aeration basins, a centrifugal blower will be a good choice. Conversely, if the system will be operated over a wide range of depths, as in an SBR, a PD blower might be a better selection.

Both discharge pressure and the weight of air vary with inlet temperature. Therefore, blowers are selected to provide the required air flow rate at the maximum discharge air temperature anticipated. Blower motors are sized to deliver the required air flow rate at the minimum inlet temperature expected.

CONTROL. Because a PD blower will deliver a relatively constant flow rate of air over a range of discharge pressures, multiple units or multiple speed motors can provide rate control. Use of variable-frequency alternating current drives allows PD blower operation as variable-capacity, variable-pressure machines.

Control of air flow rate with centrifugal blowers is typically accomplished by the use of variable-speed drives, inlet vane adjustments, or throttling valves. Engineers often select inlet throttling, which is the least complex method. Rate control can be accomplished by a manually or motor-operated valve that is controlled by some other measured parameter in the system, such as DO concentration. Air flow rate and discharge pressure can also be throttled with a valve downstream of the blower. However, this method requires more horsepower than an inlet valve. Details of blower selection, system design, maintenance, and control can be found in other publications.[54,56]

Air-piping Materials. Major considerations in piping material selection are strength and potential deterioration due to corrosion, thermal effects, and other environmental factors. Piping materials commonly used include galvanized and stainless steel, ductile iron, fiberglass reinforced plastic, HDPE, and PVC (See Chapter 8). Because blower discharge pressures are typically less than 103 kPa (15 psi), thin-walled pipe is often used. Use of such pipe requires that it be adequately protected from physical damage. Temperatures in excess of 93°C (200°F) are not uncommon in the blower discharge. Therefore, the pipe and accessories (supports, valves, gaskets, and so on) must be designed accordingly. Because thermal stresses can be significant, provisions for pipe expansion and contraction are needed. Blower discharge piping is often insulated to protect workers from burns.

Because of the potential for corrosion at the interface between the atmosphere and the liquid in aeration tanks, piping material used for the droplegs is often stainless steel or PVC rather than carbon steel or ductile iron. Inside the basin, the piping branches into a system of manifolds and headers. This piping is also usually stainless steel or PVC. The choice of material usually depends on the structural requirements of the diffuser connection and on whether a diffuser cleaning system will be provided. The designer must check the compatibility of the piping material with the cleaning gas or liquid. PVC is commonly selected due to its inert characteristics.

Stainless steel is typically selected for tube diffuser systems due to the cantilever load applied by the diffusers. However, PVC has also been used successfully. PVC is more commonly used with disc or dome diffusers because they are normally mounted on top of a header; thus, forces transmitted through the connection to the header are minimal.

Basins are periodically drained and left empty. If PVC piping is selected as the manifold and header material, it must resist sunlight deterioration. Titanium dioxide and carbon black are commonly used in PVC piping for ultraviolet (sunlight) protection. Other design considerations include the effects of freezing and thawing when tanks are empty.

Air-piping Design. Piping should be sized so that head loss in the supply lines and headers is small compared to that across the diffusers. Typically, if head loss from the blowers to the diffusers is less than 10% of the system total, the required air distribution throughout the basin can be maintained.

Basic fluid mechanics principles are used to size air piping systems. Designers usually employ standard calculation procedures such as those developed by Darcy-Weisbach. A number of handbooks describe these procedures.[57,58] Corrections for temperature rise during compression, altitude (barometric pressure), and the specific weight of air at design temperature and pressure must be included in the calculations. Head loss calculations should account for both maximum summer air temperature and the temperature rise from air compression at the maximum expected air flow rate.

Losses through fittings and valves can be calculated using head loss coefficients and velocity heads. Typical coefficients can be found in texts and handbooks.[57,58] Table 11.10 presents ranges of head losses through air filters, silencers, and check valves. Actual values selected should be verified by the manufacturer of the selected equipment.

Table 11.10   Typical head loss values.[57]

Appurtenance	Head loss, in. water column
Inlet filters	3–16
Inlet silencers	0.5–9
Check valves	1–8

Pure Oxygen Generation. On-site production of pure oxygen can be accomplished either by cryogenic means or through the use of pressure swing adsorption (PSA) equipment.[59] The cryogenic air separation process produces liquid oxygen by the liquefaction of air, followed by fractional distillation to separate the air components, mainly nitrogen and oxygen (Figure 11.23). PSA systems use a multibed selective adsorption process to provide a continuous flow of gaseous oxygen (Figure 11.24).

Cryogenic oxygen generation plants are normally more efficient than PSA units. As a general comparison, a power use of 0.15 kWh/lb O_2 generated is representative of a cryogenic plant and approximately 0.20 kWh/lb O_2 generated is typical of a PSA unit. Minimum continuous output of cryogenic units is 30 to 40 t/d, while for PSA units it is 0.5 to 1.0 t/d. Design of either system for turndown is important. With a PSA system, turndown to 20% of full generation capacity is possible. With a cryogenic system, turndown to only 70% of the full load generation rate is expected. With both systems, liquid oxygen storage is typically provided for peak oxygen demand periods and for oxygen generation equipment downtime.

MIXING REQUIREMENTS. Aeration devices must supply enough energy to mix the contents of a reactor unless it is equipped with separate mixing

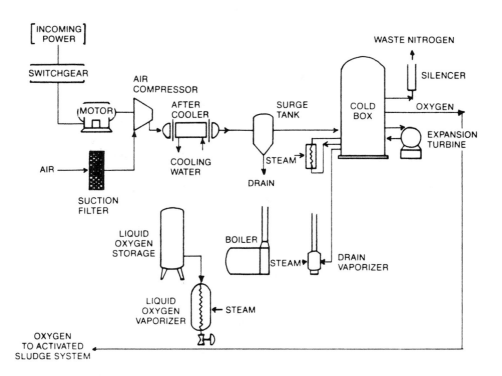

Figure 11.23   Schematic of a simplified cryogenic oxygen system.

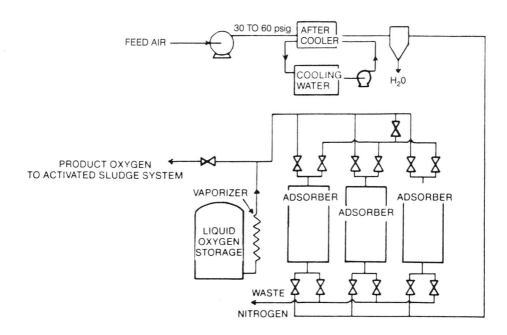

Figure 11.24 Schematic of a pressure swing adsorption system.

equipment. As a rule of thumb, mixed liquor velocities in an aeration tank should equal at least 0.3 m/s (1 ft/sec). An adequate air input for diffused air systems would range from 20 to 30 m^3 air/1000 m^3 tank volume/min (20 to 30 cu ft/min/1000 cu ft).[19] For mechanical aeration systems with a vertical mixing regime, 15 to 30 kW/1000 m^3 (0.6 to 1.15 hp/1000 cu ft) is considered appropriate. This range can be reduced somewhat for rotors or aspirators that maintain horizontal velocities. Because oxygen demand requirements normally control conventional air systems, enough energy is automatically provided for mixing. For pure oxygen systems, supplemental mixing is needed. Likewise, in extended aeration facilities, OURs are typically low and mixing requirements often control the rate of energy input.

AERATOR DESIGN AND TESTING. Historically, many methods have been used to test and specify aeration equipment. The varied methodologies, at times, have led to confusion and misrepresentation of equipment performance. Furthermore, equipment suppliers, consultants, and users often use differing nomenclature when they report equipment capabilities.

The preparation of explicit equipment specifications is vital to the eventual installation of efficient and cost-effective devices. To enable a supplier to properly specify and quote aeration equipment, the prospective user or design engineer must provide accurate and detailed information on system requirements and constraints. The supplier can then provide equipment performance information based on reliable clean water test data and on sound judgment using experience from previous field applications. In short, uniform specifications will minimize user misunderstanding of equipment quality and performance so that realistic comparisons can be made.

Evaluation of reliable clean water test data is only one step in understanding aeration system capabilities. Measuring oxygen transfer in the field with aeration equipment operating under actual process conditions is also imperative. Engineers must choose carefully among the various field test methods that are available and use them with care. The following sections discuss specification and testing protocols that will enhance the understanding of measuring aeration equipment performance. Knowledge of such protocols can simplify subsequent transactions among vendors, consultants, and users.

Aeration Equipment Considerations. As the first step in proper equipment selection, the prospective user determines the aeration system field requirements. Important elements defining these requirements are

- Site location, elevation above sea level, ambient high summer and low winter air temperatures;
- Aeration tank volume, process water depth, and basin configuration;
- Oxygen demand—minimum, average, and maximum plus spatial and temporal distributions;
- Mixing requirements—capability to maintain specified MLSS concentrations in suspension;
- Process water temperature—minimum, average, and maximum;
- Process water transfer characteristics—range of alpha and beta factors anticipated;
- Operating DO concentration (mg/L)—minimum (fifth percentile), average, and maximum (ninety-fifth percentile);
- MLSS/MLVSS concentrations (mg/L)—minimum, average, and maximum;
- Aeration equipment—desired type of system, construction materials, required performance testing, and necessary quality control for installation; and
- Penalties for not meeting performance guarantees.

Equipment suppliers should give users detailed mechanical and structural requirements and performance characteristics of their equipment, including reliable clean water performance data. Clean water test data provide the primary basis for specification of aeration equipment. These data must be reported at standard conditions and supplemented with a description of conditions under which they were derived. This information will allow the prospective user to judge the usefulness of the data for the specific application. Engineers and users should insist that basic data collection conform with protocols outlined in the ASCE Oxygen Transfer Testing Standard,[60] which is summarized in a following section.

The prospective user or design engineer must translate clean water performance results to the applicable field conditions noted above. Importance of using the most appropriate alpha values cannot be overstated; therefore, the engineer must exercise informed judgment based on the fundamentals of aeration, the specific equipment being considered, and past experiences. Based on experience with equipment applications, ask the supplier to confirm the appropriateness of the alpha values being considered.

When the above adjustments are completed, the design engineer can select the appropriate number of aeration units to satisfy design criteria. For a complete supplier response to a customer inquiry, the following information should accompany the specification and quote for aeration equipment:

- Number of units required to meet critical design conditions, including adequacy of turndown capabilities;
- Total cost of each aeration unit or that of the total system;
- Power required to operate the aeration units, if appropriate;
- Total air required to operate the aeration units, if appropriate;
- Air distribution system design, including head loss calculations, maximum blower pressure, and pipe and orifice sizes, if appropriate;
- Power required for the air blower, if appropriate;
- Inlet air filtration requirements, if appropriate;
- Clean water test data along with test conditions used for establishing standard oxygen transfer rates (SOTR) at minimum, average, and maximum oxygen demands;
- Calculations to transform the SOTR to the field condition oxygen transfer rate (OTR$_f$);
- Equipment construction materials, including detailed drawings and specifications that outline device mechanical and structural integrity;
- Quality assurance/quality control (QA/QC) programs used in equipment manufacture, shipping, storage, and installation; and
- Full-scale testing to demonstrate equipment performance guarantees (aeration and mixing at minimum, average, and maximum oxygen demand conditions, if practical).

Clean Water Testing.[60] The ASCE standard describes the measurement of oxygen transfer rate as the mass of oxygen dissolved in a unit volume of clean water by a system operating at a given aeration rate and power input condition. This method applies to laboratory-scale devices with water volumes of a few liters as well as to full-scale systems with water volumes typical of those found in a facility's aeration basins; the method is valid for many different mixing conditions and process configurations.

The method is based on DO removal from the test water volume by the addition of sodium sulfite in the presence of a cobalt catalyst, followed by transfer studies of reoxygenation to almost the saturation level. Test water volume DO inventory is monitored during the reoxygenation period by measuring DO concentrations at several points selected to best represent the tank contents. These DO concentrations can be measured *in situ* or on samples pumped from the tank. The method specifies a minimum number, distribution, and range of DO measurements at each determination point.

Data obtained at each determination point are then analyzed by a simplified mass transfer model to estimate the apparent volumetric mass transfer coefficient (K_{La}) and the equilibrium spatial average DO saturation concentration (C^*_∞).[61] Nonlinear regression analysis is used to fit the DO profile measured at each determination point during reoxygenation to the model's mathematical equation. In this way, estimates of K_{La} and C^*_∞ are obtained at each determination point. After these estimates are adjusted to standard conditions, the system SOTR is calculated from the aeration tank volume (V) and the estimates of K_{La} and C^*_∞ at each of the n determination points as follows:

$$\text{SOTR} = V \sum_{i=1}^{n} K_{La20} C^*_{\infty 20}/n \tag{20}$$

Frequently, the SAE is calculated as the SOTR divided by power input. The standard oxygen transfer efficiency (SOTE) in percent can also be estimated by

$$\text{SOTE} = (\text{SOTR}/W_{O_2}) \times 100 \tag{21}$$

The foundation and key elements of the oxygen transfer measurement test are the definition of terms used during aeration testing, subsequent data analysis, and final result reporting. A consistent nomenclature has been established with more logical and understandable terminology than the numerous and varied symbols used historically. When consistently used, this standard nomenclature will eliminate much of the difficulty in interpreting aeration literature.[60]

Transformation of Clean Water Test Data to Process Water Conditions.
The oxygen transfer rate of a particular aeration device is commonly expressed as either an SOTR or as an OTR$_f$. Calculations of a field transfer rate from a standard value can be performed as follows:[56]

$$\text{OTR}_f = [\alpha F \text{SOTR}\, \Theta^{(T-20)}/C^*_{\infty 20}] (\beta \Omega \tau C^*_{\infty 20} - C) \tag{22}$$

Where

OTR_f	=	Oxygen transfer rate estimated for the system operating under process conditions at an average DO concentration, C, and temperature, T, lb/hr.
SOTR	=	Standard oxygen transfer rate of new diffuser, lb/hr.
α	=	Average process water K_{La}/average clean water K_{La} (both with new diffusers).
β	=	Process water C^*_{st}/clean water C^*_{st}.
C	=	Average process water volume DO concentration, mg/L.
Θ	=	Empirical temperature correction factor assumed to equal 1.024, unless aeration system tests show a different factor.
F	=	Process water SOTR of a diffuser after a given time in service/SOTR of a new diffuser in the same process water.
$C^*_{\infty 20}$	=	Steady-state value of DO saturation at 20°C and a barometric pressure of 1.0 atm, mg/L.
τ	=	A temperature correction factor for DO saturation, C^*_{st}/C^*_{s20}.
C^*_{st}	=	Tabular value of DO surface saturation concentration at actual process water temperature, a barometric pressure of 1.0 atm, and 100% relative humidity, mg/L.
C^*_{s20}	=	Tabular value of DO surface saturation concentration at 20°C, barometric pressure of 1.0 atm, and 100% relative humidity, mg/L.
γ_{wt}	=	Weight density of water at process conditions, lb/cu ft.
P_{vt}	=	Saturated vapor pressure of water at process temperature, psia.
P_b	=	Barometric pressure under field conditions, psia.
P_s	=	Standard barometric pressure of 14.7 psia.
d_e	=	Effective saturation depth, ft,
	=	$1/\gamma_{wt}[C^*_\infty/C^*_{st}(P_s - P_{vt}) - P_b + P_{vt}]$.
Ω	=	P_b/P_s (for tanks under 20 ft in depth)
or		
Ω	=	$(P_b + \gamma_{wt}d_e - P_{vt})/(P_s + \gamma_{wt}d_e - P_{vt})$ (for tank depths greater than 20 ft).

Table 11.11 serves as a guide for applying equation 22 and indicates the source of information for the parameters needed to estimate OTR_f. Values of $C^*_{\infty 20}$ and SOTR are obtained from the clean water test described above. The value of C should represent the desired process water DO concentration averaged over the aeration basin volume. Temperature and atmospheric pressure correction factors (τ and Ω) are estimated as described above.

Review of the components of this equation reveals a parameter (F) recently introduced in fine-pore aeration technology analysis.[56] This parameter attempts to account for impairment in diffuser performance caused by fouling or material deterioration. This fouling factor, a dynamic term, depends on diffuser type and wastewater characteristics. Its value is still not well documented. In recent studies of fine-pore diffusers (ceramic, porous plastic, and perforated membrane), values of F ranged from 0.5 to 1.0.[56] The rate at which this value changes, also important to evaluate, is described by a fouling rate term (f_F). For further details on this parameter, refer to the EPA design manual.[56] In the past, the effects of diffuser fouling and deterioration, as well as the effects of process water on oxygen transfer, were included in the alpha term. The term apparent alpha (α') was often used to describe this dual effect.[55] With the current nomenclature, apparent alpha is replaced by the equivalent term αF.

Table 11.11 Selected factors affecting oxygen transfer field testing for estimation of K_{La}.[56]

	Oxygen transfer tests			
Factors	**Steady state**	**Nonsteady state**	**Off-gas**	**Inert gas tracers**
Sensitivity to variations in				
Flow	−	−	−	+
Uptake rate	−	−	−	+
Alpha	−	−	−	+
DO	−	−	−	+
Gas flow rate	−	−	−	−
Accurate measure of				
Uptake rate	−	+	−	+
DO	−	−	−	+
C	+	+	−[a]	+
Gas flow rate	+	+	−[a]	+
Other	+	+	−[b]	−[c]
Basin configuration				
Plug	−	−	−	+
CSTR	+	+	−	+
Aerator type	+	+	−	+
Costs				
Manpower	+	0	0	0
Analytical	+	+	0	−
Capital invest	+	+0	0	−
Calculations	+	0	0	0
Estimated precision	−	0	−	+

[a] Calculating OTE directly requires C and gas flow rate to estimate K_{La}.
[b] Requires accurate estimates of CO_2 and water vapor in gas.
[c] Requires accurate estimate of $K_{kr}:K_{La}$, especially in diffused air systems.
+ Positive response (for example, nonsensitive, less costly, more precise, and easier).
0 Intermediate response.
− Negative response.

Alpha is one of the most controversial and investigated parameters for oxygen transfer. It depends on wastewater characteristics, diffuser type, air flow rate, diffuser placement, basin geometry, system operating parameters, and flow regime. In addition, its value varies spatially and temporally. Typically, with the same wastewater, alpha will be lowest for aeration devices generating fine bubbles and highest for surface aeration systems. Recent measurements of OTR_f in municipal facilities indicated that αF values for fine-pore devices averaged about 0.4, with a range of 0.1 to 0.7.[56] These were mean weighted values for the entire aeration basin. Individual measured values were significantly lower at the influent end of the plug flow reactor. Diurnal variations of the mean weighted αF values were represented by a maximum:average ratio of about 1.2 and a minimum:average ratio of about 0.86.[56] Information on alpha for mechanical aeration equipment is more difficult to find and, when available, is less reliable than for fine-pore devices. Values from 0.3 to 1.1 have been reported in the literature.[62] Additional references discuss alpha for other systems and wastewaters.[54,56,63-66]

PROCESS WATER TESTING. Once aeration equipment is operating under process conditions, its performance should be compared with calculated design estimates. Process water testing provides the best and most reliable source of data on αF and on the effects of system design. The several methods for testing equipment during process operation are all referred to as respiring system tests.

In general, the testing methods can be categorized according to the rate of DO concentration change with respect to time in a given reactor (or reactor segment). Systems with a DO rate of change of zero are described as being in a steady-state condition; the others are classified as being in a non-steady-state condition. If influent wastewater is diverted from a reactor for testing, these tests are referred to as batch tests. The term continuous test applies where the influent wastewater flow is not diverted.

The several respiring system test methods that do not require a direct measure of OUR are broadly categorized as the mass balance, off-gas, inert-tracer, and non-steady-state methods. The mass balance method requires measurement of influent and effluent liquid flows. The off-gas method is based on the mass balance of oxygen across a system and requires measurements in both the liquid and gas streams. The inert-tracer method indirectly measures oxygen transfer rate by determining the transfer rate of a radioactive or stable inert gas tracer. For non-steady-state methods, the reactor DO level is adjusted at the beginning of the test to be either higher or lower than the steady-state DO concentration. The OUR data, though not required, are often collected to ensure that relatively constant operating conditions prevail during an evaluation. Several references include a comprehensive review of available test methods for field oxygen transfer measurements.[54,63,64] Assumptions necessary to use steady- or non-steady-state tests are given in

Tables 11.12 and 11.13[54,63,64]. Selection of the best method often depends on economics, degree of precision and accuracy required, process conditions, and other considerations.

VOLATILE ORGANIC COMPOUND (VOC) EMISSIONS. VOCs are released to the atmosphere from aerated process tanks, free falls at weirs, and other locations within a facility where air contacts the wastewater. The VOCs in wastewater originate from industrial discharges, household chemicals, and the chlorination of organic compounds such as humic substances. As a result of the potential health risk of VOCs, recent efforts have been undertaken to evaluate, regulate, and mitigate the emission to the atmosphere of some of these compounds. Properties of many VOCs of concern are presented in Table 11.14.[67]

For low concentrations of a volatile compound in water, the gas-phase concentration at equilibrium above the water surface is proportional to the liquid-phase concentration; the proportionality factor is known as Henry's law constant. The relative rate of volatilization of a compound generally increases with the value of this constant. Other factors that influence the rate

Table 11.12 Assumptions necessary to develop equations for continuous steady- and non-steady-state tests.[54,63,64]

Assumptions	Test conditions
Aeration volume DO is constant and the reactor contents are completely (uniformly) mixed.	Test time is short; they are not required for non-steady-state test.
Reactor influent flow is constant.	Test time is short; variation in influent wastewater flow is neglible during test period; recycle sludge flow is held constant.
Influent DO is constant.	Test time is short; variations in influent wastewater flow and DO are negligible during test period; recycle sludge flow is held constant.
Aeration volume oxygen uptake is constant.	Test time is short; variations in influent wastewater flow and DO are negligible during test period (note that uptake rate may be difficult, if not impossible, to accurately determine for systems with high organic loadings).
Effective oxygen transfer rate is constant.	Test time is short; alpha value remains constant during the test period.

Table 11.13   Assumptions necessary to develop equations for steady- and non-steady-state batch.[54,63,64]

Assumptions	Test conditions
Aeration volume DO is constant and the reactor contents are completely (uniformly) mixed.	Steady-state conditions have been achieved and maintained; they are required for non-steady-state test.
Recycle sludge flow is constant zero.	Recycle flow rate maintained constant or discontinued.
Recycle DO is constant.	Steady operation of recycle system if in use during the test period (for example, sludge blanket level constant).
Aeration volume oxygen uptake rate is constant.	Aeration volume biological solids and recycle sludge flow remain constant; carbonaceous and nitrogenous substrates are near zero during the test (if nitrification occurs in the test system).
Effective oxygen transfer rate is constant.	Carbonaceous substrate is near zero during the test; alpha value remains constant during the test period.

of volatilization include concentrations of the VOC in gas and liquid phases, molecular weight, diffusion coefficient characteristics of the water, and air and water turbulence and mixing. Roberts *et al.*[68] and Treybal[69] present a detailed review of the phase transfer of VOCs and the fundamentals of mass transfer.

Volatilization models for the emission of VOCs from the surface of mechanically aerated systems usually follow the two-film theory of mass transfer and employ the following function:

$$dc/dt = -K_{La}(C_L - C_L^*) \qquad (23)$$

Where

dc/dt	=	concentration rate of change, mg/L·s;
K_{La}	=	overall mass transfer coefficient, L/s;
C_L	=	liquid-phase concentration, mg/L; and
C_L^*	=	liquid-phase concentration corresponding to equilibrium with the gas-phase concentration, mg/L.

The overall mass transfer coefficient (K_{La}) is a direct function of Henry's law constant. In addition, the equilibrium liquid-phase VOC concentration

Table 11.14   Properties of volatile organic compounds.[67]

Compounds	MW	Henry's law constant, atm·m^3/mol × 10^3	Henry's law constant, dimensionless	Diffusion coefficient (water), cm^2/s
1,1-Dichloroethylene	97.00	15.00	0.62	
Dichloromethane	85.00	3.19	0.13	1.17E-05
trans-1,2-Dichloroethylene	96.94	5.32	0.22	
1,1-Dichloroethane	99.00	5.45	0.23	
Chloroform	119.40	3.39	0.14	1.00E-05
1,2-Dichloroethane	98.76	1.10	0.05	9.90E-06
1,1,1-Trichloroethane	133.40	4.92	0.20	8.80E-06
Benzene	78.10	5.55	0.23	9.80E-06
Tetrachloromethane	153.80	30.20	1.26	8.80E-06
Dibromomethane	173.85	1.00	0.04	
1,2-Dichloropropane	112.99	2.82	0.12	8.73E-06
Bromodichloroethane	129.39	2.12	0.09	
Trichloroethylene	131.39	9.10	0.38	9.10E-06
1,1,2-Trichloroethane	133.40	11.70	0.49	8.80E-06
Toluene	92.00	5.93	0.25	8.60E-06
Dibromochloromethane	208.29	0.78	0.03	
1,2-Dibromoethane	187.88			
Tetrachloroethylene	165.90	28.70	1.19	8.20E-06
Ethylbenzene	106.16	6.44	0.27	7.80E-06
m,p-Xylene	106.16	5.24	0.22	7.80E-06
Bromoform	252.77	0.53	0.02	
1,1,2,2-Tetrachloroethane	168.00	0.38	0.02	7.90E-06
o-Xylene	106.17	5.27	0.22	
Cumene	120.20	14.60	0.61	7.10E-06
Propylbenzene	120.19	6.59	0.27	
4-Ethyltoluene	120.20			
3-Ethyltoluene	120.20			
1,3,5-Trimethylbenzene	120.20	147.00	6.11	
2-Ethyltoluene	120.20			
1,2,4-Trimethylbenzene	120.20			
1,3-Dichlorobenzene	147.01	2.63	0.11	7.86E-06
1,4-Dichlorobenzene	147.00	2.72	0.11	7.90E-06
1,2-Dichlorobenzene	147.00	1.94	0.08	7.90E-06
1,3-Diethylbenzene	134.22			
1,4-Diethylbenzene	134.22			
1,2-Diethylbenzene	134.22			

(C_L^*) can be assumed to equal zero. This results in a volatilization rate that is directly proportional to the VOC concentration in the liquid phase.

For diffused aeration systems, equation 23 is invalid because the gas-phase VOC concentration varies as the air bubbles rise through the liquid. Like oxygen transfer calculations, more complex expressions are required to model VOC emissions from this type of system. Reactor configuration (plug flow, complete mixed, batch, or series) affects the extent of emissions for given conditions of flow, detention time, and influent VOC concentration.

Because measurement of the emission of VOCs from aeration tanks is difficult, few data are available. Recent work has focused on the development of models incorporating biodegradation, adsorption, and direct volatilization to the atmosphere as mechanisms of removal.[70,71] A recent study conducted by the Ontario Ministry of the Environment compared off-gas sampling results from four wastewater treatment facilities and found that actual emissions were overestimated by application of the models evaluated.[67]

SECONDARY CLARIFICATION

Secondary clarification is vital to the operation and performance of suspended-growth treatment systems. This process separates MLSS from the treated wastewater prior to discharge. It also thickens the MLSS before their return to the aeration process or their wasting. Clarifier performance is critical for meeting TSS, BOD5, and occasionally TKN discharge criteria.

In secondary clarification, mixed liquor is fed to the clarifier and the MLSS settle, forming a sludge blanket with an overlying clear-water zone. Within the clear-water zone, discrete floc particles settle, resulting in a clarified effluent. In the lower zone, the sludge blanket of MLSS thickens prior to withdrawal as clarifier underflow. Secondary clarifier process failure can involve either the clarification or thickening stages; thus, proper design must address both.

Suspended-growth system variables include process loading configuration, MCRT, HRT, DO concentration, influent wastewater characteristics, and aeration basin shear intensity. The clarification variables include hydraulic and solids loading rates, turbulence, and tank geometry. These two sets of variables together have been noted as having potential causal relationships with clarifier performance. Recent investigations have examined such relationships.[72-75] Variables including solids loading rate, MLSS characteristics, clarifier and sludge blanket depths, detention time, clarifier underflow rate, and type of sludge collection mechanism have been shown to influence the thickening performance of secondary clarifiers.

Though clarifiers have served suspended-growth processes for decades, opinions differ as to what constitutes an optimal design. Clarifiers have varied in shape, depth, and geometric detail, and methods to determine their size have changed with time. A discussion of clarifier design is given in WPCF Manual of Practice No. FD-8.[76] This reference and others can help readers develop an understanding of the underlying theories, variety of design criteria, and various geometric details used in recent years.[19,77] The following subsections present abbreviated but rather direct methods of designing activated sludge clarifiers. A dissertation covering the findings of over 300 articles is offered in an NTIS report entitled "Critical Literature Review and Research Needed on Activated Sludge Secondary Clarifiers."[78] TeKippe and Bender have prepared a brief summary of the report's findings.[79]

GENERAL DESIGN CONSIDERATIONS. The design of secondary clarifiers must be considered integrally with that of the aeration basins and return sludge pumping facilities. Area and depth requirements directly depend on settleability of the mixed liquor, average and peak wastewater flow rates, and return sludge pumping rate.

As its primary function, the clarifier provides an overflow with a low TSS concentration. Currently, federal secondary treatment standards require this value to be 30 mg/L or less on a monthly average basis for suspended-growth systems. These standards also require a comparable BOD_5 concentration of 30 mg/L or less. Effluent TSS contribute to the BOD_5 concentration of the effluent. Values of the BOD_5 associated with TSS vary with system sludge age and other factors but often approximate 50% of the effluent TSS value. It is, therefore, vitally important to remove TSS to meet not only the effluent TSS limit, but also the BOD_5 limit. To achieve the second function of the clarifier, namely sludge thickening or compaction, adequate settling time and depth are needed.

As another consideration in clarifier design, steady-state operating conditions seldom occur. Fluctuations in wastewater flow rate, MLSS concentration, and settleability occur seasonally, daily, and even diurnally. Some conservatism in design, therefore, is essential to maintain performance with these uncontrollable fluctuations.

Many states regulate clarifier design. They commonly specify a maximum allowable overflow rate for average flow as well as for peak flow, a minimum depth, and a maximum allowable weir loading rate. Some state regulations recognize differences in the type of mixed liquor settled (for example, oxidation ditch versus conventional systems). This rather simplistic approach has, in recent years, given way to more sophisticated approaches that incorporate additional parameters such as solids loading rate, initial settling velocity (ISV) or SVI, and return sludge pumping rate.

PROCESS DESIGN CONSIDERATIONS AND PARAMETERS. Approaches for clarifier design outlined herein are based primarily on the experience of engineers and operators of municipal treatment plants in the U.S.

Suspended-growth Biological Treatment   583

Most of their experience relates to circular (or radial flow) basin geometry; present data are insufficient to show whether rectangular, longitudinal flow basins require different bases for design.

The following steps summarize the historical approach to clarifier design:

1. Set the surface overflow rate for the average dry weather flow condition within 0.85 to 1.4 m/h (500 to 800 gpd/sq ft). Check the overflow rate for peak flow conditions. The allowable rate for this condition should be 1.7 to 2.5 m/h (1000 to 1500 gpd/sq ft).
2. Allow either the average dry weather or peak flow rate condition to dictate the required surface area of the clarification system.
3. Select an appropriate sidewater depth from 3.0 to 5.5 m (10 to 18 ft).
4. Choose a sludge removal mechanism (plow or hydraulic suction).
5. Choose a sloped (1:12) or nearly flat floor.
6. Specify a feedwell or other influent structure design.
7. Decide among inboard launders, a perimeter weir, or end-of-tank weir as appropriate for the tank geometry.
8. Keep the tank diameter to less than approximately 46 m (150 ft) to avoid wind disturbance problems (wind concerns exist but quantifying data have not been found in the literature).
9. Establish the minimum number of tanks required and calculate the surface area of each by dividing the total required tank surface area by the number of tanks.

This historic practice has usually resulted in success. Failures have occurred when sludge settling characteristics were unusually poor, high MLSS concentrations were employed, or stringent effluent quality limits (TSS substantially less than 30 mg/L) were required. To prevent such problems, additional degrees of design sophistication involving ISVs, MLSS concentrations, and variations in return sludge flow rates and sludge solids loading rates have been developed. The procedure outlined below represents a blend of the historical practice and the evolving design techniques.

1. Determine the operating MLSS concentration range of the plant to maintain an acceptable F:M ratio, SRT, and effluent quality under various flow and mass loading conditions. For most municipal plants, the range is 1000 to 4000 mg/L.
2. Determine the anticipated range of the MLSS SVI (or ISV). Select a statistically high value (for example, the 98% confidence interval) that would seldom be exceeded by the full-scale operating plant. This maximum design value should be based on an analysis of existing records, pilot-plant data, or information from similar full-scale plants. If no data exist, most U.S. engineers contacted in the 15-firm design practice survey said they would use an SVI value of 150 mL/g; two said 100 mL/g was acceptable and one suggested a very conservative value of 250 mL/g.

---

3. Provide for a 20 to 100% RAS pumping rate capacity (up to 150% for oxidation ditch systems or others employing high MLSS concentrations).

4. Determine the maximum theoretical solids loading rate (function of SVI) using solids flux analysis. The analysis should be based on use of the maximum expected MLSS concentration and SVI value. From such an analysis will also evolve the RAS pumping rate required under various conditions. Procedures for determining solids loading limits are presented in a later section of this chapter. This information can be converted to a limiting surface overflow rate. Then a factor of safety is applied as recommended by Riddell *et al.*[80]

5. Based on influent wastewater flow rate characteristics, select an overflow rate to achieve the required effluent quality. The overflow rate to produce a specific effluent TSS concentration has not been extensively researched, but is known to vary with the geometry of the tank inlet structure and tank depth. This rate probably will not exceed the solids limiting value determined in step 4.

6. Select a depth to provide adequate sludge clarification, thickening, and storage. Routing analysis is the best technique. Otherwise, allow 0.7 to 1.0 m (2 to 3 ft) for thickening, 1 m (3 ft) or more for buffering, and 2.4 m (8 ft) for clarification. More buffering depth is needed if diurnal flow variations or peak flow conditions are abnormally large (for example, greater than 2:1). If a flat floor is provided, an additional 0.7 m (2 ft) of depth or a decrease of 0.34 m/h (200 gpd/sq ft) in the overflow rate should be applied. About 0.7 m (2 ft) of freeboard should also be added to determine overall tank wall height.

7. Provide a reasonable weir length and place the weirs at strategic locations. Block off notches in rectangular tanks and add baffles, as needed, to eliminate problems with MLSS updraft near the effluent structure.

8. Select a mechanism for sludge removal. Plows, spiral-curved blades, or hydraulic suction systems are available.

9. Provide for other details to complete the design:

- Flocculation inlet zones, preferably separated from the rest of the tank by baffles;
- Midlength or midradius energy dissipation baffles (these may not be needed with flocculating feedwell designs);
- Full-radius skimmers for smaller tanks, full-width for rectangular tanks, and beach-type for large radial flow tanks with multiple blades or antirotation baffles;
- Rails and walkways providing access to the sludge collection drive mechanism, skimming device, and launder area;

- Launder covers or chlorine addition equipment, if necessary, for algae control;
- Hose bibs; and
- Lighting.

Principal elements of this design approach are discussed in further detail below. Special applications, such as intrachannel clarifiers and settling provisions for SBRs, are discussed in the final two subsections. When determining the number of final clarifiers, provide one unit out of service for flexibility.

Solids Loading Rate. Establishing the maximum allowable solids loading rate is of primary importance to ensure that the clarifier will function adequately. A rather simplistic approach to quickly determining limiting solids loading rates was presented in the 1977 edition of MOP-8.[81] Figure 11.25, extracted from that manual, shows the limiting rate as a function of SVI for single- and multipoint sludge removal mechanisms. The 1977 edition excludes references for the data presented and reasons for distinguishing among the various types of mechanisms. The authors of this chapter compared this approach to more sophisticated solids flux analyses and found the results to be reasonably comparable for most municipal applications where SVI values are in the 100- to 250-mL/g range. In the 1989 WPCF design practice survey of 15 major U.S. firms, about 50% of the respondents believe that the MOP-8

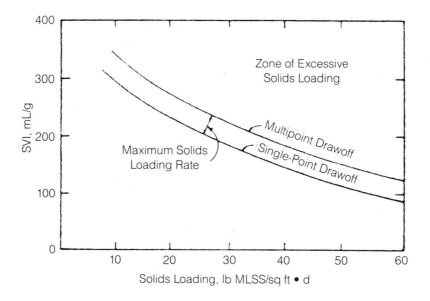

Figure 11.25   Design solids loading (lb MLSS/sq ft/day $\times$ 4.883 = kg MLSS/m^2 • d) versus SVI. The design assumes no inventory in settling basin.

Design of Municipal Wastewater Treatment Plants

curves provide reasonable approximations; the others did not use them. One respondent believes that the curves should be adjusted to be 25% more conservative.

Most design engineers prefer to keep the maximum solids loading rate in the range of 4 to 6 kg/m^2·h (20 to 30 lb/d/sq ft). Rates of 10 kg/m^2·h (50 lb/d/sq ft) or more have been observed in some well-operating plants. The curves of Figure 11.25 largely support these latter rates.

Determination of maximum allowable solids loading rate could be further refined using experimentally determined settling velocities and solids flux analysis. These methods were initially developed by Dick and Ewing[82] and advanced further by the state point concept of Keinath et al.[83,84] The concept of a factor of safety to address scale-up effects and variations in MLSS concentrations, SVI values, and flow rates was introduced by Riddell et al.[80] Data from this reference, presented in Figures 11.26 and 11.27, show

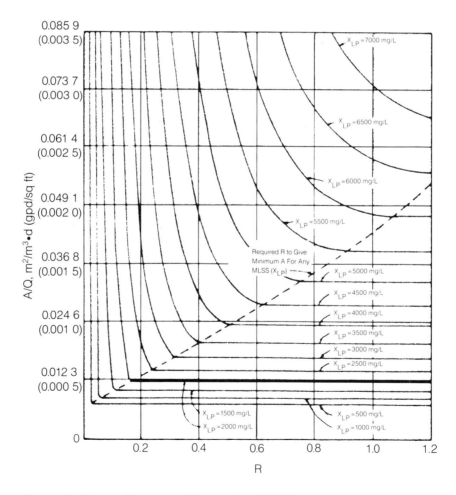

Figure 11.26 (A/Q) versus R for various MLSS concentrations.

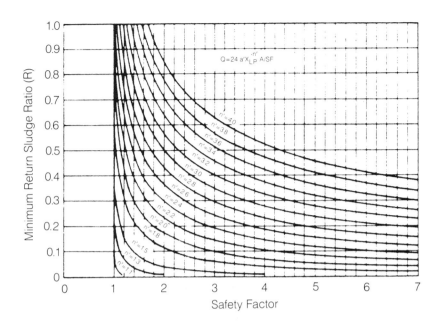

Figure 11.27   Return sludge ratio required for power model.

required RAS pumping rates and area requirements for various settled MLSS concentrations. In addition, Figure 11.27 shows various factors of safety for minimum return sludge ratios as a function of varying values of the power model coefficient.

Wilson and Lee[85] presented the following equation to determine maximum allowable hydraulic loading rate as a function of ISV at the design MLSS concentration:

$$Q = 24 \times \text{ISV} \times A/\text{CSF} \qquad (24)$$

In equation 24, the limiting hydraulic capacity [Q (m^3/d)] equals the initial settling velocity [ISV (m/h)] at the design MLSS concentration times the clarifier area [A (m^2)] and a time unit correction factor of 24, divided by the clarifier safety factor (CSF). This model calls for a high rate of return sludge pumping. Rates of 100% or even 150% would normally be used. Equation 24 includes a safety factor for scale-up. For safety factors up to three, equation 24 provides results consistent with other, more basic clarifier analyses. A CSF of two would be typical for systems known to have stable sludge settling properties, limited flow rate fluctuations, or step-feed flexibility. A minimum safety factor of 1.5 is good practice. In the above analysis, the engineer must recognize that the ISV will change with MLSS concentration and other conditions as shown in Figure 11.28. Maximum anticipated operational MLSS or the corresponding minimum ISV should be used in the equation.

Design of Municipal Wastewater Treatment Plants

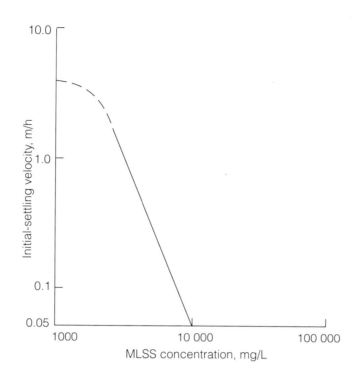

Figure 11.28 Plot of batch-settling data.

Convenient graphic design techniques using solids flux and a relationship between SVI and settling velocity have been reported by Daigger and Roper[86] and by Keinath.[87] The former authors used unstirred SVI data from two full-scale and six pilot-scale plants receiving two different influent flows treated with pickle liquor to enhance phosphorus removal; they prepared a nomograph (Figure 11.29)[86] based thereon. The latter author sought a broader data base and more consistency in SVI data. His analysis included data from 21 full-scale plants that varied considerably with respect to size, geographic location, mode of operation, method of aeration, and type and amount of industrial wastewater input. Keinath's design and operating chart, presented as Figure 11.30,[87] is based on the relationship between stirred SVI and ISV developed by Wahlberg and Keinath.[88]

Results obtained when using the nomographs of Wahlberg and Keinath[88] differ substantially from those of Daigger and Roper,[86] especially at high SVI values. Much of this difference can be attributed to the effect of stirring during the SVI test. For a single mixed liquor tested by Keinath, a stirred SVI of 122 mL/g was measured in contrast to an unstirred value of 189 mL/g. Most full-scale plant SVI data are based on unstirred test results. For such plants, sufficient stirred test data are needed to successfully use the Keinath nomograph, or a correlation between stirred and unstirred test data must be developed. Several design examples provided by Keinath illustrate the

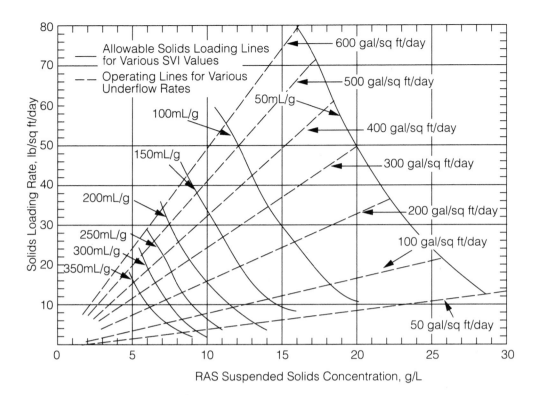

Figure 11.29 Secondary clarifier design and operation diagram
(gal/sq ft/day × 0.040 74 = m³/m² • d; lb/sq ft/day ×
4.883 = kg/m² • d).

usefulness of the nomographs. A plant with an MLSS concentration of 2 g/L, a flow rate of 4000 m³/d, a 50% RAS pumping rate, and a stirred SVI of 125 mL/g led to the prediction of a 6-g/L RAS concentration and a limiting solids loading rate of 90 kg/m²·d. This resulted in a required clarifier surface area of 133 m². A higher MLSS concentration would lead to a larger clarifier area but smaller aeration tank volume if solids loading rate was the governing criterion for tank sizing. The nomograph permits clarifier areas to be easily determined for various sets of conditions so that optimum conditions can be found.

Overflow Rate. Overflow rates used by design engineers, based on average dry-weather flow, have been observed to vary from 0.5 to 1.7 m/h (300 to 1000 gpd/sq ft). Some plants are known to operate without difficulty at the upper end of this range and produce a high-quality effluent. In many other documented cases, peak rates of 2.7 to 3.1 m/h (1600 to 1800 gpd/sq ft) do not exceed a secondary clarifier's capacity. In as many or even more cases, however, clarification efficiency suffers with much lower average and peak overflow rates.

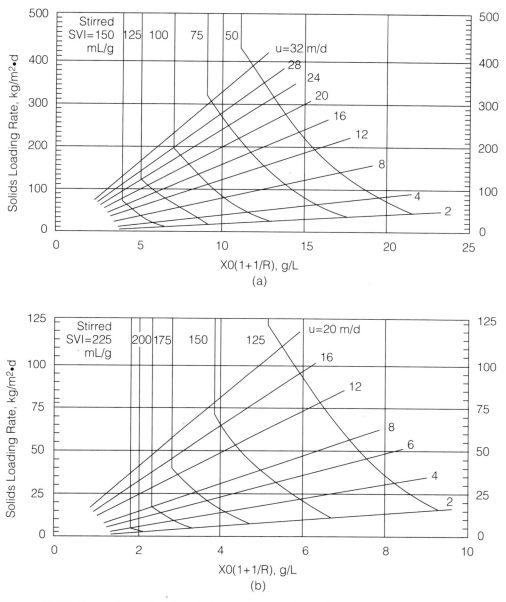

Figure 11.30 Secondary clarifier design and operations chart:
(a) low SVIs and (b) high SVIs.

The 1977 edition of MOP-8[81] suggested maximum allowable overflow rates:

- Average flow rate—1.36 m/h (800 gpd/sq ft);
- Three-hour sustained peak flow rate—2.38 m/h (1400 gpd/sq ft); and
- Two-hour sustained peak flow rate—2.72 m/h (1600 gpd/sq ft).

Within the past decade, most design engineers have chosen more conservative rates. The 1989 WPCF survey (15 major consulting firms) includes preferred overflow rates as shown in Table 11.15.

Figure 11.31 shows some correlation between effluent TSS concentration and overflow rate. The data suggest that several different designs will give an annual average value of 30 mg/L TSS if overflow rates are kept to less than

Table 11.15   Preferred overflow rates.[a]

| Flow | Circular shape | | Rectangular shape | |
	Range	Average	Range	Average
Average	0.68–1.19 (400–700)	0.95 (560)	0.68–1.19 (400–700)	0.95 (560)
Peak	1.70–2.72 (1000–1600)	2.09 (1230)[b]	1.7–2.72 (1000–1600)	2.10 (1240)[c]

[a] m/h (gpd/sq ft).
[b] 10 of 15 firms use 2.04 m/h (1200 gpd/sq ft).
[c] 8 of 13 firms use 2.04 m/h (1200 gpd/sq ft).

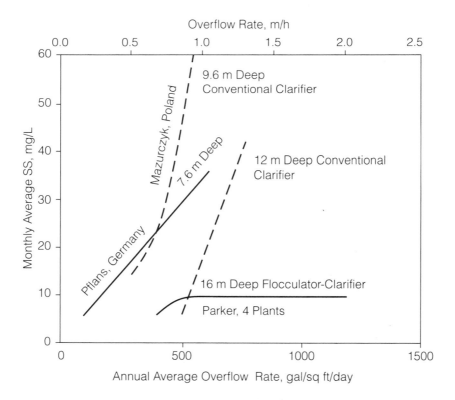

Figure 11.31   Performance response curves for conventional clarifiers and flocculator–clarifiers (ft × 0.304 8 = m).

about 0.8 m/h (500 gpd/sq ft). Such a correlation can be misleading and consequently of little value because it does not account for the effects of temperature, peaking factors, SVIs, geometrical details, RAS flow rate, and RAS concentration. Because the literature is limited in this area, designs for specific sites should be conservative or based on experimental testing.[79] Unbalanced load testing at existing plants undergoing expansion is encouraged. If such testing is not feasible, bench-scale investigations should be undertaken to provide reasonable design criteria.

Effluent TSS concentrations sometimes relate to a reactor's aeration system. Pure oxygen plants characteristically produce MLSS that are dense and settle well but have relatively high effluent turbidity values (fine particles). Most pilot plants using air employ fine-bubble diffusers. Although more data are needed, these aerators may produce fewer fine particles than mechanical or jet aeration systems. Coarse-bubble aerators, which impart more energy per unit of oxygen transferred than fine bubble systems, can also affect effluent TSS values.

Shape. Clarifier shapes include rectangular, square, and circular. A recent survey[89] led to the compilation of data shown in Table 11.16[89] and in Figures 11.32 and 11.33. No observable differences in clarification performance at average or peak hydraulic loading rates were attributed to shape alone.

Typically, chain and flight sludge removal mechanisms are used in rectangular clarifiers. Metal chains were used historically; now some are made of plastic. Because of their rotation, some circular clarifier mechanisms have mechanical seals to allow continuous sludge withdrawal. In some, the seals are near the bottom of the clarifier; in others, they are near or at the water surface. Although these seals are generally quite reliable, they require maintenance to minimize leakage.

Depth. In recent years, clarifier depth has received increasing attention. Generally, increasing depth has been observed to improve performance in terms of TSS removal and RAS concentration. Too much depth, however, increases construction cost. Another survey of major consulting engineering firms in the U.S. found that most firms are using design depths of 4 to 5 m (12 to 16 ft).[90] The trend was toward increasing depth to improve performance. Most firms agreed that larger tanks require greater depth. Table 11.17, derived from the 1977 edition of MOP-8,[81] supports this position. An overflow rate correction of 0.17 m/h (100 gpd/sq ft) for each 0.3 m (1 ft) of depth less than the minimum tabulated value has been suggested. For example, a 30-m (100-ft) diameter tank with a 3-m (10-ft) sidewater depth would have an associated overflow rate of 0.34 m/h (200 gpd/sq ft) less than that for a similar clarifier with a sidewater depth of 3.6 m (12 ft).

Table 11.16 Secondary clarifier evaluation.[89]

Plant name	Plant type	Sed type	Depth, ft[b]	Average annual flow[a]					Maximum day flow[a]				
				Overflow rate, gpd/sq ft[c]	Detention time, hours	Effluent solids, mg/L	Solids load, lb/d/sq ft[d]	SVI	Overflow rate, gpd/sq ft	Detention time, hours	Effluent solid, mg/L	Solids load, lb/d/sq ft[d]	SVI
Kenton	AAS	Ctr.	17.00	665.00	4.59	10.30	6.25	232.00	1 128.00	2.71	69.90	22.00	443.00
Spokane	AAS	Ctr.	15.00	536.00	5.07	7.00	15.39	76.00	1 054.00	2.56	6.00	29.52	70.00
Madison	AAS	Ctr.	V 12/13	NA	NA	7.00	NA	100.00	509.00	4.60	3.00	10.39	116.00
Milwaukee West	AAS	Ctr.	15.00	852.00	3.13	18.00	0.24	78.00	1 525.00	1.77	44.00	0.60	157.00
Milwaukee East	AAS	Ctr.	15.00	1 173.00	2.30	Inc. above	0.37	73.00	2 075.00	1.30	Inc. above	0.90	196.00
Vancouver	AAS	Ctr.	12.00	509.00	4.23	18.60	6.45	147.00	773.00	2.79	20.00	13.64	120.00
Anola, Minn.	AAS	Ctr.	10.00	582.00	3.09	13.00	14.02	100.00	937.00	1.92	32.00	20.17	300.00
Imperial Valley, Houston	AAS	Ctr.	9.90	111.00	16.04	4.00	4.84	90.00	283.00	6.30	3.60	22.57	NM
Chocolate Bayou, Houston	AAS	Ctr.	11.00	130.00	15.15	5.00	4.39	39.60	486.00	4.08	46.00	21.80	NM
Easterly, Cleveland	AAS	Ctr.	12.00	617.00	3.50	7.00	5.90	107.00	1 167.00	1.85	29.00	1.82	273.00
Calumet, Chicago	AAS	Ctr.	15.00	369.00	7.32	21.00	16.89	110.00	459.00	5.89	120.00	22.72	250.00
Calumet, Chicago	AAS	Rec.	12.00	443.00	4.87	21.00	20.27	110.00	550.00	3.92	120.00	21.83	250.00
Blue Plains, D.C.	AAS	Ctr.	12.00	688.00	3.16	19.90	9.94	98.00	1 210.00	1.78	45.00	41.52	205.00
Central 01, Miami–Dade	AAS	Rec.	11.00	689.00	2.87	22.00	2.62	NA	1 662.00	1.19	31.00	3.32	NA
91st Ave., Phoenix	AAS	Rec.	8.00	585.00	3.89	12.00	3.19	250.00	900.00	2.00	45.00	14.39	500.00
Metro, Minn. East	AAS	Rec.	11.00	813.00	2.44	14.00	13.27	157.00	11 102.00	11.80	116.00	128.48	1 197.00
Metro, Minn. West	AAS	Rec.	11.00	793.00	2.50	8.00	21.18	163.00	1 989.00	12.00	18.00	124.37	1 176.00
Wyandotte, Wayne Co.	HPO	Ctr.	13.50	821.00	2.95	41.00	35.37	41.80	11 020.00	12.38	168.00	136.35	175.6
Louisville	HPO	Ctr.	14.00	543.00	4.63	27.00	18.55	65.00	867.00	2.84	48.00	44.89	90.00
East Bank, New Orleans	HPO	Ctr.	12.00	522.00	4.13	25.00	19.67	71.00	900.00	2.40	246.00	101.59	250.00
Rock Creek, Hillsboro	HPO	Ctr.	12.00	455.00	4.74	17.50	18.89	72.00	908.00	2.38	453.00	59.25	131.00
Detroit	HPO	Ctr.	15.00	1 108.00	2.43	16.00	24.55	76.00	1 946.00	2.01	52.00	NM	NM
Chaska, Minneapolis	HPO	Ctr.	11.00	322.00	6.14	14.00	31.15	85.00	661.00	2.99	222.00	81.22	130.00
Central 02, Miami–Dade	HPO	Rec.	11.00	606.00	3.26	9.00	17.35	81.00	626.00	3.16	12.00	13.23	52.00
CSDLAC JWPCP, Carson	HPO	Rec.	14.00	623.00	4.05	18.00	13.35	139.00	NA	NA	NA	NA	NA
Huntington Beach, Calif.	HPO	Rec.	14.00	490.00	5.14	19.00	10.40	59.00	NA	NA	NA	NA	NA
Flat Rock, Wayne Co.	TF	Ctr.	11.00	1 896.00	1.04	36.00	NA	NA	2 780.00	0.70	59.00	NA	NA
West Bank, New Orleans	TF	Ctr.	10.00	545.00	3.30	19.00	NA	NA	NM	NM	3.00	NA	NA
Salinas 02	TF	Ctr.	8.00	495.00	2.90	26.00	NA	NA	707.00	2.03	62.00	NA	NA
Salinas 02 (filter 2)	TF	Ctr.	8.00	NA	NA	NA	NA	NA	NA	NA	NA	NA	NA
Fort Lewis	TF	Ctr.	12.00	299.00	7.21	25.00	NA	NA	582.00	3.40	100.00	NA	NA
Ellington, Houston	TF	Ctr.	11.00	12.00	158.82	11.00	NA	NA	78.00	25.48	42.00	NA	NA
Corvallis	TF/SG	Ctr.	18.00	413.00	7.83	7.70	10.63	44.00	1 895.00	13.62	18.00	145.66	156.00
Forest Grove, Hillsboro	TF/SG	Ctr.	12.00	267.00	8.07	11.80	5.93	165.00	915.00	2.36	99.00	43.29	386.00
Hillsboro, Hillsboro	TF/SG	Ctr.	V 10/15	V 232/412	V 11,644.4	10.00	V 5.62/9.99	125.00	V 796/1415	V 3.39/1.27	25.00	NA	160.00
Castroville	TF/SG	Ctr.	7.00	325.00	3.87	22.00	4.27	167.00	650.00	1.94	NM	NM	NM
Pullman	TF/SG	Ctr.	V 10/14	284.00	0.00	22.00	6.99	75.00	656.00	3.29	87.00	22.78	243.00
Central, Dallas	TF/SG	Rec.	23.75	598.00	7.13	12.30	7.69	150.00	793.00	5.45	8.00	11.68	97.00

[a] NA = data not available; NM = not measured; V = varies.
[b] ft × 0.304 8 = m.
[c] gpd/sq ft × 0.040 74 = m/d.
[d] lb/d/sq ft × 4.833 = kg/m².d

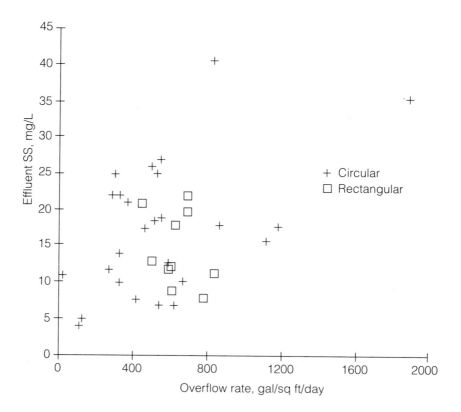

Figure 11.32 Secondary sedimentation tank survey—overflow rates (gal/sq ft/day × 0.040 74 = m/d) versus effluent SS based on annual averages.

Table 11.17 Final settling basin side water depths.[81]

Tank diameter, ft[a]	SWD, ft	
	Minimum	**Suggested**
Up to 40	10	11
40–70	11	12
70–100	12	13
100–140	13	14
>140	14	15

[a] ft × 0.304 8 = m.

For circular tanks with sludge hoppers, a 1:12 bottom slope is considered to be good practice. If a flat floor is used to accommodate a particular type of sludge withdrawal mechanism, an overflow rate correction factor of 0.34 m/h (200 gpd/sq ft) is suggested. For example, a clarifier initially designed for an

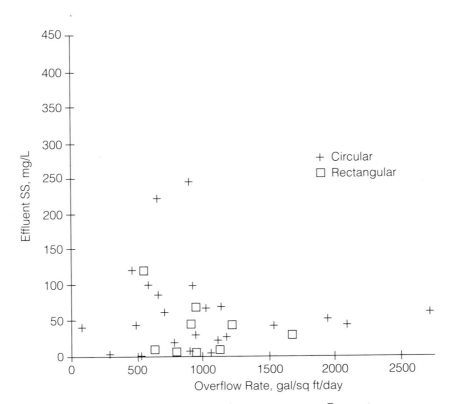

Figure 11.33 Secondary sedimentation survey—overflow rates (gal/sq ft/day × 0.040 74 = m/d) versus effluent SS based on maximum daily averages.

overflow rate of 1.4 m/h (800 gpd/sq ft) would be redesigned to reduce its overflow rate to 1.0 m/h (600 gpd/sq ft) if a flat bottom were chosen.

In recent years, additional findings about clarifier depth have been published. Data in Figure 11.34 show an increase in performance for deeper circular tanks.[91] However, a recent study comparing circular clarifiers at Renton, Washington, to rectangular clarifiers at the Los Angeles County Sanitation Districts' San Jose Creek Plant indicated very little difference in performance between the 5.5-m (18-ft) depth of the former and the 3.0-m (10-ft) depth of the latter.[92] In this study, the density current and solids flow patterns of the two tanks differed significantly. The San Jose Creek Plant, with its sludge withdrawal point at the effluent end of the clarifiers, provided a shallow, fast-moving transfer of sludge along the length of the tank. This pattern provided ample depth for clarification before the flow reached the weirs. The deep circular tanks at Renton likewise provided ample depth as the sludge flowed radially outward and was subsequently removed by a hydraulic collector system. Although additional research on clarifier depth is certain to follow, the suggested values of Table 11.17 appear reasonable for circular tanks and, perhaps, for rectangular units as well. Nonetheless, some evidence exists that rectangular tanks may not need to be as deep, especially if sludge is withdrawn at the effluent end.

Another important aspect regarding the depth requirement concerns the practice of blanket level maintenance. Studies by Crosby indicated that the top of the sludge blanket, in effect, determines the depth of clarification.[93] Relatively shallow tanks with minimal sludge blanket depths often perform as well as deeper tanks with thicker blankets. A design engineer must recognize that, if a shallow clarifier is selected, operation will be restricted to a

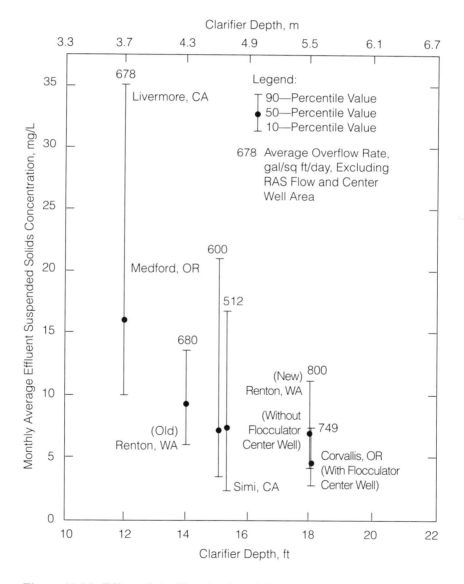

Figure 11.34 Effect of clarifier depth and flocculator center well on effluent suspended solids (gal/sq ft/day × 0.040 74 = m/d).

narrow range of blanket depths. This, in turn, might limit the maximum achievable underflow solids concentration and correspondingly increase RAS pumping rates.

Tank depth provides for short-term storage of settled MLSS during peak hydraulic loading periods. Under normal operation, the diurnal cycle creates significant changes in the solid flux on a clarifier because most plants operate with a constant RAS pumping rate. When influent solids flux exceeds the rate of sludge removal, the blanket depth increases and the surplus sludge solids accumulate in the clarifier. When the peak flow subsides, the solids inventory shifts toward the aeration tank. Therefore, ample depth for rise and fall of the blanket is necessary. For most municipal plants with peaking factors of less than two to three, a blanket depth allowance of about 1.0 m (3 ft) for this buffering should suffice. Higher peaking factors will require greater depths and might warrant a special routing analysis or even computer simulation. Crosby and others have suggested that designs be conservative enough to keep the blanket level to about 0.7 m (2 ft) at peak flows. Designing to maintain low TSS concentrations at specified frequencies can also lead to thin blankets.

Weir Loading. Numerous state regulations limit maximum allowable weir loadings to 125 m^3/m·d (10 000 gpd/ft) for small treatment plants (less than 160 m^3/h or 1 mgd) and to 190 m^3/m·d (15 000 gpd/ft) for larger plants. Experience of many operators and design engineers has led to a general consensus that substantially higher weir loading rates would not impair performance.

For radial-flow (circular or square) clarifiers, a single peripheral weir is generally considered adequate, especially if some baffling is provided to prevent an updraft wall effect that results in TSS approaching the weir. Other engineers prefer to handle this problem by locating double inboard launders at a distance of approximately 30% of the tank radius from the outer wall. The double launder concept increases construction cost but, as demonstrated by Anderson[94] more than 40 years ago, improves performance over that of simple peripheral weirs without baffling.

For rectangular tanks, launders that extend 25 to 30% of the tank length from the effluent end and are spaced at approximately 3-m (10-ft) intervals have worked well. Some engineers continue to believe that a simple full-width weir at the effluent end is sufficient. Regardless, providing extensive launder structures to meet arbitrary criteria of 125 to 190 m^3/m·d (10 000 to 15 000 gpd/ft) seems unwarranted unless necessary to meet certain state criteria.

Inlets. Inlets must dissipate influent mixed liquor energy, distribute flow evenly in vertical or horizontal directions, reduce density current impacts, minimize sludge blanket disturbances, and promote flocculation. Poor distribution or jetting of the incoming flow results in short circuiting as evidenced by concentration-time distributions in dye studies.[93] Density currents from down-flowing mixed liquor are commonly observed and are considered detrimental to clarifier performance. Some inlets are designed to reduce density current impact by discharging the mixed liquor into the lower portions of a clarifier or by distributing the mixed liquor over a large area. Small deflectors to direct inflow in an upward direction have not been successful.[95]

Since the time of Camp,[96-98] a number of designers have demonstrated that using the incoming energy to promote flocculation improves clarifier performance. Kalbkopf and Herter[99] supported this principle for rectangular tanks; Parker and Stenquist supported the same principle for center-feed circular tanks.[100] Although achieving 30/30 standards might be possible with small, center feedwells in circular tanks (10 to 15% of the tank diameter), a more effective design would employ 30 to 35% of the tank's diameter as a flocculation zone. This zone should be constructed with a cylindrical baffle that extends to approximately middepth of the clarifier. An inner chamber with a much smaller diameter and small velocity control gates is needed to optimize the feed pattern into the larger flocculating zone (Figure 11.35).[91]

For rectangular tanks, less research has been done, but the studies of Kalbkopf and Herter[99] indicated that separate flocculation at the head end of such clarifiers is effective. Paddle mixers were tested to flocculate mixed liquor entering rectangular clarifiers at the Emscher Mouth Treatment Plant, West Germany. The data show improved effluent clarity. Specially designed inlet diffusers have been used for years at the Los Angeles County Sanitation Districts' plants without the use of downstream baffles.[76] Design details recommended by studies of rectangular tanks are limited.

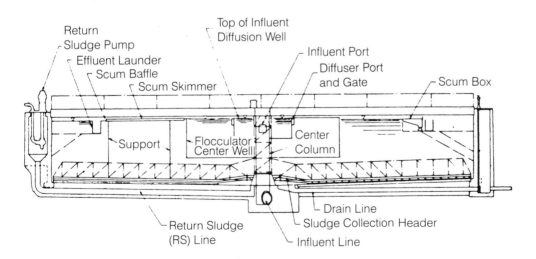

Figure 11.35 Clarifier with a flocculating center well.

Interior Baffles. Clarifier performance can often be improved by installation of interior baffles. For circular tanks, a baffle mounted on the wall beneath the effluent weir can deflect solids rising along the wall. Studies by Stukenberg *et al.*[101] and Crosby[93] showed that this solids uplift (wall effect) can be effectively mitigated by such weirs, producing an average 38% improvement in TSS removal. Figure 11.36(a) illustrates such a baffle design. Solids can deposit on top of the nearly horizontal shelf. Others have developed the design shown in Figure 11.36(b). If the former is used, an appendage of the skimming mechanism can be provided to sweep the deposited TSS from the shelf.

Crosby's work also led to development of the midradius baffle. It extends from near the floor to middepth of the clarifier and is supported by the sludge removal mechanism. This, in effect, forms a large flocculation zone, minimizes the wall effect, and dissipates additional inlet energy. A similar study, undertaken by Esler[102] for existing rectangular clarifiers, used slotted baffles located at midlength of the tanks. They were found to break up density and influent flow currents, thereby improving tank performance. In some rectangular clarifiers at the Mikawashima Treatment Plant in Tokyo, cross baffles were placed at approximately 12 and 25% of the tank length. The design engineers considered these to be near-optimum locations. Sets of cross and longitudinal sludge collectors are used to eliminate the need for large openings in the baffles to allow passage of a chain and flight mechanism.

Additional Research. Additional research in the design of final clarifiers continues under sponsorship by the ASCE, U.S. EPA, and Environment

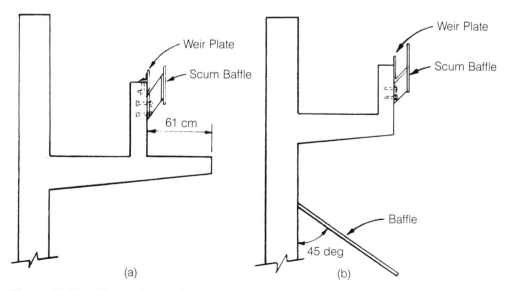

Figure 11.36   Alternative peripheral baffle designs: cm × 0.032 8 = ft.

Canada. Further refinements in tank design are expected to result within the next few years. Specifically, these will include the following: standardization of definitions, development of research protocols, recommendations on statistical practices for further research, more careful review of results to be published, and a greater effort to make full-scale clarifier research results available to practicing engineers. Although clarifiers have been designed for decades, significant improvements in performance are being realized from the findings of the past several years. Practicing engineers are advised to learn of these developments and incorporate them into the design of new clarifiers and the upgrading of existing facilities.

Intrachannel Clarifiers. A relatively new concept, known as intrachannel clarifiers (ICCs), has been developed for oxidation ditch plants. The process incorporates a clarifier installed in the channel of the ditch. Subsequent modifications by several manufacturers have included placement of the clarifier adjacent to the ditch using common wall construction. The major manufacturers of ICCs are presented in Table 11.18; system descriptions have been summarized by Zirschky[103] and, subsequently, by Montgomery.[104] A number of these systems are shown in Figures 11.37 through 11.42.

Potential advantages of ICC systems include reduced construction and O&M costs and land area requirements. Hydraulic head differences and gravity are used to force mixed liquor into the clarifier and return sludge to the ditch, thereby eliminating RAS pumping requirements. Control over sludge return rate is eliminated and sludge age is adjusted by wasting mixed liquor. Disadvantages include loss of independency of reactors and clarifiers, possible development of unanticipated sludge deposits and currents, wasting of dilute sludge, loss of system operational flexibility, and less operator control of RAS pumping rate adjustment and RAS chlorination.

Performance data for several operational systems are summarized in Table 11.19. Figure 11.43 shows effluent TSS data from 17 plants. They are compared to results from conventional oxidation ditches as reported by Ettlich.[105]

Table 11.18   Manufacturers of intrachannel clarifiers.

Manufacturer	Location	System
AEE	Kansas City, Mo.	BMTS
EIMCO	Salt Lake City, Utah	Carrousel intraclarifier
Envirex	Waukesha, Wisc.	Side-channel clarifier
Innova-Tech	Valley Forge, Pa.	Pumpless integral clarifier
Lakeside Equipment Corp.	Bartlett, Ill.	Sidewall separator
Lightnin	Rochester, N.Y.	Draft tube channel (DTC)
United Industries	Baton Rouge, La.	Boat clarifier

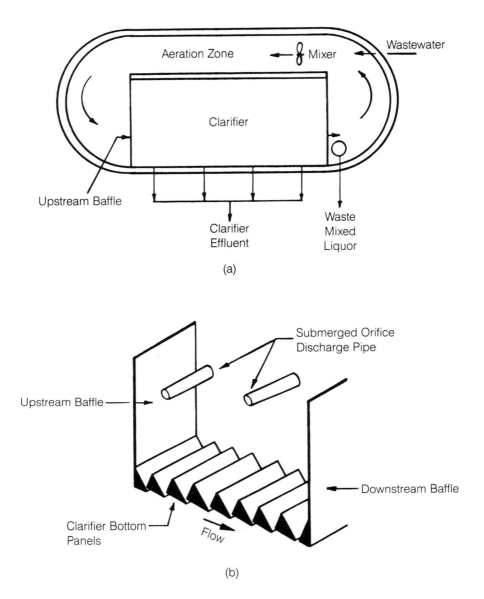

Figure 11.37 The Burns and McDonnell treatment system (BMTS): (a) plan view and (b) isometric view of intrachannel clarifier.

Most ICC systems have only recently begun operation; thus, the long-term performance capabilities of the technology cannot be determined with certainty. As with any new process, some early design and operational problems occurred.[104] Many of these, however, appear to have been corrected. Inadequate aeration or mixing was found to be a significant problem with certain systems.[104] Inadequate velocity [less than 0.2 m/s (0.7 ft/sec)] has caused MLSS settling in some ditches, resulting in sludge bulking and poor TSS

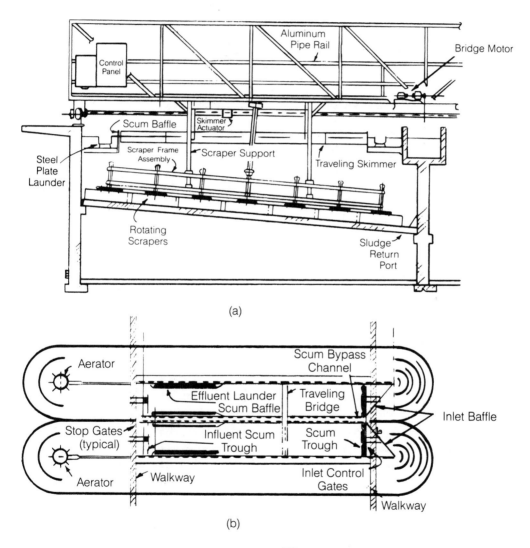

(a)

(b)

Figure 11.38   EIMCO Carrousel intraclarifier™: (a) cross section and (b) plan.

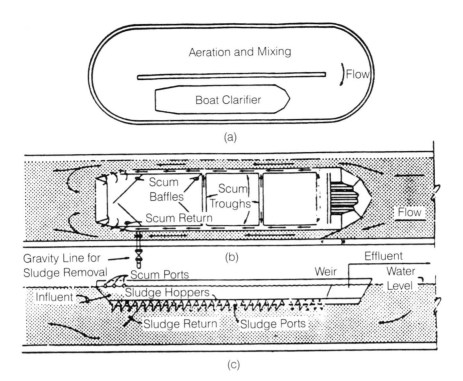

Figure 11.39   United Industries boat clarifier: (a) location in an
oxidation ditch, (b) plan view, and (c) elevation.

removal. In some designs, the ICC forms an obstacle to flow in the ditch and
requires additional energy input to maintain minimum velocity requirements.
This can be achieved by increasing the aeration system capacity or introducing separate mechanical mixers. Both require increased power input that
will correspondingly increase system capital and O&M costs.

Undersizing of sludge handling facilities is another reported problem.
Several systems have experienced difficulty in wasting sufficient volumes of
sludge to keep the MLSS concentration and sludge age at desired levels. As
ICC systems lack a conventional RAS system, conservative design would
dictate that the size of waste sludge handling facilities be governed on the
premise of wasting mixed liquor. In most cases, this premise would lead to
the inclusion of a waste sludge thickener as part of the overall waste sludge
management facilities.

To help ensure acceptable treatment performance, oxidation ditch systems
with ICCs should be designed to provide for

-   Mixing capability (ICC systems may require more mixing energy
    than conventional oxidation ditches to maintain adequate channel
    velocities);

- Aeration capacity (aeration systems that are effective in conventional oxidation ditches are generally effective in ICC systems);
- Scum removal where flow barriers occur;
- Adequate waste sludge handling capacity to keep sludge age at desired levels; and
- Adequate structural support for mixing and aeration equipment.

In addition, one manufacturer recommends that ICCs not be used if the ratio of peak to average flow exceeds 2.5.

Presently, all ICC designs are patented and, therefore, lack published design criteria. In some operating plants, the percentage of surface area

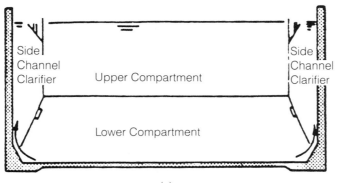

(a)

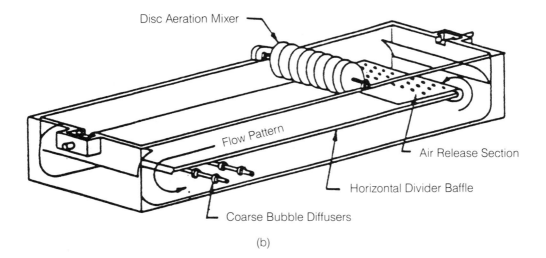

(b)

Figure 11.40 Envirex side-channel clarifier: (a) cross section and (b) isometric view.

dedicated to clarification is about 20% of the total for the ditch system. Such a value, however, will not apply consistently to all ICC designs because of variations in ditch depth and clarifier design details.

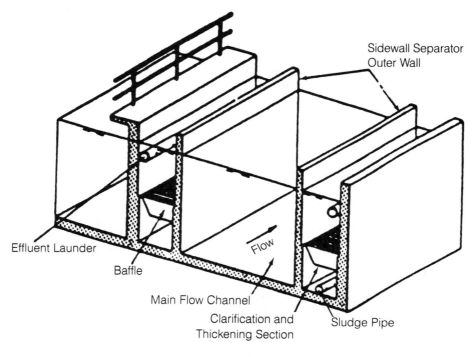

(a)

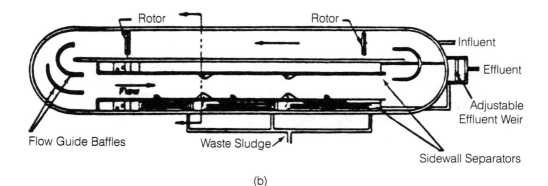

(b)

Figure 11.41 Lakeside side-wall separator: (a) isometric cross section and (b) plan view.

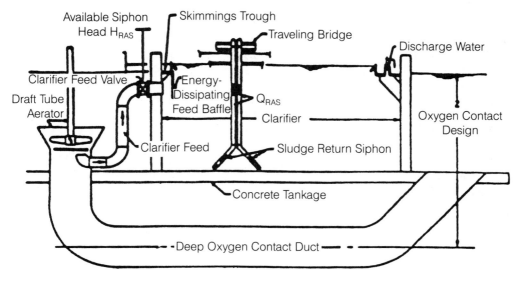

(a)

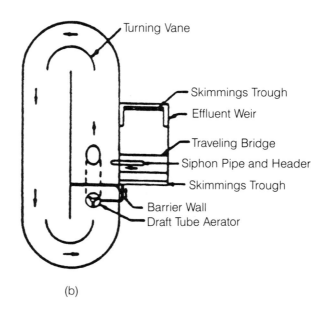

(b)

Figure 11.42 Innova-Tech pumpless integral clarifier™: (a) clarifier and ditch cross section and (b) plan view.

Table 11.19   Average performance data for select intrachannel clarifier systems.

Parameters measured	Systems and locations			
	BMTS, Sedalia, Mo.	EIMCO, Owensboro, Ky.	Lightnin, Thompson, N.Y.	Boat clarifier, Sarasota, Fla.
Months of data	9–10	3–4	4	4
Design flow, mgd[a]	2.6	6.8	1.0	1.0
Current flow, mgd	1.6	1.8	0.8	0.1
BOD, mg/L				
Influent	178	427	130	199
Effluent, %	24.4	19.8	7.25	5.8
Removal, %	86.3	95.4	94.4	97.1
TSS, mg/L				
Influent	198	270	106	214
Effluent, %	17.7	21	0.9	<2
Removal, %	91	92	91	>99

[a] mgd × 3785 = m³/d.

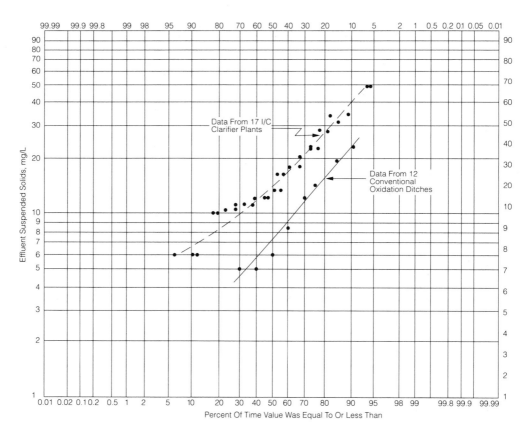

Figure 11.43   Effluent suspended solids data for intrachannel clarifiers and conventional oxidation ditches.

SBR Clarification. SBR processes are favored by some engineers because they do not require separate clarifiers. Nonetheless, proper provisions are needed to ensure a clear, high-quality supernatant when aeration and mixing are terminated. Most SBR designs involve interrupting the inflow to a basin while it is in the settling and decanting modes. One variation, however, allows flow to enter the tank at all times, relying on the distance between the inlet and outlet structures to provide adequate settling time. As an approximate rule of thumb for such a design, the tank length:width ratio should be at least 4. Even with this provision, continuous inflow SBRs in cold climates can experience density-difference short circuiting. Temperature differences of only 1 to 2°C between the influent and the mixed liquor have been shown to produce this unacceptable flow pattern.

A number of devices for decanting are illustrated in a U.S. EPA report and are shown herein as Figures 11.44 through 11.51.[106] Although most marketed decanting devices have been found to function reasonably well, some of the initial designs led to excessive TSS discharges at the onset of the decant cycle. The turbulence of aeration transferred MLSS into the decanter and the TSS subsequently left the basin when the decant cycle started. This problem has been resolved by provisions to return the initial decant to the reactor and continue the return until clarity improves to a satisfactory level. Many SBR plants have sometimes experienced significant foam buildup; the decant system should be designed accordingly. Most successful designs keep the foam out of the effluent discharge by incorporating baffles around the

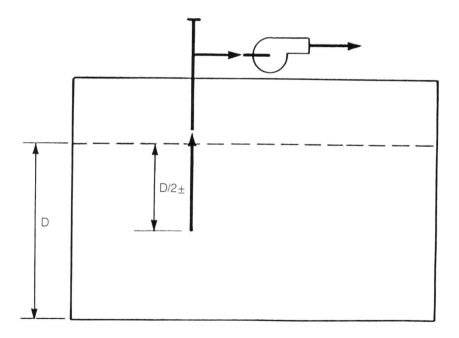

Figure 11.44   Rivercrest and Glenlea decant system.

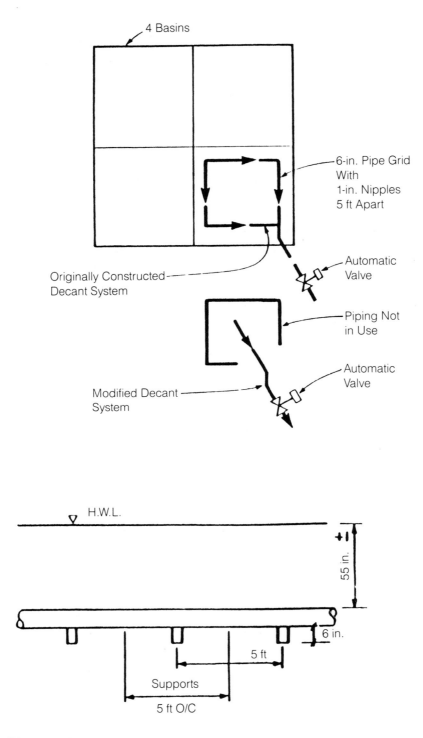

Figure 11.45   Choctaw decant system.

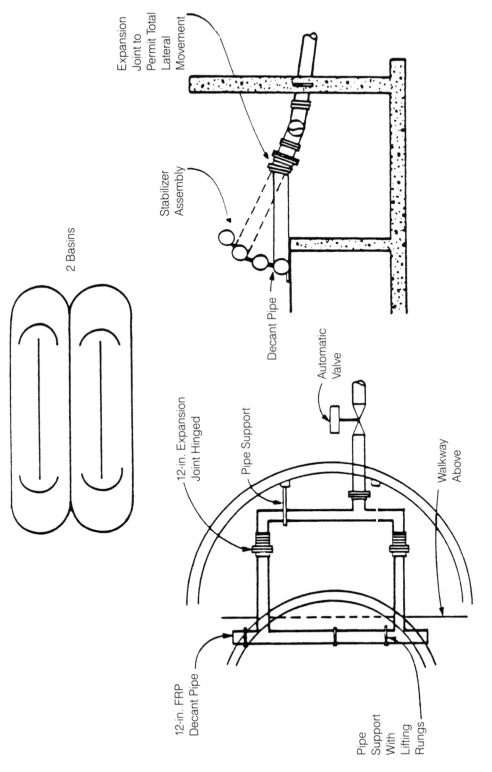

2 Basins

Expansion Joint to Permit Total Lateral Movement

Stabilizer Assembly

Decant Pipe

12-in. Expansion Joint Hinged

Pipe Support

Automatic Valve

Walkway Above

12-in. FRP Decant Pipe

Pipe Support With Lifting Rungs

Figure 11.46   Grundy Center decant system as initially built.

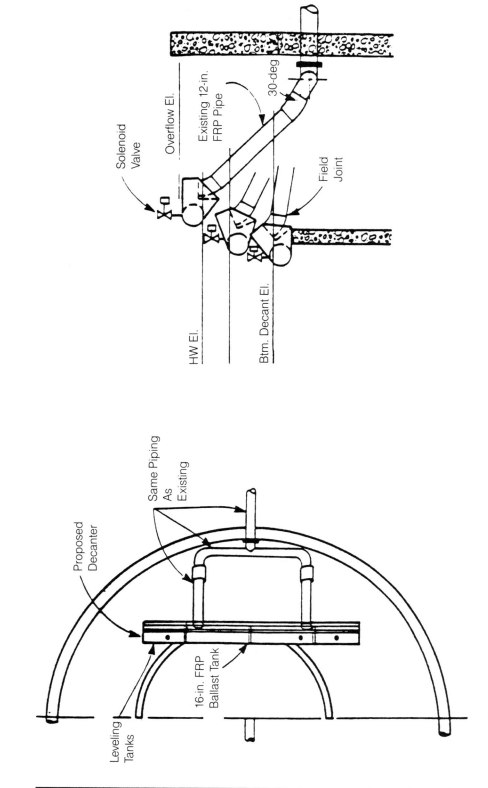

Labels in figure:
- Solenoid Valve
- Overflow El.
- Existing 12-in. FRP Pipe
- 30-deg
- Field Joint
- HW El.
- Btm. Decant El.
- Same Piping As Existing
- Proposed Decanter
- Leveling Tanks
- 16-in. FRP Ballast Tank

Figure 11.47   Retrofit of Grundy Center decant system.

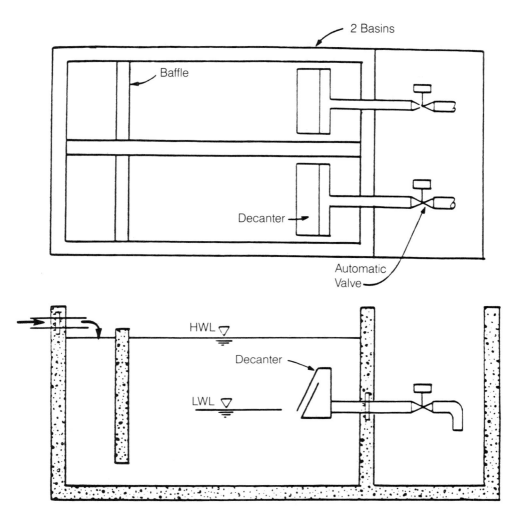

Figure 11.48   Eldora decant system in one tank.

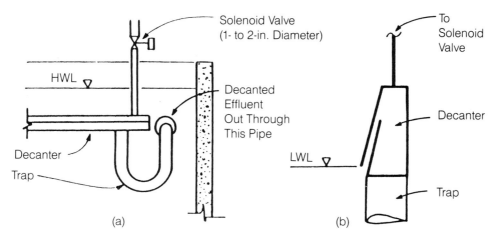

**Figure 11.49   Eldora decant system in second tank: (a) front elevation and
(b) enlarged view of decanter.**

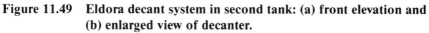

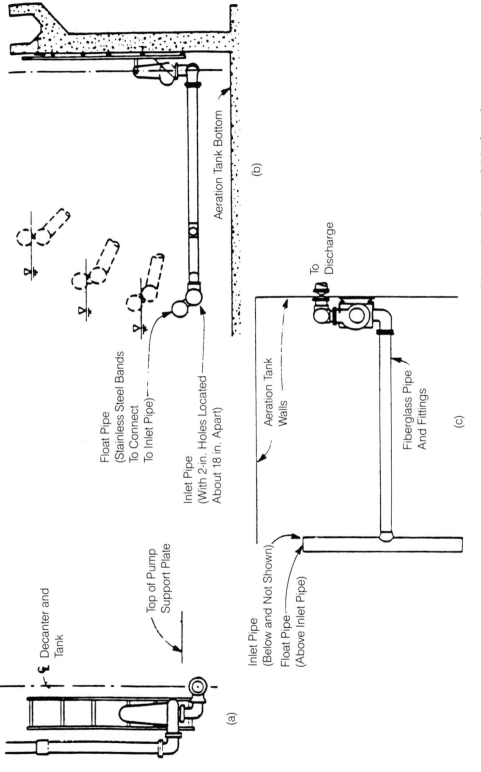

Figure 11.50 Culver decant system with well-mounted pump: (a) front elevation, (b) side elevation, and (c) plan view.

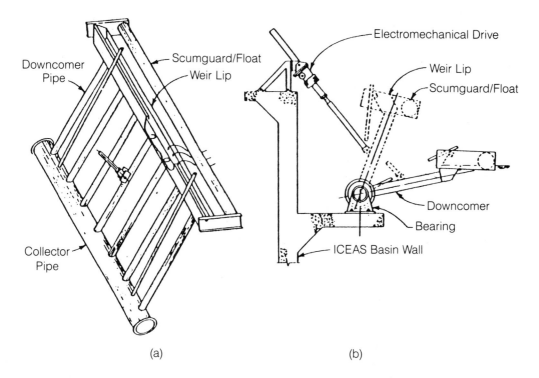

Figure 11.51 Recent decant system design: (a) isometric view and (b) section.

decanter. Foam subsequently remains in the reactor or it can be removed by separate skimming devices.

*A*ERATION BASIN CONFIGURATIONS

Configurations discussed herein include complete mix, plug flow, contact stabilization, oxidation ditch, SBRs, and deep shaft reactors. In the 1989 design practice survey, 15 engineers of major consulting firms expressed their reactor preference for a 0.2-m^3/s (5-mgd) plant as follows:

- Complete mix—4;
- Plug flow—3;
- Oxidation ditch—5; and
- Combination—3.

For smaller plants, oxidation ditches were more popular and for larger ones, conventional plug flow (some with configuration flexibility) were favored.

Most of the basin configurations can be used over a wide range of loading conditions, but the range is influenced somewhat by desired effluent quality.

Often construction requirements or the type of aeration equipment employed control the configuration selection rather than the biological reactions involved. Built-in operational flexibility for the selected configuration can enhance the ability of a plant to provide reliable performance.

COMPLETE MIX. By definition, a completely mixed reactor has uniform characteristics throughout the contents of the entire reactor. In this configuration, shown in Figure 11.52 and known as the CMAS process, influent waste is rapidly distributed throughout the basin and the operating characteristics of MLSS, respiration rate, and soluble BOD_5 are uniform throughout. Since the total body of the basin liquid has the same quality as the basin effluent, only a very low level of food is generally available at any time for the large mass of microorganisms present. This characteristic is cited as the major reason why the CMAS process can handle surges in organic loading and toxic shocks to a limited extent without producing a change in effluent quality. In recent years, many engineers have rejected CMAS systems because low food concentrations and variable DO levels can stimulate the growth of filamentous bacteria that settle poorly.

The CMAS basin is conventionally square, round, or rectangular. Tank dimensions are controlled by the mixing pattern of the aeration equipment. Surface or submerged turbine units typically provide aeration. Factors influencing the mixing regime (whether or not a basin is completely mixed) include length:width ratio, amount of air or mixing power introduced, and wastewater velocity through the tank. Achieving complete mixing in a basin is very difficult, but can be approached. From a treatment point of view, any square or circular basin with a reasonable detention time and level of mixing intensity can be considered a completely mixed reactor, regardless of the type of aeration system used.

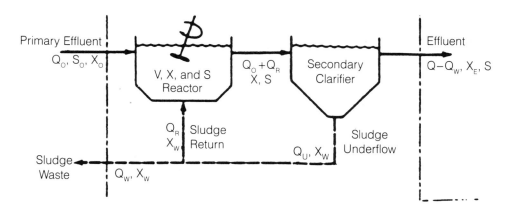

Figure 11.52   Complete mix reactor.

Experience has shown that basins with mechanical aerators can have a point feed and point draw-off and still approximate the CMAS process concept. The length:width ratio of a basin should generally be maintained at less than 3:1, although basins with 4 to 5:1 ratios can still show a very high degree of complete-mix characteristics with mechanical aerators. Multiple mechanical aeration units in a long, narrow channel (length:width greater than 5:1) create a mixing pattern similar to a number of completely mixed tanks in series. If diffused air is used, full tank width influent feed and effluent removal weir structures are normally provided as good practice. Oxidation ditches may be viewed as a complete-mix reactor even though they have some plug flow characteristics.

PLUG OR SERIES FLOW. Plug and series flow basins are discussed together since a plug flow reactor can be viewed as an infinite number of small, completely mixed basins in series. The plug flow basin is the oldest tank geometry in use since it was originally used to meet the mixing and oxygen transfer requirements of diffused aeration systems. Such basins are usually 4.6- to 9.1-m (15- to 30-ft) wide and up to 122-m (400-ft) long (L:W over 10:1). Long basins may be constructed side-by-side or in a folded arrangement. Figure 11.53 is a schematic of a plug flow reactor.

As one of its characteristics, the plug flow configuration has a high organic loading at the influent end of the basin. Loading is reduced over the length of the basin as the organic material in the wastewater is assimilated. At the downstream end of the basin, oxygen consumption primarily results from endogenous respiration. The high organic loading at the head end of this process discourages filamentous bacteria growth and often improves sludge settling beyond that realized from a complete-mix reactor.

A complete-mix reactor is noted for its ability to handle shock loads; plug flow configurations have a superior ability to avoid "bleed through" or passage of untreated substrate during peak flows. Thus, these reactors are

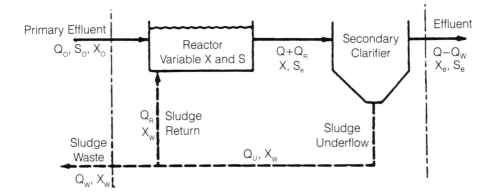

Figure 11.53   Plug flow reactor.

usually preferred for achieving nitrification. They are also often used where high effluent DO concentrations are necessary. In a complete-mix configuration, the entire tank contents would have to be maintained at the elevated DO level to achieve that objective.

CONTACT STABILIZATION. Contact stabilization describes both a system and a specific tankage configuration. As shown in Figure 11.54, it encompasses a contact basin, a clarifier, and a sludge stabilization basin. The process is best suited for smaller flows with longer target SRTs. Aerating return sludge in a separate reactor can reduce basin requirements by as much as 30 to 50% compared to that required in a conventional system. Also, sludge reaeration provides a means of achieving lower applied MLSS masses to clarifiers than those in complete-mix or plug flow systems. Lower aeration MLSS concentrations also ease attainment of nonlimiting DO concentration levels.

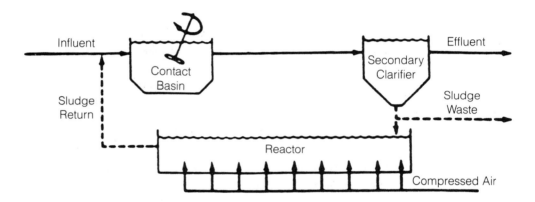

Figure 11.54   Contact stabilization configuration.

The tanks for contact and those for stabilization often have equal volumes and similar influent arrangements. Consequently, they can be operated either as a contact stabilization unit or in parallel as a conventional facility. This flexibility allows future expansion by shifting contact stabilization to a conventional system without increasing aeration basin volume. Such an expansion merely requires adding more clarification capacity. The ability to operate in either mode is particularly valuable in service areas with seasonal loading variations. Operational flexibility allows the facility to operate in the contact stabilization mode during peak flows but in a more conventional mode during off-peak periods.

Generally, contact stabilization designs provide for detention times of 20 to 60 minutes for contact and 3 to 6 hours for stabilization. Over 90% of the

BOD$_5$ can be removed from normal municipal wastewater with 1 hour in the contact basin and 3 hours in the stabilization section. Nitrification can sometimes be achieved with careful control of the operating MLSS concentration and the rate of aeration.

The design engineer must recognize that lengthy detention times may result in a process based on synthesis, sedimentation, and digestion rather than adsorption, sedimentation, and stabilization. In a contact tank with an excessive retention period for adsorption, the microorganisms will attempt to stabilize the organics by synthesizing new biomass. The stabilization tank then acts simply as an aerobic digester and storage tank for the biomass. This is not the objective of the process.

OXIDATION DITCH. In an oxidation ditch system, wastewater and mixed liquor are pumped around an oval pathway (racetrack) by mechanical aeration and pumping equipment located at one or more points along the flow circuit. Figure 11.55 shows an oxidation ditch with a rotor aerator to maintain tank motion and to aerate the ditch contents. As the ditch contents pass the aerator, the DO concentration rises and then falls while the ditch contents traverse the circuit. Oxidation ditches typically operate in an extended aeration mode with long HRTs (24 hours) and SRTs (20 to 30 days). Depending on the relative locations of wastewater input and removal, sludge return, and the aeration equipment, oxidation ditches can also achieve nitrification and denitrification (see Chapter 15). For BOD$_5$ removal or nitrification, the influent typically enters the reactor near the aerator and the effluent exits the tank upstream of the influent entrance.

Oxidation ditches have depths ranging from about 0.9 to 5.5 m (3 to 18 ft) and channel velocities from 0.24 to 0.37 m/s (0.8 to 1.2 ft/sec). Ensuring that ditch geometry will be compatible with the aeration and mixing equipment calls for consultation with individual equipment suppliers. Mechanical brushes, draft tube turbines, surface turbines, and jet devices are used to aerate and pump the liquid flow. Combinations of diffused aeration and pumping devices also have been employed, as reported by Christopher and

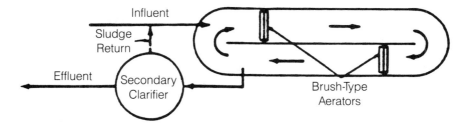

Figure 11.55 Oxidation ditch reactor.

Titus.[107] Separate or intrachannel clarifiers, as discussed above, are used to provide for separation and return of MLSS to the ditch.

A combination of oxidation ditch and SBR technologies has led to the evolution of the phased isolation ditch. In a 1986 report prepared for the U.S. EPA and subsequently published in the WPCF journal,[108] three different phased ditches—the VR-ditch, the T-ditch, and the Bio-Denitro™/ Bio-Denipho™ processes—were described. These technologies, shown in Figure 11.56 with their key features described in Table 11.20, have been used principally in Denmark, although plants have been constructed in Germany, Greece, China, and Australia. Other phased ditches are now being developed. Figure 11.57 shows a unit with three parallel runs, two used for aeration and mixing and the third used for settling. Movement of a vertically hinged wall eliminates the need to turn the aerator on and off or reverse its direction.

SEQUENCING BATCH REACTORS (SBRs). The SBR is a fill and draw system that was designated as an innovative/alternative (I/A) technology by EPA. Although SBRs have drawn attention as a new approach, Arora et al.[109] pointed out that fill and draw systems preceded the development of continuous-flow processes in Europe.

Table 11.20 Key features of Phased Isolation Ditch technologies.[108]

Ditch type[a]	Process modification	Treatment objectives		Number of ditches	Separate clarifier
VR-	–	BOD_5, SS, NH_3-N	2	Oxidation Sedimentation	No
T-	–	BOD_5, SS, NH_3-N	3	Oxidation Sedimentation	No
T-	Bio-Denitro™	BOD_5, SS, TN	3	Oxidation Denitrification Sedimentation	No
DE-	Bio-Denitro™	BOD_5, SS, TN	2	Oxidation Denitrification	Yes
DE-	Bio-Denipho™[b]	BOD_5, SS, TN, TP	2	Oxidation/ phosphorus uptake Denitrification	Yes

[a] Refer to Figure 11.57.
[b] This process modification requires an initial anaerobic tank prior to the DE-ditch system.

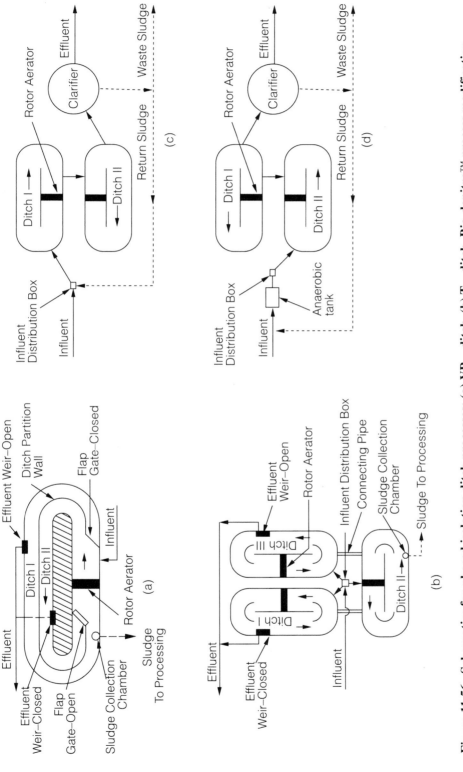

Figure 11.56 Schematics for phased isolation ditch processes: (a) VR—ditch, (b) T—ditch, Bio-denitro™ process modification, (c) DE—ditch, Bio-denitro™ process modification, and (d) DE—ditch, Bio-denipho™ process modification.

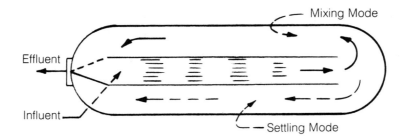

Mixing Mode

Effluent

Influent

Settling Mode

Figure 11.57   Phased ditch with three parallel runs and movable, pivoting wall aerators.

The SBR process involves a single, complete-mix reactor in which all steps of treatment occur. Discrete cycles are used during prescribed time intervals. MLSS remain in the reactor during all cycles, thereby eliminating the need for a separate clarifier. The specific treatment cycles shown in Figure 11.58 are[109]

- Fill (raw or settled wastewater fed to the reactor),
- React (aeration/mixing of the reactor contents),
- Settle (quiescent settling and separation of MLSS from the treated wastewater),
- Draw (withdrawal of treated wastewater from the reactor), and
- Idle (removal of waste sludge from the reactor bottom).

The idle cycle may be omitted by wasting sludge near the end of the react or draw cycle. As a result of the batch nature of the process, flow equalization or multiple reactors must be provided to accommodate continuous inflow of wastewater to the facility.

The intermittent cycle extended aeration system (ICEAS) was developed in Australia as a modification to the typical SBR).[110] Influent feeds continuously to the reactor during all cycles as in a continuous-flow system; but, withdrawal is intermittent, similar to the SBR system. Figure 11.59 presents a schematic of ICEAS system operation.

Advantages of SBR operation include

- Elimination of a secondary clarifier and RAS pumping,
- High tolerance for peak flows and shock loadings,
- Avoidance of MLSS "washout" during peak flow events,
- Clarification under ideal quiescent conditions, and
- Process flexibility to control filamentous bulking.

The major disadvantage is the relative lack of operational experience to date. Arora et al.[109] reported on eight installations and proposed a suggested design approach. Reported data indicated that total cycle times varied from

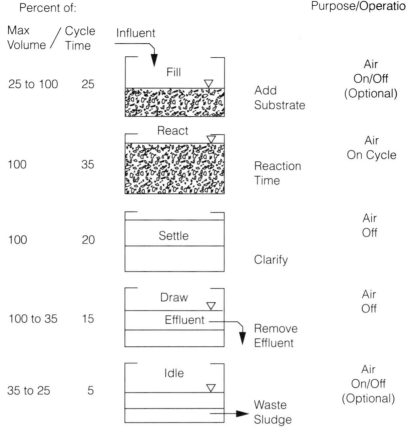

Percent of:			Purpose/Operation
Max Volume /	Cycle / Time		

Figure 11.58 Typical SBR operation.

7.6 to 49 hours, F:Ms ranged from 0.03 to 0.18 day^{-1}, and SRTs varied from 15 to 80 days. Melcer *et al.*[111] reported on the conversion of three small facilities to SBRs. Although some troubles with level control, probe fouling, and transfer pump clogging occurred, conversion resulted in successful plant operation, with minimal operator attention required. This process now seems to be an acceptable alternative to more conventional continuous-flow systems.

DEEP SHAFT REACTORS. A new form of treatment developed in England is currently marketed in the U.S. and Canada by ECO Research, Ltd. Known as the deep vertical shaft process, a deep shaft utilizing a U-tube aeration system replaces primary clarifiers and aeration basins. Drilled to depths of 120 to 150 m (400 to 500 ft), the shaft is lined with a steel shell and fitted with a concentric pipe to form an annular reactor. MLSS, wastewater, and air are forced down the center shaft and allowed to rise through the annulus. A flotation tank treats the effluent since supersaturated conditions prevail when the mixed liquor returns to the shaft outlet, causing the MLSS

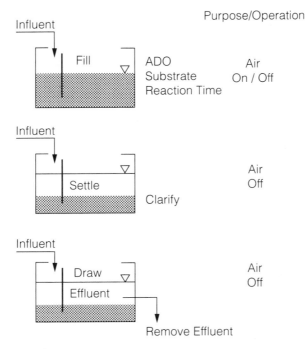

Figure 11.59 ICEAS system operation.

to float. Most of the MLSS are returned to the shaft, but some are wasted to an aerobic digestion process. Other types of sludge treatment might be satisfactory. Figure 11.60 presents a schematic diagram of this process.

Because of the high pressures involved, oxygen transfer efficiency is at least three times that of conventional air activated sludge systems. Potential advantages of the process include

- Lower capital and operational costs,
- Reduced land requirements,
- Ability to handle strong industrial wastes, and
- Immunity to climatic factors.

The process has been used in England and Virden, Manitoba, Canada. To date, no full-scale plants have started operation in the U.S.

PROCESS MODIFICATIONS

The engineering community has used a number of terms to distinguish the various activated sludge process modifications. These terms include conventional, extended aeration, high-rate, step-feed, modified aeration, tapered aeration, contact stabilization, and single-stage nitrification. Except for con-

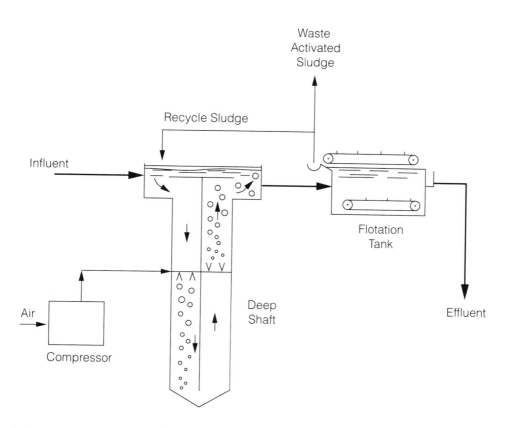

Figure 11.60 Schematic diagram of deep shaft process.

tact stabilization, single-stage system variations basically differ only in the range of the F:M ratio maintained and in the introduction points for air and wastewater. Table 11.21 provides a summary of operational characteristics for the various processes. Table 11.22 specifies typical ranges for relevant design parameters.

CONVENTIONAL. The term conventional activated sludge typically applies to a plug flow system with a F:M loading of about 0.15 to 0.40 day^{-1}. This system can yield BOD$_5$ removals in the range of 90 to 95%. Conventional system MLSS concentrations can range from 1500 to 3000 mg/L. Design MLSS concentration has increased steadily over the history of the process because of improvements in the oxygen transfer capability of aeration devices, clarifier performance, and understanding of system concepts.

As an important consideration in the design of conventional systems, nitrification might occur, even when not desired. This often happens with low loading conditions during summer months. When nitrification occurs, denitrification might begin in the final clarifiers, possibly causing rising sludge problems. As an easy remedy for limiting nitrification or denitrification, the

---

Table 11.21 Operational characteristics of activated sludge processes.[19]

Process modification	Flow model	Aeration system	BOD removal efficiency, %	Remarks
Conventional	Plug flow	Diffused air, mechanical aerators	85–95	Use for low-strength domestic wastes. Process is susceptible to shock loads.
Tapered aeration	Plug flow	Diffused air	85–95	Air supply tapered to match organic loading demand.
Step aeration	Plug flow	Diffused air	85–95	Use for general application to wide range of wastes.
Modified aeration	Plug flow	Diffused air	60–75	Use for intermediate degree of treatment where cell tissue in the effluent is not objectionable.
Contact stabilization	Plug flow	Diffused air, mechanical aerators	80–90	Use for expansion of existing systems, package plants. Process is flexible.
Extended aeration	Complete-mix	Diffused air, mechanical aerators	75–95	Use for small communities, package plants. Process is flexible.
High-rate aeration	Complete-mix	Mechanical aerators	75–90	Use for general applications with turbine aerators to transfer oxygen and control the floc size.
Pure oxygen systems	Complete-mix reactors in series	Pure oxygen with mechanical dispersion	85–95	Use where volume is limited and economical source of oxygen is available.

Design of Municipal Wastewater Treatment Plants

Table 11.22 Design parameters for activated sludge processes.[19]

Process modification	θ_c, d	F:M, kg BOD$_5$ applied/ kg MLVSS·d	Volumetric loading, kg BOD$_5$ applied/m^3·d	MLSS, mg/L	V/Q, h	Q_r/Q
Conventional	5–15	0.2–0.4	0.3–0.6	1 500–3 000	4–8	0.25–0.5
Tapered aeration	5–15	0.2–0.4	0.3–0.6	1 500–3 000	4–8	0.25–0.5
Step aeration	5–15	0.2–0.4	0.6–1.0	2 000–3 500	3–5	0.25–0.75
Modified aeration	0.2–0.5	1.5–5.0	1.2–2.4	200–500	1.5–3	0.05–0.15
Contact stabilization	5–15	0.2–0.6	1.0–1.2	(1 000–3 000)[a] (4 000–10 000)[b]	(0.5–1.0)[a] (3–6)[b]	0.25–1.0
Extended aeration	20–30	0.05–0.15	0.1–0.4	3 000–6 000	18–36	0.75–1.50
High-rate aeration	5–10	0.4–1.5	1.6–1.6	4 000–10 000	0.5–2	1.0–5.0
Pure oxygen systems	8–20	0.25–1.0	1.6–3.3	3 000–8 000	1–3	0.25–0.5

[a] Used with another biological process to achieve secondary or better treatment.
[b] Used alone to achieve secondary treatment.

MLSS concentration can be reduced. This increases the F:M ratio and lowers the SRT. Rapid sludge removal will limit denitrification in secondary clarifiers.

EXTENDED AERATION. Extended aeration plants are characterized by the introduction of screened and degritted wastewater directly to an aeration basin with a long aeration time, a high MLSS concentration, a high RAS pumping rate, and low sludge wastage. This system, typically used in the U.S. for flows of 3785 m^3/d (1 mgd) or less, incorporates complete-mix reactors. More recently, the extended aeration mode has been used in oxidation ditches of greater hydraulic capacity.

A particular advantage of using long HRTs (usually 16 to 24 hours) is that they allow the plant to operate effectively over widely varying flow and waste loadings. Secondary clarifiers must be designed to handle the variations in hydraulic loadings and high MLSS concentrations associated with this process.

One of the process goals is to maintain the biomass in the endogenous respiration phase. Since the microorganisms are essentially undergoing aerobic digestion in the aeration basin, more oxygen is required than for other single-stage systems. Many extended aeration plants experience a DO deficiency during the hours when the waste load is high. However, the long SRT and excess DO at night allow some nitrification, causing a daily, but noncoincidental, nitrification–denitrification cycle (see Chapter 15).

Some common problems with extended aeration systems include the continuous loss of pinpoint floc and the tendency to lose MLSS following short-term periods of low influent loading intensity (for example, weekends). Long aeration time combined with a long clarification time can also result in rising sludge solids in the secondary clarifiers. This condition, combined with the lack of primary sedimentation to remove floatables, requires the use of skimming devices in the final clarifiers. To avoid excessively high TSS losses, periodic sludge wasting is required. The accumulation of inert solids provides a means of estimating the rate of sludge wasting. Guo *et al.*[112] suggest that the average MLSS concentration not fall below 2000 mg/L. In cold climates, the low temperatures will likely impair performance of the extended aeration process unless heat loss is controlled. Thus, use of surface aerators would be inadvisable in cold climates. Guo *et al.*[112] noted that insufficient staffing, improper training, or both were the most common causes of poor performance, particularly in small "package" facilities.

HIGH RATE. High rate is the term applied to a CMAS system characterized by a short HRT, a high sludge recycle ratio, and a high organic loading rate. MLSS concentration generally varies from 4000 to 10 000 mg/L, and the F:M ratios are higher than those employed in conventional systems. Process integrity depends on maintaining the biomass in the growth phase.

Although high-rate systems can produce an effluent quality approaching that of a conventional system, the high-rate systems must be operated with special care. Inadequate RAS flow rates and high solids flux rates make the clarifiers of these systems more sensitive to washout. A high-rate, single-stage system can be used to partially remove carbonaceous BOD_5 as the first stage of a two-stage nitrification system.

CONTACT STABILIZATION. Contact stabilization initially was used to provide partial treatment (60 to 75% BOD_5 removal) at large coastal plants. This plug flow system is characterized by relatively short detention times (0.5 to 3.0 hours) with very high F:M loadings (1.5 to 5.0 kg BOD_5/kg MLSS·d). Compared to conventional systems, oxygen requirements are lower, but waste sludge quantities are higher. Designs may include or omit primary treatment.

The process, because of its short HRT and low MLSS concentration, can be sensitive to changes in hydraulic or BOD_5 loadings. These sensitivities have sometimes resulted in noncompliance with effluent standards, leading to diminished use of the process. This process can be effectively used as the first stage of a multistage system if loadings are not too extreme.

STEP FEED. The step feed or step aeration process, a modification of a plug flow reactor, allows entry of influent wastewater at two or more points along the length of the aeration basin. With this arrangement, the oxygen uptake rate becomes more uniform throughout the basin. Other operating parameters are generally similar to those of the conventional process. Influent flow enters the basin, usually along the first 50 to 65% of its length. Step aeration configurations generally include diffused aeration equipment. An existing plug flow reactor can be modified for step feed by simply dividing the basin into compartments and redirecting the flow so that each compartment, except the last one, receives wastewater input. A step feed tank configuration is shown in Figure 11.61.

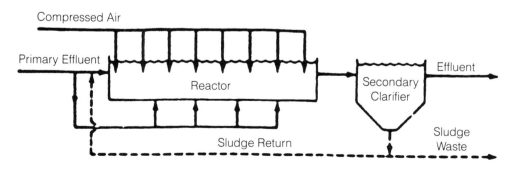

Figure 11.61   Step feed process.

Because the influent enters the reactor in stages, microbial growth can vary throughout the aeration tank. However, this has not been well documented. Torpey[113] indicated that step feeding produces a better settling sludge than does plug flow. Buhr *et al.*[114] analyzed the relationship between step-feed location and loadings on secondary clarifiers. Since RAS enters at the head of the aeration basin, mixed liquor concentration varies as a function of the number and location of influent feed points. The authors found that feeding near the head of the aeration basin increases the loading on the secondary clarifiers and consequently increases the RAS concentration; feeding near the effluent end of the aeration basin lowers the solids loading rate and RAS concentration.

TAPERED AERATION. An aeration system design that modulates the oxygen supply along the length of a plug flow reactor is called tapered aeration (Figure 11.62). This approach is commonly associated with diffused air systems. Plug flow reactor designs should incorporate tapered aeration for operational control. Mixing requirements generally govern the minimum air supply rate.

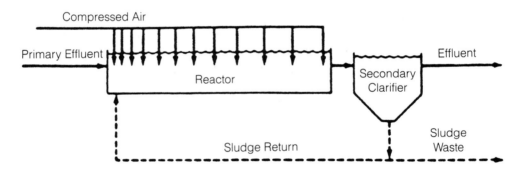

Figure 11.62   Tapered aeration process.

Design parameters for the tapered aeration process are consistent with those for conventional treatment. Adding more air at the influent end of the aeration tank than at the effluent end produces a number of beneficial results:

- Reduced blower capacity and operating costs,
- Greater operational control, and
- Inhibition of nitrification (when desired).

PURE OXYGEN SYSTEMS. Use of high-purity oxygen rather than air for aeration was evaluated in the 1950s by Budd and Lambeth.[115] The process achieved commercial status in 1970. The main advantages claimed by manufacturers of the process include reduced power for dissolving oxygen into the mixed liquor, improved biokinetics, the ability to treat high-strength soluble wastewaters, and a reduction in bulking problems from DO deficit stress.

Pure oxygen systems in the past were characterized by high MLSS concentrations (3000 to 8000 mg/L) and relatively short HRTs (1 to 3 hours). Recent practice indicates that a more appropriate MLSS concentration will likely range from 1000 to 3000 mg/L for municipal wastewater. Required MLSS concentrations range up to 10 000 mg/L for treating high-strength industrial wastewater. *Nocardia* accumulations in the reactor might cause foaming at high-MLSS concentrations.

Pure oxygen is fed concurrently with wastewater flow. For covered systems (Figure 11.63), oxygen feed can be controlled by maintaining a constant gas pressure within the tanks. A DO concentration of 4 to 10 mg/L is normally maintained in the mixed liquor. Less than 10% of the inlet oxygen vents from the last stage of the system.

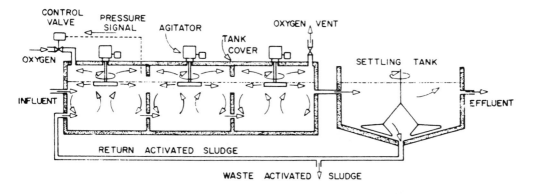

Figure 11.63 **Pure oxygen activated sludge system (covered, compartmentalized tank type).**

Data and reports on the benefits of oxygen systems have been under rigorous review. The reviews have indicated that, in some cases, a better settling sludge, greater tolerance for peak organic loadings, and reduced energy requirements have occurred. Some aspects of the process might be similar to those occurring in a well-designed and -operated system using air. In comparing air with oxygen systems, researchers have reported little or no difference in carbonaceous BOD_5 removal kinetics with levels of DO above 1 to 2 mg/L. Sludge yields and oxygen requirements seem to be essentially the same for either oxygen or air. Some studies have shown a difference in the settling properties between oxygen and air sludges, whereas others have not. The literature provides comparative information on full-scale, side-by-side air and oxygen plant operating performance.[116]

Pure oxygen systems reportedly produce lower pH values (6.0 to 6.5) due to high partial pressure of CO_2 and loss of alkalinity because of NO_3^- production. When the pH is depressed below 6.5 to 7.0, nitrification will attenuate and the system can require a longer sludge age, greater aeration tank volume, and additional settling basin area. These effects have led to consideration of separating the carbonaceous BOD_5 removal and nitrification stages when using oxygen. Some design engineers suggest that the first stage of a two-stage system receives oxygen and the second stage receives air.

Covered tanks using oxygen need provisions for warning of potential explosions that could result from the presence of combustible, volatile hydrocarbons in the influent wastewater. A detector system can be used to automatically purge the tanks with air if the volatile hydrocarbon level becomes excessive. Covered tanks capture VOC emissions; therefore, off-gas volume from pure oxygen systems is about 1% of that leaving a typical air system.

An atmosphere of high-purity oxygen and carbon dioxide in the reactor basin requires careful selection of materials used for construction. Compared with air, this atmosphere is much more corrosive and more reactive with organic compounds, such as oils and greases. Some plants have even experienced corrosion of materials used in the construction of the secondary clarifiers. Suppliers of high-purity oxygen systems have evaluated materials suitable for safe and reliable construction.

ACTIVATED CARBON ADDITION. Activated carbon was historically used as a tertiary step following biological treatment or as part of a physical-chemical process. During the early 1970s, engineers at DuPont developed and ultimately patented the PACT process in which powdered activated carbon (PAC) is added to the aeration basin of an activated sludge treatment plant. The addition of PAC forms a matrix, which has demonstrated some beneficial properties, including

- Improved sludge settling characteristics;
- Increased dewaterability of the waste sludge;
- Ability of PAC to adsorb biorefractory materials and toxic compounds to both improve effluent quality and lessen shock loading impacts;
- Reduction in odor, foaming, and bulking sludge problems; and
- Improved color and BOD_5 removal.

A key disadvantage of PAC addition is the need to regenerate the PAC for reuse or to purchase virgin PAC if the plant lacks regeneration facilities. As another disadvantage, the PACT process with PAC regeneration usually requires tertiary filtration. The ability to separate inert materials from the PAC following wet air oxidation is one of the keys to successful use of PAC in all applications. In the past, this was a problem but the patent holder has recently developed a process to accomplish such separation.

Most of the successful applications of PACT have occurred in industrial applications where the PAC has been used on a single-pass basis without regeneration. Much of the early work focused on its use for refinery waste treatment in response to stringent EPA effluent guideline limits. The oil industry considered PAC augmentation in lieu of the proposed best available technology of granular activated carbon (GAC) adsorption following conventional secondary treatment. In recent years, numerous studies have considered the kinetics of the process, the removal of specific toxic organic compounds, and the treatability of landfill leachates.

Because of the varying reasons for considering PAC treatment and the wide range of process applications, pilot-scale testing of this process in parallel with conventional systems would be advisable before proceeding with the design of such facilities.

COMBINED FIXED-FILM/SUSPENDED-GROWTH SYSTEMS. A recent redevelopment in secondary treatment entails placement of inert support media in an activated sludge reactor. This allows fixed-film biomass to grow on the media and augment the microbial population of the mixed liquor.

The concept dates back to the 1920s when the first widely accepted application had flat asbestos sheets suspended in an aeration tank (termed contact aeration).[117] More recent applications include the use of synthetic trickling filter media, polyurethane foam pads, or loops of fiber bundles as the inert material in the aeration basin.

Manufacturers of the current systems have claimed several advantages compared with conventional suspended-growth treatment processes:

- Increased biomass levels, hence lower F:M values,
- Reduced solids loading to secondary clarifiers,
- Improved sludge settling characteristics,
- Enhanced oxygen transfer efficiency, and
- Lower sludge production.

To date, combined fixed-film and suspended-growth system designs have been based on empirical relationships. A number of full-scale installations employ this concept in Germany and Japan but, currently, none in the U.S. EPA continues to evaluate this technology because of its prospective efficiency. An improved design procedure is needed before such systems will gain the acceptance of consulting firms in the U.S. Chapter 12 contains additional information on dual-type treatment systems.

PERFORMANCE HISTORY OF TREATMENT CONFIGURATIONS

The tables and figures of this section present the performance history of various suspended-growth treatment configurations. Full-scale plant performance data obtained from several EPA documents and various other sources are used to depict a cross section of municipal WWTP performance experiences in the U.S. For more detailed information on particular plants, refer to the EPA "Needs Survey Report"[118] and the "Needs Survey Tape" (raw computer database).[119]

Interpretation of performance data presented for alternate configurations must consider all of the variables affecting suspended-growth processes. Some reported performance variations likely result from variables other than reactor configuration alone. Specifically, sedimentation basin design, return sludge strategy, oxygen dissolution system, sludge age, load fluctuations, MLSS concentration, temperature, and percentage of capacity loading may differ enough to overshadow the differences in reactor configurations.

Another consideration is the quantity and quality of analytical data on which reported performance is based. For example, facilities that achieve a high level of carbonaceous stabilization often achieve partial nitrification. When this occurs and effluent ammonium nitrogen values remain above 1 mg/L, false elevated BOD_5 values are encountered if the BOD_5 tests are undertaken without nitrogen inhibition. This calls for use of the optional analytical procedure with nitrogen inhibitors.

PERFORMANCE SUMMARIES. Mean daily effluent BOD_5 values and coefficients of variation from 324 plants[120] in eight process categories are illustrated by Figure 11.64. The data for individual plants within any given category show considerable scatter about the median point with an apparent trend towards higher levels of relative variability for process categories that achieve lower mean effluent BOD_5 concentrations.

Table 11.23 summarizes actual performance of 416 well-operated treatment plants representing eight process categories.[121] Because most activated sludge plants use tapered aeration, the table omits a separate process category for this configuration. Data for this 1985 table were acquired from a wide range of geographically dispersed U.S. facilities with no design or operating deficiencies, atypical loadings, or extra unit processes. Two years of daily operating data were used for each of the representative plants. Note that several EPA reports[122] indicate that BOD_5 removals are similar for both air and pure oxygen activated sludge plants.

Table 11.24[123] summarizes small-flow plant data from inspection and technical assistance reports of the EPA regional offices. Conventional, contact stabilization, and extended aeration process configurations are

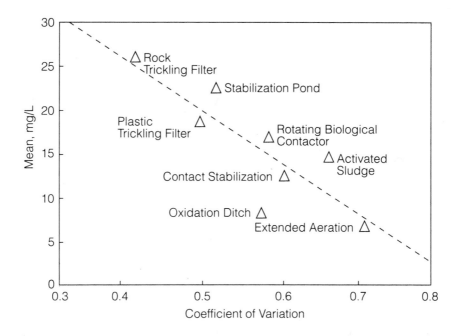

Figure 11.64   Mean and coefficient of variation of effluent BOD for the median plant from a survey of 324 facilities.

represented. Those facilities with design flows less than 0.05 m³/s (1 mgd) are primarily package plants. Most of the facilities with effluent BOD_5 and TSS values above 30 mg/L have been expanded and upgraded since 1978 to meet current effluent requirements.

Table 11.25 describes 43 facilities used in a 1981 statistical study[124] of the performance and operation of activated sludge systems. The selected treatment facilities represent a range of geographic locations and environmental conditions throughout the U.S. Plant flows ranged from less than 0.05 m³/s (1 mgd) to more than 9 m³/s (200 mgd) and included municipalities with negligible to a large percentage of industrial wastes. Data from at least 1 year of daily operation were used for each plant. None of these plants exceeded its design loadings.

Table 11.26[124] shows the average and standard deviation of the annual means, standard deviations, and coefficients of variation for each type of process. Examination of the pooled plant data suggests that, contrary to the general belief, contact stabilization and complete-mix processes do not perform in a more stable manner than do conventional activated sludge processes; however, the reasons for these results were not investigated. Possible causes include the effects of unmeasured variables such as environmental conditions, overloading, or operational and human factors.[124] Note that for conditions above design loadings, performance would be affected by the applied final clarifier loading rates. Other references showing activated

---

Table 11.23 Summary of results for secondary treatment plants $\left(\dfrac{\text{range (min − max)}}{\text{median}}\right)$ [121].

Process category	Number of plants	Effluent BOD, mg/L					Effluent TSS, mg/L					Flow, mgd		
		Mean[a]	Standard deviation[b]	Coefficient of variation[c]	S_{30}[d]	S_7[e]	Mean[a]	Standard deviation[b]	Coefficient of variation[c]	S_{30}[d]	S_7[e]	Mean[a]	Standard deviation[b]	Coefficient of variation[c]
Trickling filter, rock	64	8.49–57.58 26.03	2.41–30.72 10.06	0.11–1.02 0.40	1.71–14.58 5.88	2.02–23.27 7.67	9.18–101.3 25.25	3.24–121.1 10.77	0.11–1.65 0.50	2.12–27.00 6.28	2.91–51.09 7.46	0.03–9.42 0.65	0.004–2.36 0.23	0.06–0.97 0.27
Trickling filter, plastic	17	7.18–55.41 18.98	3.62–27.45 9.38	0.15–0.90 0.49	3.04–13.73 5.48	3.40–19.46 7.01	5.76–42.83 19.36	5.40–19.11 8.29	0.27–1.40 0.65	2.36–12.72 5.11	3.38–14.28 7.05	0.13–8.68 0.92	0.07–2.36 0.40	0.16–0.86 0.38
Activated sludge, conventional	66	3.92–41.49 14.79	2.39–34.82 8.18	0.19–1.99 0.66	1.50–21.29 5.08	1.57–27.39 6.65	5.06–52.25 14.30	2.71–78.20 12.33	0.18–2.21 0.83	1.22–39.03 6.89	1.60–54.31 8.75	0.05–82.7 1.87	0.02–18.79 0.53	0.04–1.04 0.24
Activated sludge, contact stabilization	57	3.82–45.43 12.63	1.05–43.66 8.16	0.07–1.73 0.60	1.00–32.44 5.54	0.98–38.95 6.88	3.66–46.75 13.80	1.52–75.14 9.30	0.23–2.02 0.71	1.67–39.53 6.24	2.22–50.12 8	0.002–56.23 0.36	0.002–16.22 0.13	0.06–1.35 0.34
Activated sludge, extended aeration	28	2.67–27.89 7.20	0.96–19.81 4.60	0.21–2.03 0.71	0.67–17.15 3.01	0.95–19.6 3.89	2.46–33.35 9.78	0.46–38.10 7.11	0.18–1.95 0.66	0.19–35.45 3.42	0.25–36.97 4.77	0.01–5.80 0.10	0.006–0.89 0.06	0.11–1.32 0.34
RBC	27	4.79–32.27 16.99	4.49–47.43 10.19	0.29–1.47 0.58	2.75–20.27 6.26	3.39–31.60 7.89	6.65–22.86 15.15	4.47–22.11 8.46	0.39–1.27 0.69	1.70–15.26 4.50	3.22–17.17 6.40	0.06–31.60 0.71	0.03–1.96 0.23	0.12–1.19 0.31
Oxidation ditch	28	2.93–29.20 8.40	1.13–17.95 5.88	0.14–1.59 0.57	1.01–17.22 3.98	1.23–17.88 5.58	3.77–31.55 12.26	1.88–47.65 8.61	0.21–1.86 0.68	1.46–222.12 5.98	1.92–389.8 9.86	0.01–6.33 0.18	0.004–1.15 0.07	8.09–1.16 0.31
Stabilization pond	37	8.59–20.96 22.71	5.79–46.5 9.78	0.25–1.08 0.51	4.59–38.70 8.04	5.38–46.39 8.63	12.60–133.00 39.51	5.29–156.97 22.56	0.30–1.18 0.63	3.18–128.17 17.10	4.45–135.98 19.44	0.03–2.73 0.31	0.000–1.57 0.09	0.08–0.83 0.31

a Overall facility means.
b Overall facility standard deviations.
c Overall facility coefficients of variation.
d Overall facility standard deviations of 30-day moving averages.
e Overall facility standard deviations of 7-day moving averages.

Table 11.24 Activated sludge summary, EPA region data.[123]

Location and process	Flow, mgd [a]		Influent, mg/L		Effluent, mg/L		Removal, %	
	Average	Design	BOD5	TSS	BOD5	TSS	BOD5	TSS
Conventional								
Aspen, Colo.	1.1	2.0	336	298	19	20	94	93
Missoula, Mont.	5.5	10.0	133	129	54	48	59	63
Snowmass, Colo.	0.8	0.8	185	180	34	53	82	71
Havre, Mont.	1.3	1.8	275	275	31	30	89	87
Barstow, Calif.	1.7	4.5	150	250	15	25	89	90
W. Sacramento, Calif.	3.5	2.5	140	130	20	27	86	79
Carmel, Calif.	2.5	3.0	200	300	20	30	90	90
Monterey, Calif.	3.5	4.6	250	250	35	70	86	72
Zephur Cove, Nev.	1.0	3.0	334	237	12	11	96	95
Fairfield, Ohio	1.3	1.0	173	172	16	26	85	85
Grand Rapids, N. Dak.	38	—	100	140	18	40	82	71
					Average removal		$\overline{85}$	$\overline{81}$
Contact stabilization								
Richwood, W. Va.	0.39	0.5	89	94	30	33	67	65
Pen Argyl, Pa.	1.1	0.95	162	75	16	26	90	62
Lolo, Mont.	0.15	0.25	233	172	17	12	93	93
Baden, Pa.	0.55	0.5	206	202	12	21	89	90
Lancaster, Pa.	9.24	12.0	168	227	32	35	81	85
Manteca, Calif.	1.3	3.2	200	180	30	25	75	86
Savage, Md.	5.9	—	158	206	32	41	86	80
Coralville, Iowa	0.9	—	135	150	26	24	81	84
					Average removal		$\overline{83}$	$\overline{81}$
Extended aeration								
Athens, W. Va.	0.15	0.25	129	258	14	33	88	84
Follansbee, W.Va.	0.5	0.5	200	190	15	48	92	75
Cheyenne, Wyo.	0.43	0.8	262	272	—	—	—	—
Elizabeth, Colo.	—	0.065	441	341	30	96	93	72
Elbert, Colo.	0.007	0.025	208	125	16	27	92	78
					Average removal		$\overline{91}$	$\overline{77}$

[a] mgd $\times$ 3785 = m^3/d.

Table 11.25   Activated sludge treatment plants studied.[124]

Plant number	Plant location	Year	Type of process	Average flow, mgd
1	Michigan	76	Complete mix	14.2
2	California	07-75/6-76	Conventional	100.0
3	California	76	Complete mix/contact stabilization	16.2
4	Wisconsin	76	Contact stabilization	23.7
5	Illinois	76	Step feed	69.7
6	Illinois	76	Conventional	66.2
7	Illinois	76	Conventional	69.1
8	Maryland	76	Step feed	15.2
9	New York	77	Complete mix	18.6
10	Alabama	06-76/5-77	Aerated lagoon	23.6
11	Michigan	10-76/9-77	Complete mix	0.74
12	Virginia	76	Conventional/step feed	15.8
13	Alabama	07-76/6-77	Conventional/step feed	10.5
14	Illinois	76	Conventional	2.2
15	Illinois	76	Step feed	3.3
16	California	75	Step aeration	6.3
17	California	76	Step aeration	2.4
18	California	76	Step aeration	29.0
19	California	76	Step aeration	7.1
20	Illinois	76	Conventional	209.0
21	Illinois	76	Conventional	189.0
22	Illinois	76	Conventional	188.0
23	Illinois	76	Step feed	185.0
24	Michigan	73	Conventional	1.9
25	Michigan	74	Conventional	2.1
26	Michigan	75	Conventional	2.1
27	Michigan	76	Conventional	2.3
28	Michigan	77	Conventional	2.1
29	Michigan	77	Conventional	9.5
30	Michigan	77	Conventional	9.5
31	Connecticut	76	Conventional	3.8
32	California	76	Plug flow/Kraus	47.1
33	Ohio	76	Contact stabilization	2.8
34	Nevada	75	Conventional/tapered	17.0
35	Kansas	76	Contact stabilization	10.4
36	Washington	76	Conventional/step feed	30.8
37	California	76	Complete mix/step feed	13.9
38	Michigan	10-76/10-77	Step feed	17.3
39	Michigan	77	Conventional/contact stabilization	33.3
40	Michigan	12-76/11-77	Extended aeration	0.56
41	Michigan	01-77/10-77	Conventional	26.9
42	Wisconsin	76	Conventional	81.6
43	Wisconsin	76	Conventional	54.0

Table 11.26 Statistics of the annual effluent BOD and SS concentration data for different process types.[124]

Process type	No. of plants	Annual mean		Standard deviation		Coefficient of variation[a]	
		$\bar{x}$	$S_{\bar{x}}$	$\bar{x}$	$S_{\bar{x}}$	$\bar{x}$	$S_{\bar{x}}$
				BOD			
Conventional	18	12.80	6.85	9.54	7.99	0.69	0.25
Complete mix	5	16.82	6.67	13.24	6.51	0.77	0.13
Step feed/aeration	13	10.84	7.68	8.28	7.56	0.68	0.25
Contact stabilization	4	38.38	32.08	28.17	20.92	0.76	0.14
Extended aeration	1	14.41	—	5.31	—	0.37	—
Kraus	1	24.02	—	9.85	—	0.41	—
Aerated lagoon	1	30.35	—	11.61	—	0.38	—
All plants	43	15.76	13.43	11.28	10.49	0.70	0.23
				SS			
Conventional	18	14.92	10.53	16.02	18.61	0.86	0.38
Complete mix	5	19.88	14.19	19.65	16.92	1.00	0.51
Step feed/aeration	13	16.23	16.65	16.83	23.89	0.83	0.34
Contact stabilization	4	40.88	26.75	37.66	26.18	0.90	0.14
Extended aeration	1	8.82	—	5.28	—	0.60	—
Kraus	1	24.12	—	9.26	—	0.38	—
Aerated lagoon	1	58.79	—	18.71	—	0.32	—
All plants	43	19.40	17.03	18.35	20.60	0.84	0.37

[a] Average values from all plants, not calculated from parameter averages.

sludge effluent quality for various treatment processes are included in a report by EPA.[125]

PERFORMANCE OF INDIVIDUAL PROCESSES. Performance of each of several processes is described below.

Plug and Series Flow. BOD_5 and TSS effluent quality performance as a function of normalized flow from plug flow plants are shown in Figures 11.65 and 11.66.[126] The loading relationship—normalized flow—is defined as the ratio of actual flow to design flow. Note that the overload conditions (normalized flows >1) impact conventional plants. As discussed earlier in this chapter, various engineering firms have substantially different design criteria. Thus flows above design defined by one firm may be less than those defined by another. Actual data are also shown by season of the year for demonstration of possible effluent deterioration due to cold weather. Summer and winter operation results (June through September versus December through March) did not differ significantly. Obviously, because of reduced biological activity, long-duration extreme cold conditions will adversely affect process performance. Data were collected at several locations throughout the country with the assistance of various state environmental offices, EPA offices and local operating agencies, and by actual site visits.

Contact Stabilization. BOD_5 and TSS effluent quality performance curves as a function of normalized flow are shown in Figures 11.67 and 11.68.[126] For the plants studied, overload conditions did not affect the contact stabilization plants up to normalized flows of 2.0.

Deep-shaft Biological Reactor. Table 11.27 lists design and performance data for the deep-shaft facilities cited in the 1982 EPA report entitled "Technology Assessment of the Deep Shaft Biological Reactor."[127] Since that time, additional plants have been built. As of 1989, a total of 16 municipal and 26 industrial plants were reported on-line.[128] Most of the plants are located in Japan, where some exist in the basements of large commercial buildings.

In 1979, a 0.01-m^3/s (0.2-mgd) EPA demonstration deep shaft installation was commissioned in Ithaca, New York. This installation and the 1988 full-scale plant in Homer, Alaska, are the only two installations presently known to be located within the U.S.

Pure Oxygen Systems. A summary of actual operating results from covered and staged pure oxygen processes in the U.S. is presented in Table 11.28. The data were obtained from EPA and published by Benefield and Randall.[129]

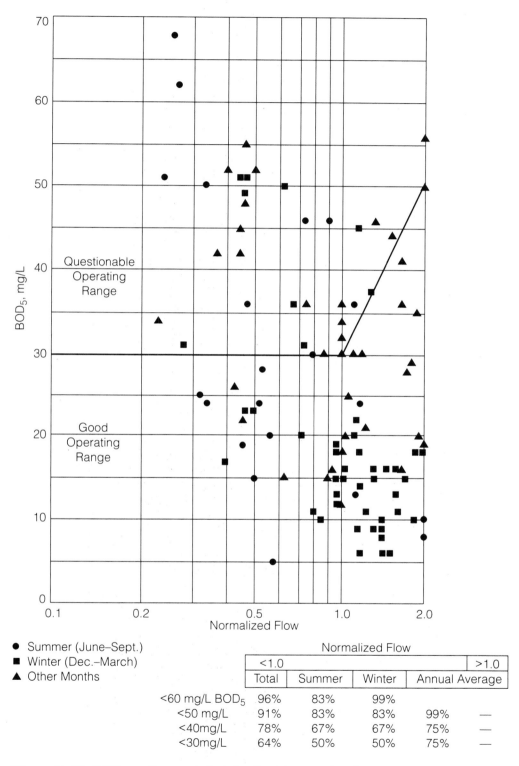

	Summer (June–Sept.)		Normalized Flow				
●	Summer (June–Sept.)		<1.0				>1.0
■	Winter (Dec.–March)		Total	Summer	Winter	Annual Average	
▲	Other Months						
	<60 mg/L BOD₅		96%	83%	99%		
	<50 mg/L		91%	83%	83%	99%	—
	<40mg/L		78%	67%	67%	75%	—
	<30mg/L		64%	50%	50%	75%	—

Figure 11.65   BOD$_5$ performance for plug flow (conventional) plants.

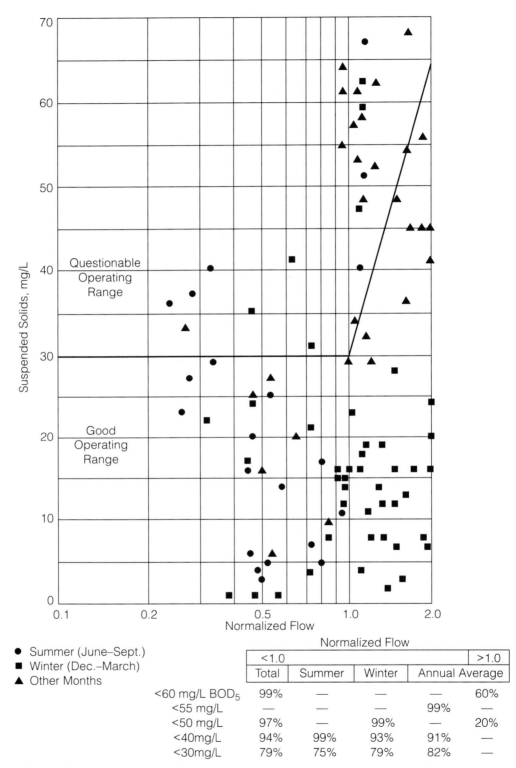

Figure 11.66 Suspended solids performance for plug flow (conventional) plants.

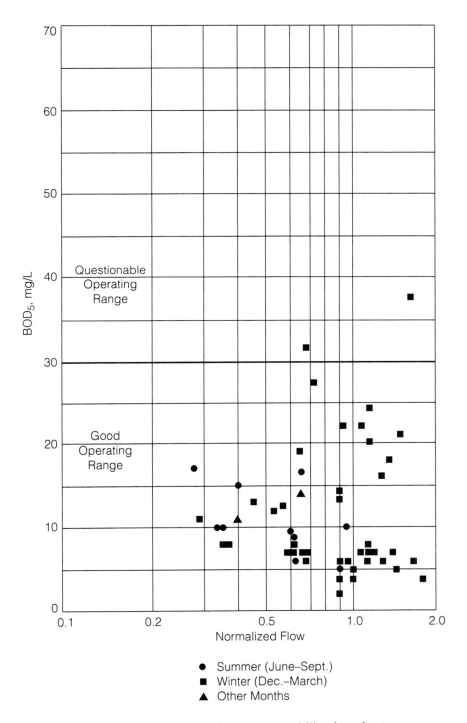

Figure 11.67   BOD₅ performance for contact stabilization plants.

Suspended-growth Biological Treatment

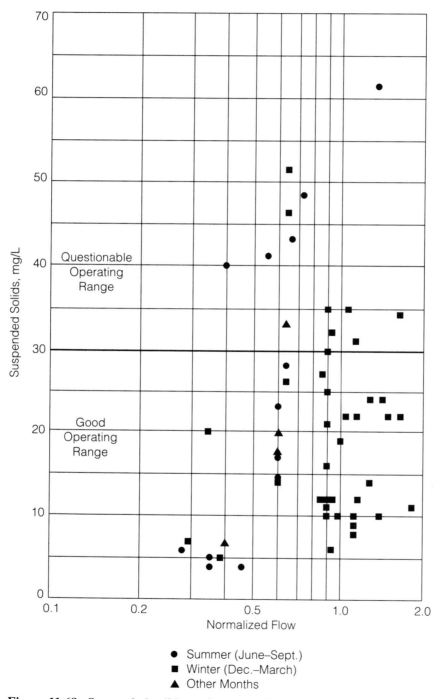

Figure 11.68 Suspended solids performance for contact stabilization plants.

Table 11.27 Design and performance data for deep-shaft facilities.[127]

Location	Design capacity,[a] m³/d	Shaft dimensions[b]	Residence time	F:M, kg/day BOD5/MLVSS	Effluent quality, mg/L BOD5	Effluent quality, mg/L SS
1. Billingham, U.K.	363	40.6 cm diameter 130 m deep	70 min	0.8	6–35	7–39
2. Mareh Farm Sewage Works Tilbury, Essex, U.K.	6 620	2 m diameter 130 m deep	ND[c]	ND	ND	
3. Emlichheim, W.Ger.	1 060	1.1 m diameter 100 m deep	120 min	1.5	60–100	290
4. Paris, Ontario, Can.	454	39 cm diameter 155 m deep	35–55 min	0.4–2.0	7–96	10–121
5. Paris, Ontario, Can.	6 246	0.01–0.91 m diameter 122 m deep	~55 min	1.35	<15	<15
6. Molson Breweries, Barrie, Ontario	125	44 cm diameter 155 m deep	288 min	2.6–5.0	43–109	47–257
7. Molson Breweries, Barrie, Ontario	2 091	21.26 cm diameter 152.4 m deep	300 min	0.86–1.3	50	67
8. Bon Conseil, Quebec, Can.	4	6.75 cm diameter 150 m deep	330 min	1.5–4	40–125	138–206
9. Ithaca, N.Y.	757	43.8 cm diameter 136 m deep	39 min	0.74	30	30
10. Virden, Manitoba, Can.	2 400	76.2 cm diameter 153 m deep	41 min	1.06	30	30
11. Portage La Prairie, Manitoba, Can.	13 615	1.32 m diameter 198 m deep	30 min	1.53	30	30[d]

[a] m³/d × 0.000 264 = mgd.
[b] cm × 0.393 7 = in.; m × 3.281 = ft.
[c] No data.
[d] Represents design criteria—not actual operation.

Suspended-growth Biological Treatment 645

Table 11.28   Actual operating data from UNOX process.[129]

Influent BOD5, mg/L	Aeration time, hours	F:M, lb BOD5/day/ lb MLVSS	Aeration MLSS, mg/L	Recycle MLSS, %	Clarifier overflow, gpd/sq ft	Effluent quality	
						BOD5, mg/L	TSS, mg/L
157	1.97	1.10	2700	0.55	456	9	22
101	1.42	1.05	2750	0.85	1595	17	31
125	3.33	0.37	3000	1.20	630	12	50
160	2.16	0.60	4000	2.40	440	12	24
357	4.70	0.33	6400	1.60	590	32	79
245	2.40	0.68	4500	1.50	541	<10	<10
114	1.29	0.70	4760	1.66	645	13	18
425	4.20	0.75	4500	1.00	502	25	70
150	1.59	0.62	5100	1.60	704	17	13
1500	9.75	0.45	9600	3.0	360	90	60

Table 11.29 contains monthly average data for five UNOX high-purity oxygen-activated sludge plants.[130] Up to 2 consecutive years of plant data are included in this table, which covers the years 1981 through 1988. Further information and several case histories are given in the EPA reference entitled "Status of Oxygen-Activated Sludge Wastewater Treatment."[131]

Extended Aeration. Performance of several types of extended aeration plants are discussed below.

CONVENTIONAL EXTENDED AERATION ACTIVATED SLUDGE. BOD_5 and TSS effluent quality performance curves as a function of normalized flow are shown in Figures 11.69 and 11.70. Note that overload conditions do not affect the extended aeration plants up to normalized flows of 2.0.[126]

Data for both extended aeration units at Alaska's Eillison Air Force Base, 1971 and 1972 data, and the 1971 data for the oxidation ditch at the University of Alaska are summarized in Table 11.30. These two facilities are located near Fairbanks, which has a subarctic climate. The mean annual temperature at Fairbanks is approximately 4°C (25°F), with minimum and maximum recorded temperatures of −55°C (−66°F) and 38°C (99°F), respectively. The area has approximately 200 days per year of temperatures below freezing. Of all the parameters affecting low-temperature performance, organic loading is the most significant. It was found that BOD_5 removals above 80% at liquid temperatures below 7°C could generally be maintained at loadings of 0.08 kg BOD_5/kg MLSS·d or less.[132]

CONVENTIONAL EXTENDED AERATION PACKAGE PLANTS. A 1978 survey of small privately operated extended aeration and contact stabilization package plants was carried out in the Cincinnati, Ohio, area (20 plants) and in Dade County, Florida (46 plants). These plants ranged in size from 0.002 to 0.04 m^3/s (0.05 to 1.0 mgd). Figures 11.71 and 11.72 show the reported reliability along with comparative data for oxidation ditch plants. These data show that the oxidation ditch plants perform better than conventional package plants.[123] The differences may be due in part to clarifier loadings as well as reactor size and configuration.

COUNTERCURRENT EXTENDED AERATION. As of the fall of 1984, 44 plants with the countercurrent aeration technology were either entirely operating or under construction in the U.S. Table 11.31 shows effluent quality observations at four WWTPs with countercurrent aeration systems. These observations were made during a period of intensive plant monitoring for a 5-year technology assessment study sponsored by EPA.[133] These four facilities had historical average BOD_5 loadings of between 22 and 62% of design values.

Table 11.29   Performance data (monthly averages) for five oxygen-activated sludge plants.[130]

Plant number	Plant parameter	High-purity, oxygen-activated sludge						Reactor influent			Secondary effluent		Month & year
		MLTSS, mg/L	MLVSS, mg/L	F:M, 1/day	HDT, hours	DO, mg/L	SRT, days	Flow, mgd[a]	BOD, mg/L	TSS, mg/L	BOD, mg/L	TSS, mg/L	
1	Average	3584	3128	0.98	2.1	10.9	1.80	4.44	241	74	21	11	January 1983 through December 1984
	Maximum	4489	3873	1.63	3.0	18.1	3.08	6.54	296	112	40	31	
	Minimum	2295	2046	0.56	1.5	3.9	0.93	3.08	159	51	9	5	
2	Average	4412		0.53	1.90		3.18	13.3	129	106	12	17	October 1987 through September 1988
	Maximum	7760		1.20	3.24		15.71	27.0	230	346	56	57	
	Minimum	2160		0.20	0.92		0.73	6.8	49	55	3	2	
3	Average	4486	3566	0.33	2.9	10.7	3.8	4.0	134	92	11	9	April 1981 through December 1982
	Maximum	7148	5953	0.49	4.1	15.0	6.1	6.3	185	161	19	17	
	Minimum	3327	2755	0.20	1.8	6.4	2.2	2.7	92	72	8	5	
4	Average	1691	1373	1.20	1.87	10.8	1.46	123.6	124	78	16	11	January 1983 through December 1984
	Maximum	2139	1668	1.57	2.61	11.6	1.99	154.8	160	110	24	27	
	Minimum	1179	982	0.78	1.25	9.8	1.06	106.9	86	53	12	5	
5	Average	3400	2633	0.7	2.75		1.39	95.7	212	176	17	18	July 1982 through March 1983
	Maximum	4600	3000	0.9	3.15		1.74	142.9	300	240	22	21	
	Minimum	2500	1900	0.5	2.53		1.03	74.5	140	150	11	15	
Averages for five plants		3515	2675	0.75	2.3	10.8	1.94	—	168	105	15	13	—

[a] mgd × 3785 = m^3/d.

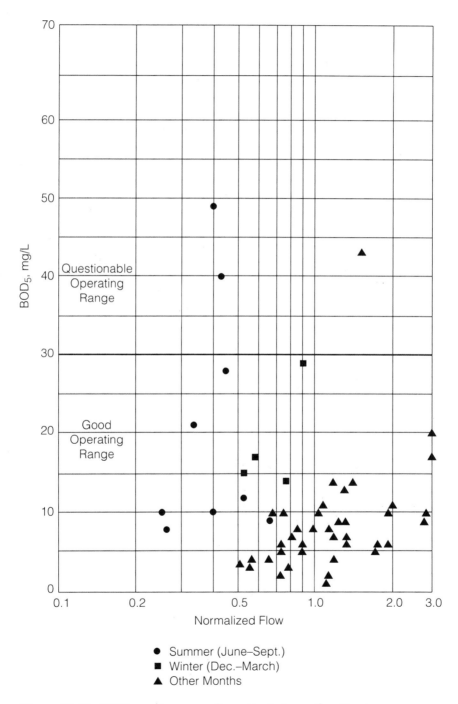

Figure 11.69 BOD$_5$ performance for extended aeration plants.

Suspended-growth Biological Treatment 649

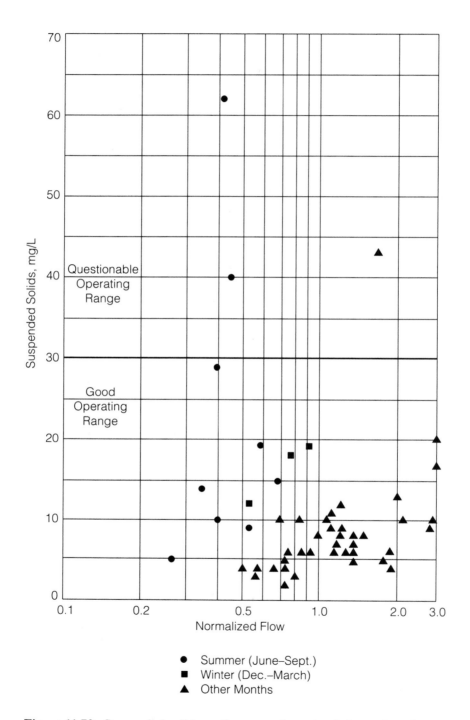

Figure 11.70 Suspended solids performance for extended aeration plants.

Table 11.30 Average performance data for subarctic Alaska extended aeration units.[132]

Period	Location	Clarifier flow pattern	Major liquid temperature range, °C	BOD removal, %	BOD/MLSS loading, day	MLSS, mg/L	Volatile, %	Overflow, $L/m^2 \cdot min$ (gpd/sq ft)		Effluent SS, mg/L
I	EAFB	Rect. Hz[a]	7-17	81	0.10	2 330	82	20	(710)	36
II	EAFB	Rect. Hz[b]	2-9	10	0.14	1 710	84	9.3	(330)	134
III	EAFB	Rect. Hz[c]	1-8	84	0.03	3 360	78	12	(420)	17
I	EAFB	Rect. Up[a]	7-19	82	0.12	2 370	80	20	(710)	99
II	EAFB	Rect. Up[b]	2-7	30	0.11	2 750	83	10	(370)	122
III	EAFB	Rect. Up[c,d]	1-8	76	0.04	2 740	79	12	(420)	50
1971	Coilege oxidation ditch	Circular up[e,f]	8-20	84	0.05	3 000	60	8.7	(310)	28

[a] From October through December 1971, period I average feed BOD = 173 mg/L.
[b] For February 1972, average period II feed BOD = 143 mg/L.
[c] For March to May 1972, average period III feed BOD = 128 mg/L.
[d] Rhodamine-B dye poisoning data excluded.
[e] Ranganathan, K.R., and Murphy, R.S., Inst. Water Resour., Univ. Alaska, Rep. #IWR-27 (1972). Septic tank sludge not included in influent BOD. Solids in quiescent sections of ditch have exceeded 20 000 mg/L.
[f] Not including data when solids were being wasted into effluent; that is, effluent BOD > influent BOD.

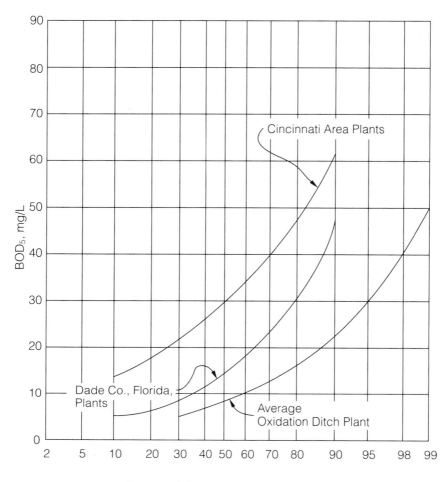

Percent of Time Value Was Less Than

Figure 11.71 Reliability of BOD$_5$ performance for activated sludge package plants.

Table 11.31 Performance of countercurrent, extended aeration plants.[133]

	Effluent BOD			Effluent SS		
	Observations			**Observations**		
Site	**Number**	**Number, >30 mg/L**	**Average, mg/L**	**Number**	**Number, >30 mg/L**	**Average, mg/L**
Carlisle	29	0	4.6	29	0	5.8
Claiborne	85	0	4.8	69	8	16.9
Hampden	63	0	1.6	62	0	1.0
Loudon	72	2	10.8	109	30	26.8

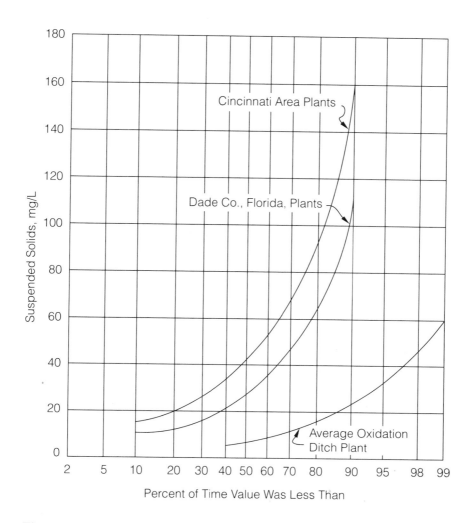

Figure 11.72 **Reliability of suspended solids performance for activated sludge package plants.**

EXTENDED AERATION OXIDATION DITCH. Performance and reliability data for 29 oxidation ditch plants were obtained from actual plant operating records for an EPA sponsored study.[123] In this 1978 study, oxidation ditch performance was developed primarily from monthly average data (Table 11.32).

Where possible, the performance for each plant was calculated for both summer and winter. Winter was arbitrarily determined to be the months of November through March. The average performance shown in the table was determined by averaging the performance of all the individual plants. Table 11.32 shows high and low individual plant performance to establish the range limits of individual plants. One of the conclusions of the study was

---

Table 11.32 Performance summary for 29 oxidation ditch plants.[123]

Parameter	Effluent, mg/L			Removal, %		
	Winter	Summer	Average annual	Winter	Summer	Average annual
BOD5						
High plant	55	34	41	87	86	87
Average	15.2	1.2	12.3	92	94	93
Low plant	1.9	1.0	1.5	99	99	99
Suspended solids						
High plant	26.6	19.4	22.4	81	82	82
Average	13.6	9.3	10.5	93	94	94
Low plant	3.1	1.9	2.4	98	98	98

that both BOD5 and TSS removal appear to be relatively independent of plant capacity.[123]

Of the 29 plants used in the performance study, 12 were further studied for reliability. Reliability curves were developed from actual daily plant data. Averaging data were not used because averaging would remove the data peaks and minimums. Reliability curves shown in Figures 11.73 and 11.74 are plotted for summer, winter, and total year for the average of all plant data. In addition, the best and worst plant reliability is also plotted. Data for the summer show a slight trend of somewhat improved BOD5 and TSS reliability compared with winter conditions. Reliability differs greatly between the best and worst plants.[123] Other sources of oxidation ditch performance data include an EPA publication.[125]

Sequencing Batch Reactors. Table 11.33 shows operational and performance summaries of eight conventional SBR and ICEAS plants in operation in Canada, Australia, and the U.S. in 1984.[109,134,135] The designs differ for several components of these plants, including inlets, aeration and mixing systems, and decanters, but their operation follows sequencing batch principles. Note that a review of the information contained in Table 11.33 indicates that no pair of the plants evaluated used the same operating strategy even though their goals were identical (removal of BOD5 and TSS).

As of 1989, over 17 full-scale municipal conventional SBR systems were either on-line or under construction in the U.S. These vary in size from 190 m^3/d (0.05 mgd) to a 23 000-m^3/d (6.0-mgd) plant in Oklahoma City, Oklahoma.[136]

Figure 11.75 is a histogram depicting the scatter of daily effluent data for the Culver, Indiana, SBR operating with the REACT treatment mode. As indicated by this case history performance summary data, effluent quality was both satisfactory and reasonably stable.[137] For the period between May 1980

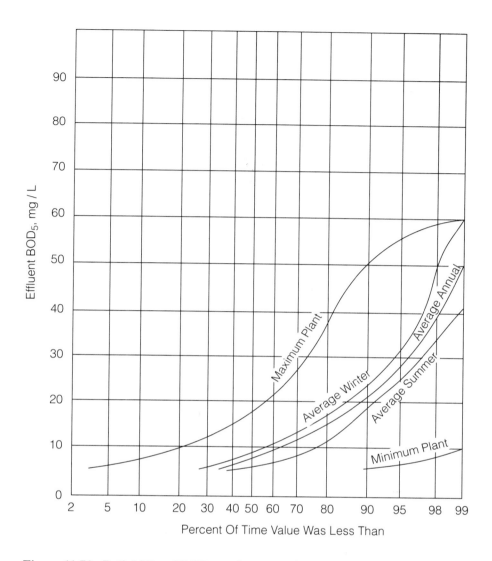

Figure 11.73 Reliability of BOD₅ performance for oxidation ditch plants.

and May 1981, daily average raw BOD$_5$ and TSS concentrations were
160 mg/L and 130 mg/L, respectively. Both daily average BOD$_5$ and TSS
concentrations in the effluent (before chlorination) were less than 10 mg/L in
both tanks. Postchlorination effluent BOD$_5$ averaged 5 mg/L.[137]

Table 11.34 shows more recent case history performance and operating
summary data for the Grundy Center, Iowa, SBR operating both with and
without the REACT treatment mode. The data were obtained between
June 10 and July 10, 1985. As shown by the table, effluent concentrations
averaged less than 8 mg/L for BOD$_5$ and about 12 mg/L for TSS. During the
test period, the influent flow rates averaged about 80% of design.[135]

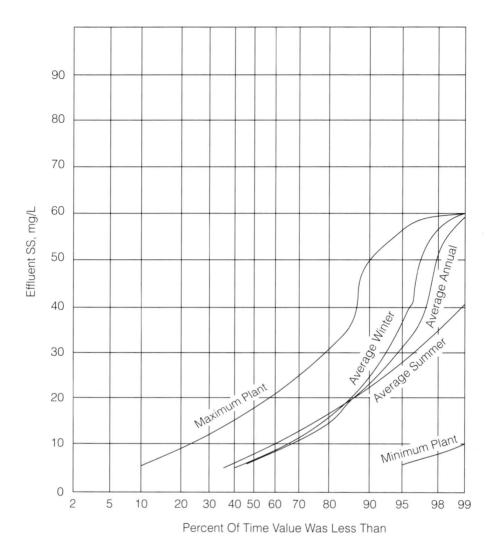

Figure 11.74 Reliability of suspended solids performance for oxidation ditch plants.

More than 100 ICEAS plants exist worldwide and, as of 1989, 39 were operational in the U.S. The latter vary in design capacity from 0.000 3 to 0.4 m³/s (0.006 to 9.2 mgd). As of 1989, the 0.4-m³/s plant located in Cleveland, Tennessee, was the largest ICEAS installation.[138]

Published performance data are limited but nonetheless include the 0.1-m³/s (2.2-mgd) Tullahoma, Tennessee, plant.[139] It is designed to produce a 30-, 30-, and 10-mg/L (BOD5, TSS, and NH4-N) winter effluent and a 20-, 30-, and 2-mg/L summer effluent. During the summers of 1986 and 1987, its effluent quality averaged 6, 6, and 1 mg/L and maximum day values were 19, 30, and 1 mg/L.

Table 11.33 Evaluation summary for SBRs (1984).[109]

Parameter	Canada		U.S.				Australia	
	Rivercrest, Manitoba	Glenlea, Manitoba	Choctaw, Oklahoma	Grundy Center, Iowa	Eldora, Iowa	Culver, Indiana[a]	Tamworth, New South Wales	Yamba, New South Wales
Mode of operation	SBR	SBR	SBR	SBR	Continuous inflow later changed to SBR	SBR	ICEAS	ICEAS
Date when operation commenced	August 1983	1978	August 1983	June 1983	April 25, 1984	May 1980	June 1983	June 1983
Design average flow, gpd	24 000	2 000	500 000	832 000	220 000	—	535 000	253 000
Design loading								
BOD, mg/L	236[b]	251[b]	260 366[b]	200	250 120[b]	170[b,c]	260	260[c]
SS, mg/L	200[b]	152[b]	260 350[b]	—	—	150[b,c]	—	—
NH3, mg/L	37[b]	55[b]	19[b]	15	25	20[b,c]	35–40	—
Current average flow, gpd	60 000	1 165	200 000 283 000 (equivalent)	800 000	220 000 106 000 (equivalent)	353 000	535 000 (est)	—
Desired effluent quality								
BOD, mg/L	TOC-40	30	20	30	30	10	30	30
SS, mg/L	30	30	20	30	30	10	30	30
NH3, mg/L	—	—	15	6 (summer), 11 (winter)	8 (summer), 10 (winter)	—	—	—
Actual effluent quality								
BOD, mg/L	11	5	8	Not being net because of decanter problems.	Data was not available. Effluent appeared to be satisfactory.	10	5 to 10	6 to 10
SS, mg/L	15	6	18			5	5 to 10	10 to 15
NH3, mg/L	10	2	—			1.0	2.2	1.0
Mode of operation at design flow								
Full time				40 minutes (without air/pumps) 120 minutes (with air/pumps)		180 minutes	Continuous	Continuous
R time	90 minutes	22 hours	18 hours		150 minutes[d]	42 minutes	120–150 minutes	150 minutes
S time	45 minutes	1 hour	3 hours	60 minutes	80 minutes	42 minutes	45 minutes	180 minutes
D time	20–60 minutes	1 hour	3 hours	40 minutes	50 minutes	42 minutes	45 minutes	45 minutes
I time	—	—	—	60 minutes	45 minutes	60 minutes Fill 30% mixed 70% aerated	—	—
Important design parameters								
DT, hours	7.6	49	48	20.4	43	16.5[b]	46	36
F:M, kg BOD/kg MLSS	0.18[b]	0.032[b]	0.037	0.078	0.05	0.08–0.16	0.04	0.05

Table 11.33 Evaluation summary for SBRs (1984) (continued).[109]

Parameter	Canada		U.S.				Australia	
	Rivercrest, Manitoba	Glenlea, Manitoba	Choctaw, Oklahoma	Grundy Center, Iowa	Eldora, Iowa	Culver, Indiana[a]	Tamworth, New South Wales	Yamba, New South Wales
Important design parameters (continued)								
SRT, days	43[b]	18.80[b]	0.028[b] Sludge wasted twice in 10 months	0.067[b] 25.30[b]	Sludge not wasted in last 2 months	15.45[b]	—	—
Power usage kWh/kg BOD applied	0.8	22.9	2.9	0.8 to 1.3	2.2	2.1	1.9	1.5
Unit processes								
Trash rack	Yes	—	Yes (bypass)	Yes (bypass)	—	—	—	Yes
Mech. screens	—	—	—	Yes	Yes	Yes	Yes	—
Comminutor	—	—	Yes	—	—	or Yes	or Yes	—
Grit removal	—	—	—	—	—	—	—	Yes
Equalization	Yes	Lift station wet well	Emergency holding pond	Yes, aerated Sideline equalization	Yes, aerated	—	—	—
Primary	—	—	—	Yes	—	Yes	—	—
Treatment SBR	Yes	Yes	Yes	Yes	Yes	Yes	Yes	Yes
Disinfection	—	—	Yes	Yes	Yes	Yes	—	—
Sludge treatment	Holding tank & land application	Agriculture farm	Holding pond & land application	Aerated sludge holding & sludge beds	Anaerobic digesters & sludge beds	Aerobic digesters & sludge beds	Polishing lagoon Sludge lagoon	Polishing lagoon Aerobic lagoon
Reasons for providing this technology	Capital cost savings & simple operation	Capital cost savings & simple operation	8.4% savings in life cycle costs	19% capital cost savings in secondary treatment process or 8% savings in overall plant cost	Capital cost savings & simple operation (100% city funding)	Full-scale study funded by EPA	Capital cost savings	Capital cost savings

[a] Culver data obtained from reference.[136]
[b] Actual operating data.
[c] Raw wastewater.
[d] Jet motive pumps on all the time, but air on and off for 40 and 10 minutes, respectively, repeated three times during the 150-minute fill & react periods.

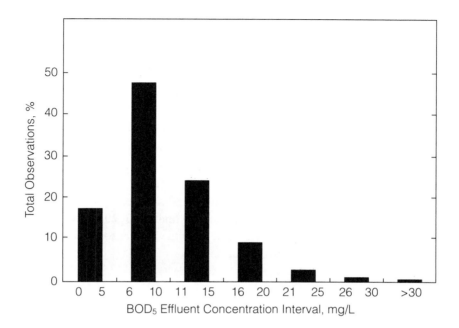

Figure 11.75 BOD5 performance histogram for SBR at Culver, Indiana (total number of observations was 407; average BOD5 effluent concentration was 10 mg/L; and observation periods were May 14, 1980–May 23, 1981, and August 19, 1981–December 31, 1981).

Table 11.34 Performance data (June 10–July 10, 1985)[135] for Grundy Center, Iowa, SBR.

Parameter	Influent	SBR with REACT	SBR without REACT
Flow, m³/d[a]	2580	—	—
BOD5, mg/L	220	8	8
SS, mg/L	170	12	12
NH4-N, mg/L	17	1.1	1.0
NOx-N, mg/L	0.7	2.1	2.6
P, mg/L	8.3	3.7	4.9
MLSS, mg/L	—	1700	1800
WS, %	—	1.2	1.2
SVI, mg/L	—	130	160
V_O, cm/min	—	4.2	2.6
Approximate detention time for both tanks		25 hours	
Approximate organic loading for both tanks		0.12 kg BOD5/ kg MLSS·d	

[a] m³/d × 0.000 264 = mgd.

PACT Process. The wet air regeneration (WAR) process incorporates wet air oxidation (WO) technology for the oxidation of excess biological solids and the regeneration of spent PAC. The combined technology of PACT plus wet air oxidation carbon regeneration has been identified by the manufacturer as the wastewater reclamation system (WRS).

Table 11.35 describes some municipal PACT plants in the U.S.[140,141] Color reduction, nitrification, limited space availability, and toxics removal were the principal reasons offered for this technology selection.

Representative performance data for five of the PACT plants, along with their current (1988) permit limits, are presented in Table 11.36.[140,141] Effluent permit limits range from typical limits of 30 mg/L each for BOD_5 and TSS (with and without nitrification), to summer limits of 7, 20, 2, and 1 mg/L for BOD_5, TSS, NH_3-N, respectively, and TP at the Kalamazoo, Michigan, WWTP (the world's largest PACT plant).

PERFORMANCE HISTORY OF SINGLE-STAGE NITRIFICATION.

Although single-stage nitrification systems are commonly used in England to obtain dependable nitrification, their use in the U.S. has only recently become widespread. Table 11.37 is a partial listing of U.S. nitrification facilities (for 1983), some of which are described in the literature.[142] If the BOD_5:TKN ratio is greater than 4, the process is classified as a combined carbon oxidation–nitrification process (single-stage nitrification). If this ratio is 4 or lower, the nitrification system is arbitrarily classified as a separate-stage nitrification process. The table also shows the distribution between BOD_5 and NOD. For separate nitrification processes, NOD accounts for at least 60% of the total oxygen demand.

Conventional Activated Sludge. The Valley Community Services District (VCSD) wastewater treatment plant at Dublin, California, treats an average daily flow of 0.16 m^3/s (3.7 mgd) from a largely residential service area. Average dry-weather flow capacity of the plant is 0.18 m^3/s (4 mgd). This plant incorporates nitrification for ammonia reduction to lower residual ammonia enough for economical breakpoint chlorination. This allows disinfection to proceed with a free chlorine residual so that stringent bacteriological requirements can be met. The aeration tank operates with the first half of the first pass serving for reaeration of the return sludge; primary effluent is step fed to the remainder of the first pass.

Table 11.38 shows case history performance data (1974) at the VCSD plant.[16] Nitrogen figures were obtained from once-monthly, 24-hour composites, while the BOD_5 and SS data are the average of daily composite samples. The performance data table shows consistent nitrification of the influent ammonia. Effluent ammonia nitrogen concentration was typically less than 1 mg/L and has remained so since August 1973. Nitrate nitrogen

---

Table 11.35 Selected municipal PACT® systems. [140,141]

Project	Current status	Current flow, mgd[a]	Design flow, mgd[a]	Application	Reason for PAC	Start-up	Comments
Rothschild, Wisc.	AS[b]	—	1.0	Municipal wastewater	Nitrification	1972–1973	Full-scale demonstration
Vernon, Conn.	PAC	4.2	6.5	Municipal wastewater, textile waste	Color	1979	Organic, color removal 50 gpm WAR
Medina Co., Ohio (Liverpool)	PAC	7	10.0	Municipal wastewater	Nitrification	1980	Nitrification 55 gpm WAR
Burlington, N.C. (East)	PAC	7.0	12.0	Municipal wastewater, textiles	Color, nitrification, toxics	1980	Color removal, nitrification 86 gpm WAR
Mt. Holly, N.J.	PAC	2.4	5.0	Municipal wastewater, textiles	Color, space	1981	Advanced treatment 28 gpm WAR
Chatsworth, Ga.	PAC	—	0.8	Municipal wastewater, carpet manufacturing	—	1982	Treatment plant upgrade
Bedford Heights, Ohio	NAC[c]	3.4	3.5	Municipal wastewater	Nitrification, space	1984	Nitrification 18 gpm WAR
Kalamazoo, Mich.	PAC	25	53.5	Municipal, industrial wastewater	Color, nitrification, toxics	1985	Nitrification, organic removal 4–110 gpm WAR
El Paso, Tex.	PAC	4.5	10.0	Municipal wastewater	Nitrification, organics	1985	Groundwater recharge 58 gpm WAR
North Olmsted, Ohio	AS[d]	6	7.0	Municipal wastewater	Nitrification, space	1985	Nitrification 18 gpm WAR
Huron Valley Hospital Milford, Mich.	PAC	—	0.05	Sanitary waste	—	1986	Aquifer recharge
Jessup, Md.	PAC	—	1.6	Prison wastewater	—	1988	Nitrification–denitrification 10 gpm skid WAR

[a] mgd × 3785 = m³/d.
[b] Converted to conventional activated sludge (AS).
[c] Activated carbon no longer needed to meet permit, converted to the use of nonactivated carbon (NAC) without regeneration.
[d] Activated carbon no longer needed to meet permit, converted to conventional activated sludge (AS) plan to convert to NAC without regeneration.

Table 11.36 Performance data for PACT® plants.[140,141]

Plant	Parameter[a]	Average flow, mgd[b]	BOD5, mg/L[c,d] Influent	BOD5 Effluent	COD, mg/L Influent	COD Effluent	TSS, mg/L Influent	TSS Effluent	Turbidity, mg/L Influent	Turbidity Effluent	Color, alpha Influent	Color Effluent	N-TKN, mg/L Influent	N-TKN Effluent	N-NH3, mg/L Influent	N-NH3 Effluent	P-total, mg/L Influent	P-total Effluent	Period
Vernon, Conn.	Design	6.5	150-300	<20	—	—	150-400	<10	—	10	50-600	0	—	—	—	—	—	—	3/79-3/80
	Actual-average	3.36	200	4	840	73	450	8	148	8	150-500	20-40	—	—	—	—	—	—	—
	Actual-average	3.5	283	7T/<4C	900	<70	338	10	—	—	—	—	—	—	—	—	—	—	7/79-12/80
	Actual-average	—	260	8.6T	—	—	390	9.0	—	—	—	—	—	—	—	—	—	—	4 years
	Permit	—	—	10C	—	—	—	20	—	—	—	—	—	—	—	—	—	—	—
	Permit-summer	—	—	—	—	—	—	—	—	—	—	—	—	—	—	—	—	—	—
Medina, Ohio	Design	10	—	—	—	—	—	—	—	—	—	—	—	—	—	—	—	—	7/88-7/89
	Actual-range	—	—	1-4	—	29-52	—	2-24	—	—	—	—	—	1.5-8.6	—	0.05-6.6	—	0.24-1.35	—
	Actual-average	—	—	3	—	38	—	10	—	—	—	—	—	3.3	—	1.4	—	0.63	7/88-7/89
	Permit	—	—	10C	—	—	—	12	—	—	—	—	—	—	—	8	—	—	—
	Permit-summer	—	—	—	—	—	—	—	—	—	—	—	—	—	—	1.5	—	—	—
East Burlington, N.C. (Annual averages)	Design	12	—	—	—	—	—	—	—	—	—	—	—	—	—	—	—	—	—
	Actual-average	—	—	6.3	—	—	—	14	—	—	—	—	—	—	—	2.5	—	—	1987
	Actual-average	—	—	5.5	—	—	—	11	—	—	—	—	—	—	—	2.6	—	—	1988
	Permit	—	—	24	—	—	—	30	—	—	—	—	—	—	—	8.0	—	—	—
	Permit-summer	—	—	12	—	—	—	—	—	—	—	—	—	—	—	4.0	—	—	—
Mount Holly, N.J.	Design	5.0	135	<5	—	—	90	<5	—	—	250	<25	40	<3	—	—	—	—	4 months/82
	Actual-average	2.51	<274	3	<532	40	<410	4.5	—	—	132	23	<46.2	2.7	—	—	—	—	—
	Permit	—	—	30	—	—	—	30	—	—	—	—	—	—	—	20	—	—	—
	Permit-summer	—	—	—	—	—	—	—	—	—	—	—	—	—	—	—	—	—	—
Kalamazoo, Mich.	Design	53	—	—	—	—	115P	4-5	—	—	—	—	—	—	25P	—	5-6P	—	—
	Actual-average	—	325P	7	600P	50	—	—	—	—	—	—	—	—	—	1-2	—	0.3	3/86
	Actual-range	—	—	1-5	—	25-125	—	0.5-15	—	—	—	—	—	—	—	0.1-12.5	—	0.1-0.8	6/86-8/86
	Actual-average	—	300	3.0	—	<75	—	1-10	—	—	—	—	—	—	—	0.2-0.8	—	<0.5	6/86-8/86
	Permit	—	—	30	—	—	—	30	—	—	—	—	—	—	—	10	—	—	0.5-1
	Permit-summer	—	—	7	—	—	—	20	—	—	—	—	—	—	—	2	—	—	—

a Influent = raw unless otherwise noted; monthly average data unless otherwise noted.
b mgd × 3785 = m³/d.
c C = carbonaceous BOD5.
d T = total BOD5.
e Underlined numbers are approximate averages of actual-range numbers.

Table 11.37 Classification of nitrification facilities.[142]

Type and location	Scale, mgd[a]	BOD5:TKN ratio	Oxygen demand distribution, %		Classification degree of separation		Pretreatment to remove BOD5
			BOD5	NOD	Combined oxidation-nitrification	Separate stage	
Suspended growth							
Manassas, Va.	0.2	1.2	20	80		x	Activated sludge
Hyperion, Los Angeles, Calif.	46	7.3[b]	61	39[b]	x		Primary treatment
Central Contra Costa Sanitary Dist., Calif.	1.0	2.4	34	66		x	Lime primary treatment
	pilot, design 30	1.0	18	82		x	Activated sludge
Livermore, Calif.	3.3	2.8	38	62		x	Roughing filter
Flint, Mich.	34	5.5	65	35[b]	x		Primary treatment
Valley Community Services District, Calif.	3.8	10.8[b]	70	30[b]	x		Primary treatment
Blue Plains, Washington, D.C.	pilot, design 309	1.3 to 3.0	22 to 39	61 to 78		x	Activated sludge
Whittier Narrows, LACSD, Calif.	12	6.6	61	39	x		Primary treatment
Jackson, Mich.	13.5	9	66	34	x		Primary treatment
Tampa, Fla.	pilot, design 60	3.0	40	60		x	Activated sludge
South Bend, Ind.	pilot	1.8	28	72		x	Activated sludge
New Market, Ontario, Can.	2.4	2.6	36	64		x	Lime primary treatment
Cincinnati, Ohio	pilot	7.2	61	39	x		Primary treatment
	pilot	1.0[c]	18[c]	82[c]		x	Activated sludge
Fitchburg, Mass.	pilot	1.0	18	82		x	Activated sludge
Marlboro, Mass.	pilot	3.6[d]	40	60		x	Trickling filter
Amherst, N.Y.	pilot	0.8 to 2.0	22	78		x	Activated sludge
Denver, Colo.	pilot	2.7	37	63		x	Activated sludge
Attached growth							
Stockton, Calif.	pilot, design 58	5.3	54	46	x		Primary treatment
Midland, Mich.	pilot	1.1	19	81		x	Trickling filter
Union City, Calif.	pilot	1.7	27	73		x	Activated sludge
Allentown, Pa.	40	1.9	30	70		x	Trickling filter
Lima, Ohio	pilot	0.79	15	85		x	Activated sludge

[a] mgd × 3785 = m³/d.
[b] Approximate, calculated from COD data.
[c] Calculated from effluent.
[d] BOD:NH_4-N ratio; BOD:TKN would be about 3.0.

Table 11.38 Performance data for single-stage nitrification plants.[a,16]

Parameters	Valley CSD, Dublin, Calif.	Jackson, Mich.	Whittier Narrows, Los Angeles, Calif.	Flint, Mich.
Flow, mgd[b]	4.0	17.0	12.0	—
BOD, mg/L	330	145	270	250
Aeration detention time, hours	8.5	6.0	6.0	—
Secondary effluent, mg/L				
BOD5	6.5–21 11.4	2.5–5.0 3.5	—	13.6
TSS	8.2–14.2 11.3	—	—	24.1
NH3–N	0.06–16 2.1	0.5–1.2 0.7	0.4–3.2 1.3	1.7
SRT, days	8.1–15.2 10.5	10.3–18.6 14.5	9.4–40.2 19.4	—

[a] Underlined numbers represent long-term average (approximately 1 year).
[b] mgd × 3785 = m^3/d.

concentration in the effluent approximated 24 mg/L, representing about 99% of the nitrogen in the effluent.

The city of Jackson, Michigan, operates a 0.75-m^3/s (17-mgd) activated sludge plant that is designed to nitrify year-round. This facility is expected to remove enough nitrogenous oxygen demand to maintain DO levels of about 4.0 mg/L in the receiving water, the Grand River. Since full nitrification started, this requirement has been consistently met. The activated sludge system was operated in the conventional or plug flow mode to gain highest nitrification efficiency, even at temperatures as low as 8°C.

Table 11.38 also shows performance data (1973–1974) for the Jackson, Michigan, plant.[16] Both the weak wastewater and the ability to maintain the mixed liquor at a high concentration allowed nitrification with HRTs of less than 8 hours, even at temperatures of 8 to 10°C.

Step Aeration. The Whittier Narrows Water Reclamation Plant is a 0.5-m^3/s (12-mgd) activated sludge plant designed and operated by the Los Angeles County Sanitation Districts. The plant reclaims water for groundwater recharge; the entire effluent of the plant enters spreading basins to recharge groundwater aquifers. This plant is an example of a step aeration activated sludge plant operating in a moderate climate.

Table 11.38 shows performance data (1973–1974) at the Whittier Narrows plant.[16] Year-round complete nitrification has been obtained. Climatic conditions for this plant favor nitrification; average monthly wastewater temperatures remained above 21°C for the year examined.

The Flint, Michigan, WWTP was upgraded to provide nitrification to prevent DO depletion in the Flint River. This plant is an example of a step aeration activated sludge plant operating in a fairly cold climate with temperatures as low as 7°C. The plant is usually operated in a step aeration mode, with 50% of the influent directed to the head ends of the second and third passes.

Table 11.38 shows performance data for the plant.[16] Nitrite nitrogen remained less than 0.1 mg/L. An appearance of high concentrations of organic nitrogen was attributed to the low rate of hydrolysis of organic nitrogen compounds.

Contact Stabilization. Contact stabilization plants with very short contact times cannot be expected to completely nitrify except under the normally impractical conditions of very high recycle rates (for example, 4.0 or higher). At a rate of about 0.1 mg NH_4-N/mg MLSS·d, about 8 mg/L of NH_3 would be oxidized during one detention time. This limits the application of contact stabilization to situations where only partial nitrification is required.[16]

Only partial nitrification can be obtained in the contact stabilization process because of an insufficient number of nitrifiers in the biological mass or contact times that are too short. Because peak hydraulic and organic loads would produce very unfavorable conditions for nitrification, this concept is discouraged. Further, high BOD_5 removals required to produce nitrifying conditions in the reactor often cannot be attained. Operating plant results for nitrifying contact stabilization plants were not found in the literature.[142]

REFERENCES

1. "Needs Survey." U.S. EPA, Washington, D.C. (1984).
2. Ardern, E., and Lockett, W.T., "Experiments on the Oxidation of Sewage Without the Aid of Filters." *J. Soc. Chem. Ind.*, **33**, 523 (1914).
3. Eckenfelder, W.W., Jr., "Industrial Water Pollution Control." McGraw-Hill, Inc., New York, N.Y. (1966).
4. McKinney, R.E., "Mathematics of Complete Mixing Activated Sludge." *J. Sanit. Eng. Div., Proc. Am. Soc. Civ. Eng.*, **88**, SA3 (1962).
5. Lawrence, A.W., and McCarty, P.L., "Unified Basis for Biological Treatment Design and Operation." *J. Sanit. Eng. Div., Proc. Am. Soc. Civ. Eng.*, **96**, 757 (1970).
6. Ramanathan, M., and Gaudy, A.F., Jr., "Steady State Model for Activated Sludge with Constant Recycle Sludge Concentration." *Biotechnol. Bioeng.*, **13**, 125 (1971).

7. Gaudy, A.F., and Kincannon, D.F., "Comparing Design Models for Activated Sludge." *Water Sew. Works,* **123,** 66 (1977).

8. Bidstrup, S.M., and Grady, C.P.L., Jr., "SSSP—Simulation of single-sludge processes." *J. Water Pollut. Control Fed.,* **60,** 351 (1988).

9. IAWPRC Task Group on Mathematical Modeling for Design and Operation of Biological Wastewater Treatment, "Final Report: Activated Sludge Model." *IAWPRC Sci. Tech. Rep. No. 1.*

10. "Summary Report: Causes and Control of Activated Sludge Bulking and Foaming." EPA-625/8-87-012, U.S. EPA, Cincinnati, Ohio (1987).

11. Pike, E.B., and Curds, C.R., "The Microbial Ecology of the Activated Sludge Process." In "Microbial Aspects of Pollution." G. Sykes and F.A. Skinner (Eds.), John Wiley and Sons, Inc., New York. N.Y. (1972).

12. Gaudy, A.F., and Gaudy, E.T., "Microbiology for Environmental Scientists and Engineers." McGraw-Hill, Inc., New York, N.Y. (1980).

13. Grady, C.P.L., and Lim, H.C., "Biological Wastewater Treatment—Theory and Applications." Marcel Dekker, Inc., New York, N.Y. (1980).

14. McCarty, P.L., "Energetics of Organic Matter Degradation." In "Water Pollution Microbiology." R. Mitchell (Ed.), John Wiley and Sons, Inc., New York, N.Y. (1972).

15. McCarty, P.L., "Stoichiometry of Biological Reactions." *Progress Water Technol.,* **7,** 157 (1975).

16. "Process Design Manual for Nitrogen Removal." U.S. EPA, Washington, D.C. (1975).

17. Water Research Commission, "Theory: Design and Operation of Nutrient Removal Activated Sludge Processes." Pretoria, S. Afr. (1984).

18. Dague, Richard R., "A Modified Approach to Activated Sludge Design." Paper presented at 56th Annu. Water Pollut. Control Fed. Conf., Atlanta, Ga. (1983).

19. Metcalf & Eddy, Inc., "Wastewater Engineering—Treatment, Disposal, Reuse." 2nd Ed., McGraw-Hill, Inc., New York, N.Y. (1979).

20. Bisogni, J.J., and Lawrence, A.W., "Relationships Between Biological Solids Retention Time and Settling Characteristics." *Water Res.,* **5,** 753 (1971).

21. Eckenfelder, W.W., Jr., "Comparative Biological Waste Treatment Design." *J. Sanit. Eng. Div., Proc. Am. Soc. Civ. Eng.,* **93,** SA6, 157 (1967).

22. "Design Manual for Municipal Wastewater Stabilization Ponds." U.S. EPA, Washington, D.C. (1983).

23. Viessman, W., Jr., and Hammer, M.J., "Water Supply and Pollution Control." 4th Ed., Harper and Row, New York, N.Y. (1985).

24. "Aeration, A Wastewater Treatment Process." Manual of Practice No. FD-13, Water Pollut. Control Fed., Alexandria, Va. (1988).

25. Boon, A.G., and Chambers, B., "Design Protocol for Aeration Systems—U.K. Perspective." In "Proceedings—Seminar Workshop on Aeration Systems Design, Testing, Operation, and Control." W.C. Boyle (Ed.), EPA-600/9-85-005, U.S. EPA, Risk Reduct. Eng. Lab, Cincinnati, Ohio, 99 (1985).

26. "Design Manual—Fine Pore Aeration Systems." EPA-625/189023, U.S. EPA, Cent. Environ. Res. Info., Cincinnati, Ohio (1989).

27. Young, J.C., et al., "Flow and Load Variations in Treatment Plant Design." J. Environ. Eng., 104, 289 (1978).

28. "Principles and Practice of Phosphorus and Nitrogen Removal from Municipal Wastewater." Soap and Detergent Assoc., New York, N.Y. (1989).

29. Hale, F.D., and Garver, S.R., "Viscous Bulking of Activated Sludge." Paper presented at the 56th Annu. Water Pollut. Control Fed. Conf., Atlanta, Ga. (1983).

30. Eikelboom, D.H., and van Buijsen, H.J.J., "Microscopic Sludge Investigation Manual." TNO Res. Inst. Environ. Hyg., Neth. (1981).

31. Jenkins, D., et al., "Manual on the Causes and Control of Activated Sludge Bulking and Foaming." Water Res. Comm., Pretoria, S. Afr. (1984).

32. Sezgin, M., et al., "A unified theory of filamentous activated sludge bulking." J. Water Pollut. Control Fed., 50, 362 (1978).

33. Buhr, H.O., et al., "Making full use of step feed capability." J. Water Pollut. Control Fed., 56, 325 (1984).

34. Air Products and Chemicals Inc., U.S. Pat. No. Re. 32,429.

35. Air Products and Chemicals Inc., U.S. Pat. No. Re. 4,488,967.

36. Lee, S.E., et al., "The Effect of Aeration Basin Configuration on Activated Sludge Bulking at Low Organic Loadings." Water Sci. Technol., 14, 407 (1982).

37. van den Eynde, E., et al., "Relation Between Substrate Feeding Pattern and Development of Filamentous Bacteria in Activated Sludge." In "Bulking of Activated Sludge: Preventative and Remedial Methods." B. Chambers and E.J. Tomlinson (Eds.), Horwood Ltd., Chichester, England (1982).

38. Casey, J.P., et al., "Non-Bulking Activated Sludge Process." U.S. Pat. No. 3,864,246 (1975).

39. Jenkins, D., "Selectors—Bulking Control Effective Contemporary Wastewater Treatment Processes." Workshop note, Dep. of Eng. Prof. Develop., Univ. of Wisconsin–Madison (1988).

40. Eikelbloom, D.H., "Identification of Filamentous Organisms in Bulking Sludge." Process Water Technol., 8, 153 (1977).

41. Eikelbloom, D.H., "Biosorption and Prevention of Bulking Sludge by Means of a High Floc Loading." In "Bulking of Activated Sludge: Preventative and Remedial Methods." B. Chambers and E.J. Tomlinson (Eds.), Horwood Ltd., Chichester, England (1982).

42. Chudoba, J., and Wanner, J., "Discussion of: The control of bulking sludge: From the early innovators to current practice." by Orris E. Albertson. *J. Water Pollut. Control Fed.,* **59,** 172 (1987).

43. Chudoba, J., *et al.,* "Control of Activated Sludge Bulking—II: Selection of Microorganisms by Means of a Selector." *Water Res.,* **7,** 1389 (1973).

44. Tomlinson, E.J., and Chambers, B., "Methods for prevention of bulking in activated sludge." *J. Water Pollut. Control Fed.,* **78,** 524 (1979).

45. van Niekerk, A.M., "Competitive Growth of Flocculant and Filamentous Microorganisms in Activated Sludge." Ph.D. thesis, Univ. of Calif., Berkeley (1986).

46. Daigger, G.T., *et al.,* "The design of a selector to control low-F/M filamentous bulking." *J. Water Pollut. Control Fed.,* **57,** 220 (1985).

47. Wheeler, M.L., *et al.,* "The Use of a 'Selector' for Bulking Control at the Hamilton, Ohio, USA, Water Pollution Control Facility." *Water Sci. Technol.,* **16,** 35 (1984).

48. Palm, J.C., *et al.,* "Relationship between organic loading, dissolved oxygen concentration and sludge settleability in the completely mixed activated sludge process." *J. Water Pollut. Control Fed.,* **52,** 2484 (1980).

49. Riddel, M.D.R., *et al.,* "Method for estimating the capacity of an activated sludge plant." *J. Water Pollut. Control Fed.,* **55,** 360 (1983).

50. Wilson, T.E., *et al.,* "Operating Experiences at Low Solids Retention Time." *Water Sci. Technol.,* **16,** 661 (1984).

51. Buchan, L., "Chapter 9—Microbiological Aspects." In "Theory, Design and Operation of Nutrient Removal Activated Sludge Process." Water Res. Commiss., Pretoria, S. Afr. (1984).

52. Wesner, G.M., *et al.,* "Energy Conservation in Municipal Wastewater Treatment." EPA-430/9-77-011, NTIS No. PB81-165391, U.S. EPA, Washington, D.C. (1977).

53. Barnhart, E.L., "An Overview of Oxygen Transfer Systems." In "Proc. of Seminar Workshop on Aeration System Design, Testing Operation and Control." EPA-60019-85-005, NTIS No. PB85-173896, U.S. EPA, Cincinnati, Ohio (1985).

54. "Aeration in Wastewater Treatment." Manual of Practice No. FD-13, Water Pollut. Control Fed., Alexandria, Va. (1988).

55. "Summary Report Fine Pore (Fine Bubble) Aeration Systems." EPA-625/8-85/-010, Water Eng. Res. Lab., Cincinnati, Ohio (1985).

56. "Design Manual—Fine Pore Aeration Systems." EPA-623/189023, U.S. EPA, Office Res. Develop., Cent. Environ. Res. Info., Risk Reduct. Eng. Lab., Cincinnati, Ohio (1989).

57. "Hoffman Air and Filtration Systems, Centrifugal Compressor Engineering." 3rd Ed. (1986).

58. Streeter, V.L., and Wylie, E.B., "Fluid Mechanics." 7th Ed., McGraw-Hill, Inc., New York, N.Y. (1979).

59. "Aeration in Wastewater Treatment." Manual of Practice No. 5, Water Pollut. Control Fed., Alexandria, Va. (1971).

60. "Standard-Measurement of Oxygen Transfer in Clean Water." Am. Soc. Civ. Eng., New York, N.Y. (1984).

61. Brown, L.C., and Baillod, C.R., "Modeling and Interpreting Oxygen Transfer Data." *J. Environ. Eng.,* **108,** 607 (1982).

62. Boyle, W.C., *et al.,* "Oxygen transfer in clean and process water for draft-tube turbine aerators in total barrier oxidation ditches." *J. Water Pollut. Control Fed.,* **61,** 1449 (1989).

63. "Development of Standard Procedure for Evaluating Oxygen Transfer Devices." EPA-600/2-83-102, U.S. EPA, Cincinnati, Ohio (1953).

64. Doyle, M., and Boyle, W.C., "Translation of Clean to Dirty Water Oxygen Transfer Rates." In "Proceedings of Seminar/Workshop on Aeration System Design, Testing, Operation and Control." EPA-600/9-85-005, NTIS No. P85-173896, U.S. EPA, Cincinnati, Ohio (1985).

65. Stenstrom, M.R., and Gilbert, R.G., "Effects of Alpha, Beta, and Theta Factors on Specification, Design and Operation of Aeration Systems." *Water Res.,* **15,** 643 (1981).

66. Mueller, J.A., and Boyle, W.C., "Oxygen transfer under process conditions." *J. Water Pollut. Control Fed.,* **60,** 332 (1988).

67. Bell, J.P., *et al.,* "Investigation of Volatile Organic Contaminants in Municipal Wastewater Treatment Systems." Ontario Ministry Environ., Can. (1988).

68. Roberts, P.V., *et al.,* "Modeling volatile organic solute removal by surface and bubble aeration." *J. Water Pollut. Control Fed.,* **56,** 157 (1984).

69. Treybal, R.E., "Mass Transfer Operations." McGraw-Hill, Inc., New York, N.Y. (1968).

70. Roberts, P.V., *et al.,* "Volatilization of Organic Pollutants in Wastewater Treatment: Model Studies." EPA-600/2-84-047, U.S. EPA, Cincinnati, Ohio (1984).

71. Namkung, E., and Rittmann, B.E., "Estimating volatile organic compound emissions from publicly owned treatment works." *J. Water Pollut. Control Fed.,* **59,** 670 (1987).

72. Chapman, D.T., "Final settler performance during transient loading." *J. Water Pollut. Control Fed.,* **57,** 227 (1985).

73. Chapman, D.T., "The influence of process variables on secondary clarification." *J. Water Pollut. Control Fed.*, **55**, 1425 (1983).

74. Mulbarger, M.C., *et al.*, "Activated sludge reactor/final clarifier linkages: success demands fundamental understanding." *J. Water Pollut. Control Fed.*, **57**, 921 (1985).

75. Cashion, B.S., and Keinath, T.M., "Influence of three factors on clarification in the activated sludge process." *J. Water Pollut. Control Fed.*, **55**, 1331 (1983).

76. "Clarifier Design." Manual of Practice No. FD-8, Water Pollut. Control Fed., Alexandria, Va. (1985).

77. Weber, W.J., Jr., "Physiochemical Processes for Water Quality Control." John Wiley and Sons, New York, N.Y. (1972).

78. Tekippe, R.J., "Critical Literature Review and Research Needed on Activated Sludge Secondary Clarifiers." Report prepared for MERL, Office Res. Dev., U.S. EPA, Washington, D.C. (1986).

79. Tekippe, R.J., and Bender, Jon H., "Activated sludge clarifiers: design requirements and research priorities." *J. Water Pollut. Control Fed.*, **59**, 865 (1987).

80. Riddell, M.D.R., *et al.*, "Method for estimating the capacity of an activated sludge plant." *J. Water Pollut. Control Fed.*, **55**, 360 (1983).

81. "Wastewater Treatment Plant Design." Manual of Practice No. 8, Water Pollut. Control Fed., Alexandria, Va. (1982).

82. Dick, R.I., and Ewing, B.B., "Evaluation of Activated Sludge Thickening Theories." *J. Sanit. Eng. Div., Proc. Am. Soc. Civ. Eng.*, **93**, SA4, 9 (1967).

83. Keinath, T.M., "Design and Operational Criteria for Thickening of Biological Sludges." NTIS No. PB-262 967, Springfield, Va. (1976).

84. Keinath, T.M., *et al.*, "Activated Sludge-Unified System Design and Operation." *J. Environ. Eng.*, **103** (EE5), 829 (1977).

85. Wilson, T.E., and Lee, J.S., "Comparison of final clarifier design techniques." *J. Water Pollut. Control Fed.*, **54**, 1376 (1982).

86. Daigger, G.T., and Roper, R.E., Jr., "The relationship between SVI and activated sludge settling characteristics." *J. Water Pollut. Control Fed.*, **57**, 859 (1985).

87. Keinath, T.M., "Diagram for designing and operating secondary clarifiers according to the thickening criterion." *J. Water Pollut. Control Fed.*, **62**, 254 (1990).

88. Wahlberg, E.J., and Keinath, T.M., "Development of settling flux curves using SVI." *J. Water Pollut. Control Fed.*, **60**, 2095 (1988).

89. Dittmar, D., "Secondary Sedimentation Evaluation Operating Data Review." Tech. Memo. Municip. Metro. Seattle, Wash. (1987).

90. Tekippe, R.J., "Activated Sludge Circular Clarifier Design Considerations." Paper presented at the 57th Annu. Water Pollut. Control Fed. Conf., New Orleans, La. (1984).

91. Parker, D.S., "Assessment of secondary clarification design concepts." *J. Water Pollut. Control Fed.,* **55,** 349 (1983).

92. Samstag, R.W., "Studies in Activated Sludge Sedimentation at Metro—Final Report." In-house Rep., Municip. Metro. Seattle, Wash. (1988).

93. Crosby, R.M., "Evaluation of the Hydraulic Characteristics of Activated Sludge Secondary Clarifiers." Rep. prepared for MERL, Office of Res. Develop., U.S. EPA, Contract No. 68-03-2782 (1984).

94. Anderson, N.E., "Design of Final Settling Tanks for Activated Sludge." *Sew. Works J.,* **17,** 50 (1945).

95. Crosby, R.M., *Clarifier Newsletter.* Newsletter, Crosby, Young and Assoc., Plano, Tex. (1983).

96. Camp, T.R., "A Study of the Rational Design of Settling Tanks." *Sew. Works J.,* **8,** 742 (1936).

97. Camp, T.R., "Sedimentation and the Design of Settling Tanks." Am. Soc. Civ. Eng. Trans., Paper No. 2285, 895 (1945).

98. Camp, T.R., "Studies of Sedimentation Basin Design." *Sew. Ind. Wastes,* **25,** 1 (1953).

99. Kalbkopf, K.-H., and Herter, H., "Operational Experiences with the Sedimentation Tanks of the Mechanical and Biological Stages of the Emscher Mouth Treatment Plant." *GWF Wasser/Zbwasser,* **125,** 200 (1984).

100. Parker, D., and Stenquist, R., "Flocculator–clarifier performance." *J. Water Pollut. Control Fed.,* **58,** 214 (1986).

101. Stukenberg, J.R., *et al.,* "Activated sludge clarifier design improvements." *J. Water Pollut. Control Fed.,* **55,** 341 (1983).

102. Esler, J.K., "Optimizing Clarifier Performance." Unpublished paper presented at the 57th Annu. Water Pollut. Control Fed. Conf., New Orleans, La. (1984).

103. Zirschky, J. "Intrachannel Clarification: State of the Art." Paper presented at the Field Evaluations of I/A Technol. Seminar Series, U.S. EPA (1986).

104. James M. Montgomery Consulting Engineers, Inc., "Intrachannel Clarifier System Evaluation Report." Rep. prepared for MERL, Office of Res. Develop., U.S. EPA (1987).

105. Ettlich, W.F., "A Comparison of Oxidation Ditch Plants to Competing Processes for Secondary and Advanced Treatment of Municipal Wastes." EPA-600/2-78-051, U.S. EPA, MERL, Cincinnati, Ohio (1978).

106. James M. Montgomery Consulting Engineers, Inc., "Technology Evaluation of Sequencing Batch Reactors." Rep. prepared for MERL, Office Res. Develop., U.S. EPA (1984).

107. Christopher, S., and Titus, R., "New Wastewater Process Cuts Plant's Cost 50 Percent." *Civ. Eng.,* **53,** 5, 39 (1983).

108. Tetreault, M.J., *et al.,* "Assessment of phased isolation ditch technology." *J. Water Pollut. Control Fed.,* **59,** 833 (1987).

109. Arora, M.L., *et al.,* "Technology evaluation of sequencing batch reactors." *J. Water Pollut. Control Fed.,* **57,** 867 (1985).

110. Goronszy, M., "Intermittent operation of the extended aeration process for small systems." *J. Water Pollut. Control Fed.,* **51,** 274 (1979).

111. Melcer, H., *et al.,* "Conversion of small municipal wastewater treatment plants to sequencing batch reactors." *J. Water Pollut. Control Fed.,* **59,** 79 (1987).

112. Guo, P.H.M., *et al.,* "Evaluation of extended aeration activated sludge package plants." *J. Water Pollut. Control Fed.,* **53,** 33 (1981).

113. Torpey, W.N., "Practical Results of Step Aeration." *Sew. Works J.,* **20,** 781 (1948).

114. Buhr, H.O., *et al.,* "Making full use of step feed capability." *J. Water Pollut. Control Fed.,* **56,** 325 (1984).

115. Budd, W.E., and Lambeth, G.F., "High Purity Oxygen in Biological Sewage Treatment." *Sew. Works J.,* **29,** 237 (1957).

116. Nelson, J.K., and Puntenney, J.L., "Performance comparison of the air and high-purity-oxygen activated sludge systems." *J. Water Pollut. Control Fed.,* **55,** 336 (1983).

117. Wilford, J., and Conlon, T.P., "Contact Aeration Sewage Treatment Plants in New Jersey." *Sew. Ind. Wastes,* **29,** 845 (1957).

118. "1986 Needs Survey Report." U.S. EPA, Office of Water, Washington, D.C. (1986).

119. "1984 Needs Survey [Computer] Tape." PB 85-172690, NTIS, Washington, D.C. (1984).

120. "A Perspective on Performance Variability in Municipal Wastewater Treatment Facilities." EPA-600/D-86/064, U.S. EPA, Washington, D.C. (1986).

121. "Technical Support Document for Proposed Regulations Under Section 304(d)(4) of the Clean Water Act, as Amended." PB-85-111397, U.S. EPA, Office of Water Program Operations, Washington, D.C. (1984).

122. "Inspector's Guide: To Be Used in the Evaluation of Municipal Wastewater Treatment Plants." U.S. EPA, Washington, D.C. (1979).

123. "A Comparison of Oxidation Ditch Plants to Competing Processes for Secondary and Advanced Treatment of Municipal Wastes." EPA-600/2-78-051, U.S. EPA, Washington, D.C. (1978).

124. "Performance of Activated Sludge Processes: Reliability, Stability, Variability." EPA-6500/2-81-227, U.S. EPA, Washington, D.C. (1981).

125. "Small Community Wastewater Treatment Facilities—Biological Treatment Systems." EPA Technol. Transfer, U.S. EPA, Washington, D.C. (1977).

126. "Estimate of Effluent Limitations to be Expected from Properly Operated and Maintained Treatment Works." PB-80-185515, U.S. EPA, Office of Water Program Oper., Washington, D.C. (1980).

127. "Technology Assessment of the Deep Shaft Biological Reactor." EPA-600/2-82-002, U.S. EPA, Washington, D.C. (1982).

128. "Manufacturer's Literature." Deep Shaft Technology, Inc., Simmons Group of Companies, Calgary, Alberta, Can. (1988).

129. Benefield and Randall, "Biological Process Design for Wastewater Treatment." Prentice-Hall, Inc., Englewood Cliffs, N.J. (1980).

130. Siralian, R., Thesis, Manhattan Coll., N.Y.

131. "Status of Oxygen-Activated Sludge Wastewater Treatment." EPA Technol. Transfer, U.S. EPA, Washington, D.C. (1988).

132. "Extended Aeration Sewage Treatment in Cold Climates." EPA-660/2-74-070, U.S. EPA, Washington, D.C. (1974).

133. "Post Construction Performance of Schreiber Counter-Current Aeration Facilities." EPA-600/2-87-089, U.S. EPA, Washington, D.C. (1987).

134. "Summary Report, Sequencing Batch Reactors." EPA-625/8-86/011, EPA Technol. Transfer, U.S. EPA, Washington, D.C. (1986).

135. "Analysis of Full-Scale SBR Operation at Grundy Center, Iowa [journal version]." EPA-600/J-87-065, U.S. EPA, Washington, D.C. (1987).

136. Irvine, R.L., and Ketchum, L.H., Jr., "Sequencing Batch Reactors for Biological Wastewater Treatment." *Crit. Rev. Environ. Control,* **18,** 225 (1989).

137. "Full-Scale Study of Sequencing Batch Reactors." EPA-600/2-83-020, U.S. EPA, Washington, D.C. (1983).

138. *Pollut. Eng.,* **24** (1989).

139. Austgen Bioget, San Francisco, Calif., Manuf. literature (1988).

140. Zimpro Passavant, Rothschild, Wisc., Manuf. literature (1989).

141. Deeny, K.J., *et al.,* "Evaluation of Full-Scale Activated Sludge Systems Utilizing Powdered Activated Carbon Addition with Wet Air Regeneration." Paper presented at the 61st Annu. Water Pollut. Control Fed. Conf., Dallas, Tex. (1988).

142. *Nutrient Control.* Manual of Practice No. FD-7, Water Pollut. Control Fed., Alexandria, Va. (1983).

Chapter 12
Attached Growth and Dual Biological Treatment

INTRODUCTION

Many years ago, it was observed that the growth of slime-producing organisms accompanied the self-cleansing of water in streams. The biological filter is one of several treatment processes developed in an attempt to identify an efficient and economical method for accelerating this naturally occurring phenomenon. Simply stated, the concept is to provide a surface on which the microbial layer can grow and expose this surface repeatedly to wastewater for adsorption of organic material and to the atmosphere for oxygen. Fixed film reactor technology and applications have been greatly expanded since 1977.[1] This experience, summarized in this chapter, will be further developed in a forthcoming WPCF design manual on fixed film reactors.

TRICKLING FILTERS. Early installations of trickling filters (TFs) used rectangular beds of rock media; presettled wastewater was sprayed from fixed nozzles onto the media surface. Automatic siphons were used with dosing tanks to provide rest periods. Circular beds with rotary distributors were later introduced to allow for continuous feeding.

Use of plastic and wood media has become standard practice to provide more surface area per unit volume and improved ventilation, thereby reducing clogging and odor problems. Synthetic media have been fabricated as assemblies of corrugated sheets that can be stacked or as small ring structures that are placed in a loose (random) packing configuration. Wood media are used in the form of stacked pallets.

Recent research suggests that adequate hydraulic flushing (rate and intensity) and forced ventilation are needed to provide maximum efficiency, control nuisance organisms, reduce odors, and eliminate excessive solids storage in the filter media. For optimum performance, TFs should be equipped with ventilation fans and electrically driven distributors. The improved hydraulic application modes and forced ventilation will provide a general improvement in TF performance.

Although the TF process is often referred to by terms such as biotowers and biofilters, this chapter mostly limits reference to the process to TFs.

ROTATING BIOLOGICAL CONTACTORS. Rotating reactors use a fixed film biomass on rotating media for biological treatment. The rotating medium, typically made from sheets of high-density plastic, provides a surface on which organisms grow and contact organic wastewater constituents and oxygen from the air. The rotating reactor carries a film of wastewater into the air. The wastewater trickles down the surfaces of the contactor and absorbs oxygen from the air. Organisms in the biomass remove both dissolved oxygen (DO) and organic materials from the wastewater film.

DUAL-PROCESS SYSTEMS. Dual or coupled biological treatment systems have a fixed film reactor and a suspended growth process. This combination results in a two-stage unit process that has unique design parameters; its treatment efficiency capabilities often exceed the individual performance of both parent systems. The activated sludge unit provides a variety of functions, including flocculation to improve clarification, removal of residual soluble 5-day biochemical oxygen demand (BOD5), nitrification, denitrification, and phosphorus removal to meet advanced wastewater treatment (AWT) requirements.

TRICKLING FILTERS

This chapter contains significant departures from the practices presented in an earlier Manual of Practice for Wastewater Treatment Plant Design.[1] The changes are a result of intervening design efforts, new data, and new process designs as well as use of TFs for AWT. TFs have historically been relegated to either roughing treatment or applications with effluent limits of 30 to 45 mg/L BOD5 and TSS. This historical categorization is reflected in today's federal regulatory standards that, as noted in Chapter 2, define secondary treatment equivalency for existing trickling filters (and stabilization ponds) at significantly lower performance levels than those allowed for suspended growth technologies described in previous chapters. However, successful conventional secondary and AWT applications are achievable with TFs but they require a better understanding of TF operation and required appurtenances. If proper design procedures are employed, TF performance equalling that of suspended growth systems can be achieved.

The following statements represent common, but inaccurate, assumptions about TF treatment capabilities:

- TFs are best suited for effluent qualities of approximately ≥30 mg/L BOD5 and TSS.
- TF effluent cannot be sparkling and bright like activated sludge effluent.
- TFs do not easily remove soluble BOD5.
- A high temperature loss through trickling filters occurs in colder climates.
- TFs are not efficient nitrification units.
- Natural ventilation is adequate.
- TFs should be dosed every 10 to 60 seconds.
- Recirculation is necessary for optimum performance.
- Sloughing cycles are normal and are not harmful to filter performance.
- TFs require more land than activated sludge treatment.

THEORY OF TRICKLING FILTER OPERATION. In principle, the TF process has the biomass attached to a fixed medium. Thus, recycling of settled biomass is generally not required. On the other hand, the activated sludge process involves aeration of wastewater and maintenance of a suspended microbiological culture by recycling settled sludge. The two

---

processes are similar because they both depend on biochemical oxidation of a portion of the organic matter in the wastewater to carbon dioxide and water. Remaining organic matter is incorporated or transformed into biomass; the energy produced is released as heat into the medium.

The settled primary treated or screened wastewater is applied to the filter medium through which the flow percolates. The surface of the media quickly becomes coated with a viscous, jelly-like, slimy substance containing bacteria and other biota. The biota remove organics by adsorption and assimilation of soluble and suspended constituents. For aerobic metabolism, oxygen is supplied from the natural or forced circulation of air through interstices in the filter media. Oxygen transfer may be direct or by diffusion through the liquid films.

After an initial start-up period, the microbial buildup may create an anaerobic interface with some of the filter media. This furthers the growth of facultative and possibly anaerobic organisms, especially if the accumulation of biomass is excessive. However, the aerobic organisms at the upper microbial surfaces provide the basic mechanism for organic removal and conversion. True anaerobic functions of hydrolysis and gasification are minimal or absent in a properly operating filter.

The quantity of biomass produced is controlled by the available food. The amount of biomass on the media surface increases as the organic load and strength increase until a maximum effective thickness is reached. This maximum thickness is controlled by physical factors, including hydraulic dosage rate, type of media, type of organic matter, amount of essential nutrients present, temperature, and the nature of the biological growth. During filter operation, a portion of the biological slime sloughs off, either periodically or continuously. Generally, accumulation of excess biomass that cannot retain an aerobic condition impairs performance. Continuous and uniform sloughing, as measured by TF effluent TSS, provides an indication of a properly operating trickling filter.

Filter flies and snails are often encountered in TF plants. Filter flies may be prevented or controlled by designing TFs to allow flooding of the media— a simple way for operators to control nuisance organisms. Low-frequency dosing also helps control nuisance organisms as well as odors.

Significant populations of snails may form, plugging pump stations, causing plugging and excessive wear of pumps, and creating problems with other equipment in both liquid and solids processes downstream from the filters. One control procedure is to provide a depressed, low-velocity channel between the TF and the final settling tank, with bypass facilities to allow cleaning of the collected snails.

Primary settling is necessary ahead of stone filters to minimize problems with clogging but may not be required in plants where comminution and fine screening of the coarser raw wastewater solids occur and the medium consists of corrugated plastic or material with large void spaces. Screening units should remove particles with sizes greater than or equal to 3 mm if primary clarifiers are not employed. Adequate final settling must be provided downstream of TFs to collect biomass sloughed from the media.

Recirculation of the TF effluent is a convenient operational tool that has been used to improve filter efficiency in numerous applications. Increasing the total hydraulic flow rate provides improved distribution and reduces the likelihood of dry or partially wetted surface areas within the filter, thereby maintaining maximum treatment capability. In addition, a higher flow rate maintains shear forces sufficient to slough excess growth, thereby reducing clogging problems associated with solids accumulation. Finally, organic matter that may have missed exposure to the slime on its first passage through the filter may be treated on the second. However, this later aspect can be insignificant if the proper dosing procedure is used and media are fully wetted.

LOADING TERMINOLOGY. TFs are classified according to applied hydraulic and organic loadings. Hydraulic loading is the total volume of liquid, including recirculation, per unit of time per unit of filter cross-sectional area. Although hydraulic loading has been expressed in various units, the current practice has been to use $L/m^2 \cdot s$ or $m^3/m^2 \cdot h$ (gpm/sq ft), but $m^3/ha \cdot d$ (mgd/ac) is sometimes used. Organic loading is currently expressed as kg $BOD/m^3 \cdot d$ (lb/d/1000 cu ft) of filter media, as well as $kg/m^2 \cdot d$ [BOD_5 and chemical oxygen demand (COD)] and $g/m^2 \cdot d$ (nitrification) (lb $NH_4-N/d/1000$ sq ft) for process loadings and performance on a unit surface area basis.

When wastewater is recycled through the filter, organic loading calculations become more complex. General practice is to ignore the recycled organic load, but recycle flow can affect treatment efficiency, an important consideration when evaluating subsequent design criteria.

Before 1936, only low-rate (also called standard-rate) TFs had been constructed in the U.S. Design hydraulic loadings ranged from 19 000 to 37 000 $m^3/ha \cdot d$ (2 to 4 mgd/ac or 0.032 to 0.064 gpm/sq ft). The first high-rate TFs were constructed in 1936. Early investigators considered an applied hydraulic loading of at least 94 000 $m^3/ha \cdot d$ (10 mgd/ac) necessary to flush organic solids from the media and prevent clogging. This translated to an average rate of at least 0.39 $L/m^2 \cdot s$ (0.16 gpm/sq ft). Many filters were also designed to operate in a so-called intermediate hydraulic range, thus accounting for the historical use of the terms "low-rate," "intermediate-rate," and "high-rate" to describe types of TFs.

Some TFs, called "roughing filters," are high-rate filters receiving high hydraulic or organic loadings. Although these filters may provide a high degree of organic load removal per unit volume, their settled effluent still contains substantial soluble BOD_5 ($SBOD_5$) and solids related BOD_5 (TSS BOD_5). Roughing filters are used to provide intermediate treatment or as the first step of multistage biological treatment.

Development of plastic trickling filter media has led to the development of "super-rate" filters. These filters, used either as roughing units for strong wastewaters or as complete secondary treatment units, do not represent a significant departure from the roughing filters discussed above. Synthetic media filters are referred to as oxidation towers, biotowers, or biofilters.

The term BOD_5 as used throughout this section is intended to represent carbonaceous or nitrification-inhibited BOD_5 ($CBOD_5$) whether referring to total BOD_5 ($TBOD_5$), TSS BOD_5, or soluble BOD ($SBOD_5$). The $TBOD_5$ of a wastewater, including partially treated or effluent wastewater, is the sum of the $SBOD_5$ and the BOD_5 of the TSS. This summation can be expressed as follows:

$$TBOD_5 = SBOD_5 + TSS\ BOD_5 \tag{1}$$

or

$$= SBOD_5 + f(TSS)$$

Where

$$f = TSS\ BOD_5/TSS = \text{respiration of TSS expressed as } BOD_5 \tag{2}$$

The value of f varies from 0.2 to 0.9 mg/mg, depending on location in the treatment process, loading, and temperature.

Although the term "$SBOD_5$" is used, it usually represents the product of filtering the wastewater at 0.45 or 1.0 μm. The filtering media used for $SBOD_5$ determinations should always be designated because the resulting $SBOD_5$ values can differ significantly depending on the respective filter pore sizes. While the "Standard Methods" procedure for TSS uses a 1.0-μm filter,[2] many laboratories use a 0.45-μm filter to determine soluble fractions.

"BOD_{5R}" refers to removal of carbonaceous BOD_5, and NOD_R refers to the removal of nitrogenous oxygen demand or the oxidation of total Kjeldahl nitrogen (TKN_{OX}).

Typical hydraulic and organic loadings for various filter classifications are shown in Table 12.1.[3] Most shallower low-rate through high-rate filters use some recirculation—usually direct recycling of unsettled filter effluent through the filter. In such cases, the ratio of recycled flow to wastewater flow is known as the recirculation ratio. If the filters are 6.1 m (20 ft) or deeper and loaded at 0.5 kg $BOD_5/m^3{\cdot}d$ (31 lb/d/1000 cu ft), recycle may be minimal during the daytime but still essential to maintain wetting during low (night) flows. If two or more methods for recirculation through a given filter are used, the recirculation ratio represents the sum of the individual recirculation ratios. Recirculation methods are discussed in another section of this chapter, "Factors Affecting Performance and Design."

Some TF plants have been built to operate with two or more TF units in series. These plants are called two-stage or multistage TF plants if

Table 12.1 Historical classification of trickling filters.

Design characteristics	Low or standard rate[a]	Intermediate rate[a]	High rate[a]	Super rate[a]	Roughing
Media	Stone	Stone	Stone	Plastic	Stone/plastic
Hydraulic loading					
mgd/ac[b]	1–4	4–10	10–40	15–90	60–180
gpd/sq ft[c]	25–90	90–230	230–900	350–2100	1400–4200
Organic loading					
lb BOD_5/d/ac-ft[d]	200–600	700–1400	1000–5000	—	—
lb BOD_5/d/1000 cu ft[e,f]	5–15	15–30	30–150	≤300	>100
Recirculation	Minimum	Usually	Always	Usually	Not normally required
Filter flies	Many	Varies	Few	Few	Few
Sloughing	Intermittent	Intermittent	Continuous[g]	Continuous[g]	Continuous
Depth, ft[h]	6–8	6–8	3–8	≤40	3–20
BOD removal, %[i]	80–85	50–70	40–80	65–85	40–85
Effluent quality	Well nitrified	Some nitrification	No nitrification	Limited nitrification	No nitrification

[a] Obsolete terminology.
[b] mgd/ac × 9353 = m^3/ha·d.
[c] gpd/sq ft × 0.040 7 = m^3/m^2·d.
[d] lb/d/ac-ft × 0.36 = kg/1000 m^3·d.
[e] Excluding recirculation.
[f] lb/d/1000 cu ft × 1.602 = kg/100 m^3·d.
[g] May be intermittent up to a total hydraulic rate of between 0.7 and 1.0 gpd/sq ft.
[h] ft × 0.304 8 = m.
[i] Including subsequent settling.

intervening clarification is included. Two filters directly coupled in series and operated at the same hydraulic rates generally perform as if they were one unit of the same total depth, especially if they have forced ventilation.

Under current practice, distinctions are made among TF applications based on the treatment provided rather than the hydraulic rate or organic loading of the application. This approach more accurately identifies the purpose of the TF operation. Hence, the general types of TFs are

- Roughing filters that provide about 50 to 75% $SBOD_5$ removal and 30 to 45% $TBOD_5$ oxidation,
- Complete treatment filters that provide the required settled effluent BOD_5 and TSS,
- Combined BOD_{5R} and NOD_R filters that provide the required settled effluent BOD_5 and NH_4-N, and
- Tertiary nitrifying filters that provide required effluent NH_4-N in a tertiary mode with a clarified secondary influent.

Adequately sized final settling tanks are required to achieve proper effluent levels of TSS and BOD_5. Application of modern deeper clarifier designs with energy-dissipating center-feed wells and moderate overflow rates are keys to good effluent quality.

PROCESS APPLICABILITY. TFs have historically been considered acceptable secondary treatment for most wastewaters amenable to aerobic biological processes. They were considered capable of providing adequate treatment of domestic waste where required effluent limits were 20 to 45 mg/L BOD_5 and TSS. Technologies currently available can produce AWT effluents of 10 mg/L BOD_5 and TSS or less and NH_4-N effluents of 1 mg/L or less. Trickling filters have historically been considered vulnerable to climatic changes because wastewater droplets must be exposed to large volumes of ambient-temperature air. However, proper engineering design can reduce temperature losses caused by wind and ventilation to less than 1.5°C (2.7°F). Improving dosing procedures and minimizing recirculation can also help control temperature loss.

Temperature effects on nitrifying trickling filters are now considered to be less significant than those on activated sludge.[4,5] Earlier observations of poor effluent quality in winter were caused by a combination of shallow filters with high surface area, low freeboard, and high recirculation ratios that caused excessive heat losses. Other conditions contributing to poor performance included poor clarifier designs and filter dosing procedures that caused excess solids accumulations.

TFs are no longer viewed only as a process to produce secondary treatment effluent. The TF process now used for AWT produces low residual BOD_5, TSS, and NH_4-N. Replacing existing TFs is often more expensive than updating and expanding units using known process technology.

In applications where more stringent effluent quality standards have exceeded the capability of existing TF designs, expanding TF capabilities often meet the requirements. The full potential of the TF has yet to be realized. The expanded treatment capabilities of new facilities, along with inherent ease of operation and low power usage, have further widened the use of TFs.

In addition to more stringent effluent limits for organics and SS, total nitrogen and phosphorus removals are now required for many effluents.[6] Nitrification can be readily accomplished in a properly designed filter. In the past 15 years, some TF plants have been designed for nitrification either in combination with BOD_5 removal or in a tertiary mode. The selected mode of operation now depends more on economics than performance capability. Space constraints now seldom rule out TFs for nitrification because TF depths can exceed 12 m (40 ft).

Limited denitrification in nitrifying filters has been observed. Because the upper reaches of TFs are generally oxygen limited, recycle flow containing nitrates can be partially denitrified in this zone. However, specific concepts[7,8] to enhance denitrification in TFs have indicated a potential to achieve 80% denitrification or even more. Chapter 15 contains a discussion of specialized denitrification filters.

Phosphorus can be removed in TF plants by mineral addition (iron and aluminum salts) to the primary or secondary settling tank. Direct addition of metal salts to the TFs has been used, but it apparently offers no advantage compared with addition to the clarifiers. Ongoing studies in British Columbia, Canada,[9] indicate that a new phosphorus removal TF concept has some potential for success. A modified TF/activated sludge (TFAS) process[10] in Chapel Hill, North Carolina, is biologically processing 263 L/s (6 mgd) of wastewater and producing effluents containing less than 1 mg/L total P. Further innovations using TF technology appear likely.

FACTORS AFFECTING PERFORMANCE AND DESIGN. Numerous variables affect TF performance and, thus, the design of TF systems. Many of these variables have been studied and, in some cases, definite trends have been established. In other cases, however, difficulties in controlling interdependent variables not under study have made it impossible to draw definite conclusions concerning performance.

Design engineers must consider effluent requirements, wastewater composition, wastewater treatability, pretreatment–primary treatment, TF media type, TF depth, recirculation, hydraulic and organic loadings, ventilation, and temperature. All of these factors are interrelated and must be considered jointly for effective design. Finally, the important step of solids–liquid separation is critical to the overall performance of the plant.

Wastewater Composition. SBOD$_5$, TBOD$_5$, and TSS determinations are the principal measurements used in BOD$_5$ removal systems to assess both the strength of the applied wastewater and the quality of the settled filter effluent. These measurements are used with influent flow to calculate the organic loading applied to the filter as well as filter efficiency or filter and secondary clarification efficiency. For nitrification designs, designers must also consider the influent TKN (not only NH$_4$-N) and effluent NH$_4$-N.

Both the volume and strength of plant influent may exhibit significant hourly, daily, and seasonal variations. However, an advantage of the TF process is its ability to handle peak conditions without upset. For this reason a roughing filter preceding activated sludge is sometimes used to smooth BOD$_5$ loading to the activated sludge unit. Further, removal of soluble BOD$_5$ in the TF can control bulking in the activated sludge (AS) system—a capability demonstrated by many industrial coupled TF/AS systems.

In special cases, such as wastewater with strong industrial contributions, organic loading used for design calculations may be based on the assumption that the wastewater strength and flow as determined for an 8- to 16-hour period remain relatively constant throughout the 24-hour day. This is necessary in situations where the TF provides the total biological treatment of the wastewater but may not be necessary for operation in a roughing mode if the activated sludge retention time provides adequate flow and load stabilization.

Like any biological process, TF influent wastewater characteristics should be carefully documented. The frequency (percent time) versus loading (kg/d) of each pollutant and the flow should be defined. While a roughing operation or a 30/30 mg/L BOD$_5$/TSS effluent requirement may permit the use of 50th percentile design values (design loading would be exceeded no more than 50% of the time), AWT requirements generally call for influent loadings in the 70th to 85th percentile range. EPA regulations require plants to meet certain effluent criteria for the maximum 30-consecutive-day average concentration and 7-consecutive-day average concentration per year. These standards dictate plant performance and a plant must have a processing reliability up to its maximum monthly effluent restriction 92% of the time and processing reliability up to its maximum weekly effluent limitation 98% of the time. State standards may be more stringent. Refer to Chapter 3 for additional information.

Wastewater Treatability. For the TF process, the treatability of a particular wastewater depends in part on the ratio of suspended/colloidal concentration to soluble organic concentration. The TF process readily removes suspended and colloidal organics by the combined processes of biological flocculation, adsorption, and enzyme-complexing—not just through biological oxidation and synthesis. Soluble, small-molecule organics, such as simple sugars, organic acids, alcohols, and so on, are readily removed from the waste stream even in the short residence times TF provide. Therefore, the process efficiently treats industrial wastes having a high percentage of soluble, small-molecule organic material.

The treatability of a waste is defined by a rate coefficient. Simple organics such as sugars are easily and quickly removed; so, they have higher rate coefficients. Complex and inhibitory compounds, as well as nonflocculant suspended solids, degrade slowly and may pass through TFs, resulting in lower rate coefficients.

Amounts of pass-through TSS have not been rigorously evaluated. These amounts depend on wastewater characteristics, hydraulic and organic loading, and dosing practice. Similar conditions occur in high-rate activated sludge processes.

Pretreatment–Primary Treatment. The degree of pretreatment affects the performance and design of TFs. Grit and screening removal, chemical treatment, clarification, equalization, neutralization, prechlorination, and preaeration are pretreatment processes that can improve the performance of TFs. Consideration of both capital expenditures and operating costs is necessary as part of the evaluation of improved treatability. For domestic wastewaters, primary clarification without equalization has provided more than adequate pretreatment. In recent years, the increase in plastic and rubber floatables passing through primary clarifiers has resulted in use of fine screens to reduce media fouling in TF installations at South Bend, Indiana; Sausalito, California; and other locations.

Although primary clarifiers should be used with rock filters, fine screens (less than 3-mm openings) have been successfully used with vertical plastic media. Fine screens will remove 15 to 20% of TSS and 0 to 5% of BOD_5. Thus, without primary clarifiers, TFs must be enlarged to handle the additional organic loading, and must be periodically well flushed to prevent solids accumulation.

TF Media Type. The introduction of synthetic media for TFs has extended the ranges of hydraulic and organic loadings well beyond those of stone media. Table 12.2 compares physical properties of various types of TF media.

Two media properties of interest are specific surface area (surface area/unit volume) and percent void space. A greater specific surface area permits a larger mass of biological slime per unit volume; increased void space allows higher hydraulic loadings and enhanced oxygen transfer. The ability of synthetic media to handle higher hydraulic and organic loadings is directly attributable to the higher specific surface area and void space of these media.

Attached Growth and Dual Biological Treatment 685

Table 12.2 Comparative physical properties of trickling filter media.

Media type	Nominal size, in. × in.[a]	Unit weight, lb/cu ft[b]	Specific surface area, sq ft/cu ft[c]	Void space, %	Application[d]
Bundle	24 × 24 × 48	2–5	27–32	>95	C, CN, N
	24 × 24 × 48	4–6	42–45	>94	N
Rock	1–3	90	19	50	CN, N
Rock	2–4	100	14	60	C, CN, N
Random	Varies	2–4	25–35	>95	C, CN, N
	Varies	3–5	42–50	>94	N
Wood	48 × 48 ×1.875	10.3	14		C, CN

[a] in. × 25.4 = mm.
[b] lb/cu ft × 16.02 = kg/m^3.
[c] sq ft/cu ft × 3.281 = m^2/m^3.
[d] C = CBOD$_{5R}$; CN = CBOD$_{5R}$ and NOD$_R$; N = tertiary NOD$_R$.

Unlike rocks or wood media, the increase in slime thickness on plastic media reduces the aerobic biological surface area.[11] Thus, as the specific (clean or unused) surface area increases, the area used for aerobic growth begins to decrease. This point is considered to be about 88 to 105 m^2/m^3 (27 to 32 sq ft/cu ft) for BOD_5 removal (BOD_{5R}) and combined BOD_{5R} and NOD_R and 135 to 150 m^2/m^3 (42 to 45 sq ft/cu ft) for tertiary nitrification.

Although use of plastic media has extended the range of application for TFs, performance of existing rock filters may often be improved by modifying the distributor operation, improving or providing forced ventilation, and otherwise improving operating conditions. Deepening the filter and replacing the rocks with plastic media is often warranted if the rock media quality is poor, space is limited, or a major increase in plant size is planned.

Figure 12.1 illustrates typical media used in TFs. The media can be designated as rock (RO), horizontal wood slats (HO), random plastic nodules (RA), vertical fully corrugated bundles (VFC), and cross-flow media (XF). Vertical semicorrugated (VSC) media with alternate flat sheets in the bundle have been widely used. VFC and VSC media were the primary synthetic media from the late 1950s to the early 1980s. Since then, XF media has become the most popular type. VFC media are used for stronger wastes and more highly loaded filters.

Studies[12] indicate that XF is superior to VFC media at low BOD_5 loadings and inferior at high loadings. However, these conclusions are based on a conventional, high dosing frequency mode of operation. There is now evidence, as discussed later in this chapter, that suggests performance varies with dosing frequency. The relative performance capabilities of XF and VFC media remain in question.

TF Depth. In the U.S., rock filters are typically 1 to 2 m (3 to 6 ft) deep and occasionally 2.5 m (8 ft) deep. This depth limitation is associated with lack of adequate ventilation produced by natural draft as well as an increased tendency to pond. In Europe, rock filters are commonly used; units in Arnheim, Holland, are typical, having been constructed 4.9-m (16-ft) deep with power ventilation. Comparative data are lacking for deep, power-ventilated rock filters and shallow, natural draft filters.

Plastic media TFs are most commonly constructed between 5 and 8 m (16 and 26 ft) deep, although units up to 12.8 m (42 ft) deep exist. The limiting depth is associated more with the tower height aesthetics, serviceability, pumping requirements, and structural design of the media than with biological treatment efficiency. Increasing the depth of the filter is generally worthwhile to reduce the minimum flow required for wetting and, thus, reduce recirculation. In taller filters that have high loadings, oxygen deficiency may occur in the uppermost layers. However, adequate ventilation and hydraulic flushing prevent odor problems from developing.

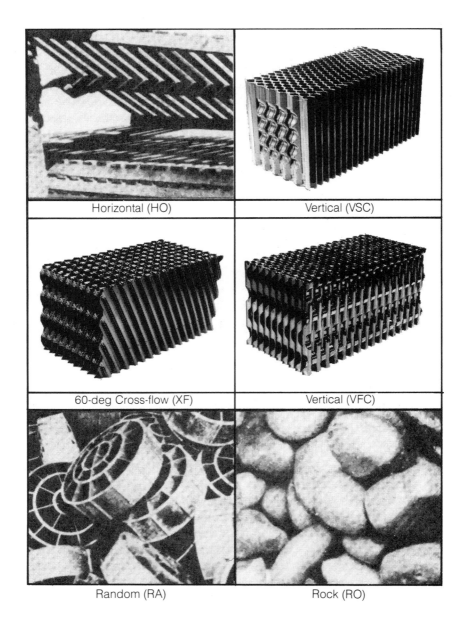

Figure 12.1 Types of commonly employed trickling filter media.

The effect of media depth on filter performance is a controversial topic. Some investigators[13-20] suggest that volume irrespective of depth controls performance, whereas others[21,22] consider that a fixed volume will perform better as depth is increased. Another investigator notes that rate coefficients vary among sites.[23] A recent model[24] predicts a minor increase in performance with deeper towers.

Most, if not all, of the improved performance with depth noted by some investigators may be a result of better hydraulic distribution. Deeper filters tend to have higher average hydraulic rates; studies[17,19] by Dow Chemical Co. indicate that the average hydraulic rate should exceed 0.51 L/m^2·s (0.75 gpm/sq ft) to ensure maximum performance. The results of these studies, illustrated in Figure 12.2, also demonstrate that the rate of BOD$_5$ removal is independent of depth from 3.2 to 12.8 m (10.5 to 42 ft). Results that indicate improved performance with increased media depth are generally based on a hydraulic rate for the shallow units below the recommended minimum wetting rates. The higher hydraulic rate typical for the deeper towers is likely the primary reason for their better volumetric performance.

The effects of hydraulics and media depth are critical to design and scale-up of results. These effects are further reviewed in subsequent discussions of design equations and process design.

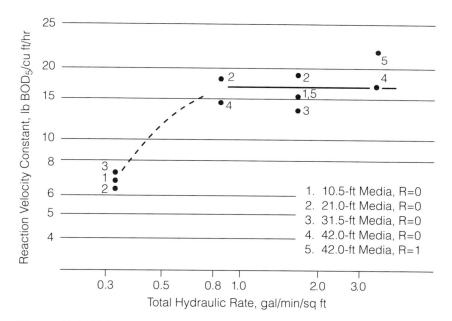

Figure 12.2   Effect of hydraulic application rate on BOD$_5$ removal rate for VFC plastic media (lb/cu ft/hr × 16.02 = kg/m^3•h; gal/min/sq ft × 0.679 = L/m^2•s).

TF Hydraulics and Loadings. Recirculation, recirculation arrangement, recirculation ratio, effect of recirculation, distributor operation, organic loading, and hydraulic and organic loading models are interrelated factors that affect TF design and performance.

Attached Growth and Dual Biological Treatment   689

RECIRCULATION. As an often important element in TF design, a portion of the TF effluent is recycled through the filter. This practice is known as "recirculation" and the ratio of returned flow to incoming flow is called the "recirculation ratio." Recirculation is an important element in stone filter design because of apparent increases in the BOD removal efficiency and is important in synthetic media filter design because it can ensure that the filter is adequately wetted. However, the instantaneous application rate resulting from modifying the distributor operation deserves equal consideration.

RECIRCULATION ARRANGEMENT. Many types of recirculation systems have been used in TF designs. Fourteen different arrangements are shown schematically in Figure 12.3, including those for both single-stage and two-stage filters. The most common single-stage flow patterns are (a) through (e).

Fluctuations in organic loading applied to filters are dampened by schemes shown in flow diagrams (a) through (d), (f), and (h). The magnitude

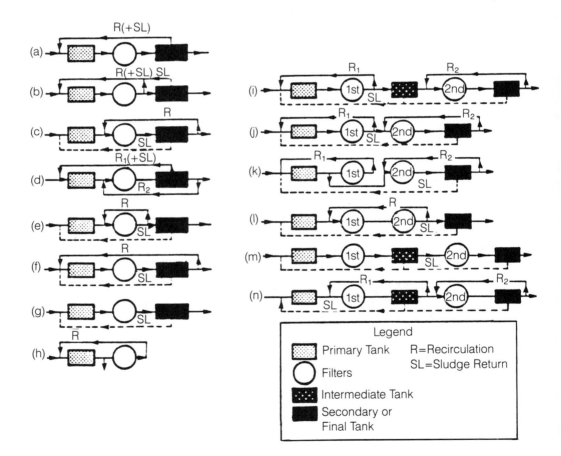

Figure 12.3   Flow diagrams of single- and two-stage trickling filter plants.

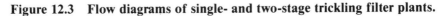

of dampening varies with the duration of organic load maxima and minima in the raw wastewater and with the retention time of the recirculated flow as it passes through the settling tanks.

In flow diagrams (a) and (d), secondary sludge combines with recirculation flow and returns to the primary tank through the same pumps. All other recirculation systems—except (h), which has no secondary settling tanks—require separate pumping arrangements for secondary sludge removal. Flow diagram (e) recycles filter sloughings to the TF, but this design provides little if any dampening of variations in the organic loading.

All flow diagrams except (e) and (g) recirculate flow through one or more settling tanks as well as through a filter. For this reason, recirculated flow must be taken into consideration in determining the capacity and other design parameters for these settling tanks. Flow pattern (g), with only sludge return, and flow pattern (h), without final settling, cannot provide the higher degrees of treatment associated with other patterns. Diagram (e) is similar to the usual flow diagram provided for low-rate filter plans. Flow diagram (d), which includes the recirculation features of both (a) and (c), has been used frequently.

A study comparing the performance of a number of single-stage plants concluded that recirculation systems (a), (b), and (c) produced equally good results.[25] Recirculation systems (c) and (e) were compared at a plant in Webster City, Iowa.[26,27] No significant differences in results were obtained when the filters operated simultaneously on the same settled wastewater under either winter or summer climatic conditions. Thus, researchers concluded that for stone media TFs, the amount of recirculation, not the arrangement, is the important variable. Because it is more cost effective, recirculation system (e) is preferable.

For two-stage filtration, flow diagrams (i), (j), (k), and (n) have been used most frequently. Flow diagrams (i), (m), and (n) may include a low-rate filter for the second-stage unit. In several instances, an overloaded low-rate filter has been relieved by pretreating the wastewater with a high-rate filter and intermediate settling tank. In such cases, the recirculation (R_2) through the low-rate filter usually has been limited to that necessary to maintain the minimum design flow.

Flow diagrams (j), (k), and (l), in addition to eliminating the intermediate settling tank, represent attempts to improve treatment by developing greater biological activity on the second-stage filter. In flow diagram (j), a part of the settleable sloughings from the first filter passes to the second-stage unit; in flow diagram (k), a part of the primary settling tank effluent is bypassed directly to the second-stage filter; and in flow diagram (l), the first-stage filter effluent is applied directly to the second filter and the recirculated flow is not settled. Studies[28,29] of TF performance on domestic wastewaters indicate that the absence of intermediate settling does not adversely affect the overall performance. However, certain wastewaters containing high proportions of industrial wastes with high concentrations of soluble organics were found to produce a large quantity of biological solids that could adversely affect the second-stage, low-rate TF if the solids were not removed by intermediate settling.

The recirculation pattern also influences the design of settling tanks in two-stage filter systems. Flow diagrams (i), (j), (k), and (n) require inclusion of the recirculated flow in the design flow for any tank through which it passes. Flow diagram (l) employs recirculation without requiring any increase in the capacity of settling tanks.

Based on the previous discussion of recommended flow diagrams, system (e) applies best for single-stage filters; (j), (k), and (l) may be considered favorably when designing a two-stage filter for domestic wastewaters.

Various recirculation pumping arrangements have been used. Probably the most frequently used consists of two or more pumps, operated by either automatic or manual control. Alternative pumping arrangements for recirculation include

- Pumping during low flows only,
- Constant rate pumping at all times,
- Pumping at rates inversely proportional to wastewater flow, and
- Pumping at two or more constant rates, either predetermined by automatic control or selected by the plant operator.

The final choice of flow pattern and recirculation arrangements is a matter of preference, taking into account the most economical method of ensuring an effluent of acceptable quality.

RECIRCULATION RATIO. Decisions on whether to use recirculation and what magnitude to use depend on comparisons of annual costs of various designs providing equal treatment. Where recirculation is used, recirculation ratios usually range from 0.5 to 4.0 but occasionally reach 10 or even higher for strong industrial wastewaters. However, a recirculation ratio greater than 4 does not materially increase the efficiency of the filters and is also generally uneconomical.[15] A formulation developed by the National Research Council (NRC) supports this conclusion. The NRC formula is discussed in a subsequent section of this chapter.

EFFECT OF RECIRCULATION. Recirculation in stone media TFs increases BOD removal efficiency for a variety of reasons:

- Organic matter in recycled filter effluent contacts the active biological material on the filter more than once. This increases contact efficiency and seeds the filter throughout its depth with a large variety of organisms.

- If the recirculated flow passes through a settling tank, it dampens variations in loadings applied to a filter over a 24-hour period. Because the strength of the recirculated flow lags behind that of the wastewater, recirculation dilutes strong wastewater and supplements weak wastewater. This helps maintain the filter in good condition during periods of fluctuations in loading.
- Recirculation through primary tanks tends to freshen stale wastewater and reduce scum formation. Also, continuous recirculation to the primary tank from the sludge hopper of the secondary settling tank removes sludge and reduces oxygen depletion in plant effluent.
- Recirculation improves distribution over the surface of filters, reduces the clogging tendency, and, if sufficiently high, aids in the control of filter flies. Therefore, providing recirculation where none exists frequently results in securing the desired degree of treatment with only a slightly higher operating cost.
- Finally, and perhaps most importantly, the increase in applied total flow increases the wetting efficiency.

The Dow Chemical Company studies summarized in Figure 12.2 demonstrate that VFC filter media require an average application rate in excess of 0.51 L/m$^2 \cdot$s (0.75 gpm/sq ft) to provide maximum efficiency. In the early years of plastic media applications, tower media depths usually exceeded 6 m (20 ft), and minimum wetting rates were often achieved without recirculation. Shallower towers required recirculation to provide minimum wetting rates. In recent years, shallow towers using XF media have used hydraulic rates of 0.11 to 0.29 L/m$^2 \cdot$s (0.16 to 0.44 gpm/sq ft); BOD$_5$ removal rates are much lower than those projected using the deeper tower data. This is reviewed further in the hydraulic design section because this significantly affects scale-up and sizing.

DISTRIBUTOR OPERATION. If the noted benefits of recirculation are mostly a result of increased wetting efficiency, the use of recycle may be unnecessary if distributor speed is significantly reduced below conventional, hydraulically driven rates[30] as discussed in more detail later in this section.

Typical distributor operation in the U.S. over the past 30 to 40 years used a rotational speed of 0.5 to 2 min/rev. With two or four arms, the filter is dosed every 10 to 60 seconds. Recent as well as historic evidence indicates that decreased dosing frequency from reduced rotational speeds can be advantageous in many situations. Such advantages were noted by some operators and researchers as long as 40 years ago but were not considered in design practices until recently.

In 1963, a summary[31] of full-scale operating results from studies[32-34] conducted from the late 1940s to the early 1950s demonstrated that rock TFs dosed every 30 to 55 minutes outperformed a more conventional operation of

---

Attached Growth and Dual Biological Treatment   693

1 to 5 min/rev. Not only did the filters perform better, but film thickness and nuisance organisms, such as filter flies, were controlled as shown in Figure 12.4. Odors were also controlled, probably because of the reduction in

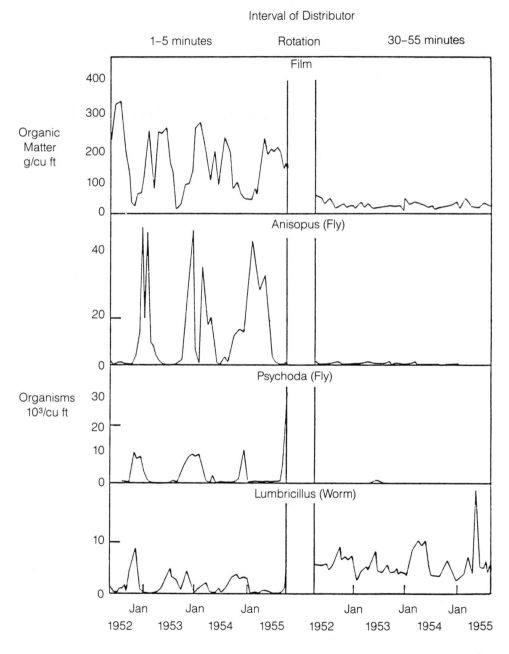

Figure 12.4 **Control of biomass growth and filter flies with low-frequency dosing** (g/cu ft $\times$ 35.31 = g/m^3; 10^3/cu ft $\times$ 35.31 = 3.5 $\times$ 10^4/m^3).

sludge accumulation. A summary[35] of work conducted in the 1960s also indicated that slowing the distributor reduces excessive biofilm storage, reduces odors, and likely improves operation of plastic media filters.

Because slowed distributor operation has benefitted facilities with both low and high flows and loadings, the reason for improvement may be associated with interrupted flow (periodicity of dosing) as well as flushing. Separating these effects is difficult.

Subsequent publications[36,37] provide further support for slowing the distributors to a fraction of the conventional speed and advocate use of the instantaneous dosing intensity (Spülkraft) approach,[30] commonly referred to as the *SK* concept. *SK* may be expressed as

$$SK = [(q + r) (1000 \text{ mm/m})] / [(a) (n) (60 \text{ min/h})] \tag{3}$$

Where

SK	=	flushing intensity, mm/pass of an arm;
q + r	=	average hydraulic rate, $m^3/m^2{\cdot}h$;
a	=	number of arms; and
n	=	rpm.

Although the average distributor in the U.S. operates in the range of 2 to 10 mm/pass, application *SK* values of 50 to over 500 mm/pass are now considered practical. This is based on the experiences of several U.S. facilities that have slowed their distributors and on long-term studies conducted in Great Britain.[32] The plants represented in Table 12.3 range from roughing filters to nitrification systems.[38,39]

Design engineers and operators of the plants cited in the table have observed better performance, a noticeable reduction of odors, and reduced or an absence of high solids sloughing cycles. The apparent conclusion is that the early work was correct but, unfortunately, was ignored for 30 to 40 years. These results also suggest that adoption of this mode of operation enhances performance of all types of media and significantly improves the general background data base for TFs. The initial 2 to 10 weeks of operating at high *SK* may be accompanied by higher TF effluent solids as the filter purges itself. BOD_5 values for this period are also generally higher as the excess (and usually anaerobic) sloughed solids are not as readily settleable.

In light of the above observations, available background data developed under less than optimum circumstances likely understate TF capabilities. This poses problems for design engineers when choosing design coefficients as discussed later.

The optimum *SK* is still to be defined and, to a degree, may be site- and application-specific. The current suggestions for *SK* are set forth in Table 12.4. These operating *SK* values may sometimes be achieved by hydraulic propulsion, but electric distributor drives offer the control necessary

Table 12.3 Wastewater treatment plants modified to high-SK operation.[38,39]

Plant name	Type of problem[a]	Plant flow, L/s[b]	BOD5 loading, kg/m²·d[c]	Filter size, number – diameter × depth, m × m[d]	Media type	Mode[e]	Averge hydraulic rate $(Q+R)$, m³/m²·h[f]	SK, mm/pass	Rotational speed, min/rev
Cedar Rapids, Iowa	1,2,3,4	1530	0.80	4 – 42.7 × 7.3	VFC	C & CN	1.8	112	15
Independence, Iowa[g]	1,2,3,4	75	0.70	2 – 16.2 × 6.7	RA	C	2.2	90	10
Huntingdon, Pa.	3,6̂	70	0.15	2 – 25.9 × 6.7	XF	CN	1.8	95	3
Clarksville, Ind.	1,3,6̂	110	0.27	2 – 20.9 × 7.3	XF	CN	1.0	40	4-5
"Baby Food," N.C.[g]	1,2,3	120	>5.0	1 – 25.9 × 6.1	VFC	RF	1.8	670	45
Cargill Inc., Iowa[g]	1,2,3,4	110	2.0	2 – 18.6 × 9.1	XF	RF	2.5	>620	>40
Hayward, Calif.	2	440	0.55	1 – 54.9 × 7.3	XF	C	0.67	47	17
Montgomery Co., Ohio	3,4	440	0.09	3 – 32.9 × 8.5	XF	CN	1.8	29	4
Auckland, N.Z.	1,2,3,4	3660	0.65	4 – 53.4 × 7.3	RA	C	2.9	230	28
Des Moines, Iowa	2,4	2060	0.72	12 – 46.3 × 2.1	RO	C	0.44	15	4
Wyoming, Mich.	1-4	570	0.21	4 – 59.5 × 2.1	RO	C & CN	0.21	27	30
Novato, Calif.	3,4,6̂	70	0.11	1 – 22 × 6.1	XF	CN	2.50	14	1.3
Central Valley, Utah	1,2,3	2190	0.53	3 – 51.8 × 4.3	XF	C & CN	2.0	100	12
Ojai Valley, Calif.[g]	1,4	80	0.65	3 – 12.8 × 4.9	RA	RF	3.2	133	5
Manteca, Calif.	3	200	1.1	2 – 16.8 × 6.1	XF	C	3.7	62	4

[a] Type of problem: 1 = fouling/plugging; 2 = odors; 3 = excessive sloughing cycles; 4 = performance reduction; 5 = filter flies; 6 = snails; and ^ = remaining problems.
[b] L/s × 0.022 8 = mgd.
[c] kg/m³·d × 62.4 = lb/d/1000 cu ft.
[d] m × 3.28 1 = ft.
[e] RF = roughing; C = CBOD5R; and CN = CBOD5R + NODR; N = NODR.
[f] m³/m²·h × 0.41 = gpm/sq ft.
[g] Electrically driven distributor.

Design of Municipal Wastewater Treatment Plants

Table 12.4   Suggestions for distributor SK rates.

BOD$_5$ loading, kg/m^3·d[a]	Design SK, mm/pass	Flushing SK, mm/pass
0.25	10–100	≥200
0.50	15–150	≥200
1.00	30–200	≥300
2.00	40–250	≥400
3.00	60–300	≥600
4.00	80–400	≥800

[a] kg/m^3·d × 62.4 = lb/d/1000 cu ft.

to identify the optimum *SK* and to provide for a high-*SK* periodic flushing of the filter media using timers and variable-speed drive arrangements as shown in Figure 12.5.

Initially, the wastewater treatment plant (WWTP) in Auckland, New Zealand,[39] used random media that fouled and caused bad odors while operating at an *SK* < 10 mm/pass. This plant is now capable of operating up to 250 mm/pass, although 125 to 150 mm/pass appears to enhance performance. Both the baby food facility (North Carolina) and the Cargill (Iowa) plant have the capability to operate at *SK*s of less than 100 mm/pass and between 500 and 1000 mm/pass. At the baby food facility, the *SK* was progressively increased until optimum performance was found at 680 mm/pass.

The electric drive unit at the baby food facility is shown in Figure 12.6. Operating at 45 min/rev ensures a high-intensity flush that minimizes solids retention in the media; maximizes aerobic surface area; and, as a consequence, probably enhances oxygen transfer.

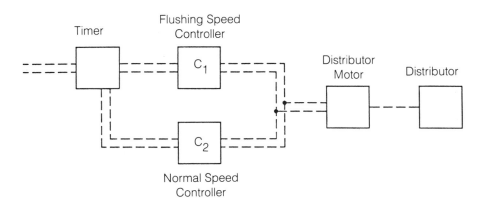

Figure 12.5   Electric drive control system for distributors using 24-hour repeat times.

Figure 12.6 **Electrically driven distributor with 4 to 70 min/rev capability at plant in North Carolina.**

In a discussion of oxygen transfer in fixed film media,[40] it was postulated that transfer was retarded by a relatively stagnant layer of water under the free-flowing surface layer. Figure 12.7 depicts this concept. If the stagnant layer were reduced by minimizing the biological mass and providing an interrupted flow, then better oxygen transfer could result. This further explains the observed improvement in performance using the infrequent high-intensity dosing of the *SK* concept.

The wetted surface area, A_w, divided by the clean surface area, A_s, is known as the wetting efficiency, A_w/A_s; this fraction is difficult to define. Media with poor wetting characteristics do not perform well. Research[41] indicates that wetting efficiencies of random media, slag, and gravel vary widely and that random media are not easily wetted. A technique to define the wetting efficiency of media has been developed[42] based on studies of the performance of lava gravel and two random media as well as literature data. The equation predicts that A_w increases asymptotically with the liquid superficial flow rate. A_w/A_s values ranged from 0.2 to 0.6, with the lowest values for a high-density random pack. These conclusions, similar to those of other studies,[41,43-46] provide a new avenue for evaluating media performance and the potential benefits of high *SK* operation.

Extensive studies[43] have been conducted in filters 4.9 m (16 ft) deep filled with 45- and 60-degree XF media with 89 m^2/m^3 (27 sq ft/cu ft) and 138 m^2/m^3 (42 sq ft/cu ft) of clean surface area, respectively. The media

---

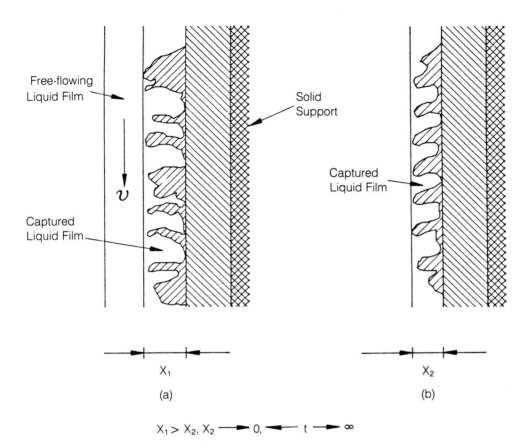

Figure 12.7   Effects of periodic dosing on oxygen transfer and slime thickness:
(a) dynamic conditions with liquid film and (b) static conditions after
flow interruption.

were dosed for 18 seconds at 60-second intervals at an average loading rate
of 0.57 L/m^2·s (0.84 gpm/sq ft). The investigators concluded that the surface
area of the media was poorly used. That is, biological rates indicated that
only 30 to 40% of the media area was being effectively wetted. The equivalent SK of these tests approximated 25 to 30 mm/pass. Earlier studies[44]
produced similar low-surface area utilization on VFC media evaluated at
Midland and Bloom Township. These findings agreed with the work of
others.[45]

Existing hydraulic propulsion-type distributors may be retrofitted with
electric drives easily, at low cost, and with minimal inconvenience. Designers of new plants should consider the value and opportunity offered by
remote variable-speed distributor drives. Benefits may include improved
performance, reduced odors, reduced power usage for recycling, reduced
nuisance organisms, and elimination of heavy sloughing cycles.

Distribution with fixed nozzles should not be used[46] because their efficiency is poor. Even distribution over the surface is not possible with fixed nozzles, and nozzle maintenance is a continuing operating problem.

ORGANIC LOADING. Trickling filters are designed for a wide range of organic loadings that sometimes, but not always, correlate with the hydraulic rate. BOD_5 loading rates may range from 0.08 to 8.0 $kg/m^3 \cdot d$ (5 to 500 lb/d/ 1000 cu ft), depending on the objective. The wastewater applied concentration is generally only a design criterion when BOD_5 concentrations exceed 400 mg/L. In such cases, dilution by recirculation is used. The minimum recycle (R) rate that has been used equals $BOD_5/400$, which results in a raw applied BOD_5 less than or equal to 400 mg/L. The origin of this design criterion is obscure, but it is included in design calculations conducted by Dow Chemical Company for plastic filter media in the early 1960s. Subsequent research[47] concluded that 420 mg/L BOD_5 is the maximum concentration based on oxygen availability and recommended the same basis for the recycle ratio.

Possibly, the above recycle requirement is a natural consequence of high-strength BOD_5 wastes producing thicker biomass.[48] The increased recycle would flush solids from the media as well as cause BOD_5 to further penetrate the media depth. Whether a high *SK* would accomplish the same objective at lower cost is unknown, but flexibility to investigate this mode of operation is warranted.

HYDRAULIC AND ORGANIC LOADING MODELS. Researchers[48-50] introduced the concept of first-order organic removal being directly proportional to filter media depth and inversely proportional to an exponential function of the hydraulic flow rate, either including or excluding recirculation. Subsequent work[51,52] indicates that recirculation, except where necessary to maintain optimum wetting rates, appears to have minimal effects on the efficiency of BOD removal in synthetic media filters.

A variety of researchers[14,15,20,24,35,53] have concluded that volumetric organic loading is the dominant criterion controlling performance. A recent design manual[54] for biological filtration supports that conclusion. On this basis, the formulations should indicate little or no effect of hydraulic loadings on performance. The conclusion of these researchers is consistent with the theory for activated sludge; that is, call retention, not liquid retention, is the dominant factor governing BOD_5 removal. The volumetric loading concept can be equated to kg BOD_5/kg MLSS·d (MLSS = mixed liquor SS) if the aerobic surface area, average aerobic biomass thickness, and average biomass solids concentration of the film are defined. Further, the solids retention time (SRT) of the filter may then be estimated based on sludge yields. In this concept, liquid flow (or hydraulic rate in the case of TFs) is not a fundamental process design factor. This concept is reviewed further in subsequent sections on design formulations and their applicability.

The significance of volumetric organic loading does not mean that hydraulic surface rate cannot be used in a model to define site-specific performance. However, it does mean that assuming that increased depth significantly reduces media volume would result in improper application of many of the design equations unless the treatability coefficient is modified for depth as discussed in the design formulations section.

The limited enhanced performance observed with increasing depth may be attributed to the general practice of having a higher total application rate, hence better wetting, in the deeper towers. Whether the SK concept could eliminate the ineffective wetting resulting from lower than optimum application rates in shallow towers remains to be determined by full-scale operations.

Ventilation. Ventilation of filters is essential to maintain the aerobic conditions necessary for effective performance. If enough passageways are provided, the difference in air and wastewater temperatures and humidity differences between ambient air and TF air provide a natural draft that often provides sufficient aeration. In many cases, however, these differences are minimal, resulting in inadequate air flow. These periods occur in the morning and evening, especially in the spring and fall. Thus, odors often occur during these periods as well as in warmer climates where temperature differences are smaller.

Good design practice requires provision of adequate sizing of underdrains and effluent channels to permit free flow of air. Passive devices for ventilation include vent stacks on the filter periphery, extensions of underdrains through filter side walls, ventilating manholes, louvers on the sidewall of the tower near the underdrain, and discharge of filter effluent to the subsequent settling basin in an open channel or partially filled pipes. However, these methods are inadequate if high performance is required or natural draft forces are low.

For municipal wastes, plastic media manufacturers have recommended 0.1 m^2 (1 sq ft) of ventilating area for each 3 to 4.6 m (10 to 15 ft) of TF tower periphery as well as 1 to 2 $\text{m}^2/1000 \text{ m}^3$ of media. The "ten states standards" require that inlet openings into the rock filter underdrains have an unsubmerged gross combined area equal to at least 15% of the surface area of the filter, and that the sizes of drains, channels, and pipe be sufficient to prevent submergence of at least 50% of their cross-sectional area under the design hydraulic loading.[53]

Currently, many designers consider forced ventilation of TFs to be unnecessary. However, optimum performance of TFs may not be attained if climatic conditions result in "turning off" the air twice a day for several hours during peak loading periods. Clearly, this is unsuitable for activated

---

sludge operation and, for the same reasons, the daily regime for TFs should avoid oxygen limitations. Reported increased odors from TFs during the more stagnant air flow periods offer ample evidence of the need for adequate ventilation, that is, powered ventilation.

Powered ventilation and enclosed filters provide other benefits: the filter may be used to destroy odorous compounds in the influent wastewater and to prevent excessive ventilation during winter, when periods of high air–water temperature differentials occur. TFs are used to deodorize gases from wastewater treatment units, composting operations, and pump stations.[55,56] The effectiveness of TFs in controlling odors in a downflow air flow mode has been documented at facilities in Cedar Rapids and Independence, Iowa. Many towers are ventilated in an upflow mode, including those at Tucson, Arizona; Stockton, California; Hayward, California; and Palo Alto, California.

A summary[57] of the literature regarding ventilation rates of TFs[40,47,58-60] concludes that TFs are generally operated in oxygen-limited environments. It has been noted that performance decreases in rock filters when air and water temperatures are identical and that the filter stagnates.[61] This stagnation occurs primarily in the highly loaded upper portion of the media and routinely depresses TF performance.

Few data exist for air flow through TFs. The differential air pressure providing air flow to the filter is minuscule as indicated by the natural drafts determined in Table 12.5. Whether the tower is 1.5 m (5 ft) or 6.1 m (20 ft) deep, the driving force is low. With temperature differentials between outside air and water of less than ±2.8°C (5°F), the tower can easily stagnate. Humidity differences also drive air flow, but ambient air values vary widely and are unpredictable. Continuous temperature measurements at Cedar Rapids, Iowa, reveal that temperature differentials between the air and water remain within ±1.1°C (2°F) for nearly 12 consecutive hours.

Recommendations for minimum air flows for design of power ventilation have been developed.[57] Downflow may reduce or eliminate biotower odors if incorporated with good flushing hydraulics. However, these recommendations indicate more information is required on the minimum air flow rates necessary to maximize TF kinetic rates. The recommendations for air flow based on O_2 demand in the tower, independent of depth, are

- Roughing TFs at 120 to 320 kg $BOD_5/100$ $m^3 \cdot d$ (75 to 200 lb $BOD_5/d/1000$ cu ft):

$$\text{Air flow (scfm)} = [\, (1200 \text{ scf/lb}) \, (0.9) \, (TBOD_5 \, SBOD_5 \text{ removed}, \quad (4)$$
$$\text{lb/d}) \, (PF) \,]/1440 \text{ min/day}$$

- Carbonaceous TFs at 40 to 80 kg $BOD_5/100$ $m^3 \cdot d$ (25 to 50 lb $BOD_5/d/1000$ cu ft):

$$\text{Air flow (scfm)} = [\, (1200 \text{ scf/lb}) \, (\text{TBOD}_5 \text{ SBOD}_5 \text{ removed}, \quad (5)$$
$$\text{lb/d}) \, (PF) \,]/1440 \text{ min/day}$$

- Combined BOD_{5R} and NOD_R TF and tertiary NOD_R:

$$\text{Air flow (scfm)} = [\, (2400 \text{ scf/lb}) \, (1.25 \text{ BOD}_{5R} + 4.6 \text{ TKN}_{OX} \quad (6)$$
$$\text{lb/d}) \, (PF) \,]/1440 \text{ min/day}$$

Where

Air volume of 1200 scf	=	21.3 lb O_2,
PF	=	peaking factor, and
$\text{TBOD}_5 \text{ SBOD}_5$ removed	=	influent TBOD_5 effluent SBOD_5.

Temperature. The effect of temperature on rock TF performance has been represented as both minimal and significant. The following conclusions have been drawn based on examination of the records of 17 plants in Michigan:[61]

- The winter efficiency was significantly lower than that during summer.
- Plants that recirculated flow during winter had more cooling and lower efficiency than those without recirculation.
- Both winter and summer performance were unaffected by organic loading.
- Filter efficiency was affected in plants without recirculation when air and wastewater temperatures were about the same. Air stagnation was the probable cause.

A combination of several conditions could explain these observations. Because of their shallow depths, rock filters have a high exposed surface area-to-volume ratio, and the low hydraulic rates increase surface heat loss. Most rock filters do not have 1 to 2 m (3 to 6 ft) of freeboard wall necessary to reduce windage effects. Also, high recycle rates are common for rock filters. Further, the observations are based on TBOD_5 values in an era when clarifier designs were not optimal for TSS removal from TF effluents. Finally, false high TBOD_5 readings might have resulted from nitrification in the BOD_5 bottle as the CBOD_5 test was not used at that time.

Most rock filters do not have air flow control. Although a 3 to 5°C (5.4 to 9°F) difference in air-water temperature is usually sufficient for natural draft, winter conditions with a 20 to 40°C (36 to 72°F) difference in the air-water temperatures can produce air flow rates more than 10 times that required. This, coupled with the other cooling factors noted, severely depresses the liquid temperature, thus causing the lowered efficiency. The inadequate clarifier designs typical of the 1960s may be another factor complicating this interpretation of temperature effects. "Ten states standards" recommend side

Table 12.5   Relative densities of tower and ambient air and natural draft forces.

Time	Location	Environmental conditions			Molecular weight of ambient air	Air density, lb/cu ft[b]	$\rho_o - \rho$, lb/cu ft[b,c]	Draft, in. H2O/ft D[d]
		Temperature, °F[a]	Saturation, %	Humidity, lb/lb				
—	Tower	70	100	0.016 2	28.69	0.074 2	—	—
0900	Ambient	60	60	0.006 6	28.85	0.076 0	+0.001 8	+0.000 35
1100	Ambient	65	70	0.009 0	28.81	0.075 2	+0.001 0	+0.000 19
1300	Ambient	70	75	0.012 0	28.76	0.074 3	+0.000 1	+0.000 02
1500	Ambient	75	80	0.015 0	28.71	0.073 5	−0.000 7	−0.000 13
1700	Ambient	80	70	0.015 3	28.71	0.072 8	−0.001 4	−0.000 27
1900	Ambient	70	80	0.012 8	28.75	0.074 3	+0.000 1	+0.000 22
2100	Ambient	60	80	0.008 8	28.82	0.075 9	+0.001 7	+0.000 33

[a] (°F − 32) (0.555) = °C.
[b] lb/ cu ft × 16.02 = kg/m^3.
[c] ρ_o = outside air density at temperature and humidity noted; ρ = saturated air density at water temperature of 21°C (70°F).
[d] Natural draft (in. H2O) per ft of tower depth $(D) = (\rho_o - \rho) \dfrac{12 \text{ in./ft}}{62.4 \text{ lb/cu ft}}$.

water depths of 2 to 2.5 m (6 to 8 ft) as well as average overflow rates of 1.7 m^3/d (1000 gpd/sq ft). Combined with the lower settling rates prevalent in colder weather, poorer performance is expected.

Researchers have determined that the temperature of the wastewater rather than that of the air was the controlling variable, particularly with high-rate filters.[62] Wastewater temperatures, except those affected by thaws and rains, rarely vary more than approximately ±3°C (5°F) during any day, and usually only from ±5 to ±7°C (9 to 13°F) throughout the year. Nonetheless, a TF in a northern state could have a wastewater temperature range of 5 to 24°C (41 to 75°F), whereas a southern facility could have a temperature range as great as 15 to 30°C (59 to 86°F) over a year. The researchers concluded that the effect of wastewater temperature becomes less important as the organic or hydraulic loadings increase.

The minimum ventilation recommendations would result in a temperature loss of only 1 to 1.5°C (1.8 to 2.7°F), even with 15°C (59°F) wastewater and air temperatures of −20°C (−4°F) if recycle is minimized and windage losses are controlled.

The effect of temperature on filter performance has been expressed by the following relationship:[63]

$$E_t = E_{20} \, \Theta^{t - 20} \tag{7}$$

Where

Θ	=	constant assumed equal to 1.035, ranging 1.015 to 1.045;
E_t	=	filter efficiency at temperature, t;
E_{20}	=	filter efficiency at 20°C; and
t	=	wastewater temperature, °C.

The range of Θ varied from 1.015 to 1.045 for carbonaceous BOD$_5$ removal; generally, the use of 1.035 produces a conservative design. Some investigators have shown[4,5] that temperature has relatively minor effects on nitrification systems; others have demonstrated significant temperature effects on nitrification rates.[64] However, the existing database is heavily influenced by natural draft results and oxygen availability. Also, first-order and zero-order nitrification occur in unknown and varying fractions of the tower volume. That is, if first-order removal occurs in a significant portion of the tower, then temperature is less significant.

DESIGN FORMULATIONS. Numerous investigators have attempted to delineate the fundamentals of the TF process by developing relationships among variables that affect TF operation. Analyses of operating data have been made to establish equations or curves to fit available data. Results of these data analyses have led to the development of various empirical TF formulations.

TF design formulations of major interest include those proposed by "ten states standards,[53]" NRC,[14] Velz,[49] Schulze,[48] a manual of practice from Great Britain,[54] Germain,[51] Eckenfelder,[65,66] Galler and Gotaas,[15] Kincannon and Stover,[20] and Logan.[24] Although these formulas represent attempts to include many of the variables that may affect TF operations, use of any one of these formulas is an approximation that does not universally predict the actual performance of TFs. On the other hand, each may model well the results of a specific installation or pilot test. Thus, designers need to assess which equation applies best to a particular situation, especially with regard to the confidence level necessary to meet discharge permit requirements.

NRC Formula. The NRC formulation[14] resulted from an extensive analysis of operational records from stone-media TF plants serving military installations. The NRC data analysis for stone-media TFs is based on two principles: the amount of contact between the filter media and that organic matter depends on the filter dimensions and the number of passes; the greater the effective contact, the greater the performance efficiency. However, the greater the applied load, the lower will be the efficiency. Therefore, the primary determinant of efficiency in a TF is the combination of effective contact and applied load. Organic loading influences filter efficiency more than does hydraulic loading.

For the 34 stone-media trickling filter plants selected for the NRC study, the efficiency curve best fitting a plot of the parameter "applied load/ effective contact area" (W/VF) is set forth below:

- First stage or single stage:

$$E_1 = 100/[1 + 0.008\ 5(W_1/VF)^{0.5}] \tag{8}$$

- Second stage:

$$E_2 = 100/[1 + (0.008\ 5/1 - E_1)\ (W_2/VF)^{0.5}] \tag{9}$$

Where

E_1 = BOD5 removal efficiency through the first-stage filter and settling tank, %;

W_1 = BOD5 loading to the first- or single-stage filter, not including recycle, lb/d;

V = volume of the particular filter stage (surface area times depth of media), ac–ft; and

F = number of passes of the organic material, equal to
$(1 + R/Q)/[1 + (1 - P)R/Q]^2$

Where

R/Q = recirculation ratio (recirculated flow:plant influent flow) and

P = a weighing factor that, for military TF plants, was found to be approximately 0.9;

E_2 = BOD$_5$ removal efficiency through the second-stage filter and settling tank, %; and

W_2 = BOD$_5$ load to the second-stage filter, not including recycle, lb/d.

Equations 8 and 9 are empirical but represent the average of the data for the rock-media TF plants, both with and without recirculation. Because of the nature of their development, the NRC formulas include several limitations and conditions:

- Military wastewater has a characteristically higher strength than average domestic wastewater.
- The effect of temperature on TF performance is not considered (most of the plant studies were in the midwest and southern latitudes of the U.S.).
- The clarifier practice of the period when the formulas were developed favored shallow units that were hydraulically loaded more than current practice allows.
- Applicability may be limited to stronger than normal domestic wastewater because no factor is included to account for differing treatability rates of lower strength wastewater.
- The formula for second-stage filters is based on the existence of intermediate settling tanks following the first-stage filters.

Figure 12.8 compares TF operational data for recirculation ratios of 0 to 2 with predicted values using the first-stage or single-stage NRC formula with a similar range of recirculation ratios.[14] This figure clearly shows that actual TF performance may deviate substantially from the removals predicted by the NRC formula. Data for loadings less than 0.3 kg/m^3·d (20 lb/d/1000 cu ft) could be biased by lack of a CBOD$_5$ test that inhibits nitrification in the BOD$_5$ bottle. Perhaps inadequate flushing, poor ventilation, or an inefficient clarifier design typical of the period could have contributed to poor performance data as well. Researchers[67,68] have demonstrated that improved clarification can significantly enhance rock filter performance. Thus, the foregoing variables should be accounted for when designing TFs based on the NRC formula curves of Figure 12.8.

British Manual of Practice. The British manual[54] provides a formulation for determining the BOD_{5R} in TFs for random media (RO and RA) and modular plastic media. The equation, produced by multiple regression analysis, follows:

$$L_e/L_o = 1/[1 + K\Theta^{(t-15)} (A_s{}^m/Q^n)]$$ (10)

Where

Θ	=	temperature coefficient;
L_e	=	influent BOD_5, mg/L;
L_o	=	settled effluent BOD_5, mg/L;
K	=	the first-order rate coefficient;
$A_s{}^m$	=	media surface area and coefficient, m^2/m^3;
Q^n	=	volumetric hydraulic rate and coefficient, $m^3/m^3 \cdot d$; and
t	=	wastewater temperature, °C.

The model was developed from data for a strong domestic wastewater with primary effluent concentrations of 360 mg/L BOD_5, 240 mg/L TSS, and

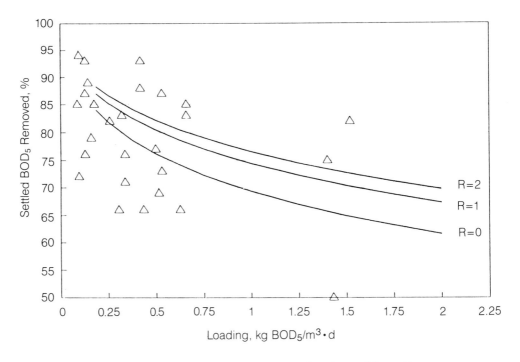

Figure 12.8 Comparison of trickling filter operating data with performance predicted by the NRC formula ($kg/m^3 \cdot d \times 0.062\ 4 = 1b/d/cu\ ft$).[14]

52 mg/L NH₄-N. The model predicts a continuous performance curve from low-rate through high-rate loadings. The test unit depths were 1.75 to 2.10 m (5.7 to 6.9 ft), surface areas were 1.0 to 5.0 m² (10.8 to 53.8 sq ft) and loadings were 0.3 to 16 m³/m³·d. The high sensitivity to temperature (high Θ value) may be due to site-specific wastewater characteristics and the data reduction procedures. The NRC equations agree with the British projection based on an influent strength of 360 mg/L BOD₅ at loadings up to 1.0 kg/m³·d (62.4 lb/d/1000 cu ft).

The British manual repeatedly mentions the need to operate the distributor at low speeds (10 to 30 min/rev) to optimize performance, control flies and odors, and minimize biofilm accumulation.

For equation 10, the reported coefficients, which account for 90% of the data variability, are set forth in Table 12.6. The equation represents a retarded form of the first-order relationship.

Table 12.6   Coefficients of the British model.

Coefficient	Random media, including stone	Modular plastics
K	0.020 4	0.400
Θ	1.111	1.089
m	1.407	0.732 4
n	1.249	1.396

Velz Formula. In 1948, Velz[49] proposed the first major formulation delineating a fundamental law, as contrasted to previous empirical attempts based on data analyses. This relationship is considered applicable to all biological beds—low-rate as well as high-rate TFs. The Velz formula indicates the TBOD remaining at depth D as follows:

$$L_D/L_O = 10^{-KD} \tag{11}$$

Where

$$
\begin{aligned}
L_O &= \text{influent BOD, mg/L;} \\
L_D &= \text{BOD}_5 \text{ removed at depth } D, \text{ mg/L;} \\
D &= \text{filter depth, ft; and} \\
K &= \text{first-order rate constant, day}^{-1}.
\end{aligned}
$$

The above formulation implies that K is constant for all hydraulic rates; however, evidence discussed later suggests that K varies with the hydraulic rate. For high-rate filters, the value of rate constant K_{20} was determined to be 0.150 5 day^{-1}. Low-rate trickling filters yielded an approximate K_{20} value of 0.175 day^{-1}.

Maximum loading of high-rate filters for any depth was estimated to be 4.9 kg/m^2·d (1 lb BOD$_5$/d/sq ft) at 30°C. Temperature was assumed to affect the rate of removal in accordance with the following expression:

$$K_t = K_{20}[1.047^{(t-20)}] \qquad (12)$$

Where

K_t	=	rate constant at any temperature (t), °C; and
K_{20}	=	rate constant at 20°C.

The temperature coefficient was higher than that generally used today. Others[11,48,69] have noted that greater biomass accumulations in the colder weather may reduce aerobic surface area and adversely affect performance. This would result in a higher Θ factor even though it is not related directly to temperature effects on biological activity. Further, the literature does not support the high Θ value of 1.047 predicted by Velz; therefore, other factors, such as poorer clarification in winter, may have influenced the determination.

The above equation does not consider hydraulic loading. For a specific surface area, such as rock media, the criterion would then translate to a volumetric or areal loading and, ultimately, to a mass-to-mass loading if the aerobic biomass quantity was known. Thus, the Velz equation followed the concept of the empirical NRC formulation.

Schulze Formula. In 1960, Schulze[48] postulated that the time of liquid contact with the biological mass is directly proportional to the filter depth and inversely proportional to the hydraulic loading rate. This is expressed as follows:

$$T = CD/q^n \qquad (13)$$

Where

T	=	liquid contact time, minutes;
C	=	constant;
D	=	filter depth, ft;
q	=	hydraulic loading rate, gpm/sq ft; and
n	=	constant, characteristic of the filter media.

Combining the time of contact with the first-order equation for BOD removal in an adaptation of the Velz theory, Schulze derived the following formula:

$$L_e/L_o = e^{-kD/Q^n} \qquad (14)$$

Where

L_e	=	BOD of settled filter effluent, mg/L;
L_o	=	BOD of filter influent, mg/L;
k	=	an experimentally determined rate constant between 0.51 and 0.76 day^{-1};
n	=	constant, characteristic of the filter media;
D	=	filter depth, ft; and
Q	=	hydraulic loading rate, mgd/ac.

This equation is similar to that proposed by Velz except that Velz's constant, K, was not formulated to consider hydraulic load, whereas Schulze's k does.

For a given wastewater strength, such as domestic primary effluent, hydraulic rate is proportional to the loading rate. Thus, volumetric loading may still be the controlling process variable. The value of k for a stone media filter with a 1.8-m (6-ft) depth at 20°C was determined by Schulze to be 0.69 day^{-1}. The dimensionless constant characteristic of stone filter media, n, was found to be 0.67. A temperature correction could be applied for k as follows:

$$k_t = k_{20} \left[1.035^{(t-20)}\right] \qquad (15)$$

Germain Formula. In 1965, Germain[51] applied the Schulze formulation to a plastic media filter as follows:

$$L_e/L_o = e^{-kD/q^n} \qquad (16)$$

Where

L_o	=	BOD5 of primary effluent fed to filter, excluding recirculation, mg/L;
L_e	=	settled effluent BOD5, mg/L;
D	=	filter depth, ft;
q	=	hydraulic dosage rate of primary effluent, excluding recirculation, gpm/sq ft;
n	=	exponent characteristic of filter media; and
k	=	treatability and media coefficient, gpm^n/sq ft.

The values of k and n are related to the media configuration, clarification efficiency, dosing cycle, and hydraulic rate; k is a function of wastewater characteristics, media depth, media surface area, and media configuration. Therefore, because a high degree of interdependency exists between k and n, this must be considered in data comparisons. Germain reported that the value of k for a plastic media filter 6.6 m (21.5 ft) deep treating domestic wastewater was 0.24 (L/s)^n/m^2 (0.088 gpm^n/sq ft) and that a value of 0.5 for n was appropriate. This VFC media had a clean surface area of 89 m^2/m^3 (27 sq ft/cu ft). Correction of the Germain k value for the high BOD$_5$ loading represented by the British Manual of Practice model, $k\left(\frac{150}{360}\right)^{0.5}$,

resulted in a close comparison at 20°C between the results of these two models for plastic media in the loading range of 0.2 to 1.5 kg/m^3·d (12.5 to 93.6 lb/d/1000 cu ft).

In tests designed to determine the effects of recirculation on BOD$_5$ removal, Germain found no statistically significant difference. However, the relatively tall 6.6-m (21.5-ft) tower resulted in high influent application rates, thus ensuring adequate wetting of the media. This conforms with the previously evolved practice of using recirculation for shallow filters where influent hydraulic rates are low and wetting efficiency would likely suffer.

The Germain equation is widely used for plastic media. The k data were developed from over 140 pilot studies by Dow Chemical and many more by other media suppliers. Most of these tests used a media depth of 6 to 7 m (20 to 22 ft).

Eckenfelder Formula. In the early 1960s, Eckenfelder,[65,66] following a previous model,[70] restated the earlier TF formulations to account for media surface area. The formula proposed[71] for SBOD$_5$ removal follows:

$$S_e/S_o = e^{(-K_sA_s^{1+m}D)/q^n} \qquad (17)$$

Where

S_o	=	influent SBOD$_5$, mg/L;
S_e	=	effluent SBOD$_5$, mg/L;
K_s	=	overall treatability coefficient based on SBOD$_5$, gpm$^{0.5}$/sq ft;
A_s	=	clean surface area, sq ft/cu ft;
D	=	depth of media, ft;
q	=	influent hydraulic rate, gpm/sq ft;
n	=	hydraulic coefficient; and
m	=	surface area modifier for surface loss with increasing area.

With recirculation, equation 17 was modified as follows:

$$S_e/S_o = (\exp - k_sD/q^n)/[(1 + R) - \exp - k_sD/q^n] \qquad (18)$$

Where

k_s	=	A_sK_s and
R	=	recycle ratio, R/Q.

Equation 18 rewritten as follows is known as the modified Velz formula:

$$S_e = S_o/[\{(R + 1)\exp k_{20}A_sD\Theta^{t-20}/[q(R + 1)]^n\} - R] \qquad (19)$$

The Velz, Schulze, Germain, and Eckenfelder equations are fundamentally the same and have similar limitations. Because the coefficients k (or K) and n must be empirically derived, background data are influenced by a host of variables, such as hydraulic rate, dosing mode, temperature, soluble waste fraction, biodegradability, media configuration, media depth, ventilation, and other unknown test-specific factors. Although the effective area, A_s^{1+m} (or A_s^m), is a valid consideration, the difficulty in accurately defining the term precludes its use.

The above equations have proven effective in modeling specific plant data, but when modifying the tower configuration the value of the so-called constant, k, generally changes for the same media and wastewater. The section on applicability of TF design equations explains this issue.

Galler and Gotaas Formula. In 1964, Galler and Gotaas[15] attempted to forecast the performance of stone media filters using multiple regression analysis of data from existing plants. Based on analysis of extensive data (322 observations), the following equation was developed:

$$L_e = [K(QL_o + RL_e)^{1.19}]/[(Q + R)^{0.78}(1 + D)^{0.67}a^{0.25}] \qquad (20)$$

Where

$\quad K \quad = \quad$ coefficient equaling the following expression:

$$[0.464(43\ 560/\pi)^{0.13}]/[Q^{0.28}t^{0.15}] \qquad (21)$$

$\quad L_e \quad = \quad$ settled filter effluent, BOD5 at 20°C, mg/L;
$\quad L_o \quad = \quad$ filter influent, BOD5 at 20°C, mg/L;
$\quad D \quad = \quad$ filter depth, ft;
$\quad Q \quad = \quad$ influent flow, mgd;
$\quad R \quad = \quad$ recirculation flow, mgd;
$\quad a \quad = \quad$ filter radius, ft; and
$\quad t \quad = \quad$ wastewater temperature, °C

The Galler and Gotaas formula recognizes recirculation, hydraulic loading, filter depth, and wastewater temperature as important variables for predicting the performance of a TF. Deeper filters performed better in their analysis. They further indicated that recirculation improves the performance of a filter but established a 4:1 ratio as a practical upper limit for recirculation. A high correlation coefficient of 0.974 was obtained using the Galler and Gotaas equation with the experimental data. The hydraulic flow rate through the filter was found to be unimportant in determining filter efficiency. The BOD5 loading correlated most closely with filter performance; that is, BOD5 loading controlled performance.

Kincannon and Stover Model. Kincannon and Stover[20] developed a mathematical model based on a relationship between the specific substrate utilization rate and the total organic loading, which followed a Monod graph. The relationship follows:

$$A_s = [(8.34 Q S_o / \mu_{max} S_o) / (S_o - S_e)] - K_b \qquad (22)$$

Where

A_s	=	total media surface area, 1000 sq ft;
Q	=	influent flow rate, mgd;
S_o	=	influent SBOD5, mg/L;
S_e	=	effluent SBOD5, mg/L;
μ_{max}	=	maximum specific substrate utilization rate of A_s, lb BOD5/d/1000 sq ft; and
K_b	=	proportionality constant of A_s, lb/d/1000 sq ft.

The biokinetic constants μ_{max} and K_b must be determined by pilot-plant tests, full-scale results, or prior experiences. They may be determined graphically by plotting BOD5 loading versus BOD5 removed^{-1}. The y-intercept is $\mu_{max}{}^{-1}$, and the slope is K_b. These investigators noted that variability in correlated data is normal, BOD5 removal is controlled by the volumetric loading and treatability, and BOD5 removal is not influenced by media depth.

Logan Model. The Logan model is an example of the new generation of TF models presently being evaluated. The computer model[24,72] was developed to predict SBOD removal in plastic media TFs as a function of media geometry. With other models, such as the Velz equation, new kinetic (k_{20} and hydraulic n) constants must be determined for each type of media used in a TF. Authors of the Logan model propose that the geometry of a single module can be measured and used in the model without the need for recalibration or performance of pilot studies for different types of plastic media.

The Logan model is based on characterizing the plastic module as a series of inclined plates covered with a thick (undefined) biofilm. The dissolved wastewater organics that exert the SBOD are equally distributed into a five-component, molecular-size distribution. The rate of SBOD removal is determined using a numerical model to solve the transport equations that describe the rate of mass transfer of SBOD components through the thin liquid film into the biofilm.

Although the model was calibrated with a single data set for only one type of plastic media, it was shown to predict SBOD removal in a variety of laboratory, pilot-plant, and full-scale TF studies. Unlike other models presented here, the equations that it is based on[24] cannot be solved in closed form, but the model is available as a computer program upon request.[72]

The model's developers determined that SBOD removal in TFs treating domestic wastewater was not limited by oxygen transfer to the biofilm.[72] Using data developed elsewhere,[73] they showed that SBOD removal rates could exceed the maximum rate of oxygen transport. Other researchers[74] have discussed the Logan model and questioned the author's premise that anaerobic activity was occurring in the biofilm when $SBOD_5$ removal exceeded oxygen transport rates. These researchers noted that a number of underlining assumptions required verification and disputed the position that no oxygen transfer limitations are present in TFs. Also, the underlying hydraulic flow modes have not been verified in the field.

TFs that exhibit anaerobic activity generally produce a poorer effluent in terms of $TBOD_5$ and TSS; they also emit odors. The ability of activated sludge biomass to remove $SBOD_5$ in the absence of oxygen is well known (that is, selectors), and this factor, if it applies to TFs, could cloud analysis of oxygen transport and $SBOD_5$ removal. This subject is worthy of further consideration.

Logan's model predicts that for a given volume of media, tower height has a minor effect on overall treatment efficiency, agreeing with the earlier conclusions that k (or K) of Velz, Schulze, and others must be modified or normalized with depth. For a specific set of operating conditions, the Logan model indicates BOD_5 removal would be slightly lower for a 3-m (9.8-ft) tower (71%) than a 6-m (19.7-ft) tower (73%).

The Logan model predicts an increase of SBOD removal proportional to an increase of wastewater temperature as a result of changes in physical parameters such as fluid viscosity, fluid film thickness, and oxygen solubility. As shown in Figure 12.9 for a 10-module tower filled with 98-m^2/m^3 (30-sq ft/cu ft) XF media at 0.68 L/m^2·s (1 gpm/sq ft), SBOD removal increases from 73 to 80% as wastewater temperature increases from 20 to 30°C (68 to 86°F). The Velz equation predicts a similar change of 73 to 84%, assuming the empirical relationship of $\Theta^{(t-20)}$, with Θ equal to 1.035.

APPLICABILITY OF TRICKLING FILTER DESIGN FORMULAS.
Design engineers may use various equations for sizing TFs. For rock filters, the NRC or Galler and Gotaas formulas are commonly employed and use of the Schulze formula also has merit. The Schulze equation is used for both rock and plastic media over wide ranges of media surface areas and depths. Thus, the coefficients k and n vary. [The word "coefficient" is used to describe k (or K) and n because they are neither constants nor treatability factors].

The Velz, Schulze, Germain, and Eckenfelder formulas are sometimes considered to be flow-rate dependent as compared to the organic loading methodology of the NRC, British, Galler and Gotaas, Kincannon and Stover, and Logan formulas. Within limits, both positions have support.

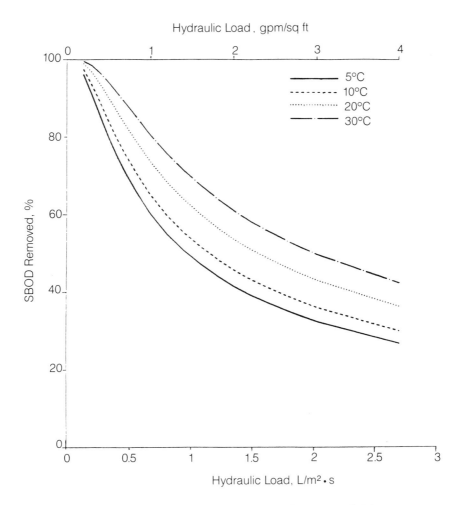

Figure 12.9 **SBOD removals predicted by the Logan model for a 10-module (6-m) trickling filter using 98-m²/m³ cross-flow (XF-98) media at various temperatures.**[24]

Researchers[13,16] have conducted simultaneous testing of two towers at the same flow and BOD5 loading but at a 4:1 ratio in the application rate because one tower was 7.4 m (24.3 ft) deep and the other was 2.1 m (6.1 ft) deep. Their data plot is shown in Figure 12.10. If these data were universally applicable, then BOD5 efficiency would be independent of depth. Thus, the value of k would be depth-dependent as follows:

$$k_2 = k_1(D_1/D_2)^{0.5} \qquad (23)$$

Initial studies[13] showed that performance varied as a function of depth. However, replacing a continuous application with intermittent dosing (higher SK) significantly improved performance and eliminated depth effects, probably by improving wetting efficiency.

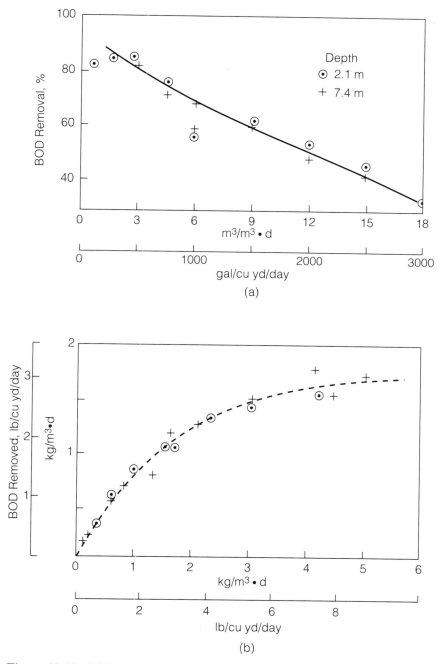

Figure 12.10 BOD$_5$ removals versus hydraulic and organic loadings of 89-m^2/m^3 VFC media at two media depths: (a) effect of hydraulic loading and (b) effect of BOD$_5$ loading.[13]

Simultaneous tests were also conducted by Dow Chemical Company[16] at Midland, Michigan, using 1.6-m (5.4-ft) and 6.6-m (21.5-ft) biotowers. The results of these studies (Figure 12.11) demonstrate that tower performance is controlled by organic loading, not hydraulic rate, and that the k value of the deep tower is exactly 50% of that for the shallow tower, that is, $(1.6/6.6)^{0.5} = 0.5$.

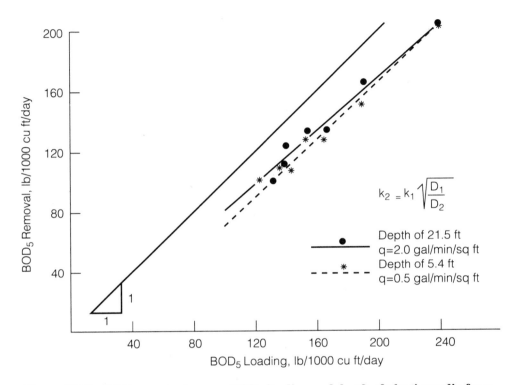

Figure 12.11 **BOD$_5$ removal versus BOD$_5$ loading and depth of plastic media from simultaneous loading studies at Midland, Michigan, plant (lb/1000 cu ft/day × 1.602 = kg/100 m^3·d; gal/min/sq ft × 0.679 = L/m^2·s).**

Similar tests conducted at plants located in Seattle,[18] Great Britain,[75] and Sedalia[19] indicated that k predictably varies with depth as set forth above. Some discontinuities in other data exist, as would be expected if inadequate wetting occurs. Perhaps the most extensive study of the depth issue was a Dow Chemical analysis of towers[16] from 3.2 to 12.8 m (10.5 to 42 ft) deep. These towers were analyzed at hydraulic rates of 0.85 to 9.8 m^3/m^2·h (0.35 to 4 gpm/sq ft). This study and supporting work derived a minimum wetting rate of 1.8 m^3/m^2·h (0.75 gpm/sq ft). As shown in Figure 12.2, depth had little or no effect on the BOD5 removal rate once the minimum wetting requirement was met. The study indicated some benefit from recycle at the highest organic loadings, probably related to oxygen supply.

The variation of k with depth is an important consideration because a k value developed for specific depth should not be employed for a different depth without modification to maintain the loading at the same value. Using data from a number of installations and simultaneous tests, research[35] showed that k could be effectively employed for any tower configuration if corrected for depth. However, this research also indicated that inadequate wetting might have produced lower k values than optimum in many studies.

In analyses of recirculation,[48,50,51] hence wetting rate, it was noted by some researchers that recirculation would theoretically offer little benefit. Nonetheless, these studies recognized the scientific consensus existing in 1960 that recirculation was beneficial due to more equalized loads, better distribution over the filter, less clogging, or possibly flaws in the theory. Benefits now observed from high-intensity dosing tend to support the 1960 consensus that recirculation removes wetting deficiencies, flushes excess solids, and improves oxygen transfer.

WASTEWATER TREATABILITY AND PROCESS COEFFICIENTS.
Variations in wastewater treatability play an important role in the design of wastewater treatment facilities. Discharge of appreciable quantities of industrial wastes to municipal sewers may call for special consideration in TF design. Problems associated with nondomestic wastewater discharges include increased organic loading, resistance of the waste to biological oxidation, and the possible presence of substances inhibitory or toxic to the biota on the filter.

Media surface area, media configuration, media depth, and the hydraulic application characteristics are all interwoven into the coefficients k and n. These values must be established to determine the required filter size.

During the early 1960s through the mid-1970s, Dow Chemical Company pilot studies of domestic and strong industrial wastes produced widely varying k values, some of which are shown in Table 12.7. The wide variations in k_{20} suggest that pilot studies are the only reliable means of determining k for industrial wastes or mixed domestic and industrial wastes. Unfortunately, much of this early work did not measure the SBOD5; thus, the k values are influenced by the settleability of the biotower effluent SS.

Stronger wastewaters, even though they are generally readily biodegradable, have lower k values, as shown in the table. The value of k_{20}, employing the Germain equation, is determined as follows:

$$k_{20} = \ln(S_o/S_e)\,(q)^{0.5}/D\Theta^{t-20} \tag{24}$$

Table 12.7 Germain k_{20} values from Dow Chemical Co. pilot studies and normalized k_{c20} values.

Type of waste/location	Flow, gpm/sq ft[a] Feed	Recycle	Depth, ft[b]	BOD5 load, lb/d/ 1000 cu ft[c]	BOD5, mg/L	BOD5R, %	Average temperature, °C	Actual k_{20}, gpm[0.5]/sq ft[d]	Normal k_{c20}, gpm[0.5]/sq ft[d,e]
Kraft mill, Ala.	0.57	0.71	21.5	51	160	76.2	37	0.028	0.030
	1.50	0.71	21.5	134	160	72.4	38	0.039	0.042
	2.00	2.00	21.5	179	160	64.6	38	0.037	0.039
	3.70	1.29	21.5	331	160	59.5	41	0.039	0.042
Cereal (dry mill), Ill.	0.29	0.71	21.5	155	956	75.2	32	0.023	0.060
	0.43	0.71	21.5	230	956	69.9	32	0.024	0.063
Meat packing (anaerobic pond effluent), N.C.	0.43	2.00	21.5	27	114	61.0	30	0.020	0.018
	3.00	2.00	21.5	204	122	41.0	30	0.030	0.028
Meat packing, N.C.	2.00	2.00	21.5	2089	1870	63.0	38	0.035	0.129
	6.43	0.00	21.5	5291	1473	42.0	38	0.035	0.112
Domestic, high industry, N.Y.	1.00	1.00	21.5	274	490	57.0	18	0.042	0.079
	2.00	2.00	21.5	547	490	39.0	18	0.035	0.065
Wet corn milling, Ind.	1.00	0.57	21.5	232	415	40.0	20	0.024	0.041
Synthetic dairy, Mich.	0.15	1.37	21.5	16	191	95.0	20	0.054	0.063
Synthetic dairy, Calif.	0.11	0.22	7.2	39	215	72.0	20	0.059	0.042
Tannery (pigskin), Mich.	0.29	1.00	21.5	118	727	84.0	33	0.029	0.067
Domestic, Ohio	1.10	1.80	9.0	364	248	47.0	21	0.071	0.062
Meat packing, Iowa	0.46	0.74	31.5	289	1645	69.0	32	0.017	0.069
	1.00	1.50	31.5	448	1175	67.0	32	0.023	0.082
	1.31	1.00	31.5	1144	2290	49.0	32	0.016	0.079
Frozen foods, Va.	0.60	0.00	21.5	488	1456	52.0	20	0.026	0.085
	0.65	0.00	21.5	453	1248	50.0	20	0.026	0.078
Pharmaceutical, Pa.	1.41	1.88	16.2	398	381	48.0	15	0.057	0.820

Table 12.7 Germain k_{20} values from Dow Chemical Co. pilot studies and normalized k_{c20} values (continued).

Type of waste/location	Flow, gpm/sq ft[a] Feed	Flow, gpm/sq ft[a] Recycle	Depth, ft[b]	BOD5 load, lb/d/1000 cu ft[c]	BOD5, mg/L	BOD5R, %	Average temperature, °C	Actual k_{20},[d] gpm$^{0.5}$/sq ft	Normal k_{c20},[d,e] gpm$^{0.5}$/sq ft
Fruit canning, Calif.	1.00	1.50	21.5	398	712	49.0	20	0.031	0.071
	2.00	1.00	21.5	1103	987	30.0	20	0.023	0.062
Sugar processing, Calif.	1.00	0.00	21.5	274	491	50.0	20	0.032	0.060
	2.50	0.00	21.5	666	477	36.0	20	0.033	0.061
Domestic, Mich.	0.50	1.50	21.5	49	175	86.0	17	0.072	0.080
	1.00	2.00	21.5	93	167	77.0	17	0.076	0.083
	2.00	0.00	21.5	162	145	71.0	17	0.090	0.092
	0.50	0.00	5.5	115	105	67.0	17	0.158	0.069
Domestic, Fla.	1.00	0.00	21.5	41	73	73.0	24	0.053	0.038
	2.00	0.00	21.5	79	71	62.0	24	0.055	0.040
	3.00	0.00	21.5	116	69	65.0	24	0.074	0.052
	4.00	0.00	21.5	179	80	54.0	24	0.063	0.048
Domestic light industry, Ga.	2.00	1.00	21.5	258	231	75.0	24	0.079	0.102
	1.00	2.00	21.5	109	196	80.0	24	0.065	0.077
Refinery, Calif.	2.00	0.00	21.5	80	72	60.0	40	0.030	0.022
	1.50	0.50	21.5	78	93	54.0	36	0.026	0.021
Potato processing, Idaho	0.72	2.00	21.5	861	2140	59.0	20	0.035	0.138
Textile mill, Va.	0.86	0.47	21.5	94	196	78.0	42	0.031	0.036
	1.50	0.71	21.5	195	233	76.0	42	0.038	0.049

[a] gpm/sq ft × 2.44 = m^3/m^2·d.

[b] ft × 0.304 8 = m.

[c] lb/d/1000 cu ft × 1.602 = kg/100 m^3·d.

[d] Corrected for temperature: $k_{20} = \dfrac{\ln\left(\frac{L_o}{L_e}\right)(q)^{0.5}}{(D)(1.035^{t-20})}$.

[e] Normalized to D = 6m (20 ft) and L_o = 150 mg/L: $k_{c20} = \dfrac{\ln\left(\frac{L_o}{L_e}\right)(q)^{0.5}}{(D)(1.035^{t-20})}\left(\frac{D}{20}\right)^{0.5}\left(\frac{L_o}{150}\right)^{0.5}$.

Note that ln (L_o/L_e) can be substituted for ln (S_o/S_e) to define TBOD5 removal. For two wastewaters with SBOD5 of 800 and 160 mg/L, each applied at the same volumetric loading and producing the same BOD5 reduction, the k value of the stronger waste will be $(160/800)^{0.5}$ or 45% of that for the weaker wastewater. The lower k values of stronger waste result from the higher organic loading of the industrial waste as affected by the 0.5 hydraulic coefficient. Therefore, industrial waste k values that are 30 to 50% of typical municipal waste values do not necessarily reflect reduced biodegradability; in fact, they often represent higher biodegradability.[76] This is reflected in the normalized k_{20} (k_{c20}) values shown in Table 12.7. The low k_{c20} values for weak domestic wastewater may be a result of a mixture of first- and zero-order rates as well as nonbiodegradable and nonremovable BOD5 because k is a direct function of ln (L_o/L_e). That is, the limiting or residual effluent value of L_e and S_e may be relatively constant with an L_o of 60 to 150 mg/L, and k must vary accordingly.

The value of the hydraulic coefficient varies widely. The empirical relationship noted in Figure 12.12 indicates that the value of n is related to the surface area of the media.[77]

Comparison of operating data is difficult if the values of n are allowed to vary. Perhaps this is the reason why engineers in recent years have tended to use a constant n value of 0.5 for plastic media, typical of the value in the NRC formulas. This recognizes the lack of any readily available means to

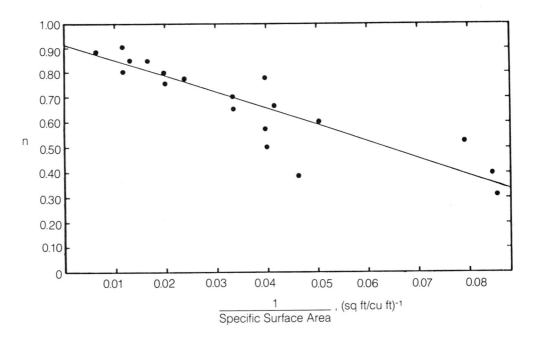

Figure 12.12   Relationship of n to the media surface area (sq ft/cu ft $\times$ 3.281 = m^2/m^3).

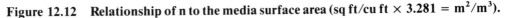

Design of Municipal Wastewater Treatment Plants

determine, or separate, individual effects of media, wastewater, and hydraulics on k and n. As demonstrated elsewhere,[143] the value of n also varies significantly with the dosing cycle. Using an n value developed from the intermittent dosing mode did not correlate with data using accepted equations. Thus, the evolved practice assumes that all synthetic media for carbonaceous BOD_5 removal with a surface area of 85 to 138 m^2/m^3 (26 to 42 sq ft/cu ft) have an n value of 0.5. This permits direct comparison of k values, wherein higher k values for the same operating conditions, waste strength, and media depth reflect better performance.

Rock media may be evaluated, assuming n equal to 0.5, even though actual n values are reportedly higher. The inherent value of n in the NRC formula is 0.5 as shown by the term $(W/VF)^{0.5}$. The performance values for BOD_5 removal determined by the NRC formula may be converted to k values in the Germain equation using a typical domestic primary effluent waste strength of 150 mg BOD_5/L. In this manner, the equivalent k values of the NRC formulation set forth in Table 12.8 are generated. This table offers some interesting insights regarding the NRC formula.

First, because of normally low wetting rates by influent flow, recirculation, as expected, produces better results, but the effects have limited significance when $R > 2$ as noted by Schulze. Increased application rates produce a higher k value. As shown earlier in Figure 12.2, this effect occurs until the average application rate exceeds 1.8 $m^3/m^2 \cdot h$ (0.75 gpm/sq ft) on plastic media. Similar effects are noted for rock media. When the filters are not effectively wetted, a lower k value is expected; this is well demonstrated in Table 12.8, which shows the projected increased NRC performance characteristic of higher flows and loadings.

Table 12.8   Germain equation k values implicit in the NRC formula.

BOD_5 loading, lb/d/1000 cu ft[a]	Hydraulic rate, gpm/sq ft[b]	NRC removals (E), %			Calculated k values,[c] $(gpm)^{0.5}/sq$ ft		
		$R = 0$	$R = 1$	$R = 2$	$R = 0$	$R = 1$	$R = 2$
15.6	0.064 7	81.9	85.3	86.7	0.072	0.081	0.086
31.2	0.129	76.2	80.5	82.2	0.086	0.098	0.103
46.7	0.194	72.3	77.1	79.1	0.094	0.108	0.115
62.3	0.259	69.4	74.4	76.6	0.100	0.116	0.127
93.5	0.388	64.9	70.4	72.7	0.109	0.126	0.135
124.6	0.518	61.6	67.3	69.8	0.115	0.134	0.144

[a] lb/d/1000 cu ft × 1.602 = kg/100 $m^3 \cdot d$.
[b] gpm/sq ft × 2.44 = $m^3/m^2 \cdot h$.
[c] $D = 1.83$ m (6 ft), $L_o = 120$ mg BOD_5/L, $n = 0.5$.

Several important points deserve consideration when developing and analyzing n and k values for TFs. These points, summarized below, allow a better understanding of the roles of n and k in the design and evaluation of TFs.

First, the value of n does vary for different media and dosing rates.[143] For data comparison and design convenience, a consistent value of $n = 0.5$ is suggested. This value conforms with the NRC, Velz, Schulze, and Germain equations and the recent Logan model. Second, without further definition, k_{20} measures neither wastewater treatability nor media effectiveness. A rock filter 1.8 m (5.9 ft) deep may have a 50% higher k_{20} value than a plastic media tower 6.1 m (20 ft) deep, but the rock filter may perform the same, better, or worse than the tower. Further, the strength of the waste impacts k because organic loading controls performance and k is a square root function of the influent application rate. Finally, the value of k is often controlled by the hydraulics of the TF, especially if the rate is less than 0.51 L/m^2·s (0.75 gpm/sq ft) and the dosing frequency is not optimized. Reported k values normalized to a media depth of 6.1 m (20 ft) are a function of total hydraulic rate as shown by Figure 12.13.[76]

The knees of the envelope curves in Figure 12.13 occur essentially where determined by Dow Chemical Company in the early 1960s (Figure 12.2); that is, at about 0.5 L/m^2·s (0.75 gpm/sq ft). This provides a further incentive to maximize hydraulic rates to enhance performance.

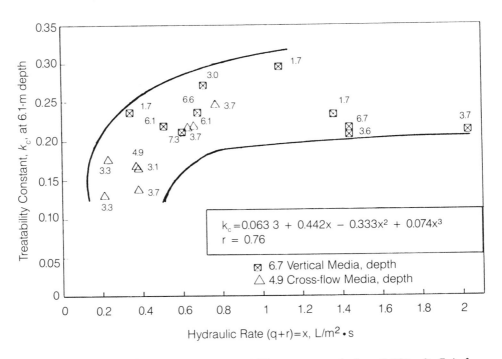

Figure 12.13   Hydraulic effect on treatability constant, k_c (m × 3.281 = ft; L/m^2·s × 1.473 = gal/min/sq ft).

Design of Municipal Wastewater Treatment Plants

As the impacts on k values have not been widely recognized, published k values have wide ranges that may not represent comparable results or even valid data. For the reasons discussed above, engineers have been advised against using tower configurations different from the test unit and have been encouraged to test industrial wastes to develop the k value. However, recognizing the effects of depth, waste strength, and hydraulics on k allows engineers to modify k if required to design towers of different configurations, as well as unify the database. Thus, the generalized or normalized[76] value of k_{c20} (represented as k_2 in the equation below) for a TF tower may be determined as follows from depth, waste strength, and k data for another tower:

$$k_2 = k_1(D_1/D_2)^{0.5}(S_1/S_2)^{0.5} \qquad (25)$$

However, this relationship is valid only where the biodegradability or treatability of waste S_1 equals that of S_2, either on a TBOD5 or SBOD5 basis. Equivalent treatability is defined as similar BOD5 removal at the same volumetric organic loading. For example, the normalized k_2 value may be determined as follows for a wastewater using a 6-m (19.7-ft) tower (D_2) and receiving a wastewater strength of 800 mg/L (S_2) based on tests conducted in a 3-m (9.8-ft) tower (D_1) with a waste strength of 350 mg/L (S_1) that produced a k_1 of 0.19 $(L/s)^{0.5}/m^2$ (0.07 $gpm^{0.5}$/sq ft):

$$k_2 = 0.19(3/6)^{0.5}(350/800)^{0.5}$$

$$= 0.089(L/s)^{0.5}/m^2 [0.033(gpm)^{0.5}/\text{sq ft}]$$

Both towers are volumetrically loaded at the same rate and have the same efficiency. The lower k_2 value represents the combined effects of a deeper tower and stronger wastewater. Optimum dosing for both towers is assumed in these calculations.

Prior conclusions in literature that compared k values from several tests must be reexamined in light of the above relationships. The absolute value of k is meaningless unless the media depth and waste strength have been defined. Further, the effluent TSS could further distort k_{20} values on a TBOD5 basis. Only after adjustments accounting for the above variables are made can the treatability and effects of media configuration be examined. The correction of k for waste strength has a limitation. At lower applied BOD5 concentrations, the removal efficiency will decrease (k decreases) as the effluent BOD5 remains relatively unchanged in the range of 15 to 20 mg/L TBOD5 and 6 to 8 mg/L SBOD5. This is reflected in Table 12.7 by the low k values for the domestic wastewater in Florida and refinery wastewaters.

Fixing n as 0.5 and establishing k_{c20} for domestic waste as equal to 0.203$(L/s)^{0.5}/m^2$ (0.075 $gpm^{0.5}$/sq ft) at a media depth of 6.1 m (20 ft) permits development of a family of curves showing k_{20} as a function of media depth and strength. At 6.1 m and 150 mg/L TBOD5, k_{c20} equals k_{20}. The curves of Figure 12.14 apply to wastewaters with biodegradability similar to that of domestic wastewater. As discussed above, the curves represent comparable loadings of towers with varying media depths and wastewater strengths. Failure to take media depth and waste strength into account could result in either oversizing or undersizing the TF tower.

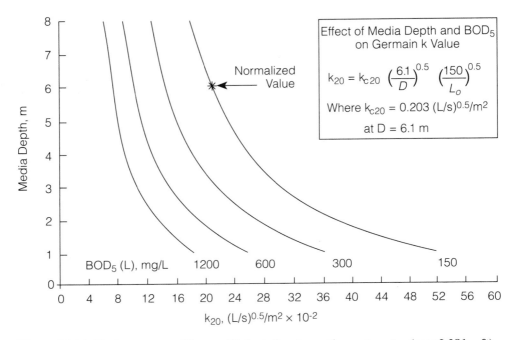

Figure 12.14   Design curves of k_{20} coefficients for domestic wastewater (m × 3.281 = ft).

One area of major concern remains. Shallow XF towers [less than 3 m (10 ft) deep] have not performed to the levels set forth above. Performance data presented for full-scale shallow XF media towers[68,78-80] have produced k values much lower than those set forth in Figure 12.14.

Although earlier studies of VFC media[13,16,17] did not show loss of efficiency with shallow media depths, the hydraulic rate, instantaneous dosing rates, or both were generally higher than those used for the XF designs. Currently, caution is advised in applying the k values set forth in Figure 12.14 for XF media depth less than 3 m (10 ft). A plot[81] of reported k data for XF media has indicated that $k_2 = k_1(D_1/D_2)^{0.3}$; this coefficient of 0.3, compared with that of 0.5 in equation 25, indicates that the shallow XF towers are less volumetrically efficient. Therefore, designers should carefully review the papers cited above before selecting or extrapolating the k value.

Recent practice of reducing the hydraulic application rate of shallow XF towers to 0.3 to 0.8 m³/m²·h (0.16 to 0.43 gpm/sq ft) might have caused the poorer performance. Better performance with higher hydraulic rates and higher SK values may be possible, but this improvement must be established by practice.

CLARIFIED EFFLUENT TOTAL SUSPENDED SOLIDS. The quantity of TSS and the endogenous respiration of these solids measured as BOD_5 influence the value of k based on $TBOD_5$. Endogenous respiration is a function of organic loading and temperature and, to a lesser degree, depends on wastewater characteristics and media configuration. Equations 1 and 2 express the components of $TBOD_5$.

The value of k_{20} in the Germain equation is derived from equation 16 as follows:

$$k_{20} = [\ln(L_o/L_e)\,(q^n)]/D\Theta^{t-20} \qquad (26)$$

Where

$$
\begin{aligned}
L_o &= SBOD_5 + f\,(TSS) \text{ of filter influent,} \\
L_e &= SBOD_5 + f'\,(TSS) \text{ of filter effluent, and} \\
n &= 0.5 \text{ for data correlation.}
\end{aligned}
$$

The $TBOD_5$ of filter systems with low loadings (low f') and good clarification (TSS <10 mg/L) is not much greater than $SBOD_5$. The k values resulting from the low TSS BOD_5 are generally high but understandable when considered in this light. Reported values of TSS BOD_5/TSS (f') as a function of BOD_5 loading were obtained from the Seattle study[18] and other reports. Figure 12.15 shows data variation without temperature corrections. The biotower efficiency, as well as the k value, depends on the clarification efficiency and the endogenous respiration of the effluent TSS. In a TF, as in an activated sludge plant, the "work" done by the TF is best described by "$TBOD_5$ in and $SBOD_5$ out."

The effluent quality of a 1.64-m (5.4-ft) TF, defined in terms of k at any temperature (k_t) versus TSS in the effluent, is shown as Figure 12.16. Because the results of Figure 12.16 were produced by a shallow, 89-m^2/m^3 VFC media TF, more significant effects of TSS on k_t are shown than would be expected from a deep tower. The $TBOD_5 - SBOD_5$ k value is 0.86 $(L/s)^{0.5}/m^2$ (0.32 $gpm^{0.5}$/sq ft). At 15 mg/L TSS, the k_t value is 0.43 $(L/s)^{0.5}/m^2$ (0.158 $gpm^{0.5}$/sq ft), or only 50% of the value at zero effluent TSS. This k_t, modified to a depth of 6.1 m (20 ft), would be 0.22 $(L/s)^{0.5}/m^2$ (0.082 $gpm^{0.5}$/sq ft)—consistent with k values for warm domestic wastewater. The equation for k_t versus TSS in Figure 12.16 differs for each facility. Until the k data are widely presented for $TBOD_5$ or $SBOD_5$ input and $SBOD_5$ output with the effluent TSS, data analysis and development of scale-up factors will be more difficult.

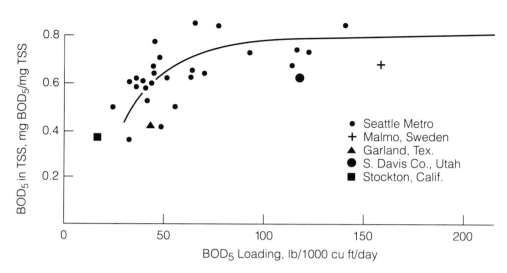

Figure 12.15 Relationship of BOD₅ contained in effluent TSS to organic loading (lb/1000 cu ft/day × 1.602 = kg/100 m³·d).

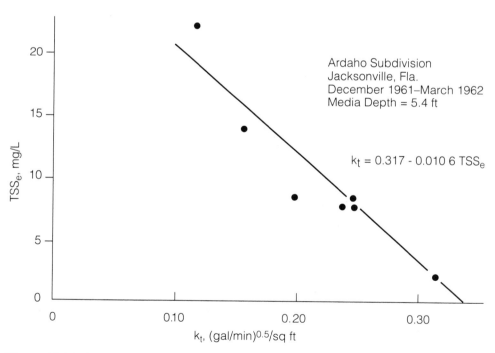

Figure 12.16 Effect of effluent TSS on k_t.

MEDIA SELECTION. General guidelines for media are set forth in Table 12.9. Design engineers should evaluate facilities similar to the plant being designed and conduct pilot studies where deemed necessary and

Design of Municipal Wastewater Treatment Plants

Table 12.9   Applications for various types of media.

Type	A_s, m²/m³ [a]	Roughing treatment	CBOD5 removal	Combined BOD5R and NODR	Tertiary NODR
Rock	40–60	[b]	X	X	
Wood	45	X	X	X	
Random	85–110		X	X	
	130–140				X
VFC/VSC	85–110	X	X	X	
XF	130–140				X
	85–110		X	X	
	130–140				X

[a] m²/m³ × 0.340 8 = sq ft/cu ft.
[b] Coarse media only, 35 to 45 m²/m³ (12 to 15 sq ft/cu ft), with depth of 1 to 1.5 m (3.3 to 4.9 ft).

beneficial. Although pilot-plant data in a number of tests have indicated that XF media are superior, full-scale operations have not always reproduced these findings.

All media that redistribute flow (RO, RA, and XF) are more prone to solids retention and fouling because of reduced flushing effects. Applications such as the treatment of strong wastes, pretreatment with fine screens, and BOD5 roughing tend to produce more biomass and a thicker biomass. Wood and vertical media types are preferred for these applications.

Media types may still evolve as the wetting and use of the media surface is better understood. Results[43,44] showing areal use of 40 to 50% for 89- and 135-m²/m³ XF media and similar levels for 89-m²/m³ VFC media may lead to a reevaluation of configurations focused on improving the wetting efficiency and, thus, surface area use. A study[42] of lava gravel at 114 m²/m³ and RA media at 95 and 220 m²/m³ found that the wetting effectiveness decreased with increasing surface area. The wetting effectiveness was found to be only 0.2 to 0.6.

A number of studies and installations have used high-density plastic media for CBOD5 removal. Analysis of the performance data[78–80] presented in the papers indicates that the k rates developed do not differ significantly from those of rock media set forth in Table 12.8. That is, many of the widely scattered data points indicate poorer performance than that predicted by the NRC formula curve for rock with $R = Q$. Use of high-density media with a surface area greater than 100 m²/m³ (30.55 sq ft/cu ft) is discouraged based on excessive levels of retained biomass and the larger surface area that generally causes poorer wetting. The increased risk of plugging is not offset by improved performance; in fact, less effective performance of 135-m^2/m^3 (42-sq ft/cu ft) media has been experienced. Technical studies support the hypothesis that wetting effectiveness decreases with increasing surface area.[42-44,63,82,83]

Generally, rock media are not used for many new installations; nonetheless, existing units may often be part of an expansion or upgrading program. Performance may be enhanced by modifying the distributor speed, power ventilation, or addition of solids contact, or by the dual biological process as described in the appropriate sections of this chapter on hydraulics, solids contact, and filter upgrading.

PROCESS DESIGN FOR CBOD$_5$ REMOVAL. With better understanding of the role of hydraulics, especially flushing intensity as defined by *SK*, a significant improvement in the database used for TF design may be anticipated in the future. Such improvement, coupled with forced ventilation and better clarifier configurations, will likely further advance TF capabilities. Although the degree of the improved performance cannot yet be determined, designs using historical coefficients (*k* and *K*) should prove to be conservative.

The need for forced ventilation is not as easily determined. However, if the dual objectives of the best possible effluent quality and the least possible odors must be achieved continuously, then forced draft ventilation is essential. Natural draft ventilation does not offer a reliable source of oxygen.

The NRC formula still provides a reliable means of designing rock filters. Improved hydraulic systems providing high-flushing intensity should control ponding, flies, and odors and likely enhance BOD$_5$ removal and solids separation. The *SK* values set forth in Table 12.4 should be used for design of rock filters, particularly if loadings exceed 0.3 kg BOD$_5$/m^3·d (20 lb/d/ 1000 cu ft) or if odors and excess sloughing become significant.

As shown in Table 12.8, the Germain equation with suitable modifications to *k* could also be employed for sizing rock filters. Although the *k* value for a constant depth increases with BOD$_5$ loading, this likely results from increased hydraulic flow and improved wetting efficiency. If the *k* values for rock media in Table 12.8 are used, the values of *k* must be adjusted for depths other than 1.83 m (6 ft) by equation 23 where D_1 equals 1.83 m (6 ft).

Properly irrigated and ventilated rock media can continue to perform as well or even better than is indicated by either the NRC or the "ten states standards" curves.

For plastic media filters, the Germain equation is used for design, recognizing that it represents a form of earlier equations. Use of the Germain coefficient *k* (for A_sK) is justified because of the lack of an adequate, properly compiled database to effectively separate the A_s and K terms. Much of the historical pilot-plant and full-scale data are impaired by lack of good hydraulics; that is, proper dosing intensity and rate. This partly explains why full-scale vertical media performance exceeds pilot results. For example, it has been reported that the volumetric efficiency of a full-scale biotower corresponded with a media volume only 60% of that predicted from a 1-year pilot study.[84]

The modified Velz equation first proposed by Eckenfelder is often used to define $SBOD_5$ removal efficiency. The beneficial effect of recirculation is reflected in both the modified Velz and NRC formulations. However, in both cases these benefits were generally derived from low application rates typical of standard-rate rock filters. Because the performance increase attributable to improved wetting is now understood, use of the modified Velz equation is unwarranted for the higher hydraulic rates common to plastic media. Researchers did not find any recirculation advantages beyond attainment of the minimum wetting level shown in Figure 12.2[26,27,51,52] and in the reported field test results of Figure 12.13.[76]

To properly and readily compare k values, the use of 0.5 for n is suggested. It has been shown that n changes with the dosing approach.[143] The literature values of n, which were derived from continuous flow studies, are likely invalid, but data will still correlate using an n of 0.5 for all types of media.

The domestic waste k_{20} values that were generated from the extensive studies by Dow Chemical Company on 89-m^2/m^3 (27-sq ft/cu ft) VFC media ranged from 0.176 to 0.244 $(L/s)^{0.5}/m^2$ (0.065 to 0.090 $gpm^{0.5}$/sq ft) when the media depth was 6.55 m (21.5 ft). This has evolved to a common design k_{20} value of 0.203 $(L/s)^{0.5}/m^2$ (0.075 $gpm^{0.5}$/sq ft) at 6.1 m (20 ft). This k_{20} value is used together with a minimum wetting rate of 0.51 $L/m^2 \cdot s$ (0.75 gpm/sq ft).

As tower depth decreases, recycle must increase to maintain the minimum wetting flow. In recent years, this criterion has been ignored in the replacement of rock media by XF media fewer than 4 m (13.1 ft) deep. In general, hydraulic rates have been 20 to 50% of the minimum wetting rate established by the earlier Dow studies. Units so designed and operated have lower efficiency[78-80] than expected. Poor wetting at the lower hydraulic rates is the possible reason, although the question of wetting effectiveness still remains to be answered.[42,44,63,82,83] Until this issue is resolved, designers must recognize that performance of XF biotowers in the range of 1 to 2.4 m (3 to 8 ft) deep will not likely exceed that of rock filters based on the NRC formula with $R = 1$ to 2.

Data available[35] for shallow VFC media towers did not reflect decreasing performance with depth. Better hydraulics (higher application rates) were present, but whether that accounts for the total observed difference between XF and VFC media is unknown.

Example 12.1. Determine the required size of an existing rock media filter expanded to meet a new BOD_5 effluent limit and, as an alternative, the required size of a new plastic media filter, given the following data:

L_o = 130 mg/L BOD_5,
L_e = 25 mg/L BOD_5, and
Q = 15 150 m^3/d (4 mgd).

Parameter	Rock filter	Plastic media filter
n	0.5	0.5
k_{20} at D, $(gpm)^{0.5}$/sq ft	NRC	0.075 at 20 ft
Diameter	150 ft	As needed
D, ft	5 ft, existing	16 ft
t, °C	15	15
Θ	1.035	1.035
q, gpm/sq ft	0.157	As calculated
R/Q	2	As needed
W, lb BOD_5/d	4337	4337
V, ac–ft	2.04, existing	As needed

SOLUTION. Step 1. Calculate required efficiency of existing rock filter to meet a limit 25 mg/L BOD_5:

$$E = \frac{(L_o - L_e)(100)}{L_o}$$
$$= \frac{(130 - 25)(100)}{130}$$
$$= 80.8\%$$

Step 2. Calculate the required volume of the rock filter at 80.8% efficiency using the NRC equation:

$$F = \frac{1 + R/Q}{[1 + (0.1)(R/Q)]^2}$$
$$= \frac{1 + 2}{[1 + (0.1)(2)]^2}$$
$$= 2.08$$

$$E = \frac{100}{1 + 0.008\,5\,(W/VF)^{0.5}}$$

$$V = \frac{W}{F}\left[\frac{0.008\,5}{\frac{100}{E} - 1}\right]^2$$

$$= \frac{4337}{2.08} \left[\frac{\frac{0.008\ 5}{100}}{80.8} - 1 \right]^2$$

$$= 2.72\ \text{ac–ft}$$

Step 3. Calculate the required depth of the expanded rock filter:

$$D = \frac{(2.72\ \text{ac–ft})\ (43\ 256\ \text{cu ft/ac–ft})}{(150\ \text{ft})^2 (0.785)}$$

$$= 6.66\ \text{ft; use 7.5 ft to allow for highly variable and non-temperature-correlated NRC data.}$$

Step 4. Select the K_{20} value from Figure 12.14 for a 4.9-m (16-ft) deep biotower using 89- to 100-m^2/m^3 plastic media:

$$K_{20}\ \text{at 16-ft depth} = (0.087\ \text{gpm})^{0.5}/\text{sq ft}$$

Step 5. Determine the hydraulic rate, q, for the biotower 16 ft deep and the corresponding tower diameter:

$$
\begin{aligned}
q &= [k_{20}\ D\ \Theta^{t-20}/\ln\ (L_o/L_e)]^2 \\
&= [0.087(16)\ (1.035^{-5})/\ln(130/25)]^2 \\
&= 0.50\ \text{gpm/sq ft}
\end{aligned}
$$

$$
\begin{aligned}
\text{Area} &= \frac{(4\ \text{mgd})(695\ \text{gpm/mgd})}{0.50\ \text{gpm/sq ft}} \\
&= 5560\ \text{sq ft} \\
\text{Diameter} &= [4(5560)/\Pi]^{0.5} \\
&= 84.1\ \text{ft (use 84 ft)}
\end{aligned}
$$

Step 6. Compare the designs for the two alternatives:

	Rock filter	Plastic media
Diameter, ft	150	84
Depth, ft	7.5	16
Volume, cu ft	132 500	88 700
Loading, lb/d/1000 cu ft	32.7	48.9
$q + r$, gpm/sq ft	0.47	0.75[a]
$q + r$, mgd	12	6
SK, mm/pass (Table 12.4)	15–300	15–300
Distributor (2 arm) speed, min/rev	1–20	1–20

[a] Minimum hydraulic rate.

The rock filter should be upgraded with an electric drive and forced ventilation, even if the existing unit did not have a history of odors or limited air distribution passages.

Rock filters can often be effectively and economically retrofitted with low-cost improvements. Powered distributors and ventilation are good practice for all TFs. In a later section, use of the flocculation/solids contact flowsheet is discussed. This process may significantly improve rock filter effluent quality without increasing media volume.[68]

PROCESS DESIGN FOR NITRIFICATION. The nitrification process is sensitive to the availability of oxygen and reportedly to high and low temperatures, the level of soluble organics, NH_4-N concentration, and perhaps media configuration and hydraulics of the TF. In recent years, TFs have been employed for nitrification, that is, oxidation of NH_4-N. This has been achieved for many years by standard-rate rock filters as well as by tertiary plastic media filters. Since 1980, use of plastic media to simultaneously remove $CBOD_5$ and TKN has expanded. These are usually single-stage units, even though the design may provide for two separate filters directly coupled in series.

The TF's inherent capabilities encourage its wider use because the process offers low energy consumption, stability, operational simplicity, and better sludge concentrating characteristics than do suspended film reactors. The TF's drawback is a lack of modern operating units providing a fully developed database. About 40 to 50 combined and tertiary nitrification facilities are in operation as of 1990. On the other hand, many more are in construction, design, and planning.

Analysis of the TF process is complex. Although it is simply a liquid plug flow-stationary biomass-reactor with backmixing as a function of recycle rate, little effort has been devoted to in-depth analysis of process reactions. Despite the several successful applications, the lack of intensive study has hampered design engineers and limited the expansion of technical understanding of the process.

Tertiary Nitrification. This chapter arbitrarily defines tertiary nitrification as the NH_4-N oxidation process in a biotower having an influent meeting two criteria: the BOD_5:TKN ratio ≤ 1.0 and $SBOD_5 \leq 12$ mg/L. Tertiary nitrification in plastic media biotowers was first reported by Duddles and Richardson[85,86] at Midland, Michigan, and later by Sampayo and Metcalf[87] at Lima, Ohio, in the period 1973–75. Parker et al. incorporated these pilot study results with those from tertiary nitrification studies at the Sunnyvale[88] plant into the EPA "Process Design Manual for Nitrogen Control."[89] Empirical relationships, developed and graphically displayed as curves in the manual, formed a design basis for several years after the manual was introduced in 1975.

Studies were also conducted at Bloom Township, Illinois,[90] using a pilot facility similar to that employed at Midland and Lima. The pilot unit was 0.91 m (3 ft) in diameter and filled with 6.55 m (21.5 ft) of 89-m^2/m^3 VFC plastic media. A rotary distributor provided nearly continuous dosing of the TF. These tests resulted in a series of empirical relationships between the rate of NH$_4$-N removal and the NH$_4$-N loading rate. Like the results of Midland and Lima, these tests demonstrated a strong rate dependency on temperature. Also, the rate of nitrification varied significantly, similar to that in activated sludge nitrification systems. However, empirical relationships developed from the early studies lacked a fundamental approach.

OXYGEN AND NH$_4$-N LIMITATIONS. Nitrification rates of suspended growth or fixed film processes are highly dependent on the available oxygen level. The rate of nitrification also depends on the concentration of NH$_4$-N. The zero-order NH$_4$-N range is reportedly the half-order region for oxygen.[59,91-93] Therefore, in the zero-order region of NH$_4$-N, a different nitrification rate will exist for each oxygen concentration. As an example, a high nitrification rate of 5 g/m^2·d was observed for an RBC operating at 29 mg/L DO in an oxygen atmosphere.[91]

The problem of establishing zero-order and first-order rates for nitrification is illustrated by data[94] shown in Figure 12.17. These data, obtained from five separate studies, indicate the range of ammonia removal expected for

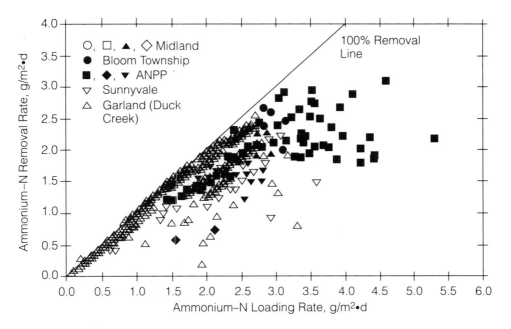

Figure 12.17   Area NH$_4$-N load versus the observed rate of NH$_4$-N removal (g/m^2·d × 0.204 8 = lb/1000 sq ft/day).

any given loading rate. The removal rate is consistent up to a loading rate of 1.2 g/m²·d, after which it becomes less predictable. These data were not corrected for temperature, and all test filters relied on natural draft ventilation. The authors concluded that oxygen availability became a dominant control variable for removal rates higher than a NH_4-N loading of approximately 1.2 g/m²·d.

The nitrification rate changes to first order when NH_4-N reaches lower concentrations and diffusion of NH_4-N, rather than O_2, into the biomass becomes the limiting parameter. A lower concentration of NH_4-N at which the rate of removal is affected is controversial,[95] but the removal rates have been observed to decrease when the residual NH_4-N is 2 to 4 mg/L. This effect is observed in TF studies,[85,88,96] including analysis of the NH_4-N profile (NH_4-N concentration versus depth).

The estimated transitional NH_4-N concentrations from zero-order to first-order nitrification were estimated by Okey and Albertson[94] as a function of DO and temperature. Figure 12.18 shows curves for the estimates that were based on extensive data. This figure relates the interferences of two limiting ammonia oxidation rate factors but does not define the impact on zero- or first-order rates.

The typical NH_4-N profile in a tertiary nitrification tower will exhibit a straight line reduction of NH_4-N at a rate controlled by the available oxygen

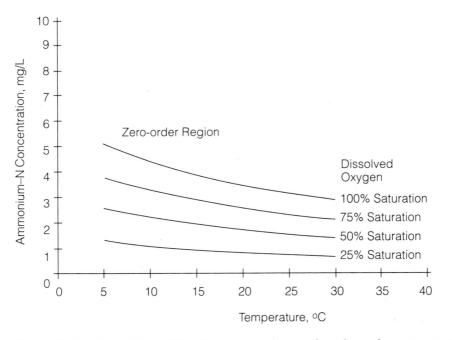

Figure 12.18 Transitional NH_4-N concentrations as functions of temperature. (The transitional region below the 100% saturation line may be either zero order or first order, depending on the oxygen concentration.)

in the upper portions of the trickling filter. The rate of removal will decrease as the controlling rate limiting factor changes from oxygen to NH_4-N diffusion. A typical tertiary nitrification profile, as NH_4-N residual versus depth, from the Midland[85] study is reproduced as Figure 12.19. This characteristic curve resembles that for plug flow nitrification in activated sludge. The point of inflection, zero to first order, is also a function of available oxygen. The poorer natural ventilation in August might have caused the differing rates shown in Figure 12.19.

TEMPERATURE EFFECTS. The effect of temperature on nitrification is similar to that reported[63] for carbonaceous filters. The effect of temperature is influenced by loading level, limiting substrates, oxygen and NH_4-N, hydraulics,[95] and wetting efficiency.[42-44] Thus, the reported effects of temperature can be variable; the higher nitrification rates are affected by temperature more than are lower rates of nitrification. As k_n approaches maximum k_n, temperature effects will increase; that is, the theta value will increase with higher nitrification rates.

Parker *et al.*[64] summarized tertiary nitrification data, shown in Figure 12.20, from several tests, which indicated a significant temperature effect on the rate. Central Valley data were developed from higher hydraulic rates than typical and excluded data for effluent NH_4-N concentrations less than 5 mg/L. Paulson,[5] reporting on the effects of temperature, showed minimal temperature effects for TFs compared with activated sludge. Okey and Albertson[94,95] found little correlation between rates and temperature and concluded that rate changes noted by others were attributable to other limiting factors such as oxygen availability, hydraulics, and NH_4-N concentration. Factors that can distort or obscure the effects of temperature and cause perturbations in the test results include oxygen availability, competitive heterotrophic activity, solids-sloughing cycles, predators, influent and effluent ammonia concentrations, and wastewater-induced effects (inhibitory). The apparent mixed response of nitrification rate to temperature changes likely results from a combination of the above factors.

pH. Huang and Hopson[97] studied biological nitrification on an inclined flat plate. The test indicated that a feed pH of 8.4 would nitrify at approximately a 50% higher rate than a feed with a pH of 7.5. Thus, pH effects on trickling filter nitrification rates appear to be significant. They concluded that the reaction was zero order at more than 2.5 mg/L. The test device was not considered to be oxygen limiting based on experiments showing the same rates of removal for varying NH_4-N influent concentrations. However, the reported DO of 5 mg/L would limit the zero-order nitrification rate and could have influenced conclusions.

Without recirculation, the TF is a liquid plug flow–fixed film reactor. Therefore, alkalinity loss due to nitrification will reduce the pH in the lower

---

Attached Growth and Dual Biological Treatment

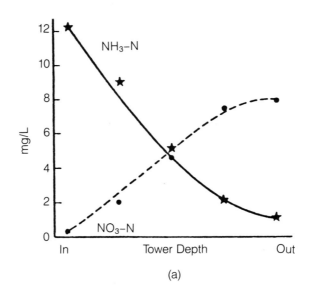

(a)

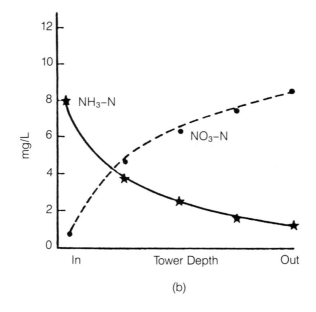

(b)

Figure 12.19 Nitrogen species profile for nitrifying
tower with influent flow of 0.71
gal/min/sq ft and zero recycle: (a) June
1972 and (b) August 1972 (gal/min/sq ft
× 0.679 = L/m² · min).

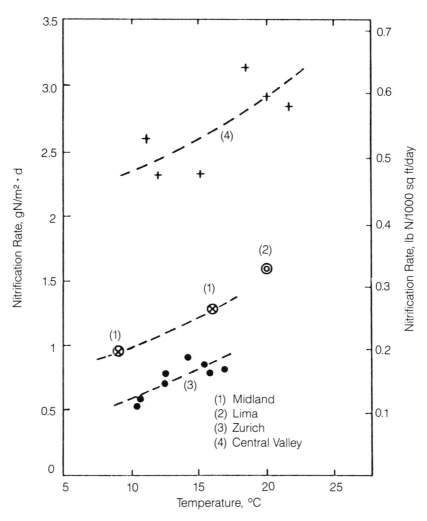

Figure 12.20 **Effect of temperature on nitrification rate of nitrifying trickling filters.**

zones where another rate-limiting factor, ammonium diffusion, also occurs. Thus, pH depression would not likely be as significant in the absence of recirculation. This suggests the use of tall towers to eliminate recirculation. Because the lowest pH would occur in the bottom of the tower where rates are already suppressed by NH_4-N limitations. Taller towers can also increase nitrification rates by minimizing first-order volume.

PREDATION. The biomass of nitrifying filters is particularly attractive to predators such as worms, snails, and fly larvae. These organisms can consume the nitrifying mass with a subsequent loss of nitrification capacity. They may also increase data variability. Control of these organisms is necessary to maximize the reliable nitrification capacity of filters. The use of high-intensity flushing (*SK* concept) effectively controls fly larvae. Whether this would also control snails is now unknown, but initial results are encouraging.[98] Because snails lay their eggs in sludge deposits, high-intensity flushing could be effective. Further, high-intensity flushing would reduce adjacent wet and dry areas that stimulate the growth and development of filter flies. One study[82] recommends periodic flooding of the biotowers for excess solids removal and elimination of snails. However, this design would be costly for deep towers, could not be readily retrofitted, and would only partially control snails. South Davis County[99] reported that weekly flooding of its rock filters does not control snails.

HYDRAULIC APPLICATION RATE. Hydraulic requirements for promoting maximum nitrification rates are still unknown. Okey and Albertson[94] and Gullicks and Cleasby[43] presented data from studies indicating that increasing the application rate ($L/m^2 \cdot s$) increased the ammonium ion oxidation rate. Application rates of more than 1 $L/m^2 \cdot s$ (1.47 gpm/sq ft) produced the best results. The authors noted that hydraulic effects were complex and might be interwoven with oxygen availability. The effects of hydraulics were found to be more significant in the zero-order range and difficult to discern in the first-order range of less than 4 mg/L NH_4-N.

Rearranged data[94,95] taken from the Arizona nuclear pilot-plant study illustrate the impact of effluent NH_4-N concentration on the rate of nitrification and hydraulic application effects. The nitrification rate depended highly on the effluent ammonia concentration as well as the available oxygen. That is, if the effluent NH_4-N exceeds 5 mg/L, rates are high. Data points and curves (Figure 12.21) were based on an operating temperature of 30°C and saturated DO of about 7.6 mg/L. The Gonenc *et al.* evaluation[91] suggests that much of the data represents oxygen-limiting conditions.

The *SK* concept's improvements of the performance of nitrifying filters (see Table 12.3) might have resulted from flushing excess solids and improved oxygenation. It is also possible that high-intensity operation keeps the media more biologically active throughout the tower's depth. Paulson[5] noted that NH_4-N may penetrate the lower depths infrequently, leading to patchy growth. However, Gullicks and Cleasby[100] reported poorer results with intermittent dosing of trickling filters.

ORGANIC LOADING. If the BOD_5:TKN ratio of the biotower influent ≤ 1.0 or $SBOD_5 \leq 12$ mg/L, design considerations presented in the section below apply. Otherwise, the engineer should use combined BOD_{5R} and NOD_R procedures explained in a subsequent section on combined BOD_5 and NOD removal. Based on the limited BOD_5 values set forth above, BOD_5 loading is not considered to be a design factor for tertiary nitrification. Poor documentation now exists for borderline systems between tertiary and combined BOD_{5R} and NOD_R ranges.

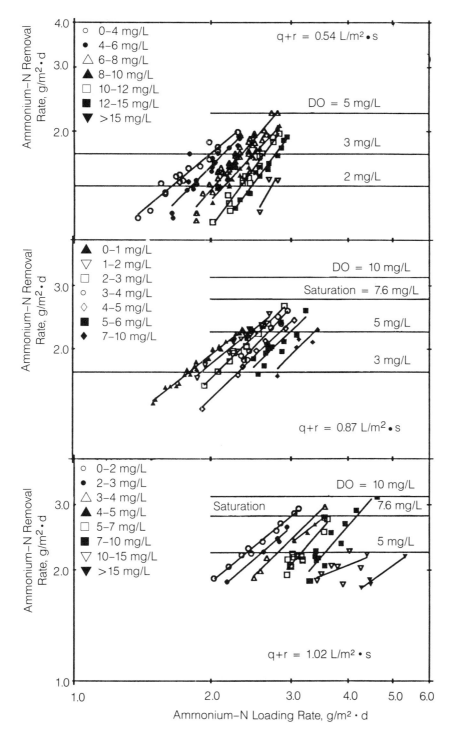

Figure 12.21 **Effects of effluent NH₄–N, hydraulic rate, and probable DO on nitrification rate (g/m² · d × 0.204 8 = lb/1000 sq ft/day; L/m² · s × 1.473 = gal/min/sq ft).**

DESIGN CONSIDERATIONS. Tertiary nitrification biotowers should be designed to maximize the media volume having zero-order removal and provide the maximum hydraulic rates without recirculation. These criteria will dictate use of a high tower with 6- to 12.2-m (20- to 40-ft) media depths. Nitrification biotowers (Buckeye Lake, Ohio) have been constructed with depths as great as 12.8 m (42 ft) with excellent results. Recirculation should be minimized to maintain the maximum NH_4-N concentration in the influent and reduce pH depression in the zero-order portion of the biotowers. For maximizing the overall nitrification rate, a biotower depth of 6 to 12.2 m will generally be optimal for producing a high hydraulic rate and maintaining a maximum zero-order volume.

Shallower towers can be employed by operating them in series. However, as Boller and Gujer[83] demonstrated, better performance will be achieved if the lead tower operation switches with the other tower every few days to prevent starvation and die-off of second-stage nitrifiers. This approach improves BOD_5 removal and resembles an alternating double filtration process developed during the late 1940s in England for improving $CBOD_5$ removal.

The optimal hydraulic application rate and frequency have not been adequately defined. Operation with the best hydraulic conditions calls for electric distributor drives with the provisions for automatic flushing cycles (Figure 12.5) during low-flow periods. Better results have been obtained from operating at slower rotational speed, but single-speed operation may not be optimal. Because hydraulically propelled units cannot reliably attain desired *SK*s, daily flushing routines are not easily instituted for such units.

Natural draft ventilation is not reliable for nitrification. Towers should be designed for forced ventilation in either an upflow or downflow mode. Nitrification towers can minimize odors and deodorize foul gases from other portions of the plant and malodorous air from carbonaceous trickling filters. For the latter purpose, the carbonaceous filter is downflowed and the nitrification filter is operated in an upflow air mode.

Ensuring air flow throughout the filter requires excess air distribution within the filter. The suggested minimum air flow rate should correspond to 50 kg O_2 supply per kg of O_2 used. During peak loading periods, the oxygen supply would be 20 to 30 kg/kg of oxygen used. Distribution of air pickup (downflow) or feed (upflow) to the underdrain area is required.

DESIGN BACKGROUND AND PROCEDURES. The difficulty in establishing design procedures stems from the variability of zero-order data as shown in Figures 12.17, 12.20, and 12.21. Therefore, a design must now be conservative, although higher zero-order rates will likely result from optimizing the tower design, including hydraulics, and providing for power ventilation.

The 1975 EPA Nitrogen Control Manual[89] indicated a zero-order rate of 1.28 g/m^2·d (3800 lb/d/sq ft) for a high-density medium (138 m^2/m^3) even though the database was derived from an 89-m^2/m^3 medium. As the surface area increases, its effectiveness decreases.[5,42,43,63,82,83] Thus, direct surface area scale-up of 89- to 98-m^2/m^3 media to 135- to 148-m^2/m^3 media is a questionable procedure.

Gujer and Boller[101] presented the following model for tertiary nitrification rates:

$$k_n = e^{0.044\,(t-10)} \cdot N/(K_N + N) \tag{27}$$

Where

k_n	=	nitrification rate at any depth in the filter, $g/m^2 \cdot d$;
t	=	temperature, °C;
N	=	bulk liquid NH_4–N concentration, mg/L; and
K_N	=	saturation constant, mg/L.

The effect of temperature [$e^{0.044(t-10)}$] on the rate equals 1.045^{t-10}, less than that reported for suspended film nitrification.

Parker et al.[82] reformulated the above model into the following design model:

$$k_n = (E \cdot jO_{2\,max}/4.3) \cdot [N/(K_N + N)] \cdot e^{-rD} \tag{28}$$

Where

E	=	media effectiveness factor;
$jO_{2\,max}$	=	maximum surface oxygen transfer rate, $g/m^2 \cdot d$;
D	=	tower depth, m; and
r	=	empirical parameter describing decrease in rate with depth.

The first term, $E \cdot jO_{2\,max}/4.3$, is the oxygen-limiting term of the maximum nitrification rate, k_n, as influenced by the effectiveness, E, of the media. The second term, $N/(K_N + N)$, represents the modifying effect of the saturation constant and residual nitrogen. The last term, e^{-rD}, adjusts the nitrification rate for its decline as a function of depth. (Note that e^{-kD} notation is used in the referenced publication.) The decline could be due to predator organisms, NH_4–N depletion, or poor wetting that leads to patchy nitrifier growth.

The term $N/(K_N + N)$ is used to adjust the nitrification rate in complete mix activated sludge systems as a function of the bulk liquid NH_4–N concentration. However, the plug flow correction term differs markedly as follows:

$$(N_o - N_e)/[N_o - N_e + K_N \cdot \ln (N_o/N_e)] \tag{29}$$

Where

N_o	=	influent NH_4–N concentration, mg/L; and
N_e	=	effluent NH_4–N concentration, mg/L.

Gujer and Boller[101] found that the correction, $N/(K_N + N)$, produced a better data fit than did the theoretical plug flow determination. Their conclusions were therefore employed in the Parker formulation.[82]

The coefficient K_N determined for the Central Valley, Utah,[82] plant was between 1 and 2 mg/L; zero-order rates and other coefficients determined from the investigations are presented in Table 12.10.

Table 12.10 Summary of nitrification profile parameters for a plant in Central Valley, Utah.

Period	Test number	k_n, maximum, g N/m²·d	r, m⁻¹	E	Temperature, °C
4/17/87–5/21/87	1	2.1	0	0.66	15.5
6/19/87–7/23/87	2	2.9	0.075	0.89	20.0
7/31/87–9/03/87	3	2.8	0	0.83	21.5
10/23/87–11/19/87	4	3.2	0.16	0.99	18
11/25/87–12/10/87	5	2.3	0	0.71	15
12/21/87–1/07/88	6	2.3	0	0.72	12
1/01/88–2/25/88	7	2.6	0	0.81	11

Table 12.10 rates are higher than those reported from other studies and shown in Figure 12.20. The rates of Table 12.10 are typical of the peak rates shown in Figure 12.17. The high rates of Table 12.10 were considered to be the beneficial result of flooding and flushing the biotower weekly and forced ventilation.

Values of r in e^{-rD} varied from 0 to 0.16. A value of 0 would imply no rate decrease as a function of media depth. This relationship needs further study. For example, a change in the r value from 0.00 to 0.164 would result in doubling the tower depth and volume[82] at $K_N = 1.0$.

Table 12.11[82] reports on nitrification rates from several studies. Predicted maximum rates were based on equation 28, which limits the rate based on the calculated oxygen transfer of each type of medium. The value of E is, therefore, the ratio of the apparent-to-predicted rates. The predicted rates are based on the calculated oxygen transfer capacity of each medium using the Logan model.

Table 12.11 indicates that apparent zero-order rates vary little among media. The Central Valley results were much higher than those of Table 12.11.

The excellent results of the Central Valley plant may be partly due to a readily nitrified wastewater and improved operating procedures, that is, good hydraulics and power ventilation. Like suspended film nitrification, TF

Table 12.11 Comparison of experimental nitrification rates with oxygen transfer-limited rates.

Plant	Media code	Temper- ature, °C	Apparent zero-order nitrification rate, g N/m^2·d [a]	Predicted maximum rate, g N/m^2·d [a]	E
Midland, Mich.	VFc27[b]	13	1.2	1.5	0.80
		7	0.93	1.3	0.73
Lima, Ohio	VFc27[b]	18	1.2	1.6	0.76
		21	1.8	1.7	1.09
		22	1.5	1.8	0.86
		22	1.2	1.7	0.68
Bloom Township, Ill.	VFc27[b]	20	1.1	1.7	0.65
		17	1.1	1.7	0.62
Zurich, Switzerland (3.9-m tower)	VFd28	17–20	1.6	2.0	0.76
	XFa68	17–20	1.2	3.4	0.29
Zurich, Switzerland (6.8-m tower)	XFa68	13	1.1	2.9	0.32

[a] g/m^2·d × 4.878 = lb/d/1000 sq ft.
[b] For example, VFc27 is vertical media with area of 88.6 m^2/m^3 (27 sq ft/cu ft).

nitrification rates will be site specific to some extent. As more TF units are brought on line, the reasons for increased nitrification rates will be better explained.

The design procedure following the formulation of Parker[64] requires determining the value of $E \cdot jO_{2\ max}/4.3$ as a function of temperature and defining the values of K_N and r. If predation is controlled and starvation (patchiness of growth) is minimized, then r approaches 0 and the term e^{-rD} approaches 1.0.

As an example of the use of the formulation, the average nitrification rate k_n is established using a maximum rate, $k_{n\ max}$, of 2.0 g/m^2·d, K_N of 1.5 mg/L NH$_4$-N, and r of 0. The average rate for an effluent of 1.5 mg/L NH$_4$-N is determined from equation 28 as follows: $k_n = (k_{n\ max}) [N/(K_N + N)] (e^{-rD})$ = (2.0) [1.5/(1.5 + 1.5)] (e^0) = 1.0 g/m^2·d.

Based on the TKN oxidized, g/d, the tower surface area (A_S) and volume could be established. The tower arrangement should encourage the maximum zero-order reaction volume.

Another design procedure[102] employs a somewhat different methodology based on analysis of operating results. This formulation, like Gujer and Bollers', was adjusted to reflect field results. The procedure has two steps:

1. Determine the zero-order medium volume using 138-m^2/m^3 media and a rate of 1.2 $g/m^2 \cdot d$ (k_n) over a temperature range of 10 to 30°C. Below 10°C, adjust the rate using $\Theta = 1.045^{t-10}$.

2. Determine the first-order medium volume using a rate (k'_n), which equals the following formulation and does not have a temperature correction between 7 and 30°C:

$$k'_n = 1.2(N_e/N_T)^{0.75} \qquad (30)$$

Where

k'_n = first-order nitrification rate, $g/m^2 \cdot d$;
N_T = the transition NH_4–N as set forth in Figure 12.18 for the coldest operating temperature, mg/L; and
N_e = effluent NH_4–N, mg/L.

The above design procedure stipulates several conditions to be met:

- BOD:TKN ≤ 1.0,
- $SBOD_5$ ≤ 12 mg/L,
- $q + r$ ≥ 0.54 $L/m^2 \cdot s$ (0.8 gpm/sq ft),
- $CBOD_5$ and TSS ≤ 30 mg/L for 138-m^2/m^3 (42-sq ft/cu ft) media,
- Forced ventilation ≥ 50 kg O_2/kg O_2 used, and
- Distributor control to provide SK of 25–250 mm/pass and flushing to ≥ 300 mm/pass.

Example 12.2. Design plastic media TF towers for tertiary nitrification using the above design procedure.

GIVEN.

Flow	=	438 L/s (10 mgd),
BOD_5 (influent)	=	20 mg/L,
$SBOD_5$ (influent)	=	8 mg/L,
TSS (influent)	=	25 mg/L,
TKN (influent)	=	28 mg/L,
NH_4–N (influent)	=	25 mg/L,
t	=	12°C, and
Design effluent NH_4–N, N_e	=	1.5 mg/L.

Tower application rate will be sufficient without recirculation. Transitional NH_4–N (N_T) concentration at 75% saturation is approximately 3.2 mg/L from Figure 12.18.

SOLUTION. Step 1. Check whether criteria for tertiary nitrification design are met.

BOD$_5$:TKN ratio	=	20/28,
	=	0.71, which is less than 1.0, and
SBOD$_5$	=	8 mg/L, which is less than 12 mg/L.

Step 2. Determine the media surface area for zero-order nitrification using the first step of the design procedure with 138-m^2/m^3 media and k_n = 1.2 $g/m^2 \cdot d$.

TKN$_{OX}$	=	[(438 L/s)/(1000 L/m^3)] (86 400 s/d) [(25 − 3.2 mg/L)/(1000 mg/g)] (1000 L/m^3)
	=	825 000 g/d
Media surface area	=	(825 000 g/d)/(1.2 $g/m^2 \cdot d$)
	=	687 500 m^2

Step 3. Determine the media surface area for first-order nitrification using the second step of the design procedure.

TKN$_{OX}$	=	(438/1000) (86 400) (3.2 − 1.5/1000) (1000)
	=	64 330 g/d
k'_n	=	1.2 $(N_e/N_t)^{0.75}$
	=	1.2 $(1.5/3.2)^{0.75}$
	=	0.68 $g/m^2 \cdot d$
Media surface area	=	(64 330 g/d)/(0.68 $g/m \cdot d$)
	=	94 600 m^2

Step 4. Determine the total media volume required.

Total media surface area	=	687 500 + 94 600
	=	782 100 m^2
Total media volume	=	(782 100 m^2)/(138 m^2/m^3)
	=	5670 m^3

Step 5. Calculate the maximum tower surface area based on the minimum flow rate of 0.54 $L/m^2 \cdot s$ stipulated by the design procedure.

Maximum tower surface area	=	(438 L/s)/(0.54 $L/m^2 \cdot s$)
A	=	811 m^2
Minimum depth	=	5670 m^3/811 m^2
	=	6.99 m

Step 6. Determine number and size of towers. Two biotowers operating in parallel with a total area of 775 m^2 (8330 sq ft) are suggested.

Diameter	=	4(775 m^2) (2)/π
	=	22.2 m (72.9 ft)

Depth	=	$5670 \text{ m}^3/775$
	=	7.32 m (24.0 ft)
q	=	$(438 \text{ L/s})/(775 \text{ m}^2)$
	=	$0.57 \text{ L/m}^2 \cdot \text{s}$ (0.84 gpm/sq ft)

As an alternative, the design could provide for two biotowers with a depth of 3.66 m and a diameter of 31.4 m in series operation, with provisions for periodically reversing the lead tower. The average nitrification rate would be

Average k_n	=	(687 500 + 94 600 g/d)/[(5670)(138)]
	=	$1.0 \text{ g/m}^2 \cdot \text{d}$

Using the two sizing procedures, projected average k_n rates, as functions of effluent NH_4-N are set forth in Table 12.12. Values of $E \cdot jO_2 \text{ max}/4.3$ and r were 2.0 $\text{g/m}^2 \cdot \text{d}$ and 0, respectively. Rates for K_N with NH_4-N effluent concentrations of 1.0 and 1.5 mg/L are shown in the table. Influent criteria employed for the American Surfpac sizing procedure provide the basis of the

Table 12.12 Average predicted nitrification rates for design procedures.

Effluent NH4-N, mg/L	Average nitrification rate, $\text{g/m}^2 \cdot \text{d}$ [a]		Example procedure
	Parker formulation		
	$K_N = 1.0$	$K_N = 1.5$	
0.5[b]	0.67	0.50	0.90
1.0	1.00	0.67	1.06
1.5	1.20	1.00	1.14
2.0	1.34	1.14	1.17
3.0	1.50	1.33	1.20[c]
4.0	1.60	1.45	1.20
			(maximum rate)

[a] $\text{g/m}^2 \cdot \text{d} \times 4.878 = \text{lb/d}/1000$ sq ft.
[b] Results < 1.0 mg/L NH_4-N are unreliable.
[c] Maximum rate allowed in procedure.

calculations.

In the most common design range of 1.5 to 3.0 mg/L of effluent NH_4-N, nitrification rates are comparable. The example procedure presently avoids use of a k_n value of more than 1.2 $\text{g/m}^2 \cdot \text{d}$ due to lack of consistent data as shown in Figure 12.17. The Parker formulation is highly sensitive to the values used for K_N as shown and more data are needed to define K_N and determine whether r approaches 0.

In the example procedure, reducing effluent NH4–N from 1.5 to 0.5 mg/L requires a tower volume increase of 31%. This reflects the uncertainty of attaining such low effluent NH4–N with tertiary filters as they are now designed.

DISTRIBUTOR OPERATION. The distributor should be able to operate over a range of 15 to $\geq$ 300 mm/pass of the arm. Therefore, the range of operating speed for a two-arm distributor in the tower design of example 12.2 follows:

q	=	$0.57 \text{ L/m}^2\text{·s}$
	=	$[(0.57 \text{ L/m}^2\text{·s}) \times (3600 \text{ s/h}) \times (\text{m}^3/1000 \text{ L})$
	=	$2.05 \text{ m}^3/\text{m}^2\text{·h}$
15 mm/pass: n	=	$[(2.05 \text{ m}^3/\text{m}^2\text{·h}) (1000 \text{ mm/m})]/[(2 \text{ arms})$
		$(15 \text{ mm/pass}) (60 \text{ min/h})]$
	=	1.14 rev/min
$1/n$	=	0.88 min/rev
300 mm/pass: n	=	0.057 rev/min
$1/n$	=	17.6 min/rev

An electrically driven distributor with automatic, dual-speed controls is suggested for example 12.2. As more *SK* data are gathered, the suggested *SK* range may change. Gear or pulley drives with a variable-speed motor would add flexibility.

POWER VENTILATION. The suggested minimum air rate provides 50 kg O_2/kg consumed. On this basis, the O_2 supply for the towers of example 12.2 can be determined as follows:

O_2 supply	=	$50(0.75 \text{ BOD}_5 \text{ applied} + 4.6 \text{ TKN}_{OX})$,
	=	$50[0.75(757 \text{ kg/d}) + 4.6(889 \text{ kg/d})]$,
	=	232 800 kg/d, and
	=	9700 kg/h (21 400 lb/hr).

The air rate to provide the oxygen is 580 m^3/min (20 500 scfm). Four fans, each with 290-m^3/min capacity, should be provided for the towers (one spare each). An air distribution (collection) system in the underdrain is suggested.

Current practice employs a 10°C zero-order nitrification rate of 1.2 $\text{g/m}^2\text{·d}$ (0.246 lb/d/1000 sq ft). However, the design engineer should incorporate data from towers built after 1989 if they have been designed for biofilm and predator control and include power ventilation.

This chapter does not includes a design procedure for natural draft systems because the performance of such TFs is considered unpredictable.

Combined BOD and NOD Removal. The design for combined removal of BOD5 (BOD5R) and nitrogenous oxygen demand (NODR) with dual-purpose biotowers using synthetic media began in the early 1980s. However, many

---

low-rate rock filters and some underloaded synthetic media trickling filters[89,103] have nitrified primary effluent in a single stage of treatment. The EPA Nitrogen Control Manual[89] notes that the design approach has been largely empirical as of 1975. Although more data are now available, the use of an empirical methodology continues because a fundamental approach to design has not yet been developed.

The kinetics of combined BOD_5 removal and nitrification are complex. Some of the reasons for this complexity stem from the process itself and the others reflect a lack of technical understanding of the influencing factors.

Unlike plug-flow suspended growth combined BOD_{5R} and NOD_R reactors where nitrification will occur throughout the reactor length, carbonaceous BOD_{5R} in a TF will occur in the upper portion of the tower, with nitrification in the lower part. Recirculation sometimes results in denitrification in carbonaceous zones.

The rate of nitrification in the combined towers will be influenced by many factors such as influent wastewater characteristics, hydraulics, ventilation, and media type. Influent wastewater characteristics include flow; $CBOD_5$, $SBOD_5$, TSS, and TKN concentrations; and peak-to-average concentration ratios. If the wastewater has been partially pretreated by chemical or biological means, its treatability has been modified. The role of temperature in combined BOD_{5R} and NOD_R designs is complex due to competing reactions for a limited substrate—namely oxygen. The effects of pH and the possible presence of inhibitory constituents can further confuse interpretation of pilot- and full-scale results.

Harremoes[92] noted that the basic phenomenon that governs reactions in fixed film reactors is the required diffusion of soluble substrates into attached biomass and products. In the case of combined treatment, the small growth of nitrifiers can be quickly submerged beyond the depth of oxygen penetration by rapid heterotrophic growth on organic substrates. Thus, nitrifier growth can only become dominant and operable after the upstream biomass removes most of the organic substrate.

BACKGROUND REVIEW. The 1975 EPA Nitrogen Control Manual[89] summarized full- and pilot-scale rock media data from TFs at Lakefield, Minnesota; Allentown, Pennsylvania; Gainesville, Florida; Corvallis, Oregon; Fitchburg, Massachusetts; Ft. Benjamin Harrison, Indiana; Johannesburg, South Africa; and Salford, England. These data show the relationship of BOD_5 volumetric loading with nitrification efficiency (Figure 12.22). The manual concluded that a loading limit of 0.16 to 0.19 $kg/m^3 \cdot d$ (10 to 12 lb BOD_5/d/1000 cu ft) was required to achieve approximately 75% nitrification. Nitrification (NH_4–N) removals at loadings above 0.4 $kg/m^2 \cdot d$ could have been synthesis removal of nitrogen. Figure 12.22 indicates that recirculation generally improved nitrification, particularly for nitrification efficiencies greater than 50%.

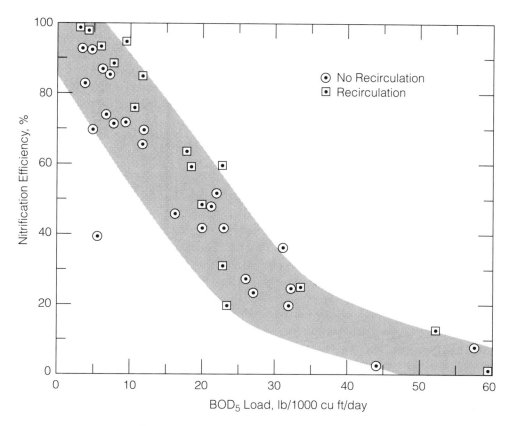

Figure 12.22 Effect of organic load on nitrification efficiency of rock trickling filters (lb/1000 cu ft/day × 1.602 = kg/100m³·d).

Stenquist et al.,[103] reporting on combined BOD_{5R} and NOD_R in synthetic and rock media TFs, related organic loading in a filter to the level of nitrification achieved. The nitrification capacity of a filter was found to be a function of the areal BOD_5 loading, kg BOD_5/m^2·d, on the media.

Bruce et al.[104] demonstrated that the effluent BOD_5 and COD had to be less than 30 and 60 mg/L, respectively, to initiate nitrification and that complete nitrification occurred with an effluent BOD_5 less than 15 mg/L. Reviewing the available work, Harremoes[92] concluded that the filtered BOD_5 ($SBOD_5$) would have to be less than 20 mg/L for the initiation of nitrification.

The removal of organic nitrogen was reported[89] as varying between 21 and 85%. Rock media studies at Gainesville and Johannesburg indicated that BOD_5 loadings must be less than 0.2 kg/m³·d to hydrolyze or otherwise remove 60 to 85% of the organic nitrogen. Studies at Bloom Township[90] showed that the removal of organic nitrogen was highly temperature dependent. Removals varied from 30 to 70% over the temperature range of 10 to 23°C.

Parker and Richards[96] reported on tests conducted at Garland, Texas, and Atlanta, Georgia. The results, reported on the basis of areal loadings (lb/d/1000 sq ft) are shown in Figure 12.23. In both studies, XF media outperformed VFC media. Results of the studies at Stockton (full- and pilot-scale) provide additional data for VFC media (Figure 12.24). Analysis of the data supported the Harremoes[92] conclusion that nitrification begins when the effluent $SBOD_5$ is less than 20 mg/L and will reach maximum rates when the effluent $SBOD_5$ is 4 to 8 mg/L.

Lin and Heck[105] reported on the successful operation of TFs designed[106] for combined BOD_{5R} and NOD_R with 1.5 mg/L effluent NH_4-N at 13°C. The tower design was based on 0.2 kg $BOD_5/m^3 \cdot d$ and a TKN loading of 0.051 kg/$m^3 \cdot d$ using 98-m^2/m^3 XF media. The $SBOD_5$ and NH_4-N removals following TF start-up are shown in Figure 12.25. Solids contact following the TFs provided further polishing of $SBOD_5$ and NH_4-N. However, performance was adversely affected by sloughing cycles, which could have been reduced by *SK* control. Complete nitrification occurred at summer BOD_5 loadings up to 0.32 kg/$m^3 \cdot d$ (19.8 lb/d/1000 cu ft).

The TF at Buckeye Lake, Ohio, employing deep media and one of two units, was operated at full-load conditions and produced 0.8 to 1.0 mg/L of ammonia nitrogen. The monthly average effluent quality with both filters on line was 2 mg/L BOD_5, 5 mg/L TSS, and 0.3 mg/L NH_4-N. Design criteria for the Wauconda, Illinois; Buckeye Lake; and Chemung Co., New York, TFs and their effluent qualities are presented in Table 12.13. The two taller units were operated at full load to satisfy performance tests. Their low average effluent NH_4-N results may be attributed to having both units on line in each plant. Updating the performance of these TFs is advisable before design of such TFs.

In addition to the facilities and pilot studies discussed, combined BOD_{5R} and NOD_R is practiced at TF plants in Ashland, Ohio; Amherst, Ohio; Montgomery Co., Ohio; Cibolo Creek, Texas; Cedar Rapids, Iowa; Sunnyvale, California; Littleton–Englewood, Colorado; and Boulder, Colorado.

DESIGN PROCEDURES. In early designs, designers selected a BOD_5 loading as a function of temperature to size the facilities. The Wauconda, Illinois, and Buckeye Lake, Ohio, facilities were designed with an empirical approach[106] that considered the organic loading and, specifically, the BOD_5 and TKN of the influent. These procedures did not fully account for the effect of the influent BOD_5:TKN ratio on the nitrification rate.

Okey[107] plotted an extensive array of data for the TKN removal rate versus the applied BOD_5:TKN ratio and sorted the data on the basis of temperature. This data array, shown in Figure 12.26, indicates a negative temperature effect. This is understandable, considering the known limitation of oxygen in tertiary nitrification biotowers. In fact, as shown, 9 to 20°C kinetic rates are the highest rates produced by the dual-purpose towers. Data

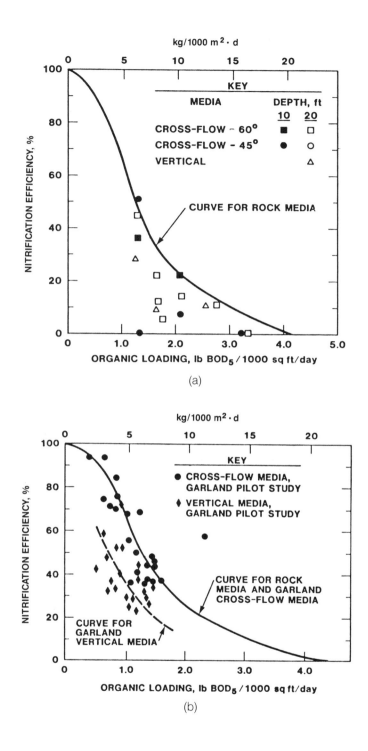

Figure 12.23　Results of nitrification pilot studies for (a) Atlanta and
(b) Garland, Georgia, plants.

Attached Growth and Dual Biological Treatment　　　753

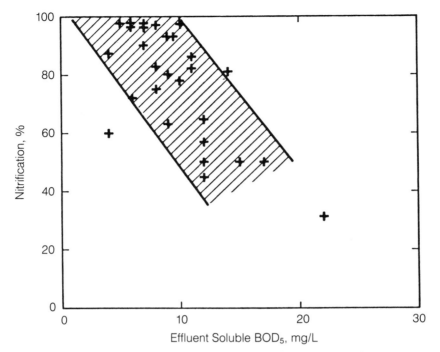

Figure 12.24 **Relationship between nitrification efficiency and soluble BOD in the effluent of a vertical media trickling filter at Stockton, California.**

and the equation of the line of best fit are shown in Figure 12.27. The standard deviation, σ, is large but not unusual for nitrification systems whether suspended growth or fixed film. TKN removal results from the combined effect of synthesis and nitrification.

The limiting effects of residual NH_4-N diffusion at low NH_4-N concentrations will be less noticeable than that at higher concentrations in terms of the average nitrification rate, k_n. The explanation is that the BOD_5 loading already suppresses the average rate to equal or below the NH_4-N limited rates of tertiary towers.

To produce the lowest NH_4-N concentrations, the biotowers need to be well irrigated and provided with biofilm control by flooding, the *SK* concept, or both. Predator control and power ventilation are also important in dual-purpose biotowers.

Example 12.3. Design a dual-purpose BOD_{5R} and NOD_R tower using the following procedure:[108]

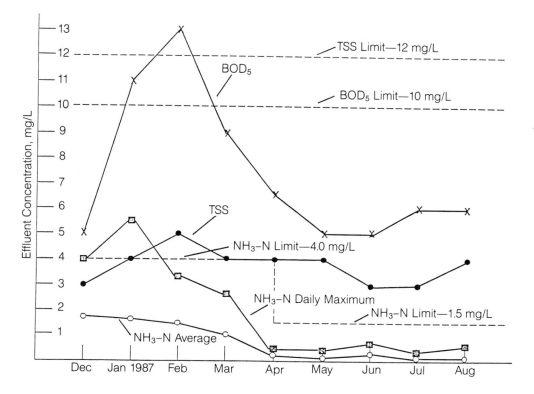

Figure 12.25 **Performance of dual-purpose tower at Wauconda, Illinois.**

GIVEN.

	Primary effluent	Clarified effluent
Flow, L/s (mgd)	219 (5.0)	—
CBOD5, mg/L	140	10
TSS, mg/L	80	15
TKN, mg/L	30	7
NH4–N, mg/L	20	2
t_{min}, °C	12	—

Tower sizing criteria are tower depth = 7 to 12 m and minimum q = 0.54 L/m²·s (0.8 gpm/sq ft). Provide biofilm control forced ventilation and minimum recirculation.

SOLUTION. Step 1. Determine the average k_n and design k_n.

BOD5:TKN	=	140/30 = 4.67,
Average k_n	=	$1.086(BOD_5:TKN)^{-0.44}$,
	=	$1.086(4.67)^{-0.44}$, and
	=	0.55 g/m²·d (8873 lb/d/sq ft).

Attached Growth and Dual Biological Treatment   755

Table 12.13   Design and performance of combined BOD$_{5R}$ and NOD$_R$ biotowers.

	Wauconda, Ill.	Buckeye Lake, Ohio	Chemung County, N.Y.
Design basis			
Number of units	2	2	2
Diameter, ft[a]	50	45	130
Depth, ft[a]	28	42	13
Volume, cu ft[b]	109 900	133 528	344 930
Media	XF98	XF98	XF98
Flow, mgd[c]	1.4	1.1	12.0
PE BOD$_5$, mg/L	117	156	73
SBOD$_5$, mg/L	80	—	—
TSS, mg/L	68	90	40
TKN, mg/L	30	36	18
NH$_4$-N, mg/L	20	26	—
Temperature, °C	7–20	10–25	10–20
Permit criteria			
BOD$_5$, mg/L	12	20	15
TSS, mg/L	10	30	—
NH$_4$-N, mg/L	1.5/4.0	3.0	5.0[d]
Design rates			
BOD$_5$, kg/m^3·d[e]	0.20	0.17	0.34
NH$_4$-N, g/m^2·d[f]	0.37	0.32	0.86
BOD$_5$:TKN	3.9	4.3	4.1
BOD$_5$ loading, kg/m^3·d[e]	0.19	0.19	0.14
Effluent quality[g]			
CBOD$_5$, mg/L	6	2	7.8
TSS, mg/L	10	5	11.7
NH$_4$-N, mg/L	<1.0	0.3	1.1

[a] ft × 0.304 8 = m.
[b] cu ft × 0.028 3 = m^3.
[c] mgd × 3785 = m^3/d.
[d] Effluent TKN, also TOD limit of 48 mg/L.
[e] kg/m^3·d × 62.4 = lb/d/1000 cu ft.
[f] g/m^2·d × 4.878 = lb/d/1000 sq ft.
[g] Solids—contact effluent following trickling filters.

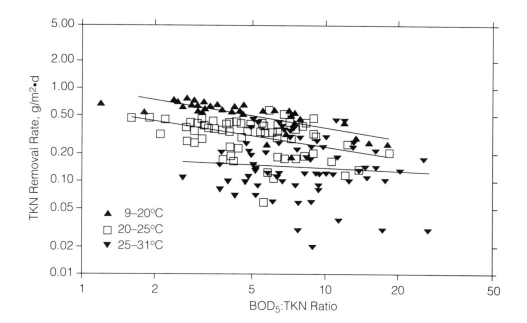

Figure 12.26 Study of nitrification in dual-purpose towers for plants in Stockton and Chino, California; Garland, Texas; and the Twin Cities Metro plant at St. Paul, Minnesota (g/m²·d × 0.204 8 = lb/1000 sq ft/day).

Average k_n	=	0.55 − 0.17 and
(−1.0 standard deviation)		
Design k_n	=	0.38 g/m²·d (12 840 lb/d/sq ft).

Step 2. Calculate media surface area, media volume, and BOD5 loading.

TKN$_R$	=	[(30 − 7 mg/L) (219 L/s) (86 400 s/day)]/(1000 mg/g)
	=	435 200 g/d (959 lb/d)
Media surface area	=	(435 200 g/d)/(0.38 g/m²·d)
A_s	=	1 145 000 m² (12 311 000 cu ft)
Media volume	=	(1 145 000 m²)/(98 m²/m³)
	=	11 690 m³ (412 500 cu ft)
BOD5 loading	=	[(140 g/m³) (219 L/s) (86 400 s/d)]/[(1000 g/kg) (1000 L/m³)]
	=	2649 kg/d (5828 lb/d)
	=	0.227 kg/m³·d (14.1 lb/d/1000 cu ft)

Step 3. Determine maximum tower surface area for 1:1 recirculation.

A	=	[(219 L/s) (2)]/(0.54 L/m²·s)
	=	811 m² (8718 sq ft)
Minimum depth	=	11 690 m³/811 m²
	=	14.4 m (47.3 ft)

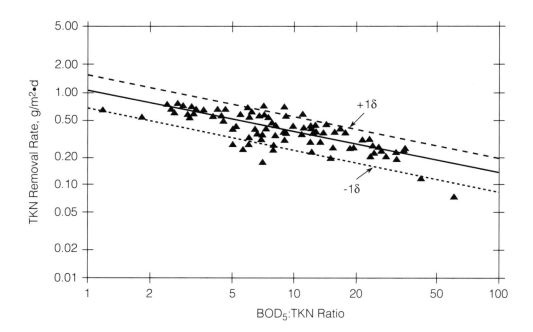

Figure 12.27 Study of nitrification at temperatures below 20°C for plants in Stockton and Chino, California; Garland, Texas; and the Twin Cities Metro plant at St. Paul, Minnesota [median $\tilde{y} = 0.460 \pm 0.175$; $\tilde{x} = 11.081$ and $\approx 15°C$ $TKN_{OX} = 1.086$ $(BOD_5:TKN)^{-0.44}$; $g/m^2{\cdot}d \times 0.204\ 8 = lb/d/1000\ sq\ ft$].

Step 4. Select biotower configuration based on above media area and volume determinations. Select two biotowers in series operation.

Diameter	=	32 m (105 ft)
D	=	7.32 m (24 ft)
q	=	0.27 L/m²·s
r	=	0.27 L/m²·s
k_n	=	0.38 g $TKN_R/m^3{\cdot}d$
A_s	=	98 m²/m³ (30 sq ft/cu ft)

Step 5. For biofilm control determine minimum speed for flushing and maximum speed for operation using *SK* range of 15 to 300 mm/pass.

- Minimum speed for flushing

$$n \quad = \quad [(q + r)\ (1000)]/[(a)\ (SK)\ (60)]$$
$$= \quad [(1.96\ m^3/m^2{\cdot}h)\ (1000\ mm/m)]/[(2\ arms)$$
$$(300\ mm/pass)\ (60\ min/h)]$$
$$= \quad 0.055\ rev/min$$
$$1/n \quad = \quad 18.2\ min/rev$$

- Maximum speed for operation (15 mm/pass)

  n = 1.1 rev/min

  $1/n$ = 0.9 min/rev

Select variable-speed distributor drive for 1 to 20 min/rev.

Step 6. Determine power ventilation of biotowers using 50 lb O_2/lb O_2 demand.

O_2 demand	=	1.2 BOD_{5R} + 4.6 TKN_R
	=	1.2 (140 − 10) + 4.6 (23)
	=	262 mg/L
	=	4957 kg/d (10 900 lb/d)
O_2 supply	=	10 300 kg/h (22 700 lb/hr)
Air rate	=	(10 300 kg/h)/(16.7 kg O_2/m^3·h)
	=	618 m^3/min (21 800 scfm)

Select two 620-m^3/min fans (one spare) that downdraft the lead tower and upflow the following tower and provide for suction and discharge air distribution in the underdrain.

The design procedure using an average −1 standard deviation k_n value produces a loading rate consistent with the fundamentals of organic loading procedure, although the procedure is still empirically based. Although the biofilm control procedures and power ventilation are expected to result in a general improvement of performance, current designs cannot take such improvements into account because of insufficient performance data. Dual-purpose systems are producing NH_4-N effluent values below 2 mg/L, but the predictability and design procedures are still not well established. However, low average k_n rates for dual-purpose towers result in a design insensitive or less sensitive to reduced nitrification rates due to NH_4-N diffusion in the range of 1 to 3 mg/L.

DESIGN AND OPERATIONAL FEATURES. This chapter will offer only a brief overview of key aspects of TF design, mostly related to operations. Valuable input is available in EPA references authored by Stenquist and Kelly,[109] Pierce,[110] and an EPA process design manual.[111]

Flow Distribution. Flow distribution is an important feature in a TF system. Flow must be evenly applied at a rate that keeps the filter unclogged. Uneven application of flow and insufficient flow rates for adequate flushing will result in poor performance. Odors will emanate as solids build up and clog the filters; growth of nuisance organisms will increase.

Most new TFs are circular to accommodate a rotary distributor. Although the ABF design (see dual biological treatment section) has employed continuous dosing with fixed nozzles, the popularity of this process has declined. Older rock filters may also employ fixed nozzles, either fed continuously or periodically dosed with siphons or a sequenced pumping arrangement.

Based on the information originally reported by Hawkes[33] and others,[32,34] fixed nozzles are not generally suitable for optimum trickling filter performance. The average rate and the instantaneous rate (*SK*) will generally be low and result in excessive solids buildup, ponding on rock filters, and odors. Also, conversion to high *SK* operation might not be feasible without extensive modification. New facilities with fixed nozzles are not advisable; existing units should be upgraded to rotating or reciprocating distributors (for rectangular filters).

If a rectangular unit is upgraded with a rotary distributor, unwetted media should be removed. The wet-dry interface area would otherwise provide a breeding area for undesirable fauna such as filter flies.

The need for and benefits of providing a means of controlling the instantaneous application rate, or *SK*, was reviewed in the discussion of hydraulics. Based on current practice, increasing the *SK* by 10 to 50 fold might improve performance. Most hydraulically propelled rotary distributors are currently operating at 0.5 to 1.5 min/rev and will thus need to be slowed at least for routine flushing to 15 to 50 minutes. More specifically, the distributor should be capable of producing *SK*s ranging up to 800 mm/pass as set forth in Table 12.4.

Hydraulic Propulsion. The conventional rotary distributor, as illustrated in Figure 12.28, is propelled by the hydraulic discharge from the trailing arms. Historically, little attention has been focused on the velocity of rotation, which in part dictates the instantaneous dosing (mm/pass of an arm). As a result, the typical dosing rate is 2 to 10 mm/pass at 0.2–1.5 min/revolution. Many believe that this practice impairs the performance of trickling filters.

The rotary distributor is generally equipped with 2 to 6 arms. The distributed flow may be staggered for full coverage per arm. That is, each arm may provide 50 or 100% coverage per revolution. The number of arms and the coverage/arm affects the value of *a* in the *SK* calculation, as shown in Table 12.14.

Thus, filters with 100% coverage/arm must operate at 50% of the rotational velocity (*n*) to maintain the same *SK* value as that for a filter with 50% coverage per pass. This is a significant consideration for a hydraulically propelled distributor.

Table 12.15 shows that operating a hydraulically propelled distributor within the desired *SK* range might be difficult or impossible. That is, the distributor might not operate as slowly as required to maintain the design range of *SK*. As this table is only an illustrative example of design *SK*, the design

Figure 12.28 Hydraulically propelled rotary distributor.

SK range should first be determined (see Table 12.4) and then the distributor design (number of arms, coverage per arm, and distributor speed) can be selected based on Table 12.14.

Providing good flushing intensity is even more difficult with rock filters operating at typical application rates of 0.2 to 0.6 $m^3/m^2 \cdot h$ (0.08 to 0.25 gpm/sq ft). Distributor speeds similar to those listed in Table 12.15 would be required to maintain the desired *SK* range and provide for routine flushing of the media.

No minimum speed has been specified for a hydraulically propelled distributor. TFs have reduced their distributor speeds with reverse jets as shown

Table 12.14 Effect of distributor design on a value in the *SK* equation.[a]

Number of arms	Value of *a*	
	50% coverage	100% coverage
2	1	2
3	—	3
4	2	4
6	3	6

[a] See equation 3.

Table 12.15 Example of distributor speed determination.[a]

| No. of Arms | Operational speed, min/rev | |
| | Flow coverage/arm, % | |
	50	100
2	1.2–12	2.4–24
3	—	3.6–36
4	2.4–24	4.8–48
6	3.6–36	7.2–72

[a] *SK* range, mm/pass of arm = 30–300; average application rate is 1.5 $m^3/m^2 \cdot h$.

in Figure 12.29 and have reported stalling speeds varying from 4 to 20 min/rev. Unless TFs with hydraulically propelled distributors receive nearly constant flow, most cannot operate at minimum speed during average to peak diurnal loading. Otherwise, the distributor would stop at low flow during the night.

Thus, even with reverse jets, the conventional hydraulically propelled distributors will often have unpredictably limited capabilities of maintaining the optimum *SK* and desired flushing intensity. To avoid these limitations of conventional hydraulic propulsion, a mechanical variable-speed drive of one of the types described below may be used.

Mechanically Driven Distributors. These include two types: water wheel distributors and electrically driven rotary distributors.

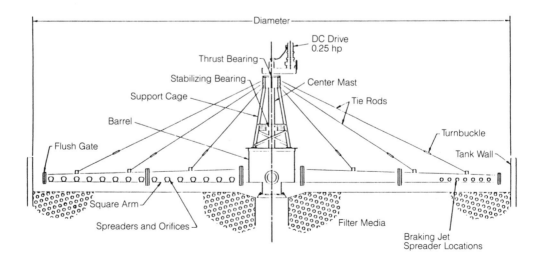

Figure 12.29 Typical rotary distributor with braking jets and electrical drive.

WATER WHEEL DISTRIBUTORS. The water wheel design is an intermediate step between the jet propulsion mode of movement and the electric drive. In its simplest form of hydraulic drive, the water wheel is energized by the influent or part of the influent flow to the trickling filter. Water wheel distributors can be driven by a wheel unit that is fixed in the center of the distributor. Providing a water wheel at the periphery of the trickling filter is also possible. Figure 12.30 shows both types.

These units can compensate for varying flows and rotate more slowly than can conventional rotary distributors. Because the water wheel needs only a portion of the flow, a varying flow affects rotation speed less than does a jet unit. However, the water wheel unit cannot be automated to provide high-intensity flushing periods, and trash can foul the unit. Thus, although the water wheel is better than the jet-propelled unit, the former still lacks the flexibility needed to optimize TF operation.

ELECTRICALLY DRIVEN ROTARY DISTRIBUTORS. The electrically driven rotary distributor may have either a center drive or peripheral drive. It generally can be easily retrofitted at low cost. The units can be programmed to operate at varying SK rates as required to optimize SK for BOD_5 removal and thoroughly flush media.

The center driven unit will be anchored to the nonrotating parts of the influent structure. Where no upper, steady bearing exists, bearing support must be installed with a stationary shaft to provide a platform for the drive unit. This can be located in the mast support for the arm guywires.

Where an upper, steady bearing exists, the stationary shaft into this bearing assembly can be extended to support the drive assembly as shown in Figures 12.6 and 12.29.

A peripherally mounted electric drive can be used instead of a center drive. The traction drive can either use the inside or top of the wall. By spring loading the drive wheel, it can operate with irregularities in the wall. A rotary union is used to transmit power. The arrangement is similar to that of a traction drive clarifier. A peripherally driven unit is shown in Figure 12.31. Hydraulic motors with equally wide speed ranges may be used instead of electrically propelled units.

Electrically propelled units with remote variable-speed controllers and timers can operate independently of flow (see Figure 12.5). This is particularly advantageous for those facilities without recycle flow or sufficient recycle flow to minimize the rotational speed. Further, with two units, the optimum SK and the flushing requirements can be readily determined. For example, the flushing would be best conducted during low-flow and loading periods, such as 1:00 to 6:00 a.m., when clarification capacity is at a maximum.

Providing two or more TFs with remote-controlled drives and timer-controllers or a microprocessor (Figure 12.5) offers significant advantages.

(a)

(b)

Figure 12.30 Two types of rotary water wheel distributors: (a) central
drive and (b) peripheral drive.

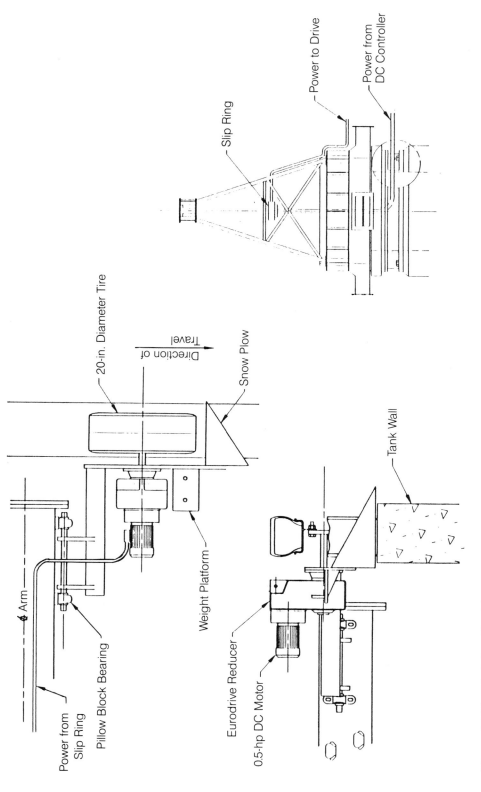

Figure 12.31 Peripherally driven rotary distributors with slip-ring arrangements.

Slip Ring

Power to Drive

Power from DC Controller

20-in. Diameter Tire

Direction of Travel

Snow Plow

Tank Wall

Weight Platform

Arm

Power from Slip Ring

Pillow Block Bearing

Eurodrive Reducer

0.5-hp DC Motor

Optimum *SK* can be determined by simultaneously operating TFs at different operating *SK*s, evaluating the SBOD5 removals, and adjusting individual speeds accordingly. Daily high-intensity flushing routines can also be programmed for the units to define optimum flushing *SK*s and durations to maximize performance. Once these two *SK* conditions are defined, the units can be controlled to maintain nearly optimum operation. Also, varying the speed (hence *SK*) as a function of the flow may result in the best performance.

Construction of Rotary Distributors. Rotary distributor arms are normally tubular in smaller units and have other shapes, such as rectangular, in larger units operating at high flows. Galvanized mild steel and aluminum are the most common materials of construction, although stainless steel may be used in more corrosive conditions. Providing flow distribution, a series of nozzles are positioned in the arm to provide either 50 or 100% coverage of the unit per pass of the arm. These nozzles will be equipped with slide gates to control the flow and splash plates to distribute water.

In many cases, distributors may be equipped with four arms in a high-low flow arrangement. Two of the arms operate at flows up to, and slightly above, average flow. The other two arms operate during peak flows. This arrangement provides maximum flushing intensity during critical low flows, containing the maximum BOD concentration and allowing better distribution because the velocity in the arms can be lower.

The hydraulic head required to operate a distributor and provide distribution is usually in the range of 400 to 1000 mm (16 to 40 in.) of water column. Maintaining the flow to the nozzle at the minimum velocity enhances distribution. High velocities will result in inadequate distribution at higher flows. Units operating at rotational velocities of 1 rpm can exert centrifugal force; the newer TF designs operating at speeds of 8 to 50 min/rev will exert insignificant centrifugal force.

A seal is necessary between the fixed influent column and the rotary section. Older designs have various types of water traps, mercury seals, or mechanical seals to prevent water from leaking between fixed and rotary parts. Modern designs employ an overflow arrangement (Figure 12.32) without a lower seal. This offers the advantages of no seal friction imposed on the mechanism and avoidance of seal maintenance. When older units are upgraded, improvements often include the sealless overflow arrangement.

One manufacturer supplies a rotary distributor with four or more Y-branched arms with nozzles that cause rainlike spray over a larger part of the filter's surface. In this type, the arms must clear the filter by at least 0.3 m (12 in.), whereas necessary clearance is usually only 0.15 m (6 in.). Where ice formation is anticipated, additional clearance of at least 0.15 m (6 in.) is frequently provided with both types. In such cases, peripheral clearance is also provided. This design has a low *SK* value that could adversely affect performance.

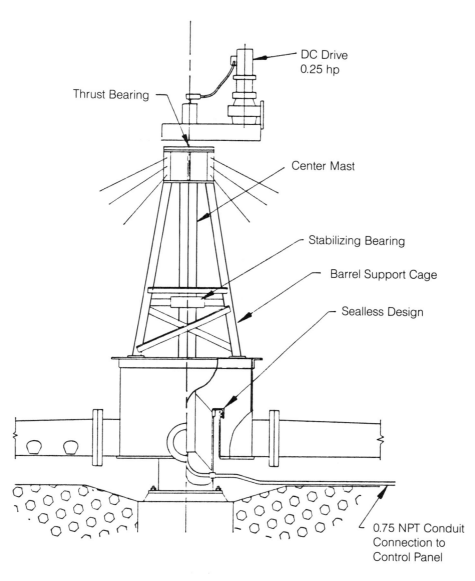

DC Drive
0.25 hp

Thrust Bearing

Center Mast

Stabilizing Bearing

Barrel Support Cage

Sealless Design

0.75 NPT Conduit
Connection to
Control Panel

Figure 12.32 Electrically driven distributor with overflow (sealless) arrangement.

Head requirements for hydraulic distributors vary with the square of the flow. The head for minimum flow is usually 0.3 to 0.6 m (12 to 24 in.) above the center line of the arms. Somewhat greater head is needed to accommodate wide flow ranges. For some distributors, this head requirement has been reduced by the use of overflow devices that result in dosing through additional arms during periods of high flow.

The general design of motor-driven, rotary-type distributors is similar to that of the reaction type. These motor-driven distributors require only minimal hydraulic head even when handling large variations in flow. A totally enclosed electric motor and gear arrangement drives rotation of the arms at a fixed rate. The motor-driven, rotary-type distributor has often proven hydraulically effective for handling extreme flow ranges. Installations of this type have been used in filters ranging in size from 7.6 to 53 m (25 to 175 ft) in diameter.

Filter Distributor and Appurtenances. Filter distributors require piping between the primary settling tank and the distributor. Where the filter is not designed for continuous dosing, a pump or dosing tank and siphon usually precedes the distributor.

PIPING TO DISTRIBUTORS. Any type of distributor requires only a single conduit to convey the liquid from the settling tank to the distributor. The hydraulic design of this conduit conforms with that for other plant piping. Design variables include head loss, provision for drainage, structural considerations, and protection against applicable climatic conditions.

RECIRCULATION PUMPS. Most high-rate filters employ recirculation pumps; recently, some low-rate filters have begun using such pumps to maintain minimum flows, control odors in the primary tank, or both. The pumps are usually constant-speed, low-head centrifugal units designed to operate against 3.7 to 6.1 m (12 to 20 ft) of head. Occasionally, two-speed motors are provided but, more frequently, two or more standard units accommodate variations in flow.

The submerged type of vertical pump, requiring only a wet well, has been used extensively, but the dry-well type with an intake pipe connection to the adjacent wet well is gaining acceptance. Because recirculated flow is generally free from clogging solids, intake screens are unnecessary. Although other types are gaining favor, pump controls are usually float-operated and include a low-water cutoff should the wet well become dry.

Head Loss Computations. Methods of supplying wastewater to the filter distributor include gravity feed, dosing siphons, and pumping. The type of feed selected depends on both the hydraulic gradient available and the distributor used. Hydraulic computations are always necessary. Computations for minimum flow are necessary to ensure adequate head to drive reaction-type distributors; computations for maximum flow indicate the head required to ensure adequate discharge capacity.

The net available head at the horizontal center line of the distributor's arm and other points may be calculated by deducting the following applicable losses from the available static head:

- Entrance loss, such as that in the primary settling tank;
- With dosing siphons, the drop in level in the dosing tank as distributor pipes are filled, friction is lost in siphon, and velocity head is imparted;
- Friction losses in the piping to the distributor, with proper allowance for head losses in elbows, valves, enlargements, contractions, and other appurtenances;
- Loss through distributor riser and center port; and
- Friction loss in distributor arms and velocity head of discharge through nozzles necessary to start reaction-type rotary distributor.

Head requirements of distributors are set by the manufacturers. The major loss of head is the difference in elevation from the distributor to the lowest water surface in the main underdrain channel. This loss alone approximates 2.4 m (8 ft) for a filter 1.9 m (6 ft) deep and can be considerably greater for a deeper synthetic-medium filter.

Head loss due to the TF usually exceeds the sum of all other head losses in the treatment plant. Unless the site offers an adequate elevation drop, the head requirement is a disadvantage of the TF process compared to the activated sludge process, which requires little drop in static head. Nonetheless, the power requirements for lifting wastewater, including that for recirculation when used, are generally less than those for aeration in the activated sludge process.

CONSTRUCTION FEATURES. The filter's shape depends on the type of distributor used. As most plants constructed since 1935 in the U.S. have rotary distributors, filters are usually circular, although some hexagonal and octagonal filters exist. Plants constructed with fixed-nozzle distribution usually have rectangular filters, but any shape may be used.

Many filters are flooded periodically to control filter flies and ponding. Such filters must have tight walls and gates for the drainage channel. An overflow pipe draining to the filter discharge pipe downstream from the gates is frequently provided to forestall the spilling of settled wastewater around the filter and the resulting nuisance conditions.

Walls. Most filters have reinforced concrete walls around their circumference, usually 0.2 to 0.3 m (8 to 12 in.) thick. Choice of materials can affect the filter's flooding capability and its ventilation and temperature. To alleviate winter conditions, wind breaks are sometimes provided either by extending the side walls several feet above the top of filter media or installing a barrier above the filter wall. Because most synthetic filter media are self-supporting, walls for TFs with such media merely serve to contain wastewater flow and maintain wastewater temperature unless the TFs are to be flooded. One of the least expensive shells is constructed of fiberglass building panels attached to an aluminum or steel frame. In deep synthetic-media filters, intermediate structural supports usually are unnecessary for media depths up to 14 m (46 ft).

Underdrains. Underdrains comprise the subfloor, filter block, and drainage channels, each discussed below.

SUBFLOOR. The floor of the filter should be able to support the under-drainage system, filter media, and water load if the filter is to be flooded. The floor commonly consists of approximately 100 to 150 mm (4 to 6 in.) of steel mesh reinforced concrete on a properly prepared subgrade. The floor slopes downward to a drainage channel, with a gradient between 0.5 and 5.0%, depending on the filter's size. Either central or peripheral drainage channels may be used for liquid removal and air passage.

FILTER BLOCK. Precast filter blocks, manufactured from vitrified clay or concrete, are used for stone-media filters. Most installations use the former. These blocks are usually rectangular, having openings in the upper face with a total area of at least 20% of the block's surface. Wastewater passing through the filter enters the blocks and flows through channels therein. These channels have sufficient capacity to carry the rated discharge with 50% or less of their cross-sectional area. In the far west, redwood timbers resting on concrete sleepers have been used instead of filter blocks, but the scarcity of redwood now prevents its use for new installations.

Filter blocks cover the entire floor of the stone-medium filter, except for the area above the drainage channel and the distributor support. These blocks are placed in parallel lines perpendicular to the drainage channel if a center drain is used or perpendicular to a diameter when a peripheral drainage channel is used. The blocks—laid on a thin layer of grout, on sand, or directly on the filter floor—slope to drain toward the drainage channel.

Where plastic or wood media are used, one of several less expensive support systems is used instead of filter underdrain blocks. Figure 12.33 illustrates three typical support systems, two using beam or column supports and the other using grating. Use of grating requires consideration of corrosion control. Plastic-coated grating best resists corrosion. Individual support systems will vary depending on the recommendations of the individual synthetic-medium manufacturer.

DRAINAGE CHANNEL. Drainage channels carry the flow from the under-drains and admit air to the underdrains for ventilation. They often extend through the filter walls to provide ventilation and should be accessible for cleaning. In a circular filter they are usually located on a diameter. If a center-feed well is used, the drainage channel may curve around the well or may parallel a diameter, slightly offset from the center. If filter media rest above

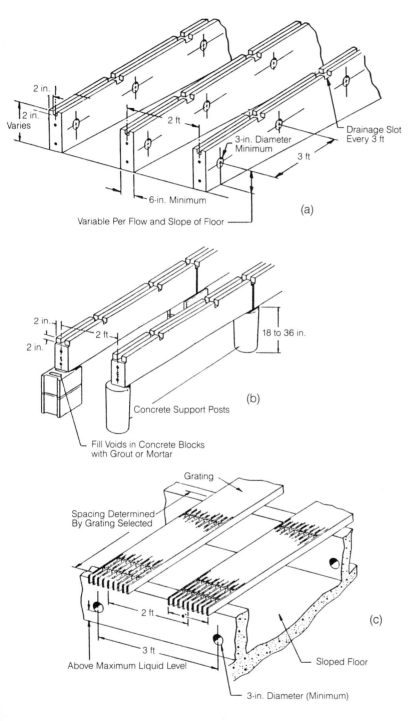

2 in.

2 in.
Varies

2 ft

3-in. Diameter
Minimum

3 ft

Drainage Slot
Every 3 ft

6-in. Minimum

Variable Per Flow and Slope of Floor

(a)

2 in.

2 ft

2 in.

18 to 36 in.

(b)

Concrete Support Posts

Fill Voids in Concrete Blocks
with Grout or Mortar

Grating

Spacing Determined
By Grating Selected

2 ft

3 ft

Above Maximum Liquid Level

Sloped Floor

3-in. Diameter (Minimum)

(c)

**Figure 12.33   Typical synthetic and wood media support structures: (a) cast-in-place
beam construction, (b) precast concrete beam supports, and (c) staggered
grating support over cast-in-place beams. Grating openings between
support bar members are not less than 1–2 in. Deflection is not more than 5%.**

the drainage channel, perforated cover blocks are used to provide support, drainage, and ventilation of that part of the filter. Small filters sometimes have semicircular, nonperforated channel covers that drain to adjacent filter blocks. Normally, channels (either rectangular or semicircular in cross section) are designed to provide velocities of 0.6 to 0.9 m/s (2 to 3 ft/sec). At maximum discharge, the water level in the drainage channel is usually below the bottom of the underdrain blocks for stone-media filters.

For plastic-media filters, the periphery wall sometimes includes multiple vents. If so, the hydraulic design required to maintain proper ventilation of the drainage channel is less critical than that for the stone-medium filter.

Forced ventilation can be in either the upflow or downflow mode. Downflow will generally result in less odor because the odor would not be stripped by upflowing air from the primary effluent. The underdrain should provide for air distribution (or suction) because head loss is so small. Axial flow fans can produce 150 to 225 m^3/m·kW (4000 to 6000 scfm/hp) of air flow. In the downflow mode, droplet entrainment will occur. Thus, droplet removal is required or, alternatively, the air can be returned to the top of the filter.

Where wide temperature extremes exist, reversing the fan direction to push air in the natural draft direction may be advisable. For example, in winter the fans would upflow and in spring, summer, and fall, they would downflow.

Media. EPA's "Upgrading Existing Wastewater Treatment Plants" [111] contains helpful information for preparing specifications for stone and synthetic filter media.

MATERIALS USED. The materials locally available or the cost of transportation often govern the choice of stone filter media. Field stone, gravel, broken stone, blast furnace slag, and anthracite coal have been used as media. Redwood blocks and inert materials molded into appropriate shapes have also been employed. Whatever material is chosen, it must generally be sound, hard, clean and free of dust, and insoluble in wastewater constituents.

SIZE AND SHAPE. Little agreement exists as to the optimum size of stone filter media. A common specification requirement is that 95% or more of the media pass 2600-mm^2 (4-sq in.) mesh screens and be retained on 1600-mm^2 (2.5-sq in.) mesh screens. Specifications usually require pieces to have a uniform size, with all three dimensions nearly equal.

Use of synthetic media for trickling filters has extended the ranges of hydraulic and organic loading beyond those of stone media. Two properties of interest are specific surface area and percent void space. Greater surface area permits a larger mass of biological slimes per unit volume, whereas increased void space allows higher hydraulic loadings and enhanced oxygen transfer. The ability of synthetic media to handle higher hydraulic and organic loadings results directly from higher specific surface area and void space of these media compared to those of stone media and blast furnace slag.

DURABILITY. Material should not disintegrate under service conditions, either by breaking into smaller pieces or by crumbling into fine material. Frequently, material is specified to be substantially sound as determined by the sodium sulfate soundness test. This test does not apply to redwood blocks and synthetic media. Synthetic media, usually constructed from polyvinyl chloride, are specified as insoluble, fire retardant, and not subject to deterioration by environmental conditions such as extreme temperatures or sunlight.

PLACEMENT. Specifications for placing stone filter media frequently include statements such as the following:

- When placing filter media, breakage and segregation of different sized particles must be prevented.
- Media shall be screened and cleaned immediately before placing to eliminate as many fines as possible.
- Media shall be placed by a method that does not require heavy traffic over them.
- Placing media by means of belt conveyor, wheelbarrow, or boxes will be acceptable.

Most synthetic media require field fabrication and hand placement of the modules within the filter shell. Generally, synthetic-media manufacturers will supervise construction to ensure proper installation.

R*OTATING BIOLOGICAL CONTACTORS*

During the late 1960s, interest in plastic media led to the development and commercialization of rotating biological contactors (RBCs), which provided many of the advantages of old rock-media TFs without some of their disadvantages. Because of new media developments and the smaller energy requirements of RBC treatment units compared with those of activated sludge units, engineers frequently selected RBC units for municipal wastewater treatment during the mid and late 1970s.

Process performance below design expectations, structural problems with shafts and media, excessive biomass buildup on media, uneven shaft rotation ("loping") for air-driven units, and other process problems have been serious concerns at many installations. These problems have resulted in the process falling out of favor with designers in recent years. Although the process may still apply in some situations, its limitations and potential problems must be understood and accounted for if it is to be used successfully.

As an estimate, more than 600 RBC plants are presently used for industrial and municipal wastewater treatment in the U.S. The first commercial RBC system was installed in West Germany in 1960. Most of the plants in the U.S. are designed and used for organic carbon removal, some for combined organic carbon removal and nitrification, and a few for nitrification of secondary effluent.[112] The first full-scale municipal RBC installation in the U.S. was in Pewaukee, Wisconsin.

The RBC consists of circular plastic disks mounted on a horizontal shaft in a tank. The shafts are rotated (1 to 2 rpm) by either a mechanical or compressed air drive. Typically, approximately 40% of the media are immersed in the wastewater. The wastewater being treated flows through the contactor by simple displacement and gravity. The rotation of disks alternately exposes the biofilm to the organic material in the wastewater and to atmospheric air. Bacteria and other microorganisms that are naturally present in wastewater adhere to and grow on the surface of rotating media. Within 1 to 2 weeks of start-up, bacteria will form a fixed biological film, covering the entire media surface. Depending on organic loading conditions, each stage will show varying slime thickness and color. Generally, the first stage shows characteristic brownish–gray color when operated within the reasonable loading range; the last (nitrifying) stages show a reddish–bronze color. Because of shearing forces, the biological film tends to slough off whenever the biomass growth on the media surface becomes too heavy. Growth and sloughing is continuously repeated. The sloughed biofilm and other suspended solids are carried away in the wastewater and are removed in the secondary clarifier.

Organic overloading, insufficient rotational speed, or other conditions may cause excessive biomass buildup on media, resulting in structural damage to shafts or media, the inability to maintain rotational speeds, and other process problems.

Figure 12.34 shows an RBC system being used for secondary treatment. The wastewater flows from a primary clarifier through a series of RBC units and then to a final clarifier where sloughed biomass is settled and removed. In addition, screens and grit removal equipment should be used ahead of the primary clarifier. If designed properly, RBCs can achieve secondary treatment, nitrification, and denitrification. Chapter 4 of Manual of Practice OM-10 "O&M of Trickling Filters, RBCs, and Related Processes"[113] provides a good overview of the RBC process and design and operational deficiencies experienced in existing RBC plants.

PROCESS CONCEPTS AND PRINCIPLES. Under most circumstances, mass transfer/diffusion of substrate and oxygen dominates organic removal in an RBC system. Mass-transfer/diffusion resistance associated with both

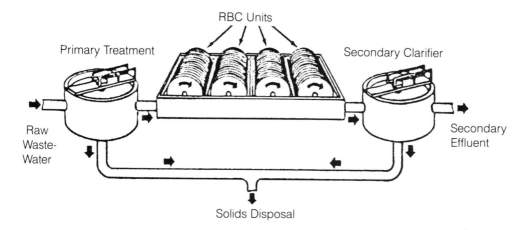

Figure 12.34   Typical RBC plant schematic for secondary treatment application.

the liquid phase and the biofilm results in significant concentration gradients from the bulk liquid to biological reaction sites within the biofilm. Oxygen transfer becomes limiting and controls the overall reaction rate in heavily loaded systems. Oxygen transfer from the air directly into the attached waste-water and biofilm represents the major oxygen source for organisms. In addition, oxygen also enters the bulk liquid as a result of turbulence generated by the rotation of media and return to the bulk liquid of wastewater that is lifted into the atmosphere and then flows freely back across the media.

The substrate removal mechanism within an attached growth process is complex. Phenomena that occur when a fixed film is brought into contact with a wastewater containing substrate and oxygen are discussed at the beginning of this chapter.

Maximum oxygen transfer and substrate removal rates are important for RBC performance; however, because of complex transfer/diffusion phenomena, substrate removal from the bulk liquid is not expected to follow any simple mathematical model.

FACTORS AFFECTING PERFORMANCE. In all RBC systems, the major factors controlling operation and performance are

- Organic and hydraulic loading rates,
- Influent wastewater characteristics,
- Wastewater temperature,
- Biofilm control,
- Dissolved oxygen levels, and
- Operational flexibility.

Organic and Hydraulic Loading. Increased flows increase hydraulic loading and decrease detention time. Sufficient detention or reaction time is

---

Attached Growth and Dual Biological Treatment   775

necessary in any biological reactor. The impact of organic and hydraulic load variations on the performance of an RBC system should be considered during the design. When the daily peak-to-average flow ratio is 2.5 or more, flow equalization merits consideration.

A limiting factor in the design of RBC systems is organic loading to the first stage(s), which should be compatible with the oxygen transfer capability of the system. With excessive organic loadings, the biofilm thickness increases, oxygen becomes limiting, and a white/gray biomass (*Beggiatoa*) frequently forms. Problems caused by overloading are septic odors, process deterioration, structural overloading, and possible equipment failure.

Stover and Kincannon[20,114,115] observed that substrate concentration and hydraulic loading are related directly to the substrate removal rate and efficiency. They indicated that substrate removal is not strictly a function of hydraulic loading or the influent substrate concentration but depends on the combination of these two variables, namely, the total organic loading to the system. They concluded, therefore, that the total organic loading concept should be used for design purposes. Dupont and McKinney,[116] Poon *et al.*,[117] and Wilson *et al.*[118] also stressed the importance of the total organic loading concept (lb BOD/d/1000 sq ft) as a design parameter. As one advantage, the total organic loading concept in design allows prediction of organic removal rates and treatment efficiency for any loading condition whether RBC units function under zero-, first-, or second-order kinetics.

An investigation of 23 RBC plants, with results summarized in Figure 12.35,[119] indicated the presence of sulfur-oxidizing organisms when the plants were overloaded organically. Figure 12.35 shows that whenever the first-stage loading limit exceeded 3.1 kg BOD_5/100 m^2·d (6.4 lb BOD_5/d/ 1000 sq ft), the system was associated with the presence of sulfur-oxidizing organisms. Higher organic loading will increase the likelihood of developing problems such as heavier-than-normal biofilm thickness, nuisance organism growth, DO depletion, and overall deterioration of process performance. Structural capacity of the equipment and other operational flexibilities require consideration if higher-end loading ranges are expected during the design life of an RBC system.

Influent Wastewater Characteristics. Municipal wastewaters are exceedingly diverse with respect to both the number of biodegradable components and the range of particle sizes. Wastewater characteristics and their impacts on biodegradability are important considerations in the design of an RBC system. High influent hydrogen sulfide (H_2S) concentrations diminish RBC performance because of the acceleration of nuisance organism growth. When higher than normal influent H_2S concentrations are anticipated, appropriate provisions in the design, such as preaeration or supplemental aeration, should

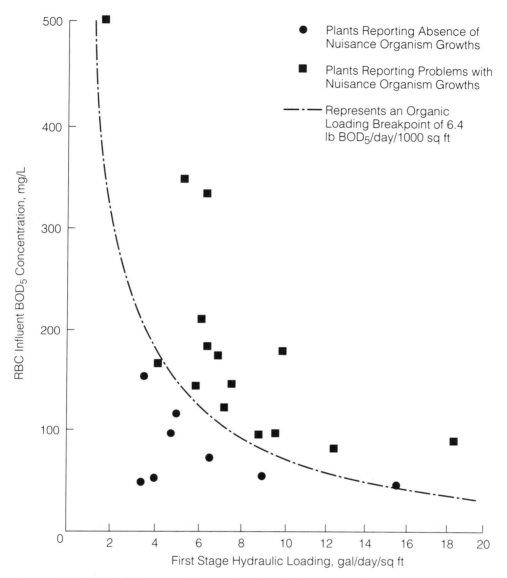

Figure 12.35 DO-limiting conditions related to influent organic concentration and hydraulic loading (lb/day/1000 sq ft × 0.483 3 = kg/100 m²·d; gal/day/sq ft × 0.040 74 = m³/m²·d).[119]

be considered. The effect of sidestreams, particularly from sludge processing, and their impact on the RBC treatment system also deserve consideration during the design.

Although many of the design curves developed by manufacturers and others using rational design approaches consider only soluble BOD, the designer should recognize that influent particulate BOD will undergo hydrolysis, thus contributing to organic loading. An RBC system must, therefore, not

---

only remove soluble wastewater components but it must also remove a substantial fraction of the particulate organic material if effective performance is to be achieved.

Wastewater Temperature. Studies have indicated that organic removal efficiency was not affected by wastewater temperatures higher than 13°C (55°F), but process performance deteriorated at lower temperatures.[120] Equipment manufacturers have developed temperature correction factors for wastewater temperatures under 13°C (55°F). Because the organic removal rate in RBC systems depends on the mass substrate and oxygen transfer/diffusion rates in addition to the biological kinetics, the use of temperature correction factors solely based on biological kinetic considerations is questionable. It is well accepted that decreasing wastewater temperature adversely affects the nitrifier's growth and, correspondingly, the nitrification capabilities of RBC systems.

In some cases, lower wastewater temperatures increase organic removal efficiencies in highly loaded systems. At lower temperatures, DO saturation increases, which promotes increased oxygen transfer.

Biofilm Control. Biofilm thickness is of great importance in the RBC process; thus, a distinction must be made between total film thickness and active film thickness. Measurements of biofilm thickness in several studies suggested that total film thickness varies from 0.07 to 4.0 mm, depending on hydrodynamic conditions.[121] Substrate removal rates within the thicker biofilm are not likely to be greater than those in thin films due to diffusional resistances within the biofilm. The active portion of the biofilm that contributes to substrate removal has been called the active biofilm thickness. This thickness has been estimated by several investigators to be between 20 and 600 μm.

Earliest observations of substrate removal with increasing biofilm thicknesses indicate that the rate of substrate utilization increased as the biofilm thickness reached 70–100 μm, after which any increases in biofilm thickness did not further increase substrate removal rates.[122] These investigators defined the active biofilm thickness as that thickness beyond which the substrate removal rate became constant after reaching a maximum value. In other words, the active film thickness is that thickness at which the substrate removal rate becomes independent of biofilm thickness.

Microprobe studies[123] to determine oxygen profiles within the biofilm indicated that respiration ceased at depths of 50–100 μm, depending on the substrate concentration. This further suggests that substrate removal is only caused by microorganisms within the active biofilm layer. Most investigators now believe that the active biofilm thickness results from oxygen and substrate transport or diffusional limitations within the biomass. Most of the studies conclude that, due to diffusional limitations of oxygen or substrate, there exists an active film thickness at which substrate removal rates reach a maximum and that removal rates do not further increase beyond that maximum active thickness.

Sufficient operational flexibility should be provided to control biofilm thickness. Because most carbonaceous material is removed in the initial stages, the first and second stages can experience the thickest biofilm growth. Load cells, devices used to "weigh" an RBC shaft in place, can be used to monitor biomass growth or accumulation. Possible biofilm thickness control methods include increasing rotational speed to enhance oxygen transfer and biofilm shearing force, periodically reversing the direction of RBC rotation to promote biofilm stripping, using supplemental aeration to increase bulk liquid DO concentration and biofilm stripping, removing baffles to reduce organic loading to a given stage, and step feeding to evenly distribute incoming flow and load. Chemical stripping of media with caustic or other chemicals may ultimately be required to control biomass buildup and shaft weight.

Dissolved Oxygen Levels. The importance of dissolved oxygen in aerobic wastewater treatment is well known as inadequate DO may be a major cause of process failure. In the RBC process, the level of available dissolved oxygen, among several operational variables, affects process performance. A minimum level of dissolved oxygen of 2 mg/L is often quoted as a requirement for proper performance.

A report noted that an RBC facility designed for carbonaceous BOD removal and nitrification failed to meet design effluent limits during both winter and summer operations.[124] Effluent limit violations in the summer were attributed to low DO levels experienced (less than 1 mg/L). However, dissolved oxygen levels increased during winter operations, contributing to sufficient carbonaceous BOD removal to achieve design expectations. Nitrification rates also improved during winter operations, but not sufficiently to achieve design effluent limits. The initial stages of RBCs had heavy *Beggiatoa* growth. Increased carbonaceous and nitrification removal rates during winter operation were attributed to increasing dissolved oxygen saturation values with decreasing temperature, which promote increased oxygen transfer.

The growth of *Beggiatoa* often suggests oxygen-limiting conditions. The low dissolved oxygen levels at high loadings promote sulfide production within a thick biofilm, which enhances the growth of sulfide-oxidizing organisms such as *Beggiatoa*. Sulfide present in influent wastewater can also stimulate *Beggiatoa* growth. *Beggiatoa*, a filamentous organism, structurally reinforces biofilm, thus complicating biomass control. This causes excess biomass and weight, and possible shaft or media failures. *Beggiatoa*, whitish autotrophic sulfur bacteria, use hydrogen sulfide and sulfur as energy sources in the presence of oxygen. These sulfide-using organisms compete with heterotrophic organisms for available oxygen and space on the RBC media surface. Their predominance can cause an increase in the concentration or thickness of biomass on an RBC unit while, at the same time, substantially reducing organic removal.

The level of dissolved oxygen is an important consideration in the RBC process because aerobic, nitrifying organisms are likely more sensitive to dissolved oxygen levels than are heterotrophic organisms. Minimum DO levels needed for nitrification vary from 0.5 to 4.0 mg/L; however, a minimum DO level of 2 mg/L is often quoted.[119]

When RBC systems are designed to produce low effluent ammonia nitrogen levels of 2 mg/L or less, generally the DO levels increase at the effluent end. Increased DO concentrations at the effluent end, combined with low levels of soluble BOD_5, can reduce nitrification rates because higher life forms (rotifiers, nematodes, and other bacterial predators), which ingest nitrifying organisms, become more numerous. To discourage bacterial predators, DO concentrations should not be allowed to increase to more than 3.5 mg/L, and a soluble BOD_5 of 6 to 8 mg/L should be maintained in the nitrification train.

RBC designs should include means of increasing dissolved oxygen levels by providing variable-speed drives, supplemental aeration, piping for effluent recirculation and step feeding of influent flow, and removable baffles, particularly in the initial stages.

Operational Flexibility. An RBC system should provide adequate flexibility for good operation and maintenance. The following measures deserve consideration:

- Possible need for supplemental aeration for mechanically driven systems to accommodate higher first-stage organic loadings;
- A necessary means for removing excess biofilm growth such as air- or water-stripping chemical additives, rotational speed control/reversal, and so on;
- Variable rotational speeds in first and second stages;
- Multiple treatment trains;
- Removable baffles between all stages;
- Positive influent flow control to each unit or flow train;
- Positive controlled alternate flow distribution systems such as step feed;
- Positive air-flow metering and control to each shaft when supplemental aeration or air drive units are used;
- Recirculation of secondary clarifier effluent;
- DO-monitoring equipment in the initial stages;

- Ease of access to shaft, media, and other mechanical equipment needing inspection, maintenance, and possible periodic removal or replacement;
- Tank drains (a necessary item);
- Load cells installed on all shafts (a necessary item); and
- Where RBCs are installed in a building, ventilation requirements (in addition to RBC oxygen requirements), humidity control, heating requirements, and provisions for removal and replacement of shaft and media.

PROCESS DESIGN. RBCs may be employed for organic removal; combined organic removal and nitrification; or separate-stage nitrification, denitrification, or both.

Numerous established design approaches can achieve the above objectives, including the use of pilot-plant studies, mathematical models, and empirical procedures. Pilot studies are suggested where economic considerations and feasibility warrant such efforts or where atypical municipal wastewater characteristics are anticipated. Whenever possible, calibration of some of the more complex mathematical models can be combined with an RBC pilot program to obtain values for model coefficients. When pilot-plant evaluations are not possible, the designer must use empirical design approaches, exercising technical judgment regarding their applicability and adaptability to specific situations and conditions.

Pilot-plant Studies. RBC design information can best be obtained by a comprehensive, on-site, full-scale pilot-plant evaluation of the wastewater. The use of full-scale-diameter media for pilot study is needed to avoid scale-up problems because studies have indicated higher removal capabilities per unit surface area when less than full-diameter units are used.[125,126] Pilot tests should be run for extended periods under diurnal- and variable-loading conditions and seasonal periods to test the performance of the system for the conditions that would be experienced by full-scale facilities. The tests should also identify any long-term changes in system performance due to solids accumulation in media or other factors. If small-diameter units are used in a pilot-plant study, stage loading below the oxygen transfer capability of a full-diameter unit is suggested to reduce scale-up problems.

Empirical Design Approaches. Few designers have used mathematical models in the design of RBC systems. In most cases, empirical approaches are used because simplistic mathematical models do not account for all of the variables that influence performance.

An empirical design equation was developed as part of a study for EPA[127] to predict the performance of RBC systems. The equation was based on

---

empirical relationships developed by Velz and Shultze for the performance of TFs and packed biological towers.[48,49] The equation predicts effluent BOD_5 based on influent BOD_5, media volume, and hydraulic loading:

$$S_e/S_i = e^{-K(V/Q)^{0.5}} \qquad (31)$$

Where

S_e	=	secondary clarifier effluent total BOD_5, mg/L;
S_i	=	RBC influent total BOD_5, mg/L;
e	=	2.718 3;
V	=	media volume, cu ft;
Q	=	hydraulic loading, gpm; and
K	=	reaction constant, 0.30.

To establish the reaction constant, Benjes reviewed data from a number of operating plants.[127] Figure 12.36 compares operating data from 27 full-scale plants with predictions from the Benjes relationship. Data represent moderate and warm operating temperatures. As indicated, the relationship provides a good estimate of the average performance of the process, although data are somewhat scattered. Note that plant data indicate a tendency for increased percent BOD removal for higher influent concentrations at the same loading

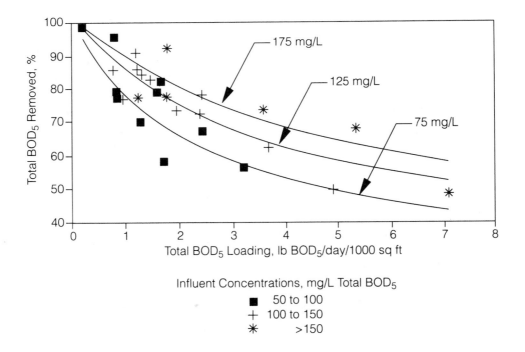

Influent Concentrations, mg/L Total BOD_5
- ■ 50 to 100
- + 100 to 150
- ✳ >150

Figure 12.36 Comparison of RBC plant data with Benjes equation results for BOD_5 removed versus organic loading (lb/day/1000 sq ft × 0.483 3 = kg/100 m²·d).

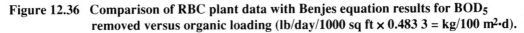

rate and that this tendency is also predicted by the equation. The scatter of data reflects differences in performance among plants and suggests a conservative approach for designing the process.

Figure 12.37 shows the trend line for total BOD_5 removed versus total BOD_5 applied, which was established from a study prepared for EPA.[128] Shown on the figure are data from 16 plants evaluated as part of the EPA study and 11 other full-scale plants.[129] Data represent moderate- and warm-temperature operation. As indicated, the trend line provides a good prediction of average performance, although the scatter of the data calls for a conservative approach in using the trend line for design.

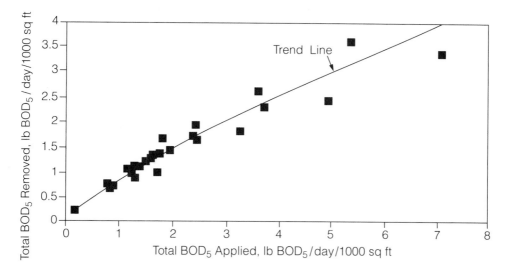

Figure 12.37 BOD_5 removed by RBC process versus BOD_5 applied (lb/day/1000 sq ft × 0.483 3 = kg/100 m² •d).

Most existing RBC plants were designed on the basis of design curves developed by equipment manufacturers. In many circumstances, actual plant performance has been well below that indicated by the curves. Figure 12.38 shows the data of Figures 12.36 and 12.37, with the predictions from design curves developed by a major supplier of RBC systems. The assumed 50% soluble BOD_5 value for RBC influent and clarifier effluent is typical of RBC systems. As indicated, the design curves substantially overestimate the performance of the process, based on the plant data shown.

Figure 12.39 shows the performance data from the 27 plants in relation to predictions from the Benjes equation for various K rates. Although the K rate of 0.30, originally proposed by Benjes, provides a good indicator of average performance based on these data, a K rate of 0.20 to 0.25 would provide a more conservative design approach, recognizing the variation in plant performance data.

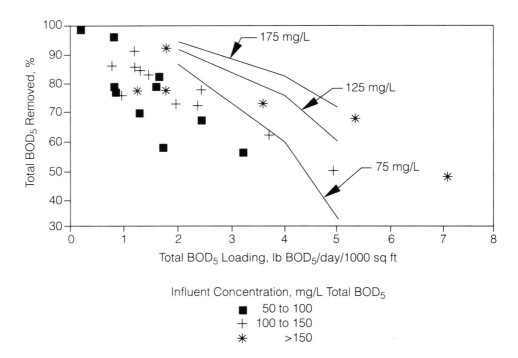

Figure 12.38 Comparison of RBC plant data with vendor curves for total BOD$_5$ removed versus organic loading (lb/day/1000 sq ft × 0.483 3 = kg/100 m^2 •d). (Curves assume 50% soluble BOD.)

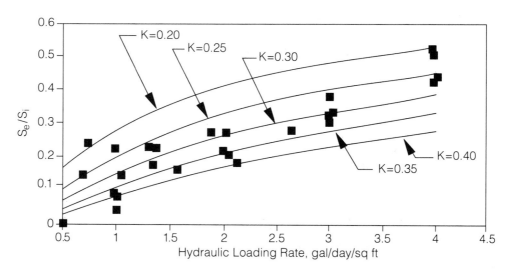

Figure 12.39 Comparison of plant data with Benjes equation for plant effluent total BOD$_5$ to plant influent total BOD$_5$ ratio, S$_e$:S$_i$ versus hydraulic loading and K rate (gpd/sq ft × 0.040 74 = m^3/m^2 •d).

The effect of cold-weather operation must also be considered in the design of RBC systems. Manufacturers of RBC equipment have developed design curves to account for reduced performance during low-temperature operation. Benjes[127] reported a reduction in K rate from a value of 0.3 above approximately 13°C to a value of 0.2 at approximately 5°C.

In designing an RBC system, the following measures merit consideration:

- Provide fine screens or primary clarification ahead of the RBC units.
- When higher H_2S concentrations are anticipated, evaluate preaeration.
- When the daily peak-to-average flow ratio exceeds 2.5, evaluate flow equalization.
- Include in-plant sidestream loadings in plant design.
- Organic loading to the first stage should not exceed 3.1 kg total $BOD_5/100$ $m^2 \cdot d$ (6.4 lb total BOD_5/d/1000 sq ft) or 1.2 kg soluble $BOD_5/100$ $m^2 \cdot d$ (2.5 lb soluble BOD_5/d/1000 sq ft).
- Evaluate appropriate staging and configuration to meet required effluent limits.
- When wastewater temperatures are less than 13°C (55°F), include operation at these low temperatures in the pilot study.
- Provide for operational flexibilities discussed earlier to improve RBC performance and properly maintain the RBC system.

Nitrification Design. Major variables influencing nitrification in an RBC system include influent organic and nitrogen concentrations, dissolved oxygen concentration, wastewater temperature, pH and alkalinity, and influent flow and load variability.

Nitrification designs for RBCs are often based on pilot-plant studies.[119] However, some equipment manufacturers have developed empirical procedures for the design of RBC nitrification systems.

Staged or plug flow configurations promote development of nitrifying organisms. Therefore, appropriate staging is essential for nitrification to occur in RBC systems. The growth of nitrifiers in any stage depends primarily on the soluble organic concentration present in that stage's wastewater. Generally, heterotrophic bacteria will dominate when the organic concentration is high. Field data indicate that nitrification is generally observed when soluble BOD_5 in stage wastewater declines to approximately 15 mg/L, and maximum nitrification occurs when soluble BOD_5 drops to 10 mg/L or less.[119]

TKN should be the design basis for nitrification systems, with use of adequate peaking factors. RBC nitrification designs based solely on influent ammonium nitrogen concentration may be undersized because of hydrolysis that can occur within the treatment system, resulting in the conversion of organic nitrogen to ammonia. Flow and mass loading variations have significant impact on the nitrification efficiency of RBCs when compared with their impact on organic removal performance. Thus, whenever flow and mass surges are experienced, the RBC system effluent generally has higher ammonia nitrogen levels unless adequate surface media are provided to compensate for influent variations. Flow equalization should be considered when flow variations are high and low effluent ammonia nitrogen levels are required.

When ammonium nitrogen exceeds 5 mg/L, the removal is claimed to proceed at a zero-order rate of approximately 0.1 kg NH_4–N/100 m^2·d (0.3 lb NH_4–N/d/1000 sq ft) at 55°F, as shown in Figure 12.40. The maximum nitrification rate of 0.1 kg NH_4–N/100 m^2·d (0.3 lb NH_4–N/d/1000 sq ft) represents an oxygen demand of 0.68 kg/100 m^2·d (1.4 lb/d/1000 sq ft), which nearly matches the maximum oxygen transfer rate of approximately 0.73 kg O_2/100 m^2·d (1.5 lb O_2/d/1000 sq ft) observed at rotational speed of approximately 1.6 rpm.[130] Thus, the maximum nitrification rate of 0.1 kg NH_4–N/100 m^2·d (0.3 lb NH_4–N/d/1000 sq ft) is limited by the oxygen transfer capabilities of full-scale RBC systems. Although some pilot studies[131] have reported nitrification rates as high as 0.3 kg NH_4–N/100 m^2·d (0.7 lb NH_4–N/d/1000 sq ft), these rates are not possible with full-scale diameter RBC units rotating at 1.6 rpm. Higher oxygen transfer rates are only possible with pilot units at increased rotational velocities and at oxygen transfer rates (per m^2) higher than those of full-scale systems.

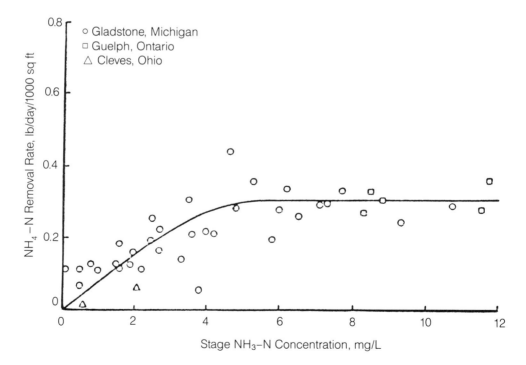

Figure 12.40 Full-scale RBC nitrification rates at design wastewater temperature of 55 ± 2°F (lb/day/1000 sq ft × 0.483 3 = kg/100 m^2 • d).

Wastewater temperature is the major variable controlling full-scale RBC nitrification below 13°C (55°F), becoming increasingly dominant when temperatures of 4°C (40°F) are approached. Wastewater temperatures higher than 13°C (55°F) do not result in higher nitrification rates in full-scale units because the oxygen transfer rate, rather than the biological growth rate, controls the rate of nitrification. Pilot-plant studies have indicated[131,132] increased nitrification rates when pH was increased from 6.5 to 8.6; however, such enhancement was seldom demonstrated with full-scale RBC units.[119]

RBC manufacturers have developed design curves to account for reduced performance during cold-weather operation. These curves typically indicate that operation at a wastewater temperature of 5°C requires approximately 2.5 times more media surface area than that for operation at 13°C achieving the same performance.

PHYSICAL DESIGN. Proper design of an RBC system includes an adequate determination of influent loadings and the impacts of sidestream loadings, selection of an overall plant layout that is compatible with other process units in the treatment system, and sufficient flexibility to promote good operation and maintenance practices. This section will discuss physical plant design factors such as layout, number of stages, equipment, and operational flexibilities.

Several manufacturers market full-scale RBC equipment in the U.S. Each manufacturer produces units with its own unique criteria and characteristics. Consequently, RBCs from different manufacturers have mostly dissimilar components, including shafts, plastic media configurations, methods of supporting plastic media, methods of separating plastic layers, bearings, and drives.

Media. Media density should be selected for the RBC system to maximize removals without sacrificing equipment life and performance because of weight overload or biomass bridging or plugging. As surface area density increases, media openings for air and water decrease in cross-section size, thus requiring a limitation of organic loading and biomass thickness, which could otherwise become detrimental to the process and equipment. Media should contain a UV inhibitor such as carbon black.

High-density polyethylene (HDPE) has been the standard for plastic media since 1972 due to its reduced cost when compared with polystyrene. Media assemblies with surface areas of 9300 to 16 700 m^2 (100 000 to 180 000 sq ft) are available for 8.2-m (27-ft) shaft length and 3.7-m (12-ft) media diameter. Manufacturers have developed various media corrugations, each with its own claimed advantages. Advantages of corrugations include increased stiffness of disks, increased available surface area, increased exposure time to air for greater oxygen transfer, constant and uniform wetting of biofilm with wastewater, and use as spacers between sheets.

Generally, low-density media are used in the lead stages of an RBC train to reduce clogging problems because of high organic loadings in the initial stages. Approximately 9300 m^2 (100 000 sq ft) is supported on 3.7-m (12-ft) diameter RBCs on an 8.2-m (27-ft) shaft. Medium- and high-density media are usually installed in the final stages of RBCs (mostly for nitrification), offering 11 100 to 16 700 m^2 (120 000 to 180 000 sq ft) of surface area for the same shaft length and diameter.

Shaft. The shaft supports and rotates the media if the RBC includes mechanical drives. For air-drive RBCs, the shaft supports the media. Typical maximum shaft length is 8.2 m (27 ft), with 7.6 m (25 ft) occupied by media.

Shaft selection, concurrent with design, must include consideration of loading, process requirements, operating environments, performance, and structural preferences. Various manufacturers use different shaft designs, which depend partly on the method of media support and attachment. Shaft shapes have included square, round, and octagonal among different manufacturers. Steel shaft thicknesses from 13 to 25 mm (0.5 to 1 in.) have been used with various synthetic coatings and galvanized coatings to reduce corrosion effects. In selecting RBC shafts, the designer should also consider attachment of stub ends to main shafts for bearings, connection of media to the shaft, maximum shaft deflection under maximum load, corrosion resistance, fatigue factors, and construction techniques for fabrication of shafts.

The length of the RBC shaft, the diameter and length of the media module, and the density of the media determine the amount of media surface area on the RBC shaft. The shaft should meet load support criteria for long life for the given application. Shaft designs should be verified by experience, and the quality of fabrication of the shaft should be verified with applicable quality assurance measures.

Drive Systems. Both mechanical- and air-drive options are available for RBC systems. Consistent rotational speed is required to provide uniform growth on media. Surface growth must be uniform to prevent the RBC from becoming unbalanced, thereby increasing torque loads on the shaft and drive of mechanically driven units or causing "loping" of air-driven units. Except for maintaining the proper rotation of the RBC, the type of drive does not directly affect the RBC treatment performance.

Generally, RBCs have mechanical-drive systems consisting of motor, speed reducer, and V-belt or roller chain components. Motor output speed is reduced to approximately 1.6 rpm through the use of various combinations of multi-V-belts, gear boxes, and chain-and-sprocket units. Electric motors used for mechanical-drive RBCs are generally high-efficiency, 3-phase, 60-hz units. The motors, designed with protective coatings for high humidity environments, are capable of providing long-term, reliable service. The motors are typically rated at 3.7 or 5.6 kW (5 or 7.5 hp) per shaft.

Manufacturers may modify the drive packages to add variable-speed capability, which allows additional operational flexibility for DO and biofilm thickness control. This capability results from changing sheaves or installing an electronic speed controller. Air-driven RBCs can provide additional oxygen to, and impose higher shear forces on, the biofilm and, reportedly, can withstand higher loadings. Air-driven shafts have previously experienced operational problems in maintaining consistent media growth and rotational speed. They are more sensitive than mechanical drives to unbalanced media loads. The RBC can develop a loping rotation, which, if uncorrected, can eventually result in the inability to maintain rotation. Actual operating experience with air-drive systems has indicated the need for significantly more air capacity than that recommended by the manufacturer.

Bearings. In the past, RBC designs experienced deflections of longer shafts, causing unequal wearing of the shaft ends and bearings. Use of self-aligning bearing units appears to have eliminated this condition. Corrosion protection by use of moisture-resistant bearings and cover plates on the idle end of the shaft has minimized other potential shaft and bearing problems. Also, the bearings can be located outside the fiberglass media covers to permit easy access for lubrication and maintenance and reduce corrosion potential. Otherwise, the cover can include adequate room inside and a door for access to facilitate maintenance. Air saturated with water and trace amounts of corrosive gaseous elements are believed to be responsible for premature bearing failures. An inside drop shield should be placed around the shaft to prevent water from running along the shaft and into the bearing.

Tapered or spherical roller bearings are commonly used in RBC applications. Spherical bearings are preferred because their design permits a higher degree of dynamic misalignment. The degree of misalignment capability depends on bearing manufacturer, seal arrangement, and bearing size.

Covers. Where RBCs are not installed in buildings, covers are provided to protect the RBCs from atmospheric and other conditions such as inclement weather and freezing temperatures, media deterioration from UV light, and exposure to sunlight that causes algae problems. Finally, covers tend to improve performance and safety. These covers are made of fiberglass or other reinforced resin plastics for durability and lightweight handling. They conform to the general shape of the media and allow sufficient access to the units for observation and minor repairs. Generally, most covers are designed in sections that can be readily dismantled for major repairs or shaft removal. Provision for easy removal of enclosures is advisable for repairs and servicing of shafts and media. In addition, the RBC design should ensure appropriate ventilation, humidity and condensation control, and control of heat loss in cold climates.

Staging. Staging, an integral part of the overall RBC design process, is a good practice for maximizing BOD and ammonium nitrogen removals. For any kinetic order higher than zero, overall carbonaceous removal for a given media surface area will be enhanced by increasing the number of stages. RBC studies[120] have reported that, for a hydraulic loading range of 41 to 204 L/m^2·d (1 to 5 gpd/sq ft), a four-stage unit produced better effluent than that obtained with a two-stage unit having the same overall surface area and treating the same wastewater. At lower loadings, below 20 L/m^2·d (0.5 gpd/sq ft), four- and six-stage systems performed similarly. As a primary consideration, when selecting the number of stages in an RBC system, ensure that the organic loading to any individual stage is less than 1.2 kg soluble BOD$_5$/100 m^2·d (2.5 lb soluble BOD$_5$/d/1000 sq ft) and 3.1 kg total BOD$_5$/100 m^2·d (6.4 lb total BOD$_5$/d/1000 sq ft). Figure 12.41 shows various staging options for an RBC system. Manufacturer's guidelines for staging vary significantly; therefore, use discretion in applying such guidelines.[119]

Generally, BOD removal efficiency is highest in the first stage and decreases through successive individual stages. For secondary treatment, designers normally select three or four stages and may add stages for nitrification. For small facilities, four stages can be installed on a single shaft by

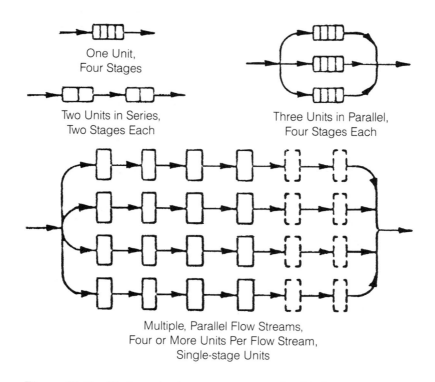

One Unit,
Four Stages

Two Units in Series,
Two Stages Each

Three Units in Parallel,
Four Stages Each

Multiple, Parallel Flow Streams,
Four or More Units Per Flow Stream,
Single-stage Units

Figure 12.41   Various staging arrangements for RBC units.

providing three interstage baffles within a tank. Overloading problems can be reduced in the first stage by removing the baffle between the first and second stages.

Flow Control. Flow distribution plays an important role in RBC operation. Normally, the RBC trains are arranged in parallel. A design including flow-measuring devices coupled with mechanisms for adjusting flow will provide the capability to maintain proper flow distribution. Balanced flow distribution will eliminate under- or overloading each train.

In addition, recirculation and step-feeding options in an RBC design can provide additional operational capabilities for flow distribution and control. Both represent an alternative method of flow and load distribution to available surface area within various stages of an RBC system. During low or intermittent flows, effluent recirculation can maintain sufficient bacteria on the media. Step feeding can reduce organic loading and oxygen demand in the beginning stages by evenly distributing incoming flow and load to the various stages within an RBC system. The design should provide for increased organic and hydraulic load due to effluent recirculation. Effluent recirculation can also improve the dissolved oxygen within the various stages of an RBC system.

Instrumentation. Electronic or hydraulic load cells are available for periodically measuring total shaft weight, which, in turn, may be translated into an estimate of biofilm thickness. The shaft-weighing device consists of a load cell bearing installed under the shaft support bearing on the idle end of a mechanically driven shaft. For the hydraulic load cell, a hand-operated hydraulic pump attached to the load cell lifts the bearing from its base after loosening bearing support bolts, while the shaft continues to rotate or is momentarily stopped. The operator reads the resultant hydraulic pressure from a gauge and converts the reading to shaft weight, which can be converted into an estimate of biofilm weight and thickness. Such measurements are useful in determining conditions that could cause excessive fatigue stress on the shaft or media. The measurements can also forewarn operators of growth of undesirable organisms such as *Beggiatoa*. The information may be used to take appropriate measures to control biofilm growth and thickeners.

The electronic strain gauge cell, a recent development, enables shaft load to be measured continuously without lifting the idle end bearing off its base. Also available, a companion converter can be connected to this type of load cell to provide a direct read-out to total shaft weight. Electronic strain gauge load cells are primarily applicable to new installations as retrofitting existing RBC installations would be difficult and costly. The designer should ensure that all shafts will have load cells or electronic strain gauge cells.

Dissolved oxygen monitoring and control is another operational tool that is important in RBC process control. Because of higher organic loadings, the

---

critical locations for DO monitoring are the first and second stages. Low DOs may indicate increased organic or hydraulic loading or both. Other instrumentation worthy of note are rpm indicators and motorized and remotely controllable air valves for speed control of air-driven systems.

DESIGN EXAMPLE. This example illustrates a method for designing an RBC system for carbonaceous BOD_5 removal.

Example 12.4. Design an RBC facility for carbonaceous BOD_5 removal (English units).

GIVEN.

Effluent BOD_5 limitation	=	30 mg/L,
Flow	=	5 mgd,
Primary effluent BOD_5	=	125 mg/L,
Temperature	=	10°C,
Shafts (low density)	=	100 000 sq ft/shaft,

Allow 15% extra surface area for 10°C operation, and
Allow 25% extra surface area as a safety factor.

SOLUTION. Step 1. Select organic loading.

BOD_5 removal, %	=	100% − (effluent BOD_5/influent BOD_5) (100)
	=	100% − (30/125) (100)
	=	76%

Using Figure 12.36 for 76% removal, select an organic loading of 1.8 lb BOD_5/d/1000 sq ft.

Step 2. Calculate required surface area.

lb/d BOD_5 loading	=	5 mgd (8.34 lb/gal) (125 mg/L)
	=	5212 lb/d
Shaft total surface area	=	(5212 lb/d BOD_5)/(1.8 lb BOD_5/d/1000 sq ft)
	=	2.9 mil. sq ft at a warm operating temperature ~ 20°C without a safety factor.
Shaft total surface area	=	2.9 mil. sq ft (1.15) (1.25) for temperature of 10°C with 25% safety allowance
	=	4.2 mil. sq ft

Step 3. Determine number of 100 000-sq ft shafts.

No. of shafts	=	(4.2 mil. sq ft)/(100 000 sq ft/shaft)
	=	42 shafts

Select 6 trains of 7 shafts each.

Step 4. Determine media area required in first stage using maximum allowable first-stage loading of 6.4 lb BOD5/d/1000 sq ft.

First-stage media area   =   (5212 lb BOD5/d)/(6.4 lb/BOD5/d/
1000 sq ft)
=   814 000 sq ft

Allow 2 shafts of each of the six 7-shaft trains for first stage.

Thus, first-stage loading   =   (5212 lb BOD5/d)/[12 shafts
(100 000 sq ft/shaft)]
=   4.3 lb BOD5/d/1000 sq ft

Step 5. Determine media area required in second stage. Assuming that approximately 50% of total BOD5 is soluble and that soluble BOD5 is removed in proportion to total BOD5, Figure 12.37 indicates approximately 60% soluble BOD5 removal in the first stage.

SBOD5 second-stage loading   =   5212 lb BOD5/d (100 − 60/100)
(50/100)
=   1042 lb SBOD5/d

Assigning one shaft per train (6 shafts × 100 000 sq ft = 600 000 sq ft) to the second stage, check the SBOD5 second-stage unit loading.

Second-stage SBOD5 loading   =   (1042 lb SBOD5/d)/[6 shafts
(100 000 sq ft/shaft)]
=   1.7 lb SBOD5/d/1000 sq ft

Because the above unit loading is less than the 2.5 lb SBOD5/d/1000 sq ft unit loading allowable, one shaft per train will suffice for the second stage.

Step 6. Determine the final arrangement of the RBC system. Use 6 trains of 7 shafts each (42 shafts) with two shafts per train (12 shafts) as the first stage. Configure remaining shafts optionally with provisions for removable interstage baffles.

*D*UAL BIOLOGICAL TREATMENT

Most dual processes use a fixed film reactor in series with a suspended growth reactor. In dual processes, the fixed film reactor usually consists of a biological tower, and the suspended growth reactor is generally an aeration basin or small contact channel. Other types of dual processes have been used where rotating biological contactors, lagoons, or other treatment processes are in series with biological towers. Dual processes have been described by various names such as two-stage, series, coupled, or combined processes.

However, in this manual, the term dual process will be used to denote the coupling in series of two different reactors of which at least one is a fixed film reactor.

This section does not repeat design guidance for individual treatment processes already described, that is, suspended growth, attached growth, or ponds. Highlighted in this section are special descriptions, sizes, or design considerations that depart from the individual or parent processes described elsewhere in the manual.

The use of dual processes has become increasingly popular, largely because designers often attempt to minimize weaknesses of individual systems by combining several processes. For example, fixed film processes are known for their ability to resist shock loads, energy efficiency, and low maintenance. By combining a shock-resistant fixed film reactor with a suspended growth process known for producing high-quality effluent and capable of operating under various treatment modes, designers have found the combination can often yield an overall biological treatment method that highlights the advantages of the two parent processes. Conversely, the weaknesses of several parent processes have been overcome by combining individual treatment steps into a dual process.

TERMINOLOGY. Much of the terminology used to describe dual processes is covered in the parts of this manual that discuss parent processes. Individual chapters can be referred to for a complete description of common terms and Table 12.16 provides a summary of common acronyms used with dual processes.[113]

ALTERNATIVES. Numerous combinations of dual processes are possible, depending on the type of parent process used, loading to the treatment units, and the point at which biological or recycle sludge is reintroduced to the main flow stream.[133] Commonly used dual processes are generally included in two groups: those that have low to moderate organic loadings to the fixed film reactor and those with high (roughing) organic loads. The dual-process schematic in Figure 12.42 illustrates alternative methods for returning secondary sludge or reaeration that are common practices and includes separate terms used by designers to distinguish process modes.

Activated Biofilter. The activated biofilter (ABF) process uses a fixed film reactor that treats moderate organic loads. High-rate plastic or redwood media must be used in the ABF fixed film reactor (rather than rock) because return sludge is incorporated with the primary effluent and recycled over fixed film media. Some designers have concluded that improved sludge settleability occurs with ABF when biological solids are incorporated with the primary effluent before distribution to the fixed film reactor. One theory of why the benefit (lower SVIs) occurs is that the high F:M ratio and plug

Table 12.16 Common acronyms used with dual processes.

Process name	Acronym
Parent (single-stage) process	
Activated sludge	AS
Aerated lagoon	AL
Biofilter	BF
Facultative lagoon	FL
Pure oxygen activated sludge	POAS
Roughing filter	RF
Solids contact	SC
Trickling filter	TF
Rotating biological contactor	RBC
Solids reaeration	SR
Solids contact and reaeration	SCR
Combined process	
Activated biofilter	ABF
Trickling filter–solids contact	TF/SC
Roughing filter–activated sludge	RF/AS
Biofilter–activated sludge	BF/AS
Trickling filter–activated sludge	TF/AS
Roughing filter–RBC	RF/RBC
Roughing filter–aerated lagoon	RF/AL
Roughing filter–facultative lagoon	RF/FL
Roughing filter–POAS	RF/POAS

flow of the filter allows heterotrophic bacteria to be more competitive than filamentous bacteria. A similar observation has been made with the use of selectors (oxic, anoxic, and anaerobic) and pure oxygen activated sludge.

Although performing well at low organic loads, the ABF process has been generally unsuccessful in consistently achieving good effluent quality (less than 30 mg/L BOD and TSS) as organics loads approach 96 to 160 kg BOD/100 $m^3 \cdot d$ (60 to 100 lb BOD/d/1000 cu ft). The ABF process without short-term aeration has also performed poorly in cold climates. To overcome these problems, the ABF tower was later modified to include a relatively small aeration basin. This will be discussed as the biofilter activated sludge (BF/AS) mode for dual processes.[46]

Trickling Filter Solids Contact. The trickling filter solids contact (TF/SC) process generally includes a fixed film that has low to moderate organic loads followed by a small contact channel. The contact channel is generally 10 to 15% the size that would normally be required in an aeration basin for activated sludge alone. By combining the fixed film reactor with the contact channel, the fixed film reactor size is reduced by 10 to 30% of that normally required if treatment had been accomplished with a TF.[134]

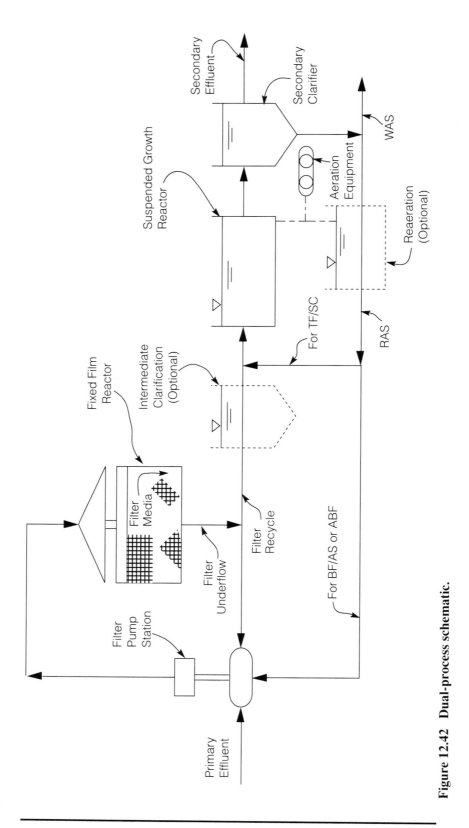

Figure 12.42 Dual-process schematic.

Benefits of the TF/SC process stem from the low power requirements due to a relatively high dependence on the TF to remove most of the soluble BOD. Other benefits include the ability to upgrade existing rock TFs through polishing the fixed film effluent by using return activated sludge as a bioflocculating agent.[68,134]

Conventional TF/SC does not include sludge reaeration. However, when both solids contact and reaeration are used, the acronym TF/SCR is used to describe the mode. When solids contact is eliminated and solids reaeration is the only type of suspended growth process used, then the acronym TF/SR describes the mode of TF/SC.

Roughing Filter–Activated Sludge. A common method of upgrading existing activated plants is to install a roughing filter ahead of the activated sludge process. As part of the roughing filter–activated sludge (RF/AS) process, the roughing filter is generally 12 to 20% of the size required if treatment had been accomplished through the use of the trickling filter process alone. Hydraulic retention time in the aeration basin is generally 35 to 50% that required with the use of the activated sludge process alone.

Both TF/SC and RF/AS have the same process schematic but with RF/AS, a much smaller TF is used so that the aeration basin must provide a significant amount of oxygen, BOD removal, and solids digestion. This differs from the TF/SC process where the TF is larger and provides almost all of the SBOD treatment, allowing the contact channel to provide only enhanced solids flocculation and effluent clarity. A deciding consideration in determining whether it is best to use the TF/SC or RF/AS process is often the availability of existing treatment units that influence the balance between capital and operating expenses.

Biofilter–Activated Sludge. The biofilter–activated sludge (BF/AS) process is similar to that of RF/AS except that with BF/AS, return activated sludge (RAS) is recycled over the fixed film reactor similar to the recycle of the ABF process. Incorporating RAS recycle over the fixed film reactor has sometimes reduced sludge bulking from filamentous bacteria, especially with food-processing wastes, which are difficult to treat. Although it has sometimes improved sludge settleability, there is no evidence that sludge recycle improves the oxygen transfer capability of the biological filter.[133]

Trickling Filter–Activated Sludge. The trickling filter–activated sludge (TF/AS) process is designed for high organic loads similar to those of RF/AS or BF/AS. However, a unique feature of TF/AS is that an intermediate clarifier is provided between the fixed film and suspended growth reactors. The intermediate clarifier removes sloughed solids from the fixed film reactor before the underflow enters the suspended growth reactor.

A major benefit of using the TF/AS mode of combined process is that solids generated from carbonaceous BOD removal can be separated from second-stage treatment. This is often a preferred mode where ammonia removal is required and the second stage of the process is designed to be dominated by nitrifying microorganisms. Another advantage in using intermediate clarification is reduced effects from sloughing of the fixed film on the suspended growth portion of the plant. However, designers generally do not believe there is evidence of significant reduced oxygen demand or improved sludge settleability from use of intermediate clarification. To eliminate the cost of intermediate clarification, most high-rate or roughing filters are designed as RF/AS or BF/AS, rather than in the TF/AS mode. Refer to Figure 12.43 for diagrams of the various dual processes currently being used.

PROCESS DESIGN. The design of dual processes is a balance between two biological treatment reactors; neither the fixed film nor the suspended growth processes can be sized separately. Good design includes consideration of the total integrated facility, including interactions among all units, especially the secondary clarifier.[46]

Numerous mathematical models are available for predicting the performance of the fixed film reactor while operating as the sole method of biological treatment. However, few rational equations are available for determining the performance of the second stage (either fixed film or suspended growth) of a dual process. Because of this, most engineers use equations only for interpolating the response of dual processes with various load conditions or during various seasons. Instead, common design practice bases reactor sizing for dual processes on either pilot- or full-scale (case history) information. Design criteria generally used by engineers for sizing dual processes are presented in Table 12.17.

The first-stage fixed film reactor in the RF/AS, BF/AS, or TF/AS process supplies from 30 to 50% of the oxygen requirements generally required for total biological treatment. However, with the TF/SC or ABF processes, essentially all of the oxygen for biological treatment is supplied in the first-stage fixed film reactor.

The second-stage suspended growth reactors with RF/AS, BF/AS, or TF/AS generally range from 15 to 50% the size that would have been required if a first-stage fixed film reactor had not been used. With the TF/SC process, the suspended growth reactor size is generally 10 to 15% of the size that would have been required if no first stage had been used. With the ABF process, the suspended growth contact is achieved through hydraulic retention as the combined recycle solids and primary effluent pass through the filter media.

The inability to control interdependent variables makes it impossible to precisely predict the performance of dual processes.[11] A number of these

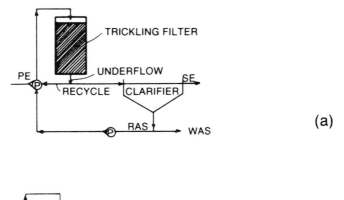

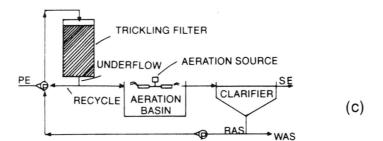

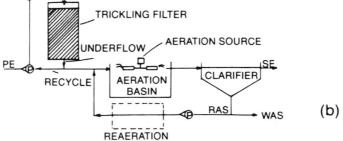

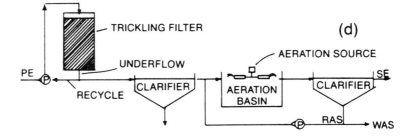

Figure 12.43   Combined process options: (a) ABF, (b) TF/SC and RF/AS, (c) BF/AS, and (d) TF/AS.

Attached Growth and Dual Biological Treatment   **799**

Table 12.17   General design criteria for dual processes.

Dual processes	Appropriate design criteria	
	Range	**Commonly used**
ABF		
Media type	High rate	High rate
BOD loading[a]	10–75	30
Hydraulic loading[b]	0.8–5.0	2.0
Filter MLSS[c]	1500–3000	2000
TF/SC		
Media type	Rock or high rate	High rate
TF BOD loading[a]	20–75	40
TF Hydraulic loading[b]	0.1–2.0	1.0
SC Basin MLSS[c]	1500–3000	2000
SC Hydraulic residence time[d]	0.2–0.5	0.4
SC Mean cell residence time[e]	0.2–0.5	0.4
SC RAS[c]	6000–12 000	8000
SC Minimum basin mixing		
Diffused air[f]	2000–4000	3000
Mechanical[g]	60–130	100
RF/AS, TF/AS, and BF/AS		
Media type	High rate	High rate
TF BOD loading[a]	75–200	150
TF Hydraulic loading[b]	0.5–5.0	1.0
AS Basin MLSS[c]	1500–4000	2500
AS Hydraulic residence time[d]	2.0–4.0	3.0
AS Mean cell residence time[e]	2.0–6.0	3.0
AS F:M[h]	0.5–1.2	0.9
AS Basin oxygen[i]		
Total available	0.6–1.2	0.9
Normally supplied	0.3–0.9	0.6

[a] BOD loading = lb BOD/d/1000 cu ft; lb/d/1000 cu ft × 1.602 = kg/100 m^3·d.
[b] Hydraulic loading = gpm/sq ft; gpm/sq ft × 2.44 = m^3/m^2·h.
[c] mg/L.
[d] Hydraulic residence time = hours based on influent Q only (no RAS).
[e] Mean cell residence time (MCRT) = days (aeration basin only).
[f] Minimum basin mixing, diffused air = scf/min/mil. gal; cu ft/min/mil. gal × 0.125 = L/1000 m^3·s.
[g] Minimum basin mixing, mechanical = hp/mil. gal; hp/mil. gal × 0.197 = kW/1000 m^3.
[h] F:M = lb primary effluent BOD/day/lb MLVSS.
[i] Basin oxygen = lb O_2/lb primary effluent BOD removed.

variables have been considered in general in the process concepts and principles section of this chapter. However, design engineers often choose both design loads and process modes for dual processes that produce the desired effluent quality and sludge characteristics under a wide range of variables.

Design Considerations. Other than environmental variables, several factors, each explained below, deserve special consideration with dual processes.

PRIMARY CLARIFICATION. Studies have shown that both the amount and nature of suspended solids in the primary effluent can greatly affect the performance and solids yield from dual processes. The effect will vary, depending on the type of filter media used in the first-stage process. The use of rock filter media tends to reduce the effects of high suspended solids in the primary effluent because produced and incoming solids are retained considerably longer in the small void space associated with rock media. The added solids retention allows anaerobic digestion and solids reduction to occur. Both the amount and characteristics of primary solids can greatly affect design choices for combined processes.

ENERGY. Designers generally consider that biological treatment with fixed film reactors requires 25 to 50% of the energy of suspended growth systems. However, with the emergence of fine bubble diffused air plus good suspended growth design, many design evaluations are proving the energy gap between fixed film and suspended growth reactors to be narrowing. With dual processes, one study indicated that the energy savings with either fixed film or dual-process operation was not realized because of the poor turndown of equipment while the units were operated at less than their full design load.[133] Generally, high power rates will tend to favor dual processes that have large fixed film reactors (ABF or TF/SC processes).

SOLIDS RECYCLE. A number of pilot studies have indicated that the recycling of biological solids over the fixed film reactor may improve sludge settleability. However, no evidence shows that sludge recycle will aid oxygen transfer or reduce the size of the second-stage suspended growth reactor. Sludge recycle has been used extensively in the treatment of food-processing waste in both the ABF and BF/AS dual-process modes. Sludge recycle has been used extensively with horizontal, redwood media as a means of increasing its relatively low surface area in comparison with plastic media. Sludge recycle use with vertical, plastic media is limited but has shown to sometimes provide the same improved sludge-settling characteristics as achieved when applied with redwood media. Sludge recycle has been reported to increase odors at some facilities.

NITRIFICATION. Requirements to achieve nitrification often sway process economics toward suspended growth rather than fixed film or dual processes. This is especially true where cold water temperatures prevail, resulting in a correspondingly greater reduction in activity for nitrifying bacteria than for BOD-removing heterotrophic bacteria. Design techniques often used to encourage nitrification for dual processes include intermediate clarification between the first and second stages. Another approach is the use of solids reaeration. Solids reaeration generally provides two to four times more mean cell residence time per unit of aeration volume than does contact through a conventional aeration basin or channel.

ODOR. Nuisance due to odors tends to be more severe with fixed film than with suspended growth reactors. Most dual processes are free from nuisance-type odors but the design engineer should, nonetheless, carefully consider the possibility of odor problems. If odor problems are occurring in the headworks or primary clarifiers, then odor control should be a significant design consideration for the fixed film reactors in dual processes. Most odor problems occur where plants have industrial loads or inorganic constituents (nitrogen or sulfur) that tend to be odor producing. Evidence indicates that odors result from waste characteristics and the presence of high percentages of certain industrial wastes. Stated differently, fixed film reactors used in dual processes with low to moderate loading can generate as much odors as can dual processes with heavily organically loaded fixed film reactors. If odors become a problem, they are often caused by (1) waste characteristics, (2) poor hydraulics, or (3) lack of ventilation.

SLOUGHING. Natural biological sloughing that occurs from fixed film reactors can be controlled (at least to a limited degree) by varying the hydraulic wetting rate to the fixed film reactor. Surveys indicate that dual processes with small suspended growth reactors (ABF or TF/SC) tend to be more susceptible to variations in MLVSS than are dual processes with large suspended growth reactors (RF/AS or BF/AS). The designer should consider the TF process design section of this chapter to obtain guidance on hydraulic load design to minimize sloughing from fixed film reactors.

SNAILS. Dual processes differ from conventional TF applications in that snails can often settle in transfer structures or aeration channels with low energy for mixing. Care should be exercised in maintaining adequate mixing and minimizing the possibility of the deposition of snail cases in unmixed areas of containment structures. Some designers include a low velocity flow transfer box between the fixed film and suspended growth reactor to encourage the settling and removal of snails.

PRODUCED SOLIDS. Deciding how to use an existing rock TF may depend on the amount of produced solids in the biological treatment process. Depending on the organic and hydraulic loading, produced solids from rock media generally range from 0.4 to 0.7 kg TSS/kg BOD (kg produced solids/kg primary effluent BOD). With the use of plastic or redwood media, the amount of solids digestion and effective mean cell residence time (MCRT) is greatly reduced because retention is less than that with rock media. Consequently, solids production with synthetic media often ranges from 0.8 to 1.0 kg TSS/kg BOD, even when a dual process is used. The amount and character of solids in the primary effluent can also greatly affect both the settling characteristics and solids production from a dual process.

EXISTING TREATMENT UNITS. Potential cost savings from using existing equipment often dictate the type of dual process selected. If existing aeration basins are already available, then economics often favor the use of dual processes with large suspended growth reactors such as BF/AS, RF/AS, or TF/AS. However, if an existing trickling filter plant is being upgraded, then economics often favor the use of process modes with small suspended growth reactors such as the TF/SC dual process.

EFFLUENT QUALITY. Actual mathematical means of predicting effluent quality with dual processes are described later in this chapter. Regarding the influence of effluent quality on the choice among various dual processes, designers generally agree that dual processes can produce better effluent quality more reliably than either of the activated sludge or trickling filter parent processes. Effluent with less than 20 mg/L BOD_5 is generally achieved with good dual-process design, and 10 mg/L BOD_5 has been achieved without advanced treatment at some facilities.

SITE CONSIDERATIONS. Dual processes often require slightly less land than do other biological treatment processes for two reasons: the ability to construct tall fixed film reactors with heights of 4.9 to 7.3 m (16 to 24 ft) and the use of slightly higher loadings of both the fixed film and suspended growth systems. However, saving space is usually not an overriding advantage in choosing dual processes because the savings are usually insignificant or other processes can be modified to realize similar space savings.

DISTRIBUTION. Dual-process performance can depend greatly on the type of distribution system chosen for the fixed film reactor. Designers concur that rotary distributors are more efficient and less troublesome to operate than are fixed nozzle distributors.[46] The use of horizontal redwood or cross-flow plastic media, which offer redistribution capability within filter media, helps to prevent problems from uneven application of pumped water to the fixed film reactor. However, the use of filter media with redistribution capabilities cannot overcome the adverse effects of major uneven application typical of fixed nozzle systems.

FILTER MEDIA. Dual-process designs for both the ABF and BF/AS processes must be limited to applications where filter media have little tendency to plug such as with horizontal redwood or vertical plastic. Cross-flow media; random; and, especially, rock media have reduced void areas or configurations that can promote solids accumulation. Therefore, media plugging can occur when recycling solids with these media.

Studies indicate that performance of vertical media can equal or exceed the performance of cross-flow media at high organic loads that occur with the BF/AS, RF/AS, and TF/AS processes.[12] At low organic loads used with the TF/SC process, cross-flow media are generally considered to outperform other types.

Compaction of filter media, media collapse, or plugging problems have occurred with nearly all types of filter media, emphasizing the need for care in selection, design, and installation of filter media.[12,135] Parallel studies of filter media, literature searches, and case history information specifically related to filter media selection with dual processes should be a major part of the design of dual processes.

TEMPERATURE. Dual processes are often selected instead of conventional fixed film processes because of temperature considerations. A second-stage suspended growth process used with dual processes is a common upgrade procedure for TFs located in colder climates.

Fixed Film Reactors. Dual-process design requires the balancing of both the first- and second-stage reactor sizing.[46] In lieu of actual pilot-plant information, standard design practice is to size the first-stage fixed film reactor based on BOD loading considered to be good practice for the particular process mode. As listed in Table 12.17, loadings to dual processes such as ABF or TF/SC can range from 16 to 120 kg/100 m^3·d (10 to 75 lb BOD/d/1000 cu ft) but generally average 48 kg/100 m^3·d (30 lb BOD/d/1000 cu ft). Dual systems with fixed film reactors for higher loads such as RF/AS, TF/AS, or BF/AS generally are designed for first-stage loadings of 120 to 480 kg/100 m^3·d (75 to 300 lb BOD/d/1000 cu ft); average design loading is usually 240 kg/100 m^3·d (150 lb BOD/d/1000 cu ft).[133]

Equations, such as those presented in the biological towers section,[23] generally do not apply for the sizing of fixed film reactors associated with dual processes. However, the modified Velz equation has been used for predicting effluent soluble BOD (SBOD) concentrations from the fixed film reactors to size the second-stage suspended growth process.[11] Studies indicate that the reaction rate constant (K) for treatment of domestic waste can vary from 0.20 to 0.33 $m^2/m^3 \cdot d$ (0.06 to 0.10 sq ft/d/cu ft).[11] Several studies also indicate that the flow exponent (n) can vary, depending on the amount of biomass present on the filter media; the n values are usually approximately 0.50 for vertical or cross-flow plastic media and 0.6 to 0.7 for random and rock media.[12]

Hydraulic loading considerations for fixed film reactors associated with dual processes are similar to those of biological towers. Because higher organic loads tend to be used with dual processes rather than with single fixed film reactors, higher hydraulic loads of 0.7 to 1.3 $L/m^2 \cdot s$ (1 to 2 gpm/sq ft) are more common in dual-process designs.[133] Many designers consider it good practice to control sloughing through operation at high hydraulic loading for short periods on a weekly basis. Another approach is to reduce the speed of the distributor by use of a series of back sprays on the front face of the filter arms. This design provides a high "instantaneous" flushing rate while allowing the filter to be operated at low "average" hydraulic loads for reduced pumping costs.[136]

Suspended Growth Reactors. Once the fixed film reactor size is selected, a common design approach is to vary the suspended growth reactor size to ensure that either the desired effluent quality or sludge settleability is maintained. Typical suspended growth design criteria to match fixed film reactors at various organic loadings are presented in Table 12.17. The most common method for sizing of suspended growth reactors in dual processes uses the mean cell residence time (MCRT) calculation. Suspended growth reactors may also be evaluated based on F:M or hydraulic residence time (HRT). Still another approach involves calculations based on soluble BOD removal. This last approach is not widely used because of the wide range of coefficients, however, a brief discussion of this approach is presented in the effluent quality section under dual processes.

FOOD-TO-MICROORGANISM (F:M) RATIO. For dual processes, the F:M ratio is calculated based on the amount of BOD (or COD) in the primary effluent (neglecting removal through the fixed film reactor). The solids inventory, M, is expressed in terms of MLVSS associated with the amount of mixed liquor in the suspended growth reactor only. In comparing F:M of suspended growth reactors (see Table 12.18) for the RF/AS, BF/AS, and TF/AS processes, values are two to three times greater than those typically used for conventional activated sludge. F:M calculations are generally not used for sizing the TF/SC process because of the small basin size.

MEAN CELL RESIDENCE TIME (MCRT). Use of an MCRT approach for sizing suspended growth reactors following fixed film systems is becoming the prevalent method for design.[46] Calculations are similar to those presented in the suspended growth chapter. For dual processes, active biomass in the

---

fixed film reactor is again neglected in the calculations. For dual processes, the calculated MCRT is an "effective" MCRT where the residence time in the suspended growth reactor may be reduced to fewer than 0.5 days for TF/SC processes; a value of only 2 to 3 days may be sufficient for dual processes with highly loaded filters such as RF/AS, TF/AS, or BF/AS (see Table 12.18). These MCRTs for dual processes compare to literature values ranging from 5 to 15 days for conventional activated sludge. Consequently, substantial savings or reduction in suspended growth reactor size is possible with dual processes but this can be partially offset by the size of the fixed film or first-stage process. Also, the type of filter media used in the fixed film reactor can significantly affect the cell yield and result in the need to increase the MCRT or size of the suspended growth reactor.

Table 12.18 Typical design criteria.

| Process | Fixed film criteria TOL[a,b] | Suspended growth criteria | | |
		F:M[c]	MCRT[d]	HRT[e]
Conventional AS	None	0.3–0.5	5–15	4–8
Conventional TF	10–20	None	None	None
RF/AS, BF/AS, or TF/AS	75–200	0.7–1.2	2–6	2–4
TF/SC	20–75	1.5–3.0	0.5–1.5	0.2–0.5

[a] TOL = a total organic loading on filter (lb BOD applied/d/1000 cu ft of media).
[b] lb/1000 cu ft $\times$ 1.602 = kg/100 m^3.
[c] F:M = lb BOD$_5$/d/lb MLVSS in the aeration basin.
[d] MCRT = mean cell residence time (day^{-1}) based on suspended growth (basin) solids only divided by the amount of solids wasted daily.
[e] HRT = hydraulic residence time (hours), based on raw wastewater flow rate only.

HYDRAULIC RESIDENCE TIME (HRT). In recent years, the use of hydraulic residence time has diminished in importance for sizing suspended growth reactors with both conventional activated sludge and dual processes. As a check, HRTs of 4 to 8 hours are commonly used with conventional activated sludge. Corresponding HRTs with dual processes for heavily loaded systems (BF/AS, TF/AS, or RF/AS) generally range from 2 to 4 hours. Dual processes with fixed film reactors receiving light organic loads (TF/SC) are often sized for less than 1 hour HRT or with little or no HRT for the suspended growth reactor similar to the ABF process.

MIXING AND AERATION REQUIREMENTS. Initially, dual processes were designed as roughing filter applications where oxygen transfer was a major consideration and minimum energy for mixing to maintain solids in suspension was of little or no concern. However, modern dual-process design may also use the TF/SC mode. The TF/SC mode is mixing limited and does not require large amounts of energy for oxygen transfer. Minimum aeration requirements for mixing will vary, depending on the dimensions of the suspended growth reactor. However, literature values of 0.61 L/m^2·s (0.12 scfm/sq ft) for fine bubble diffusion, 337 L/mil.L·s (2700 scfm/mil. gal) for coarse bubble diffusion and 11.8 W/m^3 (60 hp/mil. gal) for mechanical aeration are common minimum mixing criteria.[133]

One practical way of estimating the amount of oxygen required in the suspended growth reactor is to design based on the oxygen uptake rate (OUR). For example, typical designs for activated sludge may be based on OURs averaging 30 to 50 mg O_2/L·h, with peak days from 60 to 80 mg O_2/L·h. Corresponding suspended growth reactors following highly loaded fixed film reactors (RF/AS, BF/AS, or TF/AS) would have OURs approximately 50% that of conventional activated sludge. Suspended growth reactors following lightly loaded fixed film reactors, such as those in the TF/SC mode, are typically designed for OURs of 10 to 20 mg O_2/L·h.[46]

If a fixed film reactor precedes the suspended growth reactor, this dampens or reduces the variability in oxygen requirements. At times, the oxygen demand may be controlled by either endogenous respiration or periodic sloughing from the first-stage fixed film reactor. A survey of twelve dual-process facilities included a rational basis for calculating the amount of oxygen required in the suspended growth reactor at various fixed film organic loads preceding the second-stage process.[137] The use of rational equations or graphs, such as Figure 12.44, may be useful in predicting long-term, average oxygen requirements under various loadings. However, good design practice ensures that adequate oxygen is available to supply the needs under a wide range of uncontrollable events such as sloughing, temperature variation, or other interrelated environmental variables.

Other Combinations. Most dual processes have fixed film and suspended growth reactors in series with a biological tower as the first stage and an activated sludge reactor of reduced size as the second stage. Other combinations for a dual process are possible. For example, nearly 20 roughing filter rotating biological contactor (RF/RBC) plants operate in the U.S.[138] Also, several roughing filter–lagoon systems are in operation. The lagoon in a dual process is usually facultative, but aerated lagoons have followed fixed film reactors.

In some cases, activated sludge has been combined directly with rotating biological contactors or solids contact is used with the RBC units. These combinations are generally used in plant upgrades where varying techniques are applied to maximize the use of existing treatment equipment.

ROTATING BIOLOGICAL CONTACTORS. Oxygen deficiencies on the first stage of RBCs have been a major problem. These problems have occurred

---

Attached Growth and Dual Biological Treatment 807

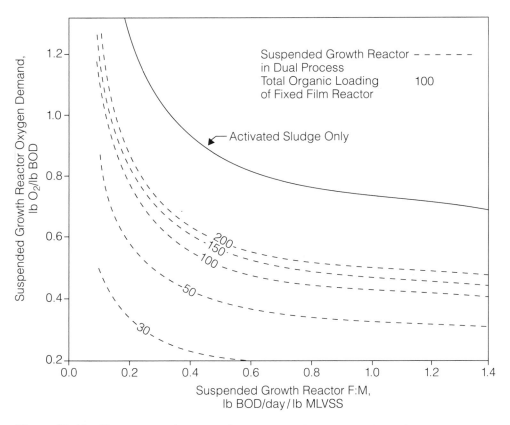

Figure 12.44 Oxygen requirements for suspended growth reactors in the dual-process mode. Numbers within the dashed lines represent total organic loading, lb BOD/d/1000 cu ft, on the fixed film reactor (lb/d/1000 cu ft × 1.602 = kg/100 m^3·d).

both at the inception of plant operation due to poor design guidance in previous times or at well-designed facilities where the design capability of the RBCs has been reached. Organic loading to RBCs is generally limited to 1.0 to 1.5 kg/SOD/100 m^2·d (2 to 3 lb SBOD/d/1000 sq ft) on the first stage and 0.5 to 1.0 kg SBOD/100 m^2·d (1 to 2 lb SBOD/d/1000 sq ft) based on all shafts in operation.

In determining the reactor sizes necessary for combined roughing filter and RBC (RF/RBC) systems, the modified Velz equation may be used to calculate the interrelationship between the first- and second-stage fixed film reactors. Limited design experience with this form of dual process indicates that a roughing filter loaded at 160 kg BOD/100 m^3·d (100 lb BOD/d/ 1000 cu ft) followed by RBC reactors sized for twice the organic load normally considered acceptable without first-stage fixed film treatment (the organic loads to both the RF and RBC are based on the kg/d in the primary effluent) should produce excellent effluent quality and avoid any organic loading of the RBCs.[138] Most RF/RBC plants in operation do not have intermediate clarification. No adverse effects or reduced efficiencies have been observed in the RBC reactors due to biological solids. Most second-stage

RBC reactors with mechanical drives have been upgraded with either air drive or air diffusers when used in the dual-process mode to prevent biological solids deposition.

LAGOONS. The use of fixed film roughing filters followed by either aerated lagoons (RF/AL) or facultative lagoons (RF/FL) can be an effective means of reducing BOD loading to the lagoons and handling biologically produced solids from the fixed film reactor. However, problems from high-effluent TSS associated with algae must be accommodated with the incorporation of lagoons for second-stage treatment. Conventional mathematical modeling or empirical estimates can be used to predict the performance of the fixed film reactor in a dual, roughing filter–lagoon system. Solids produced in the first-stage biological tower can be removed and handled in a separate digestion process or wasted to second-stage lagoons and digested through natural facultative processes.

Solids Production and Settleability. Occurrences within the fixed film reactor often dominate solids yield or production from dual-process treatment. The character and amount of inert solids in the primary effluent will also significantly influence the actual yields from the total dual process. For example, one study indicated that for dual processes with an MCRT greater than 1.0 day, little measurable benefit resulted from increasing the suspended growth reactor size to reduce sludge yield.[139]

Figure 12.45 shows various volatile solids yields for several suspended growth MCRTs and for both low and high organic loads to the first-stage fixed film reactor. The upper band in Figure 12.45 occurred during a period when the concentration of effluent suspended solids from the primary clarifier was approximately 50% higher than that from the other period. Figure 12.45 demonstrates that nonbiodegradable solids in the primary effluent can have a dominant effect on the net solids yield from a dual process. It further illustrates that significant reductions in produced solids may not occur simply by increasing the MCRT beyond 1 to 2 days.

A comparison of 43 dual-process treatment plants indicated that average solids yields were 0.7 kg TSS/kg primary effluent BOD, but that values range from 0.15 to almost 4 kg TSS/kg BOD.[133] It was concluded that sloughing and varying performances from primary clarifiers were a major cause of variability of produced solids. In addition, case history information indicates that solids from dual processes that use rock filter media may have yields 35 to 50% of those for systems with high-rate filter media.[133]

Many design engineers consider dual processes to produce better settling sludge (lower SVIs) than do conventional suspended growth (activated sludge) systems. Although dual processes are relatively resistant to shock loads, process variabilities and sludge-bulking problems can occur with dual processes in a similar fashion to those occurring in suspended growth

---

(activated sludge) processes. Therefore, no basis exists for reducing secondary clarifier sizing because of better settling sludge with dual processes. At best, designers should be satisfied that, on the average, better settling sludge can occur with the dual treatment approach. Likewise, the concentration of return sludges in a dual process will generally be similar to that of a

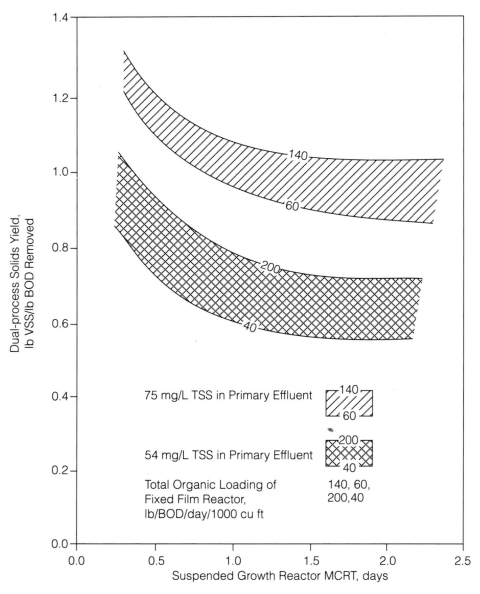

Figure 12.45 Dual-process solids yield as a function of suspended growth reactor MCRT and total organic loading of fixed film reactor with cross-flow media. Solids yield also varies with wastewater characteristics (lb/day/1000 cu ft × 1.602 = kg/100 m^3 •d).

suspended growth process. Return sludge concentrations ranging from 0.5 to 1.5% solids and averaging approximately 0.8% solids have generally been observed at plants with dual processes.

Effluent Quality. Concerns with achieving the highest effluent quality practicable often dominate process selection. An overview of numerous dual-process case histories indicates that excellent effluent quality can be achieved by any of the dual processes if designers incorporate a proper balance between the fixed film and suspended growth reactors.[133] The principal key in obtaining good effluent quality is the application of modern clarifier design with energy-dissipating center wells, low overflow rates, and depths of 3.7 to 4.6 m (12 to 15 ft).[46] For the majority of dual processes, effluent quality will be controlled not by soluble BOD, but by the amount of particulate BOD in the treated effluent.

One study comparing various organic loads ranging from 74 to 320 kg BOD/100 $m^3 \cdot d$ (46 to 200 lb BOD/d/1000 cu ft) showed that SBOD in the treated effluent decreased with increased MCRTs in second-stage suspended growth reactors.[139] However, effluent SBOD concentration was determined to be independent of organic loading to the first-stage fixed film reactor. That is, within the organic loads and MCRTs tested for domestic wastewater, varying the organic load to the fixed film reactor did not result in increased soluble BOD of treated effluents after aeration in the second-stage suspended growth step as illustrated in Figure 12.46.[139]

Numerous researchers have shown that increasing organic loads to the fixed film reactor results in increased amounts of nonsettleable suspended solids as illustrated in Figure 12.47.[139] With dual processes, second-stage treatment through a suspended growth reactor reduces the amount of effluent nonsettleable suspended solids by increasing total system MCRTs. A sharp reduction in effluent suspended solids results from providing a minimum of a 1.0-day MCRT whereas little or no additional reduction occurs with MCRTs longer than 2 days.

Several researchers have indicated that conventional Monod, first-order kinetic equations can be used to predict SBOD in treated effluent from dual processes.[12] Literature values indicate that the first-order substrate removal constant may range from 0.02 to over 0.1 L/mg MLVSS·d.[139] Because substrate removal constants can vary greatly, pilot testing is required to optimize treatment unit sizes and ensure that desired effluent quality is achievable, particularly if effluent limits are more stringent than 10 to 20 mg/L BOD.

NITROGEN OXIDATION. Nitrification through the use of dual processes is a common adaptation. One approach is to evaluate the total system (both fixed film and suspended growth reactors) through the calculation of a total MCRT.[133] Another approach is to separate the fixed film and suspended growth processes through intermediate clarification using the TF/AS process.

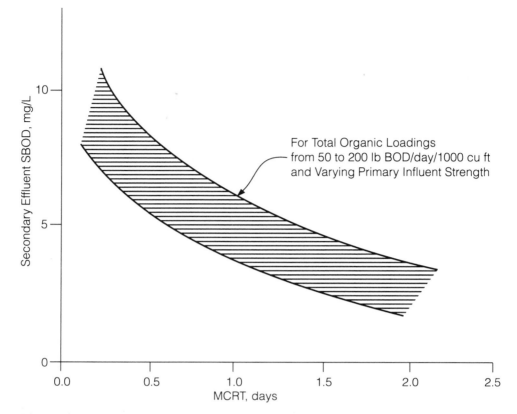

Figure 12.46 Effluent SBOD of dual process with cross-flow filter media for various MCRTs and total organic loadings. SBOD will vary with wastewater characteristics (lb/day/1000 cu ft × 1.602 = kg/100 m³ •d).

This latter approach is often preferred because it provides a more positive means of determining the actual MCRT of the second-stage suspended growth system. The TF/AS process avoids burdening the second-stage system with biologically produced solids from the first-stage system that would otherwise reduce its capability to allow nitrifying bacteria to compete for available substrate.

With dual processes, a combination of both the fixed film and suspended growth treatment steps must provide an MCRT equal to that necessary to achieve nitrification with a single-stage process such as activated sludge. One design method has been used for assigning an equivalent MCRT to the fixed film reactors so that nitrification requirements of a dual process can be determined.[46]

Except for plant upgrades, dual processes have generally not been cost effective for obtaining nitrification, compared with a single-stage activated sludge process. This is because the fixed film portion of dual processes is generally considered to have lower nitrification rates at decreased temperature[140] than does a suspended growth process.

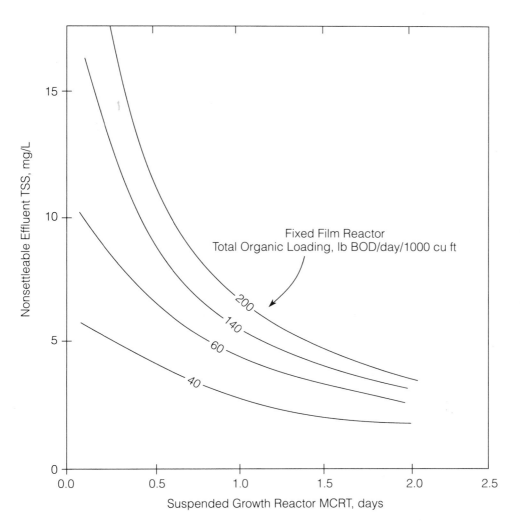

Figure 12.47 **Nonsettleable effluent TSS of dual process with cross-flow filter media for various MCRTs and total organic loadings. Nonsettleable TSS are measured in the supernatant of the settleable solids test. Secondary effluent TSS will exceed the nonsettleable TSS (lb/day/1000 cu ft × 1.602 = kg/100 m³ •d).**

PHOSPHORUS REMOVAL. Studies of phosphorus removal with dual processes indicate that chemical addition (alum or ferric chloride) does not affect the BOD removal from the biological system. Phosphorus removal by chemical addition to primary clarifiers is the most common method with dual processes. However, adaptation of biological nutrient removal by incorporating fixed film reactors and the Bardenpho™ process has also been used as a means of achieving phosphorus removal. Use of sludge fermentation to encourage the growth of phosphorus-removing microorganisms is one method of eliminating the need for chemically removing phosphorus with dual processes.[141]

Attached Growth and Dual Biological Treatment

FACILITY DESIGN. Design considerations for dual processes are similar in most respects to those for the parent processes. The designer should review appropriate parent process discussions and special design considerations for dual processes discussed below.

Fixed Film Reactors. Biological towers [Figure 12.48(a)] are the fixed film reactors most used in dual processes. Biological towers with dual processes are often operated at relatively high hydraulic wetting rates. To prevent splash or excess wind drift at the top of the tower, the walls often extend 1.5 to 1.8 m (5 to 6 ft) above the filter media. Also, for the taller towers support for jacks at the center column are commonly designed to provide support for the filter arms during a bearing replacement. For shorter TFs, a crane rather than jacks is often used to lift the mechanism.

Most dual-media fixed film filters are designed with high-rate media [Figure 12.48(d)], which is often supplied with a walking surface of high-strength media for the top layer. Many designers also specify that grating be placed on the media surface from the access ladder or walkway to and often around the center column to allow safe maintenance of the distributor.

Because fixed film reactors with dual processes are often loaded at high organic rates, designers should emphasize air ventilation. Odors are often caused by the presence of inorganic chemicals in wastewater, periodic sloughing from the filters, or constituents from industrial waste that tend to be odor producing. A study of case histories indicates that the filter underflow will have high concentrations of oxygen even at loadings of 320 to 480 kg BOD/100 $m^3 \cdot$d (200 to 300 lb BOD/d/1000 cu ft) for high-rate media. One media manufacturer recommends 0.85 m^2 of vent area per 100 m^3 of filter media (2.6 sq ft/1000 cu ft) for air ventilation. A review of operating plants indicates that natural ventilation can become limiting where less than 0.4 to 0.6 m^2 of vent area per 1000 kg of primary effluent BOD per day (2 to 3 sq ft/d/1000 lb BOD) is provided. Most biological towers used for dual processes have natural ventilation. Natural ventilation in colder climates usually includes provisions to close the louvers or vents during winter months. Likewise, if both natural and forced ventilation capabilities are provided, the vents must have provisions for closure.

Manufacturers' recommendations for forced air ventilation vary greatly depending on the literature source. One TF manufacturer recommends 0.06 m^3/kg BOD$\cdot$min (1.0 cfm/lb BOD). Another manufacturer recommends 0.16 m^3/kg BOD$\cdot$min (2.6 cfm/lb BOD) for towers loaded at 320 to 480 kg BOD/100 $m^3 \cdot$d (200 lb BOD/d/1000 cu ft). Another approach is to calculate air requirements based on the amount of BOD removed in the biological tower. For forced ventilation, regardless of the method used to calculate the air flow, air scrubbing, covers, or both might be required where extreme odor problems exist. An assessment of odor potential and provisions for odor control are mandatory with any biological tower design.

(a)

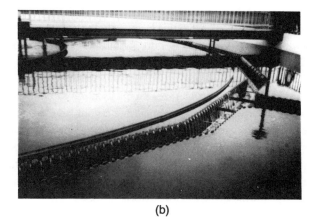

(b)

(c)

(d)

Figure 12.48 Dual-process equipment: (a) biotower, (b) clarifier, (c) contact channel, and (d) filter media.

Suspended Growth Reactors. Designers of dual processes must include or consider proper turndown capability, step feed, and various points of sludge recycle in the design of the suspended growth reactors [Figure 12.48(c)]. With dual processes, turndown capability is a key to providing good effluent quality while the plant is not fully loaded. For example, a number of dual-process designs operate with one of several basins out of service during initial operation, often because the biological towers generally outperform the designer's expectations, especially where warm temperatures and organic loading are more favorable than originally assumed. Conversely, some dual processes operate their suspended growth reactors with large basins on line, rather than using small contact channels. In at least one case, the operators have found that the larger basin could be operated with a lower dissolved oxygen concentration and less aeration power than those for operating the smaller contact channel requiring higher dissolved oxygen concentrations. Still another operator of dual processes returns sludge midway along a plug flow, second-stage suspended growth reactor so that excessive sludge age does not result in problems with nitrification or denitrification. Finally, another dual process is operated in a combined mode only when stringent effluent standards apply. When less-stringent effluent standards apply, the suspended growth reactor is taken off line or used simply as a transfer structure with no sludge recycle. For this final plant, high winter flows are accommodated with the use of the trickling filter process alone and the dual process is used during critical (low) flow conditions.

Flexibility is a key in the design of the suspended growth reactor. Many facilities with dual processes are designed to operate in various modes. Often, little extra capital expense is required to provide piping for operating in both BF/AS and RF/AS modes. Some designers prefer the use of fine bubble air diffusion in the suspended growth reactor to promote bioflocculation and prevent the shearing of biological floc. However, most dual processes have been designed with either coarse bubble or mechanical aeration devices. Plug flow reactors are preferable but not necessary for good design of suspended growth reactors with dual processes. Provisions for step feed should be considered, especially if nitrification must be controlled or solids washout during high flows is a concern. Also, the water depth in the suspended growth reactor is often not critical with dual processes as it would be with conventional activated sludge design. This especially applies where the dual-process reactor is mixing limited, such as the TF/SC process, rather than oxygen limited.

Other Combinations. Dual-process designs that have incorporated either RBC or lagoon systems differ little from their parent biological processes. However, with RBCs, mixing is generally supplied within the RBC reactor to ensure that solids deposition does not occur. Mixing is generally provided through either direct air drive or supplemental air diffusers located at the base of the biological reactor containment structure. With lagoon systems, provisions for solids deposition must accommodate the transfer of underflow from the fixed film reactor to the lagoon.

Secondary Clarifiers. Solids separation through the use of secondary clarifiers [Figure 12.48(b)] with the dual process is similar to that with the suspended growth reactors (that is, activated sludge), except for those with rapid sludge withdrawal. The use of low overflow rates is generally considered good practice for design of dual processes.[142] Some designers successfully use secondary clarifiers with scraper-type sludge collectors. Most designers of dual processes prefer relatively deep secondary clarifiers with side water depths usually from 3.7 to 4.9 m (12 to 16 ft).

Solids separation associated with combined biotowers and RBCs does not require special provisions for handling activated sludge. This is because the RF/RBC process combines two fixed film processes that produce solids more characteristic of the conventional trickling filter process.

Many designers of dual processes recommend that secondary clarifiers be operated with a less than a 0.3-m (1-ft) sludge blanket. Minimizing the blanket depth is often considered to reduce problems with denitrification.

PERFORMANCE HISTORY. A number of national surveys have gathered information on dual processes. Between 500 and 1000 dual-process plants now operate in the U.S. Use of dual processes is increasing, especially where plants are being upgraded and increased flexibility is necessary in the biological portion of the plant.

Table 12.19 lists some representative dual-process plants, including their location, process, actual dry weather flow, and typical influent and effluent BOD concentrations. In addition to the general survey information provided in Table 12.19, two exemplary plant case histories, illustrated in Figures 12.49 and 12.50, indicate the benefits of dual processes.

Figure 12.49 provides a schematic representing the case history of an activated sludge plant that has been upgraded to operate in either a TF/SC or RF/AS mode. The schematic shows how the activated sludge process at Medford, Oregon, was converted into a dual process. Electrical power costs for secondary treatment have been measured under all three modes. Power requirements are 1.9, 1.2, and 0.8 kWh/lb BOD for AS, TF/SC, and RF/AS process modes, respectively. A significant power savings has been realized through the conversion to the dual process. Table 12.20 presents comparative information for the three process modes used at the Medford facility.

Table 12.19   Dual-process plants.

Location	Process	ADWF[a]	Secondary BOD[b]		Loading[c]	
			Influent	**Effluent**	**1st stage**	**2nd stage**
Richmond, CA	RF/AS	6.0	105	30	103	9.6
Tracy, CA	BF/AS	5.0	300	30	112	6.7
Redlands, CA	TF/AS	5.3	155	14	115	9.1
Antioch, CA	RF/AS	10.0	140	13	71	3.6
San Rafael, CA	RF/AS	10.4	138	6	119	3.0
Silverton, OR	TF/SC	0.6	140	6	42	4.5
Medford, OR	RF/AS	8.4	100	6	39	2.2
Roseburg, OR	RF/AS	3.3	129	5	59	2.6
Forest Grove, OR	RF/AS	3.1	435	20	30	14.0
Eustis, FL	TF/SC	1.25	127	20	21	0.9
Chilton, WI	TF/SC	0.60	264	5	22	11.2
Muscatine, IA	RF/AS	6.1	155	34	61	9.6
Fond du Lac, WI	RF/POAS	6.4	189	22	106	4.5
Carlinville, IL	RF/AS	1.9	137	4	42	2.5
Burwell, NE	BF/AS	0.2	106	12	46	15.3
Eureka, CA	TF/SC	5.2	150	6	27	0.6
Stockton, CA	TF/FL	34	540	40	80	1159
Kirksville, MO	RF/RBC	2.3	136	10	50	0.7
Memphis, TN	BF/AS	80	400	30	290	4.0
Jamestown, NY	TF/RBC	6.5	130	30	43	3.1
Corvallis, OR	TF/SC	4.0	110	8	11	0.4

[a] ADWF = average dry weather flow, mgd (mgd × 3785 = m^3/d).
[b] BOD in mg/L.
[c] Biological tower (TF, RF, or BF) reactor-load in lb BOD/d/1000 cu ft; rotating biological contactors—load in lb SBOD/d/1000 sq ft; suspended growth reactors—load as hydraulic residence time in hours without RAS flow.

Figure 12.50 illustrates present operating conditions at the Kirksville, Missouri, plant where an overloaded RBC process was upgraded by conversion to a dual process. The RF/RBC facility at Kirksville illustrates that two parent processes can be combined successfully without adding intermediate clarification. Before installation of the biological tower, the first stages of the RBC were overloaded and dominated with *Beggiatoa*. After installation of the biological tower, growth on existing RBCs diminished, similar to the reduced growth on the last-stage nitrifying RBC discs.

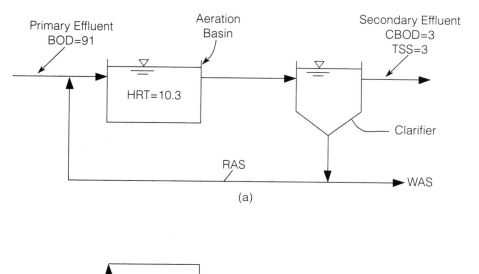

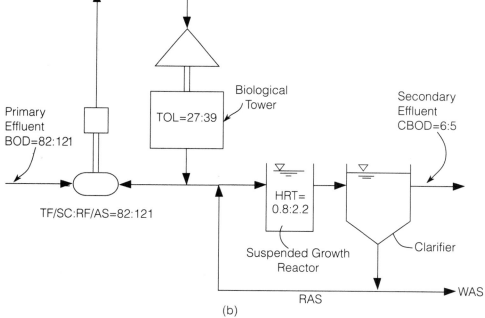

Figure 12.49 Operational modes for dual-process plant at Medford, Oregon:
(a) activated sludge mode and (b) dual-process modes, either TF/SC or
RF/AS. Table 12.20 presents data for plant.

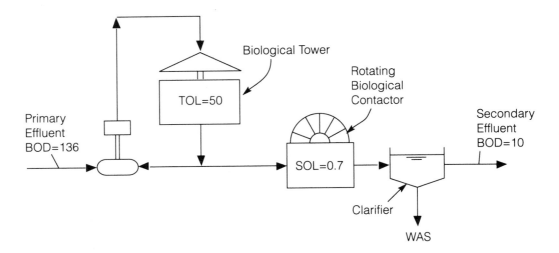

Figure 12.50 Flow diagram for RF/RBC plant at Kirksville, Missouri.
Table 12.19 presents plant data.

Table 12.20 Dual-process case history for plant at Medford, Oregon.

Item	Process mode AS	TF/SC	RF/AS
Loading			
Flow, mgd[a]	8	8.4	8.4
Primary effluent BOD, lb/d[b]	6046	5748	8494
Fixed film reactor			
TOL, lb BOD/d/1000 cu ft[c]	None	27	39.4
Suspended growth reactor			
HRT, based on influent Q, hours	10.3	0.83	2.2
MLSS, mg/L	1961	1475	1433
OUR, mg/L·h			
Average	No data	No data	21
Maximum	No data	No data	32
Final effluent quality, mg/L			
CBOD	3	6	5
TSS	3	6	4
NH₄	8.5	9.7	7.8
Power requirements, kWh/lb BOD[d]	1.9	1.2	0.8

[a] mgd × 3785 = m^3/d.
[b] lb/d × 0.453 6 = kg/d.
[c] lb/d/1000 cu ft × 1.602 = kg/100m^3·d.
[d] kWh/lb × 2.20 = kWh/kg.

REFERENCES

1. "Wastewater Treatment Plant Design." Manual of Practice No. 8, Water Pollut. Control Fed., Washington, D.C. (1977).

2. "Standard Methods for the Examination of Water and Wastewater." 17th Ed., Am. Public Health Assoc., Washington, D.C.(1989).

3. "Operation of Wastewater Treatment Plants." Manual of Practice No. 11, Water Pollut. Control Fed., Washington, D.C. (1976).

4. Okey, R.W., and Albertson, O.E., "Diffusion's role in regulating rate and masking temperature effects in fixed-film nitrification." *J. Water Pollut. Control Fed.,*, **61,** 500 (1989).

5. Paulson, C., "Nitrification in the '90s." *Water Eng. Manage.,* **57** (1989).

6. "Process Design Manual for Phosphorus Removal." EPA/625/1-87/001, U.S. EPA, Cincinnati, Ohio (1987).

7. Polprasert, C., and Park, H.S., "Effluent Denitrification with Denitrifying Filters." *Water Res.,* **20,** 1015 (1986).

8. Chen, G.H., *et al.*, "Modeling of the Simultaneous Removal of Organic Substances and Nitrogen in a Biofilm." *Water Sci. Technol.,* **21,** 790 (1989).

9. Kelly, H.G., "Bio-P Treatment Using a Fixed and Suspended Film Combination—A Demonstration Project." *Proc. Specialty Conf., Environ. Eng., Am. Soc. Civ. Eng. Proc.,* Orlando, Fla. (1987).

10. Williamson, R., "Process for Treating Wastewater." U.S. Pat. No. 4,874,519 (1989).

11. Albertson, O.E., and Eckenfelder, W.W., "Analysis of Process Factors Affecting Plastic Media Trickling Filter Performance." *Proc. Sec. Intern. Conf. Fixed Film Biol. Processes,* Washington, D.C. (1984).

12. Daigger, G., and Harrison, J., "A comparison of trickling filter media performance." *J. Water Pollut. Control Fed.,* **59,** 679 (1987).

13. Bruce, A.M., and Merkens, J.C., "Recent Studies of High Rate Biological Filtration." *Water Pollut. Control* (G.B.), **2,** 449 (1970).

14. National Research Council, "Sewage Treatment at Military Installations." *Sew. Works J.,* **18,** 787 (1946).

15. Galler, W.S., and Gotaas, H.G., "Analysis of Biological Filter Variables." *J. Sanit. Eng. Div., Proc. Am. Soc. Civ. Eng.,* **90,** 6, 59 (1964).

16. Bruce, A.M., and Merkens, J.C., "Further Studies of Partial Treatment of Sewage by High-Rate Biological Filtration." *Water Pollut. Control* (G.B.), **5,** 499 (1973).

17. Dow Chemical Co., "Surfpac Process Evaluations Report." Rep. No. AA-1925, Midland, Mich. (1964).

18. Brown and Caldwell, Seattle, Wash., "West Point Pilot Plant Study, Vol. III, Fixed Growth Reactors for Municipality of Metropolitan Seattle." (1978).

19. Dow Chemical Company, "Data Summaries on Pilot Testing Conducted at Sedalia, Mo." (1967).

20. Kincannon, D.F., and Stover, E.L., "Design Methodology for Fixed Film Reactors, RBCs and Trickling Filters." *Civ. Eng. Practicing and Design Eng., 2*, 107 (1982).

21. Poon, C.P.D., *et al.*, "Trickling Filter Renovation with Plastic Media." *Proc. 2nd Int. Conf. Fixed Film Biol. Processes*, Arlington, Va., 1167 (1984).

22. Parker, D.S., "Research Needs for Trickling Filter Design: A Consultant's Perspective." *Proc. 2nd Int. Conf. Fixed Film Biol. Processes*, Arlington, Va., 1155 (1984).

23. Parker, D.S., and Matasci, R.N., "The TF/SC Process at Ten Years Old: Past, Present, and Future." Paper presented at 62nd Annu. Conf., Water Pollut. Control Fed., San Francisco, Calif. (1989).

24. Logan, B.E., *et al.*, "Engineering implications of a new trickling filter model." *J. Water Pollut. Control Fed., 59*, 1017 (1987).

25. Rankin, R.S., "Evaluation of the Performance of Biofiltration Plants." *Trans. Am. Soc. Civ. Eng., 120*, 823 (1955).

26. Smith, R.S., *et al.*, "Direct Recirculation for High Rate Trickling Filters." Fed. Security Ag., U.S. Public Health Serv., Washington, D.C. (1951).

27. Culp, G., "Direct recirculation of high rate trickling filter effluent." *J. Water Pollut. Control Fed., 35*, 742 (1963).

28. Archer, E., and Robinson, L., "Handbook of Trickling Filter Design." Public Works Journal Corp., Ridgewood, N.J. (1970).

29. Sorrels, J.H., and Zeller, P.J., "Two Stage Trickling Filter Performance." *Sew. Ind. Wastes, 28*, 943 (1956).

30. Republic of West Germany, Arbeitsblatt A 135, "Adwassertechnische Vereinigung." E.V. (ATV), 6 (1980).

31. Hawkes, H.A., "The Ecology of Waste Water Treatment." MacMillan Co., New York, N.Y. (1963).

32. Tomlinson, T.G., and Hall, H., "The Effect of Periodicity of Dosing on the Efficiency of Percolating Filters." *J. Inst. Sew. Purif.* (G.B.), *1*, 40 (1955).

33. Hawkes, H.A., "The Effect of Periodicity of Dosing on the Amount of Film and Numbers of Insects and Worms in Alternating Double Filters at Minworth, Birmingham." *J. Inst. Sew. Purif.* (G.B.), *1*, 48 (1955).

34. Lumb, C., and Barnes, J.P., "Periodicity of Dosing Percolating Filters." *J. Inst. Sew. Purif.* (G.B.), *1*, 83 (1948).

35. Albertson, O.E., and Davies, G., "Analysis of Process Factors Controlling Performance of Plastic Bio-Media." Paper presented at 57th Annu. Conf. Water Pollut. Control Fed., New Orleans, La. (1984).

36. Albertson, O.E., "Slow Down That Trickling Filter!" *Oper. Forum,* **6,** 1, 15 (1989).

37. Albertson, O.E., "Slow-Motion Trickling Filters Gain Momentum." *Oper. Forum,* **6,** 8, 28 (1989).

38. Rockwell, D., Cedar Rapids, Iowa, Personal Communication.

39. Orr, P., and Lawtry, R., "Operating Experience with Large Random Packed Biofilm Reactors." *Proc. IAWPRC Conf.,* Paris, Fr. (1989).

40. Suschka, J., "Hydraulic Performance of Percolating Biological Filters and Consideration of Oxygen Transfer." *Water Res.* (G.B.), **21,** 865 (1987).

41. Wheatley, A.D., and Williams, I.L., "Some Comparative Studies on the Wetting and Stability of Biological Filter Media." *Water Pollut. Control* (G.B.), 433 (1981).

42. Crine, M., *et al.,* "Evaluation of the Performances of Random Media in Aerobic Trickling Filters." *Water Sci. Technol.,* **22,** 227 (1990).

43. Gullicks, H.A., and Cleasby, J.L., "Cold-climate nitrifying biofilters: design and operating considerations." *Res. J. Water Pollut. Control Fed.,* **62,** 50 (1990).

44. Gullicks, H.A., and Cleasby J.L., "Design of trickling filter nitrification towers." *J. Water Pollut. Control Fed.,* **58,** 60 (1986).

45. Honoso, Y., *et al.,* "Characteristic Evaluation of Trickling Filter Process." *Water Res.,* **14,** 581 (1980).

46. Harrison, J.R., and Timpang, P.L., "Design Considerations with the Trickling Filter Solids Contract Process." *Proc. Joint Can. Soc. Civ. Eng., Am. Soc. Civ. Eng. Natl. Conf., Environ. Eng.,* Vancouver, B.C., Can. (1988).

47. Schroeder, E.D., and Tchobanoglous, G., "Mass transfer limitations on trickling filter design." *J. Water Pollut. Control Fed.,* **48,** 772 (1976).

48. Schulze, K.L., "Load and efficiency of trickling filters." *J. Water Pollut. Control Fed.,* **32,** 245 (1960).

49. Velz, C.J., "A Basic Law for the Performance of Biological Filters." *Sew. Works J.,* **20,** 607 (1948).

50. Howland, W.E., "Flow Over Porous Media as in a Trickling Filter." *Proc. 12th Ind. Waste Conf., Purdue Univ.,* West Lafayette, Ind., 435 (1957).

51. Germain, J.E., "Economical treatment of domestic waste by plastic-medium trickling filters." *J. Water Pollut. Control Fed.,* **38,** 192 (1966).

52. Reynolds, L.B., and Chipperfield, P.N.J., "Principles Governing the Selection of Plastic Media for High-Rate Biological Filtration." *Proc. Int. Congress Ind. Waste Water,* Stockholm, Swed. (1970).

---

53. "Recommended Standards for Sewage Works." Great Lakes–Upper Mississippi River Board of State Sanit. Eng. Health Educ. Serv. Inc., Albany, N.Y. (1990).

54. Institution of Water and Environ. Manage., "Unit Processes Biological Filtration—Manuals of British Practice in Water Pollution Control." (1988).

55. Pomeroy, R., "Controlling Sewage Plant Odors." *Constr. Eng.*, **20**, 101 (1963).

56. Hicks, R., "Odor Control in Wastewater Treatment Systems." *Proc. 2nd Int. Clean Air Congress*, Washington, D.C., 867 (1970).

57. Albertson, O.E., and Okey, R.W., "Trickling Filters Need to Breathe Too." Paper presented at Annu. Iowa Water Pollut. Control Assoc. Meeting, Des Moines (1988).

58. Gonenc, I.E., and Harremoes, P., "Nitrification in Rotating Disc Systems—I. Criteria for Transition From Oxygen to Ammonia Rate Limitations." *Water Res.* (G.B.), **19**, 1119 (1981).

59. Harremoes, P., *et al.*, "Design of Fixed Film Nitrification and Denitrification Units Based on Laboratory and Pilot Scale Results." Paper presented at Europaische Abwasser Symp., Munich (1981).

60. Mehta, D.S., *et al.*, "Oxygen Theory in Biological Treatment Plant Design." *J. Sanit. Eng. Div., Proc. Am. Soc. Civ. Eng.*, **98**, 471 (1972).

61. Benzie, W., *et al.*, "Effects of climatic and loading factors on trickling filter performance." *J. Water Pollut. Control Fed.*, **35**, 445 (1963).

62. Schroepfer, G.J., *et al.*, "Temperature Effects on Trickling Filters." *Sew. Ind. Wastes*, **24**, 705 (1952).

63. Onda, K., *et al.*, "Mass Transfer Coefficients between Gas and Liquid Phases in Packed Columns." *J. Chem. Jpn.*, **1**, 56 (1968).

64. Parker, D.S., *et al.*, "New Trickling Filter Applications in the USA." *Water Sci. Technol.*, **22**, 215 (1990).

65. Eckenfelder, W.W., "Trickling Filter Design and Performance." *Trans. Am. Soc. Civ. Eng.*, **128**, Part III, 371 (1963).

66. Eckenfelder, W.W., and Barnhart, W., "Performance of a high-rate trickling filter using selected media." *J. Water Pollut. Control Fed.*, **35**, 1535 (1963).

67. Norris, D.P., *et al.*, "Production of high quality trickling filter effluent without tertiary treatment." *J. Water Pollut. Control Fed.*, **54**, 1087 (1982).

68. Matasci, R.N., *et al.*, "Trickling filter/solids contact performance with rock filters at high organic loadings." *J. Water Pollut. Control Fed.*, **60**, 68 (1988).

69. Eden, G.E., *et al.*, "Biological Filtration Using Plastic Filter Medium." *J. Proc. Inst. Sew. Purification* (G.B.), 562 (1966).

70. Ames, W.F., *et al.*, "Transient Operation of the Trickling Filter." *J. Sanit. Eng. Div., Proc. Am. Soc. Civ. Eng.*, **88**, 21 (1962).

71. Gromiec, M.J., and Maline, J.F., Jr., "Verification of Trickling Filter Models using Surfpac." Rep. EHE-70-13 (CRWR-60), Center for Res. Water Resour., Civ. Eng. Dept., Univ. Texas, Austin (1970).

72. Logan, B.E., Univ. Ariz., Tucson, Personal Communication (1989).

73. Hutchinson, E.G., "A Comparative Study of Biological Filter Media." Paper presented at Biotechnol. Conf., May, Massey Univ., Palmerston North, N.Z. (1973).

74. Hinton, S.W., and Stensel, H.D., "Discussion of: A fundamental model for trickling filter process design and engineering implications of a new trickling filter model." *J. Water Pollut. Control Fed.*, **61**, 363 (1989).

75. Chipperfield, P.N.M., "Recent Investigations of Biological Treatment Processes." *Inst. Sew. Purification J.* (G.B.), 105 (1964).

76. Albertson, O.E., "Know Your ARs from your *k*s." Submitted to *Res. J. Water Pollut. Control Fed.* (1990).

77. Roesler, J.F., and Smith, R., "A Mathematical Model for a Trickling Filter." U.S. Dept. of the Interior, Fed. Water Pollut. Control Admin., Robert A. Taft Water Res. Center, Cincinnati, Ohio (1969).

78. Deis, Gary, *et al.*, "Modular Plastic Media for Shallow Bed Trickling Filter Applications." *Proc. Spec. Conf. Environ. Eng., Environ. Eng. Div., Am. Soc. Civ. Eng.*, Boston, Mass. (1985).

79. Drury, D.D., *et al.*, "Evaluation of high density crossflow media for rehabilitating an existing trickling filter." *J. Water Pollut. Control Fed.*, **58**, 364 (1986).

80. Wood, D.K., *et al.*, "Evaluation Shows Plastic Media to be More Effective than Rock Trickling Filters." Paper presented at Calif. Water Pollut. Control Assoc., Palm Springs (1989).

81. Malmrose, P., Malcolm–Pirnie, Inc., Paramus, N.J., Personal Communication (1987).

82. Parker, D.S., *et al.*, "Enhancing reaction rates in nitrifying trickling filters through biofilm control." *J. Water Pollut. Control Fed.*, **61**, 618 (1989).

83. Boller, M., and Gujer, W., "Nitrification in Tertiary Trickling Filters Followed by Deep Bed Filters." *Water Res.*, **20**, 1363 (1986).

84. Gerlich, J.E., "Better Than the Pilot Model." *Am. City* (October 1967).

85. Duddles, G.A., *et al.*, "Plastic medium trickling filters for biological nitrogen control." *J. Water Pollut. Control Fed.*, **46**, 937 (1974).

86. Duddles, G.A., and Richardson, S.E., "Application of Plastic Media Trickling Filters for Biological Nitrification Systems." EPA-R2-73-199, U.S. EPA, MERL, Cincinnati, Ohio (1973).

87. Sampayo, F.F., and Metcalf, P.C., "Performance of Nitrification Towers at Sidney, Ohio, and Lima, Ohio." *Proc. 2nd Int. Conf. Fixed Film Biol. Processes*, Arlington, Va., **2**, 1468 (1984).

88. "Report on Tertiary Treatment Pilot Plant Studies, City of Sunnyvale, California." Brown and Caldwell, Consult. Eng., Walnut Creek, Calif. (1975).

89. Parker, D.S., *et al.*, "Process Design Manual for Nitrogen Control." Office Technol. Transfer, U.S. EPA, Cincinnati, Ohio (1975).

90. "Report Prepared for Sanitary District of Bloom Township on Pilot Studies Carried out to Study Nitrification." Baxter & Woodman, Inc., Chicago Heights, Ill. (1973).

91. Gonenc, I., *et al.*, "Nitrification in Rotating Disc Systems. I. Criteria for Transition from Oxygen to Ammonia Rate Limitation." *Water Res.*, **19,** 1119 (1985).

92. Harremoes, P., "Criteria for Nitrification in Fixed-Film Reactors." *Water Sci. Technol.*, **14,** 167 (1982).

93. Harremoes, P., "Biofilm Kinetics." In "Water Pollution Microbiology." R. Mitchell (Ed.), John Wiley & Sons, Inc., New York, N.Y. (1978).

94. Okey, R.W., and Albertson, O.E., "Evidence for oxygen-limiting conditions during tertiary fixed-film nitrification." *J. Water Pollut. Control Fed.*, **61,** 510 (1989).

95. Okey, R.W., and Albertson, O.E., "Diffusion's role in regulating rate and masking temperature effects in fixed film nitrification." *J. Water Pollut. Control Fed.*, **61,** 500 (1989).

96. Parker, D.S., and Richards, T., "Nitrification in trickling filters." *J. Water Pollut. Control Fed.*, **58,** 896 (1986).

97. Huang, C.-S., and Hopson, N.E., "Nitrification Rate in Biological Processes." *J. Environ. Eng.*, **100,** EE2, 409 (1974).

98. Dahl, R., Central Valley Wastewater Treatment Plant, Utah, Personal Communication (1990).

99. Rogers, S., So. Davis Co. Wastewater Treatment Plant, Salt Lake City, Utah, Personal Communication (1990).

100. Gullicks, H.A., and Cleasby, J.S., "Nitrification performance of a pilot-scale trickling filter." *J. Water Pollut. Control. Fed.*, **60,** 40 (1990).

101. Gujer, W., and Boller, M., "Design of a Nitrifying Trickling Filter Based on Theoretical Concepts." *Water Res.*, **20,** 1353 (1986).

102. Albertson, O.E., and Okey, R., Prepared for American Surfpac, Inc., "Design Procedure for Tertiary Nitrification." (1988).

103. Stenquist, R.J., *et al.,* "Carbon oxidation—nitrification in synthetic media trickling filters." *J. Water Pollut. Control Fed.*, **46,** 2327 (1974).

104. Bruce, A.M., *et al.,* "Pilot Studies on the Treatment of Domestic Sewage by Two-Stage Biological Filtration—With Special Reference to Nitrification." *Water Pollut. Control* (G.B.), 80 (1975).

105. Lin, C.S., and Heck, G., "Design and Performance of the Trickling Filter/Solids Contact Process for Nitrification in a Cold Climate." Paper presented at 60th Annu. Conf. Water Pollut. Control Fed., Philadelphia, Pa. (1987).

106. Albertson, O., Prepared for American Surfpac Corp., Process designs prepared for Wauconda, Ill., and Buckeye Lake, Ohio." (1984).

107. Okey, R.W., Salt Lake City, Utah, Personal Communication (1989).

108. Albertson, O.E., and Okey, R., Prepared for American Surfpac Corp., Procedure developed for sizing dual-purpose towers. (1989).

109. Stenquist, R.J., and Kelly, K.A., "Converting Rock Trickling Filters to Plastic Media—Design and Performance." EPA 600/2-80-120, U.S. EPA, Cincinnati, Ohio (1980).

110. Pierce, D., "Upgrading Trickling Filters." EPA 430/9-78-004, U.S. EPA, Cincinnati, Ohio (1978).

111. "Upgrading Existing Wastewater Treatment Plants." U.S. EPA, Office Technol. Transfer, Cincinnati, Ohio (1971).

112. Hynek, R.J., and Iemura, H., "Nitrogen and Phosphorous Removal with Rotating Contactors." *Proc. First Natl. Symp. Rotating Biol. Contactor Technol.,* Champion, Pa. (1980).

113. "O&M of Trickling Filters, RBCs, and Related Processes." Manual of Practice OM-10, Water Pollut. Control Fed., Washington, D.C. (1986).

114. Stover, E.L., and Kincannon, D.F., "One Step Nitrification and Carbon Removal." *Water Sew. Works,* **122,** 66 (1975).

115. Stover, E.L., and Kincannon, D.F., "Rotating Disc Process Treats Slaughterhouse Waste." *Ind. Wastes (Chicago),* **22,** 33 (1976).

116. Dupont, R.R., and McKinney, R.E., "Data Evaluation of a Municipal RBC Evaluation." *Proc. First Natl. Symp. Rotating Biol. Contactor Technol.,* Champion, Pa. (1980).

117. Poon, C.P.C., *et al.,* "Factors controlling rotating biological contactor performance." *J. Water Pollut. Control Fed.,* **51,** 601 (1979).

118. Wilson, R.W., *et al.,* "Scale-up in rotating biological contactor design." *J. Water Pollut. Control Fed.,* **52,** 610 (1980).

119. Brenner, R.C., *et al.,* "Design Information on Rotating Biological Contactors." EPA-600/2-84-106, U.S. EPA, Cincinnati, Ohio (1984).

120. Antonie, R.L., "Fixed Biological Surfaces—Wastewater Treatment: The Rotating Biological Contactor." CRC Press, Cleveland, Ohio (1976).

121. Atkinson, B., and Fowler, H.W., "The Significance of Microbial Film in Fermentors." *Adv. Biochem. Eng.,* **3,** 221 (1974).

122. Kornegay, B.H., and Andrews, J.H., "Characteristics and kinetics of fixed-film biological reactors." *J. Water Pollut. Control Fed.,* **40,** 11, R460 (1968).

123. Bungay, H.R., *et al.*, "Microprobe Techniques for Determining Diffusivities and Respiration Rates in Microbial Slime Systems." *Biotechnol. Bioeng.*, **11**, 765 (1969).

124. Hitdlebaugh, J.A., and Miller, R.D., "Full-Scale Rotating Biological Contactor for Secondary Treatment and Nitrification." *Proc. First Natl. Symp. Rotating Biol. Contactor Technol.*, Champion, Pa. (1980).

125. Reh, C.W., *et al.*, "An Approach to Design of RBCs for Treatment of Municipal Wastewater." Presented at Am. Soc. Civ. Eng. Natl. Environ. Eng. Conf., Nashville, Tenn. (1977).

126. Murphy, K.L., and Wilson, R.W., "Pilot Plant Studies of Rotating Biological Contactors Treating Municipal Wastewater." Rep. SCAT-2, Environ. Can., Ottawa, Ont. (1980).

127. Benjes, H.H., Sr., "Small Community Wastewater Treatment Facilities–Biological Treatment Systems." U.S. EPA Technol. Transfer Natl. Seminar, Chicago, Ill. (1977).

128. Roy F. Weston, Inc., "Review of Current RBC Performance and Design Procedures." EPA/600/2-85/033, U.S. EPA, Cincinnati, Ohio (1985).

129. Doran, M.D., Strand Assoc., Inc., Madison, Wisc., Personal Communication (1990).

130. Scheible, O.K., and Novak, J.J., "Upgrading Primary Tanks with Rotating Biological Contactors." *Proc. First Natl. Symp. Rotating Biol. Contactor Technol.*, Champion, Pa. (1980).

131. Borchardt, J.A., *et al.*, "Nitrification of Secondary Municipal Waste Effluents by Rotating Bio-discs." EPA-600/2-78/061, U.S. EPA, Cincinnati, Ohio (1978).

132. Stratta, J.M., and Long, D.A., "Nitrification Enhancement through pH Control with Rotating Biological Contactors." Pa. State Univ. (1981).

133. Harrison, J.R., *et al.*, "A survey of combined trickling filter and activated sludge processes." *J. Water Pollut. Control Fed.*, **56**, 1073 (1984).

134. Krumsick, T.A., *et al.*, "Trickling Filter Solids Contact Process Demonstration, Salt Lake City, Utah." Presented at Annu. Conf. Utah Water Pollut. Control Assoc., Salt Lake City (1984).

135. Richards, T., and Reinhart, D., "Evaluation of plastic media in trickling filters." *J. Water Pollut. Control Fed.*, **58**, 774 (1986).

136. Albertson, O.E., "Slow Down that Trickling Filter!" *Oper. Forum*, **6**, 115 (1989).

137. Harrison, J.R., "Surveys of Plants Operating Activated Biofilter/Activated Sludge." Presented at North. Reg. Conf. and Training School of the Calif. Water Pollut. Control Assoc. (1980).

138. Harrison, J.R., and Chesner, W., "Strategies for Upgrading RBC's." Presented at the Water Pollut. Control Fed. Annu. Conf., San Francisco, Calif. (1989).

139. Newbry, B.W., *et al.*, "Unit process tradeoffs for combined trickling filter and activated sludge processes." *J. Water Pollut. Control Fed.*, **60,** 1813 (1988).

140. Howland, W.E., "Effect of Temperature on Sewage Treatment Processes." *Sew. Ind. Wastes*, **25,** 161 (1953).

141. Kalb, Kathryn, *et al.*, "Nitrified Sludge—An Innovative Process for Removing Nutrients from Wastewater." Paper presented at Annu. Water Pollut. Control Fed. Conf., Washington, D.C. (1990).

142. Tekippe, R.J., and Bender, J.H., "Activated sludge clarifiers: design requirements and research priorities." *J. Water Pollut. Control Fed.*, **59,** 865 (1987).

143. Richards, T., "Evaluation of Biological Tower/Suspended Growth Systems." *Proc. 2nd Int. Conf. Fixed Film Biol. Processes*, Arlington, Va., 1584 (1984).

Notes

Notes

Notes

Notes

Notes

Notes

Notes

Notes

Notes

Notes

Notes

Notes

Notes

Notes

Notes